High Performance Liquid Chromatography in Biotechnology

High Performance Liquid Chromatography in Biotechnology

Edited by

William S. Hancock
Genentech Corporation
South San Francisco, California

A WILEY-INTERSCIENCE PUBLICATION
JOHN WILEY & SONS
New York • Chichester • Brisbane • Toronto • Singapore

In recognition of the importance of preserving what has been written, it is a policy of John Wiley & Son, Inc. to have books of enduring value published in the United States printed on acid-free paper, and we exert our best efforts to that end.

Library of Congress Cataloging in Publication Data:

High performance liquid chromatography in biotechnology
William S. Hanconck
p. cm.

"A Wiley-interscience publication."
Includes bibliographical references
ISBN 0-471-82584-0

1. High performance liquid chromatography. 2. Biotechnology-Technique. I. Hancock, William S.
TP248.25.H54H54 1990
660′.6--dc20

89-70569
CIP

Printed in the United States of America

10 9 8 7 6 5 4 3 2 1

CONTRIBUTORS

Robert A. Barford, USDA, ARS, Macromolecular and Cell Structure Research Unit, Philadelphia, Pennsylvania 19118

Irwin M. Chaiken, Department of Macromolecular Sciences, SmithKline Beecham, King of Prussia, Pennsylvania 19406

R.E. Chance, Department of Biochemistry and Biochemical Development, Lilly Research Laboratories, A Division of Eli Lilly and Company, Indianapolis, Indiana 46285

Thorkild Christensen, Nordisk Gentofte A/S, 1 Neils Steensensvej, DK-2820, Gentofte, Denmark

R.D. DiMarchi, Department of Biochemistry and Biochemical Development, Lilly Research Laboratories, A Division of Eli Lilly and Company, Indianapolis, Indiana 46285

Miral Dizdaroglu, Center for Chemical Technology, National Institute of Standards and Technology, Gaithersburg, Maryland 20899

Blair A. Fraser, Chemical Biology Laboratory, Division of Biochemistry and Biophysics, Food and Drug Administration, Bethesda, Maryland 20205

Bruno Hansen, Hagedorn Research Laboratory, Niel Steensevsin 6, DK-2820 Gentofte, Denmark

Jan Jørn Hansen, Nordisk Gentofte A/S, 1 Niels Steensensvej, DK-2820 Gentofte, Denmark

Kim R. Hejnaes, Nordisk Gentofte A/S, 1 Niels Steensensvej, DK-2820 Gentofte, Denmark

Michael P. Henry, Research and Development Laboratories, J.T. Baker Inc., Phillipsburg, New Jersey 08865

Takaharu Itagaki, Research Center, Mitsubishi Chemical Industries Ltd., Yokahama 227, Japan

Ronald D. Johnson, Process Development and Manufacturing, Chiron Corporation, Emeryville, California

Yoshio Kato, Central Research Laboratory, TOSOH Corporation, Tonda, Shinnanyo, Yamaguchi 746, Japan

Takashi Kitamura, Central Research Laboratory, TOSOH Corporation, Tonda, Shinnanyo, Yamaguchi 746, Japan

E.P. Kroef, Department of Biochemistry and Biochemical Development, Lilly Research Laboratories, A Division of Eli Lilly and Company, Indianapolis, Indiana 46285

Hiroshi Kusano, Research Center, Mitsubishi Chemical Industries Ltd., Yokohama 227, Japan

H.B. Long, Department of Biochemistry and Biochemical Development, Lilly Research Laboratories, A Division of Eli Lilly and Company, Indianapolis, Indiana 46285

Duncan Low, Pharmacia AB, Process Separation Division, S-751-2 Uppsala, Sweden

Daniel R. Marshak, Cold Spring Harbor Laboratory, Cold Spring Harbor, New York 11724

Eiji Miyata, Research Center, Mitsubishi Chemical Industries Ltd., Yokohama 227, Japan

David R. Nau, Research Speciality Products Division, J.T. Baker Inc., Phillipsburg, New Jersey 08865

Hans Holmegaard Sorensen, Nordisk Gentofte A/S, 1 Niels Steensenvej, DK-2820 Gentofte, Denmark

Hiroaki Takayangi, Research Center, Mitsubishi Chemical Industries Ltd., Yokohama 227, Japan

Johannes Thomsen, Nordisk Gentofte A/S, 1 Niels Steensensvej, DK-2820 Gentofte, Denmark

Benny S. Welinder, Hagedorn Research Laboratory, Niels Steensensing 6, DK-2820 Gentofte, Denmark

Gerald Zon, Applied Biosystems, Foster City, California 94404

PREFACE

The production of new protein pharmaceuticals by recombinant-DNA (rDNA) techniques has resulted in both governmental agencies and the pharmaceutical industry developing sophisticated new analytical and purification technologies. These more stringent requirements were introduced as a result of the perceived risk of clinical testing of new protein products that have been produced by an untried technology. High performance liquid chromatography (HPLC) has made major contributions to biotechnology as the powerful separations that can be achieved with this technique have been applied to almost all phases of research and development in the industry. The broad application of HPLC has been covered in the different chapters of this book – ranging from the major chromatographic techniques (reversed phase, ion exchange, affinity and hydrophobic interaction chromatography) to specific separations such as in monoclonal antibody and nucleic acid purification. Also the scale of HPLC use ranges from the isolation of nanogram amounts of a novel protein for characterization studies to the production of kilogram amounts in a manufacturing process.

The quality assurance of a protein pharmaceutical requires that both the recovery process and final product itself are scrutinized with a battery of analytical techniques. The use of HPLC has been most successful in the separation and quantitation of product variants which are often present in levels of 0.5 to 2%. This text will describe the different analytical techniques that are required for detection of variants and review those methods that are required for a quality control program. Furthermore a close association between the analytical chemist and process development is required for the efficient manufacture of a new product, and this book will describe the transformation of an analytical into a preparative separation. The development of high resolution methods for analytical protein chemistry has resulted in the requirement for advance technology such as mass spectrometry for structural characterization. I have therefore included a chapter on this and related technologies that allow a protein chemist to identify and then avoid unexpected side reactions in different rDNA processes.

The application of HPLC to biotechnology is an exciting and rapidly moving field and thus presents a strong challenge to the practitioner. I hope that the introductory chapter together with the specialist reports will allow the reader to get both an overview of recent developments as well as the more detailed knowledge present in the many areas of chromatography that overlap with biotechnology.

William S. Hancock

San Francisco, California
February 1990

CONTENTS

High Performance Liquid Chromatography in Biotechnology

CHAPTER 1

HPLC in Biotechnology

W.S. HANCOCK

Genentech Corporation, South San Francisco, California

The development of recombinant-DNA (r-DNA) based protein pharmaceuticals has led to a tremendous interest in new high-resolution chromatographic methods. Often, an improved analytical method will allow the detection of a previously unresolved impurity. Such a development will then generate the need for an improved preparative separation to allow removal of the impurity. An example of this interplay between analytical and preparative separations ocurred in the early days at Genentech with the development of recombinant methionyl-human growth hormone. The introduction of silver staining instead of the much less sensitive Coomassie dye staining procedures resulted in the detection of new host cell impurities (*E. coli* proteins). This observation had two important effects: one was the development of new, highly sensitive immunoassay for this class of impurities and the addition of high-resolution chromatographic steps into the recovery process. The end result is that current batches of Protropin® contain less than 10 ppm of *E. coli* proteins (12).

The purpose of this book is to bring together those chromatographic technologies that are making a key contribution to this new frontier of molecular biology and analytical chemistry. The chapters will include either updates of relevant areas of well-established methods, for example, reversed-phase chromatography, or the introduction of new technologies that are beginning to show potential, for example, capillary electrophoresis and mass spectrometry. This initial chapter will introduce the reader to examples from the author's department at Genentech (analytical chemistry), to cover areas of technology not described in the review chapters, and to describe the relevance of the review chapters to the production of protein pharmaceuticals.

Appropriately, the review chapters begin with a description of reversed-phase high-performance liquid chromatogrphy (RP-HPLC). Until as recently as 5 years ago it was generally believed that RP-HPLC was inappropriate for the analysis of proteins. Although the technique was particularly well suited to the

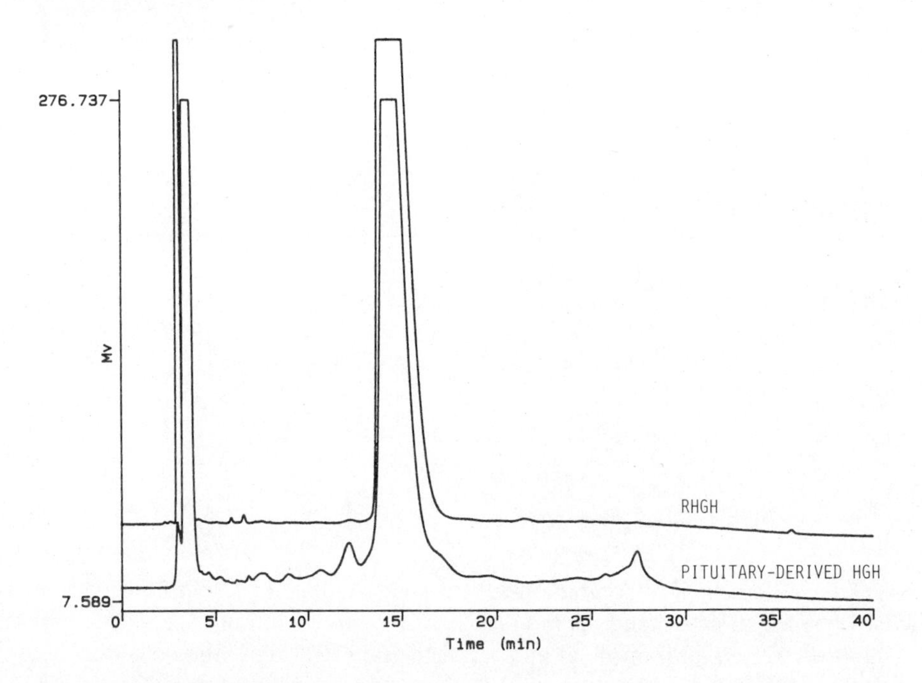

Fig. 1. The analysis of a sample of recombinant-DNA-derived growth hormone (rhGH) and a sample of pituitary-derived growth hormone by reversed-phase HPLC (RP-HPLC). The separation was carried out on a Vydac-C4 column with a mobile phase that contains 0.1% trifluoroacetic acid and acetonitrile.

separation of peptides (18), the literature described several studies where the chromatography of proteins resulted in denaturation and loss of the sample (either poor mass recoveries or low activities of the purified sample, see Refs. 3, 28, and 32). However, the early literature did give a clue to the future with a description of the successful chromatography of insulin (42) and growth hormone (4). Subsequently, Eli Lilly published a description of a highly successful manufacturing process based on RP-HPLC (29). In addition, RP-HPLC is particularly useful for the separation of samples that are insoluble in common aqueous buffers and recent studies have made use of strong acids [up to 30% trifluoroacetic acid and 50% formic acid, see Thevenon and Regnier (45) and Poll and Harding (39)] as well as organic solvents. With the potential of membrane-associated proteins as therapeutics and diagnostics, the process development scientist may need RP-HPLC to overcome the difficulty of purifying proteins with very hydrophobic transmembrane domains.

RP-HPLC played a key role at Genentech in the development of rhGH as a pharmaceutical. As can be seen in Fig. 1, this technique can be used to establish chromatographic identity between the recombinant and human-pituitary-derived material. The chromatographic analysis also demonstrates that the recombinant-DNA-derived product can be produced at a high purity level. RP-HPLC has proven to be a powerful technique for the isolation and quantitation of variants that may be present in the final product. Thus, RP-HPLC has been used to isolate a variety of degradation products of rhGH, for example, deamidated, clipped, and oxidized variants (19). It should be noted that each separation required a different mobile phase, and thus the development of the desired separation required significant effort. An even more serious pitfall is the situation where an unresolved impurity may escape detection despite the analysis of the product by a variety of chromatographic methods. In an attempt to avoid this pitfall, the principle of "orthogonal" analytical methods in biotechnology has recently been discussed (6). Thus, in the analysis of rhGH, the optimal resolution of deamidated material occurred on an ion exchange support and for the aggregated material on a gel permeation support. Owing to the lack of detailed information on the molecular properties of these variants, the selection of the different chromatographic methods could not always be predicted, for example, the separation of the desPhe variant could only be achieved by HIC in the presence of a detergent (51). It is probable that changes in the 3D structure of the individual protein species during the actual separation adds an element of unpredictability to analytical method development (37). Also, an active involvement in new technologies can improve the certainty that a new recombinant-DNA-derived product does not contain an unidentified contaminant. An example of this approach is shown in Fig. 3 where the resolution of deamidated rhGH is achieved by capillary zone electrophoresis (16).

Chapter 2 gives a careful review of the design of a series of reversed-phase packing materials that can allow an analytical separation to be scaled-up to a preparative material without dramatic differences in the separation. There are a variety of manufacturers that supply suitable packing materials [see review by Unger et al. (47)] and publications such as LC-GC® are useful for recent updates (34). This review chapter serves as an introduction to the parameters that are important in the choice of the optimal packing material for a given separation.

A number of laboratories have studied different aspects of the conformational changes that may occur during protein chromatrography (3, 14, 32, 35), and the review chapter by Barford presents a useful theoretical model that can aid to understanding of a separation based on hydrophobic interactions. Typically, an analytical separation uses a strongly denaturing mobile phase such as aqueous trifluoroacetic acid and acetonitrile, as peak shapes and recoveries are often optimal under these conditions. However, the separation of two human growth hormone variants by RP-HPLC can only be achieved under less denaturing conditions (neutral pH, substitution of propanol for acetonitrile). The sole difference of the two variants is the presence of an additional N-terminal methionine residue, which is thought to be near the surface of the molecule. Apparently this difference is more exposed to the chromatographic surface under "native-like" conditions. Thus, the use of RP-HPLC under a range of conditions can lead to differences in selectivity which can be exploited. The

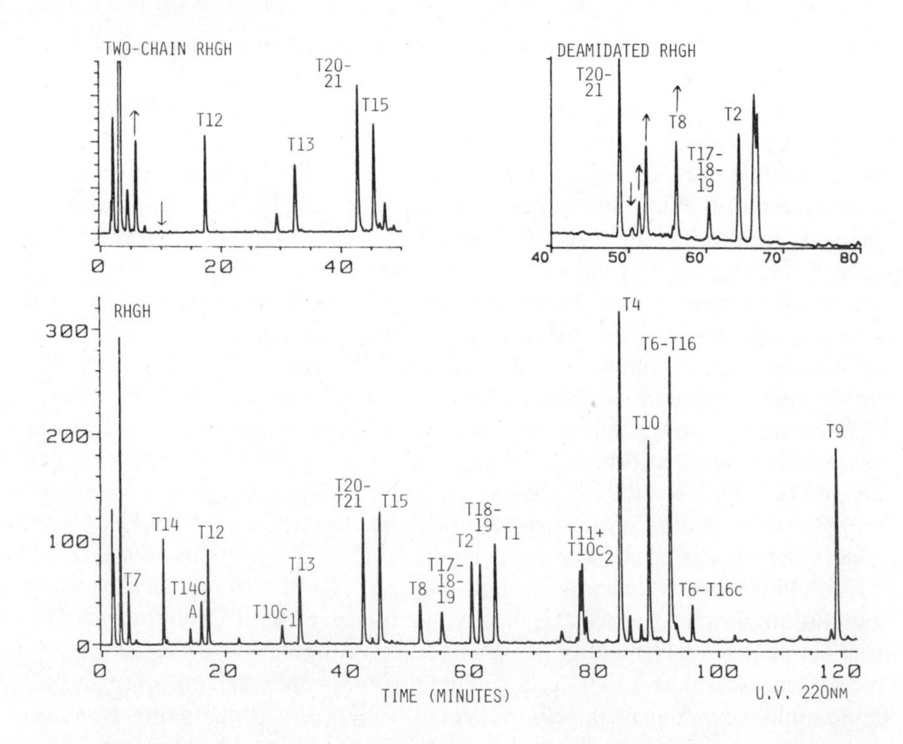

Fig. 2. The HPLC tryptic map of 100 μg of rhGH digest using 50 m*M* sodium phosphate, pH 2.85 as the aqueous mobile phase and acetonitrile as the organic mobile phase. The separation was carried out on a 15 cm, 5 μ, Nucleosil C-18 column with a linear gradient of 0–40% acetonitrile in 120 min. The flow rate was 1 ml/min, the column temperature was maintained at 40°C, and the separation was monitored at 220 nm. Tryptic maps of deamidated rhGH (upper right) and two-chain rhGH (upper left) resulted in changes in the peptide elution profile (→). Reproduced with permission from Hancock et al. (19).

ability to recover the native conformation and hence biological activity is not necessarily essential in an analytical separation but is of course essential in a preparative application. Furtunately, many of the proteins of interest in biotechnology (e.g., insulin, growth hormone, α- and γ-interferon, insulin-like growth factor, and tissue plasminogen activator) can be chromatographed on a reversed-phase column with essentially no loss of biological activity. However, these proteins are reasonably stable (most of the proteins contain disulfide bonds) and apparently can rapidly refold after the separation. In addition the problem of denaturation the biological activity may be affected by reaction between the protein and components arising from column bleed. The review by Welinder et al. looks at the activity of insulin after a RP-HPLC separation and addresses these important issues.

The characterization of rDNA-derived proteins by peptide mapping has become a very popular technique and is used by the industry to seek marketing approval from the regulatory agencies as well as for quality control of the different manufactured lots. In this application the protein is first digested at specific cleavage points with an enzyme such as trypsin or V-8 protease or cleaved with a chemical such as cyanogen bromide. Since the majority of peptides resulting from a trypsin digestion are small and without significant 3D structure, the problems of chromatography of the intact protein are largely avoided. Thus, the effect of an amino acid substitution on the retention time observed for a given peptide is usually dramatic and often follows empirical guidelines (37). Also with a volatile mobile phase such as TFA/CH_3CN, the peptides can be collected and identified by techniques such as amino acid analysis, N-terminal sequencing, and FAB mass spectrometry. If the enzymatic digestion can be carried out on an unreduced cystine-containing protein, it is often possible to assign the position of the disulfide bridges. A comparison of the reduced and nonreduced map of recombinant human growth hormone (rhGH) can allow one to estimate the amount of broken disulfides present in the product. For example, the absence of reduced T16 in the rhGH map indicated that disulfide bond in the product was completely intact (19). Also, protein degradation (deamidation or proteolysis) can be detected in a tryptic map by the disappearance of a peptide and the appearance of new fragment(s) (Fig. 2). The type of structural change can often be predicted from the shift in retention time, for example, deamidation usually results in a slight increase in retention time, while formation of methione sulfoxide and cysteic acid result in significant decreases (see Fig. 2 for examples). In a similar manner the use of an incorrect clone or the formation of mutant protein can be expected to result in a significant shift in retention time of one or more peptides. However, despite the effective and popular use of the tryptic map, it is important to remember that the technique has potential weaknesses. For example, it has been observed that the map is unable to detect low levels of variants (typically less than 2%) [see Hancock et al. (19). Also, the map is insensitive to conformational variants and proteolysis caused by trypsin-like enzymes present in the fermentation or recovery process (19). In complex maps coelutions may occur and any variability in analyst or equipment performance can easily result in unacceptable deviations in retention times and peak areas.

In addition to the characterization and quality control, (QC) aspects of protein pharmaceuticals which usually involve materials that are available in abundant amounts, the application of RP-HPLC to micropreparative, microsequencing applications and peptide mapping at the low picomole level is particularly important for the isolation of novel protein fragments. An example of the power of RP-HPLC in micropreparative applications comes from the sequencing 51 residues of cystatin (MW 14 KD) isolated from 50 μl of saliva. The separation was achieved on a RP-300 cartridge column (2.1 mm × 30 mm) with a Brownlee micropump. Recently, the potential of Micro-LC was reviewed by Verzele et al. (48). RP-HPLC using TFA/CH_3CN mobile phase is particularly useful for the isolation of samples for characterization techniques (see above).

Sample preparation is one of the major limitations to determining the amino acid sequence of polypeptide samples that have been isolated in subnanomolar quantities from their biological source material. In addition to sample concentration and desalting, RP-HPLC can be used to aid direct sequence analysis of proteins electroblotted from 2D polyacrylamide gels (41). In this approach Coomassie-Blue-stained proteins are extracted from polydivinylidene difluoride membranes, using a detergent mixture of sodium dodecylsulfate and Triton X-100. The protein sample is then isolated in a pure state by chromatography on reversed-phase columns at high organic solvent concentrations.

However, peptides isolated from complex mixtures by RP-HPLC can be contaminated with other peptides of similar hydrophobicity, and thus the use of high-performance capillary electrophoresis (HPCE) to check purity of purified peptides is becoming a popular application (see Fig. 5). Also the use of diode array detectors for the on-line measurement of the spectra (50) of peptides after separation by RP-HPLC has allowed the detection of peptide mixtures as well as the location of peptides that contain tyrosine or tryptophan residues (as predicted from cDNA sequences) in a complex map. The use of spectral as well as peak area and retention time information can allow the development of automated methods for the comparison of different tryptic maps. Such a comparison is particularly useful in QC where a typical release procedure involves the comparison of the tryptic map of a manufactured lot with that of a reference material. The development of fast separations on nonporous, microparticulate columns may have significant potential in QC applications (13, 27).

In addition to silica-based packing materials, there are available some organic-polymer-based materials suitable for protein separations. One example of these new columns is the polystyrene-based PRP-1 column (30), which has been used at Genentech for separations of growth hormone and recombinant tissue plasminogen activator (rt-PA) variants. Similar results were obtained on the polystyrene-based (PRP-1) as for the classical RP-HPLC separations based on alkyl-silica, for example, Vydac-C4. The results suggest that past problems of small pores and strong hydrophobic effects have been largely solved for these supports. Despite improvements, the current generation of organic polymers is still not completely rigid and some volume change of the matrix may occur during a gradient run. A significant advantage is the hydrolytic stability from pH 1–13 of these columns (30), which allows the development of separations at extended pH values. Many of these packing materials are too expensive and not available in large amounts, so that preparative applications are not possible. However, Chapter 5 on Sepabeads shows that a protein purification process can indeed be based on an organic polymer. Also at Genentech other packing materials such as Toyo Pearl™ have been shown to be useful. The ability to clean a packing material at high pH (e.g., 0.1 *N* NaOH) is particularly useful in process development, since such a treatment is useful both in column cleaning and ensuring sterility. Chapter 6 by Lowe from Pharmacia describes the techniques for development of preparative-scale purifications based on large columns and demonstrates the use of cross-linked polysaccharide matrices. Also Jones

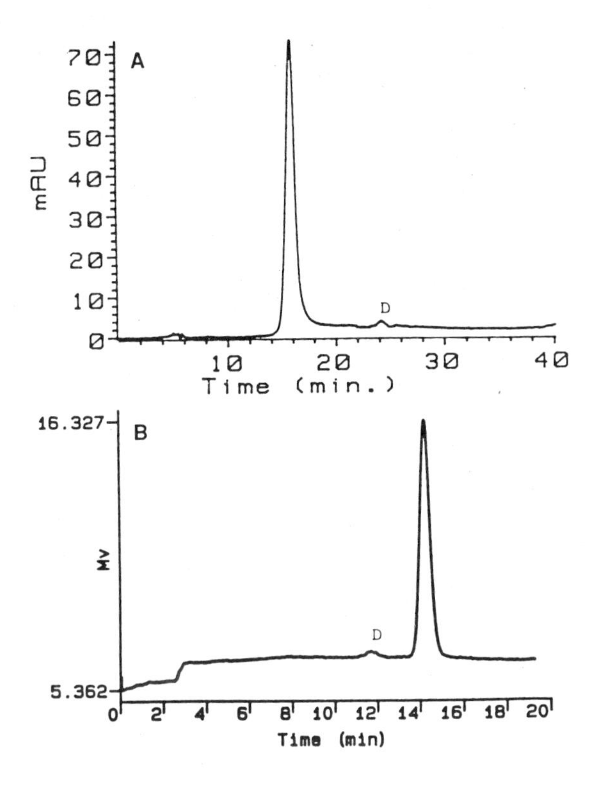

Fig. 3. The analysis of a sample of deamidated rhGH by ion exchange chromatography and by high-performance capillary electrophoresis (HPCE). (*A*) Ion exchange chromatography of rhGH stored in solution for 2 weeks. The separation was performed on a silica-based DEAE column with an acetate gradient. The more acidic peak, "D," corresponds to deamidated growth hormone, as determined by sequencing and mass spectrometry of tryptic fragments. (*B*) Capillary electrophoresis. The sample was loaded and run at 8 kV in a 20 cm coated capillary, with a pH 8.0 phosphate buffer. Detection was by absorbance at 200 nm at the anode end of the capillary, at 1 V/AU. Reproduced with permission from Frenz et al. (16).

(26) reviewed very large scale chromatography from the process development viewpoint.

In the last few years Horvath's laboratory has shown that displacement chromatography is a promising technique for large-scale protein purification. this is a nonlinear chromatographic technique that allows much higher sample loadings than is normally used in the elution mode, for example, a standard C_{18}-reversed phase column (0.4×15 cm) operated in the displacement mode can separate at least 20 mg of growth hormone vs 0.2 mg in the elution mode. In displacement chromatography the column is first equilibrated with the carrier and then loaded with the protein mixture so that the components are absorbed on the stationary phase. The proteins are recovered from the column by introduction of a displacer, which results in the mixture being separated during the displacement process as a result of competition for the binding sites at the stationary phase surface. For displacement to take place the concentrations of the displacer and the feed components must be in the nonlinear region of their respective adsorption isotherms (and hence concentrations are much higher than in traditional elution chromatography). The approach is best developed for ion exchange chromatography and a typical example was described by Liao (31) where β-lactoglobulins A and B were separated by anion exchange chromatography with chondroitin sulfate as the displacer. A new company is offering preparative systems based on displacement chromatography (J. Jacobson, Bio West Research, P.O. Box 135, Brisbane, CA 94005).

Three chapters concentrate on the analysis and purification of medically important proteins that represent different types of separation challenges. Insulin was the first product of recombinant-DNA technology and represents a small peptide-protein hormone (51 amino acid residues), whereas growth hormone at 192 amino acids is more representative of a typical nonglycosylated protein. At the other extreme is the analysis of vaccines that have traditionally consisted of complex mixtures of high-molecular-weight proteins. The chapter by Johnson describes suitable solvent systems for these mixtures and then examines the uses of different chromatographic modes. In a recent publication by Welling-Wester et al. (49) on the separation of the sendai virus integral membrane proteins, the highest yields were obtained by the use of ion exchange chromatography on either a Mono Q or a TSK DEAE-NPR column in the presence of a detergent (0.1% octylglucoside or decyl PEG-300). With the simpler insulin samples the effects of chromatographic parameters could be carefully examined and the authors showed that variables such as temperature and pH have a profound effect on the separation. Another important part of the use of RP-HPLC in biotechnology is the need to validate an analytical method for reproducibility. This aspect is described in Chapter 9 by Christiansen and Hansen on growth hormone separations.

Ion exchange chromatography has been a very popular method for the large-scale purification of proteins, since it has given efficient separations under mild conditions with retention of biological activity. These supports can typically be used early in the recovery process with crude extracts. Either a low cost, disposable material can be used, or with many of the newer matrices a harsh cleaning step can be used to allow regeneration of even a badly fouled column. Also, these columns can readily be run under sterile conditions without the need for high salt concentration or organic solvents and thus the equipment requirement is less than for HIC or RP-HPLC. Information on the large scale-use of ion exchangers can be found in Chapter 6 by Lowe. In Chapter 10 by Henry the design of silica-based ion exchange supports is described as well as operating conditions and equipment used in the separations. The silica supports provide a pressure stable medium based on small particles, which can give rise to high-efficiency separations up to moderately alkaline pH values.

The use of ion exchange supports for peptide analysis was significantly improved by the addition of a suitable organic solvent to the mobile phase. This modification has allowed for greater retention and more efficient separations of many peptides (see Chapter 11 by Dizdaroglu). The introduction of newer, more efficient ion exchange supports based on organic polymers (for example Tosohaas CM- and SP-5PW columns) can permit the use of higher pH solutions for column cleaning than can be used with silica based supports. These materials have been used at Genentech for the separation and characterization of degraded forms of proteins that arise in accelerated stability programs used for prediction of shelf life of protein therapeutics. A disadvantage of ion exchange versus reversed-phase chromatography in analytical separations has been the significantly longer times required for the reequilibration of an ion exchange column. However, the development of small-particle nonporous supports such as the Tosohaas material has allowed the development of rapid and high-efficiency ion exchange separations (see Fig. 3). Jilge et al. (24) have recently described the separation of proteins and nucleotides on 2 μm nonporous silica-based strong anion exchangers. While capillary electrophoresis is a more rapid technique, the advantage of ion exchange chromatography is that it can be readily scaled-up to semipreparative separations for characterization studies (see Fig. 3 for a comparison of the two methods). In general, RP-HPLC is less sensitive to charged variants than ion exchange chromatography, particularly at pH values that do not result in a charge difference for the variant. For example, deamidated growth hormone is not resolved by RP-HPLC with a TFA-CH_3CN mobile phase, but the pH of the mobile phase (approx. 1.9) is well below the pKa of the carboxyl group of aspartic acid (approx. 4.5), so that the native and deamidated molecule would have the same charge.

Although high-performance capilliary electrophoresis (HPCE) is not directly within the scope of this text, the exciting potential of this new analytical method is briefly described here with relevance to the needs of biotechnology. Four principal modes of HPCE have been described (17) as free-solution with either coated (5, 16) or uncoated (11) capillaries, SDS-PAGE in gel-filled capillaries (8), isoelectric focusing (22), and micellar systems (44). The principal applications of HPCE in biotechnology at this stage are in protein purity analysis and in tryptic mapping. HPCE is particularly well suited to monitoring deamidation of proteins either in the IEF or zone electrophoresis mode (see Fig. 4). The studies of Frenz et al. (16) and Grossman et al. (17) used rDNA-derived human growth hormone as an example. At this stage HPCE does not offer the degree of resolution and reproducibility of RP-HPLC for tryptic mapping, but the combination of the unique separation modes of HPCE (see Fig. 5) and further refinements in instrumentation will ensure that the new technique is widely used in biotechnology.

Hydrophobic interaction chromatography (HIC) is reviewed in Chapter 12 by Kato and Kitamura. In general, we have found that RP-HPLC gives greater resolution than does HIC in the analysis of purity of a small rDNA protein (less than 30 KD). However, HIC gives improved resolution over RP-HPLC for carbohydrate variants present in glycoproteins, as for example in the separation

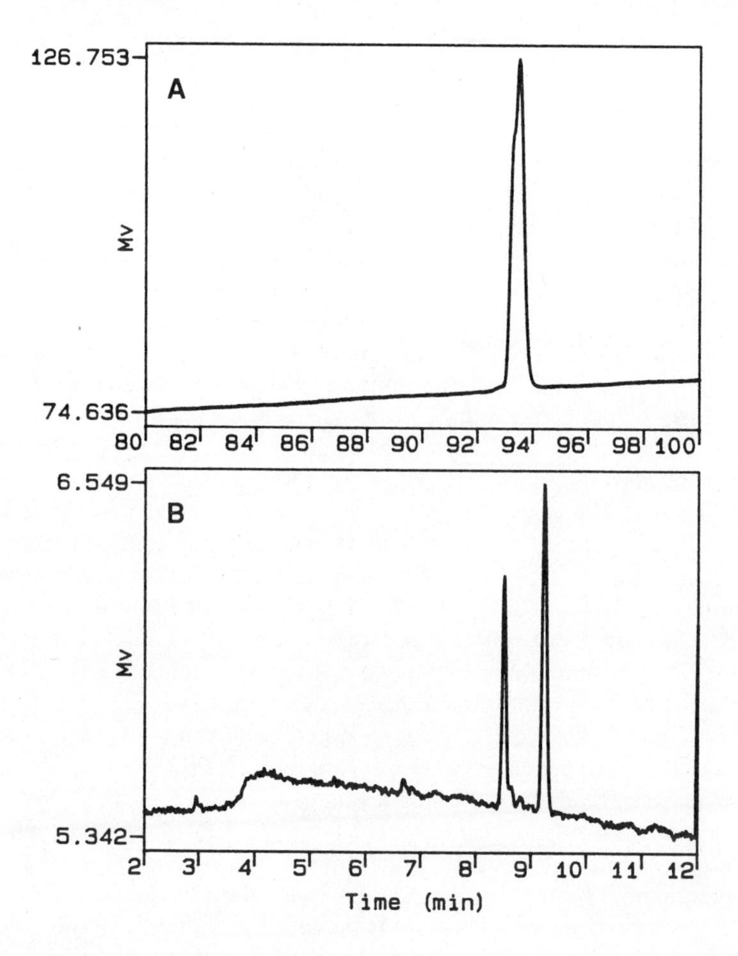

Fig. 4. (*A*) Isoelectric focusing in a 12 cm coated capillary of an rhGH sample stored 2 months at 5°C in a pH 9 buffer. Detection was at 280 nm. (*B*) Capillary electrophoresis at high pH of the aged rhGH sample shown in Fig. 3*A*. The sample was loaded and run at 8 kV in a 20 cm coated capillary, with a pH 8.0 phosphate buffer. Detection was by absorbance at 200 nm, 1 V/AU. The identitites of the peaks corresponding to deamidated rhGH variants were confirmed by electrophoresis of fractions obtained by anion exchange chromatography, and are indicated by "D" for monodeamidated rhGH and "DD" for the dideamidated species. Reproduced with permission from Frenz et al. (16).

of Type I and II forms of tissue plasminogen activator (which contain three or two carbohydrate chains, respectively). HIC may also give improved resolution of proteolytically degraded species, for example, Fig. 6 shows the separation of rhGH from a species lacking the N-terminal phenylalanine residue (51). This separation could not be achieved satisfactorily by RP-HPLC. The ability of HIC to isolate unstable proteins was demonstrated by Hyder and Wittliff (23) with

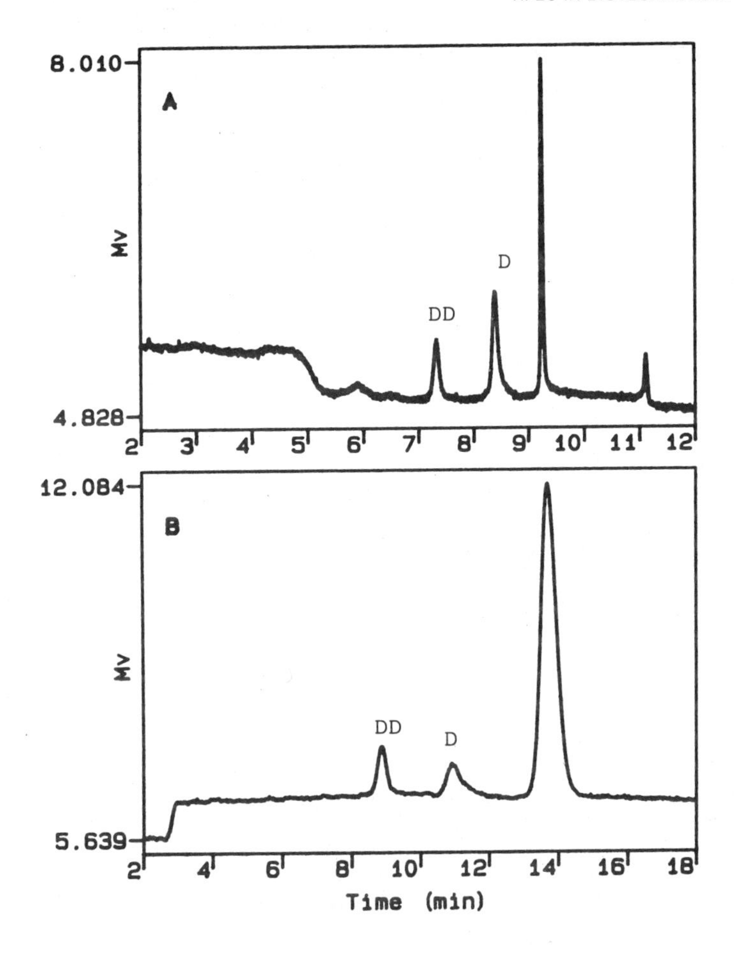

Fig. 5. The analysis of a mixture of two tryptic peptides by RP-HPLC (part A) and by HPCE (part B). The tryptic map separation was carried out using the conditions described in the legend to Fig. 2 and the HPCE separation as in Fig. 3.

the purification of the human estrogen receptor on a propyl-column. Also Alfthan and Stenman (1) described the purification of labeled antibodies on Phenyl-sepharose. In process development HIC is often preferred over RP-HPLC because of the less harsh conditions used in HIC, which can result in better retention of activity of the purified protein and in less complex equipment requirements.

Unlike the other chromatographic modes described in this volume, the application of affinity chromatography has remained primarily focused on preparative separations. Chapter 13 by Chaiken describes developments in "isocratic" affinity chromatography where the strength of interaction between the affinity support and the target protein has been modulated so that elution

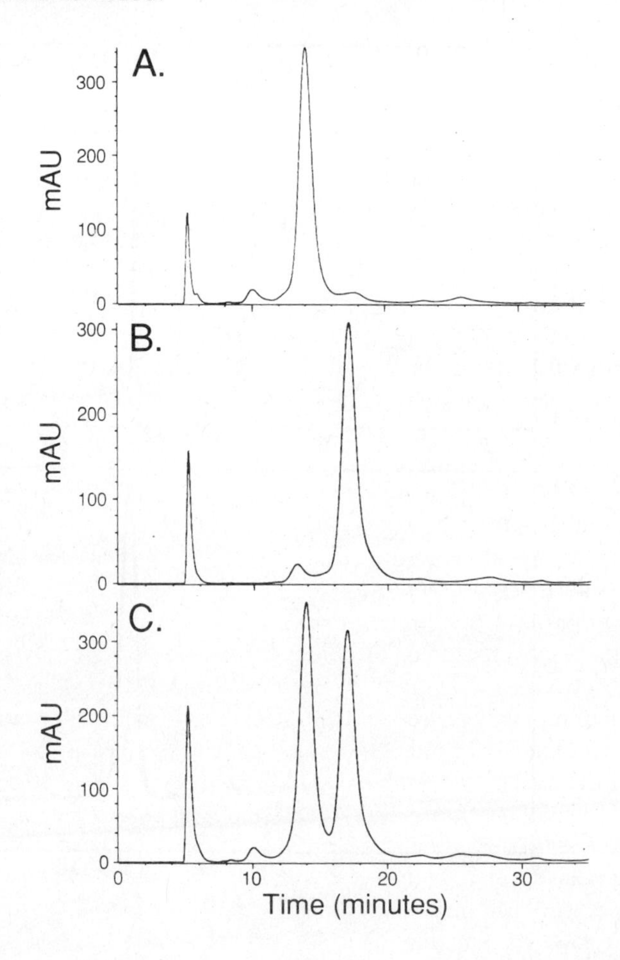

Fig. 6. The analysis of recombinant-methionyl human growth hormone and authentic sequence recombinant human growth hormone by HIC. The separation was carried out on a TSK-phenyl-5PW column at 30°C with 0.5 *M* sodium sulfate 30 m*M* Tris-HC1, 2% acetonitrile, pH 8.0. Elution with solution B [30 m*M* Tris-HC1, 5% (v/v) acetonitrile, 0.07% (v/v) Brij-35, pH 8.0] was carried out by means of the following gradient: a linear increase of 1.8% B/min for 10 min; followed by an isocratic hold for 4 min at 18% B; a linear increase of 5.125% B/min for 16 min; followed by an isocratic hold at 100% B for 5 min. Part A shows the elution profile obtained for a sample of met-hGH, part B for rhGH and part C for a 1:1 mixture.

of the bound protein and closely related variants can be achieved under mild conditions, which may allow separation of the variants. This new approach is therefore distinct from preparative separations where an on–off situation is used in the binding and elution steps. The use of affinity chromatography in protein manufacture has become very important with developments in the ability to generate a wide population of monoclonal antibodies against a given protein and improvements in the ability to scale-up production of the antibody. The major disadvantages stem from the expense of a large monoclonal column and from the time required to produce the antibody and thus develop the separation. The use of a monoclonal column in a recovery process may result in the final product being contaminated with antibody-derived impurities such as fragments or intact immunoglobulins, fetal calf serum, protein A or G, or perhaps animal retroviruses. Thus, it is important to use a stable mode of attachment of the antibody to the affinity support (see Chapter 13 by Chaiken) and to use purified monoclonal preparations (see Chapter 15 by Nau). Also Chapter 15 on antibody purification and analysis provides a wealth of detail on the practical aspects of antibody purification and on preparative separations. Another subset of this chromatographic mode is the use of IMAC (immobilized metal-ion affinity chromatography) and a publication by Maisano et al. (33) shows the potential of IMAC with the analysis of recombinant human growth hormone on either Cu^{2}, Ni^{2+}, Co^{2+}, or Zn^{2+}-iminodiacetate-agarose.

The use of mass spectrometry for characterization of polypeptides is included in the text because the ability to measure exact masses as well as in some cases fragmentation patterns has proven to be an invaluable tool for the analytical protein chemist. At Genentech we have used fast atom bombardment mass spectrometry (FAB-MS) to analyze protein variants arising from either the fermentation or recovery process or from stability programs. The most common approach is to digest the protein variant with trypsin, then measure the molecular ions of the mixture directly or of the individual tryptic peptides separated by RP-HPLC usually with the TFA-CH_3CN mobile phase (Fig. 7). This mobile phase has excellent volatility and gives a minimum of background problems in the resulting mass spectrum relative to other systems such as ammonium bicarbonate or acetate. Although we are exploring various LC/MS systems most of our studies involve the "manual interface" where each HPLC fraction is lyophilized and introduced directly onto the probe. Despite the labor involved, this approach has allowed the characterization of the sites of deamidation, proteolysis, and oxidation of rDNA products. For a polypeptide produced by chemical synthesis FAB-MS allows the characterization of deletion peptides, incomplete removal of blocking groups and side reactions such as cyclic imide formation (see Chapter 16 by Marshak and Fraser). The use of FAB-MS alone is not sufficient for the characterization of these side reactions and amino acid analysis and N-terminal sequencing are also an important part of the analytical protocol. For example, deamidation of asparagine to aspartic acid results in the gain of 1-AMU, which can be usually detected in a high-resolution mass spectrometer. However, it has been shown that deamidation occurs via a cyclic imide that hydrolyses to either an aspartic or isoaspartic acid residue (25). In this case neither amino acid analysis nor mass spectrometry was able to distinguish these isomers, while sequencing did distinguish between the two degraded forms as the sequence terminated at an isoaspartic residue. Chapter 16 by Marshak and Fraser gives an excellent description of the application of mass spectrometry to

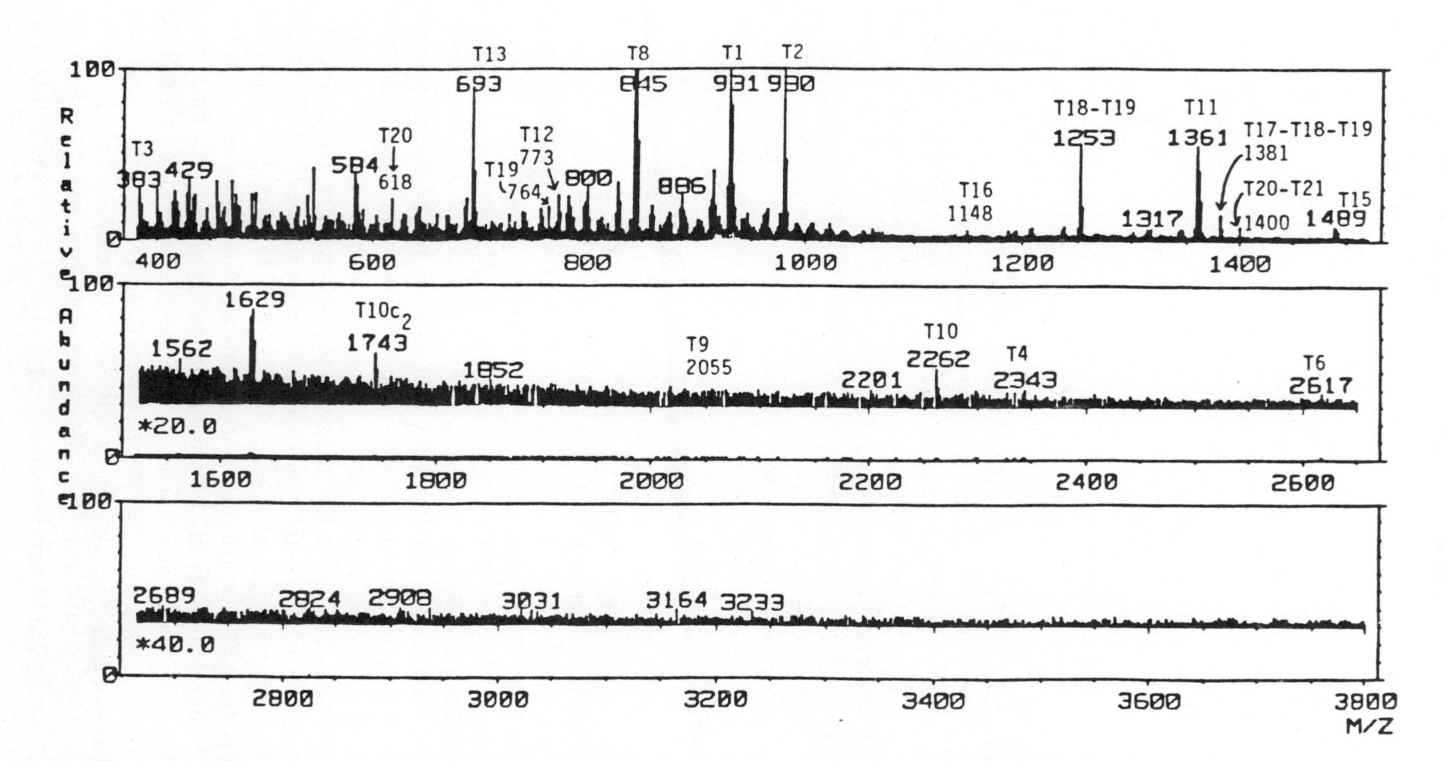

Fig. 7. FAB mass spectrum on the mixture of peptides produced by a tryptic digestion of rhGH. Spectra were obtained on either a JEOL HX100 or HX110. Fast atom bombardment was performed with 6 KeV xenon atoms using a sample matrix of either glycerolacetic acid or dithiothreitol: dithioerythritol (5:1). The table gives the result for the FAB mass spectra of the individual peptides collected after the reversed phases HPLC separations. Reproduced with permission of Canova-Davis et al. (7).

Mass spectral analysis of tryptic peptides of rhGH

rhGH peptides	Expected monoisotopic mass	Observed monoisotopic mass
T1	930.5	930.6
T2	979.5	979.6
T3	383.2	383.2
T4	2342.1	2343.4 [a]
T5	404.2	404.2
T6-T16	3761.7	3762.9 [a]
T7	762.4	762.3
T8	844.5	844.5
T9	2055.2	2056.4 [a]
T10	2262.1	2263.4 [a]
T11	1361.7	1361.8
T12	773.4	773.3
T13	693.4	693.4
T14	626.3	626.3
T15	1489.7	1489.9
T17	147.1	147.0
T18-T19	1253.6	1253.7
T20-T21	1400.6	1400.8

a Monocarbon-13 isotopic mass.

Fig. 7. *(Continued)*

the characterization of synthetic polypeptides of up to 50 residues. Recent developments have opened the possibility of mass spectrometry of intact proteins by the use of mild procedures for desorption of the macromolecule.

Electrospray ionization (ESI) is a soft ionization method capable of producing gaseous ions of ionic molecules from highly charged evaporating liquid droplets. The production of multiply charged ions of large molecules allows mass spectrometers with moderate mass/charge limit to analyze macromolecules with high molecular weights, such as lysozyme, myoglobin, and apotransferrin (15). Since capillary electrophoresis (CZE) achieves high separation efficiencies of charged species, the combination CZE/ESI-MS has allowed the analysis of protein samples of up to 80 KD (15). The development of continuous-flow FAB-MS as a liquid chromatography interface and ionization method has the advantages that ion suppression effects observed in standard FAB analysis seem to be reduced and increased signal-to-noise ratio gives better sensitivity for quantification (38). Another approach (46) uses plasma desorption via fission fragments from the decay of californium-252 to allow mass spectrometry of trypsin (23 KD), interferon α-2b (19 KD), interleukin-2 (15.5 KD), and a glycosylated domain of ovomucoid (7 KD). Also Stachowiak

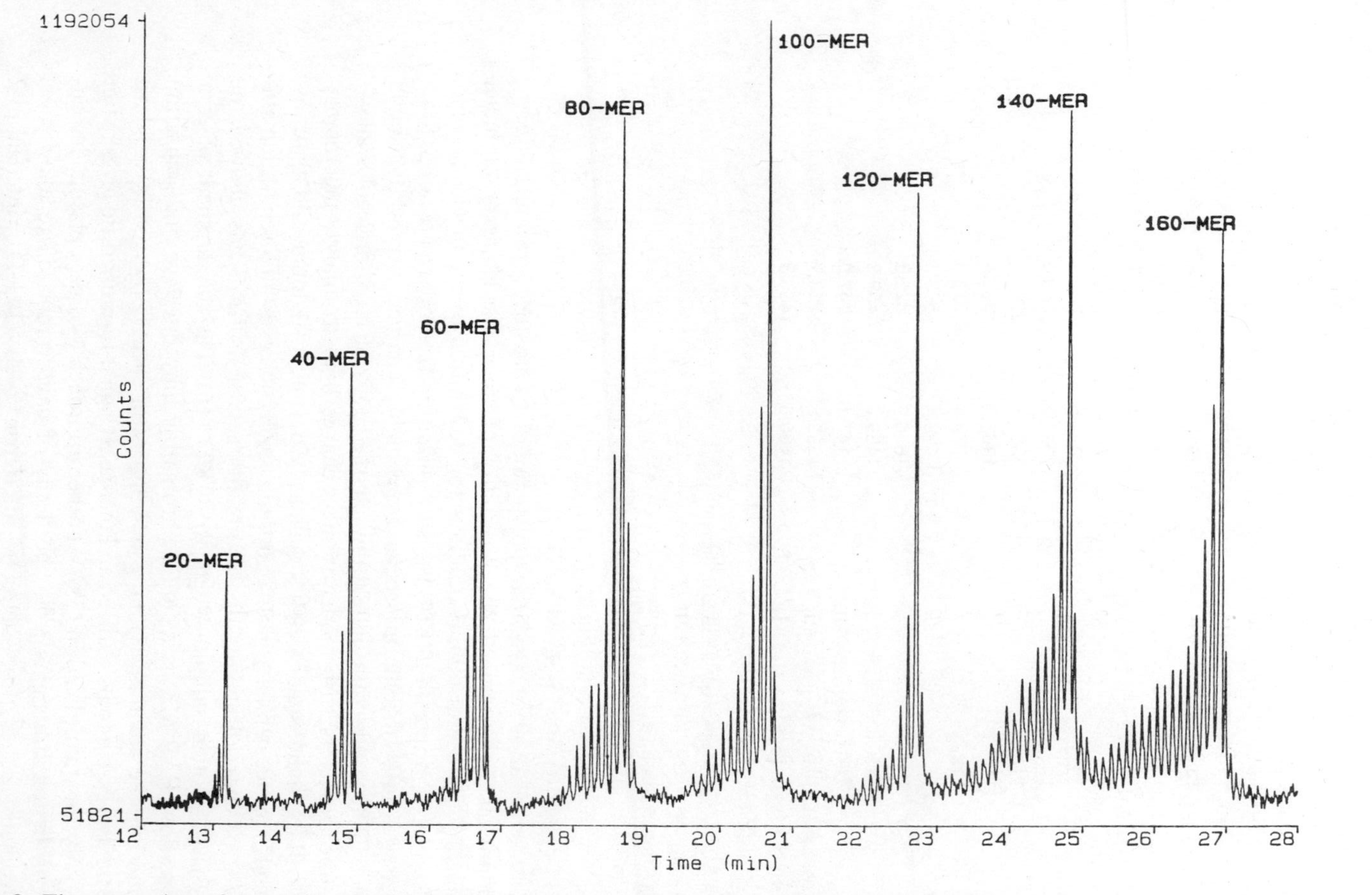

Fig. 8. The separation of a set of nucleotide standards by high-performance capillary electrophoresis. The samples were oligo-T's prepared by solid phase synthesis (N. Bischofberger, Genentech, Inc.) and the capillary electrophoresis was carried out by M. Field, Genentech, Inc. The sample was run at 400 V/cm in a 50 cm (100 micron diameter) capillary filled with polyacrylamide gel (Cohen et al., 1988). The buffer contained 90 m*M* Tris-Borate, pH 8.3 with 7 *M* urea. Detection was at 260 nm.

and Dyckes (43) used thermospray LC/MS and RP-HPLC peptide mapping for the rapid identification of hemoglobin variants.

Certain subtle differences in mass are not resolvable, for example, the observation of deamidation in rhGH would require a resolution of 1 in 20,000 or 0.0005%, which is well beyond current technologies. Also proteins that exhibit substantial microheterogeneity as, for example, a glycoprotein such as tissue plasminogen activator will not give a useful molecular ion. In the environment of the mass spectrometer, the polypeptide may undergo chemical modification such as reduction, so that a cystine peptide usually shows evidence of some of the reduced fragments. Complex samples such as mixtures of tryptic peptides will often give a mass spectrum lacking a molecular ion of some of the peptide components. For example, in the mass spectra of the trypsin digest of growth hormone, the molecular ions of the following peptides were not easily detected —T5, EQK, mass 402 D, T7, EETQQK, mass 762 D and T14, QTYSK, mass 626 D. However, each of the peptides when collected and analyzed separately after RP-HPLC gave a satisfactory molecular ion. In general, small, polar peptides are often lost in the spectrum and it recommended that FAB-MS is carried out on both the mixture of tryptic peptides and on the separated fractions to control for artifacts such as loss of individual peptides in the digestion, in the work-up step, on the HPLC column, or in the mass spectrometer.

The high-resolution separation of nucleotides has many potential applications in biotechnology ranging from plasmid mapping, quantitation of host cell DNA, analysis and purification of synthetic nucleotides to recent developments in antisense nucleotides and various analogs. The review chapter by Zon (Chapter 14) gives an excellent and comprehensive review of the use of RP-HPLC for the analysis and purification of synthetic nucleotides as well as many analogs. Recently, the use of RP-HPLC was extended to very long oligodeoxyribonucleotides (up to 143-mers) prepared by solid phase phosphoramidite chemistry (21). The separation was carried out on 5 μm ODS-Hypersil with a mobile phase containing 0.1 *M* triethylammonium acetate. Banks and Novotny (2) described the use of microcolumn HPLC for the analysis of enzymatic hydrolysis products from a 100 ng sample of rRNA-Phe. Capillary electrophoresis in both gel-filled columns and open tubes (9, 10) is capable of extremely high resolving power (see Fig. 8) and is proving to be useful in analyzing the purity of synthetic oligonucleotides. However, the low levels of DNA in the other applications such as plasmid mapping and analysis of DNA impurities in protein products will require the combination of capillary electrophoresis with laser detection to give the necessary level of sensitivity.

REFERENCES

1. A. Alfthan, and U-H. Stenman, J. Chromatogr., *470*, 385–389 (1989).
2. J.F. Banks, and M.W. Novotny, J. Chromatogr., *475*, 13–21 (1989).
3. K. Benedek, S. Dong, and B.L. Karger, J. Chromatogr., *317*, 227–243 (1984).

4. W.F. Bennett, R. Chloupek, R. Harris, E. Canova-Davis, R. Keck, J. Chakel, W.S. Hancock, P. Gellefors, and B. Pavlu, *Advances in Growth Hormone and Growth Factor Research*, Pythagora Press, Rome, 1989, pp. 29–50.
5. G.J.M. Bruin, J.P. Chang, R.H. Kuhlman, K. Zegers, J.C. Kraak, and H. Poppe, J. Chromatogr, *471*, 429–436 (1989).
6. S.E. Builder, and W.S. Hancock, Chemical Engineering Progress, *8*, 42–46 (1988).
7. E. Canova-Davis, R.C. Chloupek, I.P. Baldonado, J.E. Battersby, M.W. Spellman, L.J. Basa, B. O'Connor, R. Pearlman, C. Quan, J.A. Chakel, J. Stults, and W.S. Hancock, American Laboratory, *5* (1988).
8. A.S. Cohen, and B.L. Karger, J. Chromatogr, *397*, 409–417 (1987).
9. A.S. Cohen, D.R. Najarian, A. Paulus, A. Guttman, J.A. Smith, and B.L. Karger, Proc. Natl. Acad. Sci., *85*, 9660–9663 (1988).
10. A.S. Cohen, D.R. Najarian, J.A. Smith, and B.L. Karger, J. Chromatogr., *458*, 323–333 (1988).
11. A.S. Cohen, A. Paulus, and B.L. Karger, Anal. Chem. (to be published).
12. C.A. Dinarello, J. O'Connor, G. LoPreste, and R.L. Swift, J. Clin. Micro. *20*, 323–329 (1984).
13. M.W. Dong, R.J. Gant, and B.R. Larsen, Bio. Chromatogr., *4*, 19–34 (1989).
14. A.F. Drake, M.A. Fung, and C.F. Simpson, J. Chromatogr., *476*, 159–163 (1989).
15. C.G. Edmonds, J.A. Loo, C.J. Barinaga, H.R. Udseth, and R.D. Smith, J. Chromatogr., *474*, 21–37 (1989).
16. J. Frenz, S.-L. Wu, and W.S. Hancock, J. Chromatogr., *480*, 379–391 (1989).
17. P.D. Grossman, J.C. Colburn, H.H. Lauer, R.G. Nielsen, R.M. Riggin, G.S. Sittampalam, and E.C. Rickard, Anal. Chem., *61*, 1186–1194 (1989).
18. W.S. Hancock, *Handbook of HPLC for the separation of Amino Acids, Peptides and Proteins*, Vols. I and II, W.S. Hancock Ed., CRC Press, Boca Raton, FL, (1984).
19. W.S. Hancock, E. Canova-Davis, R.C. Chloupek, S.-L. Wu, I.P. Baldonado, J.E. Battersby, M.W. Spellman, L.J. Basa, and J.A. Chakel, in *Therapeutic Peptide and Proteins: Assessing the New Technologies*, Banbury Report 29, Cold Spring Harbor Laboratory, 1988, pp. 95–117.
20. D.H. Hawke, P.M. Yuan, K.J. Wilson, and M.W. Hunkapiller, Biochem. Biophys. Res. Commun., *145*, 1248–1253 (1987).
21. M. Hummel, H. Herbst, and H. Stein, J. Chromatogr., *477*, 420–426 (1989).
22. S. Hjerten, and M.D. Zhu, J. Chromatogr., *327*, 157–164 (1985).
23. S.M. Hyder, and J.L. Wittliff, J. Chromatogr., *476*, 455–466 (1989).
24. G. Jilge, K.K. Unger, U. Esser, H.-J. Schafer, G. Rathgeber, and W. Muller, J. Chromatogr., *476*, 37–48 (1989).
25. B.A. Johnson, J.M. Shirokawa, W.S. Hancock, M.W. Spellman, L.J. Basa, and D.W. Aswad, J. Biol. Chem., *264*, 14262–14271 (1989).
26. K. Jones, Chromatographia, *25*, 547–559 (1988).
27. K. Kalghatgi, and C. Horvath, J. Chromatogr., *398*, 335–339 (1987).
28. G.E. Katzenstein, S.A. Vrona, R.J. Wechsler, B.L. Steadman, R.V. Lewis, and C.R. Middaugh, Proc. Natl. Acad. Sci. USA, *83*, 4268–4272 (1986).
29. E.P. Kroeff, R.A. Owens, E.L. Campbell, R.D. Johnson, and H.I. Marks, J. Chromatogr., *461*, 45–61 (1989).

30. D.P. Lee, J. Chromatogr., *443*, 143–153 (1988).
31. A.W. Liao, Z. El Rassi, D.M. Le Master, and C. Horvath, Chromatographia, *24*, 881–885 (1987).
32. X.M. Lu, K. Benedek, and B.L. Karger, J. Chromatogr., *359*, 19–29 (1986).
33. F. Maisano, S.A. Testori, and G. Grandi, J. Chromatogr., *472*, 422–427 (1989).
34. R.E. Majors, LC. GC *7, 6*, 304–314 (1989); *7, 7*; 468–475 (1989).
35. W. Melander, D. Corradini, and C. Horvath, J. Chromatogr., *317*, 67–85 (1984).
36. C.T. Mant, and R.S. Hodges, J. Liquid Chromatogr., *12*, 139–172 (1989).
37. C.T. Mant, N.E. Zhou, and R.S. Hodges, J. Chromatogr., *476*, 363–375 (1989).
38. J.A. Page, M.T. Beer, and R. Lauber, J. Chromatogr., *474*, 51–58 (1989).
39. D.J. Poll, and D.R.K. Harding, J. Chromatogr., *469*, 231–239 (1989).
40. G.P. Rozing, and H.J. Goetz, J. Chromatogr., *476*, 3–19 (1989).
41. R.J. Simpson, L.D. Ward, G.E. Reid, M.P. Baterham, and R.L. Moritz, J. Chromatogr., *476*, 345–361 (1989).
42. H.W. Smith, L.M. Atkins, D.A. Binkley, W.G. Richardson, and D.J. Miner, J. Liquid Chromatogr., *8*, 419–439 (1985).
43. K. Stachowiak, and D.F. Dyckes, Peptide Research, *2*, 267–273 (1989).
44. S. Terabe, K. Otsuka, K. Ichikawa, A. Tsuchiya, and T. Ando, Anal. Chem., *56*, 113–116 (1984).
45. G. Thevenon, and F.E. Regnier, J. Chromatogr., *476*, 499–511 (1989).
46. A. Tsarbopoulos, Peptide Research, *2*, 258–266 (1989).
47. K.K. Unger, R. Janzen, and G. Jilge, Chromatographia, *24*, 144–154 (1987).
48. M. Verzele, C. Dewaele, and M. Weerdt, LC GC *6*, 966–974 (1989).
49. S. Welling-Wester, R.M. Haring, H. Laurens, C. Orvell, and G.W. Welling, J. Chromatogr., *476*, 477–485 (1989).
50. S.-L. Wu, K. Benedek, and B.L. Karger, J. Chromatogr., *359*, 3–17 (1986).
51. S.L. Wu, P. Pavlu, P. Gellerfors and W.S. Hancock, J. Chromatogr., *500*, 595–606 (1990).

CHAPTER 2

The Design of Bonded Phases for Polypeptide and Protein Chromatography Based on Wide-Pore Silicas

MICHAEL P. HENRY

Research and Development Laboratories, J.T. Baker, Inc., Phillipsburg, New Jersey

1. INTRODUCTION

Since 1980 a program of research has been carried out in our laboratories aimed at designing bonded phases for general polypeptide and protein chromatography. One result of this research is a family of chromatographic tools that is commercially available. The current status of this research is summarized in Table 1. The areas of general application of the developed bonded phases are listed in Table 2.

The approach we have taken in designing these chromatographic tools has evolved logically over the development period. The goal is to separate a polypeptide or protein of interest from a complex mixture of other components on any scale with maximum speed, recovery (mass and activity), simplicity, and economy. The following six precepts will describe the development of a family of packing materials:

1. A universal matrix was chosen for all bonded phases, in order to reduce to a minimum the variables in methodology that are required for protein and polypeptide chromatography.
2. The matrix selected was wide-pore silica owing to its numerous advantages and few limitations. The optimum pore size chosen for the widest molecular weight range, was 300 Å.
3. A minimum number of interactive surfaces was designed to solve the maximum number of purification problems. The corresponding bonded

Table 1. Current Status of J.T. Baker Research into the Design of Wide-Pore Silica-Based Bonded Phases for Polypeptide and Protein Chromatography (to End 1988)

Separation Mode	Surface Functional Group(s)	Average Particle Sizes (micron)
	A. Completed: Media for General Applications	
Reversed phase	C_{18}, C_8, C_4, diphenyl CN	5, 15, 40
Weak anion exchange	Polyethyleneimine	5, 15, 40
Weak cation exchange	Carboxyethyl	5, 15, 40
Hydrophobic interaction	Propyl	5, 15, 40
Strong cation exchange	Sulfo and Carboxyl	40

Separation Mode	Surface Functional Group(s)	Particle Sizes (micron)	Applications
	B. Completed: Media for Special Applications		
Weak Anion Exchange	Proprietary (MAb)	5, 15, 40	Monoclonal antibodies
Mixed Mode Exchange	Proprietary (ABx)	5, 15, 40	All antibodies
Affinity	Several (see Table 14)	30	Several (see Table 14)

Separation Mode	Surface Functional Group(s)	Particle Sizes (micron)
	C. In Final Development	
Strong anion exchange	Quaternary ammonium (N^+)	5, 15, 40
Strong cation exchange	Sulfo and carboxyl	5, 15
Affinity	Glutaraldehyde and hydrophilic polymer spacer	5, 15, 40

Separation Mode	Surface Functional Group(s)	Particle Sizes (micron)
	D. Experimental	
Size exclusion	Hydrophilic polymer	5, 15, 40

phases are thus:

a. Weak and strong anion and cation exchangers
b. Hydrophobic interactors, and
c. Reversed-phase packings

Table 2. Areas of Application of Bonded Phases for Polypeptide and Protein Chromatography

Property or Configuration	Appropriate Bonded Phase					
	C_4	PEI	CBX	HI-Propyl	ABx	MAb
If solute labile	No	Yes	Yes	Yes	Yes	Yes
If solute non-labile	Yes	Yes	Yes	Yes	Yes	Yes
Solute organic solvent stable	Yes	NA	NA	NA	NA	NA
Proteins, MW 10 k–1000 k	Yes	Yes	Yes	Yes	Yes	Yes
Analysis (5 micron)	Yes	Yes	Yes	Yes	Yes	Yes
Prep scale (15, 40 micron)	Yes	Yes	Yes	Yes	Yes	Yes
5 cm analysis	Yes	Yes	Yes	Yes	No	No
25 cm analysis	Yes	Yes	Yes	Yes	No	Yes
10 cm biocompatible (GOLD)	Yes	Yes	Yes	Yes	Yes	Yes
High activity recovery	Variable	Yes	Yes	Yes	Yes	Yes
High mass recovery	Variable	Yes	Yes	Yes	Yes	Yes
pI greater than 6	Yes	No	Yes	Yes	NA	NA
pI less than 7.5	Yes	Yes	No	Yes	NA	NA
pI 6–7.5	Yes	Yes	Yes	Yes	NA	NA
Prep columns (prepacked)	Yes	Yes	Yes	Yes	Yes	Yes
Prep columns (low pressure)	Yes	Yes	Yes	Yes	Yes	Yes
Bulk packings (5, 40 micron)	Yes	Yes	Yes	Yes	Yes	Yes
Minicolumns for sample preparation	Yes	Yes	Yes	Yes	Yes	Yes

4. Each bonded phase was prepared on three different particle sizes: 5, 15, and 40 micron diameter. The 5-micron particles were selected for the high efficiency they afforded and, therefore, high resolution. The 15-micron bonded phases are ideal for the medium efficiency useful in larger-scale purifications, with prepacked columns. The 40-micron packings are economical, offer good resolution, and can be easily packed by the user for small- (grams) to very-large- (kilograms) scale purification.
5. The surface chemistry was designed to provide very similar chromatography regardless of particle size. This approach simplifies the process of

scaling up from micrograms (on 5-micron columns) to kilograms (on 40-micron columns). It also means that method development and in-process monitoring of large purifications can be carried out using surface-matched analytical columns.

6. Several configurations of these bonded phases have evolved to serve a variety of purposes:
   a. Short (5-cm, 5-micron) columns for fast (5-min) analyses. These are also suitable for work requiring high sensitivity and high resolution, miniprep purifications (up to 100 mg), and guard columns.
   b. Medium length (10-cm, 5-micron) columns that operate at low (200–250 psi) pressures.
   c. Standard (4.6 mm × 250 mm, 5-micron) columns for routine analysis with maximum resolution and enhanced capacity.
   d. Larger (10 mm × 250 mm, 15-micron) semipreparative columns for up to gram quantities of protein or polypeptide.
   e. Bulk 40-micron packings for low-pressure user-packed columns or selective batch adsorption applications.

This approach has therefore resulted in the creation of an integrated system for protein and polypeptide chromatography on any scale. Although we have also prepared a number of silica-based affinity matrices for more specific applications, it is the aim of this chapter to describe in some detail, and from an industrial perspective, the rationale for the design of bonded phases for general protein and polypeptide chromatography.

2. CHOICE OF SUBSTRATE

2.1. Porous Silica

In the late 1970s the predominant separation media for proteins were the soft gels. Table 3 lists some of the major soft gel manufacturers and brand names. Soft gels had certain advantages and limitations as chromatographic packings, which are listed in Table 4. The ease of deformation of these materials, a major limitation, prompted the development of several new separation media (1–3) that were more rigid. Table 5 lists some of the more important of these packing materials with their major properties. At about the same time, porous silica was being modified to expand its applicability to large polypeptides (4, 5). After a critical assessment of the advantages and limitations of these separation media, we concluded that porous silica has the greatest potential as a substrate for protein chromatography. Table 6 lists the major advantages and limitations of silica.

Table 3. Soft Gel Manufacturers/Marketers and Trade Names

Manufacturer	Trade Names
Pharmacia	Sephadex®, Sephacryl®, Sepharose®, Sephacel®
LKB	Trisacryl®, Ultrogel®, HA-Ultrogel®, Magnogel®
Whatman	DE, CM Cellulose
Toyo Soda	Toyopearl[a]
Bio Rad	Bio-Rex®, AG®, Cellex™, Bio-Gel®, Affi-Gel®
Cuno, Inc.	Zeta Prep
Serva	Servacel®
Amicon	Matrex™ Gel, Cellufine
Phoenix Chemicals	Indion®

[a]Also marketed by E. Merck Industries, Inc., under the trademark Fractogel®.

2.2. Advantages of Silica

There is very general agreement with the advantages listed in Table 6 (6). Those that were of most interest to us were (a) the advanced state of the technology of manufacture of porous silica, (b) the rigidity of silica, (c) the control over its pore size, (d) the high surface area of silica, and (e) the remarkable surface reactivity of this material.

2.2.1. Silica Manufacturing Technology

The first choice that had to be made was whether to develop an entirely new synthetic substrate or to use an existing one. The development of a new polymer, while within the technical capabilities of our laboratories, would have been time-consuming and expensive. Silica was readily available commercially in particle sizes and economies that matched our requirements for a family of bonded phases. Most of the natural and synthetic polymer-based materials are not easily manufactured in the broad particle size range (5–60 micron) desirable. Most important, however, the manufacture and surface modification of porous silica was and is a highly developed technology that has provided the major

Table 4. Advantages and Limitations of Soft Gels

Advantages	Limitations
pH Stability	Low resolution
(2–12 in general)	Long run times
	Slow method development
	Low flow rates
	Poor mechanical stability
	Slow equilibration, regeneration
	High shrinkage, swelling
	Poor biological stability

Table 5. Semirigid[a] Packings for Polypeptide and Protein Chromatography

Manufacturer	Pharmacia		Toyo Soda	Interaction Chemicals	BioRad	Lachema	Polymer Laboratories
Trade name	Mono Beads	Superose	TSK gel PW	Interaction	High Performance hydroxylapatite	Separon	PL
Nature of substrate	—	Cross-linked agarose	Hydroxylated polyether	Divinylbenzene cross-linked polystyrene	Hydroxylapatite	(2-Hydroxy-ethylpoly-ethyleneacrylate)	Polystyrene/divinylbenzene
Particle sizes (micron)	10	10, 13, 30	10, 13, 17, 25	6	—	10, 15	8
Pressure limits (psi)	1500	450	1000	4500	250	—	1500
Pore size (Å)	1000 (est)	1000 (est)	Various	—	—	—	300
Protein capacity mg/mL)	25	—	25	—	25	—	—
pH stability	2–12	2–12	2–12	1–14	5.5–10.5	2–12	1–13

[a]Semirigid packings are defined as those that swell or shrink or have pressure or flow rate limits or both below 1500 psi or 2 mL/min.

Table 6. Advantages and Limitations of Porous Silica for Polypeptide and Protein Chromatography

Advantages	Limitations
Highly resolving	pH stability (2–8.5)[a]
High mechanical strength	Nonspecific binding[b]
High linear velocities are possible	Heterogeneous reactivity[b]
Almost unlimited potential for surface derivatization	
Controlled pore size and its distribution	
Controlled particle size and its distribution	
Controlled shape, surface area	
High capacity for binding	
No shrinking or swelling	
Compatible with a large range of solvents, buffers, and chaotropic agents	
Readily available in wide range of particle sizes	

[a] Limits extended with appropriate surface modification.
[b] Readily overcome with appropriate chemistry of modification (see Table 12).

impetus for the phenomenal development of HPLC (7). This in turn has led to the current state of excellence in the high-performance liquid chromatography of proteins and polypeptides (8, 9).

2.2.2. *Rigidity*

Rigidity has many benefits for chromatography, the major one being the ability to utilize media of small (3–5 micron) particle size with the resulting high efficiencies and resolution. Efficiency is a measure of the tendency of the thin layer of applied sample components to remain thin as the layer moves through a packed column. Expressed as theoretical plates per meter, efficiency is an inverse function of the particle diameter (10).

Column back-pressure, however, is approximately inversely proportional to particle diameter squared (11). Only rigid materials such as silica can withstand the high pressures involved with the use of efficient columns of 5- or 10-micron particles.

The great structural integrity conferred upon silica by its rigidity means that pressure increases linearly with flow rate. Soft and semirigid materials rapidly compress with increased mobile phase velocity and pressure increases exponentially and usually irreversibly. Figure 1 illustrates these phenomena for a 40-micron silica-based bonded phase and two soft gels.

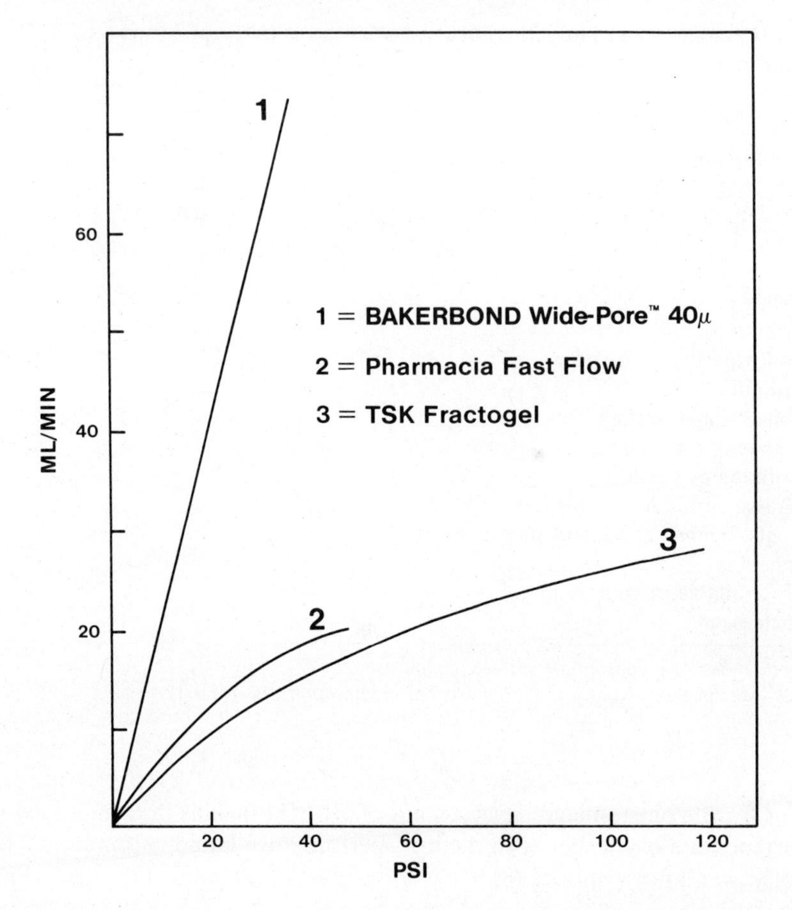

Fig. 1. Increase in pressure with flow rate for bulk media. Conditions: A. (See Appendix 1).

2.2.3. Pore Size

Another advantage of interest was the well-controlled pore size and pore-size distributions of silica. The rigidity of silica means essentially constant pore diameters and shapes during chromatography, since there is no swelling or shrinking during the wetting process or during changes in pH, ionic strength, or mobile phase polarity. Preliminary evidence was available in the late 1970s and early 1980s (4, 12, 13) which indicated that the presence of large pores (300 Å) in silica-based bonded phases correlated with substantial improvements in the resolution, capacity, and recovery of proteins.

For purely steric reasons, large pores are necessary to allow access of high-molecular-weight solutes to the maximum proportion of the interior surface area of the packing material. This access reduces the contribution of steric exclusion to the overall mechanism of separation. Furthermore, the easy access of large molecules to the wide pores improves diffusion rates and reduces band spreading.

Once inside the pore, diffusion rates are also determined by the interior geometry of the pore (14). The presence of mini- and microporous (100-Å and less than 25-Å diameter, respectively) regions may hinder diffusion and contribute to band spreading. The proportion of these smaller pores (less than 100 Å) is less in wide-pore packing materials (15). Therefore, large-pore silicas, owing to the reduced size-exclusion mechanism and the enhanced molecular diffusion rates, give increased resolution compared to small-pore silica-bonded phases.

Finally, the increased availability of the interior surface for binding of large polypeptides and proteins results in increased capacity relative to small-pore bonded phases (4, 16).

The value of 300 Å was selected for the average pore size of all the silicas used in our bonded phases, partly because of the observations described previously. We have subsequently found that this pore size is optimum for a broad range of molecular weights (up to 1 million) and general protein shapes (17).

2.2.4. High Surface Area

Modern large-pore (300 Å) silicas for chromatography have surface areas from 30 to 300 m^2/g. This variation, for a given average pore size and pore-size distribution, must be due to a different number of pores per particle. The larger the surface area, the greater the number of pores per particle and the thinner are the walls between pores. These walls maintain their integrity under pressure due to the strength of silica (see Table 12 for pressure limits).

The high surface areas of porous silicas provide the potential for good polypeptide- and protein-binding capacities. The combination of high ligand density (see Section 4.3) and high surface area allows this potential to be realized. Table 7 gives the saturation binding capacities of several bonded phases for selected proteins. These capacity values are generally four to five times larger than those for the semirigid synthetic polymer-based bonded phases (18, 19).

The importance of high binding capacities lies in the overall improvement in methodology that can be achieved. Savings are possible in spatial requirements of packed columns, buffer volumes required, analysis and purification times, and overall throughput, especially in large-scale purifications.

2.2.5. Functional Group Modification

The great potential for functional group modification of the surface of silica was a crucially important factor in the decision to choose this substrate. The chemistry involved in these modifications is the cornerstone of the resulting bonded phases.

Essentially all of the surface chemistry of silica is carried out by an initial

Table 7. Saturation Binding Capacities of Several Bakerbond Bonded Phases for Selected Proteins

Bonded Phase	Particle Size (micron)	Standard Protein	Sample Solutions	Capacity (mg/dry g)
Bakerbond HI-Propyl	5	Bovine serum albumin	2 *M* $(NH_4)_2SO_4$, pH 7.0	130
	15	Bovine serum albumin	2 *M* $(NH_4)_2SO_4$, pH 7.0	180
Bakerbond CBX	5	Lysozyme	25 m*M* KH_2PO_4, pH 6.0	200
	15	Lysozyme	25 m*M* KH_2PO_4, pH 6.0	300
Bakerbond C_4	5	Bovine serum albumin	25 m*M* KH_2PO_4, pH 7.0	80
	15	Bovine serum albumin	25 m*M* KH_2PO_4, pH 7.0	140
Bakerbond PEI	5	Bovine serum albumin	25 m*M* Tris, pH 7.0	170
	15	Bovine serum albumin	25 m*M* Tris, pH 7.0	195
Bakerbond ABx	5	Human immunoglobulin	25 m*M* MES, pH 5.6	120
	15	Human immunoglobulin	25 m*M* MES, pH 5.6	150
Bakerbond MAb	5	Human immunoglobulin	25 m*M* KH_2PO_4, pH 6.8	120
	15	Human immunoglobulin	25 m*M* KH_2PO_4, pH 6.8	150

substitution reaction at the surface silanol groups. Several such reactions are possible and some of these are shown in equations (1) to (4).

$$\equiv SiOH + SOCl_2 \rightarrow \equiv SiCl \xrightarrow{RLi} \equiv SiR + LiCl \tag{1}$$

$$\xrightarrow{RNH_2} \equiv SiNHR + HCl \tag{2}$$

$$\equiv SiOH + ROH \rightarrow \equiv SiOR + H_2O \tag{3}$$

$$\equiv SiOH + RSi(CH_3)_2Cl \rightarrow \equiv SiOSi(CH_3)_2R \tag{4}$$

The most common group of reagents used to modify the surface of porous silica is the reaction of silanes (reaction 4). Several classes of silylating reagents and their reactions are shown in Eqs. 5 and 6, and 7:

$$\equiv SiOH + R\ Si(CH_3)_2Cl \rightarrow \equiv SiOSi(CH_3)_2R + HCl \tag{5}$$

$$\equiv SiOH + R\ Si(CH_3)Cl_2 \xrightarrow{H_2O} \equiv SiOSi(CH_3)\ (OH)R + 2HCl \tag{6}$$

$$\equiv SiOH + R\ Si\ Cl_3 \xrightarrow{2H_2O} \equiv SiOSi(OH)_2R + 3HCl \tag{7}$$

When water is present during the reaction, in nonanhydrous solvents, either adsorbed to the porous silica, or deliberately added to the reaction mixture, the dichloro and trichloro silanes can undergo polymerization. This may occur outside the pore, inside the pore, or at the already covalently bound alkyldihydroxysilyl group (reaction 7) (20). One example of such a polymerization reaction is

$$\equiv SiOSi(OH)_2R + 2RSiCl_3 \xrightarrow{4H_2O} \equiv SiOSiR(OH)(-O-Si(OH)(R)-O-Si(OH)R) + 6HCl \tag{8}$$

```
                                      OH
                                      |
≡SiOSi(OH)2R + 2RSiCl3 --4H2O--> ≡SiOSiR + 6HCl
                                      |
                                      O
                                      |
                                   HO–Si–R
                                      |
                                      O
                                      |
                                   HO–Si(OH)R
```

The unreacted silanols in the polymer are then either covalently linked to the silica surface forming multiple points of attachment (21) or reacted with end-capping reagents, as shown in Eq. 9

$$
\begin{array}{lll}
\begin{array}{l}
\quad\;\; OH \\
\geqslant SiO{-}SiR \\
\qquad\; O \\
\quad HOSiR \\
\geqslant SiOH \;\; O \\
\quad HOSiR \\
\qquad\; OH \\
\\
\geqslant SiOH
\end{array}
\xrightarrow{\Delta}
\begin{array}{l}
\quad\;\; OH \\
\geqslant SiOSiR \\
\qquad\; O \\
\geqslant SiOSiR \\
\qquad\; O \\
\geqslant SiOSiR \\
\qquad\; OH
\end{array}
\xrightarrow{2(CH_3)_3SiCl}
\begin{array}{l}
\quad\;\; OSi(CH_3)_3 \\
\geqslant SiOSiR \\
\qquad\; O \\
\geqslant SiOSiR \\
\qquad\; O \\
\geqslant SiOSiR \\
\quad\;\; OSi(CH_3)_3
\end{array}
+ 2HCl
\end{array}
\tag{9}
$$

A wide range of commercial silanes are available (22), many including other functional groups capable of undergoing further reaction. Two examples are

$$
\geqslant SiO\,\underset{CH_3}{\overset{CH_3}{Si}}\,CH_2CH_2CH_2NH_2 + \text{Excess } CH_3I \rightarrow
$$

$$
\geqslant SiOSi\,CH_2CH_2CH_2\overset{\oplus}{N}(CH_3)_3I^{\ominus} + 2HI
\tag{10}
$$

$$
\geqslant SiO\,\underset{CH_3}{\overset{CH_3}{Si}}\,CH_2CH_2CH_2NH_2 + \text{(3,5-}(NO_2)_2C_6H_3\text{)}{-}CONHCH(C_6H_5)COOH \longrightarrow
$$

$$
\geqslant SiO\,\underset{CH_3}{\overset{CH_3}{Si}}\,CH_2CH_2CH_2NH\overset{O}{\overset{\|}{C}}OCH(C_6H_5)\,NHCO{-}\text{(3,5-}(NO_2)_2C_6H_3\text{)}
\tag{11}
$$

Thus, modification of silica can produce an almost unlimited variety of interactive surfaces containing groups as diverse as the extremely low polarity perfluoroalkyls, to those of very high polarity such as the fully charged sulfonic acid group.

This versatility has important consequences for the bonded phases used in polypeptide and protein chromatography. Proper control over both the surface properties of the bonded phase and the nature of the mobile phase, allows the chromatographer to select a given separation mechanism for his or her needs, which include size exclusion, anion and cation exchange, hydrophobic interaction, and reversed-phase and specific affinity binding modes.

Although the reactions described previously are subject to the predictable laws of organic chemistry, the chromatographic properties of the resulting bonded phases depend in part on the physico/chemical nature of the porous silica (23). These properties include the pore size and its distribution (16), particle size and its distribution (6), the distribution or density of surface silanols (24), and the surface concentration of water, metals, and residual organic substances (16). The last two sets of properties can exert a major influence on the density and nature of surface functional groups. The variations in surface silanol density and the presence of other materials in the surface is a major cause of bonded-phase nonreproducibility.

2.3. Pressure and Shear Effects on Proteins

HPLC of proteins and polypeptides is generally carried out under pressures varying from 100 to 3000 psi. Furthermore, relatively high linear velocities are sometimes involved, from 0.5 to 2.5 mm/sec within the column and 120 to 600 mm/sec within the sample loop and the tubing leading to the flow cell (assumed to be 0.3 mm I.D.). Shearing forces may be exerted on proteins and polypeptides where velocity gradients occur, such as at the interface between the mobile phase and the retaining frit, packing material, and walls of the column.

The effects of pressure and shearing forces on polymers, including proteins, have been studied extensively (25, 26). Pressure can only influence the properties of protein and polypeptide solutions if there is a net volume change that accompanies potential phenomena such as aggregation, denaturation, or change in the state of ionization. Johnson et al. (27) have described a pressure threshold that exists in aqueous solutions. Below about 1000 atm, pressure increases tend to stabilize the secondary and tertiary structures of proteins, whereas above this pressure proteins denature. This reversal in effect is partly due to the change in the structure of water itself that becomes significant between 1000 and 3000 atm. Thus, since HPLC is always conducted at pressures well below this threshold, pressure increases generally reduce the tendency for proteins to denature and aggregate, since both of these phenomena involve volume increases (28). On the other hand, charge formation that accompanies protonation of an amine or deprotonation of a carboxylic or phosphoric acid is accompanied by a volume decrease (29). Therefore, pressure increases tend to decrease the extent of ionization, altering conductivity, pH, and the net charge on a protein or polypeptide.

The magnitude of the pressure effect will depend on the magnitude of the pressure change, the duration of the pressure, and the nature (especially molecular weight) of the protein or proteins involved. Most evidence indicates, however, that at pressures less than 1000 atm the effects are reversible within minutes (25).

It is interesting to speculate on the possibility that proteins may be subjected to pressure jumps of up to several thousand atmospheres per second under standard HPLC conditions as they begin to enter the column inlet. Generally, they may then be held there at up to 200 atm pressure until they begin to move along the column, where the pressure decreases linearly to 1 atm at the column outlet. Changes in the extents of aggregation, denaturation, and ionization that occur during pressure increases will be reversed (although not necessarily completely) as the pressure decreases.

Although many studies have been made of the effects of shear on the properties of proteins and other biomacromolecules (26), there is no evidence yet in the literature that such a phenomenon results in any measurable permanent effects on proteins or polypeptides under HPLC conditions. Activity recoveries of sensitive enzymes, for example, after appropriate chromatography, are very high or quantitative (see Table 12).

2.4. Limitations of Silica

Before the integrated family of bonded phases could be developed, the perceived and real limitations of silica needed to be assessed and overcome, if possible.

2.4.1. Stability at High pH

The property of silica of most concern was its solubility at extremes of pH. This is not so much because of a need to do protein chromatography at extremes of pH, but that a traditional method of protein chromatographic media cleanup involves the use of sodium hydroxide (6). The usual operating range of pH for silica-based columns is normally given as 2–7.5 (6). This is the pH range over which the solubility of amorphous silica in water is understood to remain at a minimum and largely constant at 100 ppm at 20°C. The data from Alexander et al. (30), Cherkinskii and Kynaz'kova (31), and Baumann (32), however, indicate general agreement that a solubility minimum (100 ppm) exists at pH 8, slightly higher than the commonly accepted value.

The solubility characteristics of chemically modified silicas, on the other hand, will be different from those of unbonded silica. Alpert et al. (4, 33), for example, have shown that polymer-coated silica has three times the lifetime of the substrate itself. Furthermore, this column could be operated in the pH range 2–9.2 with no deterioration in 400 hr. A series of experiments in our laboratories was designed to test the stability of polymeric octadecyl-bonded silicas in the presence of pH 10 solutions. The bonded phases were suspended in water and water/methanol mixtures at several temperatures for several periods of time. Elemental analyses before and after these treatments showed only minor loss of bonded phase ligand for the polymer phases, as measured by %C, for example. The results of these experiments are given in Table 8. Clearly, the correct choice of the bonding chemistry used in preparing bonded phases is of paramount

Table 8. Bonded-Phase Ligand Stability at Various pH Values, with and without Methanol

Bonded phase: *n*-butyl on 40-micron, 60-Å irregular silica
Primary reaction: silylation with trichlorosilane
End-capping: trimethylsilyl group
Initial analysis: 8.52% C, 2.05% H
Treatment: with various solution, 50°C for 125 hr, 1 g bonded phase in 15 mL solution

Analysis[a] after Treatment:

	Aqueous Solutions		
	1% H_3PO_4 (pH 3)	HCl (pH 3)	pH 10
%C	8.36	8.28	8.32
%H	2.03	2.02	2.05
Loss (%C)	0.18	0.25	0.21
	15 mL Aqueous + 3 mL MeOH		
%C	7.97	7.90	8.35
Loss (%C)	0.56	0.64	0.18

[a]Elemental analyses are the average of duplicate measurements.

importance in improving the resistance to hydrolysis at high pH. Further discussion of this topic and its application to overall bonded phase stability is given in Section 4.

Other methods of reducing the solubility of silica include the deposition of aluminum (34) and zirconium (35) at the silica surface. Preliminary results of the research by Harrison et al. (36), and the statements by Serva Heidelberg (37), indicate that such specially treated silica-based bonded phases may be used at pH 10. However, in the last two cases, bonded-phase lifetime at pH 10 was not demonstrated with published experiments.

Regardless of pH, porous silica, whether modified or not, has a finite solubility in aqueous solvents. The extent and rate of dissolution can be substantially reduced by placing between the pump and the injector a column packed with bare silica, to saturate the mobile phase with silicic acid before it reaches the primary column. Atwood et al. (38) report on the results of such a procedure with a bare silica column using the very harsh conditions of pH 10.7 and a temperature of 60°C. The presence of the silica saturator column increased column lifetime from 50 analyses at pH 7.1 to at least 400 analyses at pH 10.7.

Thus, improvements in methods of synthesis of bonded phases and precautions taken such as the use of silica-saturator columns have extended the useful operating pH range to 2–10 in many cases.

The use of high-pH solutions for limited periods during column cleaning is a separate matter. Preliminary evidence from our laboratories indicates that it

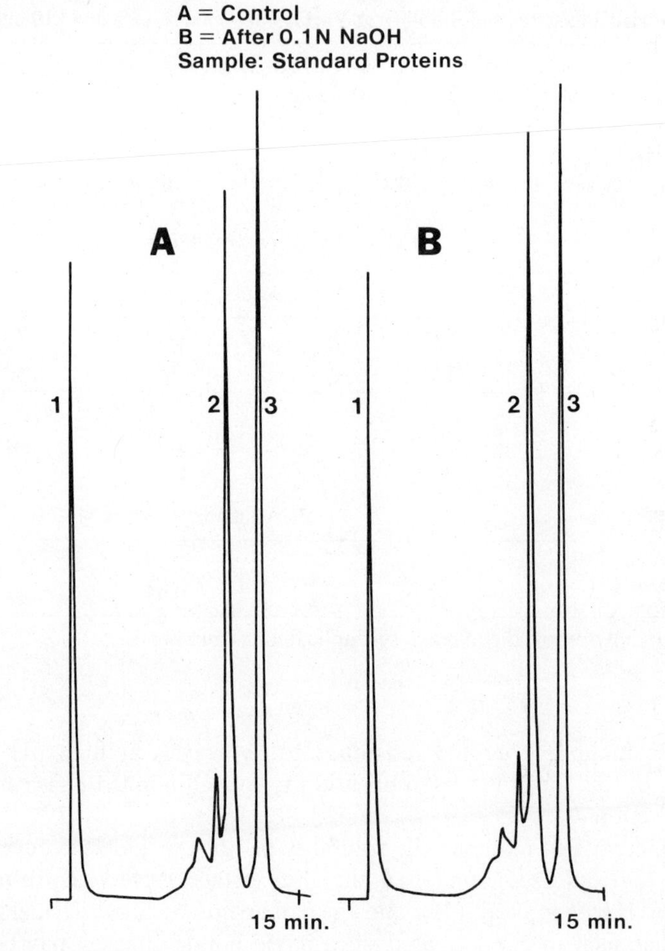

Fig. 2. Chromatographic effect of a 30 min exposure of a Bakerbond PEI HPLC column to flowing (1 ml/min) sodium hydroxide solution (0.1 *N*, pH = 13). Conditions: B. (see Appendix 1).

may be quite feasible to clean certain silica-based bonded phases with 0.1 *N* sodium hydroxide (see Fig. 2). Carefully filtered solutions of silica dissolved in this solvent may be even more effective at preserving the integrity of the bonded phase during cleanup.

2.4.2. Protein-Binding Properties of Silica

An important perceived limitation of silica is the belief that proteins tend to bind irreversibly to the matrix (39, 40). This is only partly true of even bare silica, since with the correct choice of buffers, recoveries can be greater than 80% (41). There is now sufficient evidence to indicate that with the appropriate surface chemistry, protein recoveries from bonded phases can be greater than 95% (42–44). The philosophy and description of "appropriate surface chemistry" is described in more detail in Section 4.

2.4.3. *Surface Heterogeneity*

The third limitation of silica is that its surface may be chemically heterogeneous. In other words, some regions may have higher concentrations of reactive silanols and adsorbed matter than other regions (24). This may lead to variations in bonded-phase ligand density over the pore surface. This in turn results in multiple forms of solute adsorption/desorption isotherms and generally inferior peak shapes. Reducing this phenomenon to a minimum can be achieved by pretreating silica with mineral acids. Siloxane bonds are thereby hydrolyzed to silanol groups, producing a narrower distribution of surface silanol concentrations. The subsequent bonding steps must be carried out in such a way as to minimize any introduction of ligand density variations over the surface.

One approach to minimizing surface phase density heterogeneity is to anchor covalently a well-characterized hydrophilic homopolymer to the silica surface, then graft chosen functional groups onto this polymer. This technique is the basis of most of the bonded phases developed in our laboratories for protein chromatography, and is described in more detail in Section 4.

Another commonly perceived (21, 45) source of potential nonreproducibility in bonded phases is the creation of polymeric layers by the use of difunctional and trifunctional silanes as bonding agents. Nonreproducibility from this source can also be essentially eliminated by appropriate control of reaction conditions (46); this subject is discussed in more detail in Section 4.

A more serious problem resulting in variability in silica surface activity is nonreproducibility in the manufacture of porous silica itself. For a given bonded-phase producer, there must be a protocol in place for characterizing the surface state of the adsorbent. Subsequent batches of silica purchased by that producer may then be tested according to this protocol as part of the overall control of raw material and bonded-phase reproducibility. For silica the definition of a standard state is one that can be quite complex and subtle. The subject of the standardization of silica has been carefully considered by Snyder (47), Halpaap (48), and Unger (23). However, it is beyond the scope of this chapter to address this topic in any detail.

2.5. Choice of Silica Manufacturer(s)

One of the most important advantages of silica as a chromatographic substrate (see Table 6) is its ready availability. Table 9 lists the major silica manufacturers.

J.T. Baker made the decision not to create its own unique silica, since high-quality sources of silica with an adequate range of particle and pore sizes were already available in the late 1970s (see Table 9). In addition, it was expected that well-established silica manufacturers would be developing appropriate

Table 9. Chromatographic-Grade Silica Manufacturers

Company	Major Trade Name
Amicon	Matrex™
Chemco	Chemcosorb®
Davison Chemical	Davisil™
DuPont	Zorbax®
Johns-Manville	Chromosorb®
Machery-Nagel	Nucleosil®/Polygosil®
E. Merck & Company	Lichrosorb®/Lichrospher®
Pennsylvania Quartz	PQ
Phase Separations	Spherisorb®
The Separations Group	Vydac®
Serva	Daltosil/SI
Shandon Southern Instruments	Hypersil®
Showa Denko	Shodex®
Toyo Soda	TSK-Gel®
Waters	Novo-Pak™
Woelm	Woelm®

substrates for bonded phases for protein chromatography as the technique developed.

Thus, a substantial screening program was carried out to determine the most suitable silica available, from many of the previously listed producers. The criteria used in choosing the silicas are summarized in Table 10.

A single silica source was ultimately chosen for each of the three particle sizes (5, 15, and 40 micron). In addition to the preceding criteria, a group of specifica-

Table 10. Criteria Used in Silica Selection

Property	Approximate Value/Property		
Particle size (micron)	4–6	15–20	40–50
Pore volume (mL/g)	0.5–1.5	0.5–1.5	0.5–1.5
Pore size (Å)	250–400	250–400	250–400
Surface area (m^2/g)	50–300	50–300	50–300
Shape	Spherical	Spherical or irregular	Spherical or irregular

Availability of multikilogram, single-lot batches.
Reproducibility of particle size and distribution, surface area, surface ractivity.
Cost: A major factor for spherical silicas, but less important for irregular silicas as silanes and other costs sometimes predominate in these cases.
Crushing strength under high-pressure packing conditons (up to 15,000 psi).

Table 11. Desirable Properties of Packings for Protein Chromatography

Desirable Property	Physico/Chemical Implications
High resolution, speed	Small particles, large pores (300 Å)
High recoveries (mass and activity)	Lack of nonspecific binding
High capacity	High surface area, large pores
High stability	Nonhydrolyzable
Convenience of use	High flow rates, incompressible, stored dry
Universality of methodology	Same substrate/surface characteristics
Regenerability	Range of cleaning solvents available
Sterilizability	Organic solvent compatible
Microbial inertness	Synthetic substrate

tions was set, which includes pH, extractables, trace metals content, extractable turbidity, back pressure, and chromatographic behavior.

2.6. Summary: Choice of Substrate

Over a seven-year period we have evolved a basic philosophy aimed at creating a new, carefully integrated family of separation media for polypeptide and protein chromatography.

After a study of the advantages and limitations of available substrates for these media, porous silica was the material of choice.

Several important advantages of silica (advanced manufacturing technology, rigidity, pore-size integrity, high surface area, and surface reactivity) were discussed briefly. The limitations of silica (pH stability, macromolecular adsorption properties, and reproducibility) were outlined, and techniques for overcoming these limitations were described.

3. IDEAL PROPERTIES OF BONDED PHASES FOR PROTEIN CHROMATOGRAPHY

Irrespective of the chosen material for polypeptide and protein chromatography, a number of properties are generally considered desirable for this application. These are listed in Table 11, with the physico/chemical implications of these properties.

The aim of any manufacturer of bonded phases for protein chromatography is to design the ideal family of bonded phases. We have measured most of the important chromatographic properties of the bonded phases designed in our laboratories, and some of these properties are given in Table 12. In addition, two bonded phases, having fundamentally different separation mechanisms from each other, have been designed for immunoglobulin purification. These are described in Table 13. Finally, a range of activated and nonactivated bonded

Table 12. Bakerbond™ HPLC Columns and Bonded Phases for Polypeptide and Protein Chromatography—General Characteristics

Pressure limit	5 and 15 micron—15,000 psi; 40 micron—3000 psi
Particle size	5 micron (analysis); 15 and 40 micron (preparation through process)
Pore size	300 Å (5, 15 micron); 275 Å (40 micron)
Shape	Spheroidal (5, 15 micron); irregular (40 micron)
Capacity	100–200 mg/g dry packing
Mass recovery	Greater than 95% (variable for reversed-phase)
Activity recovery	Quantitative (poor for reversed-phase)
Lifetime	Greater than 1000 hr of use
pH range for maximum stability	2–8.5
Buffer limitations	None
Organic solvent limitations	None
Efficiency	Greater than 40,000 plates per meter (ppm) (5 micron); greater than 6,000 ppm (15 micron); greater than 2,000 ppm (40 micron)
Clean up	2 *M* NaOAc (pH 8) and/or 1% acetic acid, formic acid, organic solvents, 20% HNO_3, and/or DMSO or chaotropic agents
Storage/operation temperature	Room temperature or from 4 to 80°C
Sterilization	With 70/30 ethanol/water, 0.1% azide or many other chemical sterilizing agents
Methodology of operations	Identical for all particle sizes, surface chemistries, mode of separation

Table 13. General Properties of Bakerbond MAb™ and ABx™ for Immunoglobulin Purification

Property	MAb	ABx
Substrate	Porous silica	Porous silica
Surface functional groups	Amino	Proprietary
Mechanism of binding	Anion exchange	Anion and cation exchange
Specificity	Most ascites fluid MAb's	All classes and subclasses of immunoglobulins
Application	Ascites fluid	Ascites fluid, cell cultures, and plasma
Saturation capacity (mg/g)	150 (IgG)	150 (IgG)
Mass recovery	Greater than 97%	Greater than 97%
Activity recovery	Quantitative	Quantitative

Table 14. General Properties of Bakerbond Affinity Matrices for Polypeptide and Protein Chromatography[a]

Support (Activated)	Functional Group	Ligand Specificity	Typical Applications
Glycidoxypropyl		$-NH_2$, $-OH$, $-SH$	Triazine dyes, small ligands
Glutaraldehyde		$-NH_2$	Immunoglobulins, structural proteins
p-Nitrophenyl ester		$-NH_2$	Optically active compounds soluble in organic solvents
Diazofluoroborate	RN_2^+	Phenols, aromatic amines	Proteins, peptides, nucleosides
		Support (Nonactivated)	
Amino propyl	$-NH_2$	Acid, aldehyde, epoxide (couplers)	Carbohydrates, oligonucleotides
Diol		Amine, epoxide (coupler)	Oligonucleotides
Succinoyl-aminopropyl	$-COOH$	Amine (coupler)	Proteins, oligonucleotides
Hexamethylenediamine	$-NH_2$	Acid, epoxide, aldehyde	Large molecules

[a]Particle size is 40 micron, and pore size is 275 Å in all cases.

phases have been synthesized in our laboratories for affinity chromatography. The major properties of these are given in Table 14.

Thus, we believe that with the addition of the three bonded phases currently under development (see Table 1), we shall have a complete family of bonded phases. These constitute an integrated group suitable for all protein chromatography on any scale from analytical to process. The techniques and applications for which these bonded phases are suitable are listed in Table 2.

4. THE DESIGN AND SYNTHESIS OF BONDED PHASES FOR POLYPEPTIDE AND PROTEIN CHROMATOGRAPHY

4.1. General Chemical Methods

The great majority of the bonded-phase properties given in Tables 12 and 13 were obtained through the proper functional group modification of the silica

Table 15. Approaches to the Synthesis of Bonded Phases for Protein Chromatography

Separation Mode	Synthetic Approach
Reversed phase	Conventional silanc bonding under controlled polymer-forming conditions
Ion exchange, hydrophobic interactions, size exclusion	Covalently bound hydrophilic layer modified with charged and/or polar or nonpolar groups

surface. The chemistry involved in this modification falls into two categories that depend on the mode of separation desired. The classification also distinguishes those bonded phases upon which biological (especially enzymatic) activity is often destroyed (reversed phase), and those where biological activity is largely retained (ion exchange, hydrophobic interaction, size exclusion). Table 15 describes these categories, and reactions (5), (6), (7) and (8) in Section 2.2.5 illustrate the major chemical processes involved.

4.2. Bonded Phases for Reversed-Phase Protein Chromatography

The criteria used in selecting the bonding chemistry for these packings are listed in Table 16.

It was considered most important to shield protein molecules from the silica surface, where free silanols (especially isolated silanols) may interact irreversibly with proteins. In addition, we attempted to reduce variations in ligand density to a minimum.

It was decided that trifunctional silane chemistry (see Eq. 7) had the best potential to meet the criteria we set. Independent evidence was available (49) (see also Table 8) which indicated that bonded phases made from trifunctional silanes had greater hydrolytic stability than those made using monofunctional silanes. Such evidence is still accumulating today (50). Furthermore, Nendek

Table 16. Criteria Used in Determining Bonding Chemistry for Reversed-Phase, Ion-Exchange, and Hydrophobic Interaction Packings

Strength of protein—bonded-phase interaction required
Hydrophobic nature of surface required
Ionic nature of surface required
Ligand density required
Mass and activity recoveries required
Degree of protein silanol interaction required
Stability to acid- and base-catalyzed hydrolysis required
Reaction rates and efficiency of manufacture involved
Reproducibility of bonded-phase manufacture
Cost of raw materials

and co-workers (51) have shown that the concentration of free silanols in bonded phases prepared from trifunctional silanes is approximately one-tenth that in packings made with monofunctional silanes.

The bonded layer formed from a trifunctional silane has several properties we wished to incorporate:

1. Multiple points of attachment, increasing hydrolytic stability.
2. Higher density and hydrophobicity of nonpolar groups, reducing the likelihood of rapid diffusion of polar water molecules and other bases to the silica support.
3. An "averaging" of the ligand density variations, by covalent bonding of the polymeric silane containing regular spacing of nonpolar groups.
4. A more rapid rate of equilibration with the mobile phase (52), compared with that of the bonded phase formed from a monofunctional silane.

Finally, the manufacture of reproducible nonpolar surfaces requires control of the following conditions.

1. Silica purity and water content.
2. Temperatures and reaction times.
3. Purity of silanes.
4. Nature of primary silylating reagent.
5. Nature of end-capping reagent.
6. Washing and drying protocols.

The bonded phases for reversed-phase liquid chromatography of proteins are listed in Table 1, and their general properties given in Table 12. Applications of these packings are illustrated in the applications section.

4.3. Bonded Phases for Ion-Exchange and Hydrophobic Interaction Chromatography

The bonding chemistry criteria used for these packings are listed in Table 16. In preparing bonded phases for the preceding modes of chromatography, it was again considered important to protect the protein from any silanol interaction. This would help provide maximum mass and activity recoveries. Furthermore, it was intended to produce a surface having maximum ligand density but a minimum variation in this density. Finally, we considered that hydrolytic stability was of great importance in designing the appropriate bonded phase.

The approach we have used was mentioned briefly in Section 2.4. It involves an initial covalent linking of a well-characterized hydrophilic polymer to the silica surface, followed by a chemical reaction of the polymer to introduce a chosen functional group. The general reactions are illustrated in Eqs. 12 and 13. The nature of the hydrophilic polymer and the subsequent chemical reactions are proprietary information.

$$\text{—SiOH} + \text{X} \sim\sim\sim\sim \rightarrow \text{—SiOX} \sim\sim\sim\sim \quad (12)$$

Hydrophilic Polymer Covalent Bonding

$$\text{—SiOX} \sim\sim\sim\sim + n\text{RY} \rightarrow \text{—SiOX}\sim\underset{\text{R}}{\overset{\text{R}}{\sim}}\underset{\text{R}}{\overset{\text{R}}{\sim}}\sim + n\text{YH} \quad (13)$$

Chemical Modification

R = Amino, Carboxyl, Sulfo, Propyl, Quaternary Amino
X = Hydroxyl - specific Reactive Group

The resulting bonded phases for ion-exchange and hydrophobic interaction chromatography are listed in Tables 1 and 2, and the general properties of these separation media are given in Tables 12 and 13. These soft-surface, hard-gel bonded phases are considered to have the following advantages over those prepared by conventional silane chemistry:

1. The hydrophilic layer acts to prevent close interaction between the protein and the silica backbone, thereby eliminating the possibility of strong silanol–protein binding.
2. The close spacing of the reactive functional groups on the hydrophilic layer creates a high charge density for strong and highly discriminating protein binding.
3. The soft layer causes a reduction in diffusion rates of bases from the mobile phase to the silica substructure. This phenomenon may partly explain the excellent stability of these bonded phases.
4. They combine the advantages of soft gel (good recoveries and stability) and hard silicas (high resolution and rigidity).
5. They possess a wide operational pH range.

The techniques of forming a soft polymer layer on silica are not new, of course. Alpert et al. (33) adsorbed polyethyleneimine to porous silica and partially cross-linked the primary and second nitrogen atoms to add significant hydrolytic stability to the resulting weak anion exchanger. Alpert (53) bound and then polymerized aspartic acid on the surface of porous silica to obtain a weak cation exchanger. Weak cation exchangers and hydrophobic interactors have been prepared by Regnier and his group (43, 54) by reacting polyethyleneimine-bound silica with succinic anhydride and acylating agents, respectively. The possible presence of tertiary amine groups and incompletely reacted secondary and primary nitrogens in these PEI phases, however, must be carefully considered as a potential source of positive charge and therefore unwanted interaction with the protein. Schomburg and co-workers (55) covalently attached polyalkylsiloxanes to porous silica with heat- and gamma-radiation-induced bonding. Commercial companies such as Poly LC (Polycat™), Serva Heidelberg (Si Polyol), BioRad Laboratories (Microanalyzer MAT™), Waters (Accell™), Amicon (Matrex Pel™), and The Separations Group (Vydac) have products whose chemistry is based upon the pellicular approach.

5. APPLICATIONS

All chromatographic conditions not included in figure legends are given in the Appendix.

5.1. Effect of Cleaning and Sterilizing Solutions on Chromatography

One of the most important considerations in the design of bonded phases is that they must be amenable to simple and effective cleanup and sterilization. The conditions for these processes must be such that the bonded phases are not significantly degraded during their application. We have used a number of chemicals suitable for cleanup and sterilization. Figure 3 illustrates the insignificant chromatographic effect of such treatment after limited cleanup of Bakerbond Wide-Pore CBX (Carboxyethyl) with several solutions.

5.2. High Resolution for all Surface Chemistries

From the beginning it was intended that the full capabilities of HPLC of small molecules be realized for protein chromatography. The high resolving power of reversed-phase chromatography has also been obtained with weak anion exchange, weak cation exchange, and hydrophobic interaction chromatography. The separation of four series of standard proteins through the four basic interactive mechanisms are illustrated in Fig. 4, using columns packed with 5-micron (300-Å) particles.

5.3. Duplication of Surface Chemistry

The control that was exerted over the reactions involved in the bonded-phase synthesis with three particle sizes (5, 15 and 40 micron) is illustrated in Figs. 5 and 6. The closely similar selectivities evident in these applications have a number of advantages in practice:

1. The choice of a larger-particle, more economical bonded phase for scale-up can be predicted from the chromatography obtained on a small-particle (5-micron) HPLC column.
2. Conversely, analytical HPLC of fractions from a large-scale (with 15- or 40-micron particles) purification may be carried out on a surface chemistry-matched 5-micron-particle column. The protein of interest should

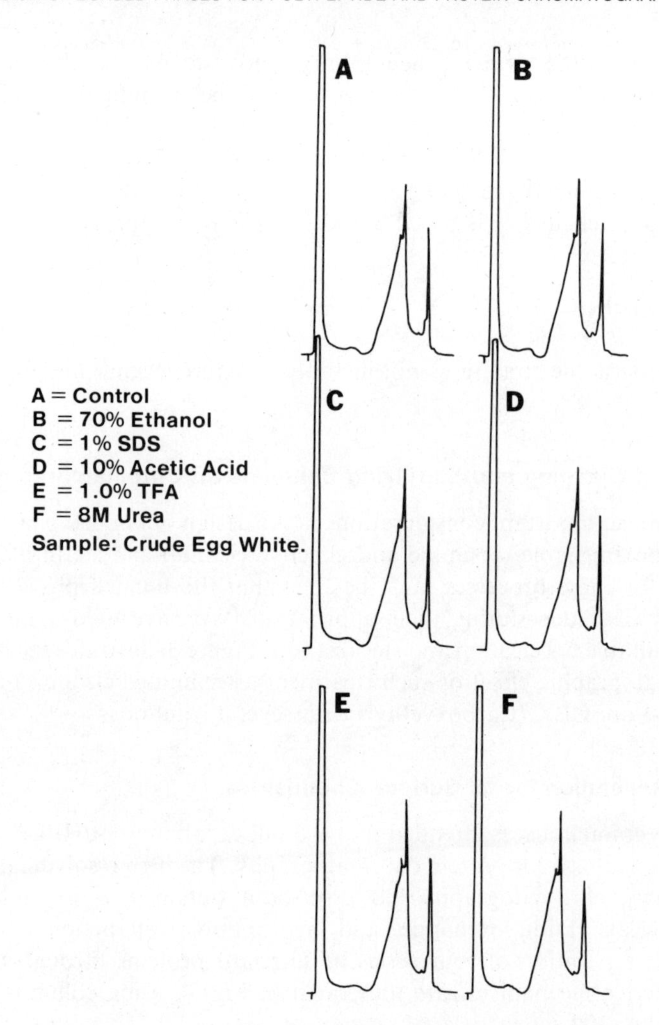

Fig. 3. Chromatographic effect of treatment of Bakerbond Wide-Pore™ CBX (carboxyethyl) with several cleaning and sterilization solutions. Conditions: C.

elute at very similar locations in the gradient on 5-, 15-, or 40-micron columns.

3. The methodology of the chromatographic techniques employed during scale-up is therefore identical, regardless of the particle size used or location in the purification scheme. In other words the flow rates, pressure limit considerations, solvent, buffer and pH compatibility, storage and handling conditions, clean-up and sterilization techniqus, product recoveries, and lifetime are essentially identical at all stages of purification.

Hydrophobic Interaction HPLC

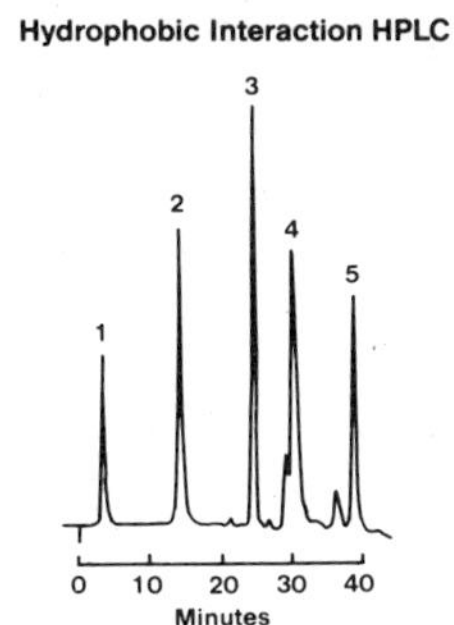

Peaks:
1. Cytochrome C, 58-μg
2. Myoglobin, 377-μg
3. Lysozyme, 203-μg
4. Ovalbumin, 580-μg
5. α-chymotrypsinogen, 232-μg

Column:	BAKERBOND Wide-Pore ™ Hi-Propyl 5-μm, 4.6-mm x 25-cm
Mobile Phase:	A = 1.8-M $(NH_4)_2SO_4$ + 25-mM KH_2PO_4; pH 7.0 B = 25-mM KH_2PO_4; pH 7.0
Gradient:	0% B → 75% B; 30 min.; 75% B → 100% B; 10 min.
Flow Rate:	1.0-ml/min.
Pressure:	1500 psi
Detection:	UV at 280; 0.32 aufs
Injection Volume:	100-μl

Reversed Phase HPLC

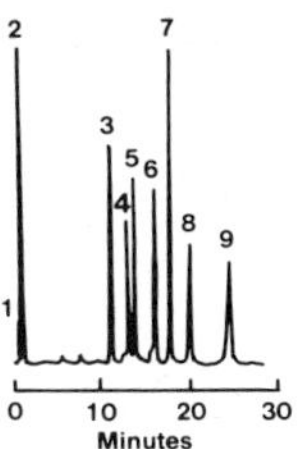

Peaks:
1. Solvent
2. Casein, 45 μg
3. Ribonuclease, 30 μg
4. Insulin, 35 μg
5. Cytochrome c, 25 μg
6. Lysozyme, 42 μg
7. Bovine Serum Albumin, 70 μg
8. β-Lactoglobulin A 28 μg
9. Ovalbumin, 80 μg

Column:	BAKERBOND Wide-Pore ™ Butyl 5-μm, 4.6 x 250 mm.
Mobile Phase:	A = 0.1% TFA in H_2O B = 0.1% TFA in CH_3CN
Gradient:	10% B to 60% B in 30 mins.
Flow Rate:	2 ml/min.
Pressure:	2000 psi
Temperature:	28°C
Detection:	UV @ 280, 0.2 aufs
Injection Volume:	20 μl

Weak Anion Exchange PEI

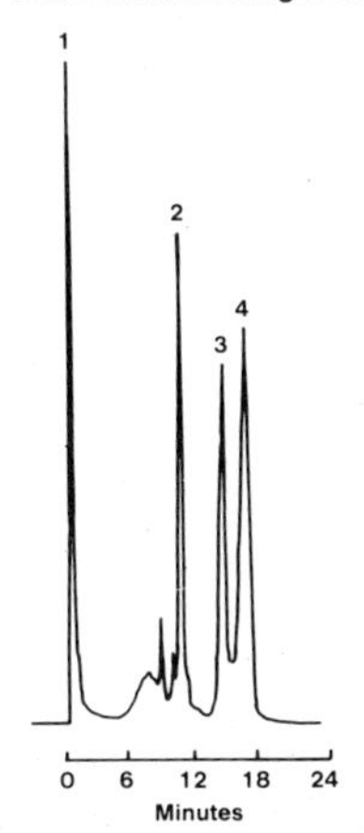

Peaks:		**Recovery:**
1. Cytochrome C, 20-μg		98%
2. Ovalbumin, 40-μg		98%
3. β-Lactoglobulin, B-chain	} 80-μg	100%
4. β-Lactoglobulin, A-chain		100%

Column:	BAKERBOND Wide-Pore™ PEI 5-μm, 4.6-mm x 5 cm
Mobile Phase:	A = 10-mM KH_2PO_4; pH 6.8 B = 500-mM KH_2PO_4; pH 6.8
Gradient:	0% B → 100% B; 60 min.
Flow Rate:	1.0-ml/min.
Pressure:	250 psi
Detection:	UV at 254; 0.08 aufs
Injection Volume:	80-μl

Weak Cation Exchange CBX

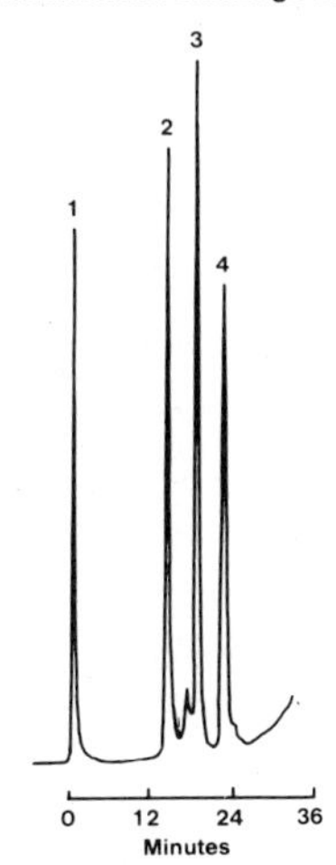

Peaks:
1. Ovalbumin, 350-μg
2. Hemoglobin, 180-μg
3. Cytochrome C 180-μg
4. Lysozyme 110-μg

Column:	BAKERBOND Wide-Pore™ CBX 5-μm, 4.6 mm x 5 cm
Mobile Phase:	A = 10-mM KH_2PO_4; pH 5.5 B = 500-mM KH_2PO_4; pH 6.8
Gradient:	0% B → 100% B; 30 min.
Flow Rate:	1.0-ml/min.
Pressure:	250 psi
Detection:	UV at 280; 0.32 aufs
Injection Volume:	35-μl

Fig. 4. HPLC of standard proteins on 5-micron bonded phases.

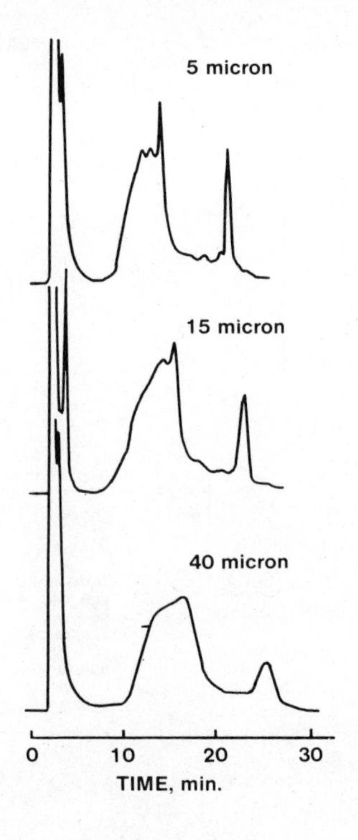

Analytical Conditions:

Column:	BAKERBOND Wide-Pore^{TM} CBX. 5 or 15 or 40 micron. 4.6 mm x 250 mm
Mobile Phase:	A = 10mM MES, pH 5.6 B = 1.0 M Sodium Acetate, pH 7.0
Gradient:	0% B to 100% B; 30 min.
Flow Rate:	1.0 mL/min.
Injection Volume:	0.05 mL
Sample:	Crude egg white

Fig. 5. Analysis of crude egg white on 5-, 15-, and 40-micron particles by weak cation exchange chromatography.

5.4. Purification of Lysozyme from Egg White

This protocol was designed in order to illustrate the power of multidimensional chromatography in achieving rapid protein purification. The general scheme is shown in Fig. 7. The lysozyme purification scheme is a tidy illustration of several well-integrated facets of the "family" approach we use in our laboratories.

1. A near-complete purification of the enzyme, from crude egg white to a greater than 99% pure (by SDS-PAGE) solution of lysozyme.

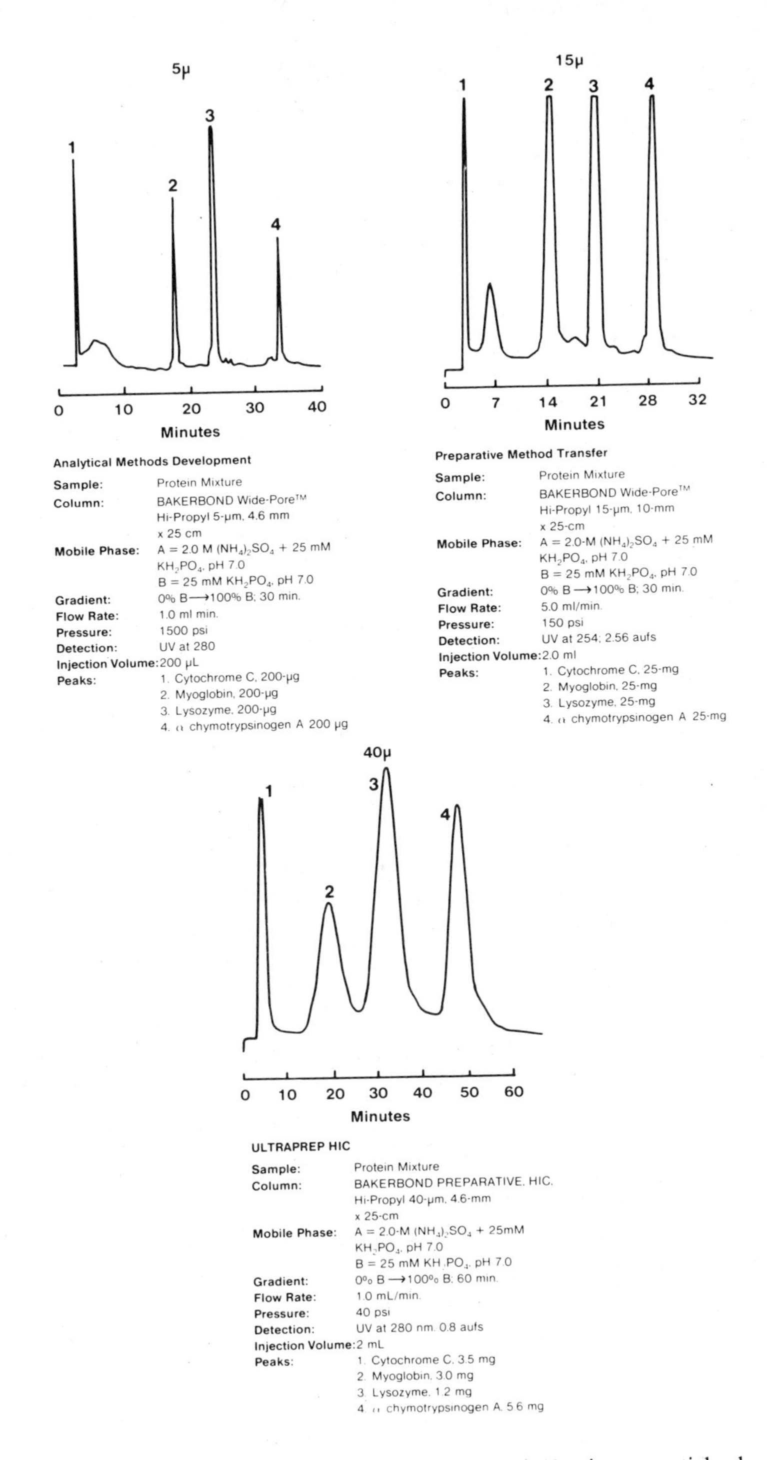

Fig. 6. Separation of standard proteins on 5-, 15-, and 40-micron particles by hydrophobic interaction chromatography.

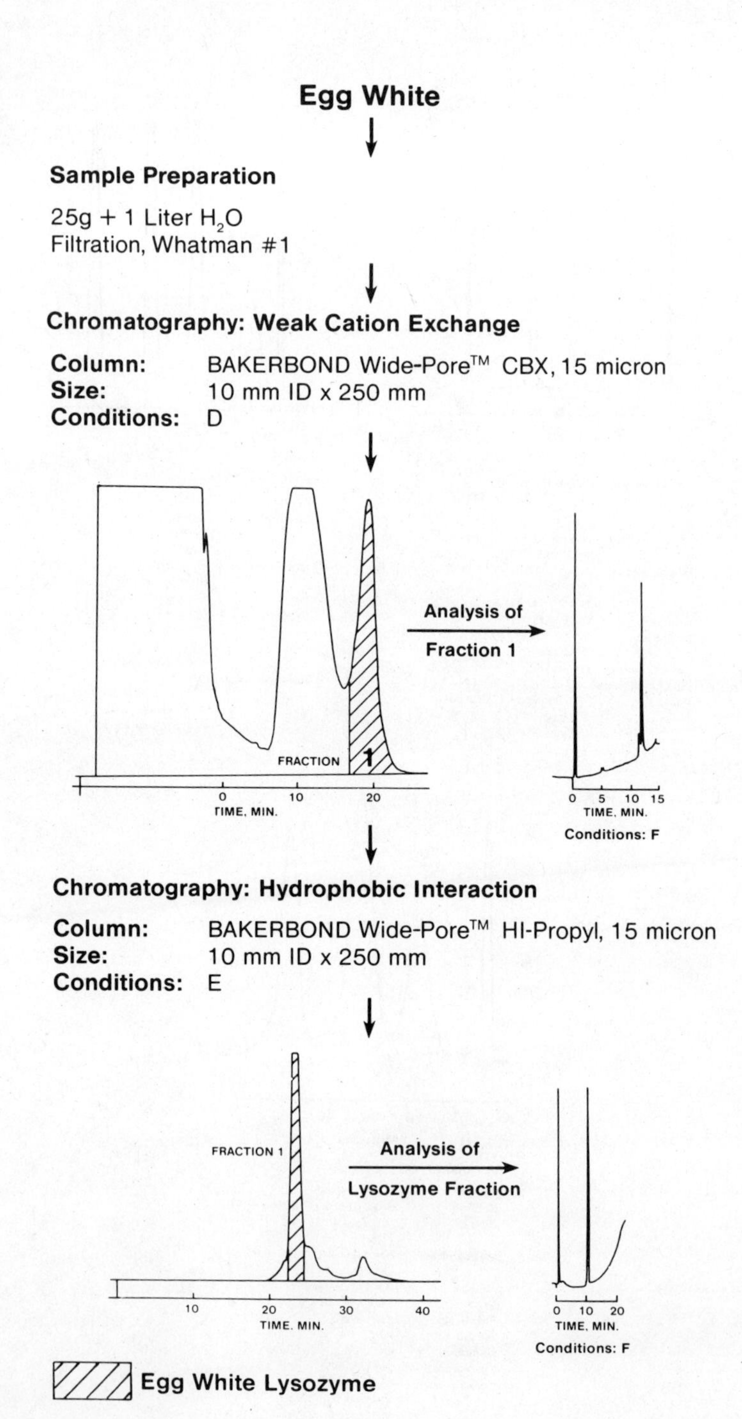

Fig. 7. Purification of lysozyme from egg white by cation exchange and hydrophobic interaction chromatography.

2. Choice of a weak cation exchanger to bind the cationic enzyme at low salt, followed by a hydrophobic interaction step that binds the lysozyme directly from the ion-exchange step, at high salt (no dilution or dialysis of collected fraction is required).
3. Moderate desalting of lysozyme fraction occurs during hydrophobic interaction chromatography.
4. Two-column approach using 15 micron packings for adequate resolution on a preparative scale, with low back-pressure, high flow rates and, therefore, rapid equilibration.
5. Fast HPLC analysis of fraction by reversed-phase SCOUT columns (4.6 mm ID × 50 mm). This technique is largely independent of the ionic strength of the fractions. This is an alternative to SDS-PAGE, in most stages of the purification.
6. 10 mm ID columns for about 0.5–5 g of protein.
7. A $1\frac{1}{2}$ hour two-step purification (not including the original sample loading) of 150 mg lysozyme from 5 g of total egg white protein.

5.5. Purification of Ovalbumin from Crude Egg White

This scheme (Fig. 8) integrates batch adsorption, preparative column chromatography, fast HPLC analysis of fractions, and SDS-PAGE. The batch adsorption technique is illustrated schematically in Fig. 9. In this particular application batch adsorption has several advantages:

1. It permits rapid, simple removal of lysozyme and phosphatin before chromatography.
2. The technique increases the capacity of the chromatographic column by initial protein removal and therefore allows greater overload with less loss of resolution.
3. It involves a useful illustration of batch method development.

5.6. Purification of Calmodulin from Bovine Brain extract

This method (Fig. 10) illustrates the potential for sequential chromatography using the same surface chemistry, but different particle size and buffers. Over 95% of the extracted protein is removed in the first step. The collected fractions containing the calmodulin (present at about 0.5% in the original extract) can be further purified on an analytical column (5 micron) whose capacity is quite adequate for the quantity of the protein of interest.

5.7. Purification of Monoclonal Antibodies from Mouse Ascites Fluids

Several applications of the bonded phase for immunoglobulin purification (Bakerbond ABx) designed in our laboratories are illustrated in Fig. 11. The major advantages of this class of bonded phase, which has been prepared on 5-,

Fresh Egg White (30g)

Sample Preparation:

Dilute with 1 liter of water
Filter, Whatman #1

↓

Batch Extraction:

Dry BAKERBOND Wide-Pore™ CBX, 40 micron
(see Figure 9 for scheme)

↓

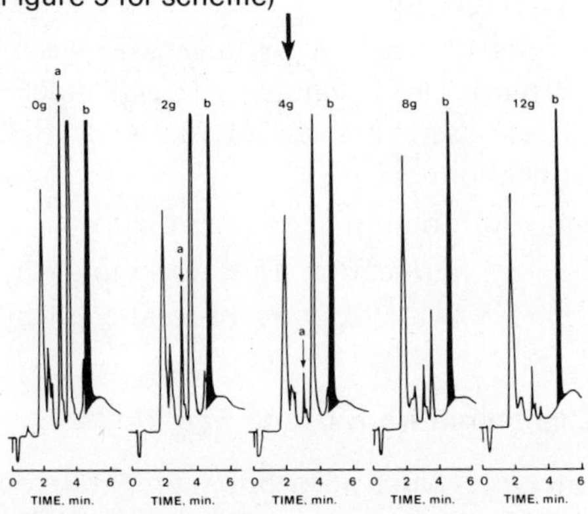

Monitoring of Supernatent by Reversed-Phase Chromatography (Conditions: G)

↓

Sample: Supernatent after treatment with 12 g of BAKERBOND Wide-Pore™ CBX 40 μ

Chromatography: Weak Anion Exchange

Column: BAKERBOND Wide-Pore™ PEI, 15 micron
Size: 21.5 mm ID x 500 mm
Conditions: H

↓

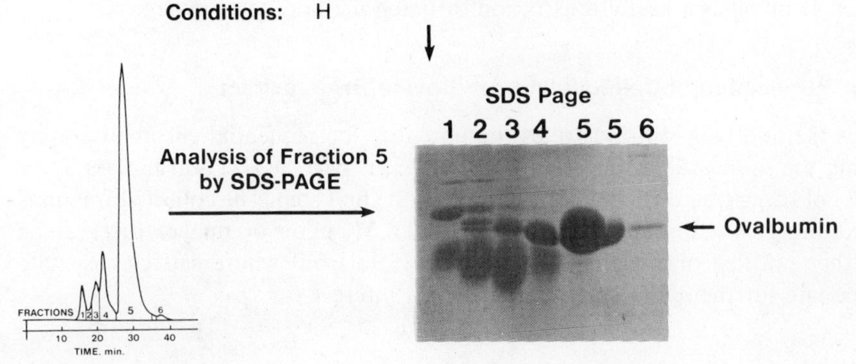

Fig. 8. Purification of ovalbumin from egg white by selective batch extraction and weak anion exchange chromatography.

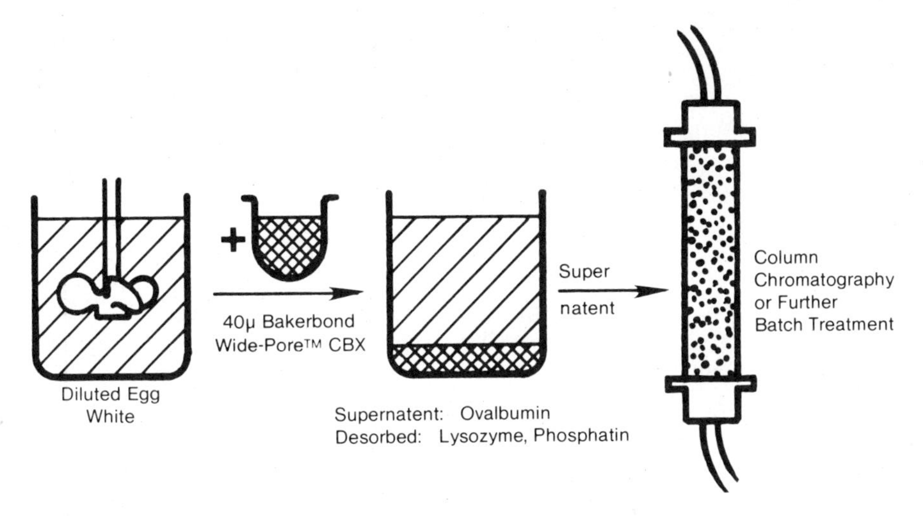

Fig. 9. Schematic for selective batch desorption of lysozyme and phosphatin from egg white.

15-, and 40-micron silica, are as follows:

1. No binding of albumins, transferrins, most proteases, and phenol red (used in cell culture as a pH indicator).
2. Binds immunoglobulins from any source.
3. High capacity, greater than 97% recoveries (mass and activity).

Further discussion of this bonded phase, its design, and applications is given in Chapter 15.

5.8. Reversed-Phase HPLC Peptide Mapping

Figure 12 illustrates the resolving power of reversed-phase HPLC in the protein sequencing studies of Hartman et al. (56), using a Bakerbond Wide-Pore Butyl column. The chromatograms contain peaks corresponding to the oxidized (OX) and reduced (RED) peptides from the tryptic digest of the hormone bovine somatotropin (bSt). The hormone was obtained from both recombinant DNA techniques (r) and pituitary (p) sources. The four chromatograms illustrate the following:

1. The difference between the oxidized and reduced reaction products of the digests of the hormone (compare especially the regions at 48 and 52 min).
2. The difference between the recombinant and pituitary hormones (compare the T_1 and T_{13} regions). The extra peaks observed with the pituitary hormones are consistent with the increased heterogeneity of this material relative to the more homogeneous recombinant-derived hormone.

Bovine Brain Extract

↓

Sample Preparation

Acetone extract dissolved in 25mM Tris, pH = 7.0
then Ultracentrifuged

↓

Chromatography: Weak Anion Exchange

Column: BAKERBOND Wide-Pore™ PEI, 15 micron
Size: 10 mm ID x 250 mm
Conditions: I

↓

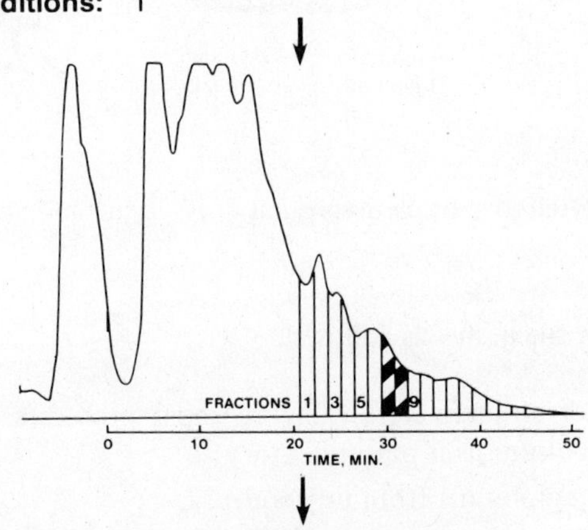

↓

Chromatography: Weak Anion Exchange

Column: BAKERBOND Wide-Pore™ PEI, 5 micron
Size: 4.6 mm ID x 250 mm
Conditions: J

↓

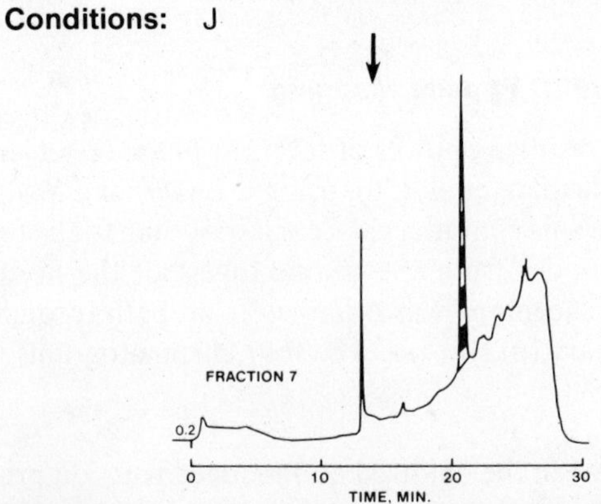

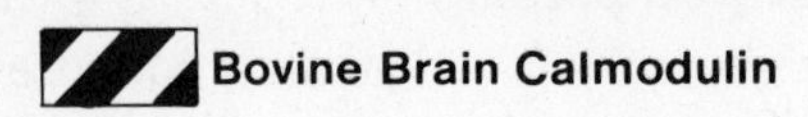

Fig. 10. Purification of calmodulin from bovine brain extract by anion exchange chromatography.

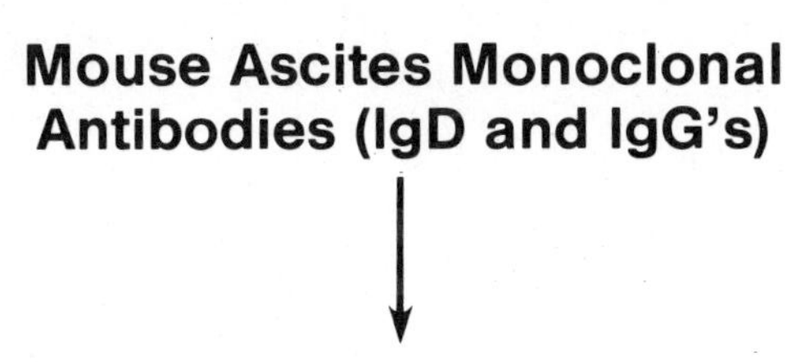

Sample Preparation

1 mL Ascites Fluid + 4 mL KH_2PO_4
(10 mM, pH = 6.0)

Chromatography: Mixed Mode

Column: BAKERBOND™ ABx, 5 micron
Size: 4.6 mm ID x 250 mm
Conditions: K

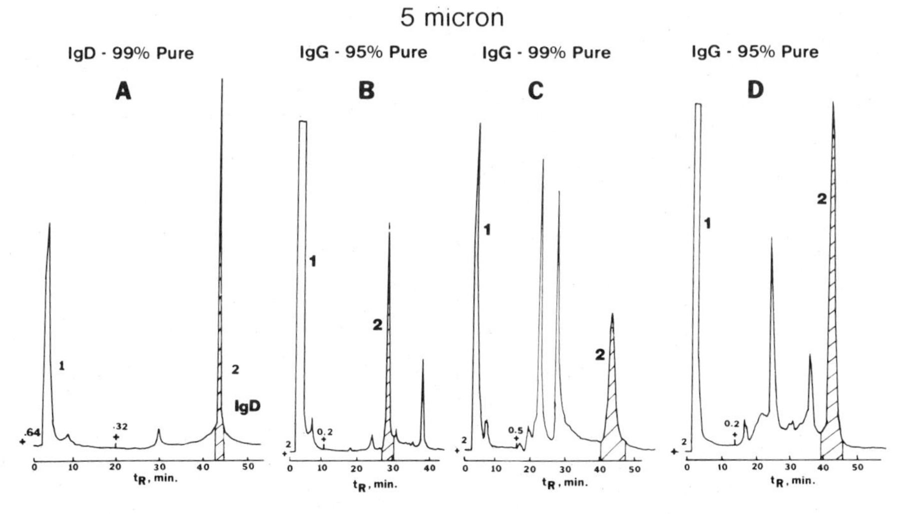

Fig. 11. Purification of monoclonal antibodies from mouse ascites fluid by ion exchange chromatography.

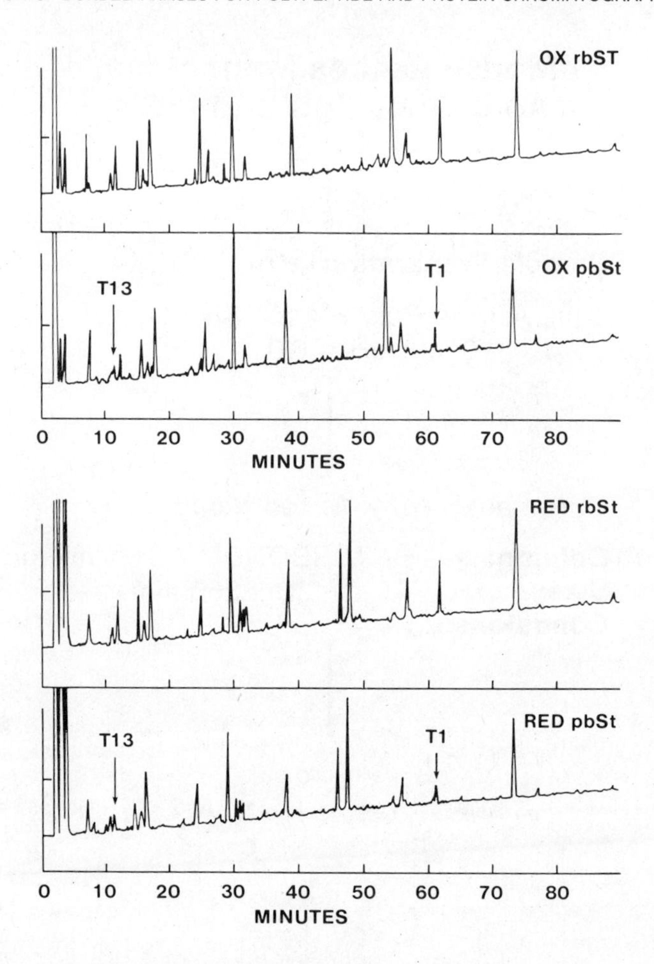

Fig. 12. Comparison of tryptic digests of rbSt and pbSt. Note the multiplicity of peaks in the T13 and T1 regions in pbSt. Gradient, 0–45% B in 90 min on Bakerbond Wide-Pore™ butyl (A = 0.1% TFA in water and B = 0.1% TFA in acetonitrile). Detection at 214 nm. Injected ca. 200 μL (200 μg of rbSt per mL). Reprinted from Ref. 56 with the permission of the publisher.

3. Bovine somatotropin contains several sites that were only incompletely cleaved by trypsin.
4. Twenty-one of the twenty-five predicted peptide fragments are resolved. The other four are two amino acids and two dipeptides that are insufficiently hydrophobic to be retained on the column. Many other retained peaks are observed, indicating the presence of impurities in both samples of the hormone.

5. Tryptic digests analyzed with wide-pore, rather than narrow-pore supports, may offer generally enhanced resolution (56).
6. The technique requires as little as 40 μg of protein (or 2 nmol assuming a molecular weight of 19,000) for full sequence analysis. It is crucial in such analyses that all peaks are sharp (giving very low limits of detection) and that mass recoveries are essentially quantitative at the very low or sub-microgram level. (All predicted peptide fragments were observed on the Bakerbond Wide-Pore Butyl column.)

6. SUMMARY AND CONCLUSIONS

Since 1980, steady progress has been made in our laboratories in designing bonded phases for peptide and protein chromatography. The quality of these packings, as reflected in the data given in Table 12, is such that the most highly developed of them combine the advantages of both soft gels and rigid substrates. The silica-based bonded phases we have designed in fact offer several advantages over soft gels and have none of the perceived limitations of silica, except that of instability at very high pH, and even this limitation has been minimized. The soft surface chemistry approach begun by Alpert and Regnier (33), and continued in our laboratories, provides increased high-pH stability. This chemistry, together with improvements in hydrolytic stability of silica itself, have effectively raised the operating pH to 10 in many cases.

Fundamental discoveries concerning the nature of the surface of silica and its modified forms continue to be made using modern spectroscopic (21, 57), thermogravimetric (24), and theoretical techniques (58). These studies offer the promise of significant improvements in the physicochemical and chromatographic properties of silica-based bonded phases. As manufacturers of silica continually improve their products, as bonded-phase producers adapt their chemistry where necessary to maintain reproducibility, new chromatographic packings will be designed to take the place of the old ones. In many fields of high-performance liquid chromatography an "improved" column offers limited immediate value, as the changed selectivity may mean a substantially new methods development. On the other hand, the HPLC of proteins is a relatively new field, toward which many of the current soft gel applications must turn in the continuing quest for speed, resolution, and throughput.

ACKNOWLEDGEMENTS

The author gratefully acknowledges the valuable discussions with Dr. Laura J. Crane of these laboratories; the elegant and precise chromatography of Drs. Steven A. Berkowitz and David R. Nau; the invaluable bonded phase synthesis by Dr. Hugh E. Ramsden; and the typing assistance of Mrs. Joanne Volkert.

APPENDIX 1

Chromatographic Conditions for Figures 1 through 12

A. Columns: Bakerbond Wide-Pore™ PEI, 40 micron (J.T. Baker Inc.
DEAE Sepharose Fast Flow (Pharmacia, Inc.)
Fractogel DEAE-650S TSK (E. Merck Industries, Inc.)
Mobile Phase: HPLC Grade Water (J.T. Baker)

B. Column: Bakerbond Wide-Pore PEI, 5 micron, 4.6 × 50 mm
Mobile Phase: A = 10 mM KH_2PO_4, pH 6.8
B = 500 mM KH_2PO_4, pH 6.8
Gradient: Linear, 0% B to 100% B over 15 min
Flow Rate: 1 mL/min
Injection Volume: 100 μL
Components:
1. Cytochrome C
2. Ovalbumin
3. Alpha-lactoglobulin A
Protocol:
(A) Initial chromatography (control), see above
(B) Same chromatography after column was cycled through a 0.1 N NaOH wash for 30 min, followed by a 0.5 M KH_2PO_4 (pH = 6.8) wash till pH of effluent reached 6, then reequilibrated to starting buffer A (see above).

C. The performance of a silica-based HPLC column after a 30-min exposure to various mobile phases that are capable of regenerating and/or sterilizing silica-based chromatography media.
(A) Initial chromatography (control) of crude egg white. The column was then washed in distilled water, followed by a 30-min wash with 70% ethanol, a rewash with distilled water, and reequilibration. A crude egg white sample was then applied to the column to generate a chromatogram for comparison purposes with control. This process was repeated for each of the remaining mobile phases tested (see Fig. 3).
(B–F) The resulting crude egg white chromatograms obtained after washing column with each indicated mobile phase. Column: 4.6 × 50 mm, 5-micron Bakerbound Wide-Pore CBX; mobile phase: buffer A = 10 mM MES, pH 5.6 and buffer B = 0.2 M sodium acetate, pH 7.0; linear gradient: 0% B to to 100% B in 15 min; detection: UV 280 nm; sample volume: 10 μL.

D. Mobile Phase A = 10 mM MES, pH 5.6
B = 1.0 M sodium acetate, pH 7.0

Gradient: 0% B to 100% B over 30 min
Flow Rate: 4 mL/min

Injection Volume: 20 mL
Concentration: 50 mg/mL total protein
Detection: UV @ 280 nm

E. Mobile Phase A = 2.0 M NH_4 sulphate + 25 mM KH_2PO_4, pH = 7.0
B = 25 mM KH_2PO_4, pH = 7.0
Gradient: 0% B to 100% B over 30 min
Flow Rate: 4 mL/min
Injection Volume: 28 mL
Detection: UV @ 280 nm

F. Column: Bakerbond Wide-Pore™ Butyl, 4.6 × 50 mm
Mobile Phase A = 0.1% trifluoroacetic acid (TFA) in water
B = 0.1% TFA in acetonitrile
Gradient: 0% B to 100% B over 5 min
Flow Rate: 2 mL/min
Injection Volume: 50 μL
Detection: UV @ 280 nm

G. Mobile Phase A = 0.1% trifluoroacetic acid (TFA)
B = 0.1% TFA in acetonitrile
Gradient: 20% B to 60% B over 5 min
Flow Rate: 2 mL/min
Injection Volume: 50 μL
Sample: Supernatant from diluted fresh egg white treated with the indicated amounts of 40-micron Bakerbond Wide-Pore™, CBX.

H. Mobile Phase A = 25 mM Tris, pH 7.0
B = 2.0 M sodium acetate, pH 7.0
Gradient: 0% B to 100% B over 45 min
Flow Rate: 20 mL/min
Injection Volume: 400 mL
Sample: Fresh egg white diluted and batch extracted with Bakerbond Wide-Pore™ CBX, 40 micron.

I. Mobile Phase A = 25 mM Tris, pH = 7.0
B = 2.0 M sodium acetate, pH = 7.0
Gradient: 0% B to 100% over 30 min
Flow Rate: 4 mL/min
Injection Volume: 28 mL
Concentration: 5.5 mg/mL total protein
Detection: UV @ 280 nm

J. Mobile Phase A = 25 mM Tris, pH 7.0
B = 2.0 M Sodium Acetate, pH 5.0

Gradient: 0% B to 100% B over 15 min
Flow Rate: 1 mL/min
Injection Volume: 5 mL of 1:4 dilution of fraction 7

K. Mobile Phase A = 10 mM KH_2PO_4, pH = 6.0
B = 250 mM KH_2PO_4, pH = 6.8
Gradient: 0% B to 50% B over 60 min
Flow Rate: 1 mL/min
Injection Volume: 0.5 mL
Detection: UV @ 280 nm

APPENDIX 2

Trademarks

Trademarks of J.T. Baker Inc.:

Bakerbond
Bakerbond Wide-pore
Bakerbond MAb
Bakerbond ABx

Where the names of products of companies other than J.T. Baker have trademark status, this has been indicated at least once in this article.

APPENDIX 3

Patents

J.T. Baker holds the following patents, which protect the chromatographic products described in this chapter: U.S. Patent Numbers 4469630, 4522724, 4540486, 4551245, 460042235, and 4606825.

REFERENCES

1. J. Richey, Amer. Lab, October, 104–129 (1982).
2. Y. Kato, K. Komiya, H. Sasaki, and T. Hashimoto, J. Chromatogr., *193*, 311 (1980).
3. R. Anderson and L. Hagel, Anal. Biochem., *141*, 461–465 (1984).
4. A.J. Alpert, "New Materials and Techniques for HPLC of Proteins," Ph.D. Thesis, 1980, Purdue University.
5. F.E. Regnier and R. Noel, J. Chromatogr. Sci., *14*, 316 (1976).
6. L.R. Snyder and J.J. Kirkland, *Introduction to Modern Liquid Chromatography*, 2nd Ed., Wiley, New York, 1979, p. 173.

7. W.R. Melander and C.G. Horvath, in *High-Performance Liquid Chromatography, Advances and Perspectives*, Vol. 2, C.G. Horvath, Ed., Academic Press, New York, 1980.
8. See, for example, the collected works from the 5th International Symposium on HPLC of Proteins, Peptides and Polynucleotides in J. Chromatogr., *359* (1986).
9. *Handbook of HPLC for the Separation of Amino Acids, Peptides, and Proteins*, Vols. 1 and 2, W.S. Hancock, Ed., CRC Press, Boca Raton, FL, 1984.
10. B.L. Karger, L.R. Snyder, and C.G. Horvath, *An Introduction to Separation Science*, Wiley Interscience, New York, 1973, pp. 135–145.
11. L.R. Snyder and J.J. Kirkland, *Introduction to Modern Liquid Chromatography*, 2nd Ed., Wiley, New York, 1979, p. 37.
12. R.V. Lewis, A.S. Stern, S. Kinura, J. Rossier, S. Stein, and S. Udenfried, Proc. Nat'l. Acad. Sci., USA, *77*, 5018 (1980).
13. K.J. Wilson, E. Van Wieringen, S. Klauser, M.W. Berchtold, and G.J. Hughes, J. Chromatogr., *237*, 407 (1982).
14. R.W. Stout, J.J. DeStefano, and L.R. Snyder, J. Chromatogr., *282*, 263 (1983).
15. A.V. Kiselev, in *The Structure and Properties of Porous Materials* (Proc. 10th Symp. Colston Res. Soc., Univ. Bristol, March, 1958), D.W. Everett and F.S. Stone, Eds., Butterworths, London, 1958.
16. B.W. Sands, Y.S. Kim, and J.L. Bass, J. Chromatogr., *360*, 353–369 (1986).
17. D.R. Nau, Biochromatography, *1*(2) (1986).
18. Compare the data in Table 5 and Table 12 for the protein capacities of Monobeads and Bakerbond Wide-Pore bonded phases.
19. Compare the data in Table 5 and Table 12 for the protein capacities of TSK gel PW and Bakerbond Wide-Pore bonded phases.
20. L.C. Sander and S.A. Wise, J. Chromatogr., *316*, 163–181 (1984).
21. E. Bayer, K. Albert, J. Reiners, M. Nieder, and D. Muller, J. Chromatogr., *264*, 197–213 (1983).
22. Silanes are available from such companies as Petrarch Systems, Inc., Bristol, PA; Silar Laboratories, Inc., Scotia, NY; and Dow Corning, Midland, MI.
23. K.K. Unger, *Porous Silica, Its Properties and Uses as a Support in Column Liquid Chromatography*, Elsevier, Amsterdam, 1979.
24. J. Kohler, D.B. Chase, R.D. Farlee, A.J. Vega, and J.J. Kirkland, J. Chromatogr., *325*, 275–305 (1986).
25. W.F. Harrington and G. Kegeles, in *Methods in Enzymology*, Vol. 27, Part D, C.H.W. Hirs and S.N. Timasheff, Eds., Academic Press, New York, 1973, p. 306.
26. E.L. Uhlenhopp and B.H. Zimm, in *Methods in Enzymology*, Vol. 27, Part D, C.H.W. Hirs and S.N. Timasheff, Eds., Academic Press, New York, 1973, p. 483.
27. F.H. Johnson, J. Eyring, and M.J. Polissar, *The Kinetic Basis of Molecular Biology*, Wiley, New York, 1954, p. 286.
28. M.T. Takahashi and R.A. Alberty, *Methods in Enzymology*, Vol. 16, K. Kustin, Ed., Academic Press, New York, 1969, p. 31.
29. B.H. Simm and D.M. Crothers, Proc. Nat. Acad. Sci., U.S.A., *48*, 905 (1962).
30. G.B. Alexander, W.M. Heston, and R.K. Iler, J. Phys. Chem., *58*, 153 (19564).
31. Y.A. Cherkinskii and X. Kynaz'kova, Dokl. Akad. Nauk SSSR., *198*, 45 (1971).

32. H. Baumann, Beitr. Silkose-Forsch., *37*, 47 (1955).
33. A.J. Alpert and F.E. Regnier, J. Chromatogr., *185*, 375–392 (1979).
34. C.M. Jephcott and J.H. Johnston, Arch. Ind. Hyg. Occup. Med., *1*, 323 (1950).
35. G.B. Alexander and G.H. Bolt, U.S. Pat. 3,007,878 (DuPont) (1961).
36. K. Harrison, W.C. Beckham, Jr., and T. Yates, paper 225, 5th International Symposium on HPLC of Proteins, Peptides and Polynucleotides, Toronto, Canada (1985).
37. Serva Fine Biochemicals, Inc., Westbury, NY.
38. J.G. Atwood, G.J. Schmidt, and W. Slavin, J. Chromatogr., *171*, 109–115 (1979).
39. T. Mizutani and A. Mizutani, J. Chromatogr., *111*, 214 (1975).
40. H.G. Bock, P. Skene, S. Fleischer, P. Cassidy, and S. Harshman., Science, *191*, 380 (1976).
41. K.C. Chadha and E. Sulkowski, Preparative Biochemistry, 11, 467–482 (1981).
42. W. Kopaciewicz, M.A. Rounds, and F.E. Regnier, J. Chromatogr., *318*, 157–172 (1985).
43. S. Gupta, E. Pfannkoch, and F.E. Regnier, Anal. Biochem., *128*, 196–201 (1983).
44. See Tables 12 and 13 in this chapter.
45. R.E. Majors and M.J. Hopper, J. Chromatogr. Sci., *12*, 767 (1974).
46. L.C. Sander and S.A. Wise, in *Advances in Chromatography*, Vol. 25, J.C. Giddings, E. Grushka, J. Cazes, and P.R. Brown, Eds., Marcel Dekker, New York, 1986, pp. 139–218.
47. L.R. Snyder, in *Principles of Adsorption Chromatography*, Marcel Dekker, New York, 1968, pp. 143–153.
48. H. Halpaap, J. Chromatogr., *78*, 63 (1973).
49. T.G. Waddell, D.E. Leyden, and M.T. DeBello, J. Amer. Chem. Soc., *103*, 5303 (1981).
50. B.D. Black, E.C. Jennings, and J.W. Higgins, paper 509, 10th International Symposium on Column Liquid Chromatography, San Francisco, 1986).
51. L. Nandek, B. Buszewski, and B. Berek, J. Chromatogr., *360*, 241–246 (1986).
52. R.P.W. Scott and C.F. Simpson, J. Chromatogr., *197*, 11–20 (1980).
53. A.J. Alpert, J. Chromatogr., *266*, 23–37 (1983).
54. J.L. Fausnaugh, E. Pfannkoch, S. Gupta, and F.E. Regnier, Anal. Biochem., *137*, 464–472 (1984).
55. G. Schomburg, A. Deege, J. Kohler, and U. Bien-Vogelsand, J. Chromatogr., *282*, 27–39 (1983).
56. P.A. Hartman, J.D. Stodola, G.C. Harbour, and J.G. Hoogerheide, J. Chromatogr., *360*, 385–395 (1986).
57. J. Wolstenholme and J.M. Walls, Research and Development, August, 58—61 (1986).
58. K.B. Lipkowitz, D.J. Malik, and T. Darden, Tetrahedron Letters, *27*, 1759–1762 (1986).

CHAPTER 3

On the Adsorption of Protein to Alkylsilica: Relevance to Reversed-Phase and Hydrophobic Interaction Chromatography

ROBERT A. BARFORD

USDA, ARS, Macromolecular and Cell Structure Research Unit, Philadelphia, Pennsylvania

1. PROTEIN ADSORPTION PHENOMENA

During the past decade, enlightenment of factors that influence the sorption of proteins to surfaces has been the thrust of much research (1). Most of the effort has been directed toward studies of blood proteins as related to biocompatibility of prosthetic devices and materials used in the medical field. Some of the general concepts and observations, however, are relevant to chromatography.

The proteins sorbed at any interface—water/air, water/oil, or water/solid—may be perceived depending on concentration and on resistance of the biopolymer to structural perturbations, as shown schematically in Fig. 1. At low surface concentrations many proteins will spread at an air or oil interface with water (2). Spreading is often accompanied by changes in secondary protein structure to permit loops of apolar and polar amino acids to partition into phases of similar polarity. At solid interfaces, the orientations may be in the surface plane or directed away from it into the solution but usually involve multiple contacts with the surface. As surface concentration increases, higher packing density inhibits spreading and promotes further looping of segments into the solution phase. Finally, with further increase in concentration multiple layers form often. These structural alterations are influenced by factors such as pH, ionic strength, and number of protein disulfide bridges (2). The last of these factors imparts rigidity in a fashion similar to increasing cross-links in organic polymers. Electrostatic interactions are important and, owing to electric double

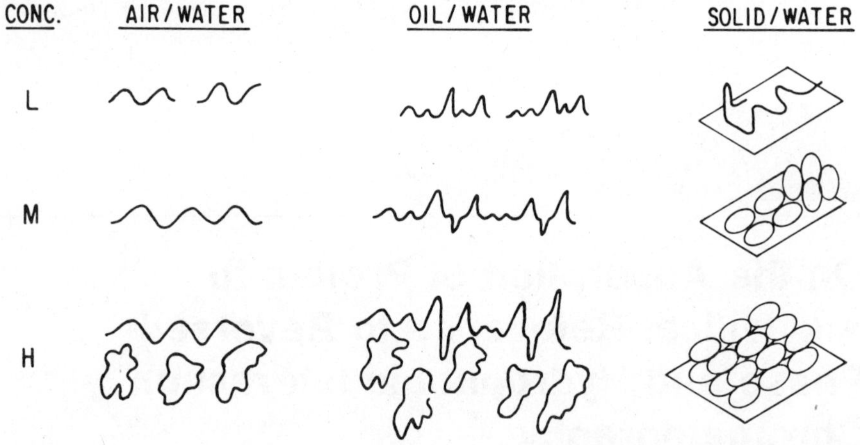

Fig. 1. Schematic of protein orientation at interfaces.

layer repulsion, tend to promote expansion of sorbed biopolymer at hydrogen ion concentrations removed from the isoionic point of a protein (2).

Sorption of proteins to ionic as well as neutral surfaces exhibit Langmuir or at least pseudo-Langmuir behavior (3, 4). Absorption that is described by the Langmuir isotherm is analogous to an ideal solution, that is, there is no interaction between adsorbed molecules. Moreover, adsorption ceases when monolayer coverage is achieved. Just as in solution chemistry, deviations from ideality are observed often in adsorption. Nevertheless, the Langmuir equation is an often used starting point.

The Langmuir equation is

$$S = S_p K C_m / (1 + K C_m) \tag{1}$$

where C_m is the protein concentration at equilibrium, S_p is the apparent binding capacity of the sorbent, S is the mg protein sorbed/g sorbent, and K is the desorption constant, which is related to the binding energy. This expression may be derived from both kinetic and statistical thermodynamic considerations and assumes a uniform surface with sorption limited to a monolayer. An example is given in Fig. 2 (5); which shows the isotherm for sorption of BSA onto highly substituted alkylsilicas and demonstrates that there was little difference in affinity of BSA for either the octyl- or octadecyl-derivatized surface. By inverting both sides of Eq. 1 and rearranging, a linear form results so that when C_m/S is plotted against C_m, the sorbent binding capacity S_p and desorption constant K can be obtained. These are shown in Table 1 for sorption of BSA from a number of different solvents. The higher capacity at pH 2 could reflect the fact that BSA is more flexible and elongated at lower pH and therefore has access to a larger fraction of sorbent pores. High affinity of the alkylsilicas for the

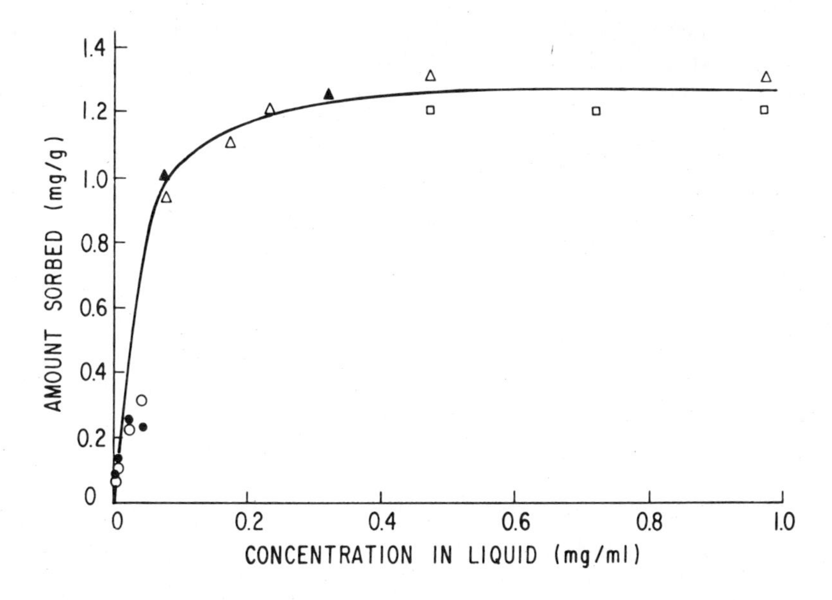

Fig. 2. Static sorption of bovine serum albumin onto alkylsilicas. Solvent: 2-propanol (40%)/0.05 *M* phosphate, pH 2.1. (▲) 500-Å pore, data using C_8 and C_{18} alkyl groups superimposable; (△) 100-Å pore, data using C_8 and C_{18} alkyl groups superimposable; (○) 100-Å pore, C_{18}; (●) 100-Å pore, C_8; (□) concentration in solvent determined by UV measurement, other data by fluorescence. $T = 23 \pm 1°C$.

Table 1. Sorption *a* of BSA to Ocytl- and Octadecyllkylsilica

Solvent	Apparent Binding Capacity (S_p) (mg/g)	Desorption Constant (K) [(g/mL) × 10^3]
	pH 2.1	
0.05 *M* Phosphate (27)[b]	3.2	5
30% 2-Propanol (29)	3.2	68
40% 2-Propanol (32)	1.9	130
0.01% Neodol (21)	2.6	11
1.0% Neodol (18)	2.6	110
	pH 7.0	
0.05 *M* Phosphate (8)	1.3	30
0.01% Neodol (13)	1.2	40
1.0% Neodol (22)	1.5	300

[a]One hour contact time at 23 ± 1°C.
[b]Number of experiments in parentheses.

protein is shown for all solvents except 1% neodol, a polyethoxylated alcohol surfact, and 40% isopropanol in phosphate buffer.

Spectroscopic techniques are used to probe the nature of species sorbed on solids (6). Infrared measurements showed, for example, that about 10% of the serum albumin molecule is in contact with the silica particle to which it is bound (7) and that this percentage is the same regardless of whether native or cross-linked protein is used in the experiment, indicating that major unfolding does not occur upon sorption. When just one disulfide of the native protein was broken before equilibration, however, the number of contacts increased by 70%. Other infrared studies demonstrated that albumin molecules attached to the surface were more ordered than molecules absorbed onto other albumin molecules (8). Studies (9) revealed that serum albumin remained immunologically active upon sorption to polystyrene. Fluorescence data support the hypotheses that ribonuclease sorbs onto this same substrate with only minor structural change (10). Sorption of fibronectin, a plasma protein consisting of two long chains of 220 kilodalton molecular weight connected by disulfide bridges at one end, on silicas has been characterized by ellipsometry (11). These studies revealed that the protein conformations were different on alkylated silica as compared to unmodified silica. Partial desorption from the hydrophillic surface was accomplished upon solution dilution but not from the hydrophobic one. Desorption upon dilution is referred to commonly as reversible desorption.

Recently it was demonstrated through the use of a novel technique for measuring surface tensions of sorbed proteins (12) that the hydrophobicity of serum albumin, immunoglobulin, and fibrinogen sorbed onto polytetrafluoroethylene (PFTE), polyvinyl chloride (PVC), and nylon (N-6,6) particles increased with decreasing solution concentration over a 1–25 mg/ml range. This change was related to alterations in protein structure that were inhibited spacially with the higher number of protein molecules on the surface. The magnitude of the effect increased with increasing hydrophobicity of the polymer particle (PFTE > PVC > N-6,6).

Not all proteins are as robust as those mentioned in the previous paragraphs and may exhibit much greater structural alterations. Studies based on chromatography, for example, have shown that papain (13) and alpha-lactalbumin (14) undergo structural alteration on hydrophobic silicas and that the degree is related to contact time. Spectroscopy also has demonstrated that changes in structure continue after sorption (8).

2. SORPTION FORCES

Having considered protein sorption at interfaces from a phenomenological perspective, let us now consider the forces acting between a protein molecule and a surface across a liquid. We assume that under the conditions usually employed in chromatography of proteins that the mobile phase ionic strength is sufficiently high that the effects due to electrostatic double layer are small

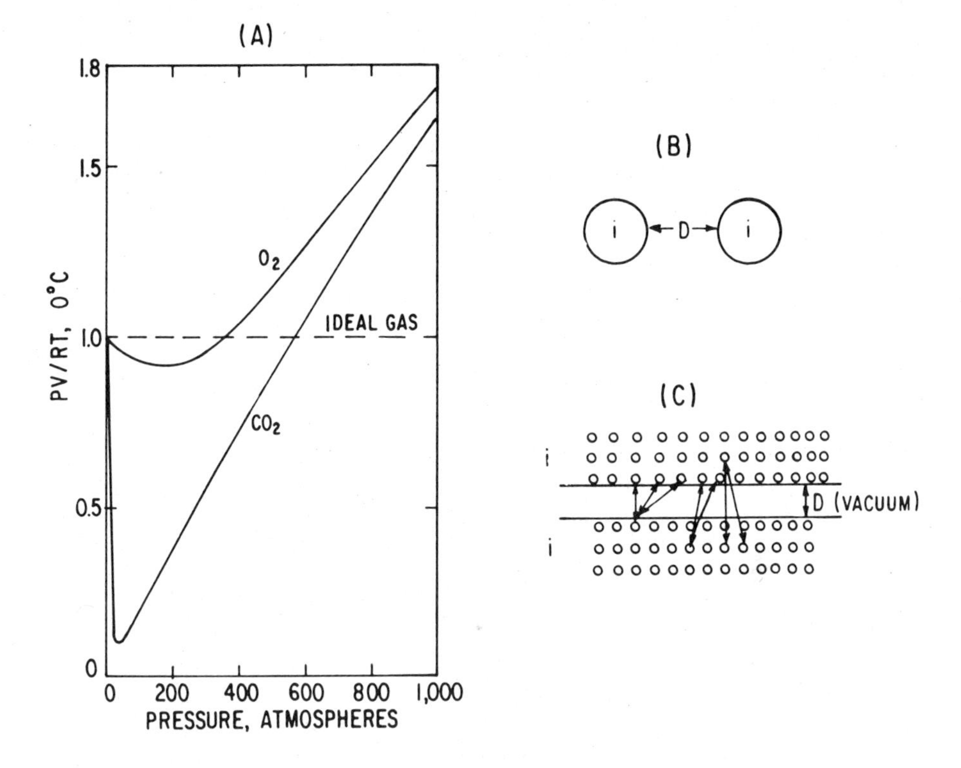

Fig. 3. Intermolecular forces: (*a*) Attraction between molecules accounts for deviation from ideality; (*b*) mutual attraction between two molecules in a vacuum (Eq. 3); (*c*) interaction of molecules in solids (Eq. 4).

when modified silicas are the sorbents. The number of ionizable silanols is small in such materials and, considering the bond lengths of covalently bound organic groups as well as their concentration, have minimal interaction with proteins. We may, therefore, consider the interaction in terms London–van der Waals forces.

The concept of cohesive forces had its origins in van der Waals attempt to explain deviations from the ideal gas law at high pressures:

$$\left(p + \frac{a}{V^2}\right)(V - b) = RT \tag{2}$$

where the correction term, a/V^2, accounts for the added contraction of volume due to the attraction between molecules. Examples of the effect show in Fig. 3*a*.

With the development of quantum mechanics, London (15) was able to quantitate van der Waals observation for molecules without permanent dipoles:

$$F = \frac{-3}{4} hf \frac{P_2}{D^6} = \frac{-B_{ii}}{D^6} \tag{3}$$

F is the mutal attraction energy between two molecules of species i under vacuum, h is Planck's constant, f is the characteristic frequency of oscillation of the charge distribution, P is the polarizability, and D is the separation distance (Fig. 3*b*).

Hamaker extended this equation to consider the attraction of assemblies of molecules in a solid with other assemblies (Fig. 3*c*). Then the interaction energy is the sum of interaction energies of all molecules present, which results in a pressure of attraction P:

$$P = \frac{-A_{ii}}{6\pi D^3} \tag{4}$$

where A_{ii} is the Hamaker constant, $\pi^2 q_i^2 B_{ii}$, is the number of atoms per cubic centimeter (16).

This approach to calculating the interaction between bodies from molecular properties has limitations in that it does not take into account the screening effect of surface molecules on interactions between molecules in secondary levels of the two bodies. It has been shown as a consequence of this effect that the predominant contribution to van der Waals forces comes from those parts of the bodies equal to the separation distance between them (17). This observation demonstrates the importance of surface layers on particles, around cells, or in structured macromolecules on the overall interaction between them.

A number of workers (18–20) have sought to overcome these limitations through consideration of macroscopic properties of materials. Such approaches lead to complex expressions for A that use measurable quantities such as dielectric constant, spectral absorbance, and surface tension. The observation that comparable values of A_{ii} are found when these various approaches are applied substantiates the validity of the concept, as well as some conclusions drawn from them. Hamaker constants for some polymers (21), normalized to that of PFTE, are given in Table 2. In general, there is good agreement between those calculated from spectral information and those determined from surface tension measurements.

The discussion in the previous paragraph and references contained in it leads us to three points for our consideration of liquid chromatography and proteins. First, van der Waals forces and the concepts evoking them are operable at distances to 100 Å. Such separation distances are not unrealistic for our stystem considering the sizes of protein molecules, high-performance supports, and the interstitial spaces between support particles. Second, the rule for Hamaker constants that describe interaction between two different species (1 and 2) is valid for many systems to within 5%. Thus,

$$A_{12} \cong A_{11} A_{22} \tag{5}$$

Table 2. Hamaker Constants

Polymer	Relative Hamaker Constant (Calculated)	Relative Hamaker Constant (from Surface Tension Measurements)
PFTE	1.0	1.0
Parafin	1.2	1.2
Polyethylene	1.3	1.8
Polystyrene	1.6	2.2
PMMA	2.1	1.9
PVC	2.6	2.2
Nylon	2.6	2.2
Polyester	2.9	2.2

for two materials interacting across a gap filled with a third, we may sum the interactions and write:

$$A_{132} = A_{12} + A_{33} - A_{13} - A_{23} \tag{6}$$

By substitution

$$A_{132} = (\sqrt{A_{11}} - \sqrt{A_{33}})(\sqrt{A_{22}} - \sqrt{A_{33}}) \tag{7}$$

Examination of this expression reveals that when A_{33} falls between A_{11} and A_{22}, A_{132} is negative. Third, from Eq. 4, we reach the often-overlooked conclusion that the van der Walls energy of interaction between two dissimilar materials in the presence of a third medium can be *repulsive*.

For this discussion, we have considered that van der Waals interactions result from forces between (a) permanent dipoles, (b) dipoles induced by dipoles, and (c) statistical dipoles resulting from random motion of electrons. These forces operate at relatively long distances from the interface. It has been proposed recently that at distances less than 2 Å beyond normal bond length need to be considered in systems consisting of polar liquids, apolar sorbents, and polymers (22). Since the supporting literature is presently small, the effect is not included in the subsequent discussion.

3. THERMODYNAMICS OF PROTEIN SORPTION

With this background we consider the Helmholz free energy of adhesion (23) of a particle (protein, p) and a solid (s) in a liquid (l):

$$\Delta F_{\text{slp}} = \gamma_{\text{ps}} - \gamma_{\text{pl}} - \gamma_{\text{sl}} \tag{8}$$

γ refers to the respective interfacial tensions. The expression therefore reflects

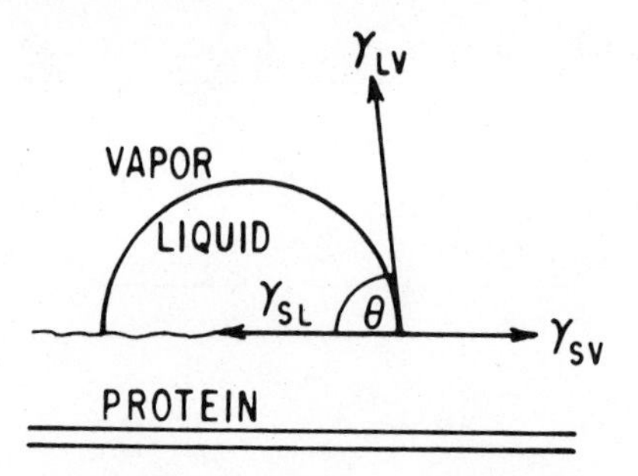

Fig. 4. Contact of liquid with thick protein layer contained on membrane. θ is contact angle, γ_{sv} and γ_{lv} are solid and liquid surface tensions, respectively; γ_{sl} is interfacial tension.

the difference in energy required to bring a protein from solution to the solid interface relative to the energy involved in returning solvent molecules to the bulk.

From the discussion in the previous section the interaction energy is described also by A_{slp} where s = 1, l = 3, p = 2 and that van der Waals repulsion is expected when $A_{ss} < A_{ll} < A_{pp}$. It has been demonstrated (24) that $A_{ii} \sim \gamma$; thus, from Eqs. (7) and (8) that sorption of protein onto the surface will be favored when the surface tension of the liquid is either above or below the surface tensions of both other components and desorption is favored when it falls between them.

3.1. Surface and Interfacial Tension Measurement

Measurement of surface tensions of liquids by DuNouy tensiometer or Wilhelmy balance is described in every text on surface chemistry and will not be discussed here. Surface tension determination for solids and proteins is less well known and will be described briefly.

Several approaches have been proposed for evaluating protein surface tension. One is to measure the incremental change as a function of protein concentration in solution. Because proteins may undergo major conformational change at the air/solution interface, low values result.

Another method uses the measurement of the angle of contact a sessile drop of liquid makes with a surface (Fig. 4). For proteins, thick layers are deposited from solution on ultrafilter membranes. The surface thus produced is in the hydrated, native state (23). Contact angles are measured with a goniometer or on enlarged photographs of the drop.

The relationships between the respective surface tensions are deduced from an equation of state (25) that is adopted from the Gibbsian dividing surface approximation. When this equation is combined with Young's equation, two basic equations result:

Table 3. Surface Tensions of Proteins

Protein	Method: Contact Angle[a] (mJ/m²)	Method: Sedimentation[b] (mJ/m²)
Human Serum Albumin	70.2	69.7
Immunoglobulin G	67.7	67.8
Bovine Serum Albumin	70.3	—
Bovine Serum Albumin[c]	35.0	—
Ovalbumin	68.8	—

[a] Liquid was saline (30).
[b] Reference (12).
[c] Liquid was EtOH/buffer, pH 2 (29).

$$\gamma_{sl} = \frac{[(\gamma_{pv})^{\frac{1}{2}} - (\gamma_{lv})^{\frac{1}{2}}]^2}{1 - 0.015\,(\gamma_{pv}\,\gamma_{lv})^{\frac{1}{2}}} \tag{9}$$

$$\cos\theta = \frac{(0.015\gamma_{pv} - 2.00)\,(\gamma_{pv}\,\gamma_{lv})^{\frac{1}{2}} + \gamma_{lv}}{\gamma_{lv}\,[0.015(\gamma_{pv}\,\gamma_{lv})^{\frac{1}{2}} - 1} \tag{10}$$

These equations may be solved through the use of published computer programs (26) or tables (27) such that a protein's surface (tension γ_{pv}) can be obtained from a contact angle determined with a liquid of known surface tension (γ_{lv}). This liquid is generally physiological saline.

The final technique to be described is based on the observation (28) that when similar masses of packing are dispersed in liquids of differing γ_{lv}, the bed volumes after settling will reach a minimum or maximum value at $\gamma_{sv} = \gamma_{lv}$, i.e., $A_{ii} = 0$. Whether a minimum or maximum is reached depends on the extent to which the particles agglomerate in early stages of sedimentation. If particles are coated with thick layers of protein absorbed from solution, $\gamma_{sv} = \gamma_{pv}$. Thinner layers may have substantially lower γ_{pv}, probably reflecting conformational change (12). Some γ_{pv} are given in Table 3.

4. SURFACE TENSIONS OF LC PACKINGS

The sedimentation technique previously described, can be used in the absence of protein solution, to determine surface tensions of LC packings (31). A table of surface tensions for chemically modified silicas is presented in Table 4. In general, the hydrocarbonaceous packings (reversed phase) are in the 32–37 mJ/m^2 range, whereas those with carbon–oxygen functionality are much higher.

5. RELEVANCE TO CHROMATOGRAPHY

We now examine the chromatography of proteins from consideration of van der Waals attraction/repulsion concepts as shown in Eq. (8) and of measured

Table 4. Surface Tension (γ_{sv}) of Derivatized Silica LC Packings[a]

Organo Group	γ_{sv} (mJ/m^2)	Organo Group	γ_{sv} (mJ/m^2)
Polyamidopropyl[b]	53	*n*-Butyl (deactivated)	41
Diether	52	*n*-Butyl	36
Diol	47	Octadecyl	37
t-Butyl	39	*t*-Butylphenyl	35
Diether (deactivated)	39	*n*-Hexyl	32

[a] Determined by sedimentation method (31).
[b] From SynChrom, Inc.; others supplied by Supelco, Inc.

surface tensions of column packings (s), proteins (p), and mobile phases (m). Plots of ΔF_{smp} against surface tensions of some modified silicas (Table 4) are given in Fig. 5 for BSA and OVAL with two hypothetical mobile phases. The higher γ_{mv} is analogous to that of water containing ~0.1 m*M* NaCl or Na_2SO_4.

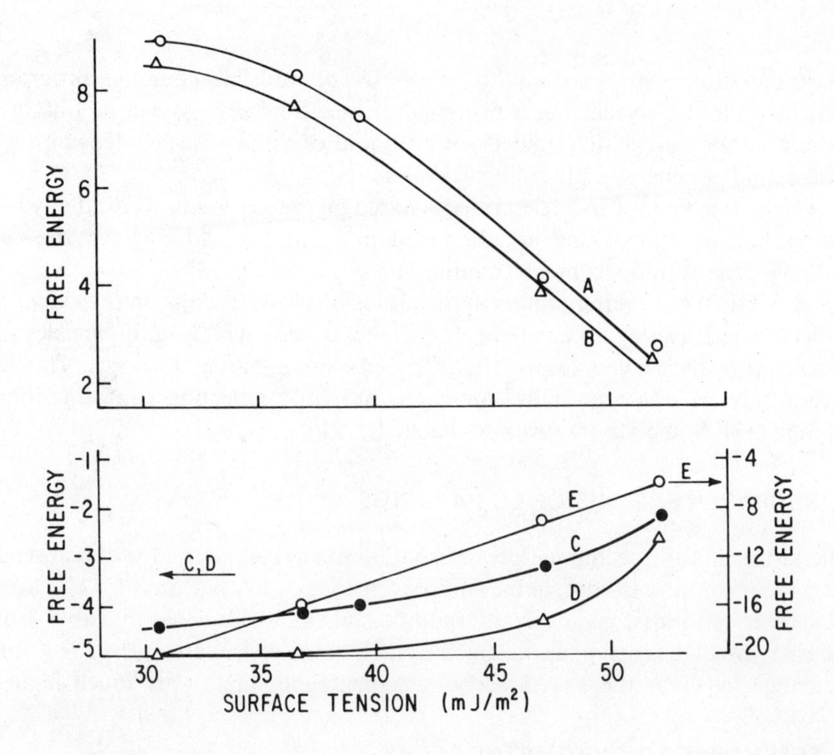

Fig. 5. Effect of support and mobile phase on protein sorption. (●–●) BSA; (△–△) Oval.; (○–○) solvent denatured BSA. γ_{mv} = 60 mJ/m^2 for curves, A, B, E; γ_{mv} = 73 mJ/m^2 for curves C and D.

With such mobile phases sorption of proteins is favored on all packings although binding affinity is reduced greatly at higher γ_{sv}. Packings with such surface tensions are used commonly for size-exclusion chromatography or "hydrophobic interaction" chromatography (HIC). The latter are similar chemically to the former but lightly alkylated. The energy of interaction becomes similar for the proteins as γ_{sv} increases but is negative unless the surface tension of the mobile phase is reduced until it is less than that of the protein. Ideally, then, determination of molecular size should be conducted with such mobile phases to prevent sorption. However, the use of buffer additives to achieve this may be precluded because the additives could alter protein size or shape or both also.

Further addition of salt increases γ_{mv} and promotes sorption. For the lightly alkylated supports, then, that protein retention can be manipulated by varying salt concentration is predicted. We may relate F_{smp} to a common measure of chromatographic retention, k':

$$k' = \frac{V_R - V_m}{V_m} = K\left(V_s/V_m\right) \tag{11}$$

and,

$$\Delta F = -RT \ln K \tag{12}$$

so that,

$$k' = e^{(-\Delta F/RT)(V_s/V_m)} \tag{13}$$

where V_s/V_m is the phase ratio, K is the distribution coefficient, and V_R is the protein retention volume. A strategy, then, for HIC is to introduce protein mixture with high surface tension (salt) mobile phase and reduce surface tension, eluting each protein as γ_{mv} passes that of the protein. Reduction of surface tension may be accomplished by reducing salt concentration or by addition of a modifier such as ethylene glycol.

Alternatively, retention may be increased by reducing the surface tension of the packing, which is accomplished by increasing the length or surface concentration of alykyl groups. This effect, as well as the salt concentration effect, were observed experimentally (32, 33). They were examined in the framework of solvophobic theory. The conclusions drawn in this report do not evoke concepts of repulsion by the solvent but concept of van der Waals attraction/repulsion of the protein and the surface that are accounted for by thermodynamic considerations. It should be noted, considering the amphiphillic nature of proteins, that the van der Waals attraction between an apolar specie and a polar one in water has been shown to be significant (34).

As the surface alkyl group density is increased further or is the only surface modification, γ_{sv} is substantially lower and ΔF becomes much more negative.

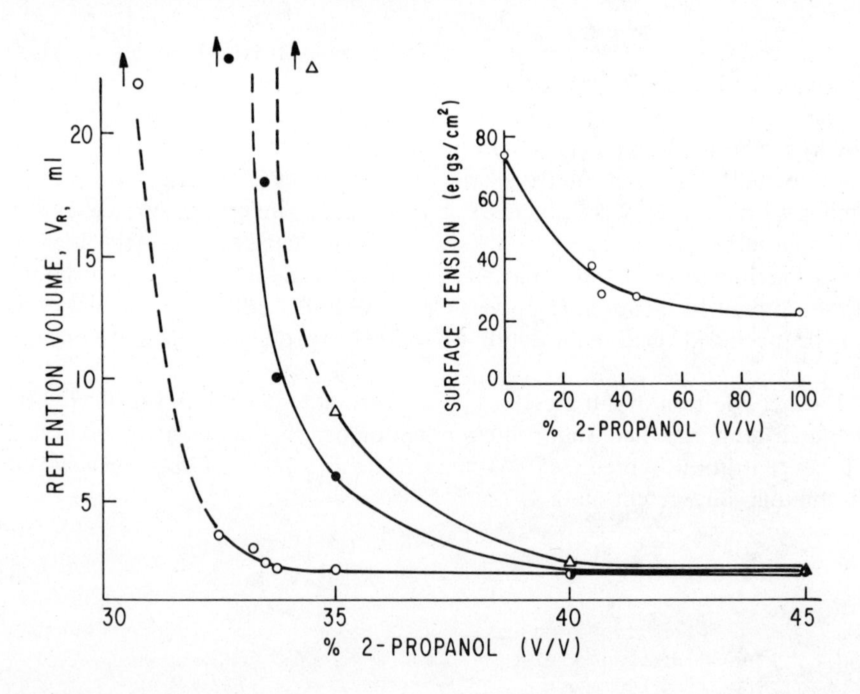

Fig. 6. Relation of surface tension to retention of (○) BSA, (●) beta-lactoglobulin, and (△) hemoglobin on a reversed-phase column. Mobile phase: phosphate buffer/ 2-propanol (pH 2.1) 1 erg/cm^2 = 1 mJ/m^2 (29).

Mobile phases containing very large proportions of organic additive in buffer are required to desorb proteins. This is shown in Fig. 6 for BSA, beta-lactoglobulin, and hemoglobin (29). The inset indicates the mobile phase surface tension that correlates with the percentages of 2-propanol that brought about elution of the three proteins from a reversed-phase column. It is seen readily that small changes in eluant composition caused large effects on retention. At 40°C, increasing 2-propranol percentage by only 3.75% reduced retention by a factor of 12 (Fig. 7). Compared to the large effect on retention due to mobile phase composition, the effect of temperature is slight but tends to decrease as temperature increases. A decrease is predicted from consideration of ΔF_{smp}, since γ_{mv} decreased with temperature increase. When ethanol replaced 2-propanol, elution of BSA occurred at the same surface tension value, although ~50% ethanol was needed to achieve 35 mJ/m^2 as compared to ~33% 2-propanol. Such proportions of organic modifier can alter the secondary structure of many proteins as observed by spectroscopic methods (35) and as evidenced by lower measured surface tensions. Solvent denatured BSA, for example, has a surface tension of ~35 mJ/m^2, whereas the value of native protein is ~70 mJ/m^2 in buffer. As seen in Fig. 5, the solvent denatured BSA has much greater attraction to a reversed-phase column than the native protein.

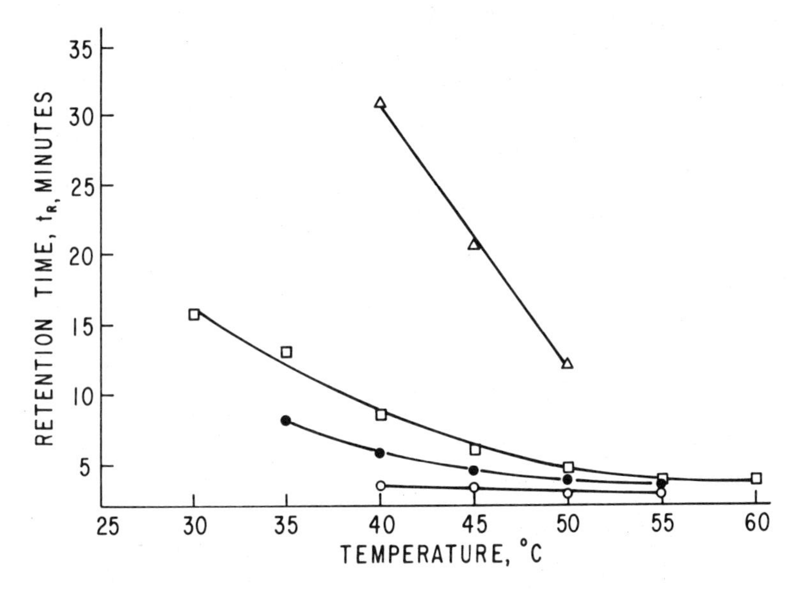

Fig. 7. Mobile phase composition and temperature effects on beta-lactoglobulin retention mobile phase and column as in Fig. 6: (△–△) 31.25% 2-propanol; (□–□) 33.25%; (●–●) 33.75%; (○–○) 35.00%.

6. CONCLUSIONS

The general view of a protein is that of a molecule of complex chemical nature with a high proportion, but not all, of the apolar amino acids arranged inside of its three-dimensional structure and a high proportion, but not all, of polar amino acids on the surface. Moreover, because proteins contain large numbers of ionizable groups so that a high electrostatic potential may occur depending on solution pH and ionic strengths. Therefore, proteins may interact with surfaces through a number of chemical functionalities. Specific interactions of proteins are mediated by many effects that influence their surface composition, structure, and the orientation of interacting groups. Furthermore, in chromatography, fluid dynamics and adsorption kinetics may be important. Nevertheless, protein adsorption and, therefore, retention in liquid chromatography may be approximated by considerations of van der Waals attractive and repulsive forces as mainifested by the free energy required to bring a protein molecule from solution to an equilibrium position at the interface. Nonpolar/nonpolar, nonpolar/polar, and polar/polar interactions are considered and arguments based on repulsion by the solvent need not be evoked. Surface energies of proteins and of chromatographic packings can be measured and used to guide development of separations by chromatography.

Size-exclusion, "hydrophobic interaction," and reversed-phase siliceous packings represent a continuum of materials with decreasing surface tension.

Retention of proteins is decreased by decreasing surface tension of the mobile phase.

For "hydrophobic interaction" packings, retention of native proteins may be influenced by number and size of alkyl groups and may be manipulated by salt concentration of aqueous mobile phase or through additon of small concentration of organic modifier. Sorption is at one or a few alkyl sites, so that protein structural changes due to sorption are minimal. Some alteration may occur, and the degree may be influenced by contact time. However, for many preparative applications where high recoveries of protein with native structure or activity or both are required, the conditions of HIC offer advantages. For highly akylated (reversed-phase) packings, high percentages of organic modifier are required to desorb protein. Such mobile phases induce structural alterations in many proteins. Moreover, multisite attachment tends to induce structural change. Alkyl chain length has little effect on sorption, but number of chains could, as they reflect unreacted silanol groups. These could act as secondary binding sites. Aryl-modified silicas have about the same surface tensions as alkyl silicas but are capable of specific π-interactions. Recoveries of protein from reversed-phase packings are improved by increasing temperature (36), use of silanophilic blocking agents (37), or protein derivatization (38) or all three. Recoveries tend to increase with sample load as expected from surface tension measurements. All of these may induce structural changes but these not likely to be important for analytical chromatography.

REFERENCES

1. L.-H. Lie, Ed., *Adhesion and Adsorption of Polymers*, Plenum, New York, 1980.
2. D.E. Graham, and M.C. Phillips, J. Colloid Interf. Sci., *70*, 427 (1979).
3. T. Suzawa, H. Shirahama, and T. Fujimoto, J. Colloid Interf. Sci., *86*, 144 (1982).
4. E.E. Graham and C.F. Fook, AIChE Journal, *28*, 245 (1982).
5. R.A. Barford, and B.J. Sliwinski, J. Dispersion Sci. Technol., *6*, 1 (1985).
6. J.D. Andrade and V. Hlady in *Advances in Polymer Science*, K. Dusek, Ed., Springer-Verlag, Berlin 1986, p. 1.
7. B.W. Morrissey, and R.R. Stromberg, J. Colloid Interf. Sci., *46*, 152 (1974).
8. R.J. Jakobsen, and D.G. Cornell, Appl. Spectrosc., *40*, 318 (1986).
9. N. Dezlic, and G. Dezelic, Croat. Chem. Acta, *42*, 457 (1970).
10. W. Norde, Commun. Agric. Univ. Wageningen, 76–6 (1976).
11. V. Jönsson, B. Ivarsson, I. Lundstroni, and L. Berghem, J. Colloid Interf. Sci., *90*, 148 (1982).
12. D.R. Absolom, Z. Policova, T. Bruck, C. Thomson, W. Zingg, and A.W. Neuman, J. Colloid Interf. Sci., *117*, 550 (1987).
13. S.A. Cohen, K.P. Benedek, S. Dong, Y. Tapuhi, and B.L. Karger, Anal. Chem., *56*, 217 (1984).
14. S.L. Wu, A. Figueroa, and B.L. Karger, J. Chromatogr., *371*, 3 (1986).
15. F. London, Z. Phys., *63*, 245 (1930).

16. H.C. Hamaker, Physica, *4*, 1058 (1937).
17. J. Visser, Adv. Collid Interf. Sci., *15*, 157 (1981).
18. I.E. Dzyaloshinski, E.M. Lifshitz, and L.P. Pitaevsku, Advan. Phys., *10*, 165 (1961).
19. J.N. Israelachvili, Proc. Royal Soc. Ser. A., *331*, 39 (1972).
20. H. Krupp, W. Schnabel, and G. Walter, J. Colloid Interf. Sci., *39*, 421 (1972).
21. J. Visser, Adv. Colloid Interf. Sci., *3*, 33 (1972).
22. C.J. van Oss, R.J. Good, and M.K. Chaudbury, J. Colloid Interf. Sci., *111*, 378 (1986).
23. C.J. Van Oss, D.R. Absolom, A.W. Newman, and W. Zing, Biochem. Biophys. Acta, *670*, 64 (1981).
24. C.J. van Oss, J. Visser, D.R. Absolom, S.N. Omenyi, and A.W. Neuman, Adv. Colloid Interf. Sci., *18*, 133 (1983).
25. C.A. Ward, and A.W. Neuman, J. Colloid Interf. Sci., *49*, 286 (1974).
26. A.W. Neuman, R.J. Good, C.J. Hope, and M. Sejpal, J. Colloid Interf. Sci., *49*, 291 (1974).
27. A.W. Neuman, D.R. Absolom, D.W. Frances, and C.V. Van Oss, Separ. Purif. Methods, *9*, 69–163 (1980).
28. E.I. Vargha-Butler, T.K. Zubovitz, H.A. Hamza, and A.W. Neuman, J. Disp. Sci. Technol., *6*, 357 (1985).
29. R.A. Barford, B.J. Sliwinski, A. Breyer, and H.L. Rothbart, J. Chromgtogr., *235*, 281 (1982).
30. C.J. van Oss, D.R. Absolom, and A.W. Neuman, Sep. Sci. Technol., *14*, 305 (1979).
31. D.R. Absolom, and R.A. Barford, Anal., Chem., *60*, 210 (1988).
32. J.L. Fausnaugh, L.A. Kennedy, and F.E. Regnier, J. Chromatogr., *317*, 141 (1984).
33. W.R. Melander, D. Covadini, and Cs. Horwath, J. Chromatogr, *317*, 67 (1985).
34. C.J. van Oss, D.R. Absolom, and A.W. Neuman, Fifty Fourth Colloid and Surface Science Syposium, ACS, Abstract 49 (1980).
35. J.M. Purcell, and H. Susi, J. Biochem. Biophys. Methods, *9*, 193 (1984).
36. N. Parris, and P.J. Kasyan, in "Methods for Protein Analysis," in J.P. Cherry, and R.A. Barford, Eds., American Oil Chemical Society Books, Champaign, IL, (1988).
37. P.C. Sadek, P.W. Carr, L.D. Bowers, and L.C. Haddard, Anal. Biochem., *153*, 359 (1986).
38. J.A. Bietz, J. Chromatogr., *225*, 219 (1983).

CHAPTER 4

Bioactivity of Insulin and Iodinated Insulin after Reversed-Phase HPLC*

BENNY S. WELINDER

Hagedorn Research Laboratory, DK-2820 Gentofte, Denmark

KIM R. HEJNAES

Nordisk Gentofte A/S, DK-2820 Gentofte, Denmark

BRUNO HANSEN

Hagedorn Research Laboratory, DK-2820 Gentofte, Denmark

1. INTRODUCTION

In spite of the extensive use of reversed-phase high-performance liquid chromatography (RP–HPLC) for the separation, isolation, and characterization of polypeptides with biological activity, only a few reports have described loss of bioactivity as a result of the chromatographic conditions used. In some cases RP–HPLC has been used to prepare polypeptides that have not been isolated using classical chemistry techniques and thus no reference compounds are available. Many biologically active proteins (such as enzymes) have so far not been purified using RP–HPLC, but have been analyzed using the less "stressful" conditions offered by high-performance ion-exchange chromatography (HPIEC) or hydrophobic interaction chromatography (HIC).

A number of reports describing partial or total loss of bioactivity after RP–HPLC of polypeptides/proteins are listed in Table 1, illustrating the potential risk in RP–HPLC such as:

1. Denaturation due to the organic solvents or acid pH or both in the mobile phase.

*Parts of this material has been presented as posters at the 4th International Symposium on HPLC of Proteins, Peptides and Polynucleotides, December 10th–12th 1984, Baltimore and the XII Congress of the International Diabetes Federation, September 23rd–28th, 1985, Madrid.

Table 1. RP–HPLC Purified Polypeptides with Partial or Total Loss of Bioactivity after High-Performance Liquid Chromatography

Reference Number	Compound Isolated	Probable Reason for Reduced Bioactivity
1	Papain	Too strong column binding
2	RNase, BSA, ovalbumin, horse radish peroxidase	Organic solvent denaturation
3	Chorionic gonadotropin	Organic solvent denaturation
4	Thyroid-stimulating hormone	Organic solvent denaturation
5	Growth inhibitory glycopeptide	Toxic compound(s) in CH_3CN
6	Lutropin	Organic solvent denaturation
7	Interleukin-2	Irreversible column binding in all but one buffer substance
8	Fibroblast growth factor	Bioactivity after HPIEC, not after RP–HPLC; solvent/organic denaturation
9	Interferon-beta	Activity lost after lyophilizing the interferon from the mobile phase
10	Somatostatin-releasing factor	Harmful binding to the stationary phase
11	Monoiodoinsulins	Column bleeding

2. Strong or irreversible binding to the stationary phase leading to reduced recoveries or large conformational changes of the sample molecule.
3. Binding of toxic substances in the mobile phase (buffer substances or acetonitrile) to the sample.
4. Column bleeding products interfering with the biological activity.
5. Denaturation during the isolation procedure where the sample is isolated from the mobile phase.

We have reported RP–HPLC purification and isolation of the four possible monoiodoinsulins and demonstrated that the purified proteins exhibit reduced binding capacity to rat adipocytes (compared to similar compounds purified and isolated using disc-gel-electrophoresis and ion-exchange chromatography) (11, 12). In order to explore possible harmful effects encountered in the chromatographical procedure, we have investigated the column bleeding (presence of degradation proudcts from the stationary phase) in the column eluate from a Lichrosorb RP-18 (5μ) column. These studies lead to the isolation of RP–HPLC purified monoiodoinsulin tracers from the column eluate in a manner that minimizes hydrophobic interactions between the polypeptide and materials liberated from the column.

Furthermore, we have purified in a preparative scale human pancreatic insulin using conventional, low-pressure chromatography (gel chromatography, ion-exchange chromatography) or RP–HPLC. We have analyzed these two human insulin preparations with respect to molecular structure, purity, and bioactivity and further compared semisynthetic human insulin against these two human insulin standards.

2. MATERIALS AND METHODS

Human crude insulin was obtained from corresponding pancreatic glands using conventional extraction procedures. This crude insulin contained 13% insulin peptide. After an alkaline precipitation the content was 40% insulin peptide, and this material was purified in two ways.

1. Conventional open-column chromatography: Gel chromatography (Sephadex G-50 SF, 3 M CH_3COOH). Ion-exchange chromatography (QAE-Sephadex A-25, Phosphate/7 M urea). Desalting (Sephadex G-25, 0.01 *M* NH_4HCO_3, pH 9).
2. RP–HPLC purification: Lichroprep RP-18 (25–40 μ), 250 × 50 mm ID 50 m*M* CH_3COONa pH 4/isopropanol. Vydac 218 TPB (15–20 μ), 350 × 38 mm ID, 1% triethylammoniumtrifluoroacetate pH 3/acetonitrile. Insulin was isolated from the eluants by zinc precipitation and desalted (Sephadex G-25, 1 *M* CH_3COOH).

All analyses were performed using the conventionally purified human insulin (Hum I), HPLC-purified human insulin (Hum II), and semisynthetic human insulin (Nordisk Gentofte A/S batch 1070, Hum III). Before analyses, all human insulin preparations were crystallized with zinc to a content of two zinc atoms per insulin hexamer.

Further details are given in the legends to figures and tables.

3. RESULTS

3.1. Column Bleeding

The optimal RP–HPLC system for separating insulin and iodinated insulins uses a Lichrosorb RP-18 (5 μ) column eluted isocratically with 0.25 *M* triethylammonium formate (TEAF) pH 6.0/21.5% iso-propanol (13). However, the monoiodoinsulins isolated after this RP–HPLC step had reduced binding affinities to isolated rat adipocytes. In order to investigate whether column degradation products could interfere with the activity of a small amount of purified proteins, the silica content in the column eluate was investigated for three new Lichrosorb RP-18 columns as well as two other silica-C18 stationary phases (Table 2).

In order to examine the effect of the organic modifier, a new Lichrosorb RP-18 (5 μ) column (No. 208810) was succeedingly eluted with 250 ml of the

Table 2. Column Eluate Silica Content[a]

	Silica Content (ng/mL)
Inlet buffer (0.25 M TEAF/iso-propanol)	11.4
Column eluate:	
Lichrosorb RP-18, 5 μ, No. 415000	216
No. 415016	30
No. 414999	188
Extraction of stationary phases:	
Lichrosorb RP-18, 5 μ, No. VV315533	204
Spherisorb ODS2, 3 μ, No. 820021	176
Vydac 218 TPB5, 5 μ, No. 270131	133

[a]Five hundred milliliter column eluate was lyophilized and analyzed for content of silica (upper part) or 1 g stationary phase was extracted in 50 ml mobile phase followed by lyophilization and silica estimation (lower part). The silica content was measured using inductive coupled plasma atomic emission spectrometry.

following solvents (followed by lyophilization and silica estimation using graphite furnace atomic absorption spectrometry):

Column Eluate	Silica Content (ng/mL)
0.25 *M* TEAF pH 6/21.5% iso-propanol	504
0.25 *M* TEAF 6/27% acetonitrile	244
0.25 *M* TEAF pH 6	88
Inital buffer	56

In an attempt to identify the products of column bleeding, 1 g of three different C18 stationary phases were extracted with 50 ml of 0.25 *M* TEAF pH 6.0/21.5% isopropanol for 24 hr. The extraction medium was lyophilized and taken up in 500 μl triethylene glycol dimethyl ether, 500 mg KOH was added, and the mixture heated at 200°C for 1 hr. The alkaline hydrolysis mixture was extracted with 500 μl *n*-pentane, which was analyzed by gas chromatography (5% silicone OV-1 on Chromosorb WHB 100/120, 185°C).

As can be seen from Fig. 1, octadecane and trace amounts of octadecanol can be demonstrated in extracts of all three RP materials if these extracts are exposed to strong base, with hydrolysis of carbon–silica bonds.

If the three RP-stationary phases are extracted with cyclohexane and this nonpolar extract is hydrolyzed as described previously, no octadecane or octadecanol could be demonstrated.

The column degradation material present in the column eluate, and which can be extracted from three stationary phases, is soluble in acid and alkaline, but not in nonpolar organic solvents. It is not sufficiently volatile for analysis by field desorption mass spectroscopy.

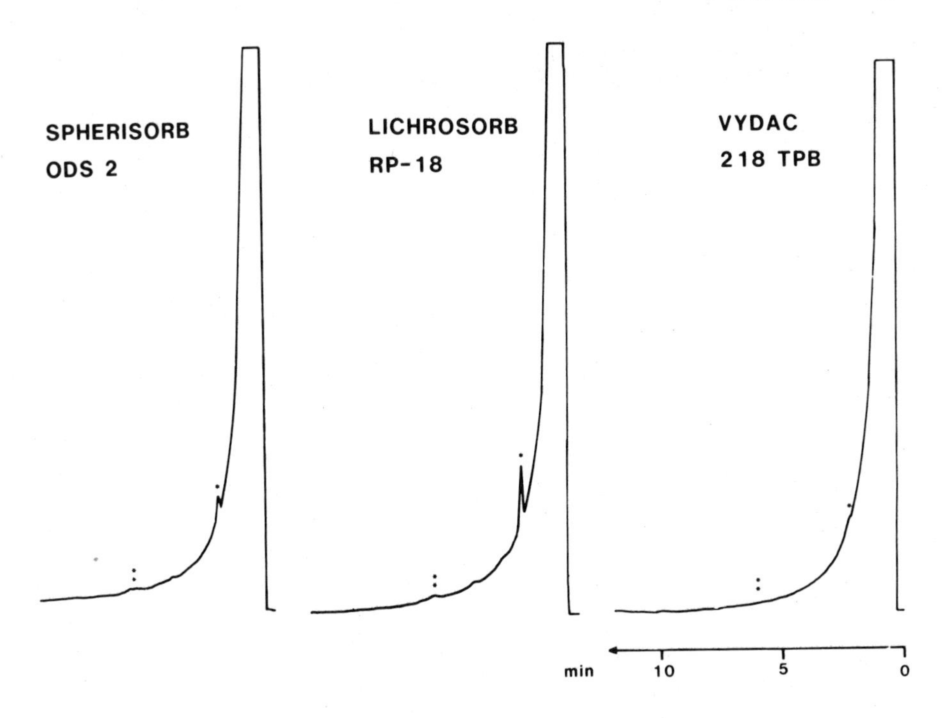

Fig. 1. Gas chromatography of 0.5 μl pentane extract of the alkaline hydrolysis mixture; * and ** indicate the elution position of octadecane and octadecanol, respectively.

3.2. Human Insulin Prepared by RP–HPLC and Open-Column Techniques

The purification of human pancreatic insulin by the two different chromatographical techniques are shown in Figs. 2 and 3.

Analytical RP–HPLC of the insulin preparations as well as NMR, CD and mass spectra of the insulins are shown in Figures 4–6.

The results of all analyses subjected to the three insulin preparations are summarized in Table 3.

4. DISCUSSION

The material present in the RP-column eluate and in the extraction medium when three different C18 stationary phases were extracted with the mobile phase used for separating insulin and the iodinated insulin derivatives are probably a silica-C18 derivative since it gives rise to octadecane and octadecanol after alkaline hydrolysis and since considerable amounts of silica can be demonstrated in the eluate and extraction medium.

Furthermore, no octadecane was detected when cyclohexane extracts of any of the three C18 stationary phases were analyzed directly or after alkaline hydrolysis of the nonpolar extracts. If silica was not bonded to the C18 moiety,

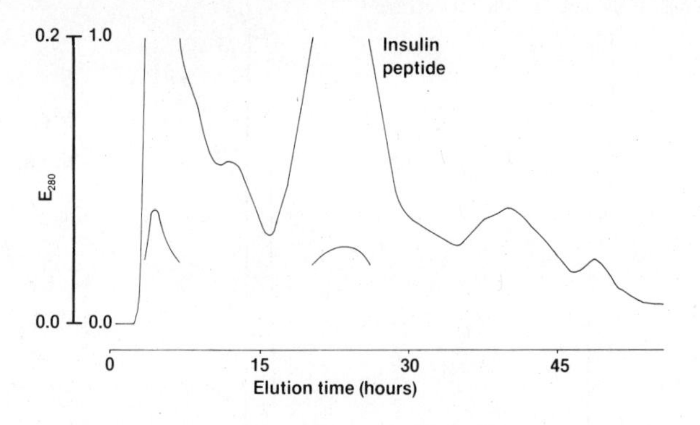

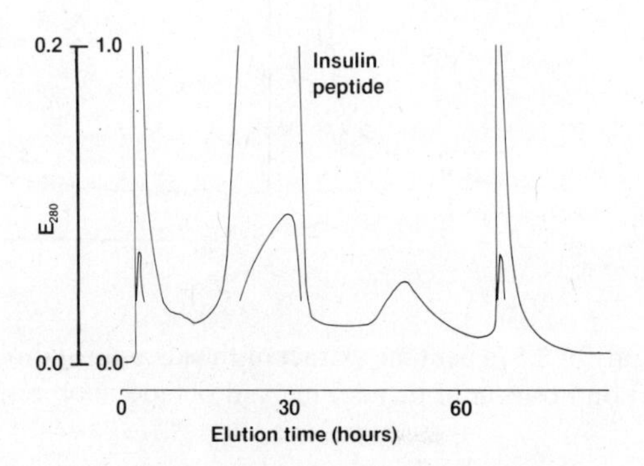

Fig. 2. (Top) conventional purification of 5-g alkaline-precipitated human crude insulin using a 85 × 9 cm ID Sephadex G-50 SF column eluted at 200 ml/hr with 3 *M* acetic acid. The fractions containing the insulin peptide where identified (analytical RP–HPLC), pooled, and lyophilized. UV detection at 280 nm. (Bottom) Anion-exchange chromatography of 1.5-g partially purified human crude insulin using a 35 × 5 cm ID QAE-Sephadex A-25 column eluted at 100 ml/hr with phosphate buffer pH 7/7 M urea. UV detection at 280 nm. Fractions containing the insulin peptide were pooled, desalted, and lyophilized.

it would be expected that the column derivative would be directly detectable by mass spectroscopy, which was not possible with the column degradation products described here.

The amounts of column bleeding products are considerable compared to the amount of monoiodoinsulins purified in a normal reversed-phase experiment: In the experiments performed so far, 1–5 ng of each of the four monoiodoisomers have been isolated in approximately 2 ml of column eluate—containing on average 100 times higher amount of column degradation products.

However, the purification of the monoiodoinsulin tracers by RP–HPLC could be performed without loss of binding affinity when the isolation procedure

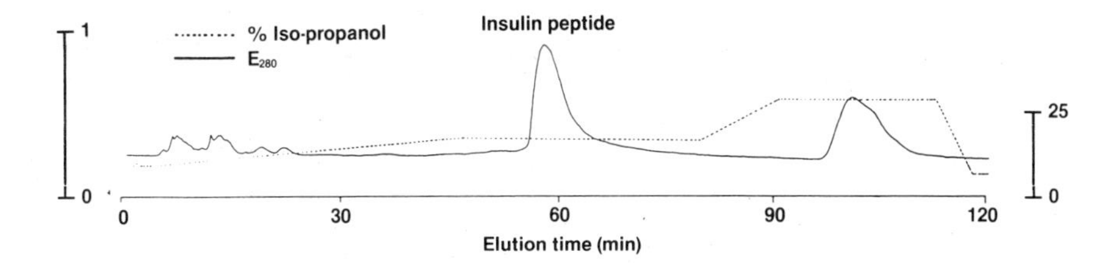

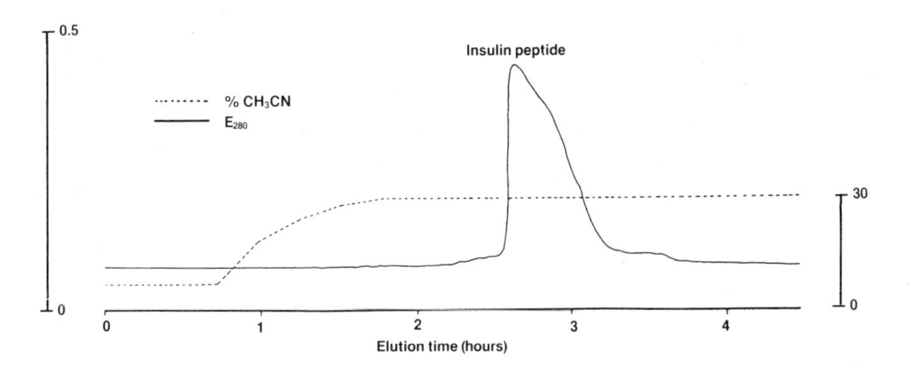

Fig. 3. (Top) RP–HPLC separation of 22-g alkaline-precipitated human raw insulin. Lichroprep C18 (25–40 μ), 250 × 50 mm ID. Eluted with 50 m*M* CH_3COONa, pH 4.0. Isopropanol gradient as indicated. UV-detection at 254 nm. Flow rate: 40 ml per min. (Bottom) RP–HPLC separation of 1.5-g partially purified human crude insulin. Vydac 218 TPB (15–20 u), 350 × 38 mm ID. Eluted with 1% triethylammoniumtrifluoroacetate pH 3.0. Acetonitrile gradient as indicated. UV detection at 280 nm. Flow rate: 10 mL per min.

was designed in a way to minimize the hydrophobic interaction between the polypeptide and possible column degradation products:

1. Larger amounts of monoiodoinsulins were purified in a single chromatographic run.
2. The tracers were isolated by gel chromatography in 40% ethanol or the RP–column eluate was extracted with cyclohexane before lyophilization.
3. Alternatively, the column eluate could be diluted with large amounts of albumin-containing buffers before use (the human serum albumin probably binds any C18 derivative present).

If the isolation of the monoiodoinsulins was performed as described above,

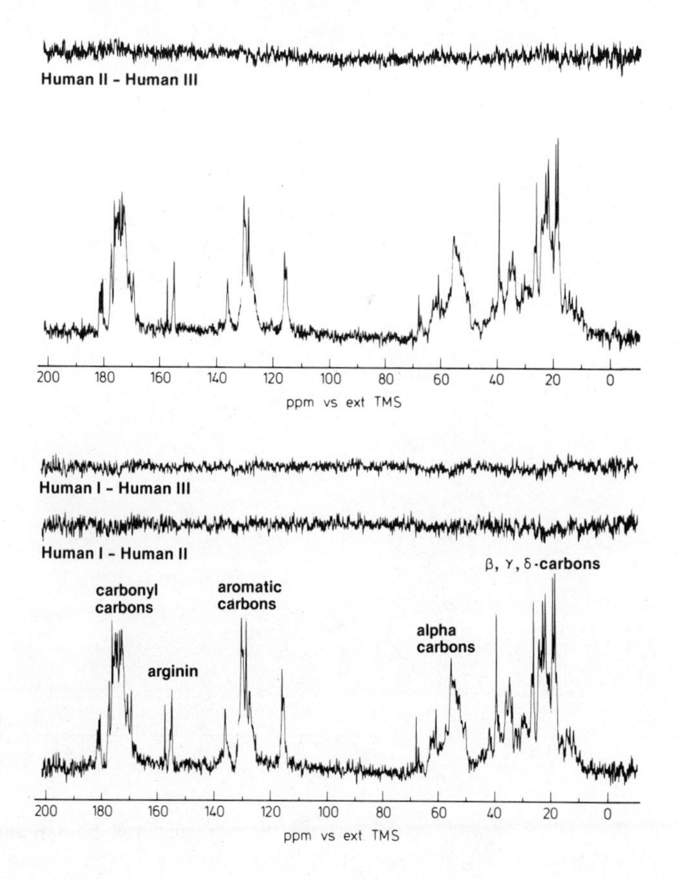

Fig. 4. NMR spectra of human insulins I, II, and III. Zn-crystallized human insulin (0.4% w/w) at 50 mg/ml, pH 7.8 was subject to analysis on a Bruckner 270. Selected parameters: 12,000 scans, delay time 6475 sec, sweep with 17,241 Hz, line broadening 5 Hz, frequency 67.89 MHz, Nuclei ^{13}C.

RP–HPLC purified tracers can be obtained (13) with similar binding affinity to tracers purified by low-pressure techniques.

It is worth noting that the extraction of the RP-column eluate with cyclohexane removes the column degradation product present in the organic modifier, which is important if the isolation procedure ends with a lyophilization step. It has been reported that residues left after lyophilization of even ultrapure acetonitrile were toxic towards 3T3 cells used by bioassaying growth inhibitors (5). We have found that A14 monoiodoinsulin, which was lyophilized after 1 hr incubation in a variety of volatile buffers and organic modifiers, displayed reduced binding affinity to isolated rat adipocytes (12). When interferon was isolated by lyophilizing, the column eluate (containing *n*-propanol as

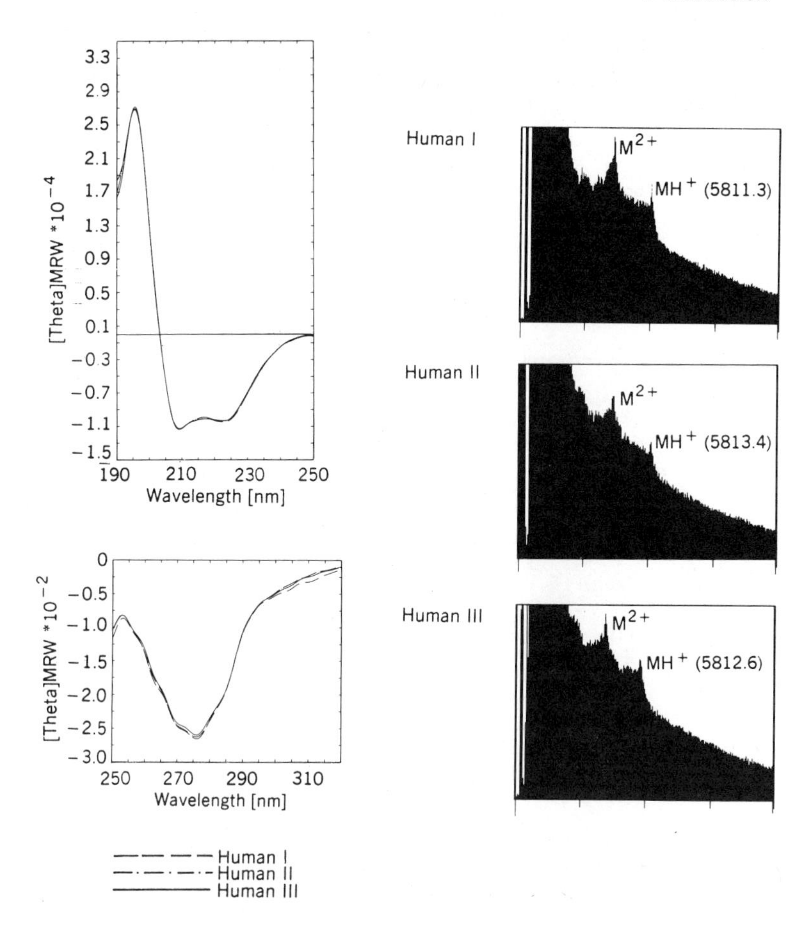

Fig. 5. Left panel: CD-spectra of Zn-crystallized human insulins I, II, and III (0.4% w/w) dissolved in 0.025 *M* Tris/HCl pH 7.8. Insulin concentration: 1 mg per ml. Right panel: MS spectra of human insulins I, II, and III. ^{252}Cf-plasma desorption time-of-flight mass spectrometry of human insulins. Samples were applied directly from nitrocellulose targets containing Zn-free insulin. The isotopically averaged mass of human insulin is calculated to 5807.6 a.m.u. corresponding to 5808.6 for the MH^+ ions.

organic modifier) the activity was completely destroyed. If the organic modifier was removed before lyophilization, the activity was preserved (9).

It therefore seems essential to separate the polypeptide from the organic phase before lyophilization in order to avoid toxic compounds in the organic modifier as well as to minimize the denaturation of the lyophilized polypeptide molecule.

In the preparative RP–HPLC purification of human insulin the polypeptide was isolated from the column eluate using zinc precipitation, redissolved in acetic acid, and desalted. Compared to highly purified human insulin prepared

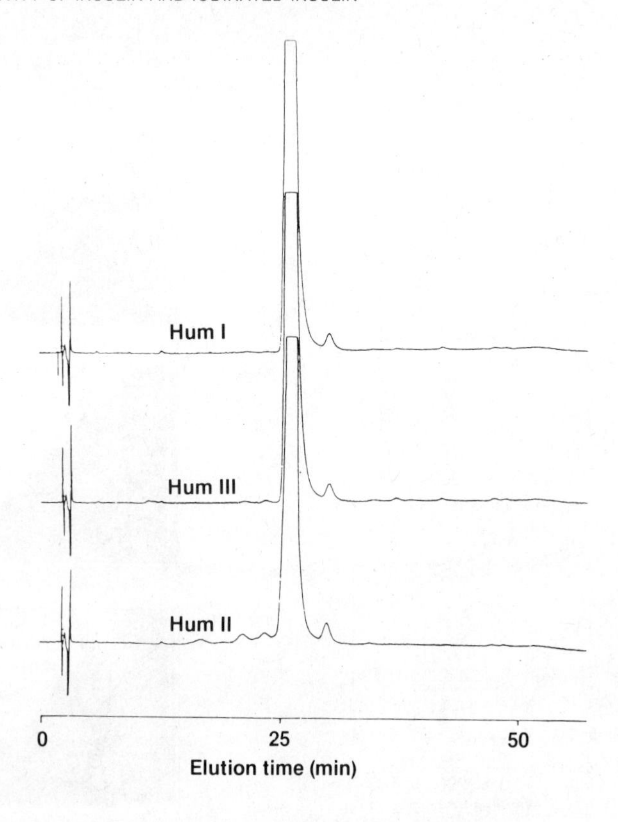

Fig. 6. RP–HPLC separation of 25 μg human insulins I, II, and III. Lichrosorb RP–18 (5 μ) 250 × 4.0 mm ID eluted at 1.0 ml per min with 0.25 *M* triethylammonium-phosphate, pH 3.0. Acetonitrile gradient from 25%–> 28% during 45 min. UV-detection at 210 nm (0.16 AUFS).

by low-pressure techniques, no differences in molecular structure could be detected (for example, superimposable CD and NMR spectra).

The chemical purity of the insulin preparations were comparable, not only by analytical RP–HPLC (Fig. 6), but also by isoelectric focusing, disc-gel electrophoresis, and amino acid analysis (data not shown). In addition, no difference in biological activity was observed within assay limits. When the two pancreatic human insulin preparations were compared to semisynthetic human insulin, identical results were obtained for purity, structure, and biopotency. It can therefore be concluded that preparative RP–HPLC can be used for purification of human insulin without damaging the molecule.

In summary, it seems essential to pay attention to the isolation step after RP–HPLC when labile molecules are subjected to reversed-phase chromatography. If the polypeptide is allowed to refold into the correct structure from a (partially or totally) denatured state induced by the mobile phase (acidic pH,

Table 3. Characterization of the Human Insulins I, II and III

Test for	Method	Human I	Human II	Human III
Species[a]	RP-HPLC	Identified	Identified	Identified
Structure	NMR	Equivalent	Equivalent	Equivalent
MW	MS	5811.3	5813.4	5812.6
Derivatives	RP-HPLC	0.2%	0.2%	0.2%
Di- and polymer[b]	GPC	0.49%	0.86%	0.59%
Des-amido insulin	RP-HPLC	0.9%	1.1%	0.9%
Potency[c]	Mouse blood glucose	28.2 IU/mg	24.5 IU/mg	28.0 IU/mg

[a]Isocratic RP-HPLC (Lichrosorb RP-18, phosphate/acetonitrile pH 3.0).
[b]Waters 1-125 columns eluted with acetic acid/iso-propanol.
[c]Fiducial limits 88.9–112.5%.

organic solvents), it should be possible to purify an increasing number of bioactive polypeptides and proteins using RP–HPLC. However, although this separation techniques is superior to others with respect to separation capacity, the chromatographic conditions limits the use for very sensitive molecules: Fibroblast growth factor could be successfully purified using HPIEC, but all attempts to purify the polypeptide using RP–HPLC as well as GPC invariably resulted in complete loss of activity or irreversible binding to the stationary phase (8).

It therefore seems advisable to consider RP–HPLC as an extremely useful purification technique for polypeptides and proteins, but not as the ultimate method of choice.

ACKNOWLEDGMENTS

We thank Ingelise Fabrin for excellent technical assistance.

REFERENCES

1. S.A. Cohen, K.P. Benedek, S. Dong, Y. Tapuhi, and B.L. Karger, Anal Chem., *56*, 217–221 (1984).
2. J. Luiken, R. van der Zee, and G.W. Welling, J. Chromatogr., *284*, 482–486 (1984).
3. J.W. Wilks, and S.S. Butler, J. Chromatogr., *298*, 123–130 (1984).
4. A.F. Bristow, C. Wilson, and N. Sutcliffe, J. Chromatogr., *270*, 285–292 (1983).
5. B.G. Sharifi, C.C. Bascom, V.K. Khurana, and T.C. Johnson, J. Chromatogr., *324*, 173–180 (1985).
6. P. Hallin, A. Madej, and L.E. Edqvist, J. Chromatogr., *319*, 195–204 (1985).
7. E.M. Kniep, B. Kniep, W. Grote, H.S. Contradt, D.A. Monner, and P. Mühlradt, Eur. J. Biochem., *143*, 199–203 (1984).

8. D. Gospodarowicz, S. Massoglia, J. Cheng, G.-M. Lui, and P. Böhlen, J. Cell. Physiol., *122*, 1985 (1985).
9. H.S. Johannsen, and Y.H. Tan, J. Interferon Res., *3*, 473–477 (1983).
10. J. Rivier, C. Rivier, D. Branton, R. Miller, J. Spiess, and W. Whale, in *Peptides: Synthesis, Structure and Function* D.H. Rich and E. Gross, Eds., Pierce, Rockford, Il, 1982, pp. 771–776.
11. B.S. Welinder, S. Linde, and B. Hansen, J. Chromatogr., *265*, 301–309 (1983).
12. B.S. Welinder, S. Linde, B. Hansen, and O. Sonne, J. Chromatogr., *281*, 167–177 (1983).
13. S. Linde, B.S. Welinder, B. Hansen, and O. Sonne, J. Chromatogr. *369*, 327–339 (1986)

CHAPTER 5

SEPABEADS FP Series: New Packing Materials for Industrial-Scale Separation of Biopolymers*

HIROSHI KUSANO, EIJI MIYATA, HIROAKI TAKAYANAGI, and TAKAHARU ITAGAKI

Research Center, Mitsubishi Chemical Industries Ltd., Yokohama 227, Japan

1. INTRODUCTION

The recent interest in biotechnology is focused on the industrial manufacturing of biopolymers such as medicines, insecticides, fertilizers, food additives, and so on. Because of the high selectivity and high yields, biotechnological methods are now recognized as very powerful and practical processes in the chemical industry. The market and manufacturing scale of biopolymers are getting larger very rapidly.

In the manufacturing of biopolymers, the efficiency, costs, and quality control of the process become very important, which is quite different from laboratory-scale work. The process should be optimized to keep the highest productivity; the productivity of an industrial process includes the total costs and time to obtain a unit amount of the target substance. We wish to obtain the target substance of the specified purity at the lowest cost within the shortest time.

Generally, the manufacturing process of biopolymers can be divided into two stages: synthesis and purification. The synthetic stage includes a variety of production processes such as biosynthesis, genetic recombination, cultivation, and so on, and thus it is very difficult to optimize this stage by a common approach. On the other hand, the purification stage consists of a small number

*SEPABEADS is a registered trademark of Mitsubishi Chemical Industries Ltd.

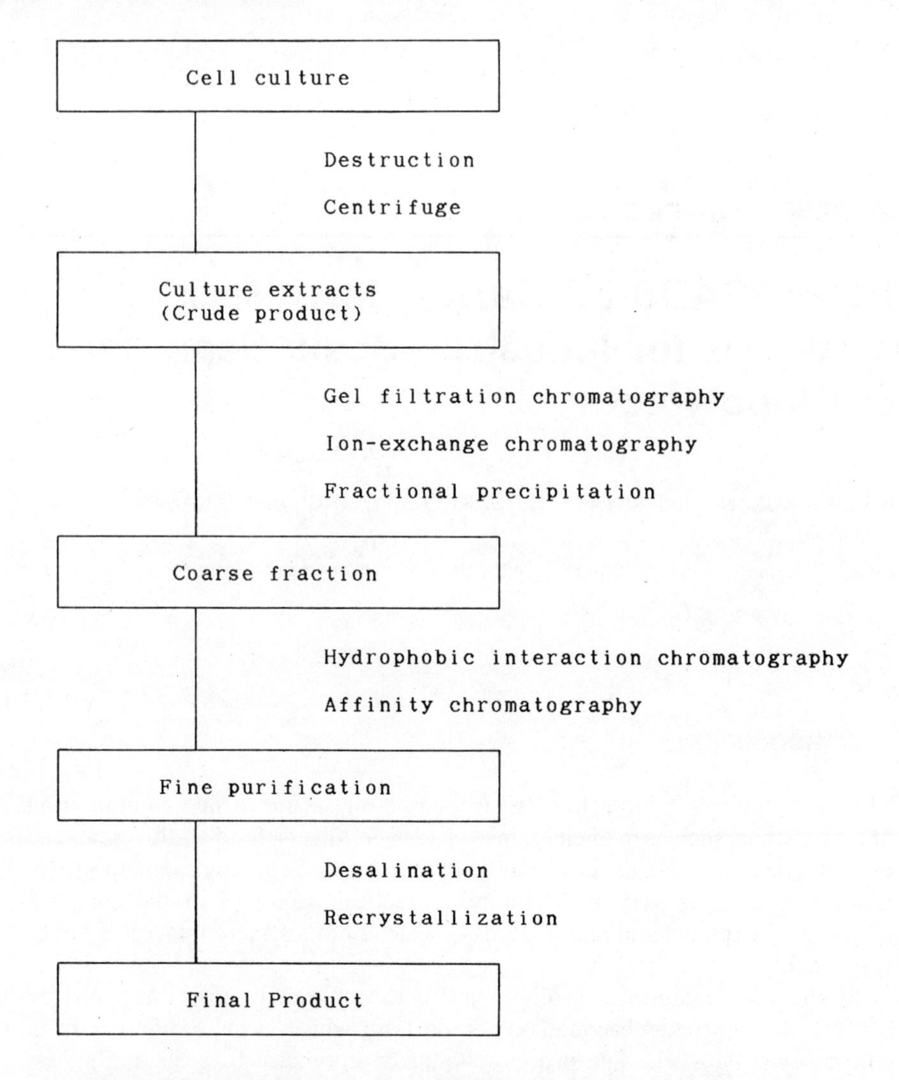

Fig. 1. Schematic diagram of a purification procedure of culture extracts.

of common techniques that are well known in analytical chemistry. Consequently, the optimization of the purification stage is effective for the productivity of the total manufacturing process.

The purification stage is composed of a series of unit operations. For instance, Fig. 1 shows a schematic diagram of a purification procedure for culture extracts. We can see from the diagram that the separation and purification operations by column liquid chromatography (LC) are the most critical in the whole purification process. LC techniques play an important role not only in analytical purposes for quality control but also in the purification stage on large scales. It should be noted here that the dimension of column beds employed in industrial-scale separations usually ranges from ~ 10 L to several kiloliters, and that little is known about packing materials suitable for such column dimensions. Although there are a number of packing materials for analytical-scale LC of biopolymers, problems still remain when industrial operations are considered. Conventional polysaccharide-based gels are somewhat expensive, and are so soft that they cannot be used in industrial dimensions. Silica-gel-based packings are not stable in acidic or basic media, and their columns usually need high inlet pressures. Polystyrene-based synthetic resins are not suited for LC of biopolymers because of their nonspecific adsorptive nature.

For the preceding reasons, there is an increasing demand for a new type of packing material. The requirements for packing materials used in industrial operations are summarized as follows:

1. *Operability*. Column pressure drop should be low to obtain high flow rates and to cut costs. For easy operations, volume change by changing conditions should be negligible.
2. *Stability*. Packings should be usable in strongly acidic or basic media as well as in organic solvents, and they should be resistant to thermal treatments.
3. *Functionality*. Many types (functional groups) of packings should be available.

This chapter will describe the fundamental characteristics of a new type of packing materials, SEPABEADS FP SERIES. SEPABEADS FP SERIES have been developed for the industrial-scale separation of biopolymers by Mitsubishi Chemical Industries Limited (MCI; Tokyo, Japan). They are designed according to the preceding requirements and are based on a new type of synthetic hydrophilic matrix. Since they have been developed for industrial uses, they have many advantages over conventional packing materials for industrial-scale operations. A number of SEPABEADS are commercially available: the assortment of SEPABEADS FP SERIES are listed in Table 1 together with their specifications and recommended uses (chromatographic modes) (1). The table also lists some of MCI GEL series (MCI) that are suitable for the high-pressure LC analysis in the same modes (2).

In the following sections, the features and use of SEPABEADS FP SERIES are described. The general properties of SEPABEADS, especially operability and stability for industrial operations, are given in Section 2. Some application examples of SEPABEADS and experimental conditions are shown in Section 3 including the evaluation procedures of the purified substances. Section 4 briefly deals with the method for scaling up the separation processes to industrial dimensions using a simple simulation model.

Table 1. List of SEPABEADS FP SERIES

Grade	Functional Group	Pore Size[a]			Chromatographic Mode					MCI GEL[g]
		Small	Medium	Large	GFC[b]	CEC[c]	AEC[d]	HIC[e]	AFC[f]	
FC–HG	–OH	FP–HG20	FL–HG13	FG–HG05	◎					CQP10
FP–CM	$-CH_2COOH$			FP–CM13		◎				CQK31
FP–QA	$-N^+(CH_3)_2C_2H_4OH$		FP–QA13				◎			
			FP–DA12							
FP–DA	$-N(C_2H_5)_2$	FP–DA20	FP–DA13	FP–DA05			◎			CQA30
FP–HA	$-NH(CH_2)_6NH_2$	FP–HA20	FP–HA13	FP–HA05			◎	○		
FP–BA	$-NH(CH_2)_3CH_3$		FP–BA13				○	◎		
FP–ZA	$-NHCH_2C_6H_5$		FP–ZA13	FP–ZA05			○	◎		
FP–BU	$-O(CH_2)_3øCH_3$		FP–BU13	FP–BU05				◎		CQH30
FP–OT	$-O(CH_2)_7CH_3$		FP–OT13					◎		
FP–PH	$-OC_6H_5$		FP–PH13					◎		CQH31
FP–CL	$-N(CH_2COOH)_2$		FP–CL13			○	○		◎	
FP–BL	Cibracron blue 3G–A		FP–BL13						◎	

[a] Details of pore size distribution are given in Section 2.
[b] GFC Gel filtration chromatography.
[c] CEC Cation-exchange chromatography.
[d] AEC Anion-exchange chromatography.
[e] HIC Hydrophobic interaction chromatography.
[f] AFC Affinity chromatography.
[g] MCI GEL series are used for analytical high pressure LC: see Ref. 2.

2. FEATURES OF SEPABEADS FP SERIES

As chromatographic packing materials for the large-scale separation of biopolymers, SEPABEADS FP SERIES have various advantages over conventionally used packings. The basic properties of SEPABEADS are briefly described in this section (3, 4). Since the operability and stability of packing materials are very important factors in industrial processes, as noted in Section 1, the first part of this section deals with the properties of the base matrix of SEPABEADS FP SERIES. In the second part, functional groups attached on the matrix are characterized.

2.1. General Properties

The properties of SEPABEADS FP SERIES, concerning the base matrix, are summarized here. SEPABEADS FP-HG and FP-DA series are mainly used as evaluation samples for simplicity, because the type or content of attached functional groups does not affect the property of the base matrix.

2.1.1. Matrix

SEPABEADS FP SERIES are hard, spherical particles based on a synthetic hydrophilic vinyl polymer. Scanning electron micrographs of SEPABEADS FP-HG13 are shown in Fig. 2. The shape and volume of the particles are maintained even when vacuum-dried. The standard diameter is 120 μm: particles of this size are practical for most industrial operations in order to keep column operability, because columns of much smaller particles (e.g., 10 μm) require a very large pressure drop and often cause plugging at the inlet. The column pressure drop should be as small as possible to lower running costs. If the solution to be injected contains a certain amount of suspended solids, much larger particles are recommended to avoid the column plugging.

In addition to the standard diameter, particles of 50 and 350 μm in diameter are also available (see Section 5). Larger particles are suitable for high-speed operations, and smaller particles are used when high efficiency is needed with a short column.

As the matrix is sufficiently hydrophilic, nonspecific adsorption of biopolymers on the surface of the base matrix is practically negligible. This is an advantageous property of SEPABEADS FP SERIES.

2.1.2. Porosity

The pore size of particles is an important factor in the separation of biopolymers. SEPABEADS FP SERIES have totally porous structure as shown in Fig. 2*b*, and have three types of structure of the base matrix. The pore size of each of SEPABEADS is indicated by two numerals in their names: 0X type (large), 1X type (medium), and 2X type (small). The pore size distribution of SEPABEADS are shown in Fig. 3. Users can select an appropriate pore size

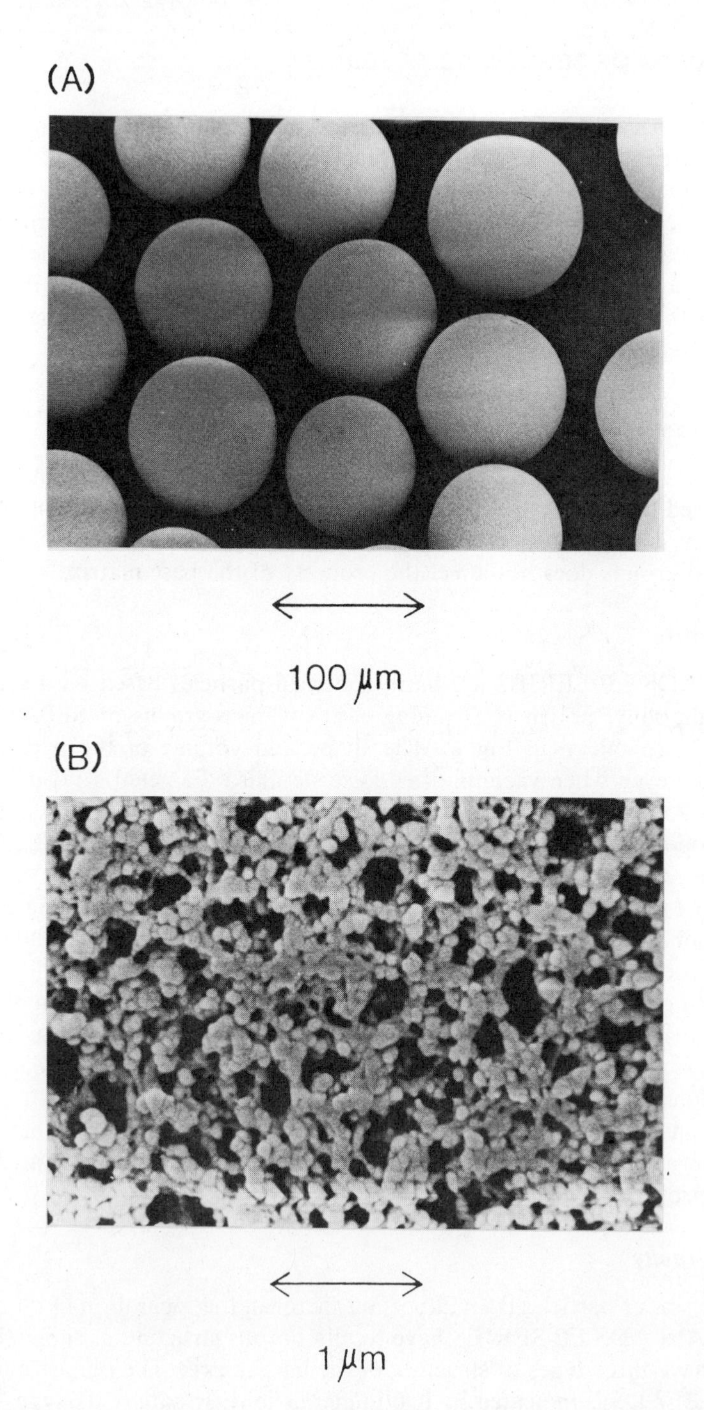

Fig. 2. Scanning electron micrographs of SEPABEADS FP-HG13. (*a*) Particle shape; (*b*) cross section.

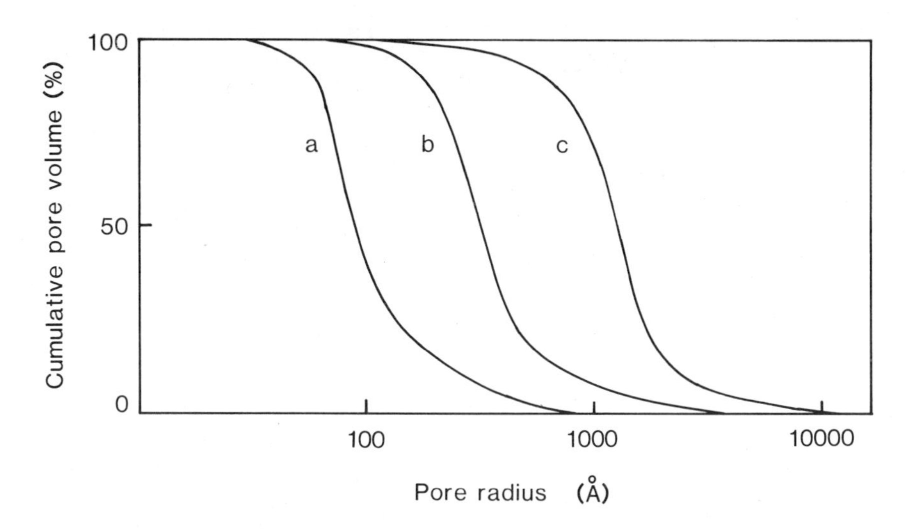

Fig. 3. Pore size distribution of SEPABEADS FP-HG SERIES samples: (*a*) SEPABEADS FP-HG20, (*b*) SEPABEADS FP-HG13, and (*c*) SEPABEADS FP-HG05.

according to the molecular size of the target substances to obtain the maximum efficiency.

The effect of porosity of packings is most apparent in gel filtration chromatography (GFC). For example, the calibration curves for dextran standards and some proteins on FP-HG columns are shown in Fig. 4, which will help the selection of pore size. In addition to GFC, the porosity of packings also affects the adsorption capacity of biopolymers (see Section 3.1).

2.1.3. Mechanical Stability

The mechanical stability of packings directly affects the operability of the process. Although high pressure (e.g., 10 atm or more) is not necessarily used in industrial processes, particles should withstand high-flow-rate operations. Figure 5 compares the flow characteristics of SEPABEADS and some conventional polysaccharide-based gels. There is a linear relationship between the pressure drop and the flow rate of the SEPABEADS FP-DA13 column in a practical operation range, and the volume change of SEPABEADS is negligible. This means that a rapid separation can be performed easily by simply increasing the flow rate with SEPABEADS columns, which is effective for the purification of labile substances. On the contrary, polysaccharide gels are severely deformed and cannot be used under the same conditions.

2.1.4. Thermal Stability

It is sometimes required to sterilize packing materials in an autoclave to avoid microbial contamination. Table 2 shows the result of the thermal stability test

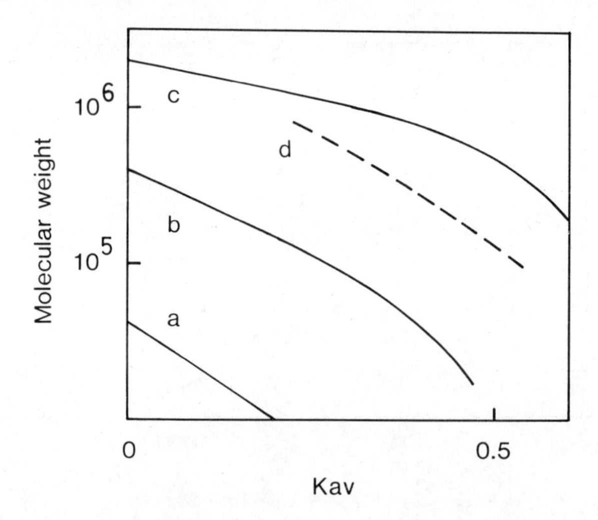

Fig. 4. Calibration curves for dextran standards and proteins on SEPABEADS FP-HG columns. For dextrans (solid lines): (*a*) SEPABEADS FP-HG20, (*b*) SEPABEADS FP-HG13, and (*c*) SEPABEADS FP-HG05. For proteins (dashed line): (*d*) SEPABEADS FP-HG13. Note: K_{av} is a parameter used for the evaluation of GFC packings: $V_e = V_0 + K_{av}(V_t - V_0$, where V_0 is void volume of the column, V_e is elution volume of solutes, and V_t is total column volume.

of SEPABEADS FP-DA13 in a buffer solution at 120°C. Neither the base matrix nor the attached functional group is affected by the thermal treatment. The result also means that SEPABEADS can be used for high-temperature separations.

Another important measure of thermal stability is long-term extraction of the packing materials. There is no detectable organics from SEPABEADS after the matrix was soaked at 40°C for weeks.

2.1.5. Chemical Stability

SEPABEADS FP SERIES are sufficiently stable for most chemical treatments usually used in industrial separations. Figure 6 shows the chemical stability of SEPABEADS FP-DA13 as a function of volume change. Figure 6*b* suggests that gradient elution operations by changing the salt concentration are easily performed without a bed volume change.

Since SEPABEADS are also stable in strongly acidic or basic madia (e.g., 1*N* hydrochloric acid or sodium hydroxide), columns of SEPABEADS can be regenerated by alkaline solutions that cannot be used for conventional packings such as silica and some polysaccharide-based gels.

Besides being stable in aqueous medium, SEPABEADS FP SERIES are stable in common organic solvents and show practically negligible volume change as listed in Table 3. Thus, they can be used in reversed-phase chromato-

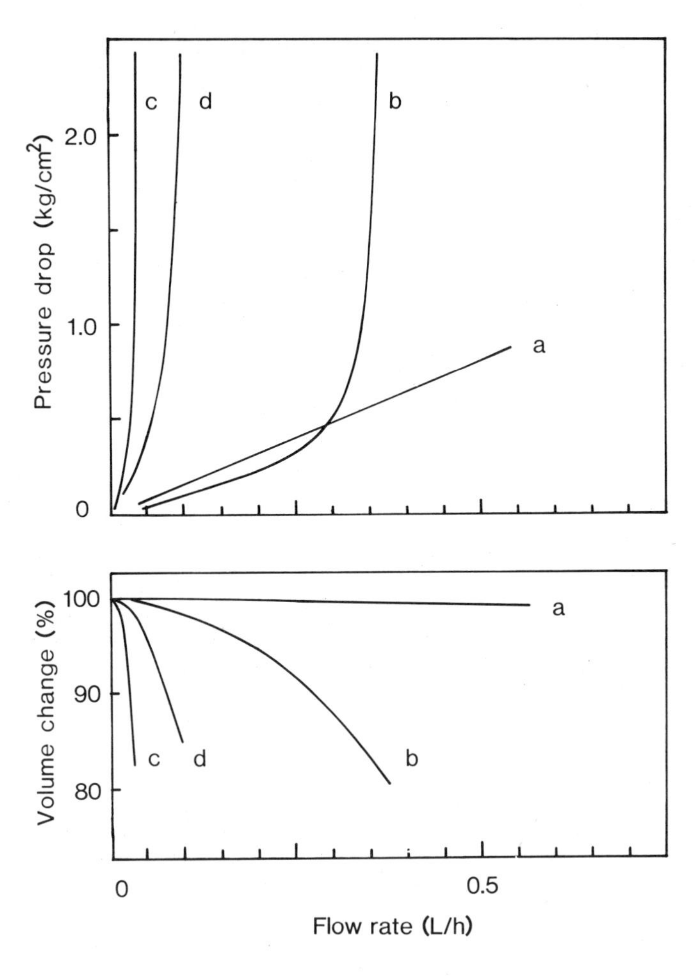

Fig. 5. Flow characteristics of SEPABEADS and some polysaccharide gel columns. Effect of flow rate on pressure drop (upper panel) and on volume change (lower panel). Columns: (*a*) SEPABEADS FP-DA13, (*b* and *c*) dextran gels, and (*d*) agarose gel; dimension, 500 × 100 mm ID for each column. Particle diameter is 100–150 μm.

graphic modes and can be washed with these solvents without shrinkage or swelling.

The chemical stability of SEPABEADS FP SERIES, especially the negligible volume change, indicates the good operability of SEPABEADS are hightly resistant to microbial degradation.

2.2. Functionality

As mentioned in Section 1, a variety of functionality is indispensable for purification processes. The characterization of some of bonded ligands is described here. For other detailed information, see Section 3.

Table 2. Thermal Stability of SEPABEADS FP–DA13

	Contact Time (hr)		
	0	1.0	2.0
Water content (%)	59.1	59.2	59.6
Ion-exchange capacity (eq/L-gel)	0.87	0.87	0.86
BSA[b] adsorption capacity (g/L-gel)	16	16	16

[a]Condition: 50 m*M* phosphate buffer: pH 7.0; temperature, 120°C.
[b]BSA, bovine serum albumin.

2.2.1. Functional Groups

As listed in Table 1, SEPABEADS FP SERIES have many types of functional groups, which are chemically bonded on the base matrices. The recommended uses (chromatographic modes) of each of the series are also included in Table 1, and most of the series are usable in different modes. For example, SEPABEADS FP-HA series are suitable for anion-exchange chromatography (AEC) and hydrophobic interaction chromatography (HIC). If two or more chromatographic modes can be accomplished in one column, the purification system will be drastically simplified (see Sections 3.2 and 3.3).

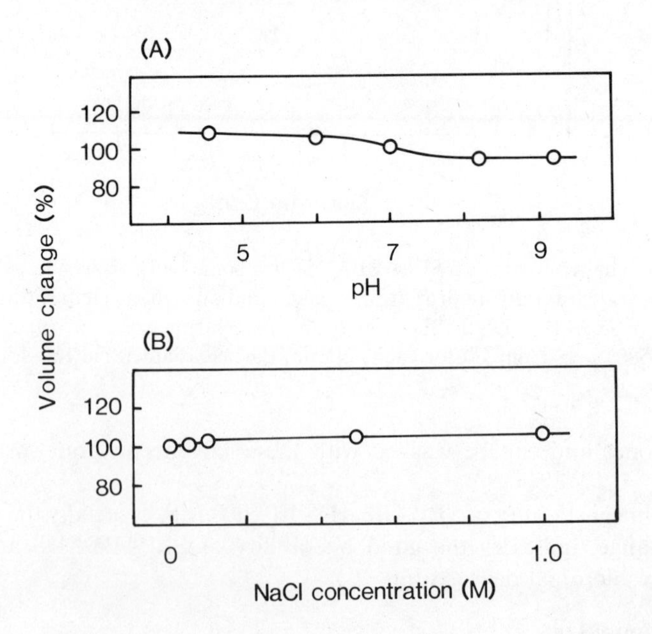

Fig. 6. Chemical stability of SEPABEADS FP-DA13: (*a*) volume change by solution pH and (*b*) volume change by salt concentration. Conditions, 50 m*M* phosphate buffer; (*b*) pH 7.0. Temperature, 25°C.

Table 3. Volume Change of SEPABEADS FP SERIES

SEPABEADS	Solvents						
	Water	Methanol	Ethanol	Ethyl Acetate	Dimethyl Sulfoxide	Dimethyl Formamide	Toluene
FP–CM13	1.00	1.03	1.00	0.97	1.13	1.10	0.93
FP–HG13	1.00	1.03	1.04	1.00	1.16	1.12	0.95
FP–DA13	1.00	1.04	1.03	0.99	1.08	1.05	0.94

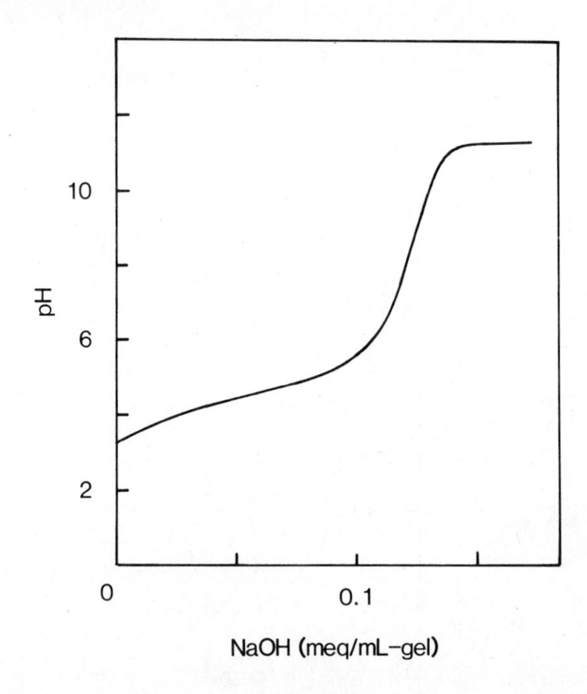

Fig. 7. Titration curve of SEPABEADS FP-CM13. Solution, 0.5 *M* sodium chloride. Temperature, 25°C.

In affinity chromatography (AFC), the efficiency of the purification depends on the affinity ligand species attached on the matrix. SEPABEADS FP SERIES have two common ligands for AFC-CL and FP-BL), and other types of AFC packings will be available from MCI as test samples (see Section 5). Since AFC is expected to have a potential effectiveness for the specific concentration of biosubstances, AFC techniques will become very significant in industrial processes.

2.2.2. *Ion-Exchange Capacity and Titration Curves*

Ion-exchange chromatography is very often used for the separation of polypeptides and related compounds. For this purpose, a number of ion-exchanger-type SEPABEADS are available (FP-CM for cation-exchange chromatography; CEC, FP-QA, FP-DA, FP-HA, FP-BA, FP-ZA for AEC). The type of the functional group or the capacity of the packing should be selected according to the property of the target substances such as pI, hydrophobicity, and so on. Figures 7 and 8 show the titration curves of SEPABEADS FP-CM13 and FP-DA SERIES respectively. This information is useful for setting the elution conditions (see Section 3.2).

In general, particles having large exchange capacity (e.g., FP-DA13) show

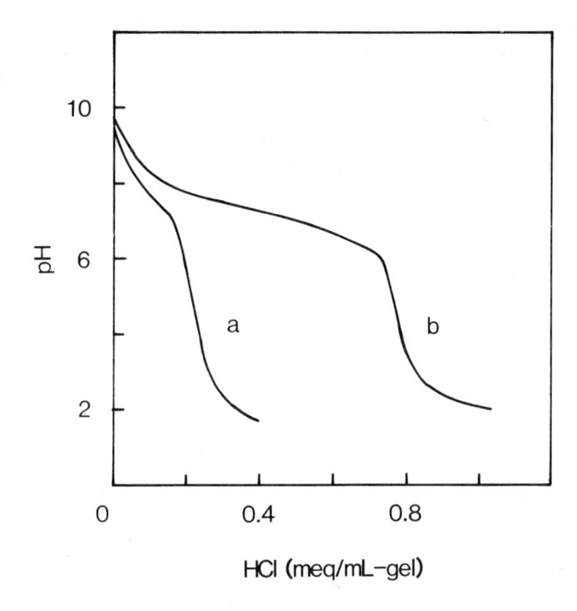

Fig. 8. Titration curves of SEPABEADS FP-DA SERIES: (*a*) SEPABEADS FP-DA12 and (*b*) SEPABEADS FP-DA13. Solution, 0.5 *M* sodium chloride. Temperature, 25°C.

large adsorption capacity for target substances and need a long regeneration time. On the other hand, particles having small exchange capacity (FP-DA12) show small adsorption capacity and need a short regeneration time. If the ion-exchange group contains a hydrophobic part (alkyl chain or phenyl group), hydrophobic adsorption of substances should also be taken into account.

2.2.3. *Hydrophobic Ligands*

FP-BU, FP-OT, and FP-PH are designed as HIC packings, and some of AEC packings such as FP-HA, FP-BA, and FP-ZA are also usable for HIC (see Table 1). It is generally accepted that particles carrying longer alkyl chain show stronger adsorption capacity, that is, they adsorb the substances at a relatively low salt concentration and need a long recovery time. Bonded phenyl groups are also known to increase the adsorption capacity. Thus, the functional group should be selected according to the hydrophobicity of the target substance and to the operation conditions (salt concentration). If the bonded ligand contains ion-exchangeable amino groups, the pH of the solution should also be taken into account.

3. APPLICATION EXAMPLES OF SEPABEADS FP SERIES

This section deals with some experimental procedures for setting the separation conditions, which must be done before scaling up the process. Two methods are given here: batch and column processes. Batch processes described in the first part in this section are used for selecting the type of packings and for obtaining quantitative information. Column processes given in the second part are carried out to set up the actual operation conditions. In the last part of this section, the evaluation of purification processes is described briefly. The concepts and methods for scaling up the system will be given in Section 4.

3.1. Batch Processes

The scale of an separation column in an industrial process depends on the amount of the product to be chromatographed and on the loading capacity of the packings. To determine the size of the column, it is necessary to know the adsorption capacity of the packing material for the target product and main impurities. Measurements in batch processes are recommended for this purpose because they are simple and sufficiently quantitative. Some examples are shown below for selecting the appropriate type of SEPABEADS and for setting the separation conditions.

3.1.1. Adsorption

The selectivity for substances to be separated can be estimated from the data of adsorption measurements except for the case of GFC. The ion-exchange adsorption of a protein (bovine serum albumin, BSA) is, for example, measured in the following manner:

> Prepare a standard solution of protein* in a properly buffered medium (0.1% w/w, 0.1 *M* Tris-HCl,† pH 7.5), and equilibrate each packing material (e.g., SEPABEADS FP-DA's) in the same buffer. Measure a portion of the equilibrated packing material in a graduated cylinder (~5 mL) and filter the material. Remove the residual liquid in the filtered material by centrifugation, and add the packing immediately into a measured volume (100 mL) of the standard protein solution in a sealed flask. Shake the mixture at constant temperature (10°C in an incubator) for a fixed time (6 hr‡). Measure the concentrations of the protein in the standard and supernatant solutions (UV 280 nm), and calculate the adsorbed amount of the protein per a unit volume of the packing material.

3.1.2. Desorption

The desorption of the substance is estimated similarly to the adsorption process. It is desirable to perform the desorption measurement just after the adsorption measurement. The information of desorption measurements is useful for setting elution conditions and for recovery tests.

*Prepare a calibration curve for protein determination.

†Tris: 2-amino-2-(hydroxymethyl)-1,3-propanediol. Tris-HCl: aqueous tris solution, pH adjustment with hydrochloric acid.

‡Equilibration time usually depends on the molecular weight of the substance investigated. As the molecule becomes larger, a longer equilibration time is needed.

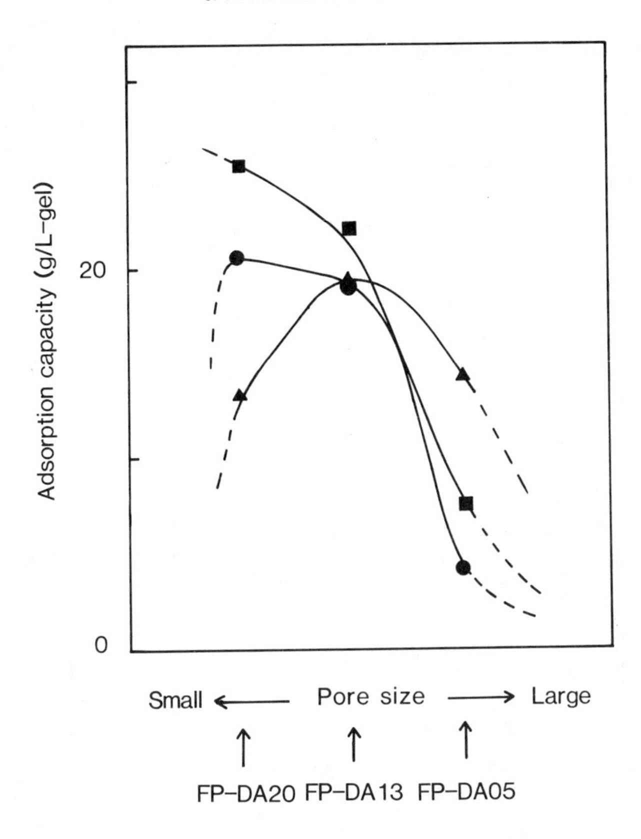

Fig. 9. Effect of pore size of SEPABEADS FP-DA SERIES on adsorption capacity for some proteins. Proteins: (●) BSA, MW ~ 67 k; (■) *a*-lactalbumin, MW ~ 150 k. Solution, Tris-HCl buffer, pH 7.5. Adsorption temperature, 10°C.

Filter the packing that has adsorbed the protein, and add the packing to the desorption solution (100 mL, 0.1 *M* Tris-HCl, pH 7.5; containing 0.5 *M* sodium chloride in a sealed flask). Shake the mixture at the same temperature (10°C) for the same period (6 hr). Measure the concentration of the protein in the supernatant, and calculate the desorbed amount.

3.1.3. Selection of Pore Size of SEPABEADS FP SERIES

The pore size of the packing is selected according to the molecular weight (MW) or molecular size of the substances to be separated. This is most significant for GFC. The calibration curves obtained on SEPABEADS FP-HG columns (Fig. 4) will help the selection of pore size.

The pore size is also effective for the adsorption capacity for biopolymers in other chromatographic modes. Figure 9 shows the effect of the pore size of

Table 4. Protein Adsorption Capacity of SEPABEADS FP SERIES

SEPABEADS	BSA Adsorption Capacity
FP-CM13	40–50
FP–DA13	15–20
FP–HA13	25–30

[a]Conditions: 0.1 *M* Tris-HCl buffer, pH 7.5 for FP–DA13 and FP–HA13; 0.025 *M* acetate buffer, pH 4.5 for FP–CM13.

SEPABEADS FP-DA SERIES on the adsorption capacities for albumin, α-lactalbumin, and γ-globulin. This figure indicates that the pore size that gives maximum adsorption is different and depends on the molecular weight of the protein. The values of adsorption capacity for other biopolymers will be obtained by the batch process measurement noted previously. These measurements will sometimes be necessary for users to select the appropriate packing material.

3.1.3. Selection of Type of Functional Group

The type of functional group also affects the separation of substances as described in Section 2. Table 4 summarizes the effect of the type of functional group on the ion-exchange adsorption capacity for BSA. In ion-exchange chromatographic modes (CEC or AEC), the capacity usually depends on the pI or hydrophobicity of the substance and on the pH and salt concentration of the solution.

In AFC, it is generally difficult to elect an appropriate ligand species by simple measurements. If the common ligands (iminodiacetate or blue dye, Table 1) do not work well, another type of ligand should be selected by the user (see Section 5).

3.2. Column Operations

The applications of SEPABEADS FP SERIES to the separation and purification of biopolymers are described here using small-scale examples. The optimum conditions for each case will be obtained by combining the information from the analytical separation data with that from batch process measurements. For the analytical separation of biopolymers by high-pressure LC, MCI GEL series (MCI) listed Table 1 are recommended [5, 6]. Details of the experimental conditions are given in the captions.

3.2.1. Column Preparation

It is not necessary to use large-scale columns for setting up the operation conditions. Columns for laboratory-scale experiments are prepared as follows:

Prepare an appropriate size (e.g., 500 × 10 mm ID) of glass tube for the column.* Equlibrate the packing material in the solution used for conditioning or elution, and pack it into the column by the slurry-packing method with gentle tapping (use a column packer if possible). Condition the packed column by pumping the eluant through the column. Set a thermostat apparatus during the conditioning process if necessary.

It is convenient to use common high-pressure LC apparatus (pump, injector, UV monitor, and so forth) for small-scale experiments.

3.2.2.–Gel Filtration Chromatography

As shown in Fig. 4, SEPABEADS FP-HG SERIES are suitable for this purpose. Salt exchange or desalting process is usually performed in this mode. Figure 10 shows an example of the desalting of an albumin solution by a SEPABEADS FP-HG13 column. Since the volume change of SEPABEADS by salt concentration is negligible as shown in Fig. 6, the packed bed is not affected by the operation and the column operability remains good.

Although GFC generally does not give good separations compared with other chromatographic modes because of GFC's short elution times, it is applicable to every stage of the purification processes.

3.2.3. Cation-Exchange Chromatography

Figure 11 shows a chromatogram of a mixture of proteins obtained on a SEPABEADS FP-CM13 column. In this operation, a stepwise gradient elution process is used because this process is more practical for industrial operations than a linear gradient system. The gradient pattern is also shown in the figure (dotted line). In CEC and AEC, gradient elution is carried out by increasing the salt concentration in the eluant.

In CEC using FP-CM columns, pH of the solution should be set so that the bonded carboxyl groups are dissociated to form carboxylate ions. The titration curve shown in Fig. 7 is useful for this purpose. Reference 5 gives some detailed analytical information for the relationship between the CEC retention and the pI of several proteins using MCI GEL columns.

3.2.4. Anion-Exchange Chromatography

A number of anion-exchanger-type SEPABEADS are available, as described in Section 2.2. Among a variety of application examples, two typical cases are described here.

Figure 12*a* shows a chromatogram of bovine serum protein (Nakashibetsu Serum, MCI) obtained on a SEPABEADS FP-DA13 column. Although experimental conditions are somewhat different, the elution pattern is in good agreement with the result of the analytical-scale separation of the same sample using an MCI GEL CQA30 column, as shown in Fig. 12*b*. It is better to use a

*Commercially available as semipreparative columns.

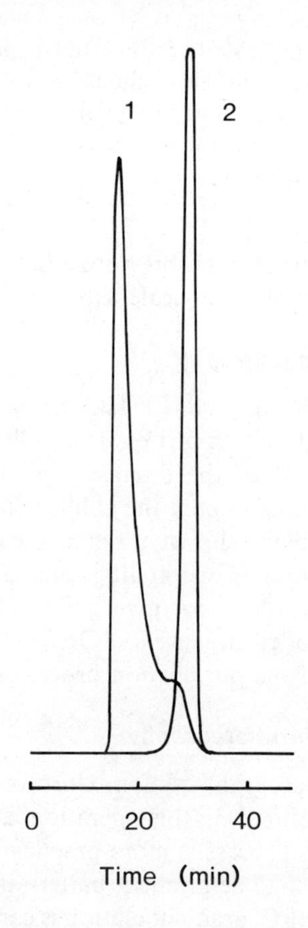

Fig. 10. Desalting of a BSA solution. Column; SEPABEADS FP-HG13; 1000 × 10 mm ID. Eluant; deionized water. Flow rate: 2.0 mL/min. Peaks: (1) BSA (UV detection, 280 nm) and (2) sodium chloride [electric conductivity (EC) detection]. Temperature; 20°C.

cheaper buffer reagent (phosphate in this case) to cut costs in industrial processes. The characterization of each fraction of Fig. 12*a* is given in Section 3.3.

Figure 13 shows a small-scale separation model of an actually working process for the purification of an acidic protease (for therapeutic use). SEPABEADS FP-HA05 is now being used successfully for the industrial process that includes harsh treatments such as washing the column with 0.5 *M* sodium hydroxide. The actual scale of this process is hundreds of times as large as shown here. Even at a high flow rate (30 mL/min; SV = 6*), the efficiency of

*SV (space velocity): a unit of flow rate often used in industrial-scale operations. SV = 1 means one column bed volume of eluant per hour.

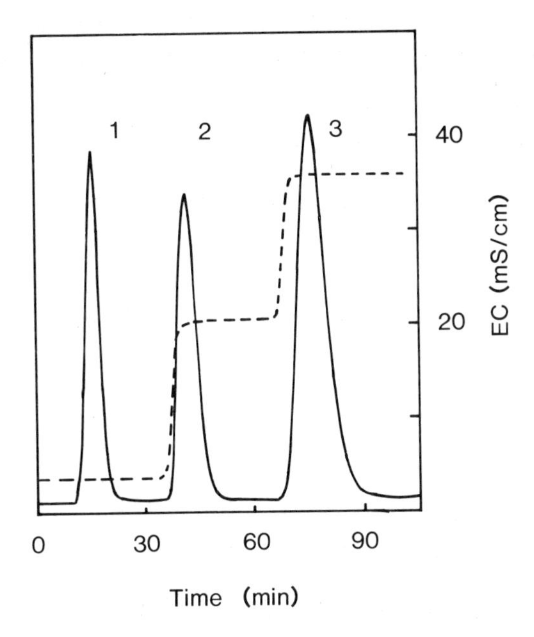

Fig. 11. Cation-exchange chromatographic separation of proteins. Column; SEPABEADS FP-CM13; 550 × 10mm ID. Solution: 0.05 *M* phosphate buffer, pH 6.6 (*A*). Step gradient elution; 0 → 0.25 → 0.5 *M* sodium chloride in *A* (dotted line). Flow rate: 2.0 mL/min. Detection: UV 280 nm. Peaks: (1) myoglobin, (2) *a*-chymotrypsinogen A, and (3) lysozyme.

the process is maintained as shown in Table 5 (see Section 3.3; the evaluation of the purification process is also given).

3.2.5. Hydrophobic Interaction Chromatography

HIC operation often needs high salt concentration at the initial adsorption stage, and the change of salt concentration during the elution process is very large. Thus, it is important for the column operability that volume change by salt concentration is negligible. SEPABEADS FP SERIES have an advantageous property for these conditions as noted in Section 2.1.

Many types of SEPABEADS FP SERIES can be used for this separation mode as noted in Section 2.2, and the type of functional group is selected according to the hydrophobicity of the substances and the operation conditions [1]. Figure 14 shows the separation of some proteins on a SEPABEADS FP-BU13 column. In HIC, gradient elution is carried out by decreasing the salt concentration in the eluant.

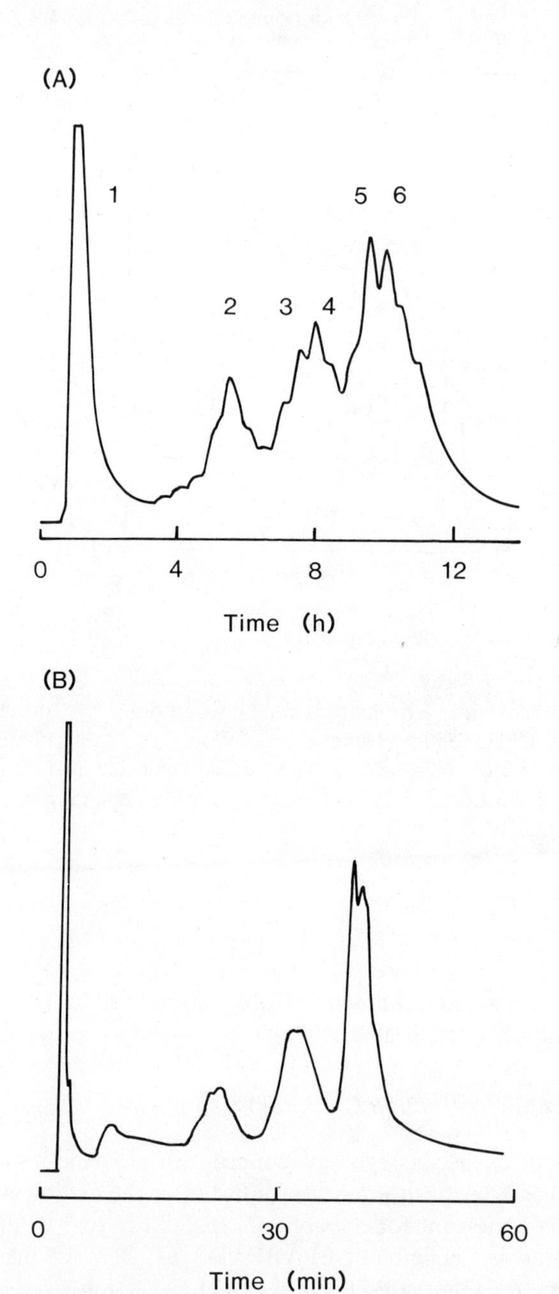

Fig. 12. Separation of bovine serum proteins. (*A*) Column: SEPABEADS FP-DA13; 1000 × 10 mm ID. Solution, 0.01 *M* phosphate buffer, pH 7.3 (*A*). Elution: 3 hr with *A*, then 0 → 0.1 *M* sodium sulfate in *A* (linear gradient, 9 hr). Flow rate: 1.0 mL/min. Detection: UV 280 nm. For fractions, see Section 3.3. (*b*) Column, MCI GEL CQA30; 75 × 7.5 mm ID. Solution: 0.014 *M* Tris-HCl, pH 7.4 (*B*). Elution: 0 → 0.5 *M* sodium chloride in *B* (linear gradient, 60 min). Flow rate: 1.0 mL/min. Detection: UV 280 nm.

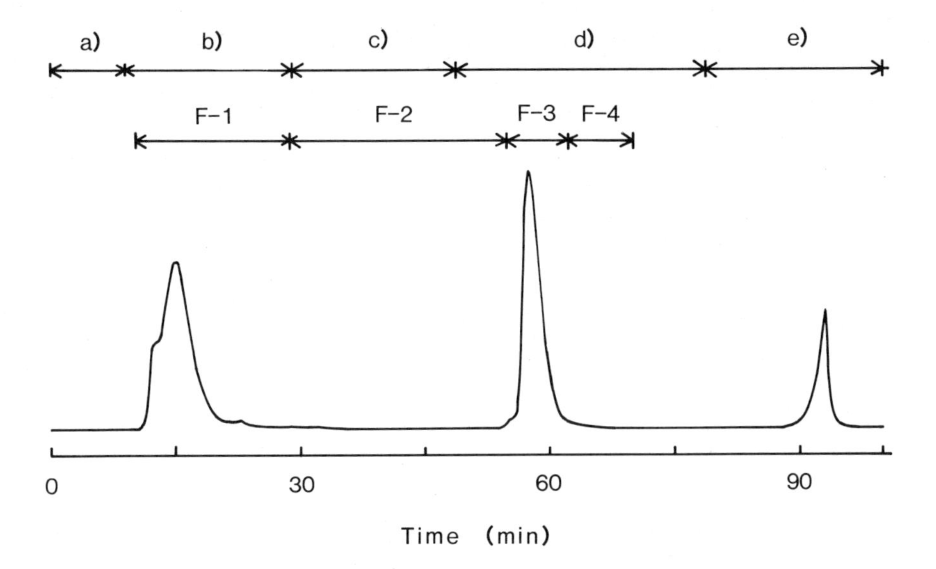

Fig. 13. Purification process of an acidic protease. Column: SEPABEADS FP-HA05; 1000 × 20 mm ID. Solution: 0.1 *M* acetate buffer, pH 4.3 (*A*). Operations: (*a*) charge at 25°C, (*b*) wash with *A* at 25°C, (*c*) wash with *A* at 30–35°C, (*d*) elution with *A* containing 0.5 *M* sodium chloride at 30–35°C, and (*e*) regeneration with 0.5 *N* sodium hydroxide at 25–35°C. Flow rate: 30 mL/min. For characterization of fractions, see Section 3.3.

3.2.6. *Affinity Chromatography*

Cibacron Blue is well known as an affinity ligand for the specific adsorption of some proteins. SEPABEADS FP-BL13 has this dye group for affinity chromatography of proteins. Figure 15 shows the separation of two proteins which are not separated by other chromatographic modes.

Chelating adsorbents are also useful for this purpose. SEPABEADS FP-CL13 in free or metal-loaded form is suited for the affinity chromatographic separation of biopolymers.

Table 5. Evaluation of Purification Process

Fraction[a]	Activity (U)	Protein (mg)	Specific Activity (U/mg)	Endotoxin (%)
Starting material	164×10^4	1870	880	(100)
F-1	3×10^4	588	—	12.8
F-2	0	25	—	1.0
F-3	141×10^4	855	1590	8.7
F-4	2×10^4	69	—	1.1

[a] For fraction number, see Fig. 12.

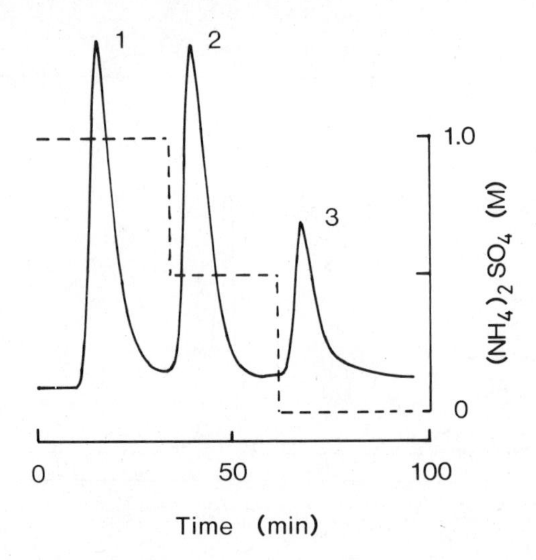

Fig. 14. Separation of proteins by hydrophobic interaction chromatography. Column: SEPABEADS FP-BU13; 550 × 10 mm ID. Solution: 0.05 *M* phosphate buffer, pH 7.2 (*A*). Step gradient elution, 1 → 0.5 → 0 *M* ammonium sulfate in *A* (dotted line). Flow rate: 2.0 mL/min. Detection: UV 280 nm. Peaks: (1) myoglobin, (2) *a*-chymotrypsinogen A, and (3) thyroglobulin.

3.3. Evaluation of Purification Processes

Before scaling up the purification system to industrial scale, it is necessary to confirm in small-scale experiments that the purity of the target substance meets the specified quality. One of the methods commonly used in biotechnology is electrophoresis, which can be performed easily. The SDS PAGE* pattern of the fractions shown in Fig. 12*a* shows zones around MW ~ 67 k in Fractions 5 and 6 which means that the main component of these fractions is albumin, while the zones around MW ~ 220 k in Fractions 1, 3, and 4 are considered to be globulins.

The specific activity of purified substance is also an important factor. The values from this method directly reflect the purification efficiency, but the evaluation process depends on the nature of the substance and can become complex. The results of the activity test of the fractions shown in Fig. 13, kindly given by the manufacturer of the medicine, are summarized in Table 5. In this case, more than 90% of endotoxin (main impurity) are removed (Fraction 3) from the target substance by a simple, inexpensive, one-column procedure. The success of this procedure is due to the fact that SEPABEADS FP-HA has two retention sites, namely, an ion-exchangeable amino group and a hydrophobic alkyl chain.

*SDS PAGE: polyacrylamide gel electrophoresis using sodium dodecyl sulfate solution.

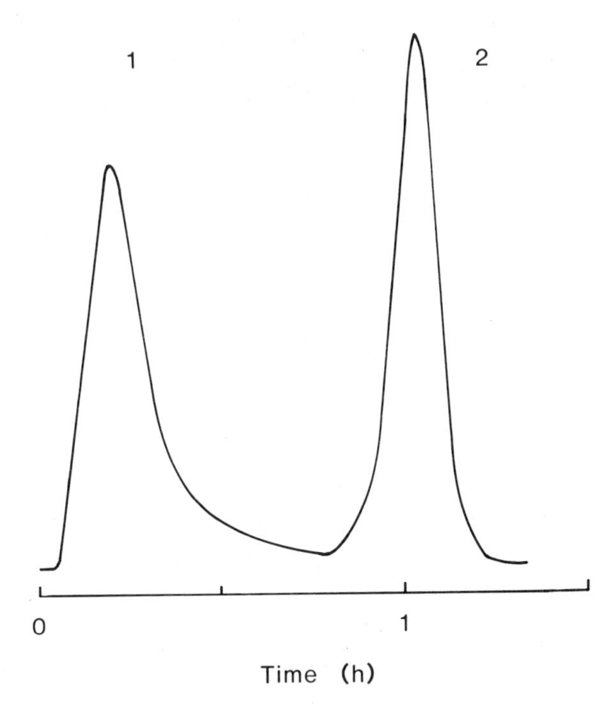

Fig. 15. Affinity chromatographic separation of proteins on SEPABEADS FP-BL13. Column dimension, 150 × 6 mm ID. Solution: 0.25 *M* Tris-HCl buffer, pH 7.5 (*A*). Step gradient elution, 0 → 1.0 *M* sodium thiocyanate in *A*, just after the elution of the first peak. Detection: UV 280 nm. Proteins: (1) *a*-lactalbumin and (2) human serum albumin.

4. SIMULATION OF LARGE-SCALE OPERATIONS

Once the operation conditions for the process are established by small-scale experiments, the next step is to scale up the system. In this step, it is convenient to use the HPLC parameters such as plate height (HETP), separation factor (α), peak resolution (Rs), and so on. Theoretical treatments of LC are given in literature [7].

It is generally known that a large injection volume of a sample gives a severe deformation of elution pattern, which is not desirable when trying to improve the total separation. In industrial-scale operations, however, we wish to load as much sample as possible on the column to cut both costs and time. A procedure for scaling up the system is given below, using isocratic elution chromatography.

Suppose that the solution of the crude material contains two main components, target product *A* and impurity *B*, and that the distribution coefficients (K_d's) of *A* and *B* are 0.3 and 1.2, respectively. If the loaded amount is sufficiently small (nearly a plug-shaped injection), the hypothetical elution pattern is similar to the analytical-scale separation of the same sample: Fig. 16*a* is the

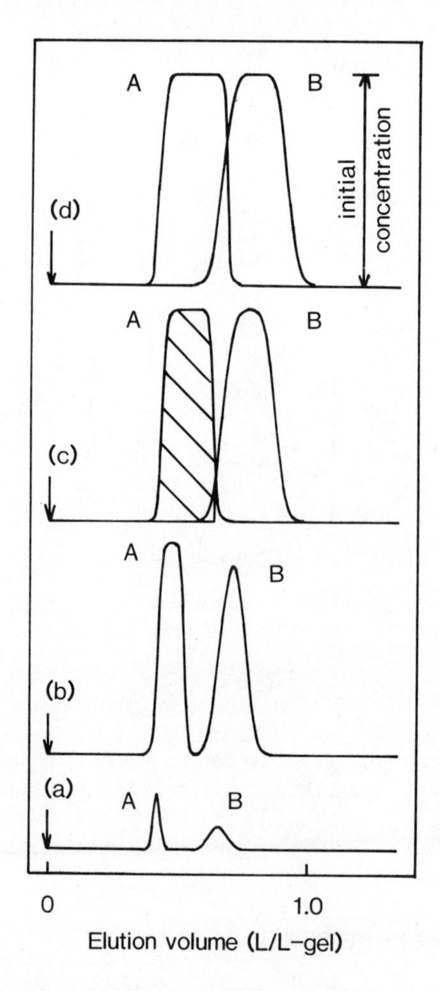

Fig. 16. Simulation of chromatograms. Loading volume: (*a*) 1% of bed volume, (*b*) 10%, (*c*) 20%, and (*d*) 25%. Details are given in Section 4.

result of simulation, where the injection volume is 1% of the bed volume. To simplify the simulation, relatively mild conditions are used here: the plate number of the column (N) is 200, and the K_d values are assumed to be independent of the concentration of the solutes, A and B.

To load a larger amount of the crude product on the column, continuous feeding of the solution is usually used in industrial operations instead of concentrating the solution. Figure 16*b* shows the result of a simulation using the same parameters when the charged volume is 10 times as large as that used in Fig. 16*a*. In this case, there is a little retardation of peaks resulting from the nonideal injection.

As expected from the above consideration, a much larger amount of the solution, that is, longer feeding time, gives a severely deformed elution pattern, as shown in Fig. 16*c* and 16*d*. These results are unsuitable for analytical purposes, but the efficiency of fractionation is not greatly affected. For example, the recovery of A in the shaded part of Fig. 16*c* is more than 95% and the contamination by B is less than 5%. thus, the loading limit is determined by the specified purity of A. When the loading capacity of the column is specified, the dimension of the column will be designed according to the total amount of A to be purified.

More accurate simulations can be made by increasing the number of parameters. the plate number will be estimated from the values of column dimension, particle diameter, flow rate, and so on. If the adsorption isotherms of A and B are known, the distribution of each substance in each theoretical plate will be estimated more accurately. The number of substances may be increased. These calculations are complicated and sometimes need the knowledge of chemical engineering, but they are not necessary in most cases.

If the purification process contains gradient elution steps, the scaling-up method becomes much simpler. Suppose that the target substance A elutes from the column rather fast (e.g., $K_d < 0.05$ with the initial solution) while the impurity B is completely adsorbed on the column under the same condition. In this case, the initial solution containing the crude product can be fed on the column until the loaded amount of B reaches the total adsorption capacity of the column for B. The gradient elution should be started when the column becomes nearly saturated with B. The actually working process shown in Fig. 13 is an example for this case.

5. CONCLUSION: FUTURE OF INDUSTRIAL SEPARATIONS

The role of biotechnology in chemical industry is becoming increasingly important. Accordingly, the technique of industrial separations by LC is widening its application range. The improvement of the methods and materials in LC is indispensable for the development of biotechnology. Especially, the costs and operability of the system are very important factors in industrial separations. SEPABEADS FP SERIES will prove their merits for various applications in industrial processes.

SEPABEADS listed in Table 1 are commercially available, and a number of other types of SEPABEADS are now being developed. Special preparations of SEPABEADS (particle diameter, functional group, or capacity) will be also available from MCI as test samples.

REFERENCES

1. Technical data sheets of SEPABEADS FP SERIES are available from Specialty

Polymers Division, Mitsubishi Chemical Industries Limited, Marunouchi 2-chome, Chiyoda-ku, Tokyo, JAPAN.

2. Technical data sheets of MCI GEL series. The mailing address is given in Ref. 1.
3. T. Itagaki, E. Miyata, and H. Kusano, Reactive Polym. (to be published).
4. T. Itagaki, H. Kusano, and E. Miyata, Ionics (Jpn.), 277 (1985).
5. H. Ouchi, H. Kato, S. Kanaya, A. Shimura, and T. Itagaki, Preprint, the 29th annual meeting of The Research Group of Automatic Liquid Chromatography, Kyoto, 1986, p. 69.
6. H. Kato, H. Ouchi, A. Shimura, and K. Kumamoto, Anal, Sci., *2*, 395 (1986).
7. There are a number of published works on the theoretical treatments of LC: for example, L.R. Snyder and J.J. Kirkland, *Introduction to Modern Liquid Chromatography*, 2nd ed., Wiley, New York, 1979.

CHAPTER 6

Scale Up of Protein Chromatographic Separations

DUNCAN LOW

Pharmacia AB, Process Separation Division, S-751-2 Uppsala, Sweden

1. INTRODUCTION

The study of biochemistry is essentially an attempt to understand the structural and metabolic aspects of life processes at the molecular level. Because of the inherent complexity of the problem, a sophisticated understanding of the nature of biomolecules and the physical and chemical reactions involved requires that suitable methods are available for their isolation and analysis.

Often, fundamental developments in biochemistry have had direct medical applications, such as an understanding of human nutritional requirements and the significance of diet in vitamin deficiencies such as pellagra, scurvy, rickets, and beriberi. However, many of the classical biochemical pathways, such as the oxidative metabolism of carbohydrates, had to be determined on soluble or cell-free systems, since no suitable methods were available for the fractionation of tissue extracts that could preserve biological activity.

Modern strategies in human and animal health care rely heavily on the identification and isolation of various biological activities and testing their potential in therapeutic applications. The availability of significant quantities of purified, active protein enables structural and functional studies to be carried out and primary sequences to be determined. Recombinant DNA technology can then be used to produce the desired biomolecule in large scale.

Computer-aided molecular design can be used to modify the molecule to make it more suitable for the intended application, and future trends could well be the use of peptide analogs to simulate the function of biological activators or repressors.

Perhaps the single most important technique for protein isolation in the last 25 years is chomatography. Since 1959, with the discovery of Sephadex® (1), ion exchange and gel filtration have formed the backbone of modern protein purification methods. New methods have been added, most significantly affinity chromatography in the 1970s and HPLC (high-performance liquid chromatography) in the 1980s, that have extended the usefulness of the technique and kept it up to date with developments in other aspects of protein purification.

The current importance of protein purification is due in part to the intensive search for new human health care (replacement therapy) products and in part due to new processes being developed for the large-scale purification of biomolecules, notably hormones and enzymes, within pharmaceutical or related industries.

Today, chromatography is well established as an essential step in the production of insulin and other hormones, interferons, monoclonal antibodies, and plasma proteins.

2. BASIC PRINCIPLES OF CHROMATOGRAPHY

Separation in chromatography is essentially a balance among resolution, capacity and speed. These three parameters are interdependent, so that a change in one will influence the other two. In a production process, all three will therefore have an effect on the overall economy. Capacity and speed combine to determine the productivity of the process, or the throughput in terms of product per unit time. Speed in itself is important in terms of product recovery, especially if the product is unstable or if proteolytic contaminants are present.

Resolution is vitally important, since it determines the quality of the product and the yield. When two adjacent peaks overlap slightly with each other, some product will have to be discarded if 100% purity is desired. Alternatively, some level of contamination has to be accepted if 100% yield is required. The balance between product quality and yield is largely determined by the application for which the product is to be used. In the case of an enzyme required for industrial use, reproducible activity is probably more important than product purity, so certain contaminants can be tolerated. A product destined for injection into humans, on the other hand, must meet completely different standards for purity, these often being set by the appropriate regulatory authorities. A product for health care purposes from a recombinant DNA source should be at least 99.5% pure, show no microheterogeneity, should have an intact 3-D structure, should be free from endotoxin and plasmids, and, interestingly, not contain any chemicals used in the purification. The impact of recombinant DNA technology on new drug development and registration is apparent from the experiences encountered in the development of human insulin by Eli Lilly (2).

2.1. Resolution

In analytical chromatography the aim of a separation is to identify and, frequently, to quantify the maximum number of components possible. Separation time should be kept as short as possible. In preparative chromatography, the

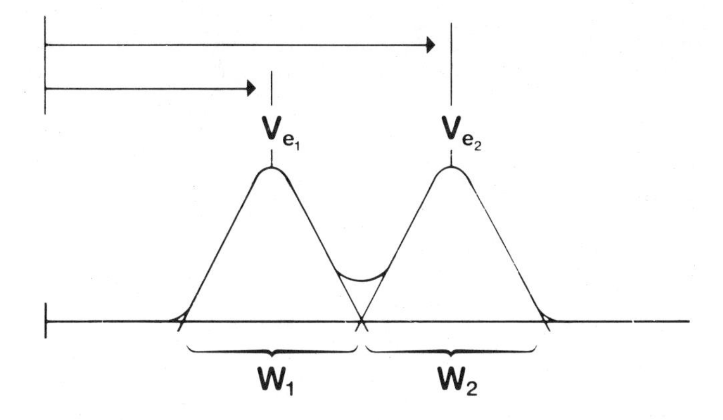

Fig. 1. Symbols used in describing resolution.

aim is to recover as much product as possible from the starting material at a defined quality. This means that what can be acceptable resolution in one case is not acceptable in the other.

Resolution can be defined in several ways. It is easiest to measure it as the distance between two neighboring peak maxima divided by half the sum of the peak's base line intercepts:

$$RS = \frac{V_{e_2} - V_{e_1}}{\frac{1}{2}(W_1 + W_2)}$$

The symbols used are described in Fig. 1.

An alternative definition is:

$$R = \tfrac{1}{4}\sqrt{N}\,\frac{\alpha - 1}{\alpha} \cdot \frac{k^1}{k^1 + 1}$$

where N = plate number
α = selectivity
k^1 = capacity

The plate number is a measure of the efficiency of the column and is defined as the length of the column divided by the height equivalent to a theoretical plate. The plate height is itself defined as the variance (σ^2) of the Gaussian concentration profile divided by the migration distance of the solute (L):

$$H = \frac{\sigma^2}{L}$$

There are a number of different ways in which column efficiency can be measured in practice (3, 4).

The plate number is dependent on the physical parameters in the chromatographic column such as quality of the packing, flow rate, column length, particle size of the packing material, temperature, and so on. One of the most important variables is the particle size of the packing material, which is very small (10 μm and less) for analytical HPLC packings, but is usually larger (30–90 μm) for large-scale applications because of economic considerations. The selectivity term is dependent on the chemical nature of the separation mechanism and will be affected by the buffers used and the relative strengths of interactions between solute molecules and the matrix. The selectivity term is extremely powerful in controlling resolution, and small changes in α can significantly improve resolution or reduce the number of plates required to achieve a defined resolution.

The capacity term is a reflection of the degree of retention available in the system and is also dependent on the chemical nature of the separation mechanism. The capacity term controls the degree of retention of the solutes. In most practical separations based on adsorption methods retention times of between 2 and 10 bed volumes are chosen. Yield usually decreases with increasing retention time, so it is normally not greatly used as a parameter to control or improve separation.

It is interesting to note that retention times in gel filtration are always less than 1 (unless adsorption effects are present), and this was thought to limit the usefulness of gel filtration as a practical separation method at the time of its introduction.

There are, therefore, two main approaches to improving resolution in chromatography. The strategy traditionally favored by biochemists has been to manipulate selectivity, achieving the desired result via a multistep purification procedure, varying the separation mechanism from step to step. The most extreme examples of the use of selectivity in chromatography come from the purification of biomolecules by affinity chromatography, where several thousand fold purification can be achieved in a single step. Selectivity is certainly a powerful parameter in large-scale, but long purification procedures increase the risk for product losses.

The strategy favored by organic chemists for the resolution and analysis of small molecules has been to manipulate the efficiency of the packing material. This has led to the development of small particle size packings (10 μm and less) and high-performance liquid chromatography (HPLC). These small particle sizes give extremely narrow zone spreading, which is exactly what is required when trying to analyze a multicomponent mixture in a single chromatogram.

Manipulating efficiency is also a successful strategy to improve resolution, although less so than selectivity (R is proportional to $\sqrt{N}$). It is probably most appropriate to the analysis of low-molecular-weight analytes that closely resemble each other.

Recent years have seen the emergence of microparticulate packings designed especially for the purification of proteins and peptides. These packings permit rapid purification of biomolecules and are available for a range of techniques, and have been widely accepted for laboratory use. However, they are expensive and require specialized equipment for optimal use, so they are not necessarily the ideal choice for large-scale operation (except perhaps as a final step). They are, however, extremely useful for rapid method development and small-scale preparative work, and some examples of this are given in Section 5.3.1.

Detailed discussions of the different factors determining resolution are available in the large number of text books that have now been published on chromatography.

2.2. Capacity and Speed

As stated earlier, capacity and speed combine to determine throughput. Once methods giving acceptable product quality have been developed, it remains to test them for their suitability for use in large scale. The key considerations at this point will be whether or not the method selected can be used to produce sufficient quantities of product to meet the market demand.

Resolution decreases with increasing sample loading, so in a case where the separation of a peak from its nearest neighbor is minimal, the only practical way to scale up is to increase the column volume by increasing the column diameter. This is not usually an effective use of the chromatographic packing, and it is well worth spending some time optimizing the separation to get as good resolution as possible in order to get high sample loading. Ideally, conditions are arranged so that only the product of interest interacts with the chromatographic packing while most of the contaminants pass straight through. In such a case the relationships described for analytical separations no longer hold true and different theoretical models are applicable (see Section 5.1.1).

Speed is a direct consequence of the quality of the packing material used. Rigid particles are required for high flow rates, and small particle sizes are required to give rapid equilibration between mobile and stationary phases to prevent excessive zone spreading. Small particles have, however, the disadvantage of resulting in high back pressure.

Speed, however, is not usually a major issue unless the objective of the step is to separate the product of interest from a degradative enzyme. The important consideration is to plan the separation so that the product is held in a convenient form from one working shift to the next. Separation steps that take less than 2–3 hr can be comfortably performed during one working day, otherwise steps may as well take several hours and be run overnight.

When choosing the dimensions of the column to be used for a large-scale separation, it is important to consider whether the whole sample has to be processed in a single cycle or whether several cycles on a smaller column can be used. Factors that determine which alternative is best will be the stability of the product in the feed solution, the time available for processing, the space available for the equipment, and the cost of automating the process. This will be dealt with further in Section 5.3.4.

3. DESIGN OF A PURIFICATION SCHEME

A protein purification method is usually designed around the known physicochemical and biological properties of the sample of interest. If these are not known, they can be determined by electrophoretic or chromatographic analysis. In the case of recombinant DNA products, the amino acid sequence is known, so it follows that the molecular weight is known. It is also possible to make some predictions as to the secondary and tertiary structure, giving an indication as to whether hydrophobic or ion exchange methods are appropriate.

The next step is to look at the methods available for the purification of proteins and, in particular, to determine which ones are suitable for large scale, and in which order they should be used.

3.1. Initial Steps

Before we can usefully use chromatography for purification, the sample has to be pretreated to make it suitable for application onto columns. The sample resulting from fermentation or extraction will be contaminated with a complex mixture of cell debris, subcellular organelles, nucleic acids, proteins, lipids, carbohydrates and low-molecular-weight substances. The major contaminant, however, will be water. The choice of initial steps will depend on whether the product is secreted extracellularly or located intracellularly, and whether it is in the soluble or insoluble form. Some or all of the following steps have to be performed before the sample can be applied to chromatographic columns:

Cell harvesting to separate the cells from the culture medium. The cells or the supernatant can then be discarded depending on where product is located.

Cell disruption, extraction, and solubilization are required if the product is located intracellularly. The product will also have to be separated from insoluble cell fragments to give a particle-free solution.

Reduction of nucleic acid content to lower the viscosity.

Concentration to reduce the total volume and give a protein content suitable for chromatography (5–7% w/v).

Obtain the product in a solution with suitable pH, ionic strength, and temperature for the first chromatographic purification step.

Various methods have been used at this stage including centrifugation, filtration and ultrafiltration, precipitation, batch adsorption and two-phase partitioning, and enzymatic digestion of contaminants. The interactions between these unit operations have been reviewed by Fish and Lilly (5). There are several recent reviews of techniques used for initial recovery covering two-phase partitioning, membrane processes, and centrifugation (6–9).

The methods used for primary recovery should not damage the product to be purified and should be economical. They should also include some form of selective purification, if possible, in order to reduce the amount of material to be handled at later stages. In practice, methods are chosen that give an effective concentration of the product with minimal losses ($< 10\%$).

3.2. The Selection of Methods

All of the methods commonly used in the laboratory for protein purification can be scaled up to large scale, however not all of them are equally suitable for large-scale use. The range of methods used are given in Table 1 together with some comments about their applicability. Methods based on adsorption, such as affinity chromatography and ion exchange, are more suitable for handling large volumes of dilute sample than partition methods such as gel filtration, where the sample volume is limited to a fraction of the total column volume.

The media used for separation steps early in the process must be resistant to strong cleaning reagents such as sodium hydroxide otherwise they will quickly become contaminated and will have to be discarded. This is particularly important to consider when using affinity chromatography or reversed phase, where the media are inherently expensive, and therefore have to be used many times to be cost effective.

3.3. Sequence of Steps

Since some techniques are better suited to the early stages of a purification while other techniques are better suited to the final stages, it follows that it is important to consider the overall sequence of steps. This is perhaps best illustrated by looking at how the nature of the separation problem varies throughout the purification scheme.

At the early stages of a purification the chief problem is recovering the product of interest from a large volume of dilute sample. The product may well be less than 10% of total protein present. It is therefore desirable to use methods that are capable of handling large volumes of sample and that can show a concentrating or enrichment effect (Fig. 2*a*) such as ion exchange or affinity chromatography. Hydrophobic interaction chromatography is also useful in the early stages, especially if the sample is in a high salt concentration after a precipitation step.

In the case where product has been concentrated and is present in roughly equal proportions to its contaminants, some form of fractionation step is required. This step need not remove *all* contaminants, and so is an intermediate step. Again, yield is important, and the degree of purification can be optimized by careful manipulation of elution conditions. Sample volumes are still likely to be high, and so methods based on adsorption are favored. As an example Fig. 2*b* shows the fractionation of plasma proteins on DEAE Sepharose® Fast Flow.

Intermediate steps should preferably use a different selectivity from the initial steps. The number of steps should be as few as possible for optimal yields, simplicity and speed, so high-resolution steps such as affinity chromatography or other high-performance techniques should be introduced as soon as possible into the process.

Table 1. Chromatographic Methods for Large-Scale Purifications of Biological Molecules

Characteristic	Separation Method	Characteristics	Applicability
Size	Gel filtration	*Resolution* is moderate in the fractionation mode, but excellent for desalting. *Speed* is low for fractionation (cycle time $\geqslant$ 8 hr) but very fast for desalting (approx. 30 min). *Capacity* is limited by sample volume.	*Fractionation* by gel filtration is best suited to the final stages of large scale purification and removing critical contaminants. *Desalting* can be used at any stage, particularly to link two steps by buffer exchange.
Charge	Ion exchange	*Resolution* is usually high. *Speed* is very fast when using the correct matrix. *Capacity* is very high. Sample volume is not limiting.	Ion exchange is most suitable during the early stages of a separation, when large quantities of material are to be handled.
Isoelectric point	Chromatofocusing	*Resolution* is very high *Speed* is high. *Capacity* is very high. Sample volume is not limited by volume size.	Chromatofocusing is more suitable towards the end of purification.
Hydrophobic nature	Hydrophobic interaction	*Resolution* is good. *Speed* can be very fast. *Capacity* is high. Sample volume is not limiting.	HIC is suitable at any stage in a separation, particularly when the sample is at high ionic strength, for example after precipitation, ion exchange or affinity.
	Reverse phase	*Resolution* is very high. *Speed* is very fast. *Capacity* is high.	RPC is most appropriate at the final stages when purifying low-molecular-weight substances.
Biological affinity	Affinity chromatography	*Resolution* is excellent. *Speed* is very fast. *Capacity* is very high. Sample volume is not limiting.	Affinity is suitable at any stage, particularly when a low concentration of sample is present in a large volume together with a lot of contaminants.

The final stages in a process can be of two types, the *removal of low-concentration contaminants* or the *removal of polymers*. Selective adsorption by affinity chromatography, ion exchange, or reversed-phase chromatography is the method best suited for the first of these two problems. Separation can be achieved by selective adsorption of the contaminant or the product. Selective adsorption of the product at this stage is likely to require a larger column. Figure 2*c* shows the selective removal of colored substances (hemoglobin and breakdown products) from albumin on CM Sepharose CL-6B.

Many biological products have a tendency to exist in polymeric forms. If these forms are to be removed, then gel filtration can be used as a final purification step. Gel filtration is slower than other techniques used in process chromatography and can only handle limited volumes (5–10% of V_t) of sample per cycle, so, if required, it can be used at the end of a purification process. It can also be used to transfer the sample to suitable buffers for packaging or prior to freeze drying. Figure 2*d* shows the removal of polymers from albumin by gel filtration.

4. CHOICE OF SEPARATION MEDIA

When choosing gels for large-scale purification of biomolecules, there are a number of conditions that have to be met.

First and foremost they should give the desired separation. They should also have excellent *chemical stability*, since this contributes to the life length of the gel and determines the type of maintenance and clean procedures that can be used with the gels. Effective cleaning procedures for gels frequently require the use of strong solutions of sodium hydroxide (0.1–0.5 *M*).

Gels used in large-scale applications also need to have good *physical* and *mechanical* stability. This is one of the more difficult aspects to meet, since many traditional macroporous matrices such as silica lack the necessary hydrophilicity and chemical stability to be used for protein separations while other media, notably cross-linked dextrans and polyacrylamide, lack the mechanical strength.

The cost of the packing material is also important. Gels should be economical in large-scale use in terms of overall cost/cycle, since a good gel will permit short cycle times and can be used for a thousand cycles or more. Small particle-size packings are inherently expensive, and it is important to consider carefully what is required before investing in high-performance media. The cost of different media relative to their performance is shown diagrammatically in Fig. 3. The different media and techniques available for large-scale chromatography have been discussed in a comprehensive review by Janson and Hedman (10) and by Cooney (11). Important developments have been made in chromatographic media for the separation of biomolecules during the last few years. The most significant of these have perhaps been improvements in the stability of silica-based media (12, 13), the development of highly stable synthetic polymer

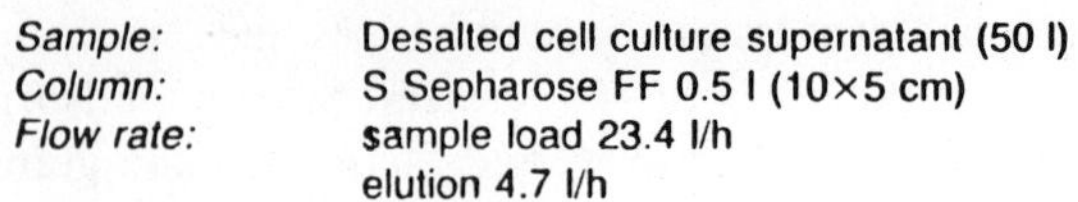

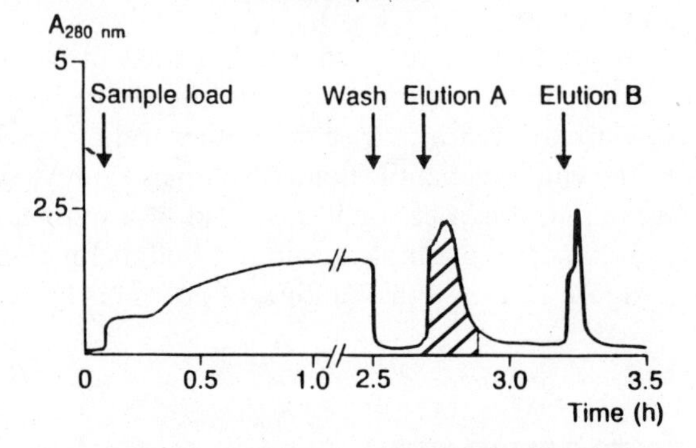

Elution profiles from DEAE-Sepharose® CL-6B of a production run

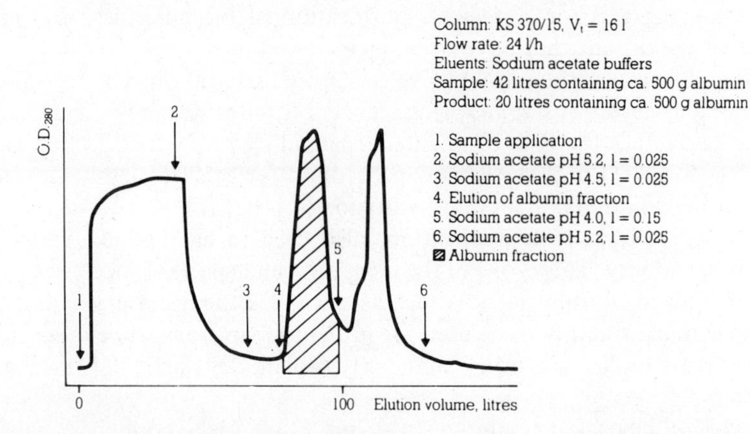

Fig. 2. Different types of separation problems in large scale chromatography. (*a*) Concentration. Recovery of monoclonal antibody from cell culture supernatant (50 L) on a 500 mL column of S Sepharose Fast Flow. Sample load in 10 m*M* citrate buffer pH 5.5, detection at 280 nm (AUFS 1.0). IgG-containing peak was eluted in the same buffer containing 0.075 *M* NaCl (AUFS 10). 81% of MAb were removed in the area indicated. (*b*) Fractionation. Separation of albumin from plasma on DEAE-Sepharose CL-6B. (*c*) Removal of contaminants. Specific removal of colored substances (hemoglobin fractions) from albumin. Sample was applied in 0.04 *M* sodium acetate buffer, pH 5.3. Colored substances were removed from the saturated column with 0.4 *M* acetate, pH 8. (*d*) Removal of polymers by gel filtration on Sephacryl[R] S-200 Superfine. Column: KS 370/15 × 4. Sample: 5 L of 6% albumin. Flow rate: $12\ \mathrm{L\,hr^{-1}}$. Buffer: 0.05 *M* sodium chloride.

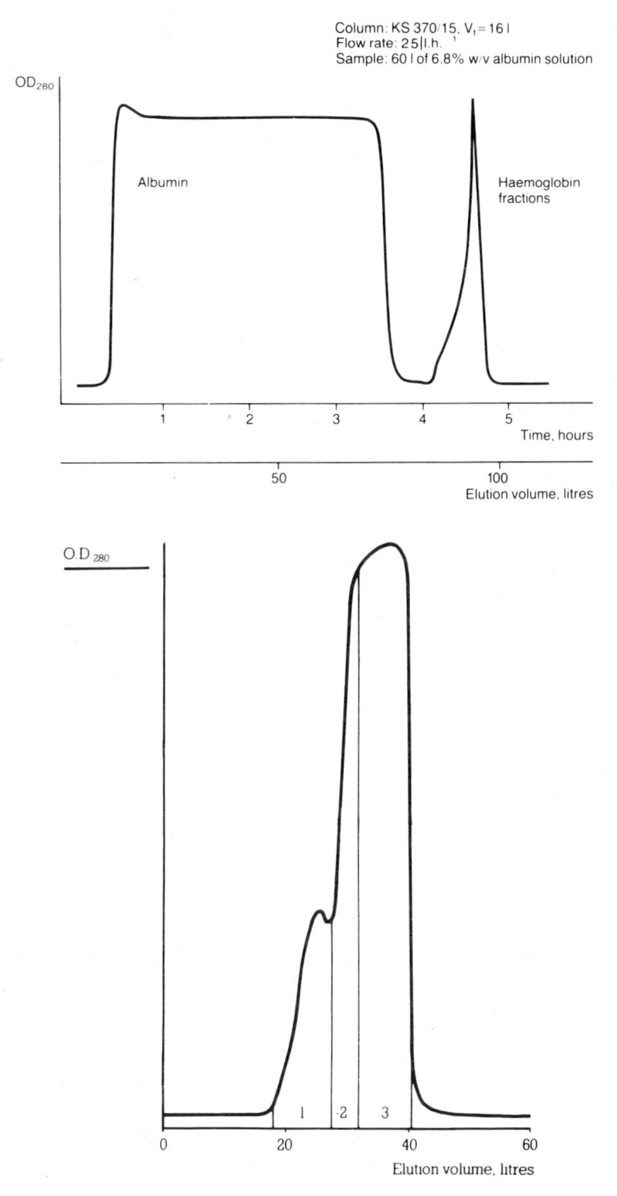

Fig. 2. (*Continued*)

matrices for various types of HPLC separations of biopolymers (e.g., MonoBead^R, Pharmacia; TSK gel PW, Toyo Soda) and the development of more rigid gel media designed for large-scale separations (Trisacryl^R, IBF; Sepharose Fast Flow, Pharmacia). Interesting new additions to this group are

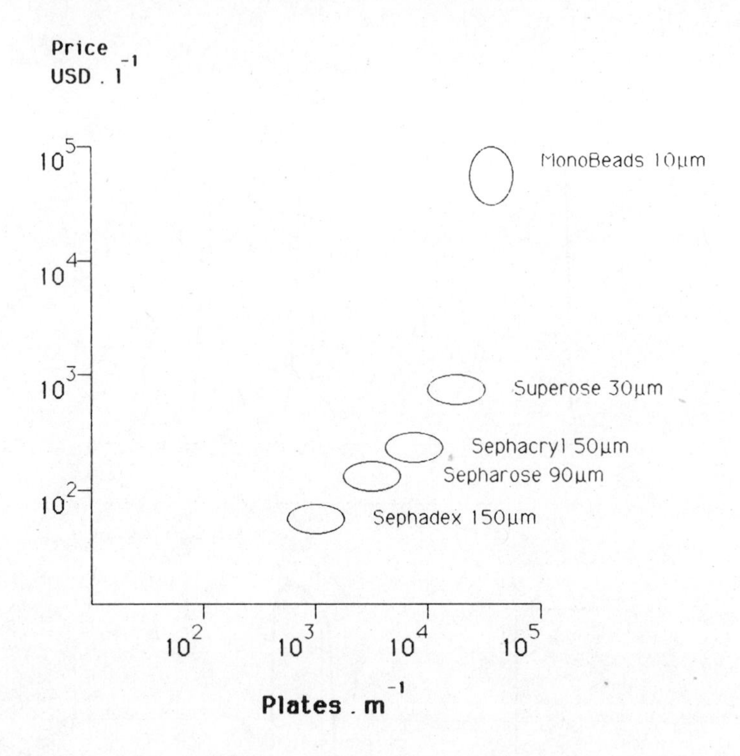

Fig. 3. Cost versus performance (expressed in plates/m) of media with different bead sizes.

fluoropolymer supports, for example, Perflex, developed by Dupont and macroporous packings for perfusion chromatography developed by PerSeptive Biosystems. All of these developments have contributed to decrease the time taken for protein purification (14).

High-resolution separation methods, using microparticular packing materials, have an important role to play in the purification of biotechnology products. There are two main areas where they are important, in method development where the short analysis time allows for rapid optimization of separation methods and in small-scale production, particularly when relatively small quantitites of product are sufficient to meet market requirements. For example, the minimal requirement of human growth hormone for the treatment of all growth deficiencies and growth-related diseases in the United States is 1 kg per annum. The requirement for the UK is around 400 g.

Other products in biotechnology, however, will need to be produced in much larger quantities, and it is probable that microparticulate packings (less than 10 μm) will be too expensive to be used in these applications for anything other than the final purification of the product or for the recovery of product from the contaminated shoulder regions of otherwise pure peaks.

The key to economy in the production of proteins in biotechnology will be finding the optimal conditions for purification and the optimal scale of the manufacturing process. High-performance methods will be an important tool for evaluating conditions and in some cases will be the preferred way to scale up.

4.1. Method Development

A great deal is known about the properties of the sample at the outset of a purification process. Nevertheless, the separation performance of each chromatographic step must be thoroughly tested and evaluated and compared with the various alternatives before it can be accepted as part of a process. HPLC is an invaluable tool for evaluation, since a large number of runs under varying conditions can be performed rapidly to determine the optimal conditions for separation.

The appropriate route for method development varies somewhat depending on the actual method being used.

4.1.1. Ion Exchange

Ion exchange methods offer considerable potential for scale up, but, because of the large number of variables that control retention, it is important to approach optimization in a rational way.

Retention in ion exchange is dependent on the nature of the protein sample, the choice of matrix, and the elution conditions (pH, ionic strength, choice of counter ions).

Proteins are amphoteric molecules, and their charge varies with pH, carrying a net positive charge under acidic conditions and a net negative charge under basic conditions. The point where the net charge is 0 is termed the isoelectric point (pI). Titration curves can be determined by electrophoresis to indicate how the charge of the sample varies with pH. Since each protein has a unique titration curve, titration curves of the sample can be run prior to chromatography to indicate a suitable pH for the run (15–19).

This approach is relatively simple, but there are drawbacks. Proteins are large molecules and do not carry a uniform charge distribution, so it is quite possible to separate two proteins that have the same pI because one has a uniform charge distribution and the other does not (16). More to the point, a protein at its isoelectric point can still bind to either anion or cation exchangers because there will still be charged groups present, as shown by Kopaciewicz et al. (16). A given protein has a preferred plane of orientation toward an ion exchange surface, and it is the structure of this plane that is important in determining retention. Structural modifications (e.g., alteration in amino acid sequence) remote from this plane have relatively little effect on retention (20).

It is important, therefore, to study ion exchange retention if a separation is to be optimized properly. This is done by plotting retention versus pH for the sample of interest and the potential contaminants.

When selecting pH, it is important to consider the stability of the protein at that pH, since yield is of great importance when the process is scaled up.

Choice of pH will also determine the choice of an anion or cation exchanger. The media used for method development should be identical to, or as close as possible to, the media selected for the final large-scale application if the investigation is to have any value. The media used for the large-scale process should fulfill the criteria described earlier, so when screening media for methods development, it is important to consider the potential for scale up.

Ion exchangers are usually available as "weak" or "strong" types, which refers to the degree of ionization of the charged groups. Strong ion exchangers are charged over a wider pH range than weak ion exchangers, and are therefore more broadly useful. However, the different types of charged group will show different selectivity effects with different proteins, and this will affect the separation between different peaks (Fig. 4). The loading capacity and recovery from different ion exchange matrices are as important as considering the degree of purification.

The buffers chosen during method development also influence the separation. Normally, a buffer is chosen that has a pKa $\pm$ 0.5 pH units from the working pH for the best results, but the choice is limited for large-scale applications because of practical considerations such as the volumes of buffers required and the final formulation of the product. Normally, buffers such as sodium acetate are used in large-scale applications because of the ease of preparation.

It may be possible to take advantage of the properties that certain counter ions and co-ions possess to improve the quality of the separation. It is therefore worthwhile to test a variety of different co-ions with a single counter ion and vice versa and then plot ion selectivity maps before choosing a specific salt.

Once the correct conditions have been chosen, the separation can be further perfected by manipulating the shape of the gradient. Linear gradients should be used for preliminary evaluation of the separation, but nonlinear gradients are more useful when trying to isolate a specific peak. Figure 5 shows how varying the gradient shape can be used to optimize the separation of enzymes present in detergents. Conditions can be adjusted so that the sample of interest is retained but most of the contaminants pass straight through the column. A small increase in salt concentration is then used to elute the sample of interest followed by a large increase to wash off any contaminants still bound to the column. Conditions should always be chosen to offer the maximum separation of the peak from the adjacent contaminants while preserving recovery of activity from the column. Theoretical models have been developed for predicting elution curves and operating conditions for ion exchange separations (21, 22). A general approach for optimizing gradients for both ion exchange and RPC has been described by Stout and co-workers (23).

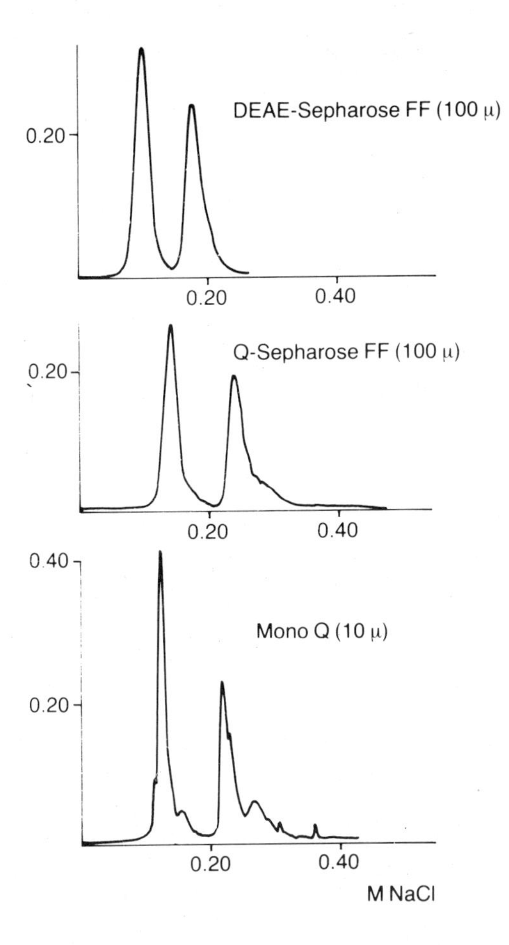

Fig. 4. Separation of protein standards on weakly and strongly basic anion exchangers. The elution profile is plotted as a function of ionic strength. Curves *a* and *b*; column, Pharmacia HR 16/10; bed height, 10 cm; eluent, 0.02 *M* Tris HCl, pH 7.5 with a linear gradient to 0.05 *M* NaCl; flow rate, 50 mL cm^{-2} hr; sample, transferrin 6 mg and albumin 15 mg. Curve *c* shows the same separation run on a Mono Q HR 5/5 column.

4.1.2. Hydrophobic Interaction Chromatography and Reversed-Phase Chromatography

Both hydrophobic interaction chromatography (HIC) and reversed-phase chromatography (RPC) separate proteins on the basis of hydrophobicity. Retention is again dependent on the nature of the sample, the choice of matrix, and the elution conditions to be used. The main difference between the two methods is that HIC is normally performed in aqueous solvents and the protein sample retains its native structure, whereas RPC is normally performed in organic

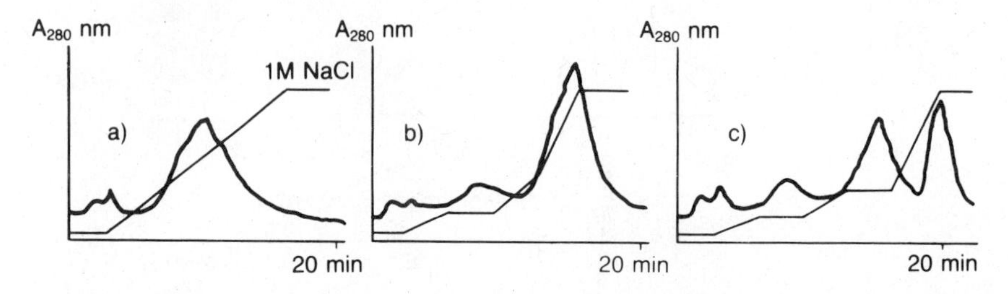

Fig. 5. Separation of bacterial proteases used in enzyme detergents on DEAE Sepharose Fast Flow in an FPLC system.

solvents and the proteins are at least partially denatured through interaction with the mobile and stationary phase (24).

The approach for optimizing separation in HIC and RPC is similar to that in ion exchange, in that retention is plotted against various parameters such as ionic strength (in HIC), pH and solvent composition and strength (in RPC). There is no simple method like electrophoresis to aid selection of conditions, but clues can be obtained from measurements of solubility and surface tension (25). Furthermore, as with charge, hydrophobic regions are not uniformly distributed over protein surfaces and the presence of different additives may affect protein conformation in different ways and thus the relative retention times on different matrices (26).

The use of organic solvents in RPC generally results in a certain amount of refolding of the protein structure, leading to the exposure of buried hydrophobic sites. Again, solvent composition and strength will be important in determining retention, and optimal conditions have to be arrived at via retention plots.

The media used for method development should be identical to, or as close as possible to, the media selected for the final large-scale application. Media used for HIC are usually alkyl or phenyl derivatives of agarose. There are very few microparticulate packing materials available for method development.

Most of the media available for RPC are based on silica supports to which a hydrocarbon phase has been bonded. Chain lengths from C_8–C_{18} have been used, but normally the longer chain lengths are too hydrophobic to be useful for the recovery of proteins in their native forms. There are limitations to the use of RPC for large-scale preparative chromatography, and the effects the mobile phase and the characteristics of the stationary phase have on protein conformation and biological activity are perhaps the most important, since they will directly affect recovery. This problem increases with increasing molecular size (and complexity) of the product to be purified, and while large-scale RPC is a useful alternative to ion exchange for the purification of peptides with molecular weights up to 10,000, ion exchange will often give superior yields for larger proteins.

Another limitation of media available for preparative RPC is the stability of the silica matrix, which is prone to attack by aqueous solvents with pH > 8. This could result in the release of leakage products (27). Effective methods for the removal of lipids and proteins from contamined silica matrices are also required, since the methods used for regenerating organic packings based on NaOH are not applicable.

New media for RPC/HIC of proteins based on organic supports have been synthesized (28, 29), but as yet they are only available as microparticulate packings. This makes them expensive to use at large scale.

4.1.3. Affinity Chromatography

Separations in affinity chromatography are achieved by taking advantage of the specific interactions between biological substances, for example, an enzyme and its substrate or cofactor, or an antibody and its antigen. Retention in affinity chromatography is mostly due to the selective binding of the sample of interest while contaminants are not retarded at all. Binding is of an all-or-none type, and elution is achieved by either a dramatic change in conditions or by adding a competitive binding substance.

Affinity chromatography methods are extremely important for large scale biotechnology because they frequently offer the maximum possible purification for the minimum number of steps. Many examples exist where over a thousand-fold purification can be obtained in a single step with extremely high yields (30–32).

The approach for optimizing affinity chromatography is mostly concerned with selecting an appropriate ligand. This can be a synthetic material with a general specificity such as a dye substance or a highly specific ligand such as a monoclonal antibody. The ligand should not have too high an affinity for the product, since this will make elution difficult, while too low an affinity makes the method uneconomical. Dissociation constants of 10^{-8}–10^{-4} M in solution are normally preferred.

The ligand should be immobilized on the gel with a suitable coupling method. The coupling method chosen should be stable, so that there is little leakage from the matrix. The ligand may or may not need to be coupled via a specific linkage or with a spacer arm. The amount of ligand coupled should give the optimal capacity for the sample of interest. Increasing the amount of ligand coupled to a matrix beyond a certain level does not result in any increase in capacity for product.

The choice of matrix to which the ligand is bound is important. It is essential that it should be inert and that there are no nonspecific interactions between the matrix and protein molecules in the sample solution, otherwise the sample will

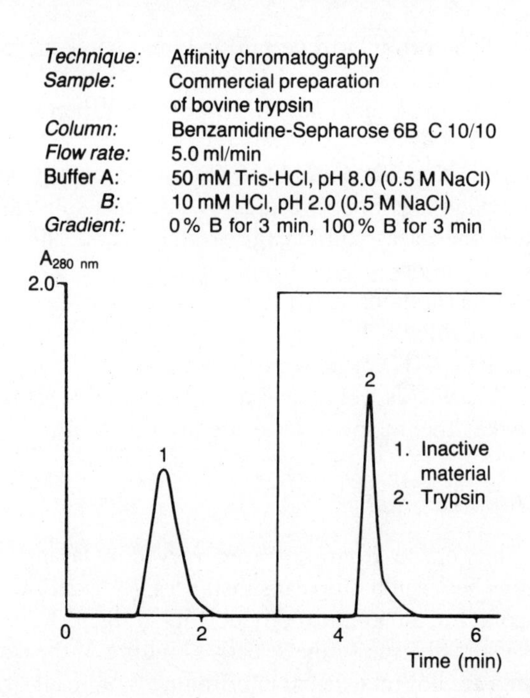

Fig. 6. Separation of active from inactive material in a commercial trypsin sample.

be contaminated. The matrix should fulfill the criteria for large-scale use described previously.

Matrices used for large-scale affinity chromatography are almost exclusively based on agarose. More recently, microparticulate packings based on silica and agarose have become available for use in HPLC (33). These media permit rapid separations primarily because the smaller particle size allows for more efficient sample loading and elution. However, mass transfer between stationary and mobile phases may be much less important than the contribution from the chemical kinetics of the binding reaction (34).

Band spreading is not normally a serious problem in affinity chromatography because of the highly selective nature of the separation. Affinity separations can be run rapidly even with relatively large-particle-size media, and flow rates in excess of 100 cm/hr have been used successfully. Figure 6 shows the rapid separation of active from inactive material in commercial preparations of trypsin. As with the previous examples, it is important to optimize the separation on the media that will be used at large scale if the physical separation parameters are to be kept constant.

Starting conditions in affinity chromatography are chosen to favor binding and are usually close to physiological conditions. Care has to be taken to include the appropriate ions and co-factors that may be required for binding. Elution is then achieved by either specific or nonspecific elution methods. Nonspecific elution is achieved by changing the pH or ionic strength or by altering structure with a denaturant such as urea or guanidine hydrochloride. Nonspecific methods are usually chosen when a highly specific adsorbent, such as a monoclonal antibody, is used.

Specific elution methods use competitive substances, are more selective, and offer gentler elution methods. However, the buffers used may be prohibitively expensive. The gentleness of the elution method permits good product recoveries and does not damage the ligand bound to the gel, which can be used for many cycles, so in the long term it may prove more economical. This should be determined in laboratory trials at small scale.

4.1.4. Gel Filtration

Optimizing gel filtration is relatively simple in comparison to the various adsorption methods, because there are fewer variables to consider. Separation in gel filtration is based on differences in molecular weight between sample molecules and the contaminants and the main factors to consider are the pore structure of the matrix, the column length, and the particle size of the packing material. Solvent effects do not contribute to the separation.

The choice of gel for gel filtration is largely governed by the application. Gel filtration is used in two main ways, for desalting and fractionation. Desalting is used to separate macromolecules from low-molecular-weight substances, while fractionation is used to separate substances of similar molecular weight. For desalting, a gel with an exclusion limit around 5000 should be chosen, whereas for fractionation a gel should be chosen whose fractionation range covers the molecular weight of the substance of interest. Information on the molecular weight can be obtained from the amino acid sequence, electrophoresis or high-performance gel filtration studies.

Desalting is widely used in a number of industries to remove low-molecular-weight substances from products and can be especially useful in separating a sensitive sample from a harsh elution buffer. Sample volumes are usually around 20% of the total bed volume and separation times are usually less than 30 min.

Fractionation by gel filtration is an important technique for production since it can be used to separate dimers and polymers from monomeric material. However, it is usually less suitable for large-scale separations than adsorption techniques, since the sample volume that can be processed in a single cycle is usually less than 5% of the column volume. This means that gel filtration is usually limited to the final stages of the purification where sample volumes are small.

The choice of matrix is determined by the steepness of the selectivity curve and the particle size of the packing material. There is a wide range of materials that show gel filtration properties and all of them will have differently shaped selectivity curves. The most suitable gels will have a steep selectivity curve in the region of interest.

The gel chosen must be sufficiently rigid to withstand the flow rates used in large-scale chromatography. Traditional media such as Sephadex G types and Biogel® P types are relatively soft and have to be packed in sectional columns (35). More rigid gels, such as Sephacryl®, can be used at flow rates up to 60–70 cm hr^{-1} in large-scale columns. Fully rigid media can be packed in 60–90 cm columns. For longer bed heights than 90 cm, it is usually wiser to link one or more columns together, or recycle, because of the practical difficulties in packing long columns.

The choice of media also affects chromatographic performance, because different media show differences in internal pore volume and the packing densities achievable with the different gels. Silica-based gels typically show higher values of V_0/V_c (void fraction) which indicates a looser packing, and lower values for the pore fraction V_i/V_c than agarose gels (36). Silica-based gels for gel filtration are frequently unstable and have to be stabilized using hydrophilic coatings (12).

Gel filtration media are available in a wide range of particle sizes, and it may well be better to compensate for a shallower selectivity by selecting medium with a narrower particle-size distribution. This can permit the use of faster flow rates. Figure 7 shows a comparison of chromatographic performance of cross-linked agarose gels with mean particle diameters of 110, 33, and 13 μm. The superior resolution using 13 μm packings meant that a very much faster flow rate could be used (36).

The use of microparticulate packings for protein purification by gel filtration is largely a question of price versus performance. For critical separations it may well be necessary to use small-particle-size materials, but with gel filtration resolution can also be increased by increasing column length or by recycling shoulder regions of the peak. Doubling column length for the separation of IgG from albumin using Sephacryl S-200 HR gives an increase in resolution sufficient to permit much greater sample loadings, or a three to four times faster flow rate (37). See also Section 5.3.3.

The other argument that has been put forward for using microparticulate packing materials in gel filtration is that throughput can be increased, but the same effect can be achieved with less expensive traditional media such as Sephadex and Sephacryl by increasing column dimensions.

Sample volumes in gel filtration are critical, and for analytical separations should not exceed 0.5% of the total bed volume since microparticulate packings are particularly sensitive to this effect (38). The larger particle-size packings used in preparative separations are less sensitive to sample volume and normally loadings of 2–5% can be used. It is important to take account of this effect when optimizing methods in the laboratory.

In summary, optimizing a separation is an essential step prior to scaling up. Factors such as choice of gel, choice of buffers and elution conditions, have to be established before sample loadings, column dimensions, and flow rates are examined in detail. The approach varies depending on the type of separation mechanism chosen, but high-performance media and methods invariably shorten the length of time required to gather all the necessary data.

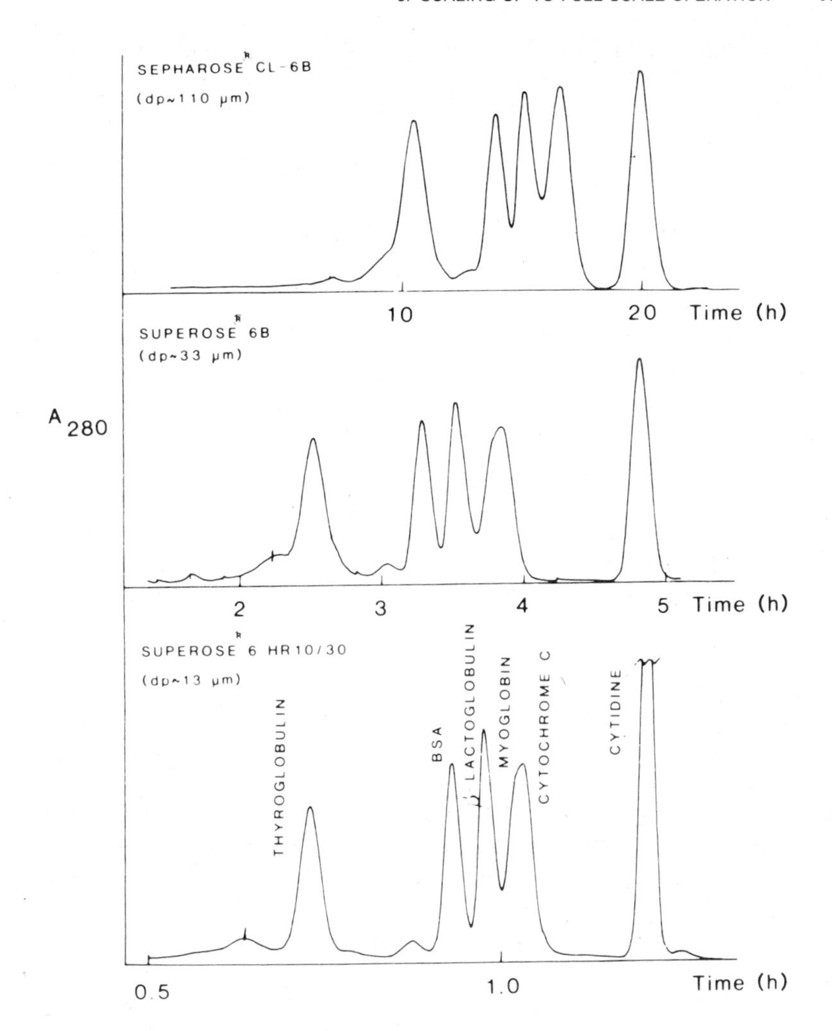

Fig. 7. Comparison of chromatographic performance of cross-linked agarose gels of different particle diameters (as indicated). Sample: 0.05 mg/mL of thyroglobulin + 0.08 mg/mL of bovine serum albumin + 0.25 mg/mL of β-lactoglobulin + 0.01 mg/mL of cytochrome c + 0.01 mg/mL of cytidine. Sample volume and linear flow rate: 1000 μL, 2.5 cm hr^{-1}; and 50 μL, 38 cm hr^{-1} from top to bottom, respectively.

5. SCALING UP TO FULL-SCALE OPERATION

The degree to which a separation is scaled up depends on economic considerations. Marketing groups will make predictions about sales volumes and selling prices for products of defined quality, while production groups will try to meet the specifications for quality while delivering sufficient product to meet market demand at an optimal price. The experienced production manager also knows that whatever marketing says about production requirements and market prices, the true answer is liable to be something different so he or she is going to have to be flexible in his or her approach. We will first consider basic principles before looking at practical considerations.

5.1.1. Basic Principles

A short examination of "large-scale-purification" methods published in the literature indicates that there are two basic approaches to scaling up. The first approach would seem to be to optimize a given separation until the required purity is reached, and then to scale up simply by increasing the column volume and sample volume in direct proportion to each other. This often results in what is a large-scale version of an analytical separation, where all the peaks have Gaussian distributions and are separated from each other by baseline separations. Only a fraction of the available capacity for the product is utilized.

The second approach is to optimize carefully the separation for purity and maximum sample loading before using larger-scale columns. This method is intended to give maximum productivity and to give optimal use of the separation media and equipment.

When larger samples are applied to the column, peak widths increase as sample loading is increased until eventually peaks will start to overlap into each other. Further increasing sample loading will mean that recovery has to be sacrificed for purity, or vice versa. It should be noted that it is quite an acceptable strategy to aim for less than 100% purity, particularly in the early stages of a separation, where recovery and speed may be of greater importance. In such a case it may be desirable to collect the contaminated shoulder regions of the peak of interest separately from the pure central portion so that the shoulders can be rechromatographed. In certain circumstances, particularly when the separation is highly selective such as in affinity chromatography, virtually all of the column capacity for the sample is used. It is important to bear in mind that adsorption isotherms will no longer be linear and that peaks will be non-Gaussian in shape, and that conventional chromatographic theory will no longer be fully relevant.

A number of workers have developed theoretical approaches to aid calculation of scale up factors in adsorption chromatography (39–47). In affinity chromatography, conditions can be arranged so that only the product of interest is adsorbed and so all of the available capacity can be used.

When working under conditions where all, or almost all, of the available capacity for sample is utilized, it is worth studying the shape of the adsorption isotherm. Under ideal conditions the breakthrough curve appears as a step function, that is, the sample is completely adsorbed from the feed solution until all of the gel is saturated. However, various factors will act to spread out the breakthrough curve such as the rate of mass transfer, the flow rate through the column, the quality of the packed bed, and the concentration of the adsorbate (41, 42). The shape of the breakthrough curve will determine the best way of applying sample to the column. If the column is switched to wash at a low effluent concentration as sample first appears in the effluent, then a portion of the bed capacity has been wasted. This effect is greatest if the shape of the breakthrough curve is shallow. On the other hand, if switching is delayed until a high effluent concentration is reached, a considerable amount of the sample may be wasted. Normally, underutilizing bed capacity is the lesser of two evils. However, the shape of the adsorption isotherm can be improved by any of a number of methods that are commonly used to improve mass transfer between stationary and mobile phases, such as decreasing flow rate, decreasing the particle size of the packing material, increasing the pore size of the matrix, or choosing a slightly different ligand or binding condition. The importance of determining which of these factors may be limiting has been discussed by Arnold et al. (41).

When calculating optimal sample loading for a less selective matrix, such as a group-specific affinity support or an ion exchanger, it is misleading simply to consider thc adsorption kinetics for the sample of interest since other compounds may also be bound as well.

Under these conditions, the breakthrough curve for the sample will give an indication of the quality of the chosen matrix in terms of rate of sample uptake as well as a measure of the capacity of the matrix for the sample of interest in the presence of contaminants. It will not, however, give an indication of the purity of the eluted product. Highly selective conditions may have to be used to separate the product of interest from contaminants.

Sample molecules tend to bind to supports in order of their affinity for the matrix, those with strongest affinity binding first. This leads to a "stacking" of the sample molecules, an effect easily seen when working with colored proteins. As more and more sites are occupied by sample molecules, competition between different proteins increases and those with a greater affinity for the matrix will displace the more weakly bound species. The point at which the sample first appears in the eluent represents the point at which the column is saturated with sample and those contaminants that bind more strongly to the support. It is therefore preferable to choose conditions whereby a minimum number of contaminants bind to the column, and the maximum number of more weakly binding species are allowed to pass through, to prevent contaminants displacing sample from the column.

When optimizing sample loading for complex separations, it is essential to remember that the retention values and separation parameters k^1, alpha, and N are normally first calculated for the analytical separation. As sample loading is increased, there will come a point when adsorption isotherms are no longer linear and these separation parameters no longer apply. This point has to be determined experimentally. The limits to sample loading are determined by the recovery required and the level of impurity that can be tolerated, while the production rate and the amount recovered per injection are determined by market requirements. Gariel and co-workers (45–47) have developed models to predict the maximal injected amount in a given preparative chromatographic system based on the experimental characterization of the peak shape obtained at high sample loading in absorption chromatography. Optimizing productivity in this way leads to production rates and recoveries per injection about 10 times higher than those obtained from a linear optimization procedure (45).

The key parameter to manipulate in the optimization of a preparative separation is selectivity. Solvent systems should therefore be chosen to give the best possible selectivity between peaks, provided of course that this does not lead to problems with solubility.

It is not normally possible to vary selectivity in gel filtration beyond choosing the specific matrix to be used, which is one of the reasons why adsorption methods are so useful for production. Increased peak separation to permit increased sample loading is therefore achieved by increasing column length, which has the double effect of increasing plate number (N) and the column volume (and therefore k^1).

5.1.2. Gradient Shape for Preparative Chromatography

In analytical chormatography, the main purpose of the separation is to obtain as much data as possible on the composition of the mixture to be analyzed. Usually sample loadings are so small that the solute components can be separated from each other with baseline separation using linear gradients.

In preparative chromatography, the objective is to separate the solute of interest as far as possible from its two nearest contaminant peaks in order to get maximal loading and recovery. This can readily be achieved by manipulating gradient shape. Gradients should be designed so that the rate of change in solvent composition is slight between the peak of interest and its contaminants, yet large through the peak itself to ensure that it is eluted in as concentrated a form as possible. It is clear, then, that while linear gradients are optimal for studying the effect of different buffer compositions on selectivity, stepwise gradients should be used as early as possible to get optimal sample loading, recovery and productivity.

Choosing the actual conditions for the steps is not always simply a matter of choosing conditions from a line drawn on the assumed elution conditions. In an ion exchange separation, various factors such as the void volume in the column, the buffering capacity of charged groups present in the column, and so on will contribute to making the conditions in the eluent slightly different from the calculated conditions in the buffer solutions going on to the column. It is necessary to monitor pH and conductivity continuously in the eluent as well as protein concentration.

There are ways of making both linear and stepwise gradients through control of pump speed or proportional valves. These offer flexible ways of manipulating gradient shape in process optimization. At production scale, however, system design is rarely changed (if ever!) from run to run, so it is convenient to use a different arrangement, for example, valve switching, to make gradients for reproducibility and hygienic reasons.

5.2. The Chromatographic Packing

HPLC packings are extremely useful for rapid method development and for small-scale production of high value products, but they are very expensive for use at large scale and indeed are not always the most cost effective solution. Mann (48) has shown that, at least for silica packings, 30 μm particles offer more theoretical plates per unit cost than 10 μm packings. Furthermore, the requirements for high pressures drop considerably for 30 and 90 μm packings compared with a 10 μm packings.

However, as discussed previously, if a separation is to be optimized for productivity, *selectivity* is the key parameter to consider. Once the desired conditions for the preparative separation have been established, including the shape of the gradient, the separation becomes essentially a process of selective adsorption and desorption, and the bead size of the packing contributes little to the quality of the separation in terms of zone spreading. The bead size of the packing is important, however, in determining the effectiveness of uptake of solutes from the sample solution passing through the column (41, 42). It is therefore important to transfer the method developed using HPLC supports to gels more suitable for large-scale use. As stated in Section 4.1, this can only be done easily if the gels to be used at large scale have similar retention characteristics to the supports used in HPLC. For ion exchange, affinity, and RPC/HIC gels this means that the nature of the substituted group must be identical and the gels must have similar surface properties (coupling density and evenness of distribution of coupling, properties of the support itself). For gel filtration this means the steepness of the selectivity curve through the range to be fractionated rather than the fractionation range itself.

These properties are very much inherent properties of the gels themselves and, because of this, the chromatographer is dependent on the different suppliers to make gels with similar selectivity in a range of particle sizes.

Examples of gels of these types are the gels Q Sepharose Fast Flow and S Sepharose Fast Flow, which have been designed to give similar retention properties to the microparticulate gels Mono Q and Mono S for FPLC®. It is therefore very easy to extrapolate results obtained with a MonoBeads column to a larger column filled with Q or S Sepharose Fast Flow.

Figures 8 and 9 show scale up from 1 mL columns (Mono Q and Mono S HR 5/5) to 50 mL columns (K 26/40) of Q and S Sepharose Fast Flow, using hen egg white as a sample. The running conditions for these columns are given in Tables 2 and 3 respectively.

Further experimentation is of course necessary in order to determine the optimal conditions for sample loading, flow rate, and gradient shape before moving on to pilot scale.

The separation characteristics for these two gel types are more or less identical, with the exception that the gradients were twice the length and the flow rates five times slower for the two Sepharose Fast Flow gels than for the two MonoBead packings.

It is also possible to obtain gel filtration media with similar selectivity curves

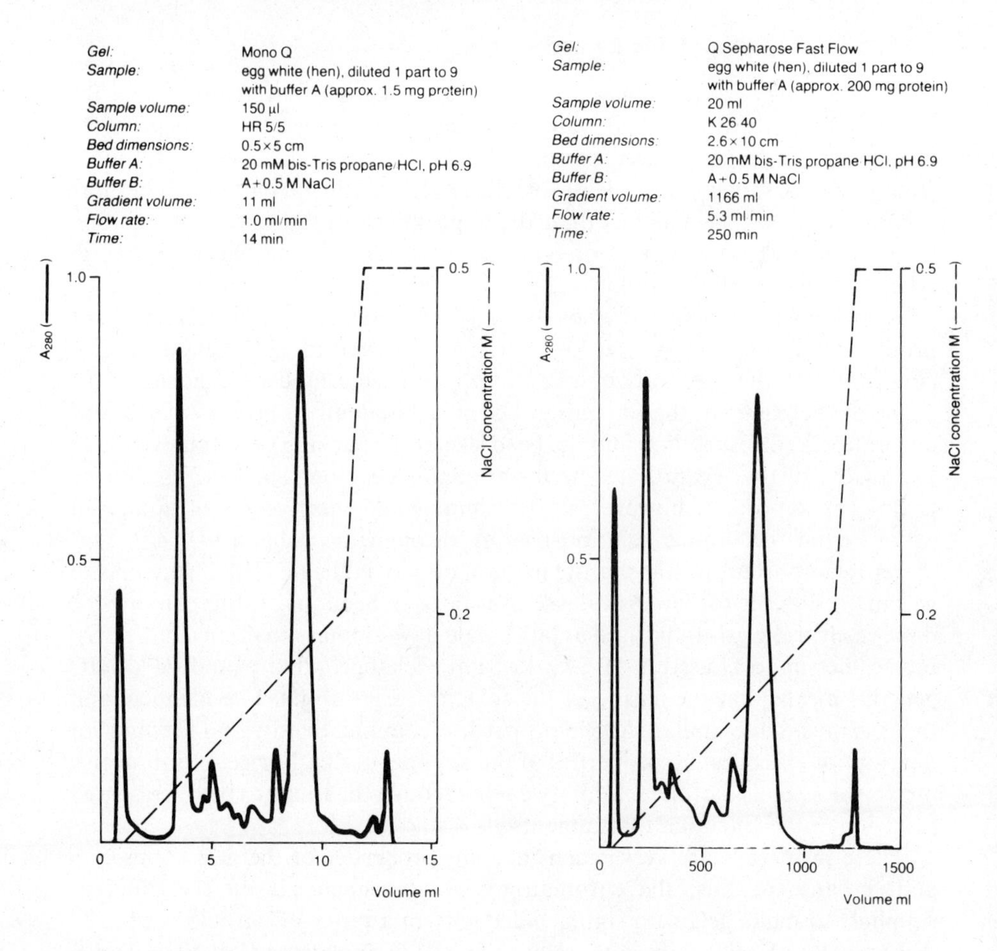

Fig. 8. Scale up of Mono Q to Q Sepharose Fast Flow. Elution profiles of hen egg white.

in a range of particle sizes. Sepharose® CL-6B, Superose 6 prep grade, and Superose 6 HR 10/30 are all based on 6% agarose with mean particle sizes of 110, 33, and 10 μm, respectively. Figure 7 showed an example of the separation that can be expected with each gel.

Different media in gel filtration differ from each other in terms of their particle size, the steepness of the selectivity curve, and the range of pore sizes. In the event that an analytical gel with a similar selectivity curve to the preparative gel is not available, or vice versa, it is possible to compromise because the change in selectivity may be compensated for by other properties of the gel such as ease of packing, efficiency, or rigidity.

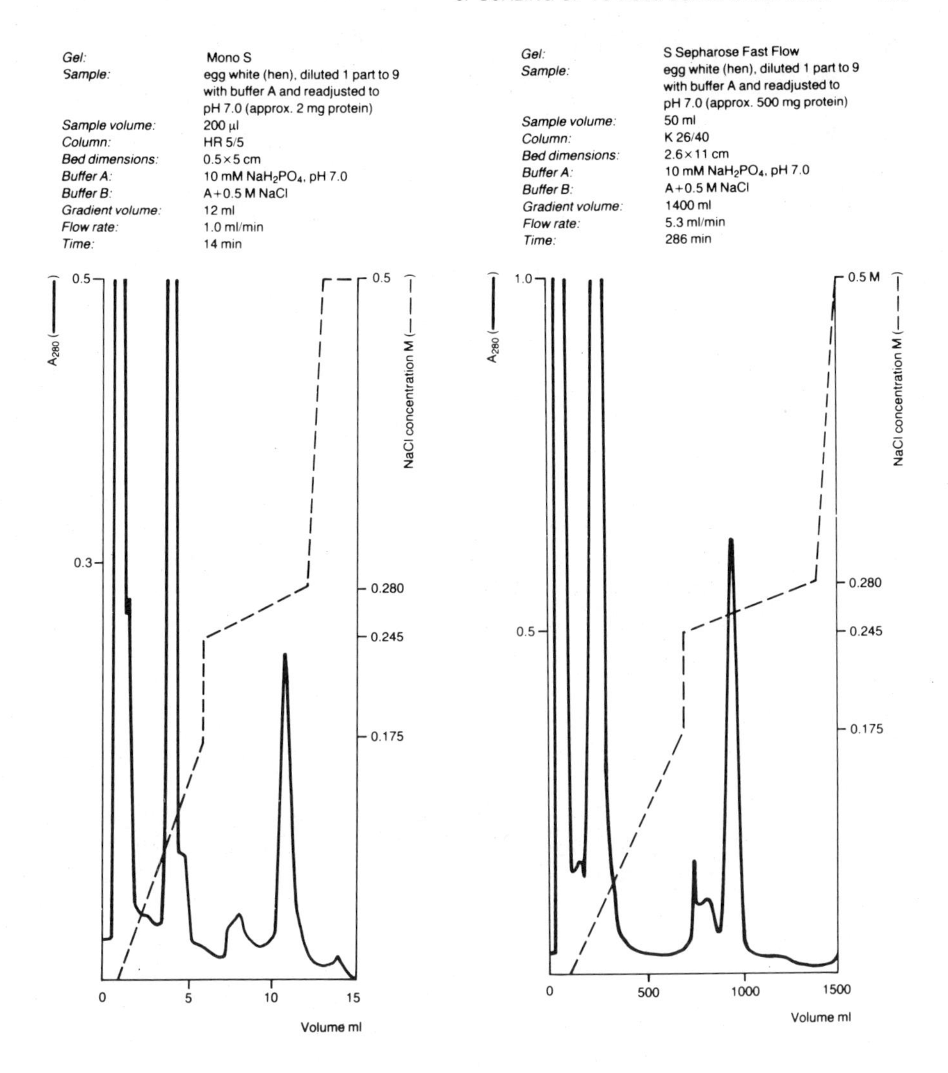

Fig. 9. Scale up of Mono S to S Sepharose Fast Flow. Elution profiles of hen egg white.

5.3. Practical Examples

5.3.1. Scaling up Methods on Microparticulate Packings

One of the principal advantages of scaling up a laboratory separation simply by increasing the column volume is that it is very straightforward and involves the minimum amount of labor. Nakamura and Kato have described preparative scale up from 1 mL analytical columns with a 10 μm packing material to preparative columns of 54 and 475 mL volumes, packed with 13 and 20 μm particles, respectively (49). Scale up was achieved without loss of resolution, but

Table 2. Scale up from Mono Q to Q Sepharose Fast Flow

	Mono Q HR 5/5	Q Sepharose Fast Flow
Cross-sectional area	0.2 cm^2	5.3 cm^2
Gel volume	1 ml	53 ml
Sample	150 μl of hen egg white diluted 1:10 in buffer (approx. 1.5 mg protein)	20 ml of hen egg white diluted 1:10 in buffer (approx. 200 mg protein)
Gradient volume	11 ml (11 × V)	1166 ml (22 × V)
Flow rate	1.ml min^{-1} 5 cm Min^{-1}	5.3 ml min^{-1} 1 cm min^{-1}

sample loadings were only in the range of 40–200 mg for the 40 mL column and 240–1000 mg for the larger column. Laboratory columns, such as the HR 5/5 column (1 mL) have a capacity for approximately 25 mg of sample per cycle. Since cycle times are normally in the region of 30–40 min, laboratory systems can easily handle several hundred milligrams of sample in an 8-hr shift. For increased throughputs it is necessary either to increase column volume or to further optimize the separation.

An example of how a separation can be scaled up reproducibly from one step to the next simply by increasing column volume is illustrated in the case of a process based on an ion exchange separation on Mono Q developed for the purification of a product synthesized by a genetically engineered bacterium.

The sample in each case is a partially purified bacterial extract. The optimal salt gradient was developed on a 1 mL HR 5/5 column (Fig. 10) prior to scaling up 200 times.

The sample concentration in all of the experiments was 200 mg mL^{-1} and the volume applied was equivalent to one bed volume. Flow rates were adjusted so that one column volume could be run in 2.5 min. The elution profiles obtained

Table 3. Scale up from Mono S to S Sepharose Fast Flow

	Mono S HR 5/5	S Sepharose Fast Flow
Cross-sectional area	0.2 cm^2	5.3 cm^2
Gel volume	1 ml	58 ml
Sample	200 μl of hen egg white diluted 1:10 in buffer (approx. 2 mg protein)	50 ml of hen egg white diluted 1:10 in buffer (approx. 500 mg protein)
Gradient volume	12 ml (12 × V)	1400 ml (24 × V)
Flow rate	1 ml min^{-1} 5 cm min^{-1}	5.3 ml min^{-1} 1 cm min^{-1}

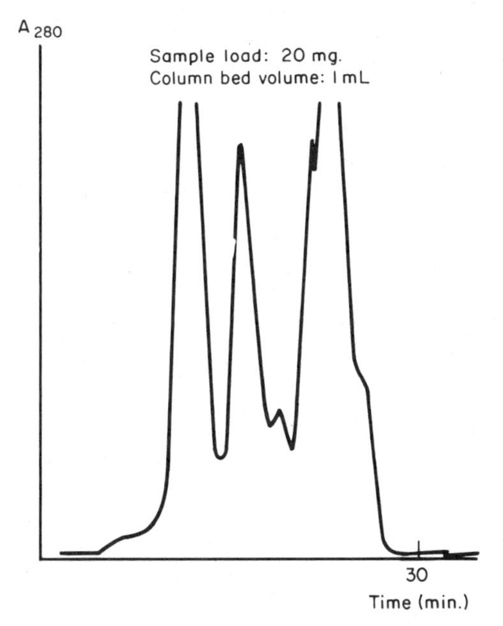

Fig. 10. Separation of a partially purified bacterial extract on Mono Q. Column: Mono Q HR 5/5. Eluent: 0.035 *M* Tris HCl, containing 30% CH_3CN with a linear gradient 0–0.14 *M* NaCl. For other conditions see text.

are shown in Fig. 11, and the running conditions are summarized in Table 4. The running time for the 200 mL column was 80 min. Note the reproducible separation of the first peak, the peak of interest.

This separation was repeated over 25 times at large scale with complete reproducibility. When the product from shoulder regions was rechromatographed, the yield was greater than 90%, and sample accountability essentially 100%.

The column was regenerated every fifth cycle with washes of 2*M* NaOH in water and 1*M* NaCl in 30% acetonitrile. Columns of this type are being used for the purification of interferon (50), interleukin (50a), growth factors and monoclonal antibodies from genetically engineered microorganisms (50).

Figure 12 shows an example of a BioPilot system capable of pumping at flow rates up to 100 mL/min. Prepacked columns of 100 and 300 mL are available for use with this system.

Careful optimization can be used to get higher sample loading. This is demonstrated in the scale up of the purification of human placental lactogen. The optimal pH and buffer system was chosen on the basis of separations achieved at different pH values before optimizing sample loading and gradient shape (Fig. 13).

Scaling up was then achieved by simply multiplying the gradient volume and sample amount (mass and volume) by the total volume of media in the larger

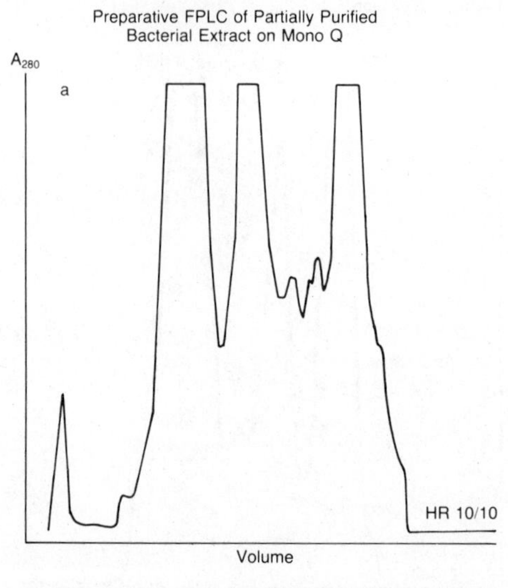

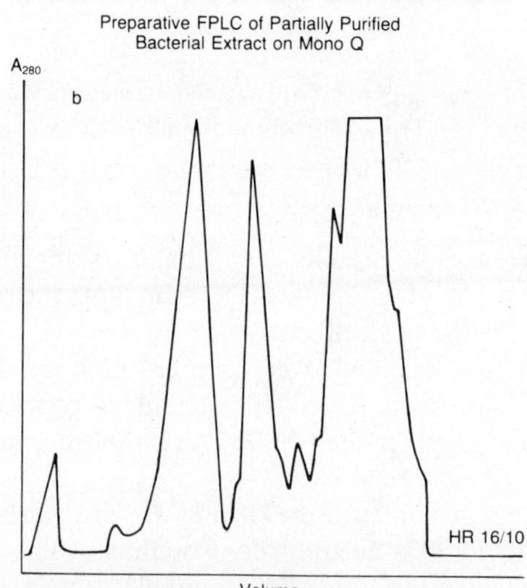

Fig. 11. Scale up of the separation in Fig. 10. (*a*) Column: HR 10/10 (8 mL). (*b*) Column: HR 16/10 (20 mL). (*c*) Column: HR 50/10 (200 mL).

column. The optimized conditions, pH, buffer and gradient shape can then be used for the preparative scale chromatography (Figs. 14 and 15).

It is worth considering the cost of using this type of purification strategy, since large columns packed with microparticulate materials are expensive (approx. 25,000–50,000 US$). Data gathered from the analytical use of these columns indicate that they can be used for over 1000 injections (51) and, while it is less likely that this type of performance can be achieved with preparative

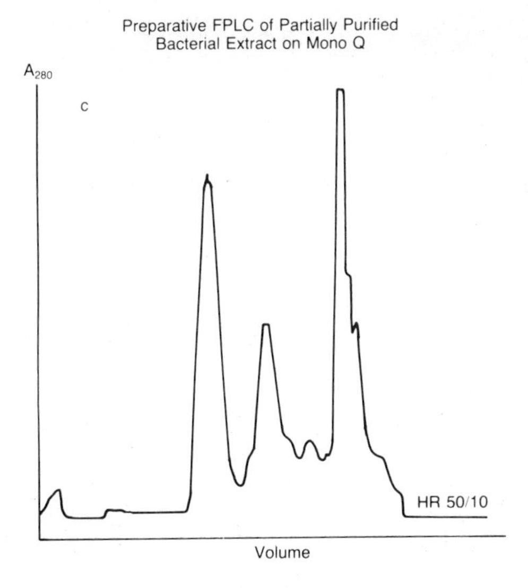

Fig. 11. (***Continued***)

loadings, particularly if the sample is relatively crude, it is reasonable to expect several hundred injections. If the dosage of a typical biotechnology product is 1 mg, then the cost of the column can be spread over a million doses, or a few cents per dose, assuming loadings of 5 g per cycle and 200 cycles.

This cost can be further reduced if the sample loading per column is fully optimized. Recent work by Hearn et al. (19, 52) has shown that 150–200 mg can be loaded onto a Mono Q HR 5/5 column and 2–3 g can be loaded onto a Mono Q HR 16/10.

Systems of this type are clearly suitable for the purification of high value products required in quantities up to 20–25 kg per annum.

High-performance methods can also be saved for the later stages of a purification process, when the sample volumes are smaller and the concentration of contaminants are less. The purification of a monoclonal antibody from a cell culture supernatant serves as an example.

Table 4. Scale Up to Preparative FPLC

Column	Bed Volume (ml)	Sample Load (mg)	Flow Rate (ml min^{-1})	Backpressure (MPa)
HR 5/5	1	20	0.4	1.0
HR 10/10	8	160	3.2	1.5
HR 16/10	20	400	8.0	2.0
HR 50/10	200	4000	30.0	0.9

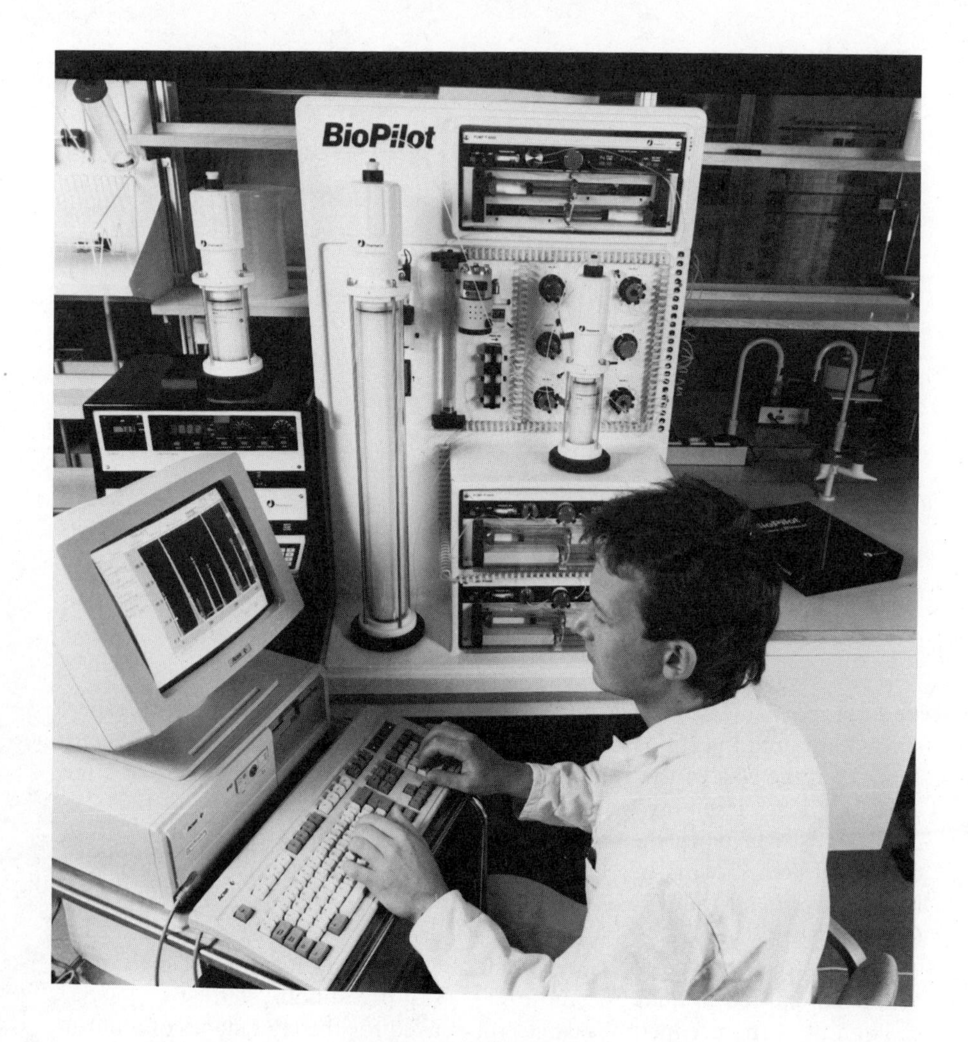

Fig. 12. BioPilot System.

Hybridoma cell cultures were grown in a spinner bottle culture system (Techne) followed by batch suspension culture in a 10-L pilot-scale fermentor. A standard DMEM (Gibco) medium supplemented with 10 μg/mL transferrin, 10 μg/mL insulin, 0.02 m*M* ethanolamine, 2 m*M* glutamine, 25 μg/mL gentamycin, and 0.2% human serum albumin was used. Culture supernatant, obtained after separation of cells and debris by centrifugation, was stored at −20°C for further studies. The process is outlined in Fig. 16.

The culture supernatant was first processed on a column (5 × 18 cm) of DEAE-Sepharose CL-6B to remove major contaminants and at the same time concentrate the immunoglobulin-containing fraction. Two litres were frac-

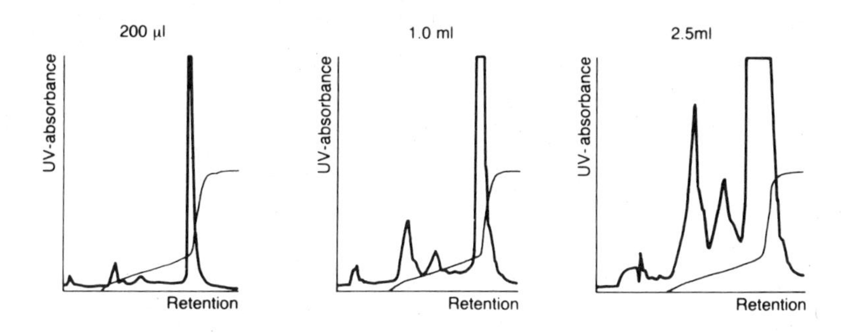

Fig. 13. Optimization of sample size in the isolation of human placental lactogen on a Mono Q HR 5/5 column. Conditions: sample, partially purified human placental lactogen; buffer A, 20 m*M* bis-Tris pH6.0; buffer B, 1 *M* NaCl in buffer A; gradient, 0–10% B in 14 mL, 10–40% B in 1 mL; flow rate, 1.0 mL min^{-1}; detection, 280 nm at 0.1 AUFS. Up to 2.5 mL protein solution was applied before losing significant resolution.

tionated in two steps to give sufficient material for purification on Sephadex G-25 M. The ion exchange separation is shown in Fig. 17.

The MCA-containing fractions from Sephadex G-25 M were finally purified on a large (20 mL) column of the cation exchanger Mono S, which removed transferrin and other unidentified components. The final chromatogram is shown in Fig. 18. The sample loading in the final step was 24.3 mg.

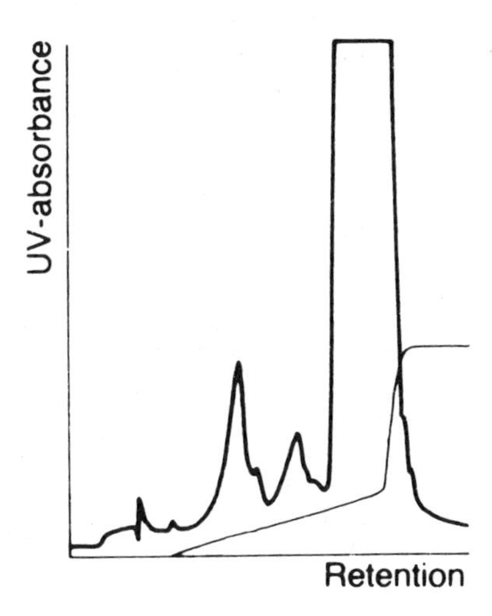

Fig. 14. Isolation of human placental lactogen on a Mono Q HR 10/10 column. Conditions: sample, 20 mL (scale-up factor 8) of partially purified lactogen, gradient, 0–10% B in 120 mL, 10–40% B in 8 mL; flow rate, 4.0 mL min^{-1}. Other conditions as described in Fig. 13 legend.

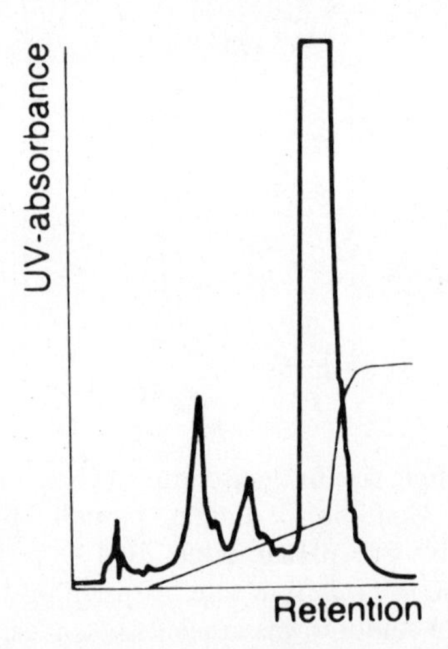

Fig. 15. Scaling up of the isolation of human placental lactogen onto a Mono Q HR 16/10 column. Conditions: sample, 50 mL (scale-up factor 20) of partially purified placental lactogen; gradient, 0–10% B in 300 mL, 10–40% B in 20 mL; flow rate, 10.0 mL min^{-1}. Other conditions as described in Fig. 13 legend.

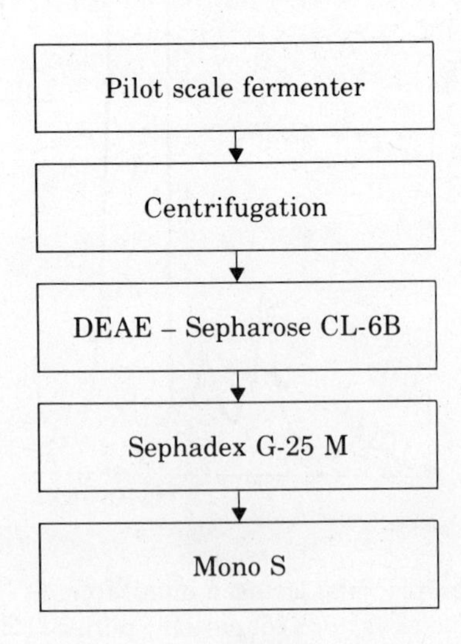

Fig. 16. Process scheme for purification of monoclonal antibodies.

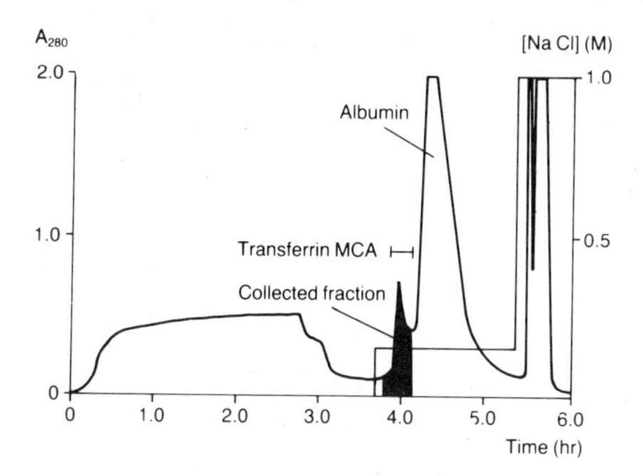

Fig. 17. The culture supernatant was processed on DEAE-Sepharose CL-6B. Two liters were fractionated in two steps (2 × 1) to give sufficient material for further purification on Sephadex G-25 M and Mono S HR 16/10.

Reversed-phase HPLC has been employed successfully in the full scale production of a genetically engineered therapeutic product for human use. In the initial process developed by Eli Lilly, recombinant human insulin is prepared by combining A and B chains and purifying by a combination of ion exchange and gel filtration. Some of the fractions prepared in this way contained significant quantities of insulin but were not sufficiently pure to be included in the mainstream. These fractions would be further purified by reverse phase (53). In the current process, insulin is prepared by reversed phase HPLC after enzymatic

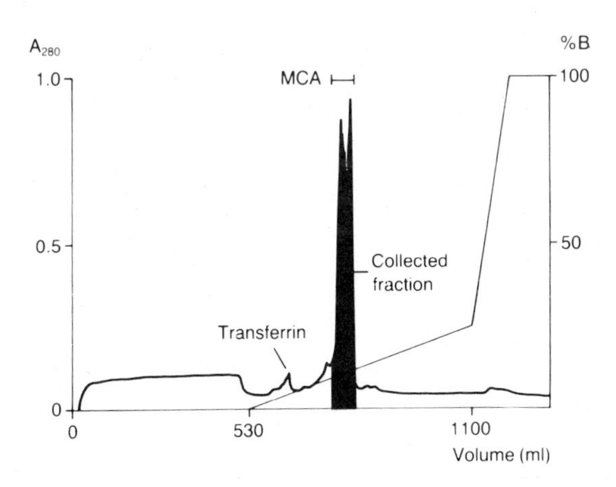

Fig. 18. Final step on Mono S HR 16/10.

treatment of proinsulin, purified by conventional ion exchange. In both cases, it was preferable to place the HPLC step late in the process when most of the contaminants had been removed to prevent fouling of the packing. Approximately 400–500 g of insulin could be run on a 30 L column with a final purity in excess of 98.5%.

Analytical HPLC separations of antibiotics have been scaled up to production scale using reversed phase. The column packing used was an octadecyl silica with a 55–105 μm particle size. Typical results were 95% yield and 93% purity with sample loadings of 200–300 g on 18.8 L columns (54). Strategic approaches to the purification of antibiotics by preparative liquid chromatography have been described recently by Sitrin and co-workers (54a).

5.3.2. Large-Scale Affinity Methods

As stated in Section 5.2, selectivity is perhaps the single most important parameter to consider in preparative separations. Affinity chromatography is the method that comes closest to giving absolute selectivity.

There are an increasing number of patents appearing for large-scale purification procedures based on affinity chromatography methods; however, there are few corresponding references reporting degree of purity and yields achieved by these methods. Some of the best documented examples come from the purification of plasma proteins (55–59).

Affinity chromatography is also used for the recovery of recombinant DNA products. Perhaps the best documented is the purification of human leukocyte interferon (60–64). The most striking feature of this purification is the very high degree of purification (over 1000-fold) reported in the original literature and the excellent recoveries (60). The current procedure consists of an immunoaffinity step, followed by copper chelate affinity chromatography, ion exchange, and gel filtration (62a).

Interferon is extracted from bacteria, and after removal of cell debris and concentration, 300 L of extract is applied to the immuno adsorbent column in 10 lots of 30 L. The column consists of antibody coupled at a density of 14 mg/mL, and has a capacity for interferon of approximately 1.7 mg/mL. Interferon was applied at a specific activity of 1.5×10^6 U/mg and eluted at 1.7×10^8 U/mg (63), representing a 10-fold concentration and 100-fold purification. This is lower than the figure reported by Staehelin et al. (60) because the specific activity of the starting material is higher. The columns are repeatedly used for several thousand cycles (61). The remaining steps are required to remove interferon fragments and interferon oligomers, which also bind to the monoclonal gels.

Monoclonals are less useful for the purification of "natural" interferon, which is present as eight subtypes (63). Monoclonals are naturally raised against a single polypeptide type.

Another substance of interest purified by affinity methods is hepatitis B antigen. Several groups have described purification procedures based on purification by monoclonal antibodies, ion exchange, and hydrophobic interaction chromatography (65–70). Hershberg (70) has compared the use of conventional purification methods (ion exchange and gel filtration) with affinity chromatography using a monoclonal antibody coupled to Sepharose. In each case the sample could be purified to homogeneity but the overall yield was 50% in the case of the conventional method compared to 96% using immunoaffinity chromatography.

Workers at Mercke Sharp and Dohme have also described a purification of HbsAg by a monoclonal antibody raised in goats in what is effectively a single-step procedure (67, 69). This vaccine was somewhat more effective than material purified by hydrophobic chromatography and gel filtration in raising antihepatitis antibodies in human trials, but was more prone to cause soreness at the injection site (68).

Hydrophobic interaction chromatography has proven to be very effective in removing hepatitis B virus from contaminated fractions of coagulation proteins (65).

High-performance media have not yet contributed significantly to affinity separations at large scale. This is mainly because decreasing particle size would have little or no affect in improving separation, which is based on a highly selective adsorption mechanism. Smaller particle sizes should, however, improve dynamic loading of the column, since the faster equilibrium between the sample and the ligand will allow for more rapid sample application or more effective uptake from the sample solution. This can, of course, also be achieved by simply increasing column dimensions, so it remains to be seen if high-performance media will provide a cost effective alternative to conventional media for large-scale affinity chromatography.

5.3.3. *Gel Filtration—A Special Case*

Scaling up gel filtration is a special case, because elution is performed under isocratic conditions. Selectivity in gel filtration is controlled by the porosity of the matrix, and it is not possible to vary porosity during the actual separation.

In scaling up it is desirable to increase throughput at a given resolution. This can be done by increasing sample size until the acceptable limit of resolution is reached and thereafter increasing column volume by increasing column diameter.

The efficiency of the separation can be increased by increasing column length, which at constant diameter means an increase in throughput. Although resolution only increases in proportion to the square root of N, throughput will be more than doubled for a doubling in column length, since this extra resolution can be used to increase sample loading further.

Efficiency can also be increased by decreasing particle size. Table 5 shows appropriate relationships for plate height and particle size as determined for analytical separations on different gel filtration media. It is important to remember that the very high plate numbers achieved with 10 μm packings occur with low sample loadings ($< 0.5\%$ of column volume), and that microparticular packings are more sensitive to sample volumes than conventional packings,

Table 5. Plate Numbers for Gel Filtration Media with Different Particle Sizes

Gel Type	Average Particle Size (μm)	Plate Number/ Meter[a]
Superose HR	10 μm	> 40,000
Superose Prep grade	30 μm	13,000
Sephacryl	70 μm	5,500
Sepharose	100 μm	4,000

[a]Plate number is determined using acetone.

which normally use loadings of 1–5% (38). Decreasing particle size also increases backpressure and makes the columns more difficult to pack, so that long columns become impractical.

The crucial limitation may well be the availability of a suitable matrix, since high-performance media are expensive to use in large volumes and traditional media such as Sephadex give low flow rates. Special types of column have been designed to surmount the handling problems associated with traditional gels (35). In addition, significant improvements have been made in media over the last few years with the introduction of composite media such as Superdex® (70a).

To obtain optimal resolution and throughput in gel filtration requires careful design of the separation, as has been demonstrated by Berglöf and co-workers (37).

Gel filtration is used in the final stage of plasma fractionation by chromatography to remove dimers and trimers from albumin (71). Good laboratory results are obtained using a bed height of about 10 cm, a sample volume of 1–2% of the total column volume, and a linear flow rate of 5 cm hr^{-1}. These conditions are not optimal for the production situation where the sample volume is increased to 8% of V_t and the flow is limited to 15 cm hr^{-1} by the rigidity of the gel.

Recently, the synthesis of Sephacryl S-200 has been optimized to improve physical rigidity several fold. To utilize this rigidity for improved resolution, the mean particle size (d_p) has been reduced from 70 to 50 μm. The improved matrix is designated Sephacryl S-200 HR (High Resolution).

To test the effect on resolution, albumin and IgG were mixed to give about equal peak heights when eluted from a column to a total protein concentration of 20 mg mL^{-1}. Sephacryl S-200 HR was packed to a bed height of 90 cm in 5-cm-diameter columns. Bed heights greater than 90 cm were achieved by coupling columns in series. At flow rates greater than 30 cm hr^{-1}, 2.6-cm-diameter columns with bed heights of 16 cm were coupled in series to give the required bed height.

Figure 19 shows the relationship between resolution and flow rate for the different columns with a sample size of 4% (37). The resolution was slightly improved by decreasing particle size, but it is particularly interesting to note the effect of doubling column length. The flow rate could be more than doubled

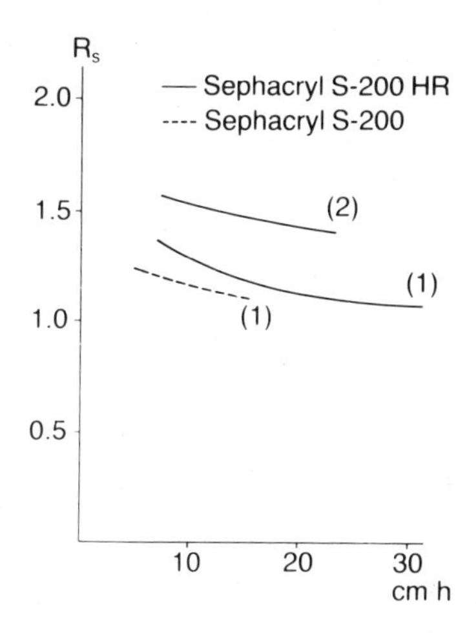

Fig. 19. R_s as a function of flow rate at (1) 90 cm and (2) 180 cm bed height. Sample size is 4% V_t.

while still giving the same resolution. In preparative columns (11.3 × 15 cm), the maximum flow rate that can be used with Sephacryl S-200 is 50 cm hr^{-1}. In practice, the former gel could not be operated at more than 15 cm hr^{-1}. Using Sephacryl S-200 HR in segmented bed heights of 16 cm and with a total bed height of 90 cm, flow rates approaching 100 cm hr^{-1} should be achievable.

In albumin production of Sephacryl S-200 a sample size of 8% V_t with a concentration of 60 mg mL^{-1} and a flow rate of 15 cm hr^{-1} has been used. Using these conditions an intermediate fraction for rerunning has to be taken out, but by reducing the sample loading to 4% of V_t this step can be omitted. Table 6 shows a comparison of the two gels. The whole peak was collected on Sephacryl S-200 HR. Rerunning of the intermediate fraction on Sephacryl S-200 was not taken into consideration in this table.

Table 6. Comparison of Sephacryl S-200 Superfine and Sephacryl S-200 High Resolution[a]

Gel	Bed Height cm	Sample (60 mg/ml) Size (% V_t)	Flow Rate (cm/h)	Cycle (1 V_t) Time (h)	Productivity	
					Column Area ($g/h \cdot 1000\ cm^2$)	Gel ($g/h \cdot 1$)
S-200 Superfine	90	8	15	6	72 (216)	0.8
S-200 HR	90	4	75	1.2	180 (540)	20

[a] The amount of protein is calculated as albumin. The figures thus represent a direct comparison of both gels' performance under optimized running conditions.

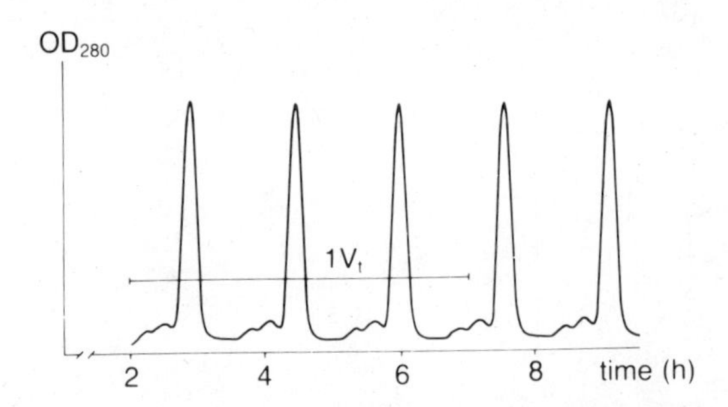

Fig. 20. Gel filtration of albumin obtained from ion exchange chromatography in repeated cycles. Sample size is 4% V_t, concentration is 15 mg/mL, flow rate is 40 cm hr^{-1}, and bed height is 12 × 16 cm. The separations are spaced so that three separations are achieved in an elution volume of one column volume (V_t), that is, removal of salt has not been considered. The distance corresponding to 1 V_t is marked.

When calculating cycle time, consideration is taken for the acetate salt eluted at 1 V_t. If desalting does not have to be considered, sample application can be spaced so that three cycles are performed within an elution volume corresponding to 1 V_t as shown in Fig. 20, giving productivity figures three times as high. These figures are given in parenthesis in Table 6.

Productivity for DEAE Sepharose Fast Flow ion exchangers run in the albumin process at 120 cm hr^{-1} is 338 g hr^{-1} per 1000 cm^2 column area. Gel filtration media with good rigidity under optimal conditions can improve productivity several fold over existing gel media and, in certain instances, approach the productivity of ion exchange media. It is not necessary to resort to microparticulate media for cost effective production.

5.3.4. Time Planning in Production Processes

There are a number of arguments favoring the use of high-performance media in production processes. Most important, they must be cost effective and lead to practical benefits, such as improved yields or purity. There is little point in using high-performance media to run a separation faster if that particular step in the procedure is not a rate-limiting step.

When planning a production step, it is important not to consider each step in isolation. Practical aspects such as how many steps can be done in a single working day and how steps can best be linked together are frequently more important.

Key parameters such as the sequence of steps, choice of gel media and buffer, bed height used, capacity of gel media, cycle volumes, linear flow rates, washing procedures, and stability of gel media all have to be established in the laboratory. When transferring the procedure into production, the production manager must consider production time and capacity and batch sizes. He or she must consider labor and maintenance costs, or whether he or she must use automation to improve running costs. He or she must consider the total level of investment, and must also consider the possibilities for expansion.

Product stability is an important factor in designing production process since products must be recovered in a stable form as quickly as possible, even though only partially pure. Once product stability has been secured, purification can proceed at a pace more suited to the convenience of the operator.

As an example of how a production process can appear, we can consider the purification of albumin from human plasma by a combination of ion exchange and gel filtration (71). This process is used in a number of centres and is described in outline in Fig. 21.

When planning the scale of the different steps, it is important to consider the total volume of material to be processed and the time available for carrying out the step in question.

The overall time plan for the process is shown in Figure 22. The first steps in this process are filtration, centrifugation, batch adsorption on DEAE-Sephadex A-50, and desalting, prior to euglobin precipitation. This point represents a convenient place to break, since precipitation takes 2–3 hr at room temperature or overnight in the cold room. Continuing with the separation process on the same day would mean an impractically long first day so it is just as well to allow precipitation to take place overnight. As a result we can allow 4 hr for processing the sample (50 L) for the initial gel filtration step.

Laboratory results indicate that a column of Sephadex G-25 Coarse with a bed height of 60 cm and a flow rate of 5 cm min^{-1} can give a satisfactory separation. The cycle time is 13 min, which means that as many as 18 cycles can be run in 4 hr. This would mean a sample volume of 2.78 L and a bed volume of 17.4 L (sample volume is 16% V_t). The column should have a minimum cross sectional area of 17,400 mL/60 cm = 290 cm^2. The most suitable column was a BP 252 with a 60 cm bed height and a 800 cm^2 cross section. Re-calculating the figures gave a process time of 2 hr 21 min.

In Day 2, material is centrifuged and filtered before being processed on DEAE Sepharose Fast Flow. The size of the columns used for each step are determined as for the example for gel filtration. In ion exchange, the column has to be thoroughly washed after a certain number of cycles, otherwise capacities start to decline. In this process, the DEAE column has to be washed every third cycle, while the CM column can be washed every 15th cycle. These washing steps have to be planned into the overall process.

The time available for the two steps is 20 hr. Laboratory runs showed that columns with a 15 cm bed height could be run at 2 cm min^{-1}. The elution volume for albumin from the DEAE column was 7 V_t (bed volumes) and the cycle was complete after 11 V_t, or 67.5 min. The sample loading was equivalent to 1 L of the original plasma (total 50 L)/L gel bed volume. These results indicate that 18 cycles can be run in 20 hr, and the bed volume required per cycle is 50 L/18 = 2.778 L. Again a BP 252 column is a suitable choice. A 15 cm bed height gives a bed volume of 7.5 L, and means that on recalculation the whole sample can be run in seven cycles. Since the column has to be washed every three

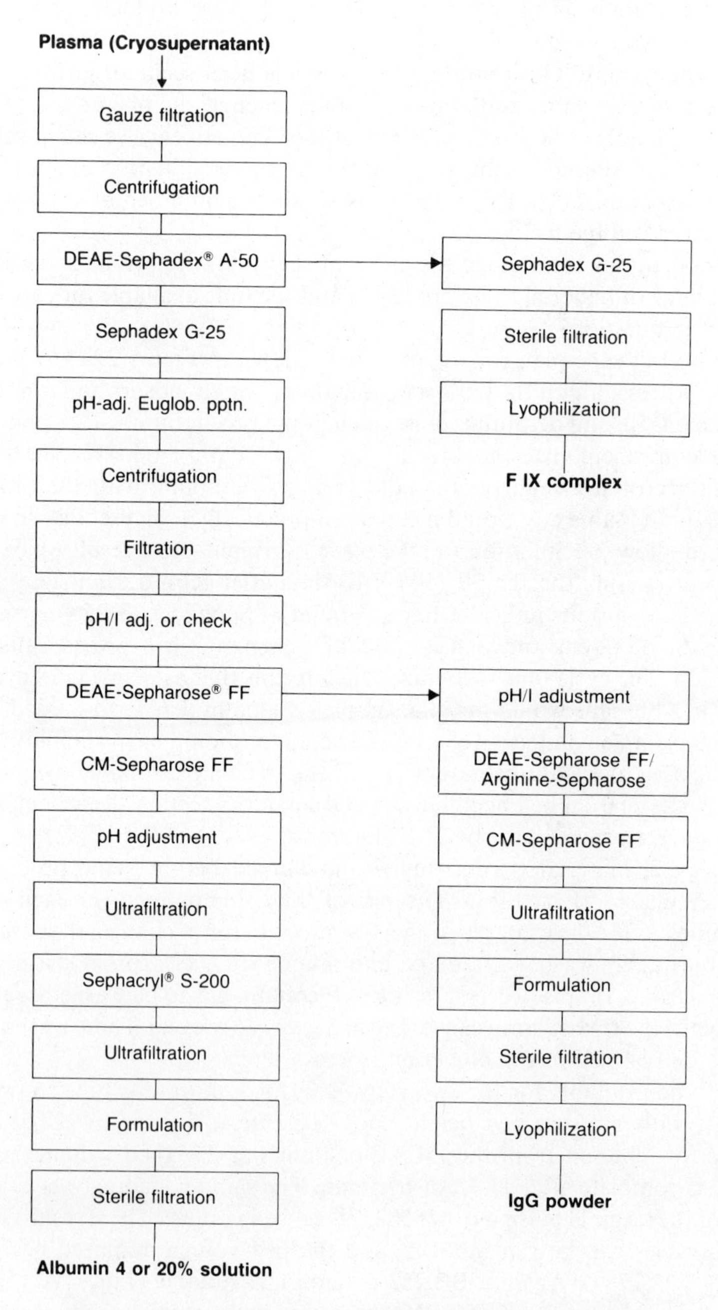

Fig. 21. Routes to prepare albumin, IgG, and Factor IX.

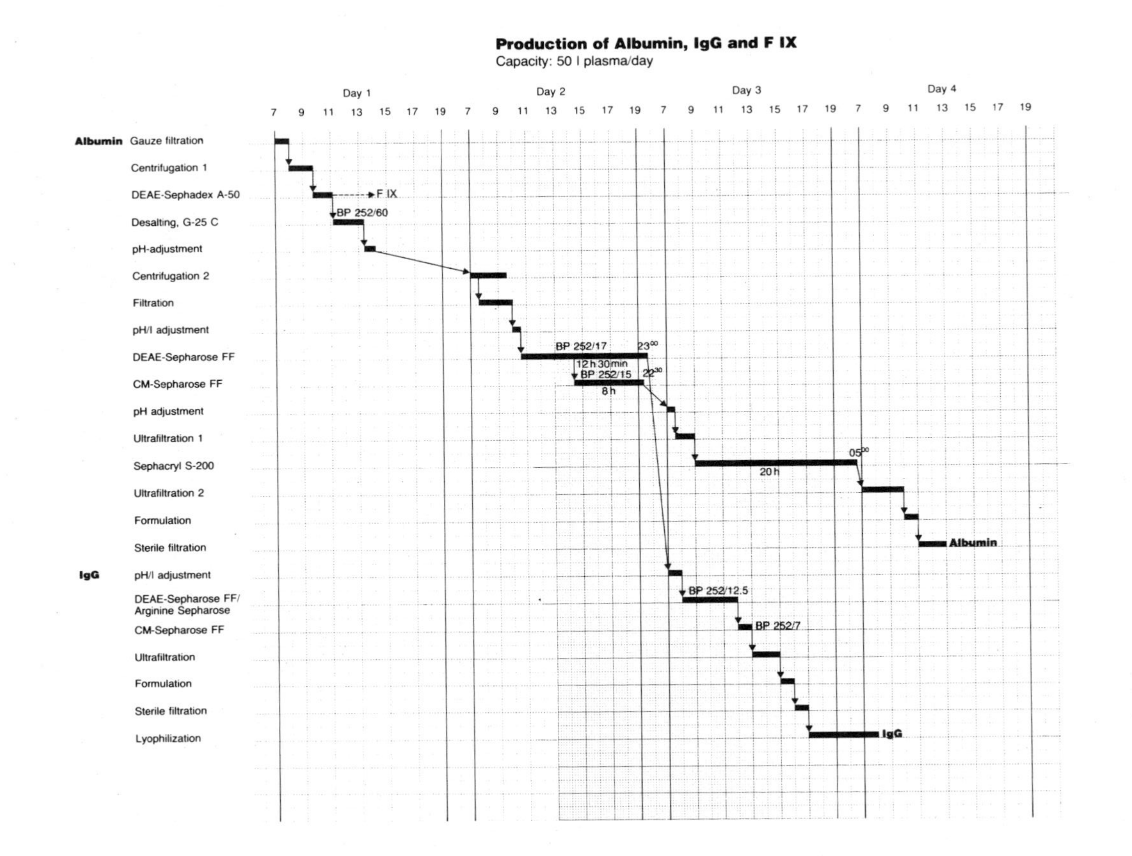

Fig. 22. Production of albumin, IgG, and Factor IX. Time schedule.

cycles, it would be better to have a cycle number which is a multiple of three. The sample (50 L of original plasma) can be processed in six cycles on a 8.33 L column, which would have a 17 cm bed height. The linear flow rate will decrease to 1.8 cm min^{-1}, but the total process time will still be under 13 hr including washes.

The CM Sepharose Fast Flow separation starts after the third cycle on DEAE Sepharose Fast Flow to shorten the overall process time.

The final step in albumin purification is gel filtration on Sephacryl S-200, while the material for IgG purification is further purified by ion exchange and affinity chromatography prior to formulation.

The columns are chosen in the same way as described previously, by working out cycle times from laboratory experiments and then calculating the size of the column required from the cross-sectional area needed to give the required throughput.

A new batch can be processed every day. It is a matter of choice whether a different production team is responsible for each unit operation or whether the same group follows the batch from start to finish.

Precise timing is clearly important in a production process, particularly when the signal for starting a step is triggered by an event in a previous process. Thus it is important that separations are highly reproducible. Automation removes the risk of human error since it enables all parameters to be monitored simultaneously.

5.4. Novel Approaches

Highly selective separations permit optimal use of the chromatographic support and simplify elution procedures. As has been described earlier, this can be achieved by affinity chromatography.

A second approach is to resort to chemical or recombinant DNA techniques to modify the sample of interest in such a way as to make it easier to purify.

An excellent example of the first of these two strategies is shown in a strategy for purification of insulin from genetically engineered *E. coli.* Insulin is first produced as a precursor, biosynthetic proinsulin (72). Before further processing it is necessary to ensure chemical homogeneity of cysteine residues and prevent random formation of disulfide bridges. This is most conveniently carried out by reversible chemical modification. S-Sulfonate derivatives have been used to regenerate native insulin from separated A and B chains by thiol mediated exchange (73), and this method has proven appropriate to protect the SH groups in proinsulin. Since 6 out of 86 amino acid residues are cysteine, these could be readily converted to the $-SSO_3$ group by oxidative sulfitolysis. The proinsulin S-sulfonate has a charge of -8 at neutral pH, so it can be purified by virtue of its strong attachment to an anion exchanger or its lack of attachment to a cation exchanger or both. The product is the last to elute from an anion exchange column and is substantially pure, ready for conversion to authentic proinsulin (53, 72). An alternative strategy could be to use sulfur chemistry to react reduced cysteine groups with SH groups immoblized to an inert matrix (e.g., Activated Thiol Sepharose) in covalent chromatography.

The second strategy is to use recombinant DNA techniques to modify the synthesized molecule to make it easier to purify. This can be done in a number of ways (74), such as by adding amino acid sequences to alter the charge (75) or hydrophobicity of the sample of interest, or by creating protein complexes where the sequence for the product of interest has been coupled to the sequence for one of a pair of interacting biological molecules, for example, Protein A-IgG (76–78), biotin and avidin, polyhistidine and chelating ligands (78a).

Sassenfeld and Brewer (75) have described a method for the purification of urogastrone using a polyarginine sequence tagged onto the C-terminal end of urogastrone. The hybrid molecule binds strongly to cation exchangers, and there is a linear relationship between the number of arginine residues incorporated and the ionic strength required for elution, which makes the separation very controllable. After separation, the polyarg terminus is removed by digestion with carboxypeptidase B and is then separated from contaminants by a second pass through the ion exchange column.

This method can be used in principal for any protein produced by recombinant DNA techniques. Sassenfeld and Brewer have reported the purification of γ-interferon using large (200 mL) columns of Mono S using this method (50).

A similar approach has been described by Uhlen and co-workers for the purification of Somatomedin C (75–78). In this method, the product of interest is linked to Protein A and the hybrid is selectively adsorbed to IgG Sepharose Fast Flow. Somatomedin C can then be split from the molecule either after elution from IgG Sepharose or while still bound to the matrix.

These strategies represent a novel approach to protein purification. While they undoubtedly simplify the overall purification scheme, expression levels of these hybrid molecules are not always as high as for simple products. Improvements in expression and recovery will certainly make these strategies an attractive alternative to conventional purification procedures.

6. NONCHROMATOGRAPHIC ASPECTS OF SCALE UP

The objective of most scale-up procedures is to arrive at a reliable and effective production method for the product of interest. When that product is intended for injection into humans, there are a number of criteria that have to be met concerning product quality and production methods which are set by the various regulatory authorities such as the Food and Drug Administration in the United States and the Department of Health and Social Security in the United Kingdom. Waife and Lasagna have described some of the regulatory problems which were encountered in the development of insulin from a recombinant DNA source by Eli Lilly and Genentech (79). HPLC methods have been very useful in the development of human insulin from rDNA by confirming structures of synthesized products (80).

6.1. Process Validation

Process validation was first used to describe the studies that are needed to give proper assurance about the sterilization process (81). More recently, it has come to mean a GMP (good manufacturing practice) assurance function providing evidence that a process is indeed capable of doing what it is supposed to do. Quality control methods are designed to ensure purity, safety, and identity of the product in question. In addition, the efficiency of the manufacturing process in removing substances that cannot be tested for in the final product should be evaluated. This is particularly important in processes used for the recovery of products from nonhuman sources, especially recombinant DNA microorganisms, where it is necessary to check that steps are present which are capable of removing nucleic acids or other products (82).

It is also important to be able to give some indication of the process operating variables, that is, the range of conditions under which a process step can be safely considered to give the required results (81, 82). This could be the upper and lower limits of pH and ionic strength for a buffer used for binding or elution in ion exchange.

Detailed information concerning concepts in validation can be found in the literature distributed by the relevant authorities. The general philosophy of validation and a useful list of references is given in Ref. 83.

6.2. Hygiene in Chromatographic Processes

Cleanliness and sterility combine to ensure the safety and integrity of the final product by removing or controlling any unwanted substances that might be present or generated in the raw material or derived from the purification system itself. This is normally not too much of a problem when working with proteins from human sources such as plasma from healthy donors, but is obviously more considerable when dealing with proteins derived from recombinant DNA bacteria. Two of the major contaminants that must be removed are bacterial DNA and endotoxins.

Chromatographic methods such as ion exchange have been reported to be effective in removing DNA (70). There are a range of methods that have been reported as useful for the removal of endotoxins (84). Whichever method is used, its effectiveness should be checked as part of the validation process.

Chromatographic systems also have to be cleaned of contaminating material picked up for the sample, and once clean, protected from any further contamination. Specific methods for handling chromatographic systems are concerned with good equipment design, protection by in-line filters, and chemical cleaning and sterilization procedures (85).

6.2.1. Equipment Design

The components used in the system must be of hygienic design. The danger zones for bacterial contamination are valves, connections, pumps, and the

Fig. 23. Large-scale chromatography system designed for hygienic use.

chromatography column itself. Hygienic design means that the fluid pathway can be cleaned easily and effectively, for example, by steam cleaning or chemical cleaning. There should be no "dead-spaces" where material can be trapped permitting microbial growth. Problems arise in threaded tubing connectors and films forming in the moving parts of valves. All materials in contact with liquid should ideally be flushed with the liquid itself.

Feed solutions should be filtered through a 0.45 or 0.22 μm filter before use. Columns are protected from particulate matter by in-line filtration with filters of pore size from 0.22 and 0.9 μm.

Materials used in the construction should be acceptable for use in pharmaceutical production, for example, stainless steel, glass, and fluoroplastics. Figure 23 shows an example of an automated system designed for hygienic production of a pharmaceutical product.

Most HPLC equipment has been designed for use in the analytical laboratory without any specific consideration for the requirements of hygienic production. Although this is quite acceptable up to pilot-scale process development, the operator should consider equipment design carefully for production scale.

The requirements on equipment automation and control vary from process development—where the key is flexibility—to production—where the key is reliability (86). Guides exist for documenting and validating computer systems used in the pharmaceutical industry for automation and control (87).

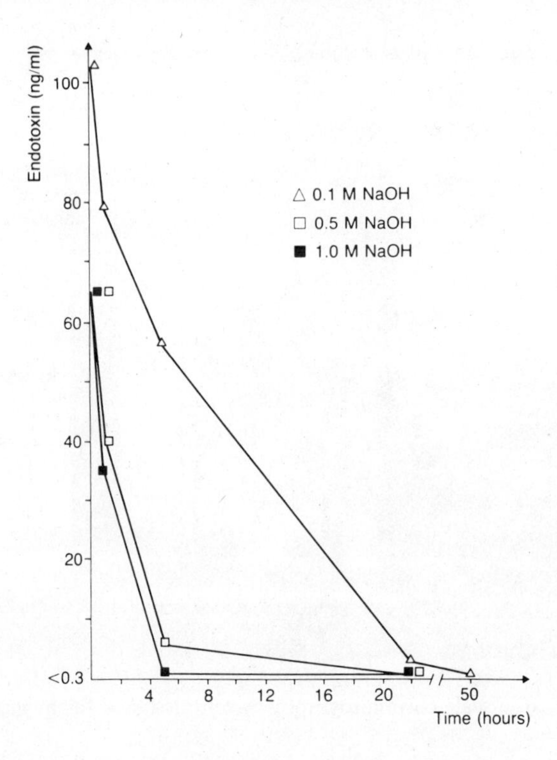

Fig. 24. Inactivation of *E. coli* endotoxin in alkaline environment at room temperature.

6.2.2. Cleaning and Sterilization of Gel Media

Whereas steam cleaning and autoclaving are suitable methods for cleaning equipment, they are normally not suitable for cleaning packed chromatography columns. Chemical cleaning methods are much more useful, and will vary according to the chemical nature of the support matrix. Sodium hydroxide is a widely used cleaning agent, and it has been shown to be effective in killing both gram negative and gram positive bacteria as well as HTLV-III/LAV virus (88–91). Sodium hydroxide also destroys *E. coli* endotoxin, as shown in Fig. 24. The cleaning effect is dependent on temperature and concentration (88).

Polymeric matrices such as MonoBeads and Sepharose Fast Flow are stable to long exposure at high concentrations of sodium hydroxide ($> 1M$ for 72 days). Table 7 shows the stability of some ion exchange and gel filtration media to sodium hydroxide. Affinity gels, particularly those with protein ligands, and gels based on silica as a matrix, are less stable to sodium hydroxide, and it is necessary to follow closely the manufacturers recommendations regarding cleaning and sterilization procedures.

Table 7. Stability of Separation Media to NaOH[a]

Separation Media	NaOH Molarity	Temp (°C)	Time
DEAE- and CM-Sepharose Fast Flow	2.0	20	72 days
DEAE-Sephadex A-50	1.0	20	2 hr
	0.1	20	30 days
DEAE- and CM-Sepharose CL-6B	1.0	40	7 days
Sephadex G-25 Coarse	0.2	20	2 years
Sephacryl S-200	1.0	20	2 hr
	0.2	20	100 hr
	0.1	20	30 days
Octyl and Phenyl Sepharose CL-4B	1.0	20	2 hr
	0.1	20	30 days

[a]The treatment shown will not change the separation properties of the gel. Maximum storage time was not investigated.

7. CONCLUSIONS

High-performance liquid chromatography is one of the most important analytical techniques available today. It is interesting to note that the majority of recent papers published using HPLC have focused on biological applications.

Traditional chromatography methods have been used for over 25 years for the purification of proteins, yet few of the hundreds of thousands of purification methods that have been published have been optimized for recovery or throughput. The main objective of the purification is purity, and it is usually easier to include a further purification step rather than to go back and optimize each step exhaustively. This is particularly true in the case of ion exchange methods, where a great deal of purification can be achieved using a starting pH fairly close to the pI of the substance of interest and eluting with a linear salt gradient. It is tedious, particularly with conventional media, to have to determine retention plots for pH, buffer type, counter ions, and co-ions when an alternative methods such as affinity chromatography or HIC will give the purity required. Purity is achieved by a series of suboptimal steps.

This approach is hardly acceptable in production, where yield and throughput at a defined purity are the key parameters.

This is undoubtedly where HPLC has played an important role, since rapid chromatographic analysis makes proper optimization feasible. As discussed earlier, it is relatively easy to improve resolution by manipulating efficiency, since this is controlled by physical characteristics such as bead size and column length. However, the key factor to optimize resolution in preparative separations with high sample loadings is selectivity. High-efficiency media allow the optimal conditions for separation to be found in a much shorter time. Many of the current studies in HPLC of proteins and peptides are designed to improve our understanding of adsorption and desorption process and their effect on protein structure and function (= recovery).

High performance in chromatography can be achieved by the use of microparticulate packings, which in turn lead to high back pressures. For many people, HPLC is high-pressure liquid chromatography, a misleading term which implies that you cannot get high resolution without high pressures.

High performance in chromatography can also be achieved by the use of highly selective separation conditions—*regardless* of bead size. Affinity chromatography is still the most powerful separation method for protein purification available today.

The objectives in process development must be to find the optimal conditions for the required separation and to select the particle size of the gel in terms of cost effective throughput, resolution, and recovery.

Large-scale "HPLC" with microparticulate packings will undoubtedly be a useful technique for the high-resolution purification of proteins, particularly for the recovery of products from contaminated peak regions. The biggest impact of microparticulates, however, will be on the rapid development of cost effective processes.

REFERENCES

1. J. Porath, and P. Flodin, Nature, *183*, 1657 (1959).
2. I.S. Johnson, Science, *219*, 632 (1983).
3. B.A. Bidlingmeyer, and F.V. Warren, Anal. Chem., *56*, 1582A (1984).
4. F.H. Arnold, H.W. Blanche, and C.R. Wilke, J. Chromatogr., *330*, 159 (1985).
5. N.M. Fish, and M.D. Lilly, Biotechnology, *2*, 623 (1984.
6. K.H. Kroner, H. Hustedt, and M-R. Kula, Process Biochem., *19*, 170 (1984).
7. H. Strathmann, Trends in Biotechnol., *3*, 112 (1985).
8. R.S. Tutunjian, Biotechnology, *3*, 615 (1985).
9. R. Datar, Process Biochem., *21*, 19 (1986).
10. J-C. Janson, and P. Hedman, in *Advances in Biochemical Engineering*, Vol. 25, A. Fiechter, Ed., Springer-Verlag Berlin, Heidelberg, 1982, p. 43.
11. J. Cooney, Biotechnology, *2*, 41 (1984).
12. R.W. Stout, and J.J. De Stefano, J. Chromatogr., *326*, 63 (1985).
13. R.W. Stout, S.I. Sivakoff, R.D. Ricker, and L.R. Snyder, J. Chromatogr., *353*, 439 (1986).

14a. D.K.R. Low, BioFair 88, Tokyo, Japan, 1988.

14. D.K.R. Low, in *Bioactive Microbial Products 3*, J.D. Strowell, Ed., Academic Press, London, 1986, p. 121.
15. L. Söderberg, in *Protides of the Biological Fluids*, Vol. 30, H. Peeters, Ed., Pergamon Press, New York, 1983, p. 629.
16. W. Kopaciewicz, M.A. Rounds, J. Fausnaugh, and F.E. Regnier, J. Chromatogr., *266*, 3 (1983).
17. W. Kopaciewicz, and F.E. Regnier, Anal. Biochem., *133*, 251 (1983).

18. F.E. Regnier, "Enzyme Purification and Related Techniques", in *Methods in Enzymology*, vol. 104, W.B. Jakoby, Ed., Academic Press, Orlando, 1984, p. 170.
19. A.N. Hodder, P.G. Stanton, and M.T.W. Hearn, *The 4th International Symposium on HPLC of Proteins, Peptides and Polynucleotides*, Baltimore, M.D., USA 1984.
20. F.E. Regnier, presented at 9th International Symposium on Column Liquid Chromatography, Edinburgh, Scotland, 1985.
21. S. Yamamoto, K. Nakanishi, R. Matsuno, and T. Kamikuko, Biotechnol. Bioengin., *25*, 1373 (1983).
22. S. Yamamoto, K. Nakanishi, R. Matsumo, and T. Kamikuko, Biotechnol. Bioengin., *25*, 1465 (1983).
23. R.W. Stout, S.I. Sivakoff, R.D. Ricker, H.C. Palmer, M.A. Jackson, and T.J. Odiorne, J. Chromatogr., *352*, 381.
24. J.L. Fausnaugh, L.A. Kennedy, and F.E. Regnier, J. Chromatogr., *317*, 141 (1984).
25. M.T.W. Hearn, "Enzyme Purification and Related Techniques", in *Methods in Enzymology*, Vol. 104, W.B. Jakoby, Ed., Academic Press, Orlando, 1984, p. 190.
26. D.F. Mann, and R.O. Moreno, Prep. Biochem., *14*, 91 (1984).
27. B.S. Welinder, S. Linde, B. Hansen, and O. Sonne, J. Chromatogr., *298*, 41 (1984).
28. Y. Kato, T. Kitamura, and T. Hashimoto, J. Chromatogr., *333*, 202 (1985).
29. Y. Kato, T. Kitamura, and T. Hashimoto, J. Chromatogr., *333* , 93 (1985).
30. J-C. Janson, in *Affinity Chromatography and Related Techniques*, T.C.J. Gribnau, J. Visser and R.J.F. Nivard, Eds., Elsevier, Amsterdam, 1982, p. 503.
31. E.A. Hill, and M.D. Hirtenstein, in *Advances in Biotechnological Processes*, Vol. 1, Alan R. Liss Inc., New York, 1983, p. 31.
32. J.C. Janson, Trends in Biotechnol., *2*, 1 (1984).
33. D. Southern, and D. Hollis, Biotechnology, *4*, 519 (1986).
34. A.J. Muller, and P.W. Carr, J. Chromatogr., *357*, 11 (1986).
35. J.C. Janson, Agr. Food Chem., *19*, 581 (1971).
36. T. Andersson, M. Carlsson, L. Hagel, P-Å. Pernemalm, and J-C. Janson, J. Chromatogr., *326*, 33 (1985).
37. J.H. Berglöf, S. Eriksson, and I. Andersson, presented at The XXI Congress of the International Society of Haematology and the XXI Congress of the International Society of Blood Transfusion, Sydney, 1986.
38. L. Hagel, J. Chromatogr., *324*, 422 (1985).
39. F.H. Arnold, and H.W. Blanch, J. Chromatogr., *355*, 1 (1986).
40. F.H. Arnold, and H.W. Blanch, J. Chromatogr., *355*, 13 (1986).
41. F.H. Arnold, J.J. Chalmers, M.S. Saunders, M.S. Crougham, H.W. Blanch, and C.R. Wilkie, *American Chemical Society Symposium Series,* Vol. 271, Amer. Chem. Soc., 1985, p. 113.
42. H.A. Chase, J. Chromatogr., *297*, 179 (1984).
43. H.A. Chase, J. Biotechnol., *1*, 67 (1984).
44. H.A. Chase, Chem. Eng. Sci., *39*, 1099 (1984).
45. P. Gariel, C. Durieux, and P. Rosset, Sep. Sci. Technol., *18*, 441 (1983).
46. P. Gariel, L. Personnaz, and M. Caude, Analusis, *7*, 401 (1979).
47. P. Gariel, L. Personnaz, J.P. Feraud, and M. Candle, J. Chromatogr., *192*, 53 (1980).

48. A.F. Mann, Int. Biotechnol. Lab., *4*, 28 (1986).
49. K. Nakamura, and Y. Kato, J. Chromatogr., *333*, 29 (1985).
50. H.M. Sassenfeld, and S.J. Brewer, Fifth International Symposium on HPLC of Proteins, Peptides and Polynucleotides, 1985.
50a. R. Tsugawa, BioFair Tokyo 86, proceedings p. 113, 1986.
51. B-L. Johansson, and N. Stafström, J. Chromatogr., *314*, 396 (1984).
52. M.T.W. Hearn, P.G. Stanton, A.N. Hodder, and M.I. Aguilar, The 9th International Symposium on Column Liquid Chromatography, Edinburgh, 1985.
53. E.P. Kroeff, R.A. Owens, E.L. Campbell, R.D. Johnson and H.I. Marks, J. Chromatogr. *461*, 45 (1989).
54. A.M. Cantwell, R. Calderone, and M. Sienko, J. Chromatogr., *316*, 133 (1984).
54a. R. Sitrin, P. de Phillips, J. Dingerdissen, K. Erhard, and J. Filan, LC GC, *4*, 530 (1986).
55. R. Eketorp, in *Affinity Chromatogrphy and Related Techniques*, T.C.J. Gribnau, J. Visser and R.J.F. Nivard, Eds., Elsevier, Amsterdam, 1982, p. 263.
56. M.J. Harvey, R.A. Brown, J. Rott, D. Lloyd, and R.S. Lane, in *Separation of Plasma Proteins*, J.M. Curling, Ed., Pharmacia AB, Uppsala, Sweden, 1983, p. 79.
57. T. Tomono, E. Sawada, and E. Tokunaga, in *Separation of plasma proteins*, J.M. Curling, Ed., Pharmacia AB, Uppsala, Sweden, 1983, p. 157.
58. E.R.A. Jeans, P.J. Marshall, and C.R. Lowe, Trends in Biotechnol., *3*, 267 (1985).
59. M. Einarsson, J. Brandt, and L. Kaplan, Biochim. Biophys. Acta, *830*, 1 (1985).
60. T. Staehalin, D.S. Hobbs, H. Kung, C-Y. Lai, and S. Pestka, J. Biol. Chem., *256*, 9750 (1981).
61. S.J. Tarnowski, and R.A. Liptak, *Advances in Biotechnological Processes*, Vol. 2, Alan R. Liss, New York, 1983, p. 271.
62. E. Hochuli, Engineering Foundation, 3rd Conference of Recovery of Bioproducts, Uppsala, Sweden, 1986.
62a. E. Hochuli, Chimia *40*, 408 (1986).
63. A.G. Laurent, J. Gruest, J. Svab, L. Montagnier, and E. Meurs, *Develop. Biol. Standard*, Vol. 57, S. Kager, Basel, 1984, p. 305.
64. K.H. Fantes, N.B. Finter, and L.L. Toy, Contr. Oncol., *20*, 98 (1984).
65. M. Einarsson, L. Kaplan, E. Nordenfelt, and E. Miller, J. Virol. Methods, *3*, 213 (1981).
66. K.G. Kenrick, S. von Sturmer, J.F. Kelly, and J.S. Stratton, in *Protides of the Biological Fluids*, Vol. 32, Peters, H. Ed., Pergamon Press, Oxford and New York, 1984, p. 929.
67. W.J. McAleer, E.B. Buynack, R.Z. Maigetter, D.E. Wampler, W.J. Miller, and M.R. Hilleman, Nature, *307*, 178 (1984).
68. E.M. Scolnick, A.A. McLean, D.J. West, W.J. McAleer, W.J. Miller, and E.B. Buynack, J. Amer. Med. Assoc., *251*, 2812 (1984).
69. D.E. Wampler, E.B. Buynack, B.J. Harder, A.C. Herman, M.R. Hilleman, W.J. McAleer, and E.M. Scolnick, in *Modern Approaches to Vaccines: Molecular and Chemical Basis of Virus Virulence and Immunogeneity*, Symposium Cold Spring Harbor, Vol. 27, R.M. Chanock and R.A. Lener, Ed., Cold Spring Harbor Laboratory, NY, 1984, p. 251.

70. R. Hershberg, Engineering Foundation, 3rd Conference on Recovery of Bioproducts, Uppsala, Sweden, 1986.
70a. H. Ellegren, B. Engström, L. Kågedal, Eighth International Symposium HPLC of Proteins, Peptides and Polynucleotides, 1988.
71. J.H. Berglöf, S. Eriksson, and J.M. Curling, J. Appl. Biochem., *5*, 282 (1983).
72. S.A. Cockle, S.M. Loosmore, M.A. Wosnick, E. James, and S-H. Shen, in *The World Biotech Report, Vol. 2: USA*, Online Publications, Middlesex, U.K., 1984, p. 393.
73. G.H. Dixon, and A.C. Wardlaw, Nature, *188*, 721 (1960).
74. S.J. Brewer, and H.M. Sassenfeld, Trends in Biotechnol., *3*, 119 (1985).
75. H.M. Sassenfeld, and S.J. Brewer, Biotechnology, *2*, 76 (1984).
76. M. Uhlén, and B. Nilsson, in *The World Biotech Report, Vol. 1: Europe*, Online, Middlesex, U.K., 1985, p. 171.
77. B. Nilsson, E. Holmgren, S. Josephson, S. Gatenbeck, L. Philipson, and M. Uhlén, Nucleic Acids Res., *13*, 1151 (1985).
78. T. Moks, L. Abrahamsen, B. Österlow, S. Josephson, M. Östling, S-O. Enfors, I. Persson, B. Nilsson, and M. Uhlén, Biotechnology, *5*, 379 (1987).
78a. E. Hochuli, H. Döbeli, J. Schacher, J. Chromatogr. *411*, 177 (1987).
79. S.O. Waife, and L. Lasagna, Regulatory Pharmacol. Toxicol., *5*, 212 (1985).
80. J.A. Galloway, and R.E. Chance, in *Proceedings of the World Conference Clinical Pharmacology and Therapeutics*, Vol. 2, L. Lemberger and M.M. Riedenberg, Eds., American Society Pharmacology Experimental Therapeutics, Bethesda, 1984, p. 503.
81. E.M. Fry, Drug and Cosmetic Industry, *137*, 46 (1985).
82. A.J.S. Jones, and J.V. O'Connor, *Develop. Biol. Standard*, Vol. 59, S. Karger, Basel, 1985, p. 175.
83. PMA's validation advisory committee, *Pharmaceutical Technol.*, Sept. 1985, 78 (1985).
84. G. Sofer, Biotechnology, *2*, 1035 (1984).
85. J.M. Curling, and J.M. Cooney, J. Parent. Sci. Techol, *36*, 59 (1982).
86. H. Johansson, in *The World Biotech Report 1985, Vol. 1: Europe*, Online, U.K., 1985, p. 193.
87. N.R. Kuzel, Pharmaceutical Technol., *9*, 60 (1985).
88. *Downstream* No. 2, 1 (1986), Pharmacia Process Separation Division, Uppsala, Sweden.
89. R.L. Whitehouse, and L.F.L. Clegg, J. Dairy Res., *30*, 315 (1963).
90. A.M. Prince, B. Horowitz, and B. Brotman, Vox Sang., *46*, 36 (1984).
91. L.S. Martin, J.S. McDougal, and S.L. Loskowski, J. Infect. Diseases, *152*, 400 (1985).

CHAPTER 7

High-Performance Liquid Chromatography of Antigenic Proteins and Vaccines

RONALD D. JOHNSON
Process Development and Manufacturing, Chiron Corporation, Emeryville, California

1. INTRODUCTION

Vaccination has been defined as the prevention of infectious diseases by the introduction into the body of reasonably safe preparations made from whole live or killed microbial disease agents, or from chemical components of bacteria or viruses, or from toxins (poisons) produced by such microbial pathogens (1). The mechanism of action of a vaccine is defined by the science of immunology, which deals with the various biological and chemical processes of resistance or the impairment of resistance to infectious agents or foreign bodies and conditions (1).

The introduction of these whole live or killed microbial agents or impure preparations of microbial chemical components, including toxins, into the body has saved countless lives since the conception of this technique. However, many vaccines developed severe and sometimes lethal reactions to impurities in these crude vaccines. A more complete understanding of antigen-antibody reactions, the isolation and purification of antigens responsible for immunization, and a better understanding of the biochemical and physicochemical nature of such antigens have allowed the generation of safer and more potent vaccines. An ideal vaccine antigen could well be an individual molecule or complex mixture biosynthesized from a source unrelated to the disease-causing organism. The purification and characterization of a single entity, even macromolecules such as proteins, glycoproteins and polysaccharides, is clearly superior to the production of crude, complex vaccines that contain many components, some of which have nothing to do with the generation of immunity. The concept of an ideal vaccine is being pursued via recombinant technology (2) and (bio) chemical synthesis (3), with the advantages of safety to operators and other staff, purity of preparation, and ease of production. These advantages are exemplified by the FDA approval, in 1986, of a Hepatitis B vaccine manufactured by Merck Sharp and Dohme, West Point, PA from a recombinant yeast strain licensed from the Chiron Corporation, Emeryville, CA. However, most vaccines are still isolated from a form of the causative organism and many are live or attenuated organisms. The greater our understanding of the physicochemical and biological properties of the antigen and the better our analytical detection and quantitation techniques, the easier it becomes to generate a controlled manufacturing process that produces a safer vaccine.

High-performance liquid chromatography (HPLC) is a very powerful analytical technique that has been applied extensively to peptides and soluble proteins (2). Unfortunately, many antigenic proteins are very hydrophobic and separate poorly because of complications associated with insolubility, denaturation, aggregation, or strong adsorption to packing matrices (5). The application of HPLC to hydrophobic proteins, such as membrane proteins and structural polypeptides of viruses, has been difficult. However, recent advances in solubilization methods and solvent systems have allowed the generation of RP-HPLC separations (5, 6, 12), which can be used as rapid, quantitative measures of the antigen(s) of interest.

2. SPECIFIC APPLICATIONS

2.1. Reversed-Phase HPLC

One of the older and most biologically applicable HPLC technique is reversed-phase HPLC (RP-HPLC). It is particularly applicable to peptides and soluble proteins, but is less applicable to higher-molecular-weight or more hydrophobic proteins or both due to the strong denaturating effect of protein-packing interactions and of organic solvents (7).

One of the earlier examples of an RP-HPLC separation of viral antigenic proteins was reported by Henderson et al. (8). A sample of Gross Leukemia virus was disrupted by 6 M guanidine-HCl at pH 2.0 in trifluoroacetic acid (TFA) and chromatographed over a 0.4 × 30 cm microBondapak phenylalkyl column. An eight minute isocratic elution with 0.05% TFA in water eluted the polar components, such as viral RNA, buffer salts, and guanidine-HCl. A 40 min gradient to 60% acetonitrile in 0.05% TFA (1 mL/min) eluted the following viral proteins with baseline separation: P-10 (MW = 10,000), P-12 (MW = 12,000), P-30 (MW = 30,000) and P-15 (MW = 15,000) (10). In this case, a standard peptide HPLC solvent, dilute TFA and organic solvent, was sufficient to provide an excellent separation of Gross Leukemia viral proteins. Unfortunately, more hydrophobic viral proteins do not elute from RP-HPLC columns using conventional solvent systems. Heukesoven and Dernick (6, 9–11) have conclusively shown the applicability of 60% formic-acid-based elution solvents for the solubilization and elution of the structural polypeptides of poliovirus strains. The separation was executed on Baker widepore C18 column at room temperature and a flow rate of 1 mL/min. TCA-precipitated virus was dissolved in 70% formic acid, injected and eluted with an acetonitrile gradient in 60% formic acid. The four major structural polypeptides were separated for each of three poliovirus strains (type 1 = strain Mahoney; type 2 = strain MEF-1; type 3 = strain Saukett). Slight changes in the elution times for the peptides from different strains were indicative of slight structural variations (7).

Van Der Zee and co-workers (12–14) have studied the separation of detergent-extracted, DTT-reduced Sendai virus proteins by RP-HPLC using ethanol–*N*–butanol–HCl–water systems as well as TFA–acetonitrile–water systems with success. Sendai virus contains three envelope-associated proteins, HN (MW = 66,000), F or fusion protein [composed of two disulfide-linked polypeptides, F1 (MW = 52,000) and F2 (MW = 13,500), and M or matrix protein (MW = 38,000) (12). Triton X-100 was removed by treatment with XAD-2 and the solubilized viral proteins were filtered in the presence of 20 m*M* DTT. A 4-cm, 300 Å Supelcosil LC-318 (C18) column eluted with a linear gradient from 10–60% ethanol–butanol (4:1, v/v) in 12 m*M* HCl was used to separate several of these proteins (12, 14). Protein F2 (peak 1) was recovered quantitatively in pure form. However, the other three proteins (F1, NH and M), were poorly recovered (5–50%), but microgram quantities of moderate purity were obtained (12). A superior separation of these proteins was achieved on a 7.5 cm TMS-250 C1 column, monitored at 205 nm, with a 25 min linear gradient from 25–75% acetonitrile in 0.05% TFA in water at 1 mL/min. This yielded considerably higher recoveries (60–100%) for all four of the proteins (13).

Influenzae virus components (16) and peptides resulting from proteolytic digestion and CNBr digestion (15) have been separated by RP-HPLC. By isolating the membrane (M) protein, a water-detergent insoluble, organic insoluble, hydrophobic protein from Influenzae A virus, LeComte et al. (15) showed that RP-HPLC can resolve I-125 labelled peptides from proteolytic or CNBr digestion in > 85% yield. Separation was achieved on a Waters microBondapak C18 column eluted with linear acetonitrile or isopropanol gradients in triethylamine phosphate (15). Comparison of the *S. aureus* V8 protease digestion of the I-125 labeled peptides from two influenzae viruses (N and PR8) showed distinct peptide profiles. The M protein from the N virus had two peptides that were not found in the M protein from the PR8 virus, while the PR8 virus showed a peptide not found in the M protein of the N virus (15).

Phelan and Cohen (16) used RP-HPLC to separate the proteins of A/Bangkok 1/79 × 73 influenza virus. Dissociation of the virus was achieved in 2 m*M* DTT and 8 *M* guanidine-HCl. The separation was achieved on a 25 cm Aquapore RP-300 column at 1 mL/min using a linear gradient from 5% to 75% acetonitrile in 0.05% TFA (16). Three major peaks were noted: Peak I = HA1 (Hemagglutinin monomer subunit); Peak II = NP (Nucleoprotein); Peak III = M (Matrix Protein). The recoveries of these proteins were 32%, 13%, and 22%, respectively, while purities ranged from > 95% to 88% (16). Only the M protein recovered following chromatography was still reactive with monoclonal antibodies directed against it, suggesting that the higher-order structure of M is preserved (16).

Welling-Wester et al. (17) have investigated the use of RP-HPLC in the analysis and purification of a 57,000 dalton Herpes simplex virus (HSV) polypeptide, which is likely involved in the infection process. Virion envelope proteins were extracted by the detergent Nonidet P-40 (NP-40). Separation of the envelope extract components was achieved on a 30 cm Nucleosil 10 C18 column at 1 mL/min monitored at 215 nm. Elution was achieved by a complex gradient from 100% solvent A (3% *n*-butanol, 10% 2-methoxyethanol in 0.1% TFA) to 60% solvent B (70% ethanol, 20% *n*-butanol, 10% 2-methoxyethanol in 0.025% TFA) (17). The 57 KD polypeptide eluted at 33 min and was collected and studied by SDS-PAGE, amino acid analysis, and gas-phase microsequencing. Monoclonal antibody studies indicated that the 57 KD protein is related to the VP16 HSV protein (a 65 KD phosphorylated glycoprotein), probably as the nonglycosylated portion (17). Generally, only small quantities of virus proteins are available, therefore the combination of HPLC and microsequencing supplement each other (17).

The structural proteins of murine type C retroviruses, which result from the proteolytic cleavage of two different precursor polyproteins coded for by the viral *gag* and *env* genes, were quantitatively isolated from Rauscher and Moloney strains of type C murine leukemia viruses (R-muLV and M-MuLV) using RP-HPLC by Henderson et al. (18). Components isolated from R-MuLV (homologous in M-MuLV) included p10, p12, p15, p30, p15(E), gp69, gp71, and three unknown components designated p10′, p2(E), and p2(E)*. The virus was disrupted with 83% saturated guanidine-HCl in pH 7 phosphate buffer (0.1 *M*). The sample was acidified to pH 7 with TFA and injected onto a Waters microBondapak phenyl column at 23°C with UV detection at 206 nm. Separation was achieved with linear gradients of acetonitrile in 0.1% TFA and isopropanol at 50°C (18). The stoichiometry of *gag* cleavage products to *env* cleavage products was shown to be 4 to 1, and the data were consistent with the proposal that proteolytic processing of precursor polyproteins occurred after virus assembly (18).

RP-HPLC has also been used to investigate the structural proteins of Hepatitis A virus, a picornavirus with one antigenic type (Dr. V. Gauss-Mueller, verbal communication, 1986). The method of Heukeshoven and Dernick (11, 12), which included a 4 *M* guanidine-HCl dissociation and separation in 60% formic acid-based elution solvents, was used. A Baker Wide Pore, 300 Å, C18 column was eluted with a 30 min gradient from 20–40% acetonitrile in 60% formic acid. Proteins VP0 (composed of VP2 + VP4), VP1 (30–33 K daltons), VP2 (24–27 K), VP3 (21–23 K), and VP4 (7–14 K; no 280 nm absorption) were isolated and characterized.

Separations by RP-HPLC techniques of the trypsin-digested peptides from viral proteins have been achieved with more conventional elution solvents. Basak and Compans (19) have demonstrated the RP-HPLC separation of the tryptic peptides of the hemagluttin (HA) glycoprotein from influenzae virus. A microBondapak C18 column, eluted with a 0–30% gradient of *n*-propanol in 0.1% phosphoric acid at 2 mL/min, was used to separate eight components in 150 min at 90–95% recovery (19).

Kemp et al. (20) have demonstrated the separation of the tryptic glycopeptides from influenza HA glycoprotein (strain A/WSN/HON1) using a LiChrosorb RP-18 column at 2 mL/min using acetonitrile gradients in 0.1 *M* phosphoric acid at pH 2.85. Elution was achieved with an initial 10 min isocratic, 0% acetonitrile elution followed by a linear gradient from 0% to 40% acetonitrile for 150 min (20). The separation of the HA1 tryptic glycopeptides was achieved with a modified gradient from 12.5% to 35% acetonitrile gradient following a 20 min isocratic elution at 1 mL/min (20).

Rosner and Robbins (21) have demonstrated the separation of the tryptic peptides from the envelope glycoprotein (G) of the vesicular stomatitis virus using a 0.46 × 25 cm Ultrasphere-ODS RP-HPLC column at 0.5 mL/min. Elution was achieved with a linear gradient from 0% to 50% acetonitrile in 0.1 *M* phosphate buffer, pH 2.2 allowing the determination of the two major glycosylation sites. HPLC analysis of appropriately treated glycoproteins can be used to determine the nature and extent of glycosylation at individual sites (21).

Clark et al. (22) have used RP-HPLC to isolate and characterize the major structural glycoprotein (gp64; 64 K daltons) of human cytomegalovirus (HCMV). HCMV grown in the presence of [^{35}S]methionine and [^{3}H]glucosamine was dissociated in 6 *M* guanidine-HCl and chromatographed over a 0.41 × 25 cm Synchropak C18 (300 Å) column at 1 mL/min (22). The major component eluted at 50–52 min and was shown by SDS-PAGE analysis to have a molecular weight between 64 and 69 kilodaltons. This example amply demonstrated the applicability of RP-HPLC to the preparative purification of the major structural glycoprotein of HCMV and allowed the rapid generation of abundant, highly purified material for clinical investigations (22).

2.2. Ion-Exchange HPLC

High-performance ion-exchange chromatography (HPIEC), a less widely used HPLC technique for antigenic proteins and vaccines, is a valuable complement to RP-HPLC. Although often used in conjunction with RP-HPLC as a fractionation step, HPIEC can be used to purify these types of components directly. Welling et al. (23) have purified the F protein from Sendai virus using anion exchange HPIEC. Virions were treated with 2% Triton X-100 in Tris buffer (pH 7.2). The detergent extract contained the Fusion protein (F) and the hemagglutinin-neuraminidase protein (HN) and was loaded onto a 0.5 × 5 cm Mono Q column at 1 mL/min with UV detection at 260 nm (23). Elution was achieved with a 0.15 *M* to 1.5 *M* NaCl (23). The HN protein was not adsorbed to the column; however, the F protein eluted as several peaks that, after reduction with mercaptoethanol, migrated in SDS-PAGE analyses as the F1 protein, the 50,000 molecular weight disulfide cleavage product of F (23). The heterogeneity of the F protein was probably due to the differences in charge resulting from acidic oligosaccharide chains attached to the protein (24).

Welling-Wester et al. (25) have demonstrated the use of HPIEC in a multiple HPLC purification scheme involving RP-HPLC and high-performance size exclusion chromatography (HPSEC). Herpes simplex type 1 virions were extracted with 0.5% NP-40 in pH 7.4 phosphate buffer and subjected to chromatography on a 0.5 × 5 cm Mono Q column at 1 mL/min with UV detection at 280 nm. Elution was achieved with a 15 min linear gradient from 0.1 *M* Tris-HCl (pH 7.5) to 0.5 *M* sodium acetate in the same buffer (25). Peak 2 was shown by SDS-PAGE analysis to be predominantly an 80–85 K protein, while peak 3 contained a highly purified 60 K protein and peak 4 was predominantly a 20 K polypeptide (25). It is interesting to note that separation was achieved without the addition of detergent to the elution buffer.

Calam and Davidson (26) have purified viral proteins from X-49 influenzae virus and from three influenzae vaccines by HPIEC or HPSEC. Unfortunately, the protein recovery from an Aquapore AX300 column (0.4 × 25 cm), eluted with a NaCl gradient in phosphate buffer, was poor, although the peaks were sharp. The resolution of components was poor when a TSK IEX-645 DEAE column (0.75 × 75 cm) was used, even when eluted with an organic solvent-containing buffer (26). It is likely that resolution and recovery would have improved if a nonionic detergent or chaeotrope was present in the elution buffer. Green and Brackman (27) have used HPIEC and RP-HPLC to separate and purify several human adenovirus proteins (apparent molecular weights of 53 K, 19 K, and 20 K with smaller amounts of 21 K, 22 K, and 23 K). RP-HPLC on an Ultrasphere C-8 or a SynchroPak RP-P column (0.46 × 25 cm) at 0.33 mL/min, using a linear gradient from 0% to 60% 1-propanol in 0.5 *M* pyridinium formate, pH 4.1, allowed the purification of five of these proteins (19 K, 20 K, 21 K, 22 K, and 23 K) with retention of their immunological activity (27). HPIEC, on a SynchroPak AX-300 column (0.46 × 25 cm) at 1 mL/min using a 70 min linear gradient from 0.01 *M* to 1 *M* NaCl in a complex pH 7.6 buffer containing detergent, a protease inhibitor, DTT, glycerol, and HEPES, was used to separate the same five proteins with improved resolution of the 19 K protein (27).

In general, HPIEC of viral antigens has not been thoroughly investigated, but remains a viable complement and alternative to RP-HPLC.

2.3. High-Performance Size Exclusion Chromatography

Although the resolving capabilities and capacity of HPSEC are much lower than those of RP-HPLC and HPIEC, it has often been applied to viral antigens. The high recoveries and nondenaturing mechanism of action in HPSEC make it an ideal candidate for purification of native, biologically active antigens.

Calam and Davidson (26) have used HPSEC extensively in the purification of the constituent antigen proteins from X-49 influenzae virus as well as several influenzae vaccines. A TSK-4000SW column (0.75 × 30 cm) was eluted at 0.5 mL/min with 0.1 *M* phosphate buffer, pH 7.0, containing 0.1% of one of the following detergents: SDS, Brij 35, or Lutensol ON 70D. These conditions yielded the separation of one whole virus vaccine and two "surface antigen" vaccines containing varying amounts of HA, NA, and M proteins. HPSEC can also be used to separate the viral proteins from disrupted influenzae virus or vaccines and recovered with activity in immunodiffusion assays using samples as small as 4% of a single human dose (26).

Flavivirus membrane and capsid proteins have been purified by a multistep HPLC process that was optimized by immunological monitoring as reported by Winkler et al. (28). The three structural proteins, E (55 K), C (15 K), and M (7.5 K), of encephalitis virus were separated by a two step HPLC process involving HPSEC and RP-HPLC. The hydrophobically associated membrane proteins, E and M, were successfully separated by HPSEC on a TSK-3000 SW column in the presence of SDS, while the separation of M and C, and the desalting and removal of SDS, was achieved by subsequent RP-HPLC on an Ultrapore RPSC C3 column using a linear gradient of acetonitrile-isopropanol in TFA (28). Immunological monitoring with a highly sensitive dot immunoassay with polyclonal or monoclonal antibodies revealed peak tailing and cross contamination not readily apparent from the UV trace or SDS-PAGE analyses (28). This observation allowed the optimization of the conditions of chromatography for maximum separation.

Welling et al. (29) have also used combinations of HPSEC and RP-HPLC for the separation and purification of crude envelope proteins obtained by Triton X-100 extraction of Sendai virus virions. HPSEC on a TSK-4000 PW column (0.75 × 60 cm) at 1 mL/min in 45% acetonitrile in 0.1% HCl yielded several partially purified components. Subsequent purification by RP-HPLC on a TMS-250 C1 column, eluted with a linear 25 min gradient from 25% to 75% acetonitrile in 0.1% TFA, yielded pure F2 protein (15 K) and nearly pure M protein (38 K), F1 protein (50 K), and HN protein (66 K) as measured by SDS-PAGE analyses (29). The previous examples illustrate the suitability of combinations of HPLC modes for the rapid purification of small amounts of antigens.

Another combination HPLC purification of viral antigens was described by Ricard and Sturman (30) in which coronavirus envelope proteins were purified by HPSEC and hydroxyapatite HPLC (HPHAC). The coronavirus virion contains three major structural proteins, E1 (23 K membrane glycoprotein), E2 (180 K glycoprotein), and N (50 K phosphoprotein). Virions were solubilized with Triton X-114 and phase fractionated. Trypsin treatment hydrolyzed E2 into two 90 K subunits designated A and B, which copurify upon HPSEC on a TSK-4000 SW column (0.75 × 60 cm) eluted at 0.5 mL/min with pH 6.8 phosphate buffer in 0.1% SDS (30). Separation of A and B was achieved on a BioRad HPTH column (0.75 × 10 cm), eluted at 1.0 mL/min with a linear gradient (0.15–0.5 *M*) of phosphate buffer, pH 6.8, containing 0.1% SDS (30). The two 90 K subunits were shown to have a methionine ratio of 2.0/1.2. Radiolabeling studies showed the palmitic acid acylation site to be on subunit B (30).

2.4. High-Performance Hydrophobic Interaction Chromatography

Welling et al. (31) have shown the applicability of using high-performance hydrophobic interaction chromatography (HPHIC) in the purification of small quantities of relatively hydrophilic viral proteins for amino acid analyses and other chemical and biological characterization. An ideal vaccine should mimic the immunological stimuli of the natural infection, evoke minimal side effects, be readily available, stable, inexpensive, and easily administered (3). Vaccines based on (bio) synthesized peptides that represent only a part of the pathogenic entity could potentially meet these criteria (31). Studies using a TSKgel Phenyl-5PW column (0.75 × 7.5 cm), eluted with a 15 min linear gradient from 1.5 *M* to 0 *M* ammonium sulfate in 0.1 *M* phosphate buffer, pH 7.0 at 1 mL/min, yielded high recovery of native proteins. However, highly hydrophobic proteins, such as those from a detergent extraction of Sendai virus under reducing conditions, were not eluted from the column (31). It is likely that the introduction of detergent or a small amount of organic solvent or both would have eluted these proteins. HPHIC would be especially applicable to the purification of relatively hydrophilic viral proteins or to the smaller molecular weight peptide vaccines (bio) synthesized based on characterization data obtained on the native antigen.

3. FUTURE CONSIDERATIONS

3.1. Analytical Applications

The advent of quality HPLC separations of viral protein antigens opens the path to the development of HPLC assays for purity determination and quantitation of these antigens during process development, pilot-scale production, and bulk manufacturing. The use of HPLC as a rapid in-line process monitor for product purity and quantitation is an accepted practice in the manufacture of antibiotics, peptides, and small MW recombinant proteins and can be applied similarly in vaccine manufacturing. Of course, the application of HPLC for quality control release testing would also be advantageous.

3.2. Preparative Applications

Preparative HPLC is less widely used as a manufacturing process step. However, a few examples do exist in production facilities around the world including the manufacture of recombinant human insulin at Eli Lilly (32). With the development of elegant membrane techniques for the removal of salts, detergents, and chaeotropes, the use of preparative HPLC in the development and manufacture of vaccines is almost a foregone conclusion.

REFERENCES

1. A. Chase, *Magic Shots*, William Morrow and Company, Inc., New York, 1982, pp 41–83.
2. J.S. Emtage and N.H. Carey, "The Production of Vaccines by Recombinant DNA Techniques", in *New Developments with Human and Veterinary Vaccines*, A. Mizrahi, I. Hertman, M.A. Klingberg, and A. Kohn, Eds., Alan R. Liss, Inc., New York, 1980, pp. 367–376.
3. M.W. Steward and C.R. Howard, Med. Lab. Sci., *42*, 376–387 (1985).
4. F.E. Regnier, Science, *222*, 245 (1983).
5. B.G. Sharifi, C.B. Bascom, V.K. Khurana, and T.C. Johnson, J. Chromatogr., *324*, 173–180 (1985).
6. J. Heukeshoven and R. Dernick, J. Chromatogr., *252*, 241–254 (1983).
7. R. Johnson, Dev. Ind. Microbiol., *27* (Suppl. 1), 77–83 (1987).
8. L.E. Henderson, R. Sowder, and S. Oroszlan, Dev. Biochem., *17*, 251–260 (1981).
9. J. Heukeshoven and R. Dernick, J. Virological Methods, *6*, 283–293 (1983).
10. J. Heukeshoven and R. Dernick, Chromatographia, *19*, 95–100 (1985).
11. J. Heukeshoven and R. Dernick, J. Chromatogr., *236*, 91–101 (1985).
12. R. Van Der Zee, S. Welling-Wester, and G.W. Welling, J. Chromatogr., *266*, 577–584 (1983).
13. G.W. Welling, J.R.J. Nijmeijer, R. Van Der Zee, G. Groen, J.B. Wilterdink, and S. Welling-Wester, J. Chromatogr., *297*, 101–109 (1984).
14. R. Van Der Zee and G.W. Welling, J. Chromatogr., *237*, 377–380 (1985).
15. A. Darveau, N.G. Seidah, M. Chretien, and J. LeComte, J. Virol. Meth., *4*, 77–85 (1982).
16. M.A. Phelan and K.A. Cohen, J. Chromatogr., *266*, 55–66 (1983).
17. S. Welling-Wester, T. Popken-Boer, J.B. Wilterdink, J. van Beeumen, and G.W. Welling, J. Virol., *54*, 265–270 (1985).
18. L.E. Henderson, R. Sowder, T.D. Copeland, G. Smythers, and S. Oroszlan, J. Virol., *52*, 492–500 (1984).
19. S. Basak and R.W. Compans, Journal of High Resolution Chromatography and Chromatography Communications, *4*, 302–304 (1981).
20. W.C. Kemp, W.L. Holloway, R.L. Prestidge, J.C. Bennett, and R.W. Compans, J. Liq. Chromatogr., *4*(4), 587–598 (1981).
21. M.R. Rosner and P.W. Robbins, J. Cell. Biochem., *18*, 37–47 (1982).
22. B.R. Clark, J.A. Zaia, L. Balce-Directo, and Y.-P. Ting, J. Virol., *49*(1), 279–282 (1984).
23. G.W. Welling, G. Groen, and S. Welling-Wester, J. Chromatogr., *266*, 629–632 (1983).
24. M.C. Hsu, A. Scheid, and P.W. Choppin, Virology, *95*, 476–481 (1979).
25. S. Welling-Wester, T. Popken-Boer, and G.W. Welling, Protides Biol. Fluids, *32*, 1053–1056 (1984).
26. D.H. Calam and J. Davidson, J. Chromatogr., *296*, 285–292 (1984).
27. M. Green and K.H. Brackmann, Anal. Biochem., *124*, 209–216 (1982).

28. G. Winkler, F.X. Heinz, F. Guirakhoo, and C. Kunz, J. Chromatogr., *326*, 113–119 (1985).
29. G.W. Welling, G. Groen, K. Slopsema, and S. Welling-Wester, J. Chromatogr., *326*, 173–178 (1985).
30. C.S. Ricard and L.S. Sturman, J. Chromatogr., *326*, 191–197 (1985).
31. G.W. Welling, R. Van Der Zee, G. Groen, J. Van Beeuman, and S. Welling-Wester (unpublished results).
32. Eli Lilly and Company (unpublished results).

CHAPTER 8

Utilization of Analytical Reversed-Phase HPLC in Biosynthetic Insulin Production

R.D. DIMARCHI, H.B. LONG, E.P. KROEFF, and R.E. CHANCE

Department of Biochemistry and Biochemical Development, Lilly Research Laboratories, A Division of Eli Lilly and Company, Indianapolis, Indiana

The ability to achieve biosynthesis of foreign proteins in genetically altered microorganisms has fostered numerous advancements in analytical and preparative liquid chromatography of peptides and proteins. Insulin, being recognized since 1922 (1) as a hormone of vital importance in the treatment of diabetes mellitus, was immediately identified as a desirable target for biosynthesis (2). In 1982, human insulin produced by Eli Lilly and Company was approved by regulatory agencies for general distribution and thereby became the first human health product of this recombinant DNA technology. Through similar methodology proinsulin, the natural human insulin precursor, has also been produced (3). It is currently being clinically evaluated (4, 5). Much of what was rapidly accomplished by a group of scientists, engineers, and other corporate personnel, too numerous to be individually acknowledged, now appears routine. However, the initial formulation of a commercial process for the production of a two-chain disulfide-containing peptide was by no means obvious. The poorly efficient disulfide pairing of insulin A and B chains and the large-scale utilization of cyanogen bromide were just two of the many obstacles to practical insulin production that were recognized from the start. High-performance reverse-phase chromatography proved instrumental in the optimization of these chemical procedures and the preparative chromatography utilized in insulin purification. Numerous reports on reverse-phase chromatography of insulin have appeared throughout the course of our studies (6–18). The purpose of this chapter is not to review but instead to focus on the practical aspects of high-performance reverse-phase chromatography, as employed in the production of biosynthetic human insulin.

The initial production of biosynthetic human insulin utilized separate bacterial expression of the respective insulin A and B chains and purification as their S-sulfonates. The chains are combined by a disulfide-interchange reaction to yield insulin, which is subsequently purified to near homogeneity (2, 19). As development and optimization of each step in the process proceeded simultaneously, there was an immediate need to analyze three peptides at varying purity levels. The additional requirement for analysis of proinsulin and its S-sulfonate derivative, two peptides that represent intermediates in an alternative process for the biosynthesis of insulin (3), increased the need for a highly efficient and rapid method of analysis.

Reversed-phase chromatography of peptides at ambient temperature on a silica-based support at low pH is an established high-resolution analytical technique (20). However, chromatographic analysis of proteins produced in various genetically engineered microorganisms can present difficult quantitation problems, as protein concentrations and purities are initially quite low. Alternatively quantitation by radioimmunoassay (RIA) has been used. Analysis by RIA requires the handling of a radiolabeled peptide and the availability of a suitable antibody. In general, when chromatographic analysis can be employed, it has not only proven to be faster, but more accurate and discriminating than RIA.

Separation of peptides by reversed-phase chromatography is most commonly achieved through elution with a linear gradient of increasing acetonitrile concentration, in an aqueous acidic buffer consisting of dilute trifluoroacetic or phosphoric acid. A frequent substitute for acetonitrile is l-propanol, owing to its enhanced strength of elution. However, its inherent effect to increase the chromatographic operating pressure and often diminish peak shape makes the former preferable. The large variety of columns currently available can render the proper selection a tedious, expensive, and long process. Changes in resin matrix, ligand type and density, bonding chemistry, degree of end-capping, particle size, shape, and pore diameter can markedly alter the degree of peptide resolution, capacity, and recovery. Several reports have appeared that indicate distinct preferences regarding these chromatographic variables that can significantly minimize the process of chromatographic optimization (20–26). For the peptides we studied, a spherical silica particle of 5–7 micron size having a mean pore diameter of either 60 or 150 Å yielded the best results. These particles had been derivatized with an octyl-ligand (C8) and appropriately end-capped. In particular, Zorbax® and IBM C8 columns were most commonly used. A positive feature of Zorbax® was the early ability of the supplier to produce and reserve large quantities of resin, as a means to increase chromatographic reproducibility from column to column. More recently, similar comparative analysis with other synthetic and biosynthetic peptides has occasionally indicated an advantage for columns from other suppliers. In particular, chromatography of growth hormone releasing factor and its analogs in aqueous trifluoroacetic acid has required the greater eluting strength of propanol for analysis with these same Zorbax® C8 columns. Similar analysis using Vydac wide-pore C18 columns revealed quantitative recovery of growth hormone releasing factor, with enhanced resolution from structurally related derivatives, through elution with acetonitrile.

Silica-based resins are used in the vast majority of reverse-phase chromatographic separations. With limited exceptions acidic buffers are employed due to the alkaline instability of the silica resin matrix and the adsorptive effects of unprotonated silanols at elevated pH values. If only for practical reasons, the availability of a comparably performing alkaline-resistant reversed-phase resin could prove revolutionary. By analogy, occasional washings of Pharmacia Mono Bead® ion exchange columns with dilute sodium hydroxide provides a simple, inexpensive, and effective method for column regeneration. In this way routine analyses can be achieved with a single column for more than 1 year, thereby far surpassing the longevity of similar silica-based columns utilized under analogous conditions. Chromatography of insulin or proinsulin on alkaline-resistant polystyrene–divinylbenzene copolymers, such as PRP-1® Hamilton columns, did not provide the peak shape, resolution, or recovery that is achievable on silica-based supports at any pH studied. Most recently, a new alkaline-stable support for reverse-phase chromatography has been reported and are worthy of serious consideration (27). Our initial interest in alkaline-stable reversed-phase supports resulted from the inability to perform repetitive analyses of insulin-related peptides at low pH. While reasonably pure insulin and proinsulin chromatographed efficiently in an assortment of acidic buffers containing acetronitrile, the respective three S-sulfonate peptides proved less amenable. Chromatographic analyses of these peptide S-sulfonates begins early in the biosynthetic process among a large array of acid-insoluble contaminating *E. coli* natural products. These impurities constitute the major obstacle to reproducible low-pH analyses. To eliminate the insoluble contaminants, analytical samples were pretreated with acid or organic solvent or both, and centrifuged. Unfortunately, this proved to be a significant source of analytical variance, as each S-sulfonate peptide exhibited a variable degree of precipitation, which appeared dependent on its initial purity. Consequently, alternatives to low-pH reversed-phase analysis needed to be considered.

Throughout the two insulin processes there is a total of five steps where an interconversion between one or more of the insulin-related peptides exists. The disulfide interchange of A- and B-chain S-sulfonates to yield insulin, and the analogous conversion of proinsulin S-sulfonate to proinsulin, are two such examples. The respective sulfitolyses of insulin and proinsulin polymers, in addition to the enzymatic transformation of proinsulin to insulin, represent the remaining three interconversion reactions. It was deemed most desirable to identify a chromatographic system where through a single analysis, the reaction substrates, intermediates, and products could be singly quantitated. This approach would serve to eliminate experimental variance and diminish the total number of samples to be analyzed. Consequently, the objective was to identify a chromatographic buffer where high-resolution quantitative analysis of insulin and all related peptides at varying levels of purity could be achieved.

Chromatography in a buffer of 0.1 M $(NH_4)_2HPO_4$ at a pH 7 or above with a linear gradient of acetonitrile was observed repeatedly to elute all five insulin and related peptides at near quantitative levels ($>95\%$), with insignificant carryover ($<1.0\%$). The effect on chromatography of temperature in the range of 25–55°C was examined, and three representative points are shown in Fig. 1.

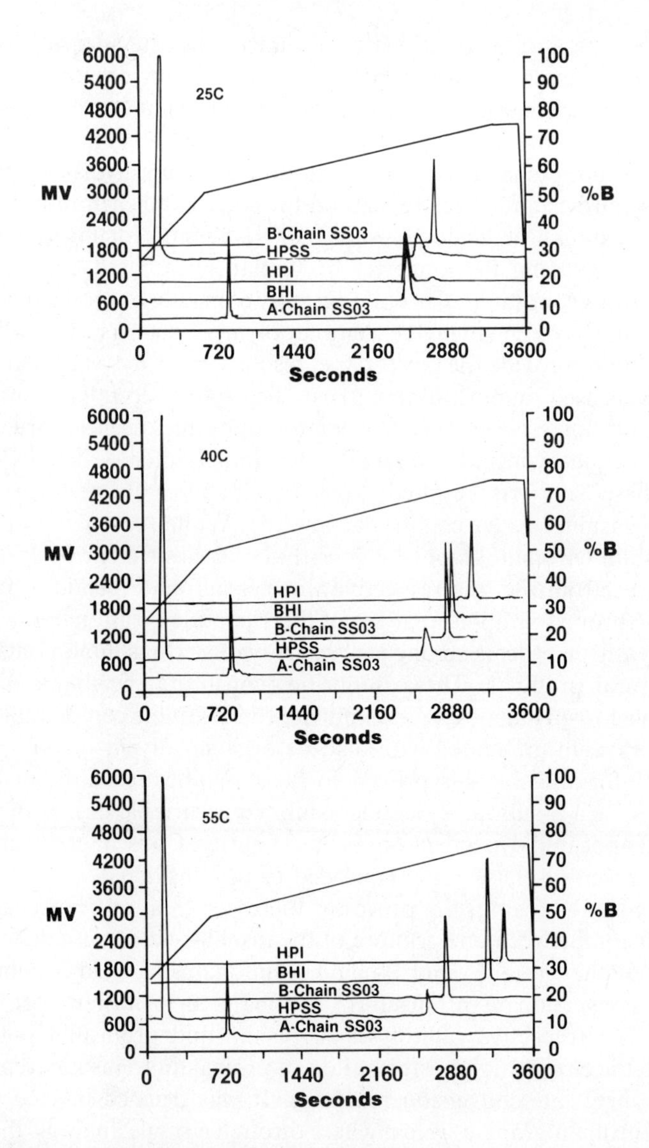

Fig. 1. Reversed-phase chromatographic analysis of insulin and related peptides at pH 7 in 0.1 *M* $(NH_4)_2HPO_4$, at three differing temperatures. Elution is enacted at 1 mL/min through application of a linear gradient of acetonitrile concentration from 10% to 35%, followed by 5 min of chromatography at 35%. A buffer: 0.1 *M* $(NH_4)_2HPO_4$, pH 7; B buffer: 0.1 *M* $(NH_4)_2HPO_4$, pH 7/50% CH_3CN. A (4.6 × 25 cm) column of Zorbax C8, 5 μm particle size, 150° Å pore size was exployed with detection at 214 nm. The following abbreviations are used: BHI; biosynthetic human insulin; HPI, human proinsulin; B-chain SSO_3, human insulin B-chain S-sulfonate; A-chain SSO_3, human insulin A-chain S-sulfonate; HPSS, human proinsulin S-sulfonate.

The increase in retention displayed most dramatically by proinsulin and insulin with elevated temperature is quite anomalous (16, 28–31), and offered a degree of selectivity in resolution. In particular, note the inversion in elution order of proinsulin and its sulfonate with changing temperature from 25 to 40°C. Analogous inversion in retention is seen for insulin and B-chain S-sulfonate. Even proinsulin and insulin display small differences in temperature sensitivity under gradient elution with acetonitrile. Where at low temperature there is no apparent resolution of these two peptides, at elevated temperatures they are efficiently separated. While column longevity at elevated pH was an initial concern, it has been observed that through mobile phase pretreatment with silica, an average of greater than 400 analyses of highly impure samples can be performed at pH 8 and 45°C. Appropriate treatment of the mobile phase is most easily achieved by passage through a column (2.5 × 15 cm) containing activated silica (100–200 μm). In gradient analysis this precolumn is positioned between each solvent reservior and the respective pump, where for isocratic analysis a single column is placed between the solvent mixer and sample injector, preferably maintained at the temperature of chromatography.

The utilization of this alkaline phosphate system in the site assignment of chemically modified insulins has proven most useful. Following sulfitolysis a relative change in the retention of the A- or B-chain S-sulfonate provides a rapid identification of which peptide possesses the site of derivatization. Likewise, the inability to generate either peptide has proven hightly indicative of a insulin sulfhydryl/disulfide irreversible modification. Application of this approach of sulfitolysis followed by chain analysis to the assessment of final insulin, provides an added dimension in characterization and assurance of purity. In the generation of insulin through A- and B-chain combination the yield is highly dependent on an exact ratio of these two peptides (19). Finally and of significant importance is the ability when utilizing this chromatographic buffer to repetitively analyze fermentation samples. These samples, assayed immediately following cyanogen bromide cleavage and sulfitolysis, contain as little as 1% of the solids as the desired peptidyl S-sulfonate. The quantitation of A- and B-chain S-sulfonates is shown in Figs. 2*a* and 2*b* to be nearly linear, by measurement of peak height or peak area, in the range of 20–500 μg/mL.

Certain precautions are taken to maintain efficient chromatography of these highly impure peptides, applied directly from sulfitolysis solutions containing 7 *M* urea. All samples are initially centrifuged at a modest speed for a short period of time to remove insoluble substances. A guard column filled with an appropriate reversed-phase resin is used to diminish the irreversible binding of highly hydrophobic natural products to the analytical column. Elution of the S-sulfonate peptides is performed isocratically to maximize the resolution of the substances exhibiting similar relationships of retention to acetonitrile concentration (32), and to facilitate subsequent analysis by eliminating equilibration time. Initially, guard columns dry-packed under atmospheric pressure with pellicular C_{18} resin (30–45 μm) were employed. This approach was not successful, since operation under high pressure generated voids that markedly reduced

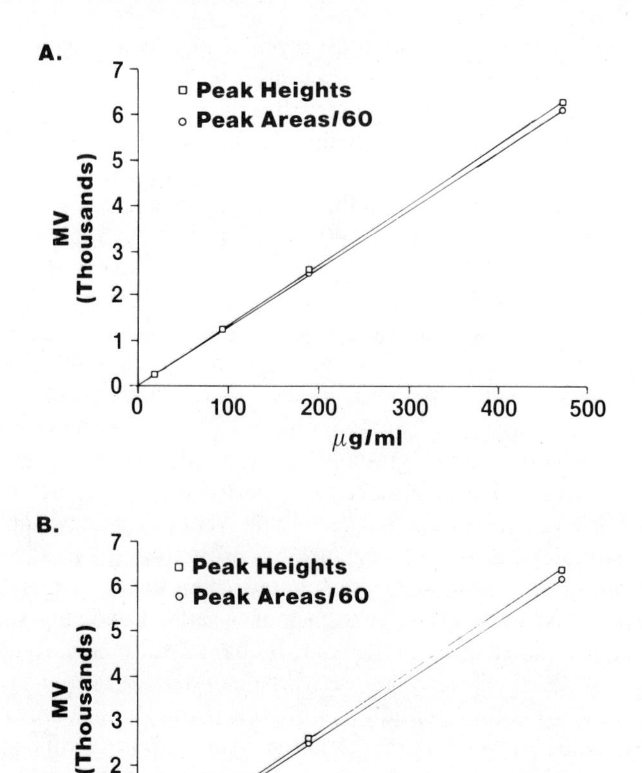

Fig. 2. (*a*) Graphical representation of the relationship between the measured peak area and peak height versus human B-chain S-sulfonate concentration. (*b*) Graphical representation of the relationship between the measured peak area and peak height versus human A-chain S-sulfonate concentration.

the efficiency of isocratic analysis. An alternative to this problem was the use of replaceable cartridge guard columns (33). This system has proven effective, but not without some reservations. In particular, the use of cartridges is of appreciable expense, and the generation of cartridge voids cannot be established without cartridge destruction. Finally, with frequent replacement it has not been uncommon for a newly inserted cartridge to prove more retentive than an aging but still usable analytical column. A simple and inexpensive solution has been achieved through the use of shortened sections (5 cm) of discarded analytical columns. Having been packed and operated at elevated pressures, these shortened columns are slow to void, and by their inherent nature are less retentive than a properly functioning analytical column. Application of this last approach to analysis of the least pure samples has shown baseline absorbance to return within 10 min of injection, at a flow rate of 1 mL/min. By appropriate adjustment of the isocratic acetonitrile concentration, elution of the desired peptide is routinely set at 15–30 min following injection. Under these chromatographic conditions, more than 50 samples of a highly impure nature can be analyzed in a single day through automated injection.

The high degree of resolution achievable through reversed-phase insulin analysis has been exemplified by the separation of insulins from different species in low-pH buffers (6–18). A comparable separation of five insulins in phosphate buffer, pH 7, is shown in Fig. 3*a*. Within a 10 min analysis all of the differing insulins are efficiently resolved. A more difficult test of chromatographic performance has proven to be the separation of human insulin analogs that differ in structure only at the NH_2-terminal residues. The chromatography of insulins separately modified by carbamylation, acylation with formic acid, and isopropyl alkylation at the respective A- and B-chain NH_2-terminal amines was studied. With the exception of the N^aisopropyl-Bl derivative which elutes at 45 min, the separation of the remaining derivatives and insulin is shown in Fig. 3*b*. Respective modification of the A and B chains with identical substituents yielded derivatives that were easily separated. In particular, where the isopropyl alkylation of A chain increased the retention of insulin by approximately 30%, the same substituent on the B chain results in more than a doubling of elution time. In similar fashion the carbamylation and formylation of the B chain has a more pronounced effect on the chromatographic retention than these same modifications of the A chain, suggesting an enhanced degree of solvent exposure for this site (17). The resolution of the insulins modified, respectively, on a single chain by formylation or carbamylation can be quite difficult. While the A-chain derivatives exhibit near baseline separation in 15 min, the analogous B-chain derivatives are only half-height resolved in the same time period. Improvements in separation of this partially separated pair can be achieved through reductions in temperature, flow rate, or acetonitrile concentration or all three.

The identification and development of this alkaline phosphate buffer for reversed-phase analysis of insulin and related peptides facilitated the biosynthetic proinsulin and insulin process development through the rapid analysis of highly impure intermediates. The utilization of elevated analysis temperatures offers an added dimension in resolution while serving in a practical sense to lower operating pressures and thereby extend the life of chromatographic columns and hardware. In the assessment of final insulin purity alkaline analysis has complemented acidic reversed-phase analysis and peptidyl mapping strategies by offering an additional approach in chromatographic characterization.

ACKNOWLEDGEMENT

The support of Lilly Laboratories is gratefully acknowledged. The authors are grateful for the assistance provided by Ms. Gale Kautzman in the preparation of this manuscript.

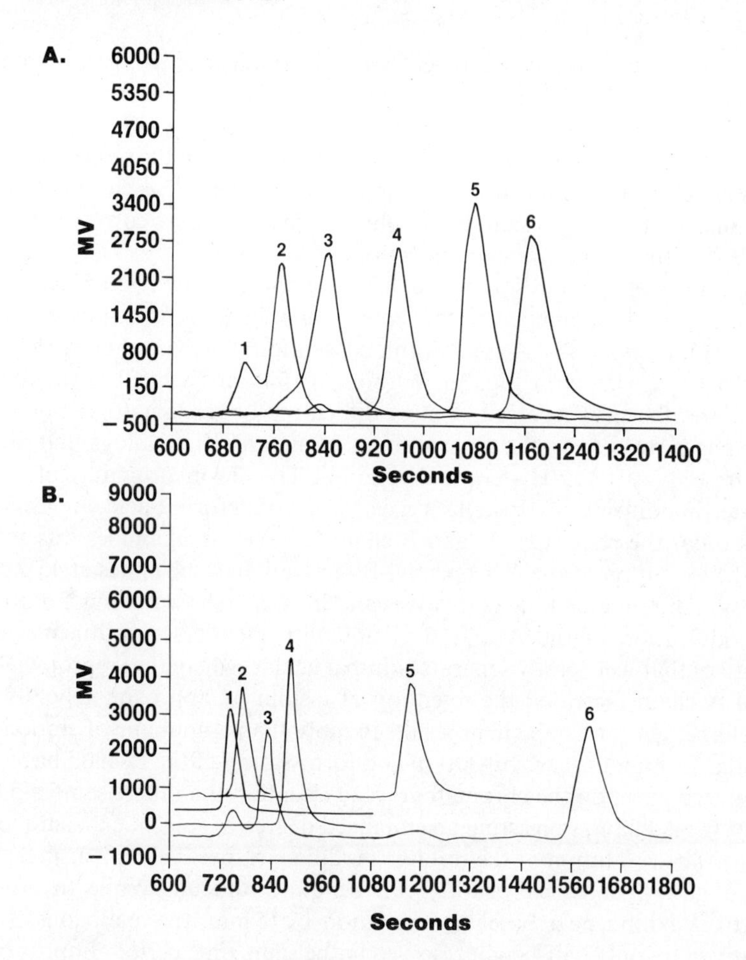

Fig. 3. (*a*) Isocratic separation of insulins from different species in 0.1 *M* $(NH_4)_2HPO_4$, pH 7.0/30% CH_3CN at 1 mL/min and 55°C. 1: Monodesamido bovine insulin, Asp-A21; 2: bovine insulin; 3: ovine insulin; 4: Rabbit insulin; 5: porcine insulin; 6: human insulin. (*b*) Separation of insulin NH_2-terminal analogs under identical conditions as in (*a*) 1: N^acarbamoyl-Bl insulin; 2: N^aformyl-Bl insulin; 3: N^aformyl-Al insulin 4: N^acarbamoyl-Al insulin; 5: human insulin; 6: N^aisopropyl-Al insulin. Not shown is N^aisoproyl-Bl insulin, which elutes at 45 min (2700 sec).

REFERENCES

1. F.G. Banting, C.H. Best, J.B. Collip, J. Hepburn, J.J.R. MacLeod, and E.C. Noble, Trans. Roy. Soc. Can., *16*, Sect. V, 1–18 (1922).
2. I.S. Johnson, Science, *219*, 632–637 (1983).
3. B.H. Frank, J.M. Pettee, R.E. Zimmerman, and P.J. Burck, Proc. Am. Pept. Symp., 7th, 1981, pp. 729–738.

4. R.M. Bergenstal, R.M. Cohen, E. Lever, K. Polonsky, J. Jaspan, P.M. Blix, R. Revers, J.M. Olefsky, O. Kolterman, K. Steiner, A. Cherrington, B. Frank, J. Galloway, and A.H. Rubenstein, J. Clin. Endo. Metabol., *58*, 973–979 (1984).
5. R.R. Revers, R. Henry, L. Schmeiser, O. Kolterman, R. Cohen, R. Bergenstal, K. Polonsky, J. Jaspan, A. Rubenstein, B. Frank, J. Galloway, and J.M. Olefsky, Diabetes, *33*, 762–770 (1984).
6. A. Dinner, and L. Lorenz, Anal Chem., *51*, 1872–1873 (1979).
7. U. Damgaard, and J. Markussen, Horm. Metab. Res., *11*, 580–581 (1979).
8. G. Szepesi, and X. Gazdag, J. Chrom., *218*, 597–602 (1981).
9. Z. Varga-Puchony, E. Hites-Papp, J. Hlavay, and G. Vigh, Hungarian J. Chem. Veszprem, *9*, 339–346 (1981).
10. Y. Pocker, S.B. Biswas, J. Liq. Chrom., *5*, 1–14 (1982).
11. J. Rivier, and R. McClintock, J. Chrom., *268*, 112–119 (1983).
12. A. McLeod, and S.P. Wood, J. Chrom., *285*, 319–331 (1984).
13. K. Hayakawa, and H. Tanaka, J. Chrom., *312*, 476–481 (1984).
14. S. Seino, A. Funakoshi, Z. Zhe Fu, and A. Vinik, Diabetes, *34*, 1–7 (1985).
15. D.J. Smith, R.M. Venable, and J. Collins, J. Chrom. Sci., *23*, 81–88 (1985).
16. R.D. DiMarchi, and H.B. Long, Proc. Natl. Acad. Sci. USA (to be published).
17. E.P. Kroeff, and R.E. Chance, in *Hormone Drugs*, The United States Pharmacopeial Convention, 1982, pp. 148–162.
18. R.E. Chance, E.P. Kroeff, J.A. Hoffman, and B.H. Frank, Diabetes Care, *4*, 147–154 (1981).
19. R.E. Chance, J.A. Hoffmann, E.P. Kroeff, M.G. Johnson, E.W. Schirmer, W.W. Bromer, M.J. Ross, and R. Wetzel, Proc. Am Pept. Symp., 7th, 1981, pp. 721–728.
20. F. Regnier, in *Methods in Enzymology*, Vol. 91, C.H.W. Hirs, and S.N. Timasheff, Eds., Academic Press, New York, 1983, pp. 137–190.
21. J.D. Pearson, W.C. Mahoney, M.A. Hermodson, and F.E. Regnier, J. Chrom., *207*, 325–332 (1981).
22. R.V. Lewis, and D. DeWald, J. Liq. Chrom., *5*, 1367–1374 (1982).
23. M.J. O'Hare, M.W. Capp, E.C. Nice, N.H.C. Cooke, and B.G. Archer, Anal. Biochem., *126*, 17–28 (1982).
24. J.D. Pearson, N.T. Lin, and F.E. REgnier, Anal. Biochem., *124*, 217–230 (1982).
25. K.J. Wilson, E.V. Wieringen, S. Klauser, and M.W. Berchtold, J. Chrom., *237*, 407–416 (1982).
26. N.H.C. Cooke, B.G. Archer, M.J. O'Hare, E.C. Nice, and M. Capp, J. Chrom., *255*, 115–123 (1983).
27. Y. Kato, T. Kitamura, and T. Hashimoto, J. Chrom., *333*, 93–106 (1985).
28. W.C. Mahoney, and M.A. Hermodson, J. Biol. Chem., *255*, 11199–11203 (1980).
29. K.J. Wilson, A. Honegger, R.P. Stotzel, and G.J. Hughes, Biochem. J., *199*, 31–41 (1981).
30. R.A. Barford, B.J. Sliwinski, A.C. Breyer, and H.L. Rothbart, J. Chrom., *235*, 281–288 (1982).
31. K.A. Cohen, K. Schellenberg, K. Beneder, B.L. Karger, B. Grego, and M.T.W. Hearn, Anal. Biochem., *140*, 223–235 (1984).
32. B. Grego, and M.T.W. Hearn, Chromatographia, *14*, 589–592 (1981).

CHAPTER 9

RP-HPLC of Biosynthetic and Hypophyseal Human Growth Hormone

THORKILD CHRISTENSEN, JAN JØRN HANSEN, HANS HOLMEGAARD SØRENSEN, and JOHANNES THOMSEN

Nordisk Gentofte A/S, 1 Niels Steensensvej, DK-2820 Gentofte, Denmark

1. INTRODUCTION

The major component of human growth hormone (hGH) contains 191 amino acid residues in a single protein chain with a molecular weight of approximately 22K (1) (Fig. 1). Several isoforms of hGH have been described and characterized (2), including a 20K variant that differs from 22K hGH by deletion of amino acid residues 32–46.

RP-HPLC is a versatile and efficient method for the determination of the identity and purity of peptides and proteins. However, relatively few RP-HPLC investigations appear to have been carried out on growth hormone. Possibly, this is because most RP-HPLC studies have emphasized silica packing materials with a pore size of 60–100 Å and a C_{18} bonded phase, using acetonitrile as organic modifier. We find such systems unsuitable for RP-HPLC analysis of hGH.

In 1981 Nice et al. (3) made a comparison of short-chain and ultrashort-chain alkylsilane-bonded silicas for RP-HPLC of a number of proteins, including bovine growth hormone (bGH). Using acetonitrile as organic modifier at pH 2.1, the retention time of bGH increased with increasing chain length from C_1 to C_8. Recovery was optimal at C_3, and it was suggested that further improvements in chromatographic properties might be obtained by using wide-pore packing materials with a pore size of 300–500 Å and propanol as organic modifier (4).

In 1982 Kohr et al. (5) demonstrated RP-HPLC of biosynthetic methionyl-hGH. Elution was done with a linear gradient between 0.1% trifluoroacetic acid (TFA) in water and 0.1% TFA in 1-propanol. The column, Synchropak RP-P,

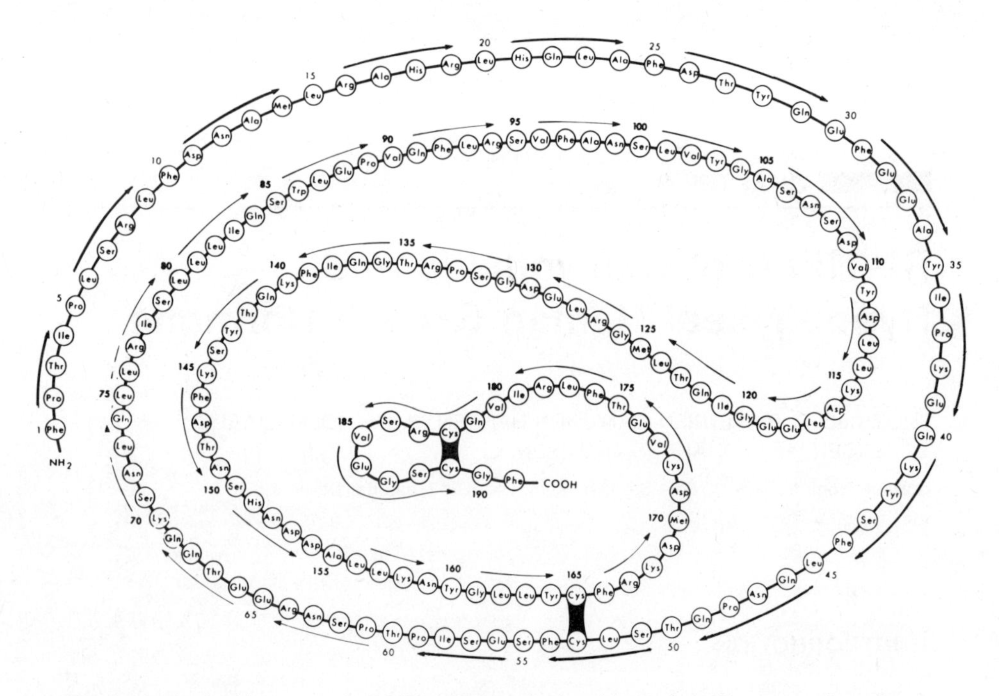

Fig. 1. Amino acid sequence of 22K hGH as depicted in Ref. 1 (Courtesy of professor C.H. Li). Note that residues 74, 107, and 109 should be Glu, Asp, and Asn, respectively.

contained a wide-pore packing material with a pore size of 300 Å and, presumably, a C_{18} bonded phase. No information was given with respect to recovery and reproducibility.

In 1983 Baldwin et al. (6) described RP-HPLC of phosphorylated hGH on a C_4 column (LiChrospher) with a particle size of 10 μm and 500 Å pore diameter. The chromatography was carried out in 0.1 *M* ammonium bicarbonate and a mixture of acetonitrile and 2-propanol was used as organic modifier. A similar system was described in 1984 by the same authors (7) and stated to give high recoveries (95–97%) of native and S-carboxymethylated hGH. It was further mentioned that the 20K hGH variant could not be readily separated from 22K hGH by RP-HPLC under a variety of conditions.

In 1984 Grego and Hearn (8) compared different RP-HPLC systems with respect to resolution and recovery of a number of proteins, including several pituitary hormones. A system using 15 m*M* orthophosphoric acid and acetonitrile as organic modifier was compared with the systems described previously (6, 7). On μBondapak C_{18} columns the recovery of radioiodinated hGH from the ammonium bicarbonate system was superior to the recovery obtained with the orthophosphoric acid system. However, recoveries appeared to be load related and sample loads in the picogram range gave considerably lower recoveries than did sample loads on the micro- to milligram scale.

Also in 1984, Chang et al. (9) showed that hGH isolated from pituitary tumors of acromegalic patients eluted at the same position as a reference hypophyseal hGH when subjected to RP-HPLC. These authors used an RP-18 column and elution was performed with a linear gradient of acetonitrile in 0.1% TFA.

In 1985 Patience and Rees (10) described an additional RP-HPLC system with TFA in the mobile phase. An Ultrapore RPSC column containing a 5 μm column packing material with 300 Å pores and a C_3 bonded phase was used, and elution was performed with a 1-propanol gradient in 0.1% TFA. RP-HPLC analysis of fractions from gel filtration of a pituitary extract showed that both the dimer fractions and the aggregate fractions contained considerable amounts of monomeric hGH. The monomer could be separated by RP-HPLC into two immunoreactive peaks, of which the more abundant was considered to be 22K hGH. The other was suggested to be a chemically modified form of 22K hGH.

Tryptic digestion of hGH has been carried out in several cases and the fragments separated by RP-HPLC on C_{18} or alkylphenyl columns (6, 11, 12). RP-HPLC of tryptic digests can be used to demonstrate the correct position of disulfide bridges in proteins produced by recombinant DNA technology by comparing the tryptic digestion maps (fingerprints) for the biosynthetic and the authentic native protein. In this way Kohr et al. (5) compared biosynthetic methionyl-hGH with authentic pituitary-derived hGH. Only one pair of non-coincident peaks was found and shown to be caused by the N-terminal methionine residue.

As part of our studies on the preparation of biosynthetic hGH, we have developed an RP-HPLC analysis for biosynthetic and hypophyseal hGH. The procedure uses a wide-pore column with 300 Å pores and a C_4 bonded phase. Elution is performed with a convex gradient of 1-propanol in 0.05% TFA. These conditions are similar to those described by Patience and Rees (10). Furthermore, we present a comparison between the tryptic digestion maps of authentic 22K hGH isolated from pituitaries and our biosynthetic hGH (13) with an amino acid sequence identical to that of 22K hGH.

2. MATERIALS AND METHODS

Three preparations of hGH, all formulated in sodium bicarbonate–glycine–mannitol buffer, were used in this investigation. Two of the preparations contained pituitary-derived hGH, either as a commercial preparation (Nanormon^R, Nordisk Gentofte) (P-hGH) or as a further purified fraction, essentially consisting of 22K P-hGH. The third preparation contained biosynthetic hGH (B-hGH) prepared by recombinant DNA technology (13) with an amino acid sequence identical to that of authentic 22K hGH (Fig. 1). Before RP-HPLC, lyophilized samples of P-hGH and B-hGH were dissolved in water or buffer A and added buffer B to the same composition as used for (initial) elution.

RP-HPLC of P-hGH and B-hGH was performed on a 250 × 4.6 mm Vydac 214TP54 column connected to Waters instrumentation, consisting of two 6000A pumps, a 710B WISP autosampler, and a 720 System Controller with a 730 Data Module. Detection was performed with a Pye Unicam LC-UV detector at 277 nm, unless otherwise noted. The column was maintained at 40°C in a thermostatted column heater (Mikrolaboratoriet, Århus) and eluted at 1.0 mL/min with a 1-propanol gradient constructed with the two buffers A and B:

A: 0.05% TFA in 10 m*M* aqueous ammonium sulfate.

B: 0.05% TFA in 1-propanol mixed with buffer A in the ratio of 10:1.

The gradient used was set up on the Waters 720 System Controller as follows:

Initial: 1.0 mL/min, 80% A, 20% B
15 min: 1.0 mL/min, 65% A, 35% B, gradient No. 7
25 min: 1.0 mL/min, 65% A, 35% B, gradient No. 6
35 min: 1.0 mL/min, 80% A, 20% B, gradient No. 6
45 min: 1.0 mL/min, 80% A, 20% B, gradient No. 6

Usually this was followed by a 5 min equilibration delay before the next injection.

Peak area determinations were performed by the standard "dropped-vertical-lines" procedure of the Waters 730 Data Module.

Isocratic elution was performed at 1.0 mL/min with an A:B/65:35 ratio of the buffers described previously. Unless otherwise noted, the column was maintained at 40°C and detection performed at 277 nm. These conditons albeit at 20°C, were also used for determination of recovery where injections of P-hGH on the column were compared to similar injections through 2000 × 0.8 mm of fluoroplast tubing.

Tryptic peptide mapping of hGH was carried out according to published procedures (11, 12), using DPCC treated bovine trypsin (Sigma T-1005). To 1.0 mL of a 1 mg/mL dialyzed solution of B-hGH or 22K P-hGH in 50 m*M* Tris-HCl, containing 2 m*M* $CaCl_2$ at pH 7.8, was added 10 μL 1 mg/mL trypsin solution in 2 m*M* $CaCl_2$ with 1 m*M* HCl. The digests were kept at 37°C for 6 hr after which 25 μL or 100 μL aliquots were subjected to analytical or preparative RP-HPLC.

These RP-HPLC separations were performed on a 250 × 4 mm LiChrosorb RP-18 column with a 5 μm particle size (Merck 50333) connected to Waters equipment consisting of two 6000A pumps, a 710B WISP autosampler, a 480 LC spectrophotometer, and an 840 Data and Chromatography Control Station. The column was maintaind at 45°C and eluted at 1.0 mL/min with a linear gradient of acetonitrile using the following buffers:

A: 0.05% TFA in water

B: 0.05% TFA in 50 vol% acetonitrile in water

and gradient:

Initial: 1.0 mL/min, 100% A, 0% B
60 min: 1.0 mL/min, 0% A, 100% B, gradient No. 6
70 min: 1.0 mL/min, 0% A, 100% B, gradient No. 6
75 min: 1.0 mL/min, 100% A, 0% B, gradient No. 6

Detection was performed at 215 nm and peaks appearing in the chromatograms were collected and identified by amino acid analysis (14) and gas-phase sequencing (15).

3. RESULTS AND DISCUSSION

3.1. RP-HPLC of hGH

The RP-HPLC analysis of hGH can be performed with detection at both 213 and 277 nm. Detection at 213 nm is inherently more sensitive, but the gradient system used here gives a strongly sloping baseline at this wavelength. Therefore, detection at 277 nm was chosen for our analytical procedure.

Figure 2 shows the chromatograms obtained after injections of 250 μg of pituitary-derived hGH (P-hGH) and biosynthetic hGH (B-hGH), respectively. The gradient system in both cases shows the presence of several small impurity peaks in addition to the dominating hGH peak. These impurity peaks are common to both samples and are likely to be hGH degradation products. Apart from a putative dimer eluting approximately 2 min after the hGH peak, none of the impurity peaks amounts to more than 0.2% of the area of the hGH peak.

The RP-HPLC procedure is also applicable to crude mixtures containing growth hormone. Figure 3 shows gradient elution of a pituitary extract, and the arrow indicates the hGH peak. The identity of the other peaks in the chromatogram is unknown.

3.2. Linearity

The detector response linearity was investigated by injecting varying volumes of a fixed concentration of hGH. Figure 4 shows that an approximately linear response is obtained within the range of 33–500 μg B-hGH. Other series of injections (not shown) at lower protein loads showed that linearity is upheld down to approximately 1 μg hGH.

During these variations in protein load it was observed that increasing the amount of protein gives a slight but steady and reproducible decrease in the retention time for the hGH peak. This effect within a series can be exemplified with a 20 μg injection of P-hGH giving a retention time for the hGH peak at 24.2 min, decreasing to 23.7 min for a 400 μg injection.

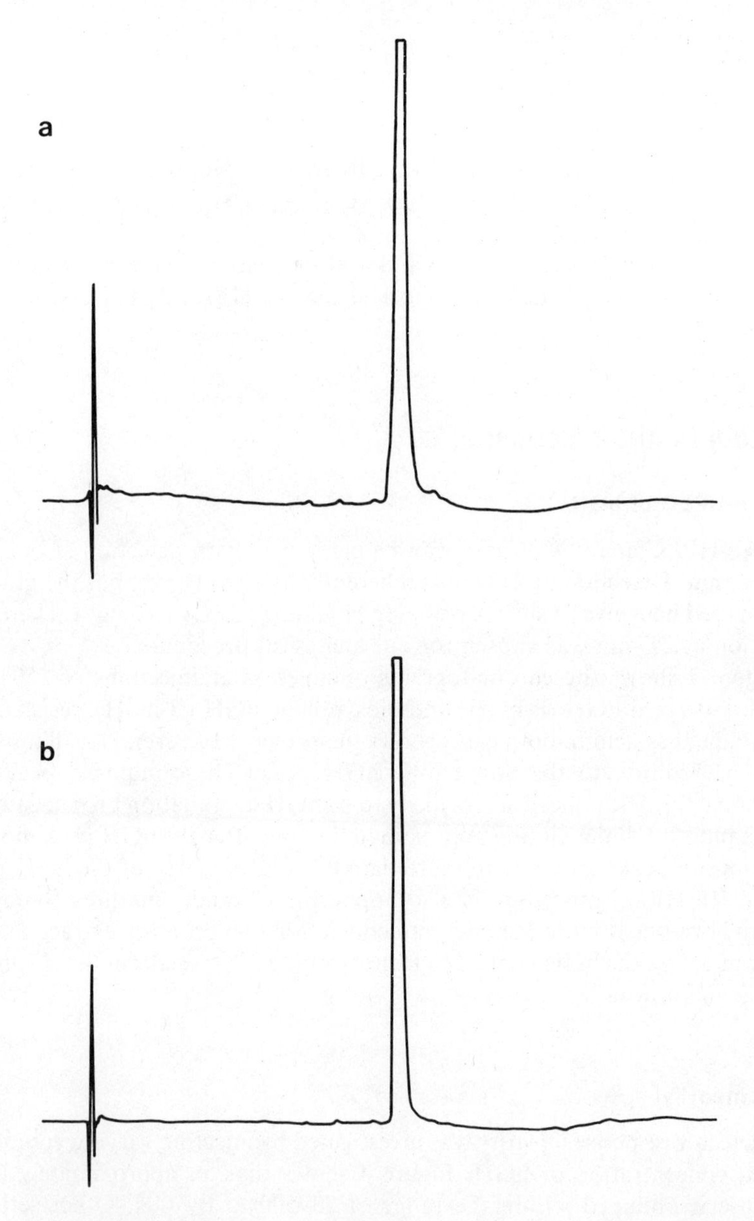

Fig. 2. Gradient elution of (*a*) 250 μg P-hGH and (*b*) 250 μg B-hGH.

3.3. Detection Limit

The detection limit may be defined as the amount of protein corresponding to an area of three times the standard deviation (SD) obtained by repeated injections of a protein load five times as high as the estimated minimum detectable

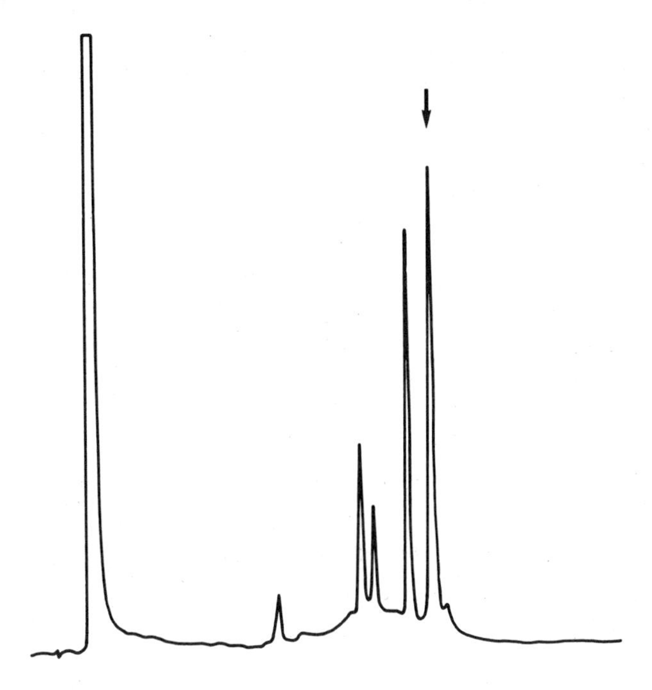

Fig. 3. Gradient elution of a crude pituitary extract. Pituitary glands were extracted 3.5 hr with 0.1 *M* sodium phosphate, pH 8.7. Solid urea was added to the supernatant to a concentration of 7 *M* and, after standing overnight, the extract was filtered through a 0.45 μm filter.

amount. A 0.8 μg injection of P-hGH gives a barely discernible peak for hGH, while no detectable hGH peak is obtained with a 0.32 μg injection. Thus, the minimum detectable amount of hGH in this procedure is not much below 0.8 μg.

Eight 4 μg injections of each of two different solutions of 0.4 mg/mL P-hGH gave a mean of 811 and 902 area units and an SD of 66 and 76 area units, respectively. Thus, both series of injections gave an SD of approximately 70 area units, corresponding (3 $\times$ SD) to a detection limit of 0.7 μg.

3.4. Precision and Reproducibility

The area variation within a series of consecutive 4 μg injections is relatively large owing to the low amount of protein injected. Thus, the previously listed SD amounts to approximately 8% of the mean area. In order to determine the precision (within-day-variation) with a higher amount of protein, a series of six consecutive 40 μg injections from a single solution of P-hGH was carried out, giving a coefficient of variation (CV) of 1.7%.

Six injections of different solutions of 0.8 mg/mL P-hGH, separately diluted from a common stock solution and injected (50 μL) on different days, were used

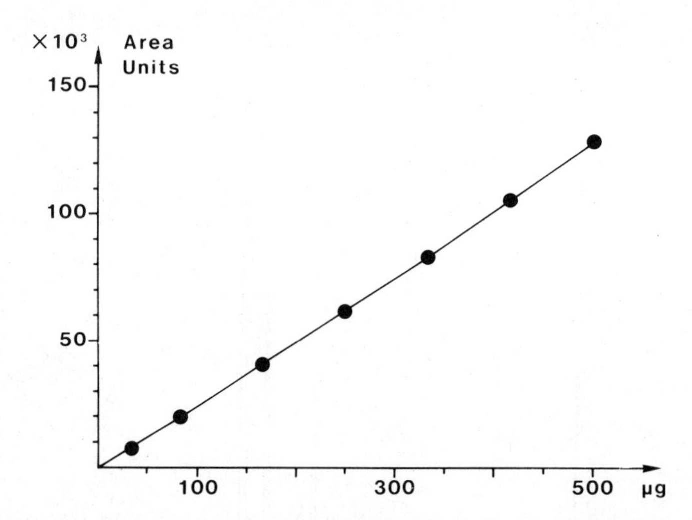

Fig. 4. Detector response at 277 nm obtained by injecting 10–150 μL of 3.3 mg/mL B-hGH.

to determine the reproducibility (day-to-day variation) of the procedure. Two different preparations of eluants were used for the first two and the last four injections, respectively, without significantly affecting area or retention time. The CV obtained for this day-to-day variation was 1.1%, which is approximately the same as found for the within-day variation, showing that any dilution variations are small compared to variations due to the RP-HPLC procedure.

3.5. "Column Memory"

An injection of 400 μg P-hGH and the two subsequent injections of 0 μL and 100 μL eluant A:B/80:20, respectively, gave the chromatograms shown in Fig. 5. Although the latter two injections did not contain any hGH, a small peak eluting with the gradient at the position of hGH is clearly detectable in the first of these injections.

This phenomenon has been generally observed whenever a blank gradient is run after the gradient elution of a hGH sample. The area of the ghost peak usually corresponds to 2–3 μg hGH, but has varied between 0.8 and 5 μg. That these peaks are due to "column memory" and not to carryover from the injection system is supported by the previously mentioned experiment and by the fact that ghost peaks never appeared after isocratic runs. Thus, injections of small amounts of hGH should be preceded by a blank gradient in order to give reproducible quantification.

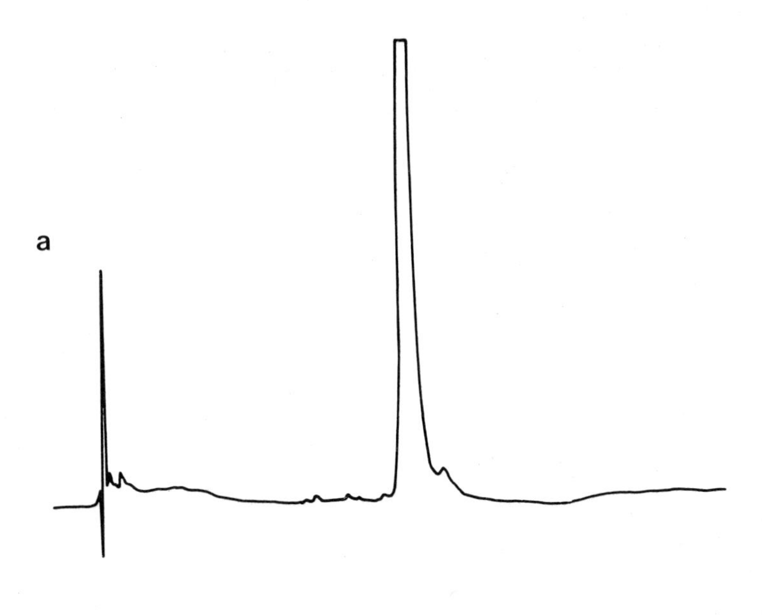

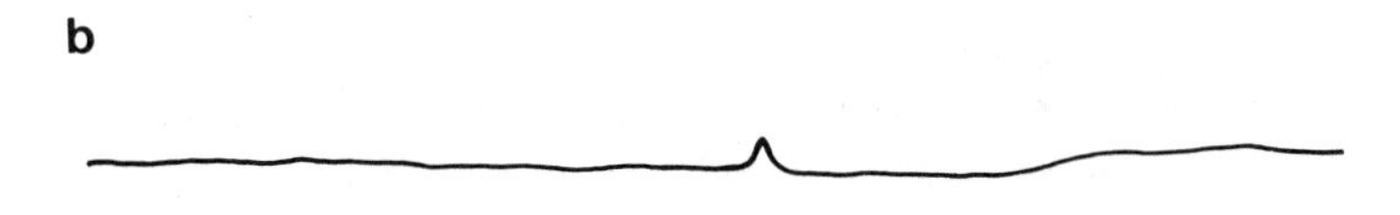

Fig. 5. Gradient elution of (*a*) 400 μg P-hGH and the two subsequent injections of (*b*) 0 μL and (*c*) 100 μL eluant A:B/80:20.

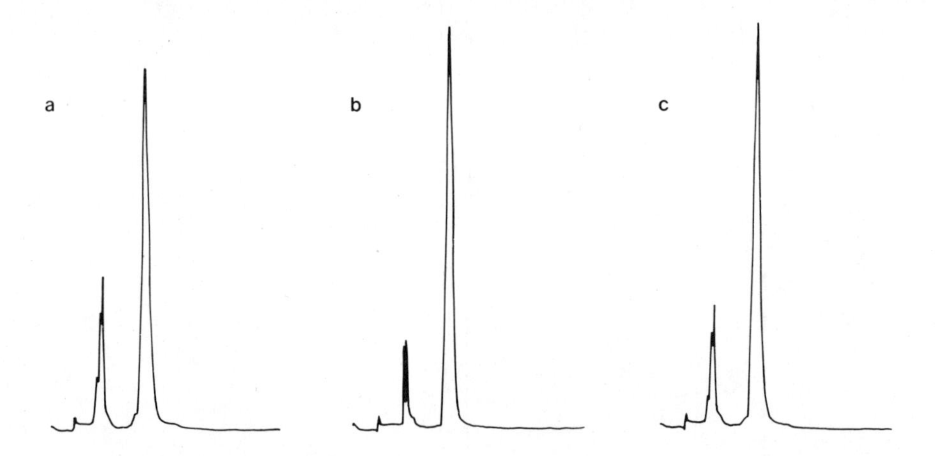

Fig. 6. Isocratic elution of (*a*) 70 μg P-hGH, (*b*) 83 μg B-hGH, and (*c*) a (1:1 mixture) of P-hGH and B-hGH.

3.6. Isocratic Elution

Figure 6 shows isocratic elution of P-hGH and B-hGH with 35% B at 40°C. Both still elute as essentially one peak and Fig. 6*c* shows that the two forms of hGH coelute. The efficiency of the column is a relatively low ~3000 plates per meter for the hGH peak, but B-hGH has consistently shown a slightly sharper peak than P-hGH under various isocratic conditions. The systems described here do not resolve 20K hGH and 22K hGH, nor do they separate hGH from its deamidated forms. This lack of resolution appears to be common for all described RP-HPLC separations of hGH. B-hGH contains no 20K fraction and considerably smaller amounts of deamidated forms than P-hGH. This might explain the small difference in column efficiency observed betweeen B-hGH and P-hGH.

3.7. Recovery

The recovery of hGH was determined at 20°C under isocratic elution with 35% B. The chromatogram obtained under these conditions by injection of 78 μg P-hGH on the column is shown in Fig. 7. Compared to Fig. 6*a* the hGH peak, is retained considerably more owing to the lower temperature. The UV trace from a corresponding injection of 78 μg P-hGH after the column has been substituted by 2000 × 0.8 mm fluoroplast tubing is shown in Fig. 7*b*. This set of injections was performed in triplicate and supplemented by blank injections of 50 μL eluant A:B/65:35 (Figs. 7*c* and 7*d*). The total peak area obtained on the column (including the void volume peaks due to the additives of P-hGH) relative to the peak area obtained with the fluoroplast tubing, when corrected for the blanks, gives a recovery of 90%. The area of the hGH peak in Fig. 7*a*

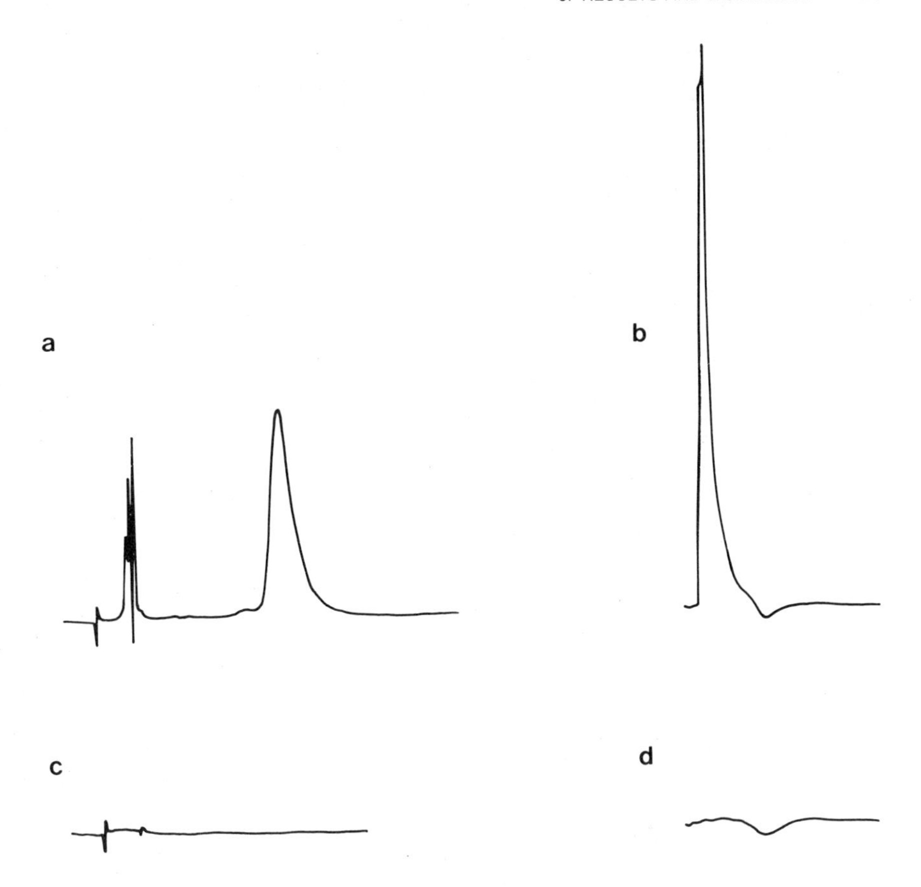

Fig. 7. Recovery of hGH from 78 μg P-hGH injected through (*a*) column and (*b*) through 2000 × 0.8 mm fluoroplast tubing. Also shown are the UV traces obtained at 277 nm from 50 μL blank injections of eluant A:B/65:35 through (*c*) the column and (*d*) the fluoroplast tubing.

agrees within 2–3% with the area obtained in the gradient system for a similar injection of 78 μg P-hGH. Therefore, we consider 90% also to be a reasonable estimate of the recovery in the gradient system.

3.8. Tryptic Peptide Mapping

RP-HPLC of tryptic digests of B-hGH and 22K P-hGH are shown in Fig. 8. In both cases at least 21 fragments are obtained, all of which have been identified by amino acid composition and gas phase sequencing as listed in Table 1. The essentially identical peptide maps strongly support the identity of B-hGH and 22K P-hGH. In particular, the identical position and size in the two chromatograms of peaks 7, 17, and 20, containing disulfide-linked peptides, confirm the correct native position of the disulfide bridges in B-hGH. A similar peptide

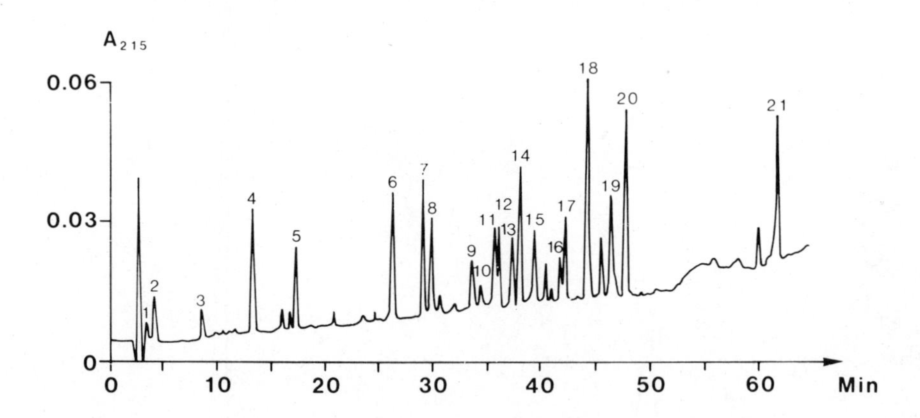

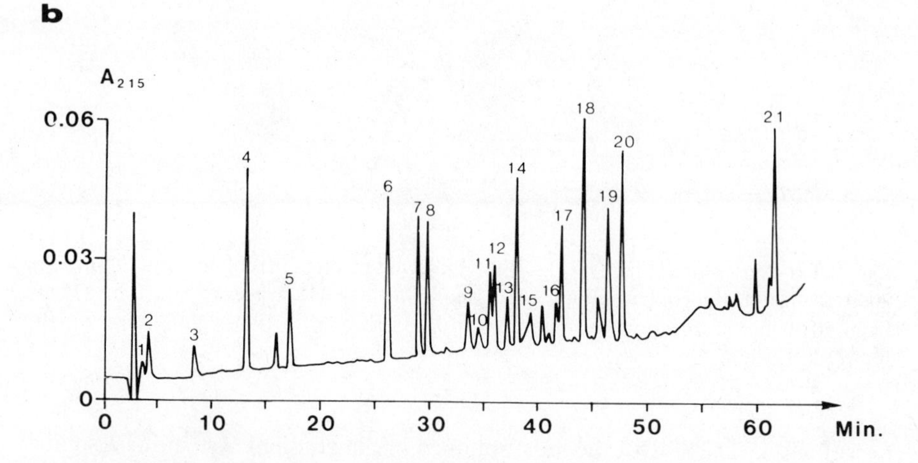

Fig. 8. Tryptic peptide mapping of (*a*) 22K P-hGH and (*b*) B-hGH.

mapping of methionyl-hGH (not shown) gave a chromatogram where the retention time of peak 14 had increased by 2.7 min. Apart from this change, the chromatogram was identical to those obtained with B-hGH and 22K P-hGH.

In contrast to earlier reported sequences (1, 7, 12), we find for both B-hGH and 22K P-hGH that the residues at position 74, 107, and 109 (Fig. 1) should be Glu, Asp, and Asn, respectively. This result has been predicted from hGH cDNA sequencing (16, 17) and was also found by tryptic peptide mapping of biosynthetic methionyl-hGH (5).

Table 1. Identity of Tryptic Fragments Appearing in Fig. 8

Peak	Trypsin Fragment[a]	Residues (See Fig. 1)
1	5	39 – 41
2	3	17 – 19
3	7	65 – 70
4	14	141 – 145
5	12	128 – 134
6	13	135 – 140
7	20/21	(179 – 183) / (184 – 191)
8	15	146 – 158
9	8	71 – 77
10	17–18–19	168 – 178
11	18–19	169 – 178
12	2	9 – 16
13	16A[b]	159 – 164
14	1	1 – 8
15	10A[b]	95 – 111
16	11	116 – 127
17	6/16B[b]	(42 – 64) / (165 – 167)
18	4	20 – 38
19	10	95 – 115
20	6/16	(42 – 64) / (159 – 167)
21	9	78 – 94

[a] Fragment 17, Lys(168), elutes in the void volume.
[b] Chymotryptic fragments.

Note added in proof: Since the completion of the manuscript, RP- HPLC separations have been reported by R.M. Riggin, G.K. Dorulla, and D.J. Miner, Anal. Biochem., *167*, 199–209 (1987) and by B.S. Welinder, H.H. Sørensen, and B. Hansen, J. Chromatogr., *398*, 309–316 (1987).

REFERENCES

1. C.H. Li, Mol. Cell. Biochem., *46*, 31–41 (1982).
2. R.K. Chawla, J.S. Parks, and D. Rudman, Ann. Rev. Med., *34*, 519–547 (1983).
3. E.C. Nice, M.W. Capp, N. Cooke, and M.J. O'Hare, J. Chromatogr., *218*, 569–580 (1981).
4. M. Rubinstein, Anal. Biochem., *98*, 1–7 (1979).
5. W.J. Kohr, R. Keck, and R.N. Harkins, Anal. Biochem., *122*, 348–359 (1982).
6. G.S. Baldwin, B. Grego, M.T.W. Hearn, J.A. Knessel, F.J. Morgan, and R.J. Simpson, Proc. Nat. Acad. Sci. USA, *80*, 5276–5280 (1983).
7. B. Grego, G.S. Baldwin, J.A. Knessel, R.J. Simpson, F.J. Morgan, and M.T.W. Hearn, J. Chromatogr., *297*, 21–29 (1984).
8. B. Grego and M.T.W. Hearn, J. Chromatogr., *336*, 25–40 (1984).

9. W.-C. Chang, A.L.-Y. Shen, C.-K. Chou, and L.-T. Ho, Int. J. Peptide Protein Res., *23*, 637–641 (1984).
10. R.L. Patience and L.H. Rees, J. Chromatogr., *324*, 385–393 (1985).
11. M.T.W. Hearn, B. Grego, and G.E. Chapman, J. Liq. Chromatogr., *6*, 215–228 (1983).
12. B. Grego, F. Lambrou, and M.T.W. Hearn, J. Chromatogr., *226*, 89–103 (1983).
13. H.D. Andersen, H.-H.M. Dahl, J. Pedersen, J.W. Hansen, and T. Christensen, Bio/Technology, *5*, 161–164 (1987).
14. B.A. Bidlingmeyer, S.A. Cohen, and T.L. Tarvin, J. Chromatogr., *336*, 93–104 (1984).
15. R.M. Hewick, M.W. Hunkapiller, L.E. Hood, and W.J. Dreyer, J. Biol. Chem., *256*, 7990–7997 (1981).
16. J.A. Martial, R.A. Hallewell, J.D. Baxter, and H.M. Goodman, Science, *205*, 602–607 (1979).
17. W.G. Roskam and F. Rougeon, Nucleic Acids Research, *7*, 305–320 (1979).

CHAPTER 10

High-Performance Ion-Exchange Chromatography of Proteins: A Review of Methods and Mechanisms

MICHAEL P. HENRY

Research and Development Laboratories, J.T. Baker Inc., Phillipsburg, New Jersey

1. INTRODUCTION

1.1. History

The history of the development of ion-exchange chromatography of proteins is characterized by the now familiar technological acceleration experienced by other fields of science. It was 100 years from the discovery of the phenomenon of ion exchange by Thompson (1) to the first experiments by Margoliash (2) with synthetic ion-exchange resins to separate proteins chromatographically. Soft gels based on ionically derived cellulose have been known since 1941 [Yackel and Kenyan (3)], but Peterson and Sober (4) in 1956 were the first to separate proteins with these materials. The first commercialization by Serva Fine Biochemicals, about 1956, of celluloses bearing charged functional groups was followed by the development of other ion exchangers based on agarose, dextran, and polyacrylamide. Improvements in the physico/chemical properties and preparation of these soft gel ion-exchangers continue to this day with the relatively recent invention of chromatographic media based upon regenerated celluloses (rayons). Table 1 lists some currently available soft gel ion exchangers.

In 1976, Chang, Gooding, and Regnier (5) reported the preparation and applications of high-performance silica-based ion exchangers for protein chromatography. At that time the phenomenal growth of high-performance liquid chromatography (HPLC) in many application areas was well underway (6). Some of the methods of HPLC were applied by Chang et al. (5) to separation problems that were then solved using soft gels. It had, therefore, been 20 years from the first application of soft gel ion exchangers in protein separations to the first invention of high-performance chromatographic media for this area.

Table 1. Soft Gel Manufacturers/Marketers and Trade Names

Manufacturer	Trade Names
Pharmacia-LKB Biotechnology AB	Sephadex®, Sephacryl®, Sepharose®, Sephacel®, Trisacryl®, Ultragel®, HA-Ultrogel®, Magnogel®
Whatman, Inc.	DE, CM Cellulose
TOSOH (formerly Toyo Soda)	TOYOPEARL®
Bio-Rad Laboratories	Bio-Rex®, AG®, Cellex™, Bio-Gel®, Affi-Gel®
Cuno, Inc.	Zeta Prep
Serva Fine Biochemicals, Inc.	Servacel®
Amicon	Matrex™ Gel, Cellufine
Phoenix Chemicals	Indion®

Over the decade following the work of Chang and co-workers (5), the technique of ion-exchange HPLC of proteins developed rapidly, although its commercialization lagged behind for several reasons. The 1980s saw the first dramatic and widespread increase in the use of HPLC in protein chromatography. Chromatographic analysis of protein mixtures became a reality. High-performance gel filtration, reversed phase, strong and weak ion-exchange, hydrophobic interaction, and bioaffinity modes of chromatography were commercialized from research work originating largely in universities in Norway, Japan, and the United States. Tables 2–4 list companies who have commercialized high-performance materials for ion-exchange chromatography of proteins and peptides.

1.2. Definitions

High-performance liquid chromatography is generally characterized by the use of microparticulate (less than 20 micron) chromatographic media and advanced instrumentation. Together these achieve a high level of resolution, speed, and precision in the analysis and purification of complex protein and polypeptide mixtures.

This review of high-performance ion-exchange chromatography will include separations where components are not collected for further study (analysis) and those where they are (preparative).

Ion-exchange chromatography is defined as that separation technique where the predominant binding process involves the substitution of one mobile ionic species for another. The exchange occurs at the surface of a solid matrix bearing fixed (nonexchanged) charged groups. In true ion exchange there is no significant change in the structure of this matrix. While the predominant attractive force is ionic, other forces and mechanisms may play a role in achieving the final separation. These forces include dispersive and dipole–dipole attraction, solvophobic repulsion, and hydrogen bonding. In addition, the size-exclusion process will normally operate to achieve a slightly more rapid elution of large ions

Table 2. Commercially Available Wide Pore HPIEC Columns for Anion Exchange of Proteins and Peptides

Manufacturer	Packing	Strong	Weak	Pore Size (Å)	Particle Diameter (microns)
J.T. Baker Inc.	BAKERBOND Wide-Pore*	Yes	Yes	300	5, 15
Dupont	ZORBAX®	Yes	Yes	300	7
SynChrom	SynChropak®	Yes	Yes	300, 1000	6.5, 10
Supelco	SUPELCOSIL®	Yes	No	300	5
Pharmacia	MonoBeads®	Yes	No	approx. 800	10
BioRad	Microanalyzer™	No	Yes	No Pores	7
Rainin	Dynamax®-300A AX	No	Yes	300	12
Polymer Labs, Inc.	PL-SAX	Yes	No	1000	8, 10
Brownlee	AQUAPORE®	No	Yes	300	7
TOSOH (formerly Toyo Soda)	TSK® IEX	No	Yes	1000	5, 10
Serva	Daltosil, Si	No	Yes	300, 100	5, 10
Poly LC	PolyWAX LP	Yes	Yes	300	5, 7
Separations Industries	Nugel™	Yes	Yes	300	5, 10
Interaction	Hydrophase™	No	Yes	approx. 1000	10
Chemco	Chemcosorb	Yes	No	300	7
Amicon	Matrex™	No	Yes	450	10
Shimadzu	Shim-pack	No	Yes	300	5
Mitsubishi	MCI Gel	Yes	Yes	300	5

Table 3. Commercially Available Wide Pore HPIEC Columns for Cation Exchange of Proteins and Peptides

Manufacturer	Packing	Strong	Weak	Pore Size (Å)	Particle Diameter (microns)
J.T. Baker Inc.	BAKERBOND Wide-Pore*	Yes	Yes	300	5, 15
Dupont	ZORBAX®	Yes	Yes	300	7
Synchrom	SynChropak®	Yes	Yes	300, 1000	6.5, 10
Supelco	SUPELCOSIL®	Yes	No	300	5
Pharmacia	MonoBeads®	Yes	No	approx. 800	10
BioRad	Microanalyzer™	No	Yes	No Pores	7
Brownlee	AQUAPORE®	No	Yes	300	7
TOSOH (formerly Toyo Soda)	TSK® IEX	Yes	Yes	1000	5, 10
Serva	Daltosil, Si	No	Yes	300	3, 5, 10
Poly LC	PolyCAT A™	Yes	Yes	300	6.5
Separations Industries	Nugel™	No	Yes	300	5, 10
Chemco	Chemcosorb	Yes	No	300	7
Shimadzu	Shim-pack	No	Yes	300	5
Mitsubishi	MCI Gel	Yes	Yes	300	5

Table 4. Commercially Available Wide Pore Mixed and Multimode HPIEC Columns for Proteins and Peptides

Manufacturer	Product Name	Application Area	Mechanisms
		Mixed Mode	
J.T. Baker Inc.	BAKERBOND ABX*	Monoclonal antibodies	Weak anion and weak cation exchange
Alltech Assoc., Inc.	Alltech Mixed-Mode	Acidic peptides	Weak anion exchange and reversed phase
		Basic peptides	Weak cation exchange and reversed phase
		Multimode	
Rainin	Dynamax®-300A AX	Acidic proteins (ion exchange mode) General proteins (hydrophobic interaction mode)	Weak anion exchange and hydrophobic interaction

relative to that occurring in the absence of any steric effect. There is no consensus to the question of when a peptide becomes a protein. For the purpose of this chapter, however, it is useful to consider a peptide as an amino acid polymer with a molecular weight of about 5000 or less. Such a peptide can generally be successfully chromatographed under a wide range of pH values, ionic strengths, buffer types, organic solvents, temperatures, and matrix types. In other words, good peak shape and recoveries (mass and activity) are largely maintained regardless of the chromatographic protocols employed.

1.3. Aim and Scope

The aim of this chapter is to describe current methods in analytical and preparative high-performance ion-exchange chromatography of peptides and proteins.

2. SAMPLE CHARACTERIZATION, HANDLING AND PREPARATION

For the purposes of handling and preparing samples prior to ion-exchange HPLC, it is useful to know or measure the properties of the sample (Table 5) and the component(s) (Table 6) of interest.

Details and reviews of sample characterization, handling, and techniques are

Table 5. Important Sample Properties for Sample Preparation and Column Choice

Sample Property	Utility
Toxicity	Safety precautions and disposal procedures required
pH	Determines charge for polypeptide
Ionic strength	Usually needs to be low for efficient binding
Cell content	Filtration may be necessary
Colloid content	Ultrafiltration may be necessary
Viscosity	Potential sample application problems
Assay methods	Important factor in throughput
Matrix constituents (e.g., lipids, proteins, nucleic acids)	Source of interferences, adsorption problems, aggregation, enzyme-catalyzed degradation
Pyrogen content	May require removal or use of pyrogen-free buffers or both

described in Refs. 7–11. Selected aspects of these topics are discussed in Sections 4.1 and 4.2.

3. INSTRUMENTATION

Parris' book (12) on the requirements for suitable instrumentation for HPLC is one of several modern, thorough treatments of the subject. Detailed descriptions of several commercial liquid chromatographs have also appeared outside of the normal industrial technical marketing literature. McNair (13) has published an objective review of current instrumentation in this area. Wehr (14) has reviewed some commercially available HPLC columns suitable for ion exchange of peptides and proteins.

Table 6. Important Component Properties Useful for Column Choice

Property of Component(s) of Interest	Utility
pI (isoelectric point)	Determines choice of anion or cation exchanger
Molecular weight	Determines desirable pore size
Hydrophobicity	Extent of possible hydrophobic binding to ion exchanger
Reactivity (oxidation/reduction, catalysts)	Inhibitors may be required
Degree of aggregation	Oligomers may need to be removed; further aggregation may need to be prevented, disaggregation may be required
Concentration	May determine ease and scale of purification
Solubility	Must be soluble in ion-exchanger buffers
Stability	May determine temperature and speed of operation nessary
Microheterogeneity	May require highly resolving techniques

There are a number of aspects of the technology of high-performance liquid chromatographs that induce uncertainty in the minds of many chromatographers.

1. The sometimes perplexing variety of manufacturers of such instruments that are available. It has recently (January, 1988) been estimated (15) that 65 instrument companies compete in the field of HPLC. Distributors of columns for these instruments are suggested to number 200. There are far fewer basic manufacturers of HPLC columns, however.

2. Any high-performance liquid chromatograph can be used for peptides or proteins despite a current trend toward so-called "biocompatible" systems. The latter instruments use mobile phase delivery systems and detectors in which there are no wetted steel-based surfaces.

Pharmacia was the first company to commercialize a suitable chromatography system that contained only glass, ruby, and fluoropolymer wetted surfaces. The "FPLC" (Fast Protein Liquid Chromatography) system as it is known has all the usual components of a standard HPLC. It has a maximum pump pressure limit of 600 psi and, consequently, can only be used with low-pressure HPLC columns that contain 10 micron particles or are short (usually less than 10 cm). The cross-linked polystyrene-based MonoBead® ion-exchange columns (Mono Q, Mono S, and Mono P) were released simultaneously with the chromatograph and are widely used in the HPLC of peptides and proteins.

The Dionex 4000i, Wescan 2851, and Waters™ 650 liquid chromatographs use no glass or metal wetted components, but rely on a strong fluoropolymer to achieve pressures of 4000 psi (Dionex and Wescan) and 600 psi (Waters). Perkin-Elmer, Scientific Systems, Pharmacia-LKB, and Shimadzu have titanium components instead of stainless steel. With titanium there is far less leaching of ferrous and heavy metals into the mobile phase, where they may interact with sensitive proteins or the surface or the substructure of the column packing material. Several companies have introduced titanium HPLC column components.

3. The use of high salt concentrations [up to 3 M $(NH_4)_2SO_4$ for example] and alkali metal halides in mobile phases should be attempted with confidence. However, the seals in conventional piston pumps will in general begin to leak in a matter of days as the evaporating buffers leave abrasive salt crystals on the piston. A number of companies have attempted to overcome this problem by using continuously washed or "floating" pistons. The former are designed to prevent crystal formation. The latter induce less stress in the seal and consequently extend its life. Buffers in general should be thoroughly filtered through 0.2 micron filters before use to prevent wear on seals. The salt solutions should be completely flushed from the interior of the instrument when not in use. Manufacturer instructions concerning the storage conditions for the columns should be followed.

Table 7. Five Major Steps in Column Selection

1.	Characterize sample [matrix + component(s) of interest]
2.	Characterize component(s) of interest
3.	Choose type of ion exchanger/pH range
4.	Choose column configuration
5.	Examine quality, cost, service

4. BONDED PHASE SELECTION

A complete listing of all companies that provide bonded phases that can be used for peptide and protein chromatography is beyond the scope of this chapter. However, the sources of those chromatographic packings that contain wide pores are relatively few in number. A list of these is given in Tables 2, 3, and 4.

A general approach to be used in choosing a bonded phase for the ion-exchange HPLC of peptides and proteins is given in Table 7. A detailed discussion of each of the major steps follows.

4.1. Sample Characterization (see also Table 5)

The properties of the sample that are most important in bonded phase selection for ion-exchange HPLC are as follows: pH, ionic strength, pyrogen content, indicator dyes, volume, matrix constituents (lipids, proteins, nucleic acids, for example).

The pH and ionic strength will normally need to be adjusted before ion-exchange chromatography, principally in order to achieve binding. Unwanted pyrogens may be removed by dialysis after chromatography or by the choice of a bonded phase that binds the component(s) of interest but does not bind the pyrogen. Dyes (such as phenol red) that indicate pH may be added to biological samples. These must be removed during or after chromatography. The sample volume, when taken in conjunction with the total mass of bindable material, the capacity of the bonded phase (see Section 4.3) and whether purification or analysis is intended, will determine the size of the column. Many ion exchangers exhibit sufficient hydrophobic character to bind lipids even at low ionic strength. Immissible solvents such as LIPOCLEAN™ may be used to extract nonpolar constituents before chromatography and without harming the activity or solubility of the peptide or protein of interest.

Association among sample constituents and the component(s) of interest may occur in such a manner as to alter the binding and other properties of the latter. For example, a strong association between a nucleic acid and a basic peptide or protein may give misleading information concerning the nature of the species of interest. Techniques to dissociate such complexes must be applied, preferably to the sample prior to chromatography.

pK_a ⦚ pH ⦚ pI

pK_a of fixed charge group

pI of protein/peptide

Fig. 1. A schematic guide to the use of cation and anion exchangers. The operational pH of the mobile phase will in general lie between the isolectric point (pI) of the protein or peptide and the pK_a of the ion-exchanger acidic groups. The acidic groups may be $-SO_3H$ (sulfo), $-COOH$ (carboxy), $-OPO_3H_2$ (phospho), $-\overset{+}{N}H$ (primary, secondary, tertiary ammonium). In anion exchange, $pI < pK_a$, and in cation exchange, $pI > pK_a$.

4.2. Component Characterization (see also Table 6)

The properties of the peptide or protein of interest that are most important in bonded phase selection are as follows: pI, molecular weight, hydrophobicity, and reactivity.

The importance of these properties is mentioned briefly in Table 6 and amplified in Section 4.3.

4.3. Choice of Ion Exchanger

The properties of the ion-exchange medium that must be considered are as follows:

4.3.1 Nature of surface (especially charge as a function of pH)

4.3.2 Nature of substrate (particle size, pore size, chemical nature, surface area, resistance to deformation).

4.3.3 Nature of bonded phase (scale-up capability, stability, ease of cleaning, regeneration and sterilization).

4.3.1. Nature of Surface

4.3.1.1. Charge. When the range of pH values over which the peptide or protein of interest is stable is known, bonded phase surface charge may be chosen to bind the component via electrostatic forces only (if possible). The isoelectric point (pI) of the constituent should be known, and the type of ion exchanger and operational pH range selected according to the guide in Fig. 1.

At pH values greater than the pI, a peptide or protein is generally generally negatively charged and at pH values less than the pI, these materials are usually positively charged. Thus, a given peptide or protein may be chromatographed on both an anion exchanger or a cation exchanger, depending on the pH of the sample and mobile phase and whether the surface is appropriately charged at this pH. Kopaciewicz and co-workers (16) have provided retention (vs pH) maps of several proteins that can bind to both negatively and positively charged ion exchangers depending on the pH. In general, however, acidic proteins (pI less than 6) are chromatographed when they are negatively charged; basic proteins (pI greater than 8) are chromatographed when they are positive; and proteins whose pI lies between 6 and 8 can be chromatographed as either negative or positive species on anion exchangers or cation exchangers, respectively (see Fig. 2).

The order of elution is approximately related to pI but Kopaciewicz and co-workers (16) have shown that this is an oversimplification for the following reason: Owing to a possible asymmetrical charge distribution over the protein's surface, binding may still occur via a region of the molecule that may have a charge opposite to that of the molecule as a whole. Isoelectric points are measured for the whole peptide or protein, of course.

It is important to know or appreciate the manner in which the surface charge of an ion exchanger varies as a function of pH and ionic strength. The last parameters are those most commonly chosen to control efficient binding and elution. For weak and moderate ion exchangers, the lower the pH, the more positive or less negative the surface becomes; and the higher the pH, the more negative or less positive the surface becomes. For true strong ion exchangers, however, surface charge is independent of pH within the range 2–12. A knowledge of the magnitude of the surface charge of the ion exchanger and that of the polypeptide or protein at the same pH and ionic strength, permits a rational choice of an ion exchanger to bind a given biopolymer. In general, conditions must be chosen to optimize the strength of binding between oppositely charged protein and ion exchanger. Total binding energy may include contributions from electrostatic, dispersive, and dipole–dipole attractive forces, depending on the nature of the protein surface and ion-exchanger surface. Further discussion on this subject is given later in this section. The aim in resolving a component biopolymer of interest from a complex mixture is to control its binding strength so that it is different from those of all other components.

The functional groups that determine the variation of surface charge with pH, can be used to define the various classes of ion exchangers, as given in Table 8. A qualitative representation of the manner in which surface charge varies with pH is given in Fig. 3.

The actual shape and position of the curves in Fig. 3 will depend on the total environment in which the functional groups are located, which in turn will usually depend on the manner of synthesis, the substrate, and the manufacturer.

The use of the terms "strong," "moderate," and "weak" still appears to be a subject of significant confusion and controversy, when applied to ion-exchange chromatography. The origin of these terms lies clearly with the definitions of strong, weak, and moderate Bronstead acids and bases (17). Such are the functional groups that are bound to the ion-exchange surface. The average pK_a and pK_b values of the various acid and basic groups, set the boundaries beyond which there is more charge or less charge. The strength of attraction between the fixed charges on the ion exchanger and mobile ions being exchanged is not implied in these terms. In any case, in practical ion-exchange chromatography the total binding strength as measured by retention times, is determined by charge density (16), surface area, and other attractive forces (see Sections 4.3.1.2, 4.3.1.3, and 4.3.1.4), as well as the nature of the functional group of a given charge (18).

SUMMARY. Sample and component characterization will provide information to determine the surface charge best suited to bind the peptide or protein of interest. The isoelectric point of the component of interest (pI) and the pK_a (or pK_a range) of the surface functional groups, are numerically the limits of pH than can be used in principle to achieve binding (Fig. 1).

The variation of surface charge with pH (Fig. 3) can be used to select a surface pK_a range that is most suitable to bind the component of interest.

Commercially available ion exchangers are listed in Tables 2, 3, and 4.

4.3.1.2. Ligand Density and Charge Density. Ligand density is usually measured in terms of the number of interactive functional groups present per unit area of a surface. For example, the ligand density of fully hydroxylated silica is 8–9 micromoles of silanol groups per square meter (19). Polypeptides and proteins that have tertiary structures can also be considered to have ligand density. The interactive groups on both the chromatographic support and the protein will include fixed charges such as amino and carboxyl groups. In addition, the ion exchanger may bear other functional groups such as a polyamine, phospho, and sulfo, and the protein may have phenolic and phospho chargeable groups. The protein surface will also contain nonpolar and polar uncharged regions such as carbohydrate, lipid, and hydrocarbon functional groups. These are all capable of binding to the surface of the ion exchanger.

The importance of the charge distribution, in other words, the variation of ligand density, over the surface of a protein, is well established for high-performance ion-exchange chromatography (HPIEC). However, the influence of the variation of ligand density at the ion-exchange surface upon HPIEC has rarely been considered or measured. A great deal of research remains to be done in this area, since ligand density variations on the support surface may contribute to band broadening.

It is important to differentiate between the terms "ligand" and "charge" density. The latter refers to a property measured in terms of electronic charge per unit area. The former (ligand) may include any interactive functional groups: charged, dipolar, or nonpolar. Binding forces may be primarily electrostatic in nature, but because of the close proximity of protein with the support surface, may also involve dipole–dipole and dispersive interaction.

Charge density can be defined as the number of charged groups per unit area of surface on a bonded phase. Charged groups in high-density ion exchangers are placed as close as every two or three atoms apart, and can theoretically exhibit charge densities of 1 per 20 $Å^2$, for example. This figure is estimated from the diagram shown in Fig. 4. The polymeric anion exchanger polyethyleneimine

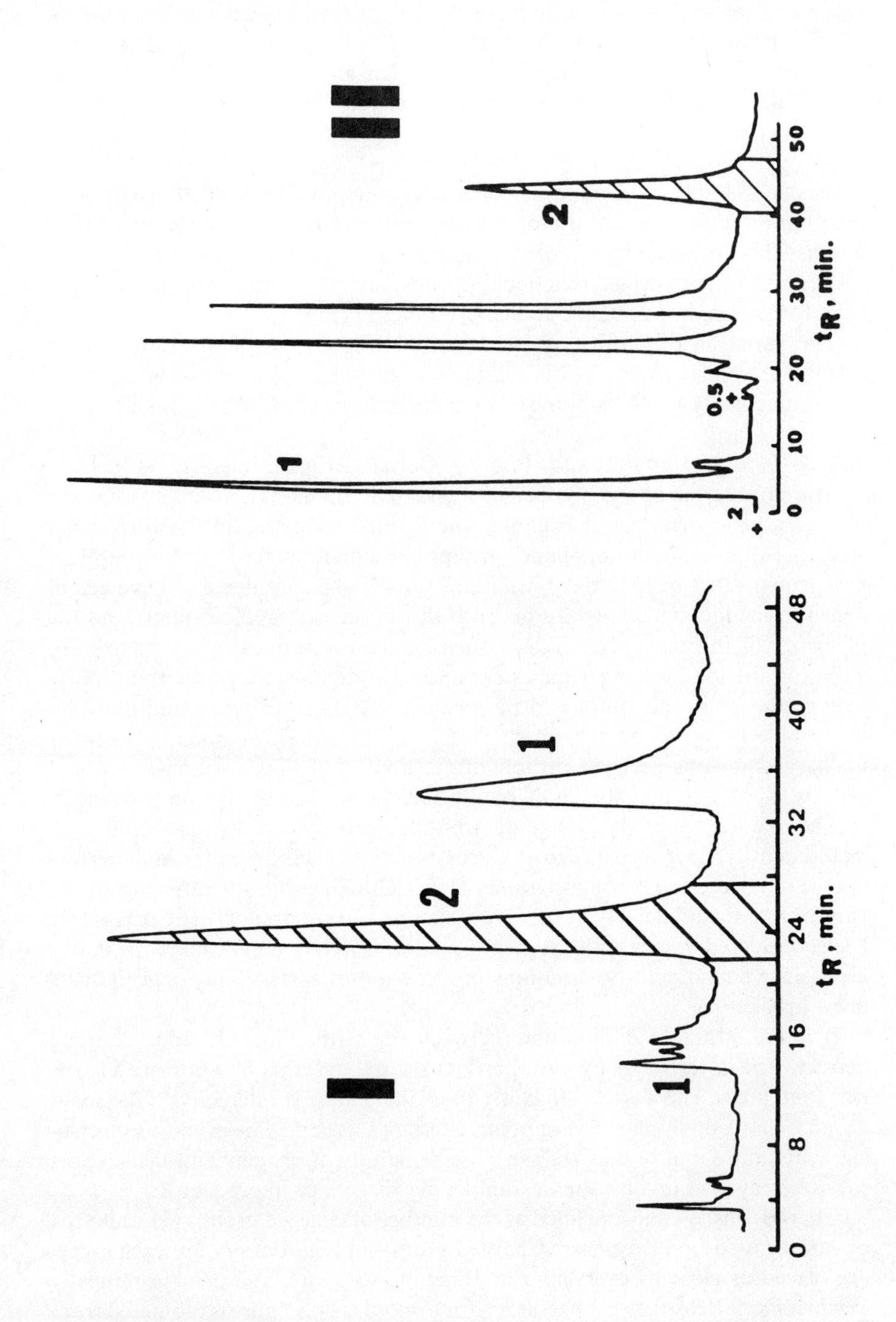

I
1
2
1
0
8
16
24
32
40
48
t_R, min.
II
1
2
0.5
0
10
20
30
40
50
t_R, min.

Table 8. Grouping of Ion Exchangers Based on Functional Group Type

Anion Exchangers	Typical Functional Group	Cation Exchangers	Typical Functional Group	Mixed-Bed Exchangers
Weak	Polyamino	Weak	Carboxymethyl	Amino + carboxy
Moderate	Diethylamino-ethyl	Moderate	Phospho	Carboxy + sulfo
Strong	Quaternary ammonium	Strong	Sulfo	Quat + sulfo

in this example is considered to consist of a branched polymer network with each chargeable nitrogen placed at the center of four ethylene groups. Assuming C–C and C–N bond lengths of 1.54 and 1.51 Å, respectively, this is equivalent to a charge density of 1 positive nitrogen per 20 $Å^2$.

The concept of charge density in surface-polymer-bound ion exchangers, however, may be an elusive one. Alpert (20) has explained the high capacity of a silica-based polyaspartic acid-bonded ion exchanger as being due to long "polymer strings" extending into the pore, creating extra binding sites for proteins. This phenomenon may be a general one for those ion exchangers that consist of a chargeable polymer bound to a surface at several points along the molecular chain. The rest of the polymer may reach into the mobile phase in the pore creating extra binding surface. The description by Kopaciewicz and co-workers (16) of observed increases in ligand density may be increases in "effective" surface area, causing stronger protein binding. Such ion exchangers have also been prepared by Ramsden (21) and Kitigawa (22). Analogous behavior is

Fig. 2. (*Opposite*) Separation of a monoclonal antibody from mouse ascites fluid using weak anion-exchange (I) and mixed-mode exchange (II) chromatography. Conditions for chromatogram I: Column: 4.6 mm × 250 mm, 5 micron BAKERBOND MAb. Mobile phase: Initial buffer (A) 10 m*M* potassium dihydrogen phosphate, pH 6.80; final buffer (B) 500 m*M* potassium dihydrogen phosphate, pH 6.4. Gradient: 0% B to 25% B over 60 min. Flow rate: 1 mL/min. Detection: UV (280 nm), 0.2 AUFS. Sample: 0.2 mL (mouse ascites fluid, 40 μL, diluted with 160 μL of buffer A). [Chromatogram from M.P. Henry, (84).] Conditions for Chromatogram II: Column: 4.6 mm × 250 mm, 5 micron BAKERBOND ABx. Mobile phase: Initial buffer (A) 10 m*M* potassium dihydrogen phosphate pH 6.0; final buffer (B) 250 m*M* potassium dihydrogen phosphate, pH 6.8. Gradient: 0% B to 50% B over 60 min. Flow rate: 1 mL/min. Detection: UV (280 nm) 2.0 AUFS (flow through) and 0.5 AUFS (bound peaks). Sample: 0.5 mL (mouse ascites fluid, 100 μL, diluted with 400 μL of buffer A). Peaks: 1. albumins, transferrin; 2. monoclonal immunoglobulins (IgG class). [Chromatogram courtesy of Nau (70).]

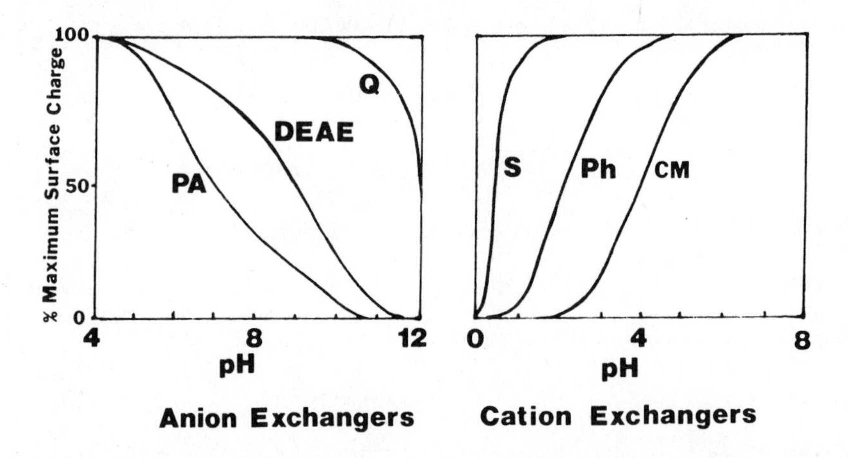

Fig. 3. Idealized representation of the variation of surface change with pH for various acid and basic groups on ion exchangers. PA = Polyamine. DEAE = Diethylaminoethyl. Q = Quaternary ammonium. CM = Carboxymethyl. Ph = Phospho. S = Sulfo.

exhibited during the well-established mechanism for the flocculation of colloidal particles by polyelectrolytes (23).

The generally accepted models constructed to describe aspects of the binding mechanism between proteins and anion-exchange surface, shed some light on the concept of charge density. In such a model it is postulated that a protein binds to a surface at Z points of attachment. There must, therefore, be a reasonable spatial relationship between the attachment points on each surface (protein and ion exchanger). If the charge density across the surface of the latter is too great (charges too close together), there may not be optimal binding with the binding sites on the protein. Conversely, where charges are widely spaced (low charge density) the protein may be attached by too few binding sites. There

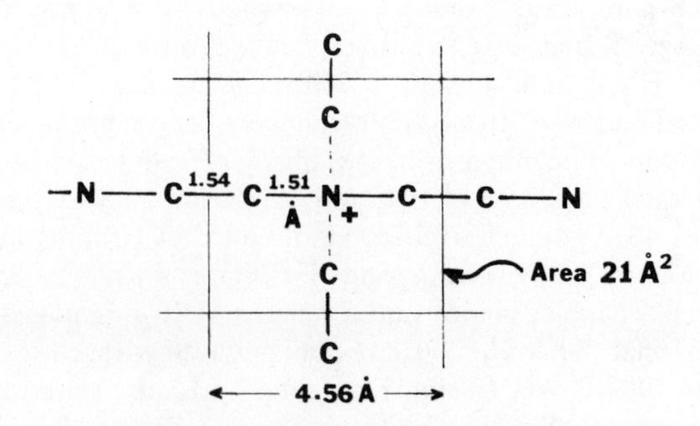

Fig. 4. Schematic model illustrating one method of calculating an approximate theoretical charge density on a surface.

should be an optimum spacing of charges on the ion exchanger to maximize the Z value for a given protein.

Kopaciewicz and co-workers (24) have obtained evidence illustrating this optimization approach. Three weak anion exchangers of different ligand densities were prepared on the same silica substrate, using different amounts of polyethyleneimine and the same cross-linking agent. Here ligand densities were measured in terms of the ion-exchange capacity for picric acid. Retention of ovalbumin (OVA) and soybean trypsin inhibitor (STI) at pH 8.0, increased with increasing ligand density. Myoglobin (MYO), on the other hand, was only retained on the ion exchanger of intermediate ligand density. It is likely in these cases that as ligand density increases, so does hydrophobicity, since the hydrophilic polyethyleneimine is cross-linked with a hydrophobic diepoxide. Myoglobin, with a pI of 6.9 and 7.3, is not bound at a low or high ligand density. However, at an intermediate hydrophobicity and ligand density, the configurations of the functional groups are optimal for binding of MYO to the weak anion exchanger. Myoglobin is also bound to the strong anion exchanger, Mono Q, at pH 8, with the same strength (16).

In the same series the large-molecular-weight protein, ferritin, is only recovered from the low-ligand-density ion exchanger. It is likely that the higher hydrophobicity of the higher-ligand-density ion exchangers, causes the nonelution of this protein. The hydrophobic attraction is greater at higher ionic strength (see Section 4.3.1.4). The increasing hydrophobicity of the three ion exchangers may also account in part for the increasing retention of OVA and STI at higher ligand densities.

In practice, however, the spacing of chargeable atoms such as nitrogens or oxygens is fixed for a given ion exchanger. There will be a degree of molecular flexibility in both the protein and the ion exchanger, which will provide limited assistance to optimal binding. Binding strength will thus be a compromise among the number and distribution of opposite charges on the protein capable of binding to the ion-exchange surface, dispersive interaction between nonpolar regions of the protein and ion-exchanger, and repulsive forces between surface charges and similar charges on the macromolecule. Further discussion of these effects upon selectivity is given in Section 4.3.1.4.

SUMMARY. Proteins interact with the surface of ion exchangers via ligands on both substances. The sum of the attractive and repulsive interactions of various kinds (ionic, dispersive, electrostatic, and dipolar) will determine the net binding strength. Changes in mobile phase pH and ionic strength that usually occur during peptide and protein chromatography, may result in variations in effective charge density during separation.

The spatial arrangement of interactive fixed ligands will influence optimum binding to a given peptide or protein.

Attempts to vary independently properties of bonded phases such as ligand and charge density may not be possible. Simple structure-retention correlations, for example, are not normally obtained, indicating that multiple changes occur during attempts to synthesize well-characterized bonded phases in a predictable manner.

4.3.1.3. Nature of Functional Group. Chargeable functional groups include amino (primary, secondary, tertiary, quaternary), phospho (phosphate, phosphite), sulfo, and carboxy. Virtually all anion exchangers are amines, whose degree of alkylation determines the pH range over which the nitrogen is charged. The effects of the alkyl groups upon protein and polypeptide binding are many and little evidence has been obtained to separate these effects.

Gooding and Schmuck (18) were among the first to compare the chromatographic properties of a weak (largely tertiary) ion exchanger with a strong (largely quaternary) one. Since the strong exchanger was prepared from the weak one, most of the physical and chemical properties of both will be identical. These authors found that the weak anion exchanger, in general, bound proteins more tightly than the strong anion exchanger. These observations were explained in terms of a steric inhibition of close approach by anions, and an enhanced hydrophobicity of the weak over the strong anion exchangers. The latter property increases hydrophobic attractions between protein and ion exchanger, resulting in longer retentions. It seems unlikely, however, that the preceding authors claims for great exposed hydrophobicity of the polyethyleneimine backbond can be substantiated. The chromatography of proteins at high salt, on a noncrosslinked polyamine-bonded phase (210), indicates negligible hydrophobic character (see Fig. 5). It is more likely that steric effects of the methyl group (if they occur) on protein binding to a quaternary ion exchanger, are of two kinds:

1. Possible steric inhibition of close approach, mentioned above.
2. Steric protection of the hydrophobic cross-links by the bound methyl group. The second effect may reduce the hydrophobic attraction component in the quaternary ion-exchanger, leading to reduced retention.

Clearly, then, it is difficult to alter systematically the structure of an ionized surface without producing multiple effects upon protein retention. Kopaciewicz and co-workers (24) have attempted to do this, and their work is reviewed in the previous section.

Charge density has been determined (25) to be a major factor in permitting isocratic elution of proteins of widely differing isoelectric points within a reasonable time. According to Yao and Hjerten (25) the other requirements are small particles, negligible nonspecific (hydrophobic) interactions, and symmetrical peaks (linear absorption isotherms). Small particles are required for sharp peaks and symmetrical peaks maintain resolution. Low ligand density compresses the range of k' values for proteins, by reducing the number of points of attachment any one protein can have with the ion-exchange surface. The fourth requirement, minimal nonspecific interactions, has not been established by Yao and Hjerten (25).

The retention model construed by Kopaciewicz and co-workers (16) for ion-exchange chromatography of proteins can in fact be used to predict the relation between low charge density and facile isocratic elution. The preceding authors derived Eq. 1—relating capacity factor and salt (here NaCl concentration:

$$\log k' = 2Z \log (1/[\mathrm{NaCl}]) + \log k_z \quad (1)$$

where Z is the number of charges associated with the absorption–desorption process, k' is the capacity factor, and k_z is the capacity factor when [NaCl] is unity. By deliberately keeping ligand density low, Z values for all proteins will decrease, and for a given value of [NaCl], all k' values will be small. Thus, isocratic elution can separate a group of proteins over a small range of capacity factors.

It is also possible to change the charge density of "weak" ion exchangers by altering their state of ionization. At low pH the charge on carboxyl groups is neutralized; and at high pH the charge on amino groups decreases. Unfortunately, this approach to facilitating isocratic separations has no general validity. Changes in pH simultaneously alter the net charge on the protein and the sum of both changes is often unpredictable.

Perkins and co-workers (26) have demonstrated this phenomenon in the separation of several basic proteins by weak cation-exchange chromatography. Isocratic elution at a higher pH of 6.5 (where charge density was high) produced weak binding of proteins. At a pH of 5.85 (where charge density was lower) protein binding was stronger. The ionic strength was the same for both pH values. In this example the charge density on the protein surface changed more rapidly than that on the ion-exchanger surface. In the preceding case, order of elution remained constant at both pH values, but the selectivity changed significantly.

In deriving Eq. 1, Kopaciewicz et al. (16) have shown that Z is constant under conditions of fixed pH, but may vary at different pH values. For example β-lactoglobulin A, chromatographed on a Pharmacia Mono Q column, showed no variation in its Z value of 4.10 ± 0.05 from pH 8 to 6, but dropped to 1.8 at pH 5. The ionic strength of binding also decreased at this pH, although retention (measured as $\log k'$) could be increased by using a sufficiently low salt concentration.

The fact that β-lactoglobulin A could be just as retained on a positively charged (Mono Q) surface at pH 8 (where it is negatively charged) as at pH 5 (where it is slightly positively charged) illustrates several important aspects of the nature of the protein–stationary-phase interaction:

1. β-lactoglobulin A must contain regions of overall negative charge on its surface, even at a pH where it is electrophoretically positive. In other words the molecule has a nonuniform surface charge distribution. (The use of the term "charge asymmetry" by Kopaciewicz and co-workers (16) is not strictly correct

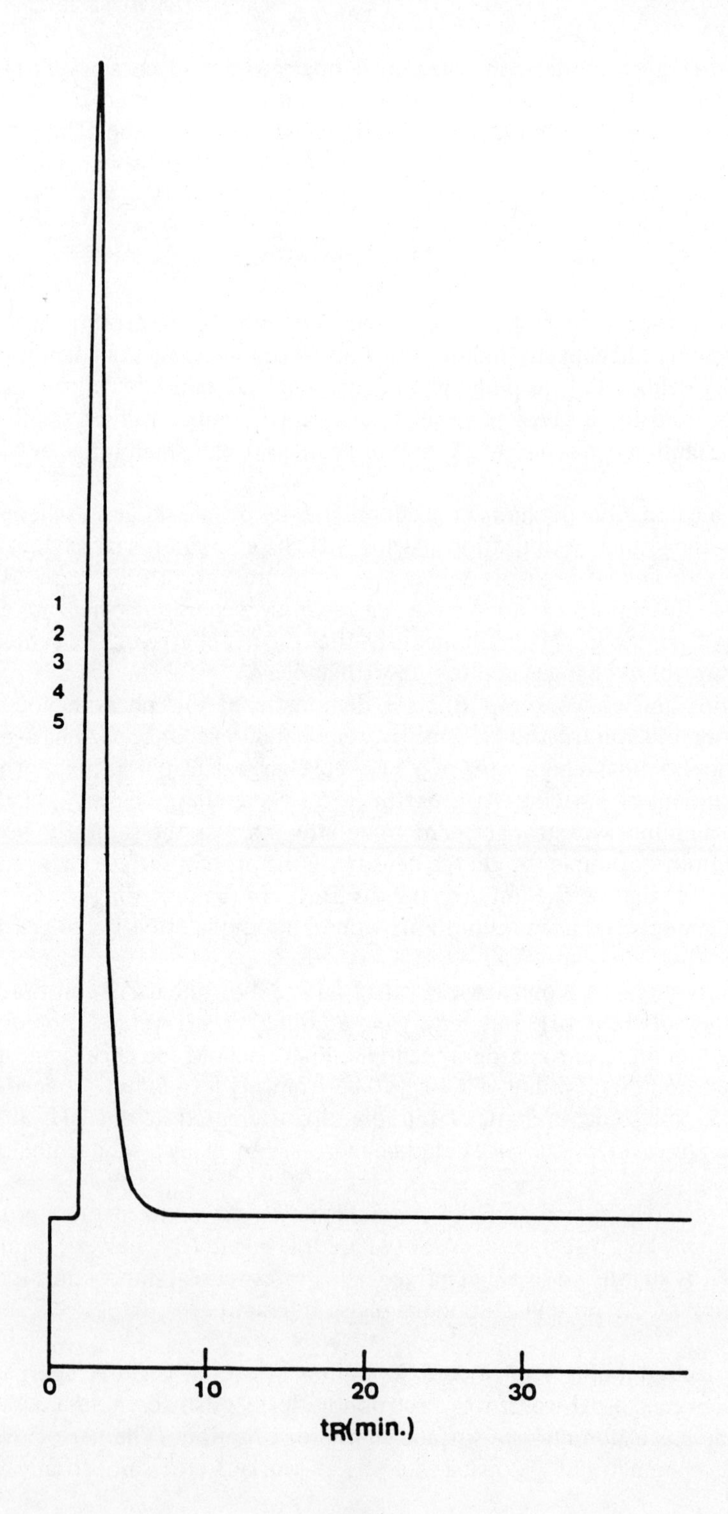
1
2
3
4
5
0
10
20
30
t_R(min.)

since charge may be localized at regularly spaced regions on the surface of the protein.)

2. Since the Z value and binding strength remain substantially constant over the pH range 8 to 6, the functional groups (amino, phenol) on the β-lactoglobulin molecule that are being protonated at the lower pH's must be located away from the protein absorption site. At pH 5, where Z is 1.8—and by implication corresponds to a two-point binding site (at pH 6 to 8 there is a four-point binding site)—two of the original four points of attachment must have been eliminated. To achieve significant retention, the eluting buffer must then be lower in ionic strength.

3. Above an ionic strength of about 0.15 M, Kopaciewicz and co-workers (16) showed that the double-layer thickness is constant at about 2 Å. Thus, provided that measurements of k' at various buffer concentrations are made above this value (0.15 M), calculations of Z will not be influenced by changes in double-layer thickness. On the other hand, the linearity that the preceding authors observed for the plots of log k' against log 1/[salt], indicates that small changes of several angstroms in double-layer thickness do not influence binding strengths. It would be interesting, however, to observe the linearity of these plots as ionic strength approached 0.002 M where the double-layer thickness is a substantial fraction of the diameter of the protein (i.e., 10–12 Å).

SUMMARY. Functional group types may be changed by further, chemical, permanent modification (as in the derivatization of weak amines to quaternary amines); or by a change in pH (which may alter the state of ionization).

Correct manipulation of the latter process allows the chromatographer to carry out isocratic elution of mixtures of proteins. This occurs when pH and ionic strength are optimized.

Isocratic experiments can be used to develop mechanistic theories of adsorption of proteins. Information can be obtained concerning charge asymmetry over the surface of a protein; and the nature and location of the adsorption sites on the protein.

4.3.1.4. Hydrophilic Properties. It is generally considered desirable to use bonded phases whose surfaces are hydrophilic, in other words, that are polar

---

Fig. 5. (*Opposite*) Absence of hydrophobic interaction under typical HIC conditions on both 40 micron BAKERBOND PREPSCALE* PEI and CBX. Column: 4.6 × 250 mm 40 micron BAKERBOND PREPSCALE PEI, or 4.6 × 250 mm 40 micron BAKERBOND PREPSCALE CBX. Mobile Phase: A = 2 M $(NH_4)_2SO_4$ plus B; B = 25 mM KH_2PO_4, pH 7.0. Gradient: 100% A–100% B over 30 min. Flow rate: 1.0 mL/min. Pressure: 30 psi. Detection: UV at 280 nm, 1.0 AUFS. Sample: Protein standards. Peaks: 1. cytochrome c (horse heart, type VI); 2. myoglobin (sperm whale skeletal muscle); 3. lysozyme (hen's egg); 4. α-chymotrypsinogen (bovine pancreas); 5. calmodulin (bovine brain). Recovery: > 99% of a total protein mass. Chromatography is identical on both columns. [From Nau (85).]

and will attract water molecules. Such surfaces, when bearing covalently fixed charges, will bind mobile ions and water molecules via largely electrostatic forces that form and break under conditions of high surface tension. Functional groups on the ion-exchange surface that are less polar will undergo dispersive or so-called nonspecific interactions with similar regions on the peptide or protein. These short-range attractive forces are difficult to break under the conditions of high surface tension that exist in aqueous salt solutions. Consequently, portions of the sample may be indefinitely bound to the ion exchanger, causing low mass recoveries and unwanted build-up of material on the bonded phase (27). The tendency of a peptide or protein to bind to nonpolar or hydrophobic surfaces, increases with the increasing surface tension that usually accompanies increasing ionic strength. Consequently, a peptide or protein bound tightly to an ion exchanger with some hydrophobic character at low ionic strength may experience two opposing forces as ionic strength increases: the displacing effect of the increased number of ions in solution and the increased binding forces brought about by the greater surface tension of the more concentrated buffer. In some cases a solute may not elute at all (27).

It should be appreciated that all peptides and proteins are bound to ion exchangers via dispersive forces as well as electrostatic forces (27–29). It is the aim of the research worker and the manufacturer to be able to control the relative importance of these non-specific interactions.

Kopaciewicz and co-workers (24) have published several detailed examinations of the effects of the stationary phase surface on protein retention in high-performance ion-exchange chromatography. These investigations are unique insofar as the structure and properties of the bonded phase itself are altered systematically, allowing optimized packings to be made for different classes of proteins. Although there was no attempt to quantify the separate contributions of electrostatic, hydrophobic, and steric interactions to overall protein retention on Regnier's bonded phases, it is clear that all three play a more or less important role. This does not mean, of course, that such interactions occur with all types of high-performance ion exchangers.

Kopaciewicz et al. (24) for example, have correlated protein retention, binding capacity, and mass recoveries on ion exchangers, with the nature of epoxides used to cross-link and derivatize PEI-coated silicas. A more hydrophobic cross-linker or derivatizer increased retention and resolution, and for large proteins such as ferritin (FER), recovery decreased. Conversely, a more hydrophilic cross-linker epoxide reduced capacity, retention, and resolution of small proteins but allowed FER to be eluted. Certain cross-linkers shut off access of proteins to a proportion of the ionized groups, causing a possibly sterically induced reduction in hemoglobin binding capacity. Figure 6 illustrates the sometimes dramatic changes in selectivity Kopaciewicz and co-workers (24) obtained by introducing varying kinds of hydrophobic groups into a PEI coating, via cross-linking. Clearly, a large number of high-performance ion exchangers can be prepared having different selectivities depending on the degree of hydrophobicity of the surface.

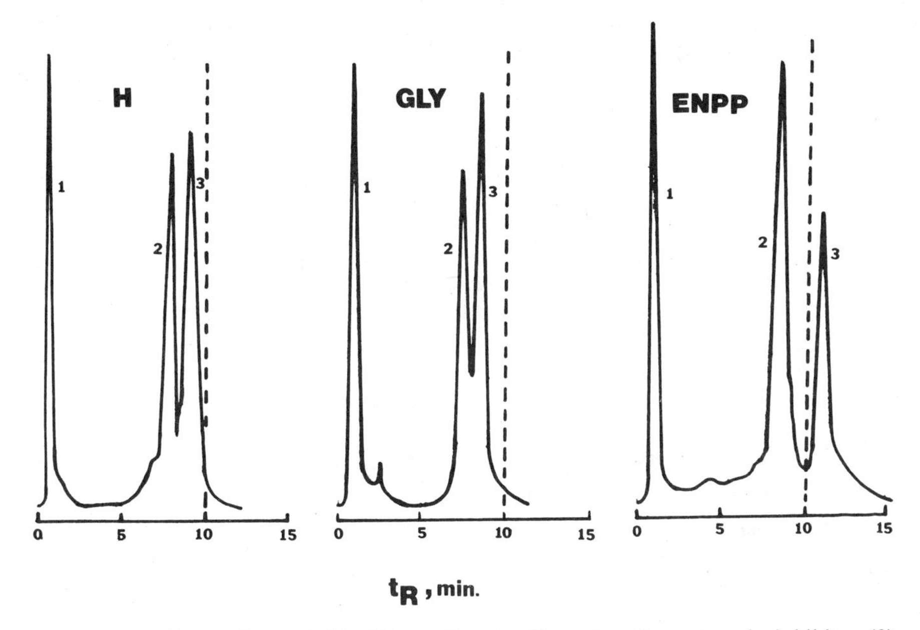

Fig. 6. Separations of myoglobin (1), ovalbumin (2), and soybean trypsin inhibitor (3) on three PEI-coated stationary phases of equal ligand density but varying hydrophobicity. The hydrophobic character was obtained by cross-linking the base coating (H) with 1,2-epoxy-3-hydroxypropane (GLY) and 1,2-epoxy-3-(p-nitrophenoxy)-propane (ENPP). Chromatographic conditions follow. Column: 4.1 × 50 mm. Mobile phase: A = 10 mM Tris hydrochloride, pH 8.0; B = 500 mM NaCl + A, pH 8.0. Gradient: 0: 0% B to 100% B over 20 min. Flow rate: 1 mL/min. Detection: 254 nm [From Kopaciewicz et al. (24).]

Kopaciewicz and Regnier (30) have suggested the presence of cooperative hydrophobic–ionic interactions in the retention properties of a weak cation exchanger. The latter was one of several bonded phases made by reacting a cyclic anhydride with an adsorbed, cross-linked PEI silica. The reaction scheme is shown in Fig. 7.

Where the anhydride was glutaric (largest nonpolar group), the corresponding bonded phase exhibited the greatest pH sensitivity with regard to protein retention. Furthermore, at pH 5 the longest retentions of lysozyme were observed for the anhydrides with the largest CH_2 to COOH ratio. Kopaciewicz and Regnier (30), suggested that the pH-induced retention variation and the longer retention were both due to increased hydrophobic interaction contributions.

In an earlier related work, Gupta, Pfannkoch, and Regnier (31) demonstrated a possible hydrophobic effect in a weak cation exchanger. This was prepared by reacting covalently bound PEI with glycolic anhydride. At pH values of 5 and below, where carboxylic groups are beginning to lose their charge, retention

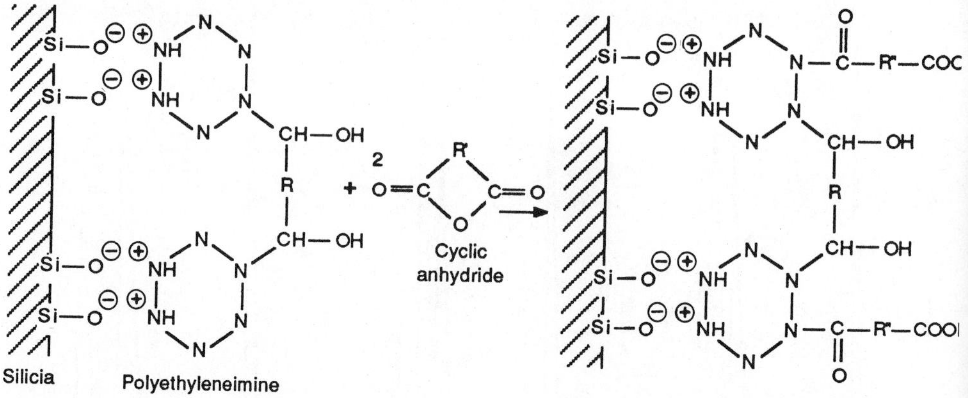

Fig. 7. Reaction of cyclic anhydrides with adsorbed, cross-linked, polyethyleneimine [From Kopaciewicz et al. (30).]

of lysozyme increased rapidly as resolution and recovery decreased with pH. Alpert (20) has prepared a weak cation exchanger exhibiting little or no apparent hydrophobic character. The polyaspartic-acid-bound silica consists of a polymer of aspartic acid covalently linked to aminopropyl silica via side chain carboxyl groups. The acidic polypeptide nature of the surface probably accounts for its hydrophilic nature, as exhibited by the high activity (and therefore mass) recoveries of several enzymes.

In another investigation, Kennedy, Kopaciewicz, and Regnier (32) set out to prepare mildly hydrophobic surfaces from coated PEI silicas. The extensively cross-linked bonded phases showed the ability to separate proteins when used as both weak anion exchangers and hydrophobic interactors. The degree of overlap of the two modes of interaction was estimated using Eq. 2, where H is the ionic strength that produces a k' of unity in the hydrophobic interaction mode and I is the value that produces a k' of unity in the ion-exchange mode; S is a measure of the segregation of the two mechanisms:

$$S = H - I \tag{2}$$

Elution problems arise when a protein is tightly bound to an ion exchanger and is also quite hydrophobic. The value of S may be very small or negative and the protein will not elute.

Synthetic control over the values of H and I can be exercised in several ways, in order to maximize the magnitude of $H - I$. For a given ligand density, for example, and therefore a given value of I, a more hydrophobic cross-linker or a great degree of cross-linking may increase $H - I$. In fact, as was discussed above, the decreased steric access of proteins and peptides to the ionized surface obtained after cross-linking will reduce *I*, producing an even more favorable value of *S*.

Kennedy and co-workers (32) showed that an exhaustive cross-linking reaction of an already cross-linked coated-PEI silica produced a reduction in ion-pair capacity of approximately 25%; and a reduction in hemoglobin binding capacity of 15%. On the other hand, acylation of the amine groups, which neutralizes the primary and secondary nitrogen atoms, reduce ion-pair capacity by 40% and the hemoglobin binding capacity by 80%. Values of *I* decrease as fewer basic nitrogens are available for binding and consequently *S* values increase. Kagel of Rainin Instruments Co. Inc. (33) has commercialized a multi-mode PEI-based anion-exchanger/hydrophobic interactor in which the *S* value was maximized by keeping the ion-exchange capacity (and therefore *I*) quite low. Such a procedure results in reduced protein binding capacity, however, for the column, when operated in the anion-exchange mode. Figure 8 shows how this ion-exchanger (Dynamax-300A AX) can be operated in both modes. The advantage of such a material is that chromatography can be carried out in both modes using the same column.

Using a substantially different method of covalently attaching polyethyleneimine to silica, Ramsden (21) produced a suitable bonded phase without the need to cross-link the polyamine and that exhibited no hydrophobic properties. Ramsden and co-workers have also prepared a group of substantially nonhydrophobic packing materials including a weak cation exchanger (carboxyethyl) (21), quaternary ammonium strong anion-exchanger (34), and a mixed-mode carboxylic/sulfonic weak/strong cation exchanger (35). Figure 5 illustrates the chromatography carried out on packed columns of one of the above four materials with high salt mobile phases. No binding occurs in the latter circumstances of any protein selected. The major advantages of such ion exchangers is that they combine high capacity (high surface area) and retention (high ligand density) with high recovery of proteins and peptides of a broad range of hydrophobicities and molecular weights (minimal nonspecific binding).

The degree to which other commercially available ion exchangers show hydrophobic character is not known from any published information with the exception of the Dynamax 300A AX. Work in our laboratories, with products from several other major manufacturers of ion exchangers, indicates different levels of hydrophobicity for different products.

For example, Fig. 9 shows separations of standard proteins on a high-performance anion exchanger from a major commercial source and on a BAKERBOND Wide-Pore* hydrophobic interaction column under conditions designed to promote hydrophobic binding in both cases. The bonded phase used in the commercially available weak-anion exchange column (A) is prepared by cross-linking polyethyleneimine adsorbed to silica. The cross-links impart a significant degree of hydrophobic character to the surface; the column binds proteins at high salt, and therefore can be used in the hydrophobic interaction mode. The hydrophobic interaction chromatography obtained with the BAKERBOND* column specifically designed for this mode, is illustrated in B for comparison.

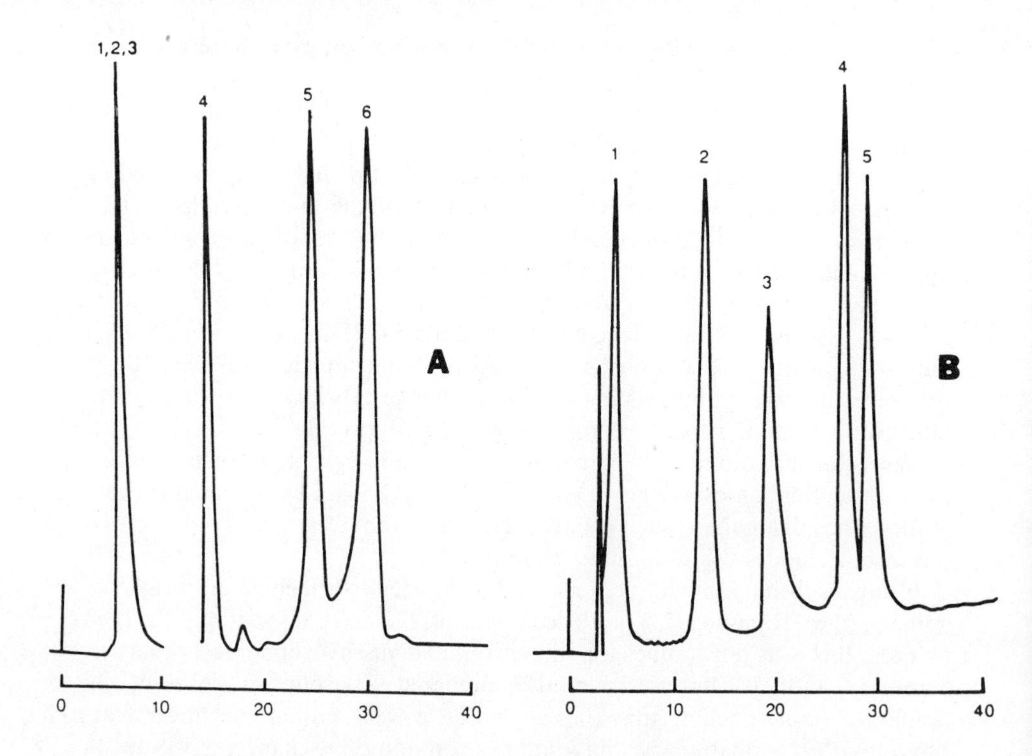

Fig. 8. Separation of standard proteins on a multimode bonded phase under conditions for ion exchange (A) and hydrophobic interaction (B). Chromatographic conditions follow. Column: Dynamax®-300A AX, 4.6 × 250 mm. For Chromatogram A, Mobile phase: A = 5 m*M* KH_2PO_4, pH 7.0; B = 300 m*M* KH_2PO_4, pH 7.0. Linear gradient: 0–100% B over 40 min. Flow rate: 0.6 mL/min. Detection: UV at 280 nm. Peaks: 1. α-chymotrypsinogen; 2. cytochrome c; 3. ribonuclease A; 4. α-lactalbumin; 5. β-lactoglobulin B. For Chromatogram B, Mobile phase: A = 100 m*M* KH_2PO_4 plus 3 *M* $(NH_4)_2SO_4$, pH 7.0; B = 100 m*M* KH_2PO_4, pH 7.0. Linear gradient: 0–65% B over 40 min. Flow rate: 1.0 mL/min. Detection: UV at 280 nm. Peaks: 1. cytochrome c; 2. ribonuclease A; 3. β-lactoglobulins A and B; 4. α-chymotrypsinogen; 5. α-lactalbumin. [From Kagel (33).]

There have been several qualitative methods devised to determine the true hydrophobic nature of a charged chromatographic surface:

1. One or two percent of an organic solvent such as 2-propanol is added to the buffered mobile phases involved in the ion-exchange chromatography. The retention times of standard proteins such as bovine serum albumin, ovalbumin, lysozyme, and cytochrome c, chromatographed within ±1 pH unit of their isoelectric point were measured with and without the 2-propanol. Retention time decreases of no greater than 5% indicate the absence of a significant hydrophobic effect. This procedure has been followed by Gupta and co-workers

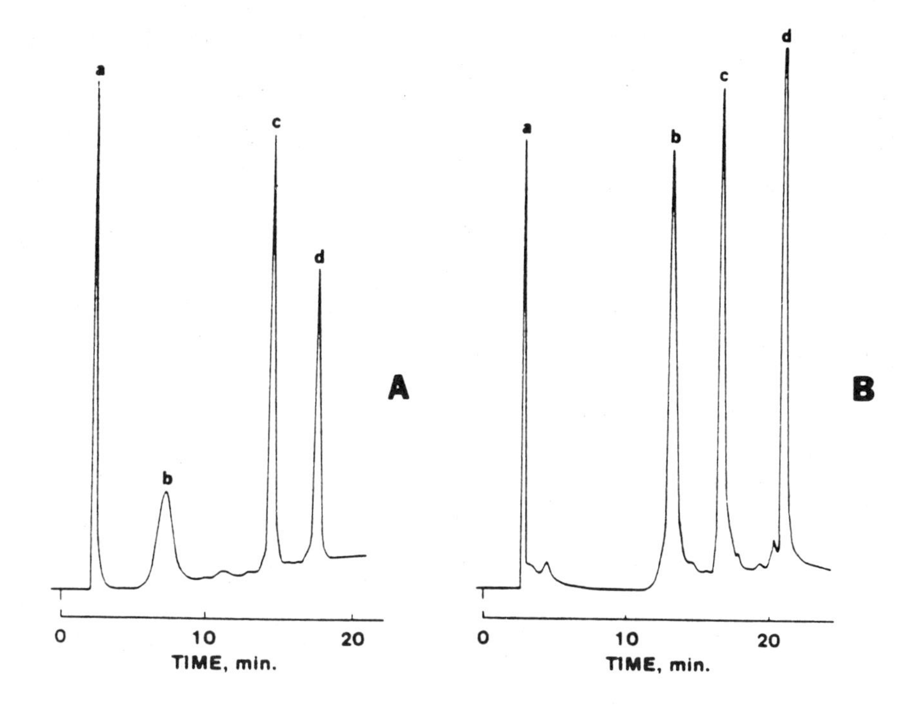

Fig. 9. Separation of standard proteins under typical hydrophobic interaction conditions on a commercially available weak anion exchanger (A) and a BAKERBOND hydrophobic interactor (B). Chromatographic conditions follow. Column A: weak anion exchange (commercially available) 4.1 × 250 mm. Column B: BAKERBOND Wide-Pore HI-Propyl*, 4.6 × 250 mm, 5 micron mobile phase in both cases: A = 2 *M* ammonium sulfate + B, pH 7.0; B = 25 m*M* KH_2PO_4, pH 7.0. Linear gradient: 0–100% B in 15 min. Flow rate: 1 mL/min. Detection: UV at 280 nm. Peaks: a. cytochrome c; b. myoglobin; c. lysozyme; d. α-chymotrypsinogen. [From S.A. Berkowitz (27).]

(36), Kopaciewicz and co-workers (14), and Stout et al. (37). The last authors however criticized this approach as being invalid, although the basis for their disagreement was not provided.

2. A mixture of proteins is adsorbed at the inlet of a column of packing material for various lengths of time (dwell times) up to several hours. Elution of the components is then carried out in the usual manner, and the chromatography for each dwell time is compared. Similarity among the chromatograms is an indication of negligible hydrophobic character of the ion exchange material.

Hofstee and co-workers (28, 19) have provided a description of phenomena that may occur to support the above conclusions: Initial ionic attraction that binds a protein close to a charged surface may lead to a time-dependant increase in nonspecific (hydrophobic) attractive forces. Under these circumstances, samples chromatographed with different dwell times will exhibit different elution patterns. Yao and Hjerten (25) have demonstrated the application of this method in the anion-exchange chromatography of proteins using a new cross-linked microparticulate diethylamino agarose. No change in the chromatography was observed after a 2 hr dwell time, indicating a negligible hydrophobic effect on the elution profiles.

The preceding technique is only sufficient to demonstrate the absence of hydrophobicity, but not its presence. Strong ionic attractions may also be sufficient to cause slow unfolding of the proteins on the surface of the ion exchanger, leading to dwell-time-related changes in chromatographic elution patterns.

It is certainly true, on the other hand, that the more hydrophobic a surface is, the more sensitive is the protein chromatography to the dwell time.

For example, Hearn and co-workers (38) have examined the effects of dwell time upon the chromatographic elution profile of proteins eluted under reversed-phase conditions. Here surfaces are quite hydrophobic, resulting in very strong non-specific attractive forces. Proteins undergo a degree of time-dependent unfolding both on the surface and in the mobile phase. Consequently, the elution pattern will vary with dwell time. Benedek and co-workers (39) have shown that rapid but incomplete denaturation occurs at the moment of absorption of a protein to a nonpolar surface. This denaturation is not as extreme when a hydrophilic modifier such as 1-propanol is present to solvate the stationary phase ligand. Longer dwell times generally alter the chromatography profiles even further, as a slower unfolding process occurs. Substantial loss of biological activity occurs. With the more hydrophilic ether phases of Wu et al. (40), kinetic unfolding is substantially reduced as evidenced by the high activity and mass recoveries of various proteins. Gradual loss of mass recovery, however, is still observed with increased dwell times.

3. Chromatography is attempted at high ionic strength [for example, 2 *M* $(NH_4)_2SO_4$] where ion exchange is largely absent, but hydrophobicity of the surface may be revealed. High concentrations of most inorganic salts increase surface tension of aqueous solutions. If the surface tension is raised high enough, an energetically favorable adsorption of a peptide or protein occurs to a weakly hydrophobic surface. Where there is a negligibly small degree of hydrophobic character in the ionized surface, no binding will occur. The greater the degree of hydrophobicity in the ion-exchanger the lowest surface tension needed to bring about adsorption. Thus, for a given protein chromatographed isocratically on a series of ion exchangers of differing hydrophobic character, the probable general characteristics of plots of k' vs ionic strength surface tension are shown in Fig. 10.

SUMMARY. The presence of nonpolar groups in an ion exchanger results in a hydrophobic component of the total binding interaction that increases in importance at high ionic strengths (see Fig. 9). In some cases, this may result in nonelution of more hydrophobic proteins. Studies have been made correlating

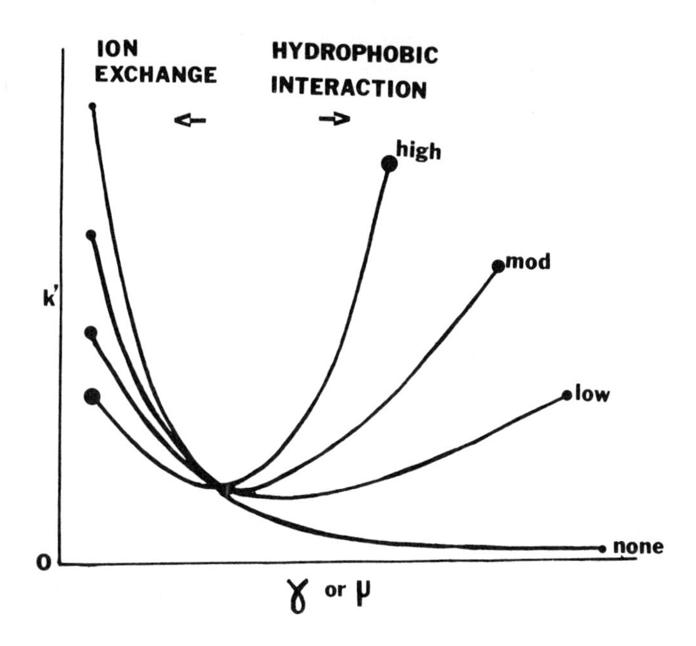

Fig. 10. Schematic illustration of the probable variation of k' with ionic strength (μ)/surface tension (γ) of a single protein chromatographed isocratically on ion exchangers of different hydrophobic character.

retention, capacity, and mass recoveries, with the nature of cross-linker used to stabilize coated polymers. These studies emphasize the difficult of obtaining simple changes in these properties as a function of hydrophobicity of the cross-linker. There are clearly cooperative ionic/hydrophobic forces that produce the resultant chromatographic effects.

Bonded phases that have hydrophobic character, may often be used as both ion exchangers and in the hydrophobic interaction mode. Careful synthesis of such bonded phases is required to control degree of mixing of the two mechanisms.

The ionized bonded phases prepared by Ramsden show very little hydrophobic character due to their unique method of synthesis, which employs no cross-linkers.

Three tests for the absence of hydrophobicity were described including the retention changes using alcoholic solvents, dwell-time-related effects, and high salt effects.

4.3.2. *Nature of Substrate*

Relatively few substrate materials have been developed for the HPLC of peptides and proteins.

Mikes, Strop, and Sedlackova (41) have specified several physico/chemical

Table 9. Physico/Chemical Requirements for High-Performance Ion-Exchange Substrates

Property	Necessary	Desirable
Structural integrity (resistance to deformation)	Yes	Yes
Microparticulate nature	Yes	Yes
Porosity	No	Yes
Range of pore sizes	No	Yes
Lack of solubility in mobile phases	No	Yes
Surface hydrophilicity	No	Yes
Chemical reactivity	Yes	Yes
Chemical resistance	Yes	Yes
Microbial resistance	No	Yes

characteristics required of ion exchangers suitable for high-performance chromatography of technical enzymes. These characteristics and others are listed in Table 9.

It is only relatively recently that the technology has been developed to produce particulate materials that meet these requirements. The first of these substrates was silica (6), followed by polystyrene highly cross-linked with divinylbenzene (42), hydroxylated polyether (43), polymethacrylates cross-linked with glycerol and glycols (44), and alumina (45).

The major requirement that spurred the development of the preceding materials for HPLC was that of structural integrity. Columns packed with small particles, necessary for high efficiencies, must generally be able to withstand high pressures with little or no deformation of the particle's structure. The high-pressure requirement is less of a concern for the actual analysis or purification itself, since a short column (5 cm) packed with 5 micron particles may exert less than 50 psi at a useful linear velocity of 0.3 mm/sec with the aqueous mobile phases used in ion-exchange chromatography. High pressure (several thousand psi) are required, however, during the slurry packing of such columns. Furthermore, rapid reequilibration of HPLC columns is best accomplished at high linear velocities and, consequently, high pressures. The previously mentioned short column, for example, could be routinely reequilibrated at half the maximum flow rate of a standard analytical HPLC pump (5 mL/min) if the particles could withstand 1000 psi. At the corresponding linear velocity of 5 mm/sec, the reequilibration could be achieved at 20 times the flow rate used in the actual analysis.

The requirement for small diameter particles in high-resolution chromatography is well established. The great majority of HPLC columns for analysis contain particles that are 10 micron or less in diameter. HPLC columns used in preparative work contain particles that are rarely greater than 25 micron, which may be considered the upper limit of the diameter of so-called "high-performance" packings. Beyond this figure, it is generally agreed that the term "medium performance" should be used. It is also at about 25 micron that columns containing such packings may be efficiently dry-packed.

The great majority of HPLC bonded phases are prepared on porous particles. The major advantage of such substrates is their high specific surface area. Binding capacity of a given peptide or protein is dependent on the proportion of this surface that can be accessed by the biopolymer. Ion exchangers prepared from porous substrates can bind up to 40% of their own weight of an average sized (MW = 40,000) protein (20). Furthermore, such a high capacity means that complex samples containing low concentrations of a protein of interest may be safely analyzed and purified with a low risk that this protein will not bind.

There is far less control over pore-size distribution than over particle-size distribution in bonded phases. Consequently, a range of pore diameters and shapes is unavoidable, even in so-called controlled-pore substrates such as glass and alumina. Is a narrow pore-size distribution preferable to a wide distribution? This question has not been satisfactorily answered in the published literature (19, 46). Comparison of pore-size effects upon chromatographic properties such as efficiency, capacity, recovery, and retention has been the subject of several investigations (47, 48). The problem with most of these comparisons, however, is that as pore size (diameter) and its distribution changes, so do other parameters as fundamental as silica type, pore volume, and surface area.

Several bonded phases suitable for the ion-exchange HPLC of peptides and proteins have been developed using nonporous substrates (49). The main advantage of these is the speed with which separations occur. This is because the large polymers do not need to diffuse into and out of pores. The main limitation of such supports is their very low capacity, which results in rapid overloading and peak-broadening unless due care is taken. Thus porosity is not necessary, but is desirable.

However, there is general agreement that pore diameter must be greater than a given size to be suitable for chromatography of a given protein. Figure 11 indicates the pore-size limits of bonded phases as a function of molecular weight. This diagram delineates approximate limits only and will not necessarily give optimum pore size for chromatography of larger proteins. In general, it is advisable to choose a pore size nearer the upper limit (total inclusion) for a peptide or protein of a given molecular weight.

Low solubility of the ion-exchanger substrate in mobile phases and cleaning/regenerating/sterilizing solutions is, of course, desirable. It is not necessary, however, since subsequent chemical modification of the substrate produces derivatives whose solubility characteristics may be significantly different from those of the substrate itself. In this respect, most substrates that have been developed for ion-exchange chromatography have one or more limitations. Silica dissolves rapidly at pH values above 8; polystyrene derivatives cannot be used with moderately polar organic solvents such as tetrahydrofuran; polyhydroxyethers should not be used with high-ionic-strength solutions (greater than 0.5 *M*) and polyacrylates are hydrolyzed in alkaline solvents.

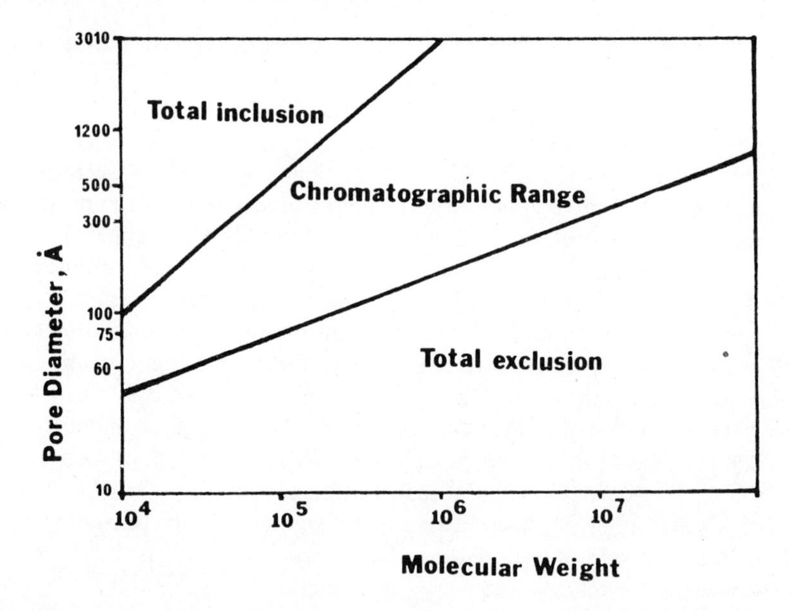

Fig. 11. Pore diameter limits for total exclusion and total inclusion of polymers as a function of their molecular weights.

Hydrophilic (readily hydrated) surfaces are generally considered necessary for ion exchange, since such environments are energetically favorable for the formation and breaking of only ionic bonds. The presence of nonpolar regions in an ion exchanger may promote dispersive and dipole–dipole attractive forces with nonpolar regions on the peptide or protein. The simultaneous operation of several disparate forces to bind a polymeric ion, has undesirable consequence for ion-exchange chromatography (see Section 4.3.1.4).

The substrate must be chemically reactive in order to produce the fixed charged groups that are bonded covalently or electrostatically to the surface. Unfortunately, little detailed chemistry has been published in the patent or other literature concerning the reactivity of substrates for commercially produced high-performance ion exchangers (21, 34, 35).

Resistance of the substrate to chemical reaction (inertness) is important to determining the lifetime, reusability, regenerability, cleaning, and sterilizability of the ion-exchange material. Naturally, the chemical inertness required is that existing under normal chromatographic conditions:

pH values from 2 to 10
Ionic strengths from zero to up to 3 molar
Possible presence of organic solvents

Table 10 gives a list of the major high-performance substrates, together with

Table 10. General Nature of Major Substrates Used in HPIEC of Proteins and Peptides

Name	Structure	P max (mPa)	pH Range	Organic Solvents Used
Silica	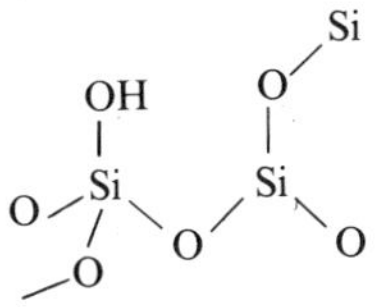 (silica structure: Si–OH, Si–O–Si network)	100	2–8	All
Polystyrene/ Divinylbenzene	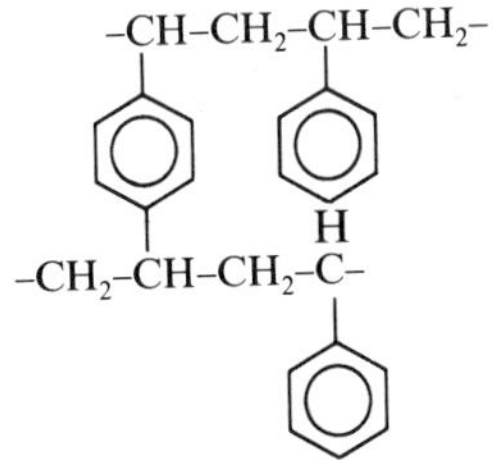 $-CH-CH_2-CH-CH_2-$ (phenyl rings); $-CH_2-CH-CH_2-C(H)-$ (phenyl)	40	2–12	Most
Hydroxylated Polyether	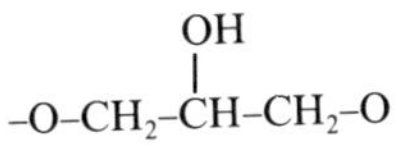 $-O-CH_2-CH(OH)-CH_2-O$	10	2–12	Most
Polymethyl methacrylates	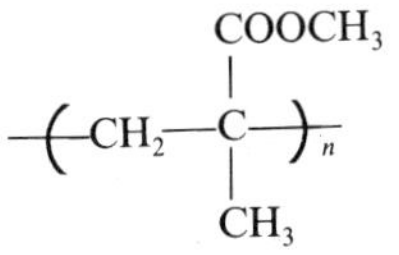 $\left(CH_2-C(COOCH_3)(CH_3) \right)_n$	10	2–10	Most
Alumina	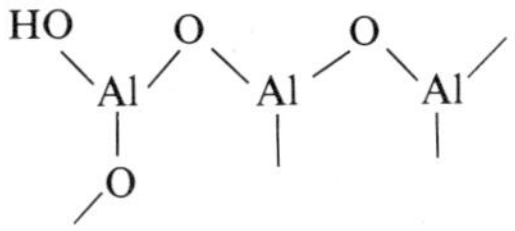 HO–Al–O–Al–O–Al network	100	2–14	All
Polyester/ polyamine	$\left(CH_2-CH_2-NH \right)_n$	40	2–12	Most

general conditions under which the matrix is stable. Microbial and enzymatic resistance is desirable for a number of reasons:

1. Most biological fluids contain enzymes that catalyze various reactions of reactive substrates such as polysaccharides. These processes not only tend to degrade the structural integrity of the substrate, but where microorganisms are involved, may cause the release of endotoxins into the product stream.

2. Accumulation of microorganisms within a packed column often causes increased back pressure, preventing further separation and purification.
3. Concerns about the preceding reduce column lifetime and increase maintenance of high performance columns. Currently, all but one (SUPEROSE™) commercial high-performance substrates are synthetic materials, exhibiting minimal sensitivity to microbial or enzymatic attack.

4.3.3. Nature of Bonded Phase

The properties of a high-performance ion exchanger are comprised of the combined properties of the substrate and surface. Surface charge and its density, the factors determining surface hydrophobicity, and structural requirements for the substrate, have been described in Sections 4.3.1 and 4.3.2. Other important factors in HPIEC that relate to the nature of the bonded phase as a whole are

1. Scale-up capability.
2. Stability.
3. Ease of regeneration (reequilibration).
4. Ease of cleaning.
5. Ease of sterilization.

4.3.3.1. Scale-Up Capability. HPIEC for protein and polypeptide purification on a large scale (100 g per injection) is a rapidly developing field. Although this area will not be reviewed in detail in this chapter, it is useful to examine the systematic philosophies that exist for approaching this subject. It is assumed that attempts to carry out large-scale purifications will automatically involve small-scale trials to optimize operating conditions:

1. Determine most suitable mode of chromatography in relation to its influence on sample preparation, buffer removal, analysis of fractions, mass and activity recoveries, degree of purification obtainable, capacity, throughput rates and costs.
2. The above determination may be done in several ways:
   a. Using HPIEC with 5 or 10 micron particles in an analytical column, then using larger particles with similar properties at the preparative step.
   b. With small columns of the larger particle preparative material, determine the best chromatographic conditions. The scale-up step to larger columns would then be carried out on the same ion-exchange packing as that used for analysis.
3. When the appropriate separation made has been determined using a small system, the chromatographic conditions are finally optimized (pH, buffer types, buffer ionic strengths, linear velocities, and gradient slope).
4. Transfer these conditions to a system in which column diameter is in-

creased by 5–10 times, keeping the column length unchanged. Scale-up flow rates, sample volumes in proportion to the preparative column volumes. For example, if analyses were optimized with a 4.6 mm × 25 cm column at 0.5 mL/min flow rate, scale up to a 21.2 mm × 25 cm column (21 × volume) at about 10 mL/min flow rate. A 500 μL analytical sample volume can be increased to 10 mL per injection.

5. Large columns generally contain packings of large particle diameter (25–65 micron) for reasons of economy. Large high-performance liquid chromatographs exist that can operate at 4000 psi, and, therefore, in principle may accommodate columns packed with 5 micron bonded phases. But these columns are usually very expensive and will place great stresses upon the equipment. Furthermore, since the current trend in large-scale preparative chromatography is to operate systems under various conditions of overload, the advantages of high efficiencies will be lost in many cases.

The capability for scale-up of a given bonded phase will need to include an assessment of the degree of chromatographic similarity between a small-particle and a large-particle ion exchanger. Both materials should ideally have similar polypeptide and protein adsorption isotherms. This ensures that changes in peak shapes and positions during loading remain similar for the small-particle methods development column and the preparative column containing large particles. Such similarity is also important where high-performance ion exchangers are used for rapid in-process monitoring.

The scale-up process should exploit the advantages, wherever possible, of high-performance packings. The large-particle ion exchangers should be operated with the same methodology as the high-performance packings. (In other words, similar physical and chemical characteristics, similar limits regarding pH ranges, buffer types, pressure, and linear flow rates.) Naturally, larger-particle packings must be available and economically priced. Large-scale high-performance and medium-performance liquid chromatographic equipment is available (52–55) to effectively handle column packing, sample loading, gradient elution, peak detection, and fraction collection from columns up to 6–12 in. internal diameter.

4.3.3.2. Stability. The chemical and physical stability of a high-performance ion exchanger is determined by the same properties of the support matrix, the firmness with which the surface layer (when used) is attached to this support, and the chemical inertness of the surface layer itself. Aspects of the stability of the substrate and the bound surface layer were discussed in Sections 4.3.2 and 4.3.1, respectively. Ideally, maximum bonded phase stability would be achieved where the charged groups are as integrated as possible with the entire structure of the packing material, and where no groups exist that can react with water, oxidants, reductants, or enzymes capable of catalyzing such reactions.

Table 11. Clean-Up Reagents for BAKERBOND Wide-Pore Columns for HPIEC

Reagents (in order of priority)

1. Sodium acetate (2 *M*, pH 8)
2. Acetic acid (10%), formic acid (10%), or trifluoracetic acid (0.2%)
3. DMSO/water (50/50)
4. Chaotropic agents such as aqueous urea, guanidine hydrochloride (3–6 *M*) is necessary

Protocol for use of above reagents

1. Wash the column with 10 column volumes of water
2. Run a gradient with 2 *M* sodium acetate, pH 8 (10 volumes)
3. Follow this with water (5 volumes); then acetic acid (10%, 10 volumes)
4. Repeat steps 1 to 3 if a single clean-up is insufficient

It is well established that synthetic polymers used in HPIEC, such as derivatized polystyrenes and polyhydroxyethers, are generally stable to hydrolysis over the pH range 2–12. In addition, Chicz and co-workers (45) have demonstrated that ion exchangers made from high-performance alumina and zirconia-coated silica have negligible solubilities at pH 14. Zirconium-treated silica, patented by Dupont, is known to withstand pH values up to 8.5 with little increase in solubility. The BAKERBOND series of polymer-bound silica-based ion exchangers and hydrophobic interactors has general stability over the pH range 2–10. Serva Fine Biochemicals has claimed a stability of pH 2–8.0 for the Daltosil and SI range of bonded phases for HPIEC of biopolymers.

Stability of an ion exchanger at high pH may be important when alkaline clean-up washes are used to remove final traces of polypeptides or proteins or the lipopolysaccharides that are the major class of endotoxins. However, alternate, less harsh methods have been developed, to achieve this clean-up (see Section 4.3.3.3).

4.3.3.3. Clean-Up. Modern packings for HPIEC should show mass recoveries of greater than 90% for a wide range of peptide and protein classes. Even in those cases where recoveries approach 100%, there will usually be small quantities of biopolymer that need to be removed either after each separation or after a given number of cycles. The efficiency with which a clean-up step may be performed is an important consideration in the choice of a bonded phase. Several commercial suppliers recommend clean-up procedures as part of the use and care of their products. Table 11 lists the reagents that can be used for clean-up of the BAKERBOND Wide-Pore family of bonded phases. Figure 12 illustrates the chromatographic elution profile obtained from a sample of standard proteins before and after clean-up with 0.1 *N* NaoH. The use and care of HPLC columns for biopolymer analysis has been reviewed by Wehr (7).

4.3.3.4. Sterilization. Sterilization is the process by which living cells or viruses are killed or prevented from reproducing. Table 12 lists the several agents that

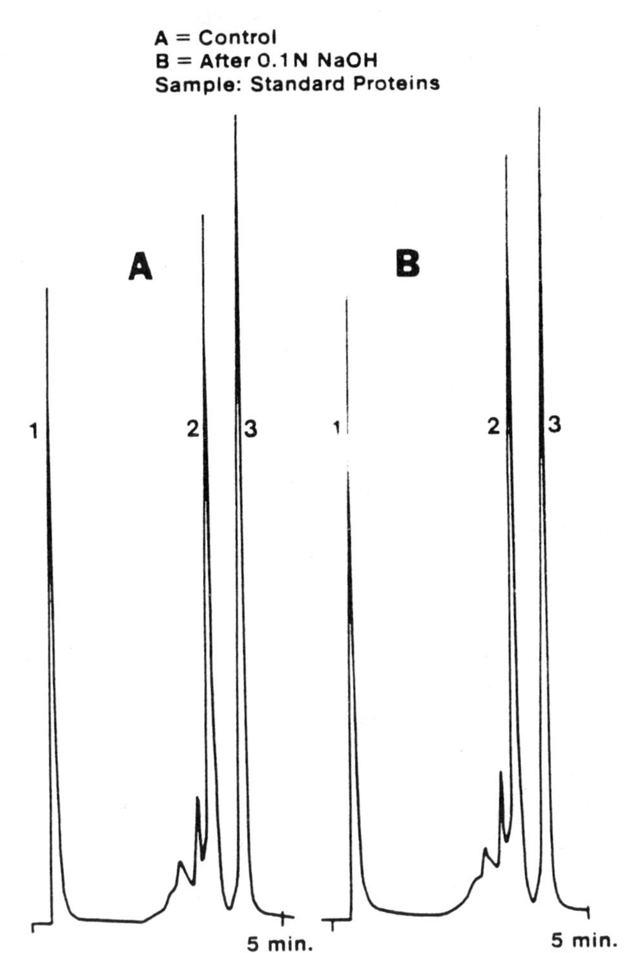

Fig. 12. Separation of standard proteins using weak anion exchange, before and after treatment with 36 column volumes of 0.1 *N* NaOH. Column: 4.6 mm × 50 mm, 5 micron BAKERBOND Wide-Pore PEI. Mobile phase: Initial buffer (A) 10 m*M* potassium dihydrogen phosphate, pH 6.8; final buffer (B) 500 m*M* potassium dihydrogen phosphate, pH 6.8. Gradient: 0% B to 100% B over 2 min. Flow rate: 1 mL/min. Detection: UV (280 nm) 0.1 AUFS. Peaks: 1. Cytochrome c; 2. Ovalbumin; 3. β-Lactoglobulin A. Chromatogram A: initial chromatography. Chromatogram B: same chromatography after 30 mls (36 column volumes) of sodium hydroxide (0.1 *N*) was cycled through column, neutralized with buffer B to pH 6.8 and reequilibrated. [From S.A. Berkowitz et al. (79).]

Table 12. Sterilizing Agents[a] for BAKERBOND HPIEC Columns

Acetone	Trichloroacetic acid (10%)
2-Propanol	Hydrogen peroxide (10%)
Hexane	Sodium dodecyl sulfate (1%)
Dichloromethane	Nitric acid (10%)
Ammonium Hydroxide (1 *N*)	Ethanol and 70/30 ethanol/water
Sodium Hydroxide (0.1 *N*)	

[a]Agents giving 100% prevention of the growth of microorganism after at least 84 hr, in a tryptic soy digest agar medium. Details of method are given in Ref. 56.

are suitable for this process. Some of these solutions cannot be used with certain high-performance ion exchangers, since the latter interact physically or chemically with the solutions recommended. Often the column or packing material is washed or stored or both in the sterilizing solution. The manufacturers recommendations should be followed in all cases.

4.3.3.5. Reequilibration. At the end of an elution in which pH and ionic strength may be different from the starting conditions, reequilibration will be necessary. A high-performance ion exchanger can be said to be reequilibrated when the ionic strength and pH of the solution within and without the pores of the packing (if porous) are the same as those at the beginning of the separation. Furthermore, the state of ionization of the charged surface groups must be the same as that reached initially.

During the process of reequilibration of an HPIEC column, from say a higher ionic strength and pH to a lower ionic strength and pH, it is postulated that the following processes occur:

1. Ionic strength and pH outside the pores drop immediately at the inlet of the column.
2. Ions diffuse out of the pores of the ion exchanger, tending to counteract the decrease.
3. Protons diffuse into the pores and begin titrating the basic groups at the ion-exchanger surface. This will tend to counteract the higher concentration of protons in the reequilibrating solution.
4. As fresh solution enters the column, processes 2 and 3 will continue but at a slower rate, since the concentration gradient from inside to outside the pore will be shallower.
5. The new rate of exchange of ions and protons will slow down and reach zero at equilibrium.

The time taken for reequilibration will thus depend on the linear sweep velocity of the initial solution, the rates of diffusion of ions into and out of the pores, the void volume inside the column, and the speed at which the "titration" occurs. For strong ion exchangers such as those bearing a quaternary

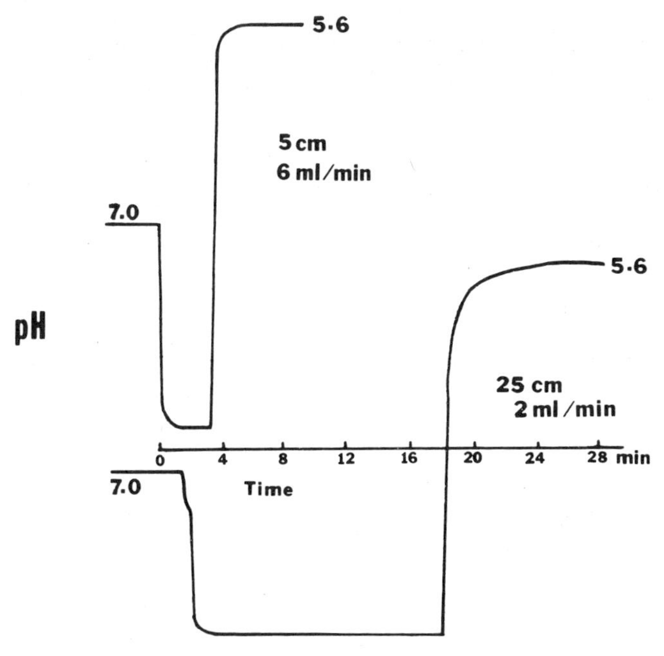

Fig. 13. Reequilibration speed of short columns. Columns: BAKERBOND Wide Pore CBX, 5 micron, 4.6 × 50 mm and 4.6 × 250 mm. Both columns were equilibrated with 1.0 *M* NaAcetate, pH 7.0 at the flow rates indicated, while monitoring the effluent pH. The buffer was changed to 10 m*M* MES, pH 5.6 and the pH monitored as a function of time. [From Berkowitz (86).]

ammonium group and the sulfonic acid group, no titration (or interaction with protons) will occur. The reequilibration process will thus be faster for strong exchangers than for weak exchangers, all else being equal. Diffusion of protons to the titratable groups in weak ion exchangers will be controlled by the accessibility of the chargeable groups. Those buried deep in a bound polymer layer, for example, will be slow to titrate and reequilibration will be slow. Diffusion rates will be larger for larger cross-sectional pore areas, and larger pores are an advantage here. If diffusion and titration rates are high enough, the rate of reequilibration will depend only on void volume and linear velocity. For a given flow rate, therefore, smaller columns will equilibrate faster than larger columns. Figure 13 illustrates this process for the reequilibration of a weak cation exchanger in a 5 cm and a 25 cm column of the same internal diameter. Naturally, higher flow rates will also reduce reequilibration time, although not necessarily in proportion to column volume. High flow rates are possible with shorter columns, since back pressures are lower.

4.4. Column Configurations and Construction

The choice of column configuration is dictated by the manufacturer of column hardware or by commercially available prepacked columns. The first columns

Table 13. Major Commercial Column, Configurations Advantages, Limitation and Application Areas

Size (ID × length, mm)	Advantages	Limitations	Application Areas
4.6 × 30	Low pressure/high speed	Low capacity/ decreased resolution	Fast analysis
4.6 × 50	Low pressure/high speed	Low capacity/ decreased resolution	Fast analysis
5.0 × 50	Low pressure/high speed	Low capacity/ decreased resolution	Fast analysis
4.6 × 100	Moderate pressure and speed	Decreased capacity	Moderate speed analysis
4.6 × 150	Moderate pressure and speed	Fair capacity	Moderate speed analysis
4.6 × 250	High resolution and capacity	High pressure	High resolution analysis, semi-preparative
7.75 × 100	Low pressure/high capacity		Moderate speed analysis, semi-preparative
10.0 × 100	Low pressure/high capacity		Analysis, semi-preparative

for analytical HPIEC were 25–30 cm long and 4–5 mm ID. More recently, shorter columns (3.3 cm long) have become commercially available (50). Table 13 lists the major column sizes vailable with a description of their advantages, limitations, and important areas of application.

The construction of an HPIEC column is important when considering chemical inertness, biological compatibility, adaptability to various chromatographs, pressure tolerances, and, occasionally, ease of disassembly for column repair when necessary. In addition, column hardware design will influence its chromatographic efficiency somewhat, although this latter parameter depends far more on the nature of the packing medium and the sophistication of the column packing technique. Table 14 lists the major materials of construction of HPIEC columns together with brief comments on their physical, chemical, and biological compatibilities with mobile phase components, instrumentation, and peptides and proteins of interest.

Tables 13 and 14 can be used to select a general configuration and construction of a column to fit the requirements of the separation. Naturally, the choice will be limited to the column types offered by the company whose bonded phases appear best suited for a particular separation, although some companies offer a custom column service.

Table 14. Major Column Construction Materials

Material	Physical Strength	Chemical Compatibility	Biological Compatibility
		End-Fitting	
316 stainless steel	Highest	Avoid halides	Generally compatible
Titanium	High	Noncorrodible	Compatible
Fluoropolymer	High	Noncorrodible	Compatible
		Frits	
316 stainless steel	High	Avoid halides	Compatible, adsorbs protein slightly
Titanium	Moderate	Noncorrodible	Compatible, adsorbs protein slightly
		Column Wall	
316 stainless steel	Highest	Avoid halides	Compatible, adsorbs protein slightly
Borosilicate glass	Low	Noncorrodible	Compatible, adsorbs protein slightly
Fluoropolymer	High	Noncorrodible	Compatible, slight adsorption of protein

4.5. Quality, Cost, and Service

These aspects of bonded phase selection are, of course, very difficult to assess. However, a brief, necessarily incomplete general guide to this topic can be given:

4.5.1. Quality

Bonded phases must be selected that are reproducible, sufficiently resolving, stable, easy to clean, regenerate, and sterilize and give high mass and activity recoveries. Some or all of this information may be obtainable from the manufacturer. Each manufacturer's weak anion exchanger, for example, will often be different from the others. This may be useful when two components, poorly resolved on one ion exchanger, may successfully separate on another.

4.5.2. Cost

Take into account the specific binding capacity, since several available HPIEC columns can be used for both analysis and small prep-scale work. Cost should, of course, be related to quality. Catalogs and price lists from all major suppliers should be obtained and kept conveniently together.

4.5.3. Service

Determine the depth of technical knowledge available in a company to support their bonded phase products. This knowledge is usually found in their customer service departments, technical sales or marketing specialists, and research and development scientists. The name of the marketing people or scientists of most value in this respect will be found on papers published on the products in which you are interested. These people will be able to give detailed information on column use and care, applications, and in most cases can help in solving new separation problems and troubleshooting a range of technological difficulties. Timely delivery is important and assurances on this matter should be obtained from customer service and other independent users of a product.

5. MOBILE PHASE SELECTION (57)

The full range of buffers, salts, and other reagents that have been used over the past 30 years in soft-gel ion-exchange chromatography is available for HPIEC (Section 9, Applications). In addition, many high-performance packings can be used with organic solvents. The following questions should be asked in selecting a mobile phase for HPIEC:

1. What has been used in the past for the ion-exchange chromatography of the protein?
2. What are the properties and components of the buffers with which the protein of interest is most compatable (stable)?
3. How volatile is the buffer? How expensive is the buffer?
4. Will the buffer react with the packing material or any components of the liquid chromatograph?
5. Can sources of sufficiently pure buffers and salts be found? How reliable are those sources?
6. What are the solubilities of the protein of interest in various buffers at what pH values?
7. What temperature is necessary and is the buffer functional under this condition?

A list of buffers, salts, and other reagents used in HPIEC is given in Table 15, along with the most appropriate pH range over which they can best be used. The choice of pH depends on the isoelectric point of the peptide or protein, and was discussed in Section 4.3.1.1.

Much information needed for mobile phase selection will be obtained as the sample and components of interest are characterized (Sections 4.1, 4.2) and the appropriate ion exchanger and column construction are chosen (Section 4.3).

Table 15. Common Buffers and Salts Used in HPIEC

Buffer	pH Range	Buffer	pH Range	Neutral Salts
H_3PO_4	1–3	Na citrate	3–5	NaCl
KH_2PO_4	6–8	Na glycinate	8–10	Na_2SO_4
NH_4 acetate	6–8	Acetic acid	4–5	$(NH_4)_2SO_4$
Na acetate	4–5	2[*N*-morpholino] ethanesulfonic acid (MES)	5–7	
Tris hydrochloride	7–9			

6. BINDING AND ELUTING

In HPIEC the stationary and mobile phases are chosen so that either the protein of interest is bound and as many of the other components are not bound, or the protein of interest is not bound and the other components are bound to the ion exchanger. In general, the former is usually practiced, since a degree of concentration can be achieved in addition to purification.

Peptides and proteins are bound to ion exchangers at low ionic strength and at a pH that maximizes the difference in charge between the surface of the bonded phase and the binding site of the protein. Under these conditions there is the least competition from other ions in the buffer with the macromolecular ion, and there is the strongest ionic binding. When the competition becomes greater, in other words ionic strength increases and when the pH changes such that the charge on the protein's surface approaches that on the ion exchanger, elution will occur. The correct choice of mobile phase is important in order to achieve optimal binding and release under conditions of least interference from other components in the mixture (Section 7).

Most HPIEC is carried out under gradient (rather than isocratic) elution conditions, where ionic strength and pH may change during chromatography. There is great scope for controlling the binding and elution patterns with this system. The nature of the sample (temperature, ionic strength, pH, concentration of peptide components, volume) is also important in achieving efficient binding, since perturbations of the equilibrated column at the inlet by the sample solution may change the binding ability of the ion exchangers. The diffusion of proteins to the interior binding surface of porous ion exchangers is a time-dependent process, and, consequently, the linear velocity at which the sample is applied will be important. In general, best binding is achieved when the flow rate for sample application is at most one-half of the flow rate that will be used for elution.

The dwell time—the time that the protein of interest remains bound to the ion exchanger before elution—can influence the state of denaturation or aggregation in which the protein of interest is obtained after separation (Section 4.3.1.4). Short dwell times will ensure the minimum of undesirable changes in these properties. Dwell times may be minimized in some cases by binding the protein of interest at ionic strengths just below the elution value, then rapidly desorbing the protein by slightly increasing ionic strength. Unfortunately this technique is not generally applicable, since many proteins are most effectively adsorbed at ionic strengths significantly below the elution value. The reasons for this are not clear. Table 16 lists the general binding and elution buffers for selected proteins and sample types on various BAKERBOND Wide-Pore ion-exchange columns.

It is instructive to monitor both pH and ionic strength (by conductivity) of the column eluent using flow cells. These techniques will enable the determination of the point when the column has been equilibrated and how the pH and conductivity changes during elution. Figure 14 shows how pH changes during elution of standard proteins from BAKERBOND PREPSCALE* CBX (40 micron weak cation exchanger) with different buffer systems. The pH of the solution in which the protein elutes can be controlled to some extent by the correct choice of elution buffer. More important, however, this choice will determine the selectivity of the ion exchanger (Section 7).

7. RESOLUTION CONTROL

Resolution is the primary aim of HPIEC, and its control is consequently of major importance. The tools that a chromatographer has available to him in this regard are as follows:

1. Choice of bonded phase(s).
2. Choice of binding and elution buffers, gradient slope, and flow rates.
3. Temperature of sample, column and buffers.

7.1. Choice of Bonded Phase(s)

The most important choice that can be made is that of the bonded phase. Aspects of this procedure were described in Section 4.

7.2. Choice of Binding and Elution Buffers

The nature, pH, and ionic strength of these buffers can often influence the resolution of the protein of interest from other components. This is illustrated in Figure 15. In this case, calmodulin, a calcium-binding protein, may be more or less resolved from the other components in bovine brain extract, depending on the buffer properties.

The application of these observations to the prediction of optimum binding and elution conditions has no general validity, unfortunately. Each peptide and protein in a mixture will exhibit its own buffer-dependent retention characteristics. It is useful initially, however, to gain a general idea of the retention

Table 16. Common Binding and Eluting Buffers and Sample Types for BAKERBOND HPIEC Columns

Sample	Major Protein Bound	Column Type	Binding Buffer (A)	Eluting Buffers (B)
Ascites fluid	Monoclonal antibody	ABx[a]	10–25 m*M* MES, pH 5.6	1 *M* Na acetate, pH 7.0 or 500 m*M* $(NH_4)_2SO_4$ + A or 250 m*M* KH_2PO_4, pH 6.8
Serum	Immunoglobulins	ABx	As above	500 m*M* $(NH_4)_2SO_4$ + A
Cell culture	Monoclonal antibodies	ABx	As above	As above
Muscle extract	Actin	PEI[b]	25 m*M* Tris, pH 7.0	2 *M* Na Acetate, pH 7.0
Protein standards	Acidic proteins	PEI	10 m*M* KH_2PO_4, pH 6.8	250 m*M* KH_2PO_4, pH 6.8
	Basic proteins	CBX[c]	10 mM KH_2PO_4, pH 6.0 or 10 mM MES, pH 5.2	500 mM KH_2PO_4, pH 6.0
Bovine brain extract	Various, calmodulin	PEI	25 m*M* Tris, pH 7.0	2 *M* Na acetate, pH 6.0
Rabbit muscle extract	Aldolase	CBX	20 m*M* MES, pH 5.4	500 m*M* NaCl plus 50 mM KH_2PO_4, pH 6.7
Yeast extract	Glucose-6-phosphate dehydrogenase	PEI	25 m*M* Tris, pH 7.0	2 *M* Na acetate, pH 6.0
Crude egg white	Lysozyme	CBX	10 m*M* MES, pH 5.6	1 *M* Na acetate, pH 7.0

[a] BAKERBOND ABx, Mixed-Mode (see Table 4).
[b] BAKERBOND Wide-Pore PEI (polyethyleneimine, weak anion exchanger).
[c] BAKERBOND Wide-Pore CBX (carboxyethyl, weak cation exchanger).

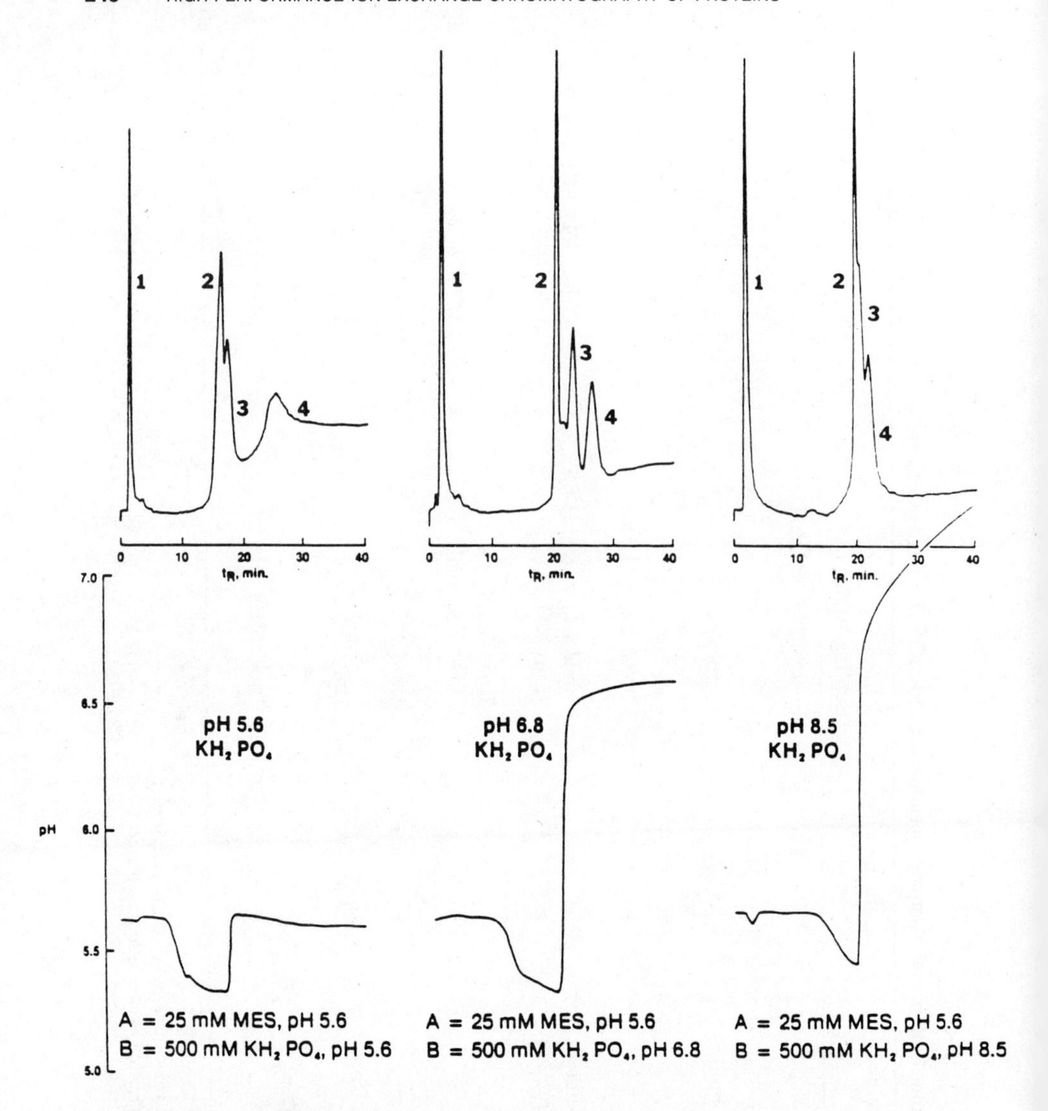

Fig. 14. Effect of B buffer pH upon protein selectivity and eluent pH profile. Column: BAKERBOND PREPSCALE CBX, 4.6 × 250 mm, 40 micron. Mobile phase: A and B, see figure for details. Gradients: 0–100% B over 30 min. Flow rate: 1 mL/min. Peaks: 1. ovalbumin; 2. hemoglobin; 3. cytochrome c; 4. lysozyme. [From Nau (85).]

properties of a protein of interest in its isolated state (in other words, as a standard, purified material). Location of the protein in the chromatogram of the real sample will then be easier; and the correct choice of mobile phase buffers giving the best resolution may be simplified.

There are many empirical approaches that could be made in peptide and

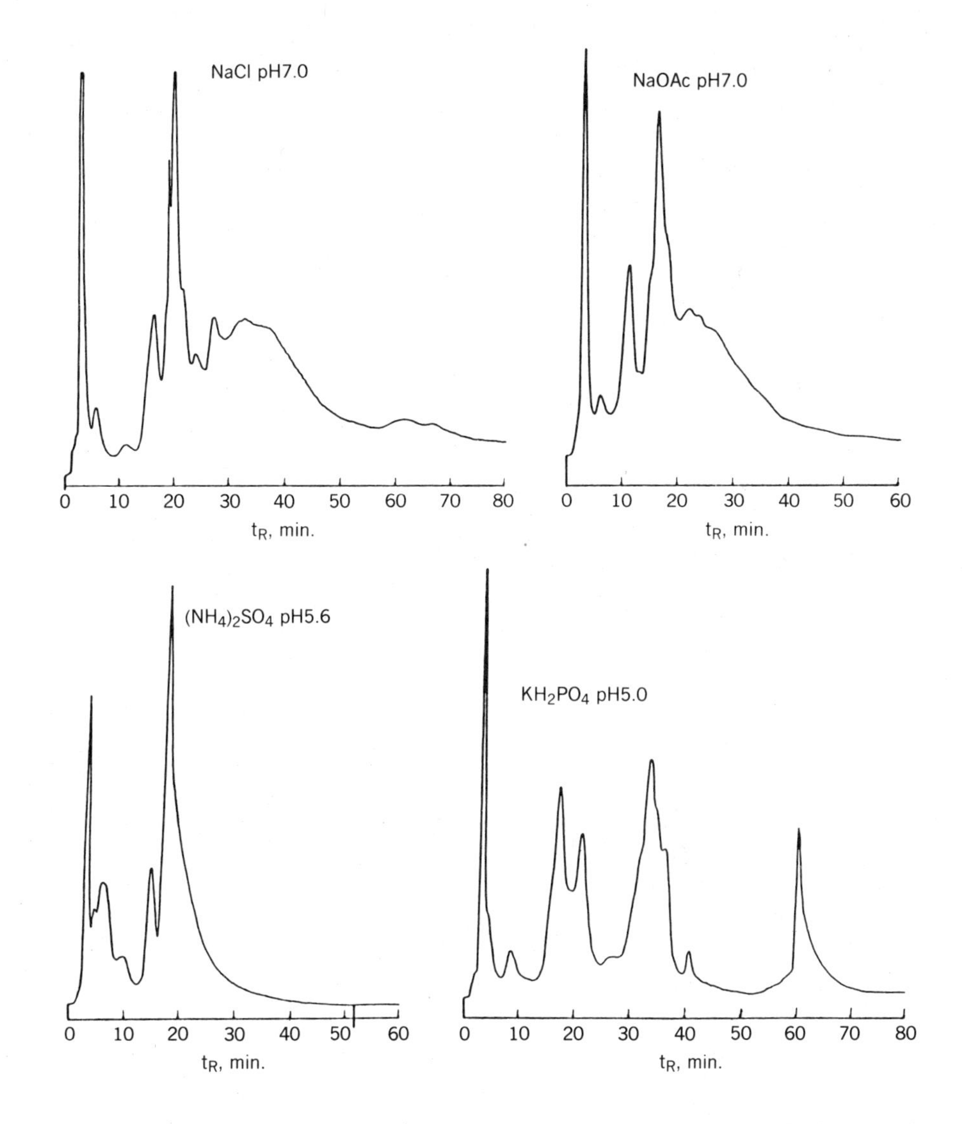

Fig. 15. Dependence of resolution in a complex protein mixture (bovine brain extract) upon elution buffer. Column: BAKERBOND PRESCALE QUAT*, 4.6 × 250 mm, 40 micron. Mobile phase: A = 25 m*M* Tris, pH 9.0; B = see above. Gradient: 0% B to 100% B over 60 min. Flow rate: 1 mL/min. Pressure: 25 psi. Detection: UV at 280 nm. Sample: bovine brain extract, 5 mg protein. [From Nau (85).]

protein retention information once the bonded phase has been selected. The following procedure is one such approach using a standard protein or proteins, with gradient elution:

(i) Choose linear velocity for sample application (binding) and elution.

Initially, the linear velocity for sample application should generally be within the range 20–40 mm/min. For chromatography the linear velocity range may be double this value, or 40–80 mm/min. These velocities correspond to 0.5 mL/min (sample application) and 1 mL/min (elution), for a standard 4.6 × 250 mm analytical column. These values are those typically used in the author's laboratories and also as reported in the literature. (See Section 9, Applications.)

(ii) Choose sample properties such as pH, ionic strength, peptide and protein concentrations, and volume. Some of these parameters will depend on the solubility and the binding strength of the protein of interest, the binding capacity of the bonded phase, and the column configuration. For simplicity, it will be assumed that the sample solution properties will be the same as those of the binding buffer (buffer A). [See Step (iii) for choice of binding buffers.] This can be achieved by dialysis of the sample against buffer A, in order to avoid any precipitation upon sample application.

(iii) Choose buffer A to maximize the binding of the protein of interest. (See Section 6.) Ionic strength is usually kept low, at about 25 m*M*. The pH is selected to lie between the isoelectric point of the protein of interest (if known) and the pK_a of the functional groups on the ion-exchanger surface. In those cases where the surface charge of an ion exchanger varies continuously over a wide range of pH values (see Fig. 3), there are many pK_a values and this choice of pH is consequently much wider.

The effect of pH variation of buffer A upon retention of hemoglobin on a strong cation exchange column is illustrated in Fig. 16. In this case, the pI of hemoglobin is about 7.2, and the pK_a of the functional group is about 1. Provided pH lies between these limits (1 and 7) hemoglobin will bind. As the pH increases, hemoglobin at first bound, is seen to bind less and less tightly.

The buffer type is chosen according to the pH of buffer A, and a guide is given in Table 15.

(iv) Choose the elution buffer (B) with a fairly high ionic strength (see Table 15). Adjust the pH to be the same as buffer A initially. Figure 17 shows the influence of the type of buffer B upon retention and selectivity of a group of standard proteins, chromatographed upon a strong cation exchanger.

The variation in chromatographic selectivity among a group of proteins as pH and ionic strength of buffer B are changed, is a function of the nature of the ion-exchanger. (The physicochemical contributions of pore size, hydrophobicity, charge density and homogeneity, and function group type to retention was discussed in Section 4.3.)

(v) Choose the gradient steepness, assuming in the first instance that the full range of buffers A and B will be used (100% A to 100% B). Typically, the increase in buffer B should be at rates between 1.5 and 3.0 volume percent per minute. Figure 18 illustrates the changes in resolution between standard proteins observed in varying the gradient steepness from 7.0 to 3.5 to 1.8 volume percent per minute.

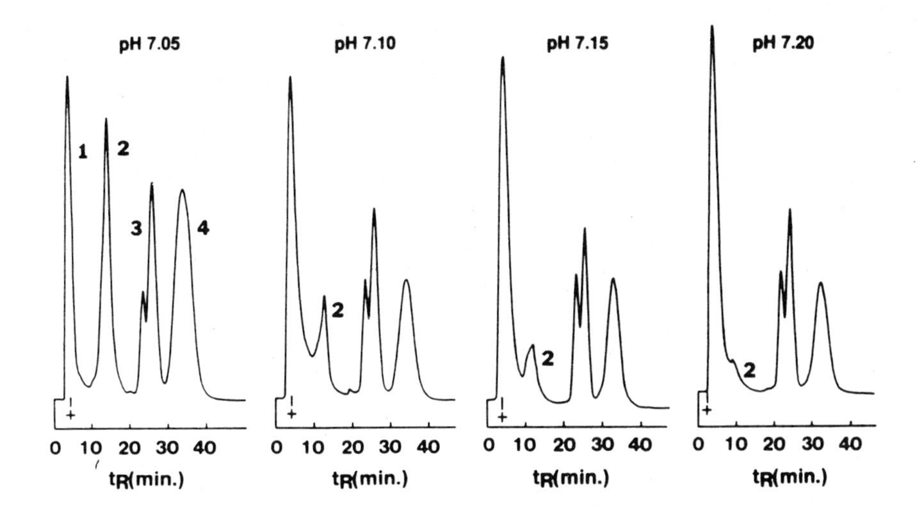

Fig. 16. Effect of binding buffer pH on elution profile. Column: 7.75 × 100 mm, experimental strong cation exchanger, 40 micron. Mobile phase: A = 12 m*M* KH_2PO_4, pH (see above); B = 500 m*M* $(NH_4)_2SO_4$ plus 200 m*M* Na acetate, pH 7.0. Gradient: 0% B to 100% B over 30 min. Flow rate: 1.0 mL/min. Detection: UV at 280 nm, 0.5 AUFS. Peaks: 1. ovalbumin; 2. hemoglobin; 3. cytochrome c; 4. lysozyme. [From Nau (87).]

7.3. Temperature of Sample, Column and Buffers

Choice of temperature will be determined by the lability of the peptide or protein under chromatographic conditions. The effects of temperature upon denaturation, oligomerization, and chemical degradation of a biopolymer should be studied before chromatography is carried out. These properties may also depend upon the condition of purification or concentration of the protein.

Much work has been done in HPIEC to maximize resolution, and some of this has been published. Section 9 (Applications) reviews some of this work, in addition to providing selected sources of detailed information on HPIEC of peptides and proteins.

8. COLUMN USE AND CARE

Instructions for column use and care are generally supplied with each HPIEC column. It is worthwhile requesting a copy of these from the manufacturer before purchasing since the instructions usually contain much useful information which will help in choosing a column.

The following guideline is necessarily brief and describes the most important aspects of column care. The subject has been reviewed by Wehr (14).

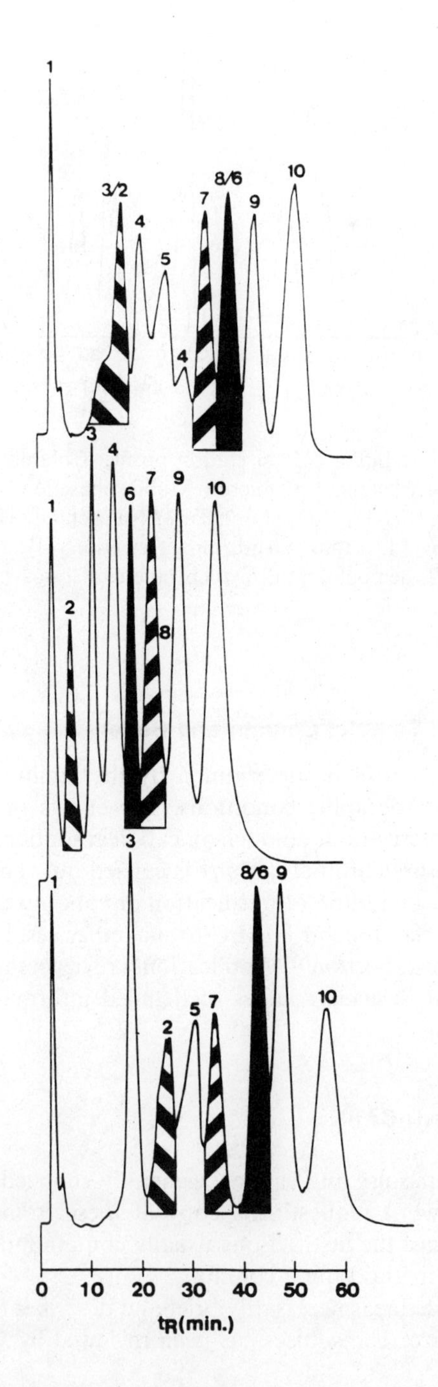

1
3/2
4
5
4
7
8/6
9
10
3
4
1
6
7
9
10
2
8
3
1
8/6
9
2
5
7
10
0 10 20 30 40 50 60
t_R(min.)

Step 1. Examine the column and its container for any signs of damage. The column end caps should be screwed in tight. Do not subject the column to any shock.

Step 2. Check the contents of the column box, which normally contains the column, extra nuts, ferrules and frits, instructions and a QC certificate. Keep the contents inside the box for safekeeping.

Step 3. Read Column Use and Care Instructions.

Step 4. Read the QC certificate carefully and note the column shipping solvent.

Step 5. Ensure that the mobile phase in your HPLC is compatible with this solvent.

Step 6. Follow the manufacturer's instructions for equilibrating the column. This normally includes a fairly prolonged wash with a low-ionic-strength buffer at a slow flow rate. This is followed by a gradient elution from low to high ionic strength and back.

Step 7. Monitor eluent spectroscopic properties and column back-pressure during the above column conditioning.

Step 8. Separate the components of the prepared sample according to the determined procedure.

Step 9. At the conclusion of the work, store the column according to the manufacturer's instructions.

Step 10. Further extended equilibration is normally unnecessary.

9. APPLICATIONS

The technique of HPIEC of proteins and polypeptides has developed at an accelerating pace since the first applications in 1976. The number of applications is extraordinarily large, and the list given in Table 17 is consequently very selective. The references given in this table describe more or less detailed procedures and results in HPIEC of proteins and polypeptides. They cover a range of applications of varying complexity, several classes of proteins and polypeptides, and illustrate the use of the major commercially-available columns. Figures 2, 15, 19, and 20 illustrate the HPIEC of a number of naturally

---

Fig. 17. (*Opposite*) Effect of elution buffer on protein selectivity on a strong cation exchanger. Column: 10 × 100 mm, BAKERBOND PREPSCALE CARBOXY-SULFON*, 40 micron. Mobile phase: A = 10 m*M* KH_2PO_4, pH 7.15; B = 500 m*M* KH_2PO_4, pH 7.0 (a), 500 m*M* NaCl plus 12.5 m*M* KH_2PO_4, pH 5.0 (b), 2 *M* NH_4 acetate, pH 7.0 (c). Flow rate: 2.0 mL/min. Pressure: 15 psi. Detection: UV at 280 nm, 0.5 AUFS. Peaks: 1. ovalbumin; 2. aldolase; 3. hemoglobin; 4. trypsinogen; 5. trypsin; 6. ribonuclease A; 7. α-chymotrypsinogen; 8. cytochrome c (oxidized); 9. cytochrome c (reduced); 10. Lysozyme. Sample: Protein standards as above (20 mg total). [From Nau (85).]

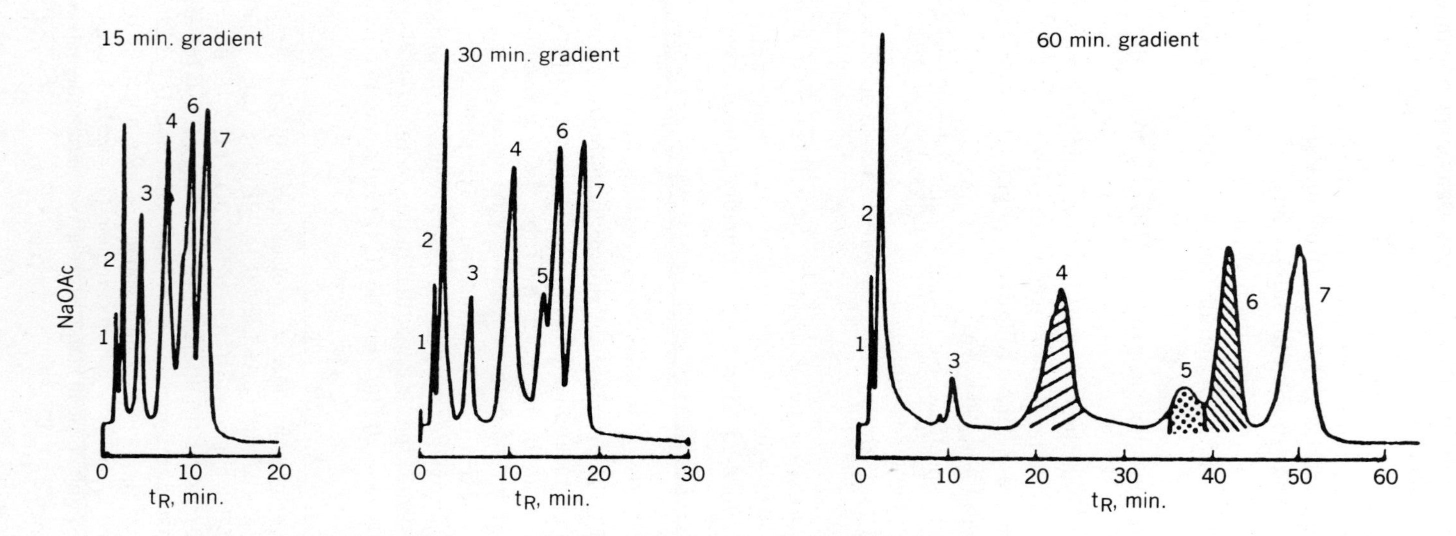

Fig. 18. Effect of gradient steepness upon the resolution of standard proteins on a weak cation exchanger. Column: 10 mm × 10 mm, BAKERBOND PREPSCALE CBX, 40 micron. Mobile phase: A = 25 m*M* MES, pH 5.4; B = 1.0 *M* Na acetate, pH 7.0. Gradient: 0% B to 100% B over 15, 30, and 60 min. Flow rate: 2 mL/min. Pressure: 5 psi. Detection: UV at 280, 0.2 AUFS. Peaks: 1. impurities; 2. ovalbumin (monomer); 3. ovalbumin (dimer); 4. hemoglobin; 5. reduced cytochrome c; 6. oxidized cytochrome c; 7. Lysozyme. Sample: Protein standards as above, 20 mg each. [From Nau (85).]

Table 17. Applications of HPIEC to the Separation of Mixtures of Peptides and Proteins

Protein/Peptide or General Class	Reference Number
Apolipoproteins	58
Basic peptides	59
Calmodulin	60
Cereal proteins	61
Creatine kinase	62
Serum proteins	63
Hemoglobins	64–68
Growth factors	69
Monoclonal antibodies	70–72
Ribosomal proteins	73
Plant globulins	74
Hormones	75
Membrane proteins	76
Complex proteins mixtures	77
Sample preparation	7
General strategies	78, 79
Insect venoms	80
Viral proteins	81
Alkaline phosphatase	82
Red cell membrane proteins	83
General proteins	20, 77, 79

occurring sources of proteins. The set of chromatographic conditions given with each figure is an accurate guide to obtaining reproducible chromatography with the respective samples. Sample preparation is important for protein-containing samples, and guidelines for this technique are given in Sections 2, 4.1, and 4.2.

TRADEMARKS

BAKERBOND ABx, BAKERBOND, BAKERBOND Wide-Pore, PREP-SCALE, CARBOXY-SULFON, HI-Propyl and QUAT are Trademarks (*) of J.T. Baker Inc.

PARAFILM® is a registered trademark of American Can Company.

LIPOCLEAN™ is a trademark of Behringwerke AG.

SUPEROSE™ is a trademark of Pharmacia LKB Biotechnology.

FPLC® is a registered trademark of Pharmacia LKB Biotechnology.

IsoPURE™ is a trademark of the Perkin-Elmer Corporation.

ZORBAX® is a registered trademark of E.I. duPont de Nemours & Co., Inc.

SynChropak® is a registered trademark of Synchrom, Inc.

MonoBeads® is a registered trademark of Pharmacia LKB Biotechnology.

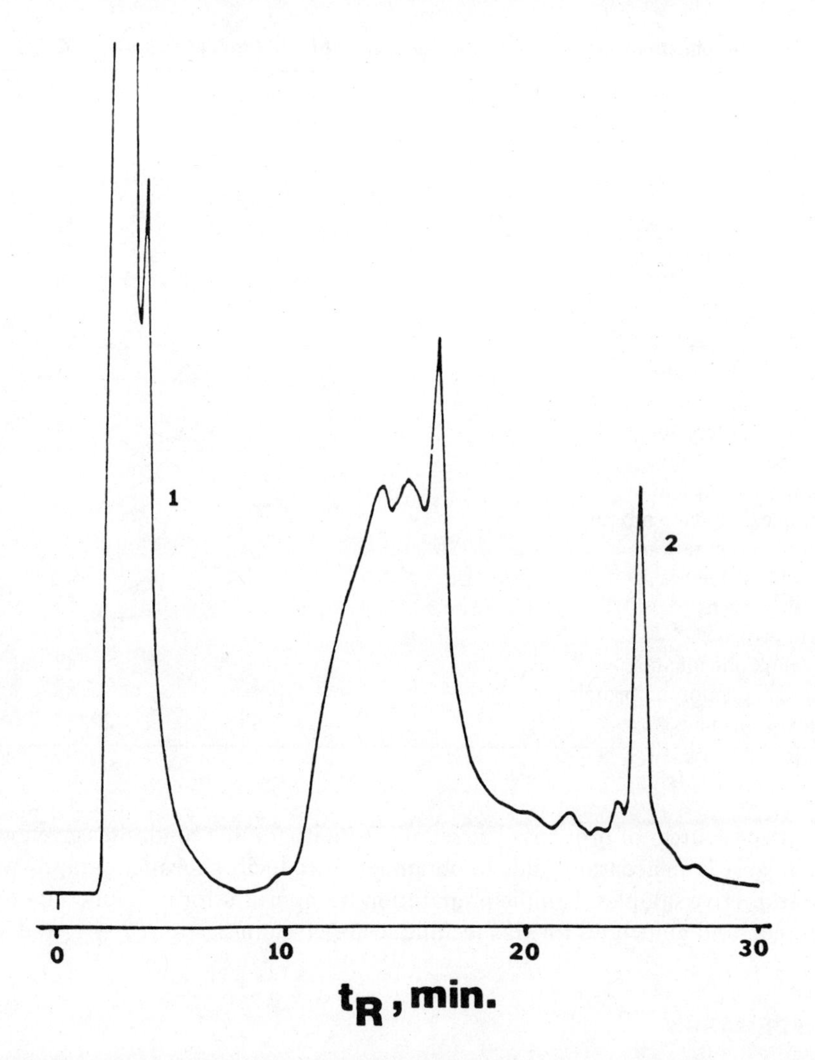

Fig. 19. Purification of lysozyme from crude hen egg white using a weak cation exchanger. Column: 4.6 mm × 250 mm, 5 micron BAKERBOND Wide-Pore CBX. Mobile phase: Initial buffer (A) 10 m*M* MES, pH 5.6; final buffer (B) 1.0 *M* sodium acetate, pH 7.0. Gradient: 0% B to 100% B over 30 min. Flow rate: 1.0 mL/min. Detection: UV at 280 nm, 1.0 AUFS. Sample: 0.05 mL [fresh egg white, 30 mg dissolved in buffer A (100 mL) and carefully filtered]. Peaks: 1. ovalbumin; 2. lysozyme. (From Berkowitz (86).]

Microanalyzer™ and Cellex™ are trademarks of Bio-Rad Laboratories.

Bio-Rex®, AG®, Bio-Gel® and Affi-Gel® are registered trademarks of Bio-Rad Laboratories.

ServaCel® is a registered trademark of Serva Fine Biochemicals Inc.

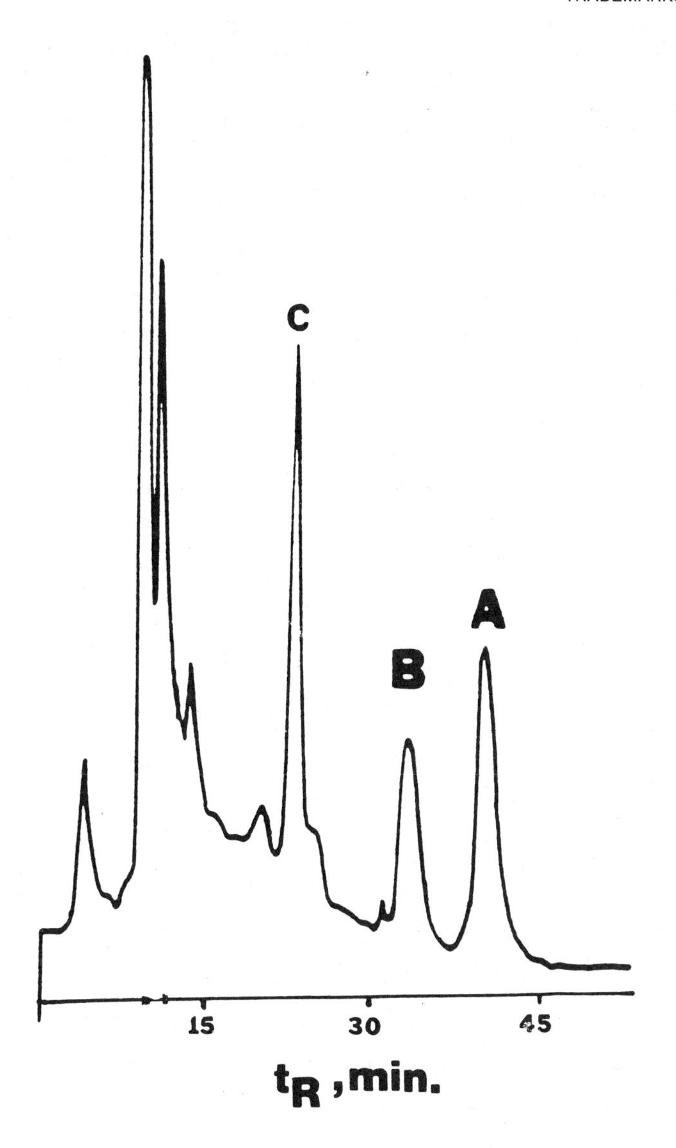

Fig. 20. Separation of milk proteins using a weak anion exchanger. Column: 7.75 mm × 100 mm, 15 micron BAKERBOND Wide-Pore PEI. Mobile phase: Initial buffer (a) 10 m*M* Tris, pH 7.0; final buffer (B) 2 *M* sodium acetate, pH 6.0. Gradient: 0% B to 100% B over 60 min. Flow rate: 1 mL/min. Detection: UV at 280, 0.5 AUFS. Sample: 0.5 mL (LIPOCLEAN™-extracted skim milk, diluted three times with buffer A). Peaks: C = Conalbumin, B = β-Lactoglobulin B, A = β-Lactoglobulin A. [From Nau (85).]

Dynamax® is a registered trademark of RAININ INSTRUMENT COMPANY, INC.

AQUAPORE® is a registered trademark of Brownlee Labs, Inc.

TSK® and TOYOPEARL® are registered trademarks of the Toyo Soda Manufacturing Company.

Nugel™ is a trademark of Separations Industries Inc.

Hydrophase™ is a trademark of Interaction Chemicals Inc.

Matrex® is a registered trademark of WR Grace & Co.

SUPELCOSIL® is a registered trademark of SUPELCO, Inc.

PolyCAT A™ is a trademark of PolyLC Inc.

Waters™ is a trademark of Millipore Corp.

Sephadex®, Sephacryl®, Sepharose®, Sephacel®, Trisacryl®, Ultrogel®, HA-Ultrogel®, Magnogel® are registered trademarks of Pharmacia-LKB Biotechnology.

Indion® is a registered trademark of Phoenix Chemicals.

ACKNOWLEDGEMENT

I would like to thank Drs. Laura J. Crane and Harold A. Kaufman for their practical comments on this manuscript; Drs. Steven A. Berkowitz and David R. Nau for the chromatography; Ms. Joanne Volkert and Mary Boncik for the excellent production of this chapter.

REFERENCES

1. H.S. Thompson, J. Roy. Agr. Soc. Engl., *11*, 68 (1850).
2. E. Margoliash, Nature, *170*, 1014 (1952).
3. E.C. Yackel and W.O. Kenyan, J. Am. Chem. Soc., *64*, 121 (1942).
4. E.A. Peterson and H.A. Sober, J. Am. Chem. Soc., *78*, 756 (1956).
5. S.H. Chang, K.M. Gooding, and F.E. Regnier, J. Chromatogr., *125*, 103 (1976).
6. L.R. Snyder and J.J. Kirkland, *Introduction to Modern Liquid Chromatography*, Wiley, New York, 1974.
7. C.T. Wehr, J. Chromatogr., *418*, 27–50 (1987).
8. C.T. Wehr and R.E. Majors, LC-GC Magazine, *5*, 548–552 (1987).
9. C.H. Suelter, *A Practical Guide to Enzymology*, Wiley, New York, 1985.
10. J.M. Brewer, A.J. Pesce, and R.B. Ashworth, *Experimental Techniques in Biochemistry*, Prentice-Hall, New Jersey, 1974.
11. *Methods in Enzymology*, see Cumulative Subject Index, Volumes 81–94, 96–101, Academic Press, Orlando, FL, 1986.
12. N.A. Parris, *Instrumental Liquid Chromatography*, 2nd rev. ed., Elsevier, Amsterdam, 1985.
13. H.M. McNair, J. Chrom. Sci., *25*, 564–582 (1987).

14. C.T. Wehr, *Methods in Enzymology*, Vol. 104, Part C, W.B. Jakoby, Ed., Academic Press, New York, 1984, pp. 133–169.
15. *American Laboratory, 1988 Buyer's Guide Edition*, International Scientific Communications, Inc., Shelton, CT, 1988.
16. W. Kopaciewicz, M.A. Rounds, J. Fausnaugh, and F.E. Regnier, J. Chromatograph., *266*, 3–21 (1983).
17. J.N. Bronstead, Chem. Rev. *3*, 231 (1928).
18. K.M. Gooding and M.N. Schmuck, J. Chromatogr., *327*, 139–146 (1985).
19. K.K. Unger, *Porous Silica—Its Properties and Use as a Support in Column Liquid Chromatography*, Elsevier, Amsterdam, 1979.
20. A.J. Alpert, J. Chromatogr., *266*, 23–37 (1983).
21. H.E. Ramsden, U.S. Patent Number 4,540,486.
22. N. Kitigawa, LC-GC, *6*, 250–262 (1988).
23. T.W. Healy, in *Polymer Flocculation*, T.W. Healey, Ed., Royal Australia Chemical Institute, Melbourne, Australia, 1973.
24. W. Kopaciewicz, M.A. Rounds, and F.E. Regnier, J. Chromatogr., *318*, 157–172 (1985).
25. K. Yao and S. Hjerten, J. Chromatogr., *385*, 87–98 (1987).
26. R.V. Perkins, V.J. Nau and A. MacPartland, Chromatography Review (Spectra-Physica), *14*, 13–15 (1987).
27. S.A. Berkowitz, J. Liquid Chrom., *10*, 2771–2787 (1987).
28. B.H.J. Hofstee, in *Methods of Protein Separation*, Vol. 2, N. Catsimpoolas, Ed., Plenum Press, New York, 1976.
29. B.H.J. Hofstee and N.F. Otillis, J. Chromatogr., *159*, 57 (1978).
30. W. Kopaciewicz and F.E. Regnier, J. Chromatogr., *358*, 107–117 (1986).
31. S. Gupta, E. Pfannkoch, and F.E. Regnier, Anal. Biochem., *128*, 196–201 (1983).
32. L.A. Kennedy, W. Kopaciewicz, and F.E. Regnier, J. Chromatogr., *359*, 73–84 (1986).
33. R. Kagel, The Retention Times (Rainin Instrument Company, Inc.), *1*, 3–5 (1987).
34. H.E. Ramsden and D.R. Nau (patent applied for).
35. H.E. Ramsden and D.R. Nau, U.S. Patent Number 4,721,573.
36. S. Gupta, E. Pfannkoch, and F.E. Regnier, Fed. Proc., *41*, 875 (Abstract 3541) (1982).
37. R.W. Stout, S.I. Sivakoff, R.D. Ricker, H.C. Palmer, M.A. Jackson, and T.J. Odiorne, J. Chromatogr., *352*, 381–397 (1986).
38. M.T.W. Hearn, M.I. Agiular, T. Nguyen, and M. Fridman, J. Chromatogr., *435*, 271–284 (1988).
39. K. Benedek, S. Dong, and B.L. Karger, J. Chromatogr., *317*, 227 (1984).
40. S.L. Wu, K. Benedek, and B.L. Karger, J. Chromatogr., *359*, 3 (1986).
41. O. Mikes, P. Strop and J. Sedlackova, J. Chromatogr., *148*, 237–245 (1978).
42. J. Ugelstad, P.C. Mork, K.H. Kaggerud, T. Ellingsen, and A. Berge, Adv. Colloid Interface Sci., *13*, 101–140 (1980).
43. Y. Kao, K. Nakamura, and T. Hashimoto, J. Chromatogr., *266*, 358–394 (1983).
44. P. Vratny, O. Mikes, P. Strop, J. Coupek, L. Rexova-Benkova, and D. Chadimova, J. Chromatogr., *257*, 23–35 (1983).

45. R.M. Chicz, Z. Shi, and F.E. Regnier, J. Chromatogr., *359*, 121–130 (1986).
46. Y.S. Kim, B.W. Sands, and J.L. Bass, J. Liquid Chromatogr., *10*, 839–851 (1987).
47. K.J. Wilson, W. Van Wieringer, S. Klauser, and M.W. Berchtold, J. Chromatogr., *249*, 19 (1982).
48. R.V. Lewis, A.S. Stern, S. Kinura, S. Stein, and S. Udenfriend, Proc. Natl. Acad. Sci. U.S.A., *77*, 5018 (1980).
49. Y. Kato, T. Kitamura, A. Mitsui, and T. Hashimoto, J. Chromatogr., *398*, 327–334 (1987).
50. D.J. Burke, J.K. Duncan, L.C. Dunn, L. Cummings, C.J. Siebert and G.S. Ott, J. Chromatogr., *353*, 425–437 (1986).
51. J. Thayer, W.T. Edwards, and C.A. Pohl, *New Polymeric Columns for Ion Exchange and Reversed Phase Separations of Biomolecules*. Paper 1143 presented at the 39th Pittsburgh Conference, New Orleans, LA, February 22–26, 1988.
52. Separations Technology, Inc., Wakefield, R.I.
53. Water Associates, Milford, MA.
54. Amicon Corporation, Danvers, MA.
55. YMC, Inc., Mt. Freedom, NJ.
46. D.R. Nau and J.G. Guenther, *Effective Antimicrobials for the Sterilization of Silica-Based Chromatographic Supports: Solutions to an Ongoing Controversy*. Paper 1010 presented at the Sixth International Symposium on HPLC of Proteins, Peptides and Polynucleotides, Baden-Baden, West Germany, October 20–22, 1986.
57. W. Kopaciewicz and F.E. Regnier, Anal. Biochem., *133*, 251–259 (1983).
58. G.S. Ott and C.G. Shore, J. Chromatogr., *231*, 1–12 (1982).
59. P.J. Cachia, J. Van Eyk, P.C.S. Chang, A. Taneja, and R.S. Hodges, J. Chromatogr., *266*, 651–659 (1983).
60. S.A. Berkowitz, Anal. Biochem., *164*, 254–260 (1987).
61. J.A. Bietz, Cereal Chem., *62*, 201–212 (1985).
62. A.H.B. Wux and T.G. Gornet, Clin. Chem., *31*, 25–31 (1985).
63. T.D. Schlabach and S.R. Abbott, Clin. Chem, *26*, 1504–1508 (1980).
64. S.M. Hanash and D.N. Shapiro, Hemoglobin, *5*, 165–175 (1981).
65. T.H.J. Huisman, J. Chromatogr., *418*, 277–304 (1987).
66. J.B. Wilson and T.H.J. Huisman, in *The Hemoglobinopathies*, H.T.J. Huisman, Ed., Churchill Livingstone, Edinburgh, 1986.
67. C.Y. Ip and T. Asakura, Anal. Biochem., *156*, 348–353 (1986).
68. C.N. Ou, G.J. Buffone, and A.J. Alpert, J. Chromatogr., *266*, 197–205 (1983).
69. R.C. Sullivan, Y.W. Shing, P.A. D'Amore, and M. Klagsbrun, J. Chromatogr., *266*, 301–311 (1983).
70. D. Nau, Biochromatography, *1*, 82–94 (1986).
71. L. Crane, *Monoclonal Antibody Production Techniques and Applications*, Chap. 9, L.B. Schook, Ed., Marcel Dekker, New York, 1987.
72. P. Dorfman, *Gen. Eng. News*, May/June 15–17 (1984).
73. M. Capel, D. Datta, D.R. Nierras, and G.R. Craven, Anal. Biochem., *158*, 179–188 (1986).
74. N. Lambert, G.W. Plumb, and D.J. Wright, J. Chromatogr., *402*, 159 (1987).

75. J.D. Pearson, M.C. McCroskey, and D.B. DeWald, J. Chromatogr., *418*, 245–276 (1987).
76. G.W. Welling, R. Van Der Zee, and S. Welling-Webster, J. Chromatogr., *418*, 223–243 (1987).
77. F.E. Regnier, J. Chromatogr., *418*, 115–143 (1987).
78. M.T.W. Hearn, J. Chromatogr., *418*, 3–26 (1987).
79. S.A. Berkowitz, M.P. Henry, D.R. Nau, and L.J. Crane, Amer. Lab., 33–42, May, 1987.
80. R. Einarsson and B. Rench, Toxicon, *22*, 154–160 (1984).
81. G.W. Welling, J. Chromatogr., *297*, 101–109 (1984).
82. V.J. Britton, Liq. Chromatogr., *1*, 176–179 (1983).
83. P. Lundahl, J. Chromatogr., *297*, 129–137 (1984).
84. M.P. Henry, in *Liquid Chromatography in Pharmaceutical Development: An Introduction*, Chapter 2, I.W. Wainer, Ed., Aster, Springfield, OR, 1985.
85. D.R. Nau, *J. Chromatogr.* (to be published).
86. S.A. Berkowitz (personal communication).
87. D.R. Nau (personal communication).

CHAPTER 11

The Purification of Polypeptide Samples by Ion-Exchange Chromatography on Silica-Based Supports

MIRAL DIZDAROGLU

Center for Chemical Technology, National Institute of Standards and Technology, Gaithersburg, Maryland

1. INTRODUCTION

Since its introduction by Moore and Stein (1), ion-exchange liquid chromatography has played an important role in the separation and purification of peptides. This technique has been used extensively for the separation of peptide fragments of chemically or enzymatically cleaved proteins prior to sequence analyses (2–5). Later, automated ion-exchange chromatography provided highly reproducible separations of peptides (6–8). Various anion-and cation-exchange resins have been used as stationary phases. Most of these ion-exchangers have been prepared from divinylbenzene cross-linked polystyrene and functional groups have been attached to the polymeric matrix (9). The introduction of volatile buffers was an important improvement in ion-exchange chromatography, since this permitted the isolation of salt-free peptides to be used directly in sequence analysis (10).

The development of high-performance liquid chromatography (HPLC) during the past decade or so contributed greatly to the improvement of peptide separations by liquid chromatography. The reversed-phase mode of HPLC has been the most popular and broadly used technique for many separation problems in peptide chemistry.

The purpose of this paper is to review the recent applications of ion-exchange HPLC on silica-based supports to peptide separations and purifications.

2. ION-EXCHANGE HPLC OF PEPTIDES

In the past, ion-exchange HPLC has been also used for peptide separations, although to a lesser extent than reversed-phase HPLC. New developments have

Table 1. Peak Identification and Sequences of Peptides in Figures

	Figure 1
Vasopressin,	Cys-Tyr-Phe-Gln-Asn-Cys-Pro-Arg-Gly-NH_2;
Oxytocin,	Cys-Tyr-Ile-Gln-Asn-Cys-Pro-Leu-Gly-NH_2;
Substance P,	Arg-Pro-Lys-Pro-Gln-Gln-Phe-Phe-Gly-Leu-Met-NH_2;
Angiotensin,	Asp-Arg-Val-Tyr-Ile-His-Pro-Phe-His-Leu.
	Figure 4
A and B,	Asn-Gln-Lys-Leu-Phe-Asp-Leu-Arg-Gly-Lys-Phe-Lys-Arg-Pro-Pro-Leu-Arg-Arg-Val-Arg-Hse;
C,	Gly-Gly-Phe-Lys-Arg-Pro-Pro-Leu-Arg-Arg-Val-Arg-Amide;
D,	Ac-Gly-Lys-Phe-Lys-Arg-Pro-Pro-Leu-Arg-Arg-Val-Arg-amide;
E	Ac-Gly-Gly-Phe-Lys-Arg-Pro-Pro-Leu-Arg-Arg-Val-Arg-amide;
	Figure 5
1	Ser-Asp-Asn-Ile-Pro-Ser-Phe-Arg-Gly-amide;
2	Ac-Ser-Asp-Gln-Glu-Lys-Arg-Lys-Gln-Ile-Ser-Val-Arg-Gly-Leu-amide;
3	Leu-Lys-Ala-Leu-Leu-Gly-Ser-Lys-His-Lys-Val-Cys-Hse;
4	Leu-Lys-Ala-Leu-Leu-Gly-Ser-Lys-His-Lys-Val-Cys-Hselac;
5	Ac-Gly-Lys-Phe-Gly-Arg-Pro-Pro-Leu-Arg-Arg-Val-Arg-amide;
6	Ac-Gly-Lys-Phe-Lys-Arg-Pro-Pro-Leu-Arg-Arg-Val-Arg-amide;
7	Asn-Gln-Lys-Leu-Phe-Asp-Leu-Arg-Gly-Lys-Phe-Lys-Arg-Pro-Pro-Leu-Arg-Arg-Val-Arg-HseLac.
	Figure 7
CN1	Leu-Gly-Ile-Ala-Ala-Thr-Glu-Leu-Glu-Lys-Glu-Glu-Gly-Arg-Arg-Glu-Ala-Glu-Lys-Gln-Asn-Tyr-Leu-Ala-Glu-His-Cys-Pro-Pro-Leu-Ser-Leu-Pro-Gly-Ser-HseLac;
CN2	Asp-Leu-Arg-Ala-Asn-Leu-Lys-Gln-Val-Lys-Lys-Glu-Asp-Thr-Glu-Lys-Glu-Arg-Asp-Val-Gly-Asp-Trp-Arg-Lys-Asn-Ile-Glu-Glu-Lys-Ser-Gly-HseLac;
CN3	Ala-Glu-Val-Gln-Glu-Leu-Cys-Lys-Gln-Lys-His-Ala-Lys-Ile-Asp-Ala-Ala-Glu-Glu-Glu-Lys-Tyr-Asp-HscLac;

Table 1. (*Continued*)

	Figure 7
CN4	Asn-Gln-Lys-Leu-Phe-Asp-Leu-Arg-Gly-Lys-Phe-Lys-Arg-Pro-Pro-Leu-Arg-Arg-Val-Arg-HseLac;
CN5	Ac-Gly-Asp-Glu-Glu-Lys-Arg-Asp-Arg-Ala-Ile-Thr-Ala-Arg-Arg-Gln-His-Leu-Lys-Ser-Val-HseLac;
CN6	Glu-Ile-Lys-Val-Gln-Lys-Ser-Ser-Lys-Glu-Leu-Glu-Asp-HseLac;
CN7	Leu-Lys-Ala-Leu-Leu-Gly-Ser-Lys-His-Lys-Val-Cys-HseLac.
	Figure 11
1, somatostatin (Ala-Gly-Cys-Lys-Asn-Phe-Phe-Trp-Lys-Thr-Phe-Thr-Ser-Cys);	
2, proctolin (Arg-Tyr-Leu-Pro-Thr);	
3, neurotensin (pGlu-Leu-Tyr-Glu-Asn-Lys-Pro-Arg-Arg-Pro-Tyr-Ile-Leu);	
4, Met-enkephalin (Tyr-Gly-Gly-Phe-Met);	
5, bradykinin potentiator c (pGlu-Gly-Leu-Pro-Pro-Gly-Pro-Pro-Ile-Pro-Pro);	
6, Lys-Glu-Thr-Tyr-Ser-Lys;	
7, α-endorphin (Tyr-Gly-Gly-Phe-Met-Thr-Ser-Glu-Lys-Ser-Gln-Thr-Pro-Leu-Val-Thr);	
8, EAE-peptide (Phe-Ser-Trp-Gly-Ala-Glu-Gly-Gln-Arg);	
9, glucagon (His-Ser-Gln-Gly-Thr-Phe-Thr-Ser-Asp-Tyr-Ser-Lys-Tyr-Leu-Asp-Ser-Arg-Arg-Ala-Gln-Asp-Phe-Val-Gln-Trp-Leu-Met-Asn-Thr);	
10, ribonuclease s-peptide (Lys-Glu-Thr-Ala-Ala-Ala-Lys-Phe-Glu-Arg-Gln-His-Met-Asp-Ser-Ser-Thr-Ser-Ala-Ala);	
11, IgE-peptide (Asp-Ser-Asp-Pro-Arg).	
	Figure 12
A I,	Asp-Arg-Val-Tyr-Ile-His-Pro-Phe-His-Leu;
A II,	Asp-Arg-Val-Tyr-Ile-His-Pro-Phe;
A III,	Arg-Val-Tyr-Ile-His-Pro-Phe.
	Figure 13
NT,	pGlu-Leu-Tyr-Glu-Asn-Lys-Pro-Arg-Arg-Pro-Tyr-Ile-Leu.

made this technique a powerful tool for this purpose (for reviews, see Refs. 11–13). A large number of reports have demonstrated the use of polymeric ion-exchange resins for peptide separations (for a review, see Ref. 12). Silica-based, ion-exchange-bonded stationary phases have also found broad application in various areas of ion-exchange HPLC of peptides. Strong cation exchangers and strong anion exchangers, carrying the labels SCX and SAX, respectively, have usually been employed (11).

2.1. Cation-Exchange HPLC

Radhakrishnan et al. (14) first demonstrated the application of a silica-based strong cation-exchange column (Partisil SCX) in the separation of peptides using volatile pyridine–acetic acid buffers and an automated fluorescamine column monitoring system. Figure 1 shows the separation of some synthetic peptides by this method. Retention times of chromatographed peptides were

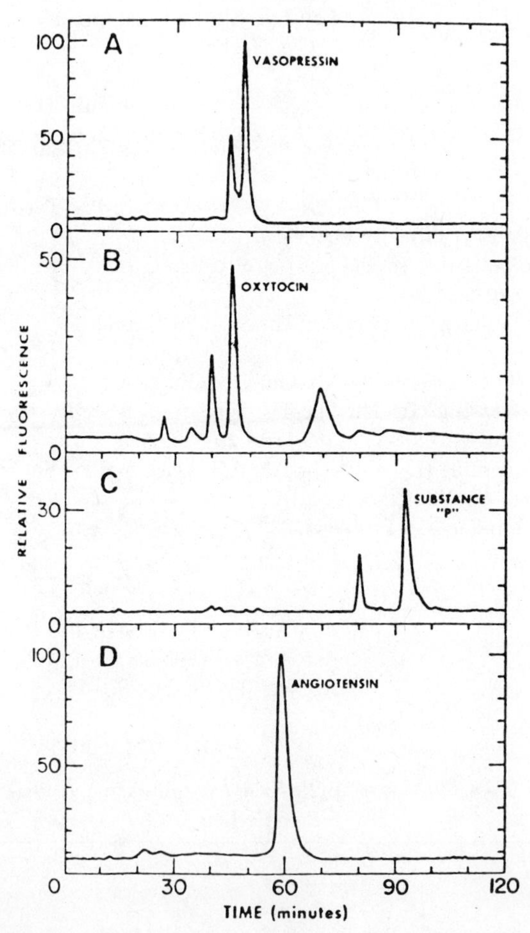

Fig. 1. Chromatography of some synthetic peptides on a Partisil SCX column using a 50 min linear gradient from 5×10^{-3} *M* pyridine, pH 3.0 to 5×10^{-2} *M* pyridine, pH 4.0 followed by a 60 min linear gradient to 5×10^{-1} *M* pyridine, pH 5.0. Approximately 10 nmoles of each sample were applied to the column and 8% (800 pmoles) was utilized for detection. (From Ref. 14 with permission.) For peak identification see Table 1.

found to increase with both size and basicity of the peptide. Successful separation of some polyamines was also reported in the same paper. In a subsequent study, this method has been used for the purification of some biologically active peptides such as Leu-enkephalin and insulin with a slight modification of the elution system (15). As an example, Fig. 2 illustrates the elution profile of Leu-enkephalin and the removal of sodium dodecyl sulfate from this peptide by a simple two-step gradient elution. The use of an SCX column and sodium phosphate buffers has also been reported for the separation of Met-enkaphalin from Leu-enkephalin, β-endorphin, and enkephalin metabolites (16). Nakamura et al. (17) have reported analysis of histidine-containing dipeptides, polyamines, and related amino acids on an SCX column with lithium citrate buffers and fluorescence detection. Figure 3 shows the separation of some histidine-containing dipeptides. The SCX column was noted to be less adsorptive of peptides than polystyrene–divinylbenzene resins. On the other hand, it's relative instability has been cited as a disadvantage.

Recently, a silica-based weak-cation exchanger has been utilized in the separation of a number of highly basic peptides with the use of KH_2PO_4 buffer containing KCl as the eluting salt (18). Figure 4 shows a chromatogram of five

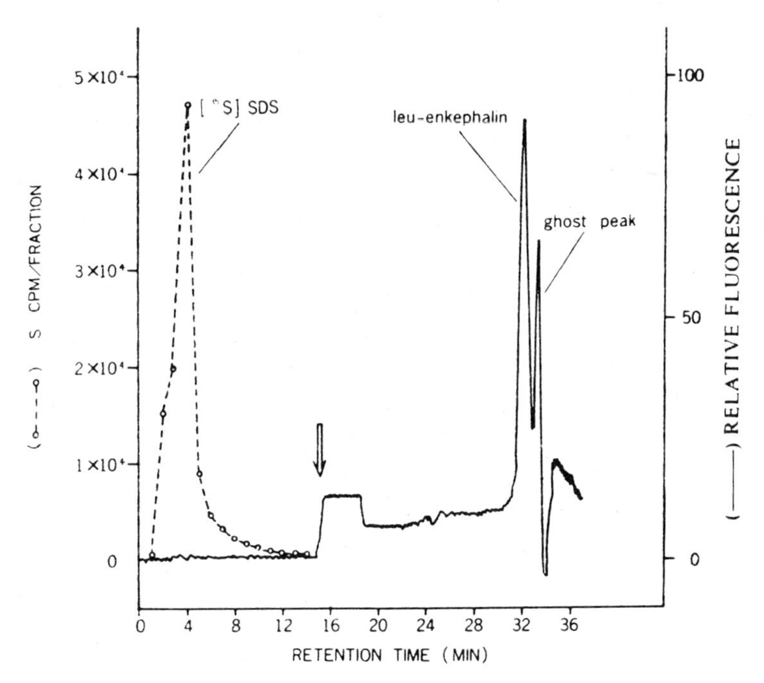

Fig. 2. Elution profile of Leu-enkephalin (Tyr-Gly-Gly-Phe-Leu) dissolved in a solution containing SDS and [^{35}S] chromatographed on a Partisil SCX column with a stepwise gradient elution. Mobile phase: (1) water at a flow rate of 1.5 mL/min, 15 min; (2) 3 *M* pyridine–0.5 *M* acetic acid at a flow rate of 0.5 mL/min, 30 min. The arrow indicates the beginning of the elution with phase 2. (From Ref. 15 with permission.)

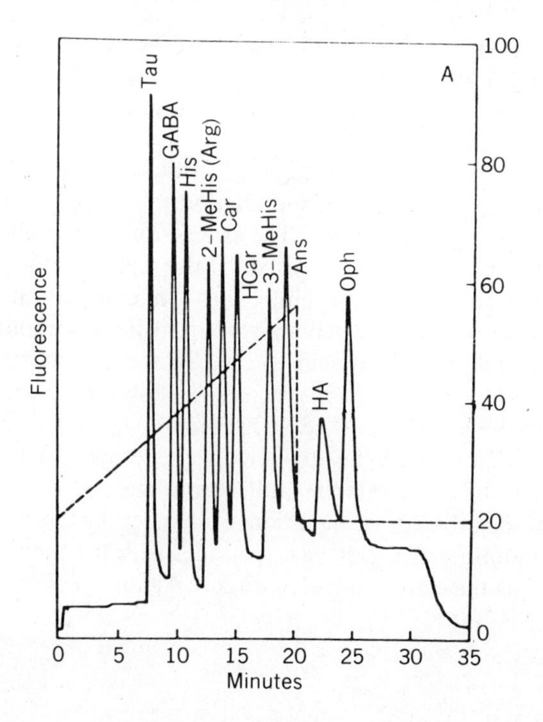

Fig. 3. Separation of some histidine-containing dipeptides on a Partisil SCX column. Mobile phase: buffer A, 0.2 *M* citric acid (pH 1.9); buffer B, 0.2 *M* lithium citrate buffer (pH 4.5). Gradient elution from 20% B to 54% at 1.8% B/min. Flow rate, 0.7 mL/min. Temperature, 50°C. Abbreviations: Tau, taurine; GABA, α-amino-*n*-butyric acid; Car, carnosine; HCar, homocarnosine; HA, histamine; Oph, ophidine. (From Ref. 17 with permission.)

basic peptides. Changes in the pH and the ionic strength of the eluent were found to have a large effect on the resolution and the retention times of the peptides examined. Similar results have been obtained with basic peptides on a silica-based strong cation-exchange column using the same eluting buffer and salt (19). Peptides varying in net charge from +1 to +9 were well separated as Figs. 5 and 6 illustrate. The elution time was demonstrated to increase with an increase in the net charge of a peptide or in the pH of the eluent. Separations of several peptides on strong-cation- and strong-anion-exchange columns were also compared. The same strong-cation-exchanger has been used in another study, which demonstrated the combined application of size-exclusion, cation-exchange, and reversed-phase HPLC techniques to a complete resolution of a protein digest (20). In the first step, a cyanogen bromide cleavage mixture of rabbit skeletal troponin I was fractionated by size-exclusion HPLC. Individual fractions were then separated and purified by cation-exchange HPLC followed by reversed-phase HPLC. As an example, Fig. 7 demonstrates such a separation. It was concluded that the combined use of these three techniques can

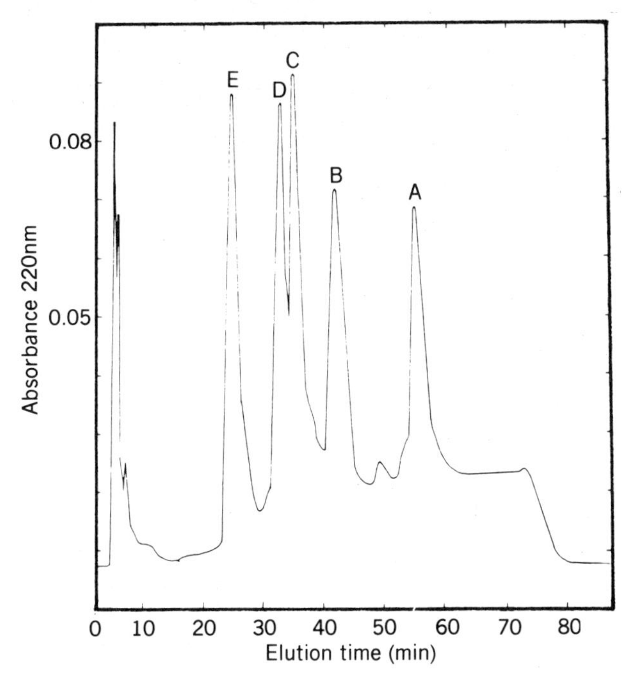

Fig. 4. Separation of some basic peptides. Column, CM300. Buffers: A, 50 m*M* KH_2PO_4 (pH 4.5); B, 50 m*M* KH_2PO_4, 1 *M* KCl (pH 4.5). A linear KCl gradient (8 m*M* B/min) was applied with the following compositions at the times indicated (min): (0) 80% A; (50) 40% A; (60) 40% A; (65) 80% A. Flow rate, 1 mL/min. (From Ref. 18 with permission.) For peak identification see Table 1.

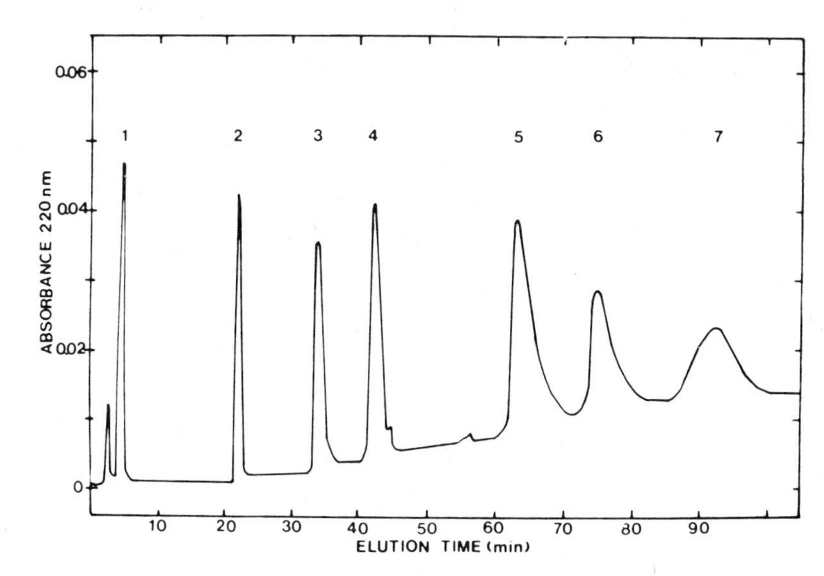

Fig. 5. Separation of some basic peptides. Column, S300. Buffers: A, 5 m*M* KH_2PO_4 (pH 6.5); B, 5 m*M* KH_2PO_4–1 *M* KCl (pH 6.5). A linear KCl gradient (5 m*M* B/min) was applied, following a 10 min isocratic elution with buffer A. Flow-rate, 1 mL/min. (From Ref. 19 with permission.) For peak identification see Table 1.

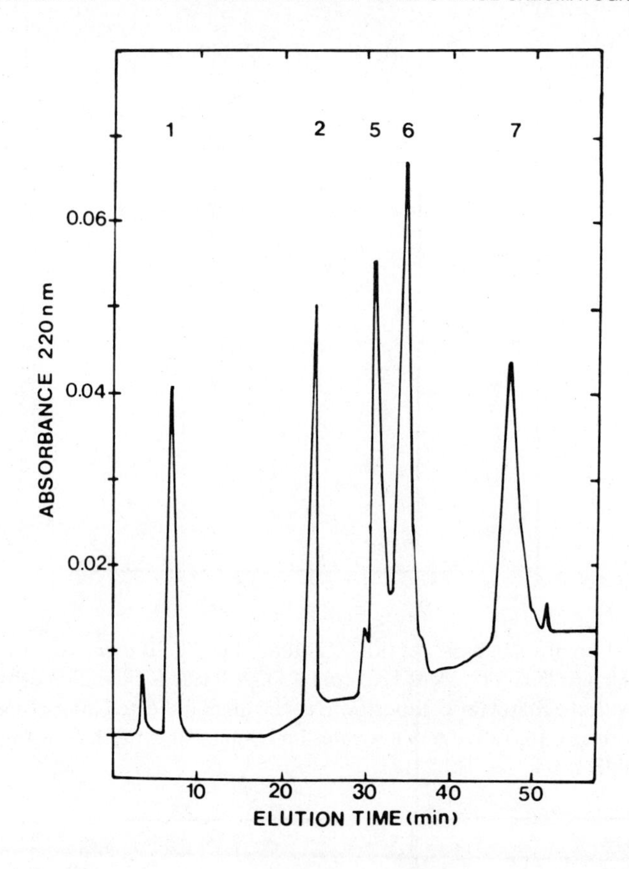

Fig. 6. Separation of some basic peptides (for peak identification see Table 1 and Fig. 5). Column and buffers as in Fig. 5 except for pH 3 for both buffers. Gradient and flow-rate are as in Fig. 5. (From Ref. 19 with permission.)

provide an optimal separation of complex peptide mixtures. The separation of troponin subunits from bovine and rabbit skeletal troponin on a weak cation-exchanger has also been demonstrated (21).

2.2. Anion-Exchange HPLC

Recently, a method has been introduced for the separation of dipeptides, including sequence isomeric and diastereomeric dipeptides, by weak-anion-exchange HPLC (22). For this purpose, a difunctional weak-anion-exchange bonded phase prepared on porous silica has been used. This commercially available stationary phase, that is, MicroPak AX-10 (Varian), had been introduced previously for simultaneous analysis of nucleotides, nucleosides, and nucleobases (23) and has also been applied to the separation and sequencing of deoxypentanucleotide sequence isomers (24). Peptides usually have little or no

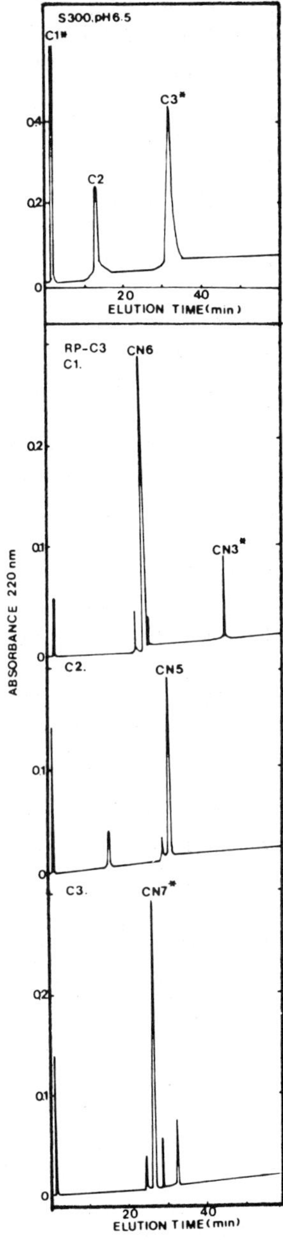

Fig. 7. Cation-exchange and reversed-phase chromatography of a peptide fraction obtained by size-exclusion chromatography from a cyanogen bromide cleavage mixture of rabbit skeletal troponin I. Cation-exchange column, buffers, and flow-rate are as in Fig. 5. A linear gradient of KCl (5 m*M* B/min) was applied. Reverse-phase column, C3. Solvents: A, 0.1% aqueous TFA; B, 0.05% TFA in acetonitrile. A linear gradient of 0.5% B/min was applied. Flow rate, 1 mL/min. (From Ref. 20 with permission.) For peak identification see Table 1.

retention on MicroPak AX-10 when only aqueous buffers are used as the eluent; however, the addition of an organic solvent such as acetonitrile to the buffer increases the peptide retention. Mixtures of triethylammonium acetate (TEAA) buffer and acetonitrile were used as the eluent for separation of dipeptides (22). This elution system allowed gradient elution and sensitive detection of peptides in the wavelength range of 210–225 nm. The TEAA buffer is also volatile and thus enabled easy recovery of peptides. Gradient elution was carried out by increasing the amount of the buffer in the eluent.

Figure 8 illustrates the separation of some dipeptides by this method. Dipeptides containing acidic amino acids do not elute under those conditions. Such dipeptides could be eluted and separated by reducing the pH of the eluent (22). An excellent resolution of sequence isomeric and diastereomeric dipeptides has also been achieved by weak-anion-exchange HPLC as Figs. 9 and 10 illustrate. All DL, DL-dipeptides examined were resolved into two components (Fig. 10). Since the individual four diastereomers of each dipeptide were not available, peak assignments were based on the elution order of the available four diastereomers of Ala-Ala. In this case, the D, L- and L, D-isomers eluted first as a single peak (peak 7 in Fig. 10), whereas the L, L- and D, D-isomers coeluted later (peak 11 in Fig. 10).

The weak-anion exchange HPLC method introduced by Dizdaroglu and Simic (22) has been applied later to the separation of a variety of peptides.

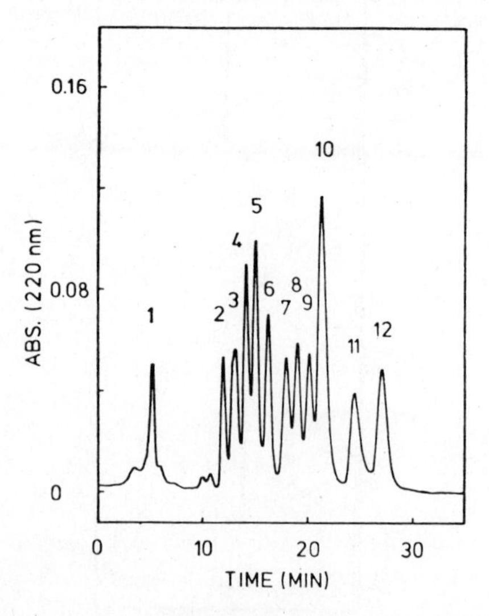

Fig. 8. Separation of some selected dipeptides. Column, MicroPak AX-10 (10 μm), 30 × 0.4 cm. Temperature, 40°C. Eluent, mixture of 32% 0.01 *M* triethylammonium acetate (pH 4.3) and 68% acetonitrile. Flow rate 1 mL/min. Peaks: 1, L-Arg-L-Phe; 2, L-Leu-L-Leu; 3, Gly-L-Ile and L-Leu-L-Trp; 4, L-Ala-L-Ile; 5, L-Trp-Gly; 6, L-Trp-L-Phe; 7, L-Val-L-Val and L-Ala-L-His; 8, L-Trp-L-Ala; 9, L-Ala-L-Thr and L-Met-L-Met; 10, Gly-Gly and L-Phe-L-Phe; 11, L-Ser-L-Phe; 12, L-Tyr-L-Tyr. (From Ref. 22 with permission.)

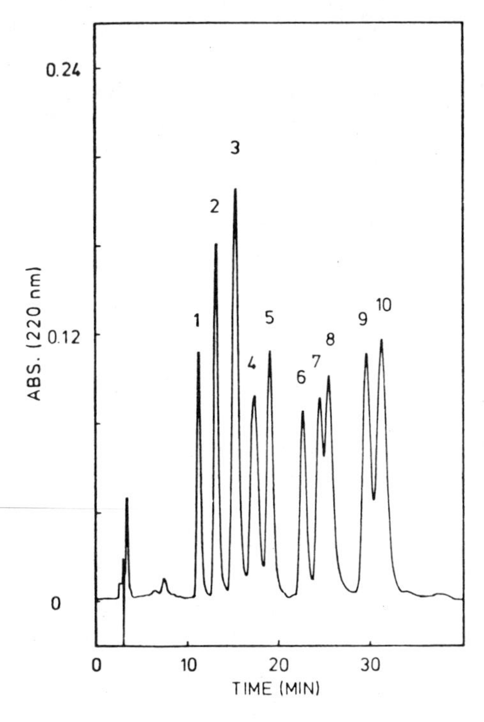

Fig. 9. Separation of some sequence isomeric dipeptides. Column details as in Fig. 8. Peaks: 1, DL-Leu}-DL-Ala; 2, Gly-L-Phe; 3, L-Ala-L-Leu, Gly-L-Met, and L-Ala-L-Phe; 4, Gly-L-Tyr and DL-Leu-DL-Ala; 5, L-Ala-L-Tyr; 6, L-Phe-Gly; 7, L-Tyr-Gly; 8, L-Met-Gly; 9, L-Phe-L-Ala; 10, L-Tyr-L-Ala. (from Ref. 22 with permission.)

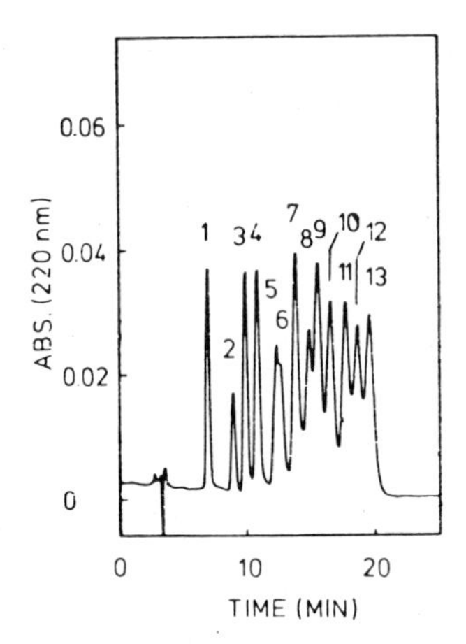

Fig. 10. Separation of some diastereomeric dipeptides. Column as in Fig. 8. Temperature, 45°C. Eluent, mixture of 35% 0.01 *M* triethylammonium acetate (pH 4.3) and 65% acetonitrile. Flow rate, 1 mL/min. Peaks: 1, DL-Leu-DL-Phe; 2, DL-Ala-DL-Phe; 3, DL-Leu-DL-Ala; 4, DL-Ala-DL-Val; 5, DL-Leu-DL-Phe; 6, L-Ala-L-Phe; 7, DL-Ala-DL-Ala and DL-Ala-DL-Val; 8, DL-Leu-DL-Ala; 9, DL-ALa-DL-Ser; 10, DL-Ala-DL-Asn; 11, DL-Ala-DL-Ala; 12, DL-Ala-DL-Ser; 13, DL-Ala-DL-Asn. (From Ref. 22 with permission.)

Figure 11 illustrates the separation of a multicomponent peptide mixture (25). Acidic peptides with no compensating basic residues did not elute from the column under the conditions described in Fig. 11; however, such peptides could be chromatographed using isocratic elution with dilute formic acid (25). Closely related peptides with important biological activities such as bradykinins, angiotensins, and neurotensins have been also separated by this method (25–27) Figures 12 and 13 demonstrate the separation of 12 angiotensins (26) and of some diastereomers of neurotensin (27), respectively. In both cases, a significant effect of the column temperature on the peptide retention was observed. For instance, two coeluting angiotensins (peak 6 in Fig. 12) could be separated from each other by elevating the column temperature (12, 26). More recently, separation of some diastereomers of angiotensin I and their purification from impurities have been demonstrated (28).

The weak-anion-exchange HPLC method has also been applied to the separation of peptides resulting from the tryptic digestion of some proteins (25, 29). As an example, Fig. 14 shows the separation of a tryptic digest of rat small myelin basic protein (29). Fragments represented by peaks 1 to 16 were found to cover the total sequence of this protein. An aliquot of this particular digest was also separated by reversed-phase HPLC in order to compare the two HPLC

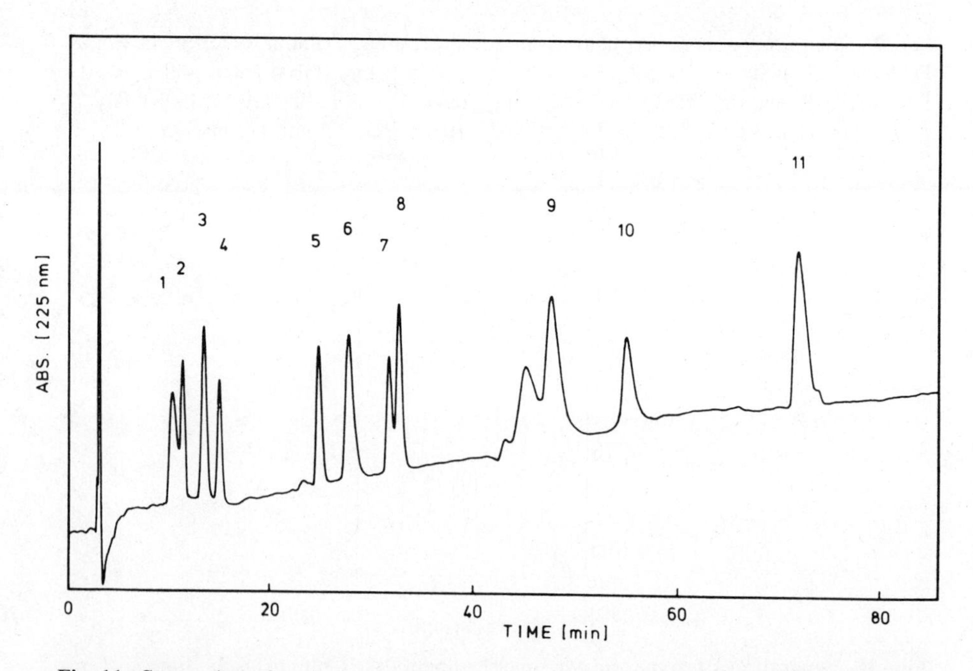

Fig. 11. Separation of various peptides. Column as in Fig. 8. Temperature, 30°C. Eluent: A, acetonitrile; B, 0.01 *M* triethylammonium acetate (pH 6.0), gradient program: linear starting from 25% B with a rate of 1% B per min. Flow rate, 1 mL/min. Amount of injection, 0.5–5 μg per peptide. For peak identification see Table 1. (From Ref. 25 with permission.)

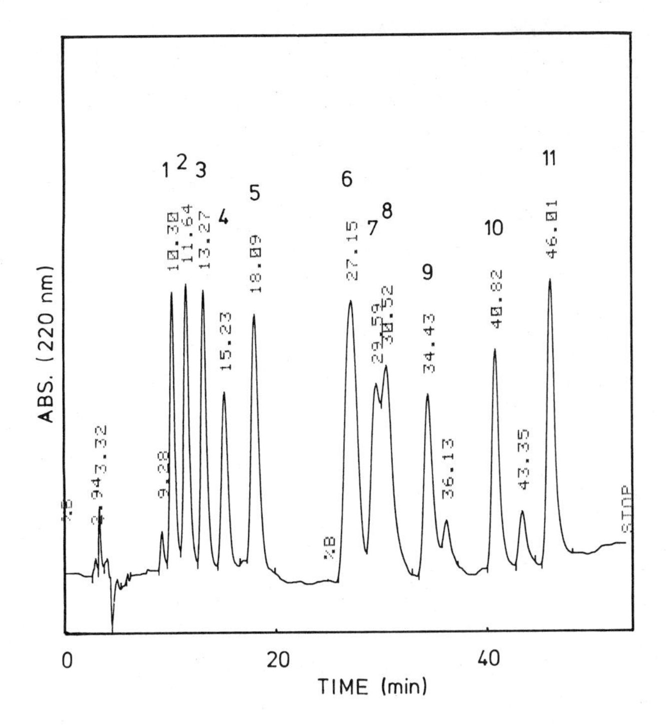

Fig. 12. Separation of some angiotensins (A) by HPLC. Column as in Fig. 8. Eluent: A, acetonitrile; B, 0.01 *M* triethylammonium acetate (pH 6.0), gradient elution starting from 24% B with a rate of 1% B per minute for 25 min then 0.5% B per minute; column temperature, 26°C; flow rate, 1 mL/min; amount of injection per peptide, approximately 1 μg (1 nmol based on A II); AUFS: 0.1 at 220 nm. Peaks: 1, A III; 2, (Val^{4})-A III; 3, A III inhibitor; 4, (Asn^{1}-Val^{5})-A II; 5, (Sar^{1}-Ile^{8})-A II; 6, (Sar^{1}-Ala^{8})-A II and (Sar^{1}-Gly^{8})-A II; 7, (Sar^{1}-Thr^{8})-A II; 8, (Sar^{1}-Val^{5}-Ala^{8})-A II; 9, A II; 10, A I; 11, (Val^{5})-A II. For sequences of A I, A II, and A III see Table 1. (From Ref. 26 with permission.)

methods that utilize different separation principles. Comparison of the results suggested that the combined use of these two methods of separation can provide even more resolving power for a given mixture of peptides.

The combined use of reversed-phase HPLC and weak-anion-exchange HPLC has been also demonstrated for the purification of peptides contained in a subtilisin digest of purple membrane protein bacteriorhodopsin (30). Furthermore, the separations of some diastereomers of angiotensin I from their impurities by weak anion-exchange HPLC has been compared recently with those obtained by reversed-phase HPLC (28, 31). This comparison strongly suggested that these two HPLC methods are complementary; hence, their combined use leads to a more confident assessment of the purity of a given peptide preparation.

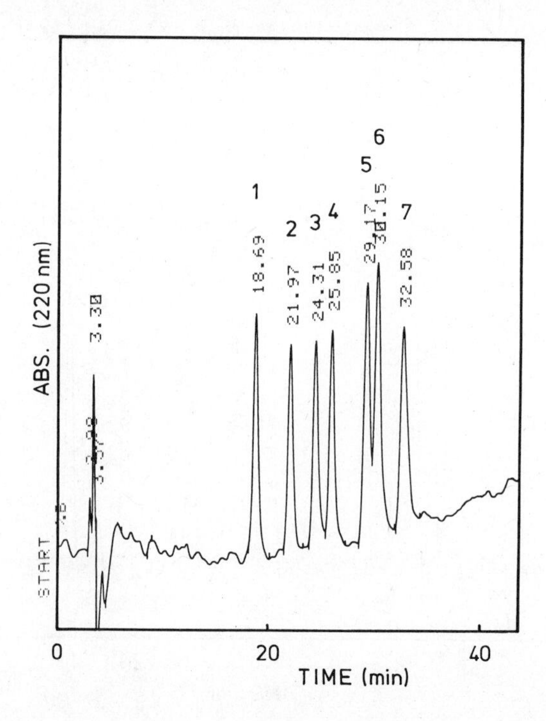

Fig. 13. Separation of some diastereomers of neurotensin (NT). Column as in Fig. 8. Temperature, 50°C. Eluent: A, acetonitrile; B, 0.01 *M* triethylammonium acetate (pH 6.0), gradient program: linear starting from 23% B with a rate of 0.3% B/min. Flow rate, 1 mL/min. Amount of injection per peptide, ca. 1 nmol. AUFS, 0.1 at 220 nm. Peaks: 1, (D-Phe^{11})-NT; 2, (D-Tyr^{11})-NT; 3, (D-Pro^{10})-NT; 4, (Phe^{11})-NT; 5, (D-Arg^{9})-NT; 6, NT; 7, (D-Glu^{4})-NT. For sequence of NT see Table 1. (From Ref. 27 with permission.)

3. CONCLUSIONS

Recent developments have demonstrated that ion-exchange HPLC is a powerful technique for peptide separations and purifications. Both the cation- and anion-exchanger methods which have been developed appear to be extremely useful for many separation problems in peptide chemistry. Moreover, studies indicate that, in some instances, ion-exchange HPLC methods are complementary to those of reversed-phase HPLC. This means that their combined use can provide the optimal separation of a given peptide mixture or the confident assessment of the purity of a peptide.

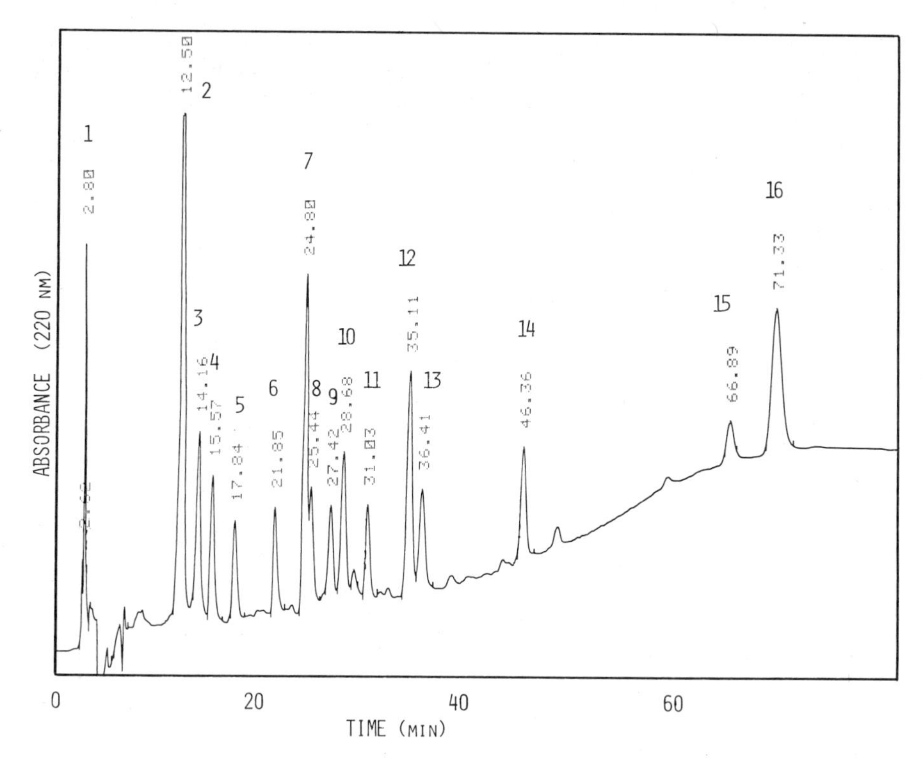

Fig. 14. Separation of a tryptic digest of rat small myelin basic protein. Column as in Fig. 8. Eluent: A, acetonitrile; B, 0.01 *M* triethylammonium acetate (pH 6.0), gradient program: linear starting from 23% B with a rate of 0.7% B/min for 40 min then 1% B/min to 100% B. Flow rate, 1 mL/min. For peak identification, see Ref. 29. (From Ref. 29 with permission.)

REFERENCES

1. S. Moore, and W.H. Stein, J. Biol. Chem., *192*, 663 (1951).
1. C.H.W. Hirs, S. Moore, and W.H. Stein, J. Biol. Chem., *219*, 623 (1956).
3. A.B. Edmundson, and C.H.W. Hirs, J. Mol. Biol., *5*, 683 (1962).
4. E. Margoliash, and E.L. Smith, J. Biol. Chem., *237*, 2151 (1962).
5. A. Light, and E.L. Smith, J. Biol. Chem., *237*, 2537 (1962).
6. R.T. Jones, Cold Spring Harbor Symp. Quant. Biol., *29*, 297 (1964).
7. J.V. Benson, R.T. Jones, J. Cormock, and J.A. Patterson, Anal. Biochem., *16*, 91 (1966).
8. R.L. Hill, and R. Delaney, Methods Enzymol., *11*, 339 (1967).
9. J.R. Benson, Methods Enzymol., *47*, 19 (1977).
10. W. Machleidt, J. Otto, and E. Wachter, Methods Enzymol, *47*, 210 (1977).
11. J.A. Smith, and R.A. McWilliams, Am. Lab., *12*, 25 (1980).
12. M. Dizdaroglu, "Separation of Peptides by High-Performance Ion-Exchange Chro-

matography," in *Handbook of HPLC for the Separation of Amino Acides, Peptides and Proteins*, Vol. 2, W.S. Hancock, Ed., CRC Press, Boca Raton, FL, 1984, p. 23.
13. M. Dizdaroglu, J. Chromatogr., *334*, 49 (1985).
14. A.D. Radhakrishnan, S. Stein, A. Licht, K.A. Gruber, and S. Udendriend, J. Chromatogr., *132*, 552 (1977).
15. H. Mabuchi, and H. Nakahashi, J. Chromatogr., *213*, 275 (1981).
16. T.P. Bohan, and J.L. Meek, Neurochem. Res., *3*, 367 (1978).
17. H. Nakamura, C.L. Zimmerman, and J.J. Pisano, Anal. Biochem., *93*, 423 (1979).
18. P.J. Cachia, J. van Eyk, P.C.S. Chong, A. Taneja, and R.S. Hodges, J. chromatogr., *266*, 651 (1983).
19. C.T. Mant and R.S. Hodges, J. Chromatogr., *327*, 147 (1985).
20. C.T. Mant, and R.S. Hodges, J. Chromatogr., *326*, 349 (1985).
21. P.J. Cachia, J. van Eyk, W.D. McCubbin, C.M. Kay, and R.S. Hodges, J. Chromatogr., *343*, 315 (1985).
22. M. Dizdaroglu, and M.G. Simic, J. Chromatogr., *195*, 119 (1980).
23. E.H. Edelson, J.G. Lawless, C.T. Wehr, and S.R. Abbott, J. Chromatogr., *174*, 409 (1979).
24. M. Dizdaroglu, M.G. Simic, and H. Schott, J. Chromatogr., *188*, 273 (1980).
25. M. Dizdaroglu, H.C. Krutzsch, and M.G. Simic, J. Chromatogr., *237*, 417 (1982).
26. M. Dizdaroglu, H.C. Krutzsch, and M.G. Simic, Anal. Biochem., *123*, 190 (1982).
27. M. Dizdaroglu, M.G. Simic, F. Rioux, and S. St-Pierre, J. Chromatogr., *245*, 158 (1982).
28. S.A. Margolis, and M. Dizdaroglu, J. Chromatogr., *322*, 117 (1985).
29. M. Dizdaroglu, and H.C. Krutzsch, J. Chromatogr., *264*, 223 (1983).
30. H.D. Lemke, J. Bergmeyer, and D. Oesterhelt, Methods Enzymol., *88*, 89 (1982).
31. S.A. Margolis, and P.L. Konash, Anal. Biochem., *134*, 163 (1983).

CHAPTER 12

The Purification of Polypeptide Samples by Hydrophobic Interaction Chromatography

YOSHIO KATO and TAKASHI KITAMURA

Central Research Laboratory, TOSOH Corporation, Tonda, Shinnanyo, Yamaguchi 746, Japan

1. INTRODUCTION

Hydrophobic-interaction chromatography has been used extensively for the purification of proteins. Alkyl or aryl derivatives of agarose have been used mainly for these purifications. Since this type of column packing material cannot withstand high pressure, hydrophobic-interaction chromatography had to be performed at low pressure. However, microparticulate rigid supports have been developed during the last few years (1–7), and rapid purifications have become possible. Many proteins could be purified rapidly with high resolution without denaturation on these supports. Some examples of the protein purifications are described here.

2. HYDROPHOBIC INTERACTION CHROMATOGRAPHY EXPERIMENTS

Preparative columns (150 × 21.5 mm ID) of TSKgel Phenyl-5PW (2) and Ether-5PW (8) (TOSOH, Tokyo, Japan) were used. These columns were packed with particles of 13 μm in diameter. Protein purifications on these columns were carried out at 25°C with a high-speed liquid chromatograph Model SP8700 (Spectra-Physics, San Jose, CA) equipped with a variable-wavelength UV detector Model UV-8 (TOSOH) operated at 280 nm. Elutions were usually performed with linear gradients of decreasing ammonium sulfate concentration in 0.1 *M* phosphate buffer (pH 7.0) at a flow rate of 4 mL/min.

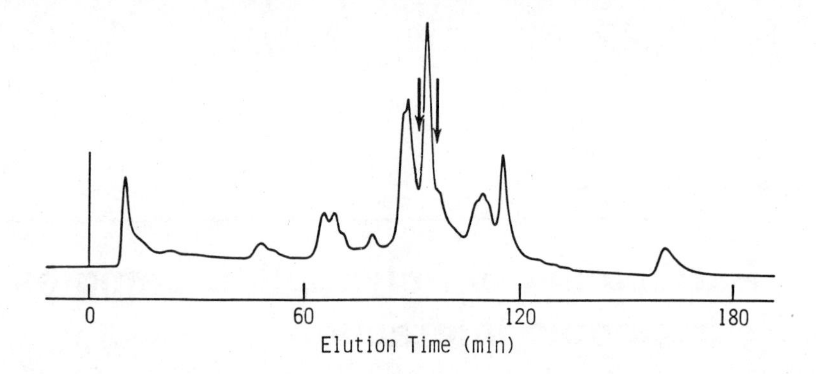

Fig. 1. Purification of 200 mg lipoxidase on the Phenyl-5PW preparative column. (Data from Ref. 9. With permission.)

3. PURIFICATION OF LIPOXIDASE

Lipoxidase was purified on the Phenyl-5PW preparative column with a 120 min linear gradient of ammonium sulfate from 1.5 *M* to 0. A crude sample of lipoxidase was purchased from P-L Biochemicals (Milwaukee, WI). The crude sample was separated with various loadings. Almost identical patterns were obtained with sample loadings up to 200 mg, and the resolution decreased gradually with increasing sample loading above 200 mg. Figure 1 shows the separation of a 200 mg sample. The peak corresponding to lipoxidase between the two vertical lines was fractioned. The recovery of enzymatic activity in the fraction was 86%. The degree of purification based on the specific activity was 5.6-fold. The purity of the fraction was tested by high-performance liquid chromatography (HPLC). Hydrophobic-interaction chromatography (HIC) was performed on an analytical Phenyl-5PW column (75 × 7.5 mm ID) with a 60 min linear gradient of the same eluents as in the preparative purification. Reversed-phase chromatography (RPC) was performed on a TSKgel Phenyl-5PW RP column (75 × 4.6 mm ID) with a 2 min linear gradient of acetonitrile from 5% to 20% followed by a 48 min linear gradient of acetonitrile from 20% to 80% in 0.05% trifluoroacetic acid. Gel filtration (GF) was performed on a TSKgel G3000SW column (600 × 7.5 mm ID) in 0.05 *M* phosphate buffer containing 0.2 *M* sodium chloride (pH 7.0). Ion-exchange chromatography (IEC) was performed on a TSKgel DEAE-5PW column (75 × 7.5 ID) with a 60 min linear gradient of sodium chloride from 0 to 0.5 *M* in 0.02 *M* Tris-HCl buffer (pH 8.0). The flow rate was 1 mL/min in all these separations. The results of the purity tests are shown in Fig. 2. One major peak and several very small peaks are seen in each chromatogram of the fractions. Because the major peaks showed enzymatic activity except in the case of reversed-phase chromatography, they must correspond to lipoxidase. This means that lipoxidase of high purity was obtained from commercial lipoxidase containing large amounts of impurities.

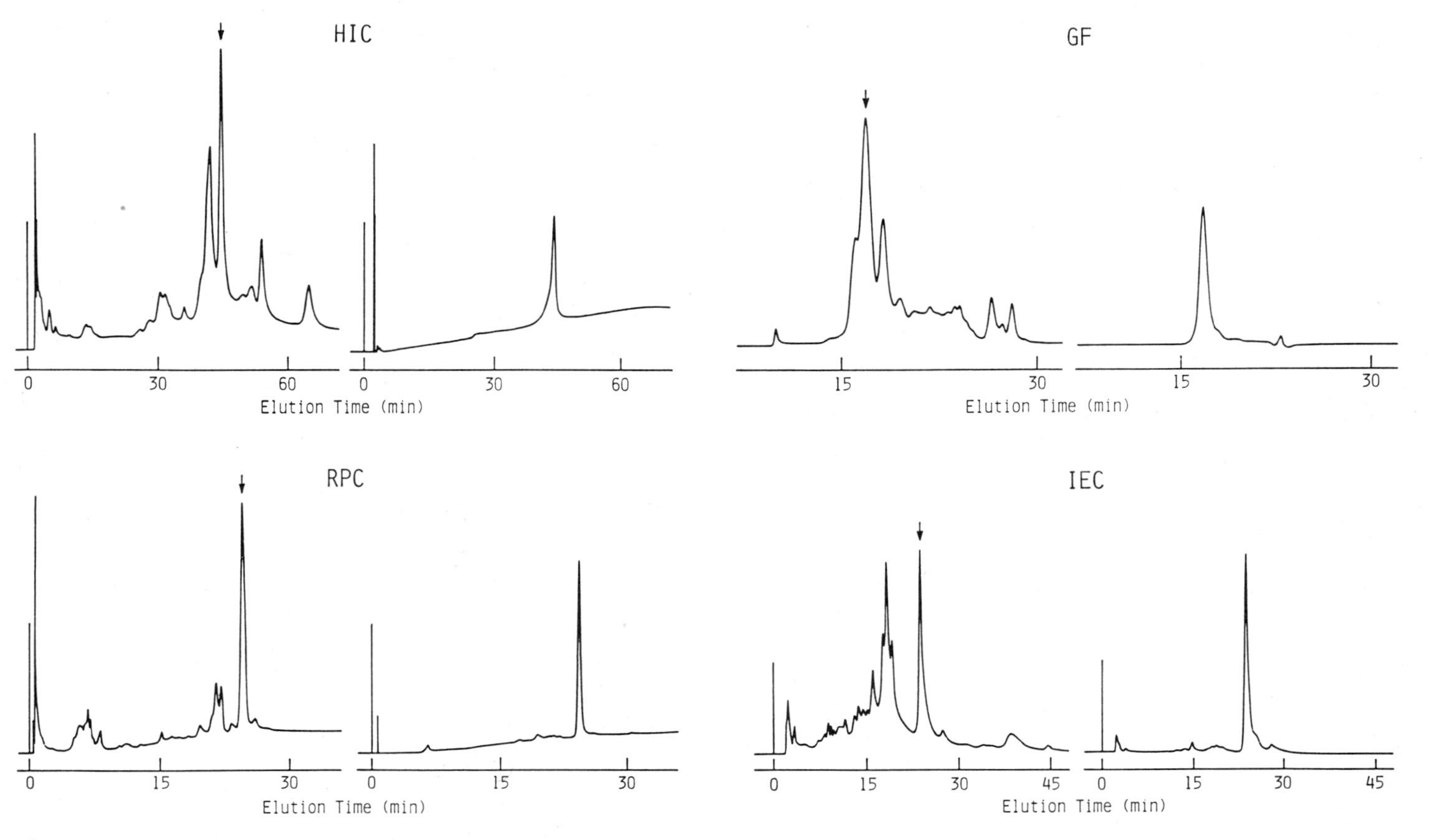

Fig. 2. Chromatograms of the crude sample of lipoxidase (left) and the fraction in Fig. 1 (right) obtained by HPLC. Peaks indicated by arrows in the chromatograms of the crude sample correspond to the main peaks of the fraction. (Data from Ref. 9. With permission.)

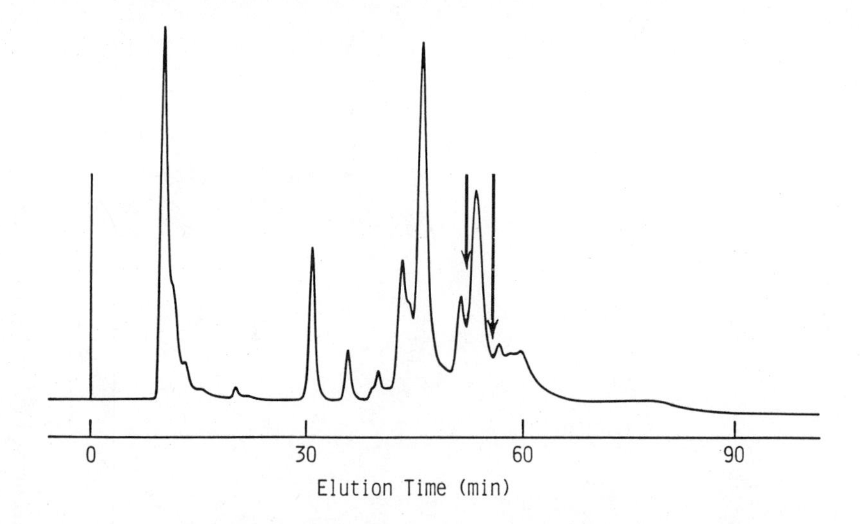

Fig. 3. Purification of 100 mg phosphoglucose isomerase on the Phenyl-5PW preparative column. (Data from Ref. 9. With permission.)

4. PURIFICATION OF PHOSPHOGLUCOSE ISOMERASE

Phosphoglucose isomerase was purified on the Phenyl-5PW preparative column with a 60 min linear gradient of ammonium sulfate from 1.5 *M* to 0. A crude sample of phosphoglucose isomerase was purchased from Sigma (St. Louis, MO). In the separation of the crude sample with various sample loadings, almost the same results were obtained up to 100 mg. The separation of a 100 mg sample is shown in Fig. 3. In this separation, enzymatic activity was found in several peaks. Of the applied activity, 70% was found in the peak between the two vertical lines and 26% was recovered in three small peaks eluted between 55 and 60 min. The degree of purification was 3.7-fold for the fraction between the two vertical lines. Figure 4 shows the results of purity tests of this fraction by HPLC. Conditions of HPLC were the same as in the case of lipoxidase. The main peaks of chromatograms of the fraction were confirmed to correspond to phosphoglucose isomerase by the enzymatic activity test. This indicates that rather pure phosphoglucose isomerase was obtained.

5. PURIFICATION OF HUMAN SERUM ALBUMIN

Human serum albumin was purified on the Ether-5PW preparative column with a 60 min linear gradient of ammonium sulfate from 1.7 *M* to 0. Human serum was purchased from Miles Labs. (Elkhart, IN). Before the serum was applied to the column, ammonium sulfate was added to give the final concentration of 1.7 *M*. The human serum could be applied up to 1.2 mL, which contains

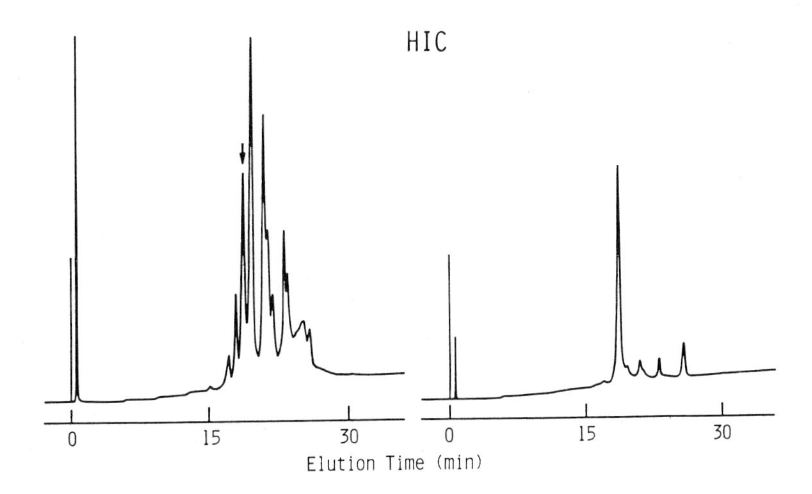

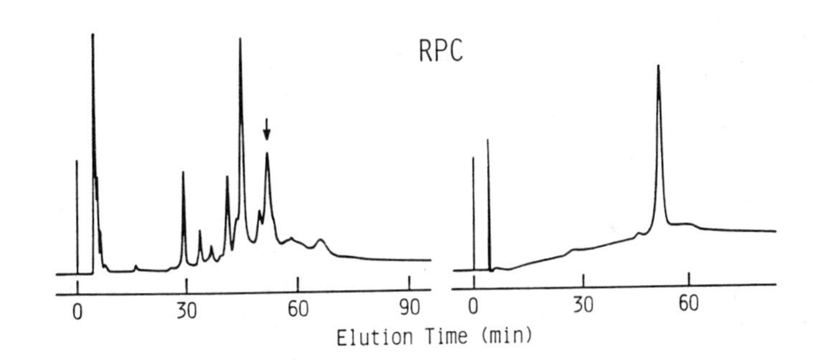

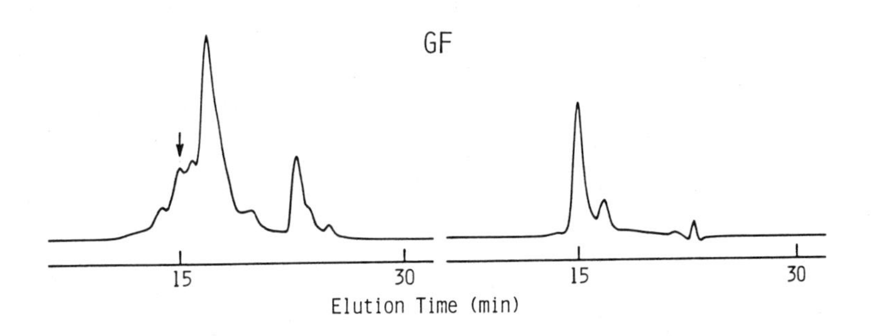

Fig. 4. Chromatograms of the crude sample of phosphoglucose isomerase (left) and the fraction in Fig. 3 (right) obtained by HPLC. Peaks indicated by arrows in the chromatograms of the crude sample correspond to the main peaks of the fraction. (Data from Ref. 9. With permission.)

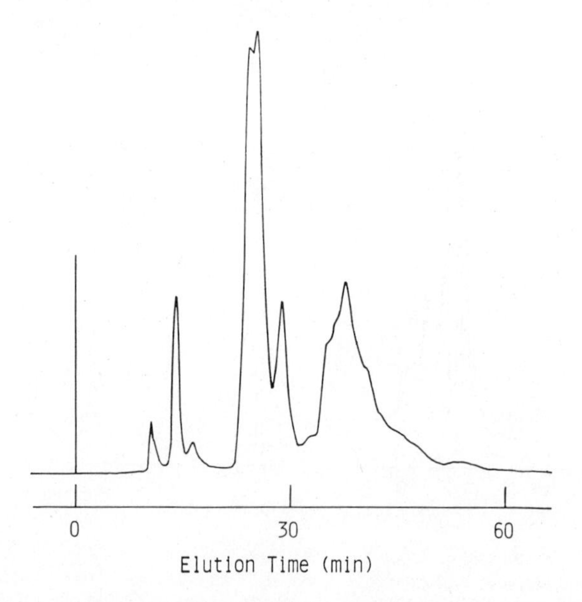

Fig. 5. Purification of human serum albumin on the Ether-5PW preparative column. A 1.2 mL sample of human serum containing ~90 mg of proteins was applied to the column.

~90 mg of proteins, without loss of separation efficiency. The separation of 1.2 mL of human serum is shown in Fig. 5. Column effluents were fractioned and subjected to immunoelectrophoresis. Two peaks appeared at elution times of 20–27 min were identified as albumin, a peak around 28 min was transferrin, and a broad peak at 34–50 min was confirmed to be γ-globulin. The broad peak of γ-globulin should be indicative that the components of γ-globumin were partially separated. The albumin peaks were collected between the two vertical lines. The recovery of albumin was determined by immunodiffusion. A 89% yield of albumin was recovered in the fraction. The purity was examined by immunoelectrophoresis (Fig. 6). Only a single band corresponding to albumin is seen in the pattern of the fraction, indicating that very pure albumin was obtained.

6. PURIFICATION OF α-AMYLASE

α-Amylase was purified on the Ether-5PW preparative column with a 60 min linear gradient of sodium sulfate from 1.1 *M* to 0. A crude sample of α-amylase (type VI-A from porcine pancreas) was purchased from Sigma. Almost identical patterns were obtained with sample loadings up to 100 mg. Figure 7 shows the separation of 100 mg sample. The peak corresponding to α-amylase between the two vertical lines was collected. The purity of the fraction was examined by HPLC as in the case of lipoxidase. In HIC, however, an analytical Ether-5PW

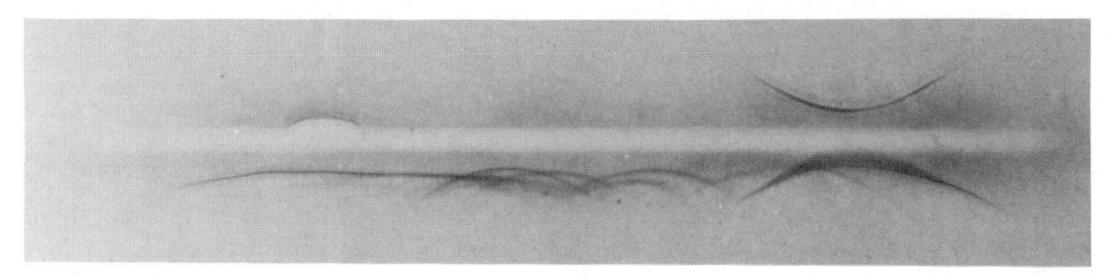

Fig. 6. Immunoelectrophoresis patterns of the albumin fraction in Fig. 5 (upper) and the original human serum (lower).

column (75 × 7.5 mm ID) was used and elution was performed with a 60 min linear gradient of the same eluents as in the purification. The results are shown in Fig. 8, indicating that α-amylase was purified to a great extent.

7. CONCLUSION

High-performance hydrophobic interaction chromatography is very useful to purify proteins rapidly with high resolution and recovery. Proteins are separated according to their hydrophobic properties, just like in the case of reversed-phase

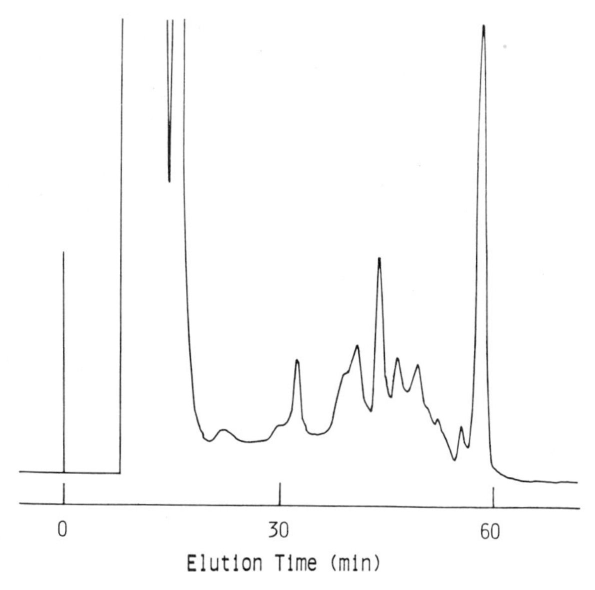

Fig. 7. Purification of 100 mg α-amylase on the Ether-5PW preparative column.

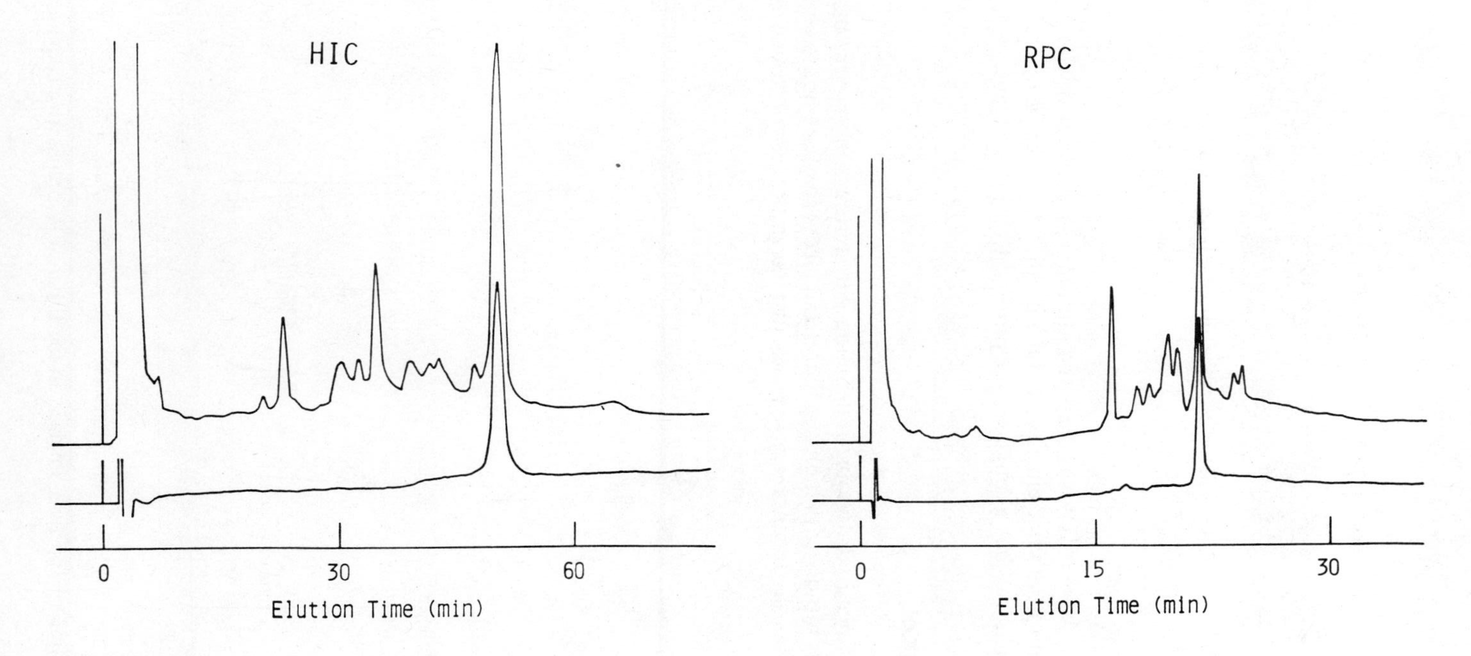

Fig. 8. Chromatograms of the crude sample of α-amylase (upper) and the fraction in Fig. 7 (lower) obtained by HPLC.

chromatography. However, proteins are not denatured in hydrophobic interaction chromatography. This is a great advantage of hydrophobic interaction chromatography over reversed-phase chromatography.

REFERENCES

1. Y. Kato, T. Kitamura, and T. Hashimoto, J. Chromatogr., *266*, 49 (1983).
2. Y. Kato, T. Kitamura, and T. Hashimoto, J. Chromatogr., *292*, 418 (1984).
3. D.L. Gooding, M.N. Schmuck, and K.M. Gooding, J. Chromatogr., *296*, 107 (1984).
4. J.L. Fausnaugh, L.A. Kennedy, and F.E. Regnier, J. Chromatogr., *317*, 141 (1984).
5. J.L. Fausnaugh, E. Pfannkoch, S. Gupta, and F.E. Regnier, Anal. Biochem., *137*, 464 (1984).
6. N.T. Miller, B. Feibush, and B.L. Karger, J. Chromatogr., *316*, 519 (1985).
7. J.-P. Chang, Z.E. Rassi, and CS. Horváth, J. Chromatogr., *319*, 396 (1985).
8. Y. Kato, T. Kitamura, and T. Hashimoto, J. Chromatogr., *360*, 260 (1986).
9. Y. Kato, T. Kitamura, and T. Hashimoto, J. Chromatogr., *333*, 202 (1985).

CHAPTER 13

High Performance Affinity Chromatography: Isolation and Analysis of Biological Macromolecules

IRWIN M. CHAIKEN

Department of Macromolecular Sciences, SmithKline Beecham, King of Prussia Pennsylvania

1. INTRODUCTION

During the past decade, biomolecular separations increasingly have been carried out in the high-performance liquid chromatography mode. For pre-HPLC chromatographic processes such as gel filtration and ion-exchange chromatography, adaptation to HPLC has offered precision, automation, microscale usage, and scale-up applications. And, more recently developed separation modes, including reverse phase and hydrophobic interaction, have been devised predominantly for HPLC and with this technology have become increasingly commonplace for a wide range of biomolecules.

Affinity chromatography has entered the HPLC field more slowly, in spite of the availability of increasingly sophisticated sample delivery, elution, and monitoring instrumentation (Fig. 1). The lag has been due in large part to the separation vehicle itself, namely, the affinity matrix. The most commonly used affinity support, agarose, has been eminently successful for many preparative needs, has required only simple instrumentation, and has been employed in a compressible porous gel form not amenable to the mechanical forces of HPLC (1–7, and references therin). Moreover, the intrinsic specificity of a particular immobilized ligand for only a limited number of mobile molecules makes it difficult to produce a small set of high-performance affinity matrices for a wide range of chromatographic needs, the hallmark of most other HPLC development.

In response to the compulsion to develop matrices with a wide range of

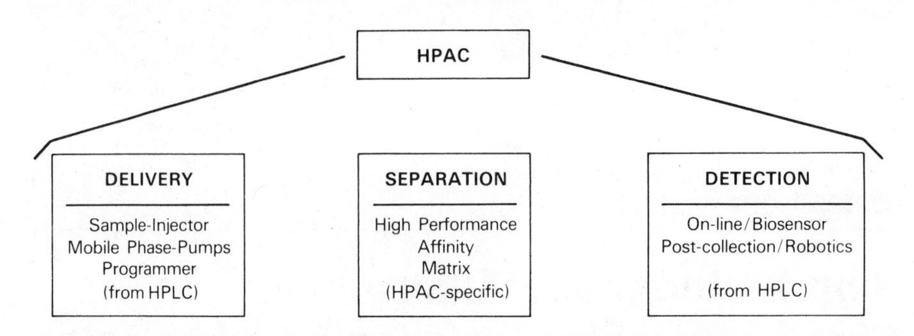

Fig. 1. Scheme of major components of a prototypic HPAC isolation and analysis system.

specificity, general ligand affinity supports have been described with such immobilized interactors as protein A, lectins, boronic acid, and triazene dyes (4, 8, 9). These have been produced with high-performance (noncompressible) matrices and used to isolate immunoglobulins, saccharides, and glycosylated macromolecules, and a variety of dye-binding (often nucleotide-binding) proteins (Refs. 10 and 11, and References therein). The general ligand matrices can bind rather large families of proteins and other macromolecules but still can be used for selective isolation based on differential molecular affinities within the respective families.

Nonetheless, broadly speaking, it is the versatility of affinity chromatography, using a wide variety of specialized affinity supports with immobilized molecules tailored to isolate an equally wide range of soluble biomolecules, that is the special appeal of the method. And it is here that soft gel technology historically has been so successful. With such a methodology already in place, one may ask, why bother with high-performance affinity chromatography (HPAC or the more pronounceable acronym HPLAC for high-performance liquid affinity chromatography)?

The answer to this question rests largely on advantages of HPAC and noncompressible affinity matrices for separating micro and macro amounts of biomolecules. For isolation of large amounts of molecular species, often from very large volumes of extracts or biological fluids, noncompressible affinity supports suitable for HPAC offer an important opportunity to scale-up separations. Such a possibility is more difficult with compressible gel supports, which cannot be packed in large beds and subjected to high flow rates. At least as important, matrix compactness and overall miniaturization of HPAC can be advantageous for microscale separation needs. These latter include isolation of an increasing array of biologically interesting molecules present in only small amounts, and often as minor components of complex mixtures. The forces of both micro- and macroscale separation needs have helped stimulate the current development of HPAC. The early results suggest that, at least in principle, micro and macroscale HPAC can be used in much the same way that reversed phase and other HPLC modes are, for both isolation and analysis.

2. ANALYTICAL AFFINITY CHROMATOGRAPHY THEORY AND ANALYTICAL HPAC

Previous experience with conventional affinity chromatography has shown that simple theoretical relationships can be used to describe affinity chromatographic elution behavior of macromolecules (11–14). Beyond allowing the evaluation of macromolecular interaction properties chromatographically, a useful research goal in itself, the quantitative analysis allows the specificity of affinity matrices to be measured and compared to solution interactions, thus providing an evaluation of the degree of biospecificity with an affinity support and therein of its reliability as a separatory vehicle. Several high-performance affinity matrices have been subjected to this analysis and found to be faithful biospecific interactors. These have been used increasingly for microscale isolation and characterization of peptides and proteins. The progress of this work has emphasized the benefits of HPAC for isolation generally. This includes the potential to carry out microscale isolations from crude extracts diagnostically by analytical HPAC, in order to obtain profiles of the functional molecular species present in a particular biological source (tissue extract, cell culture, biological fluid).

The early development of affinity chromatography involved predominantly preparative, two-step fractionations (Fig. 2*a*), with a binding step to allow selective retention of a desired macromolecule on an affinity support and a subsequent chaotropic elution step to break the noncovalent interaction to the immobilized ligand and recover the bound macromolecule of interest. Such separations were simple but nonetheless often quite selective. Eventually, the biospecificity inherent in affinity chromatographic interactions stimulated more analytical elution approaches (Fig. 2*b*) in order to describe mobile macromolecule-immobilized ligand interactions quantitatively. The quantitative potential is likely to have a continuing strong impact on HPAC usage.

When a zone of mobile macromolecule is eluted on an affinity matrix under binding conditions but adjusted to allow moderate retardation instead of functional retention (very strong retardation), the extent of retardation during isocratic elution can be used as a measure of the binding of mobile and immobilized interactors (for full derivations, see Ref. 16). And, the effect on elution volume of soluble molecules competing with immobilized ligand can be used to evaluate binding properties of the mobile macromolecule fully in solution. In the prototype monovalent case, the equilibria that occur are

$$M + P \overset{K_{M/P}}{\rightleftharpoons} MP \tag{1}$$

$$L + P \overset{K_{L/P}}{\rightleftharpoons} LP$$

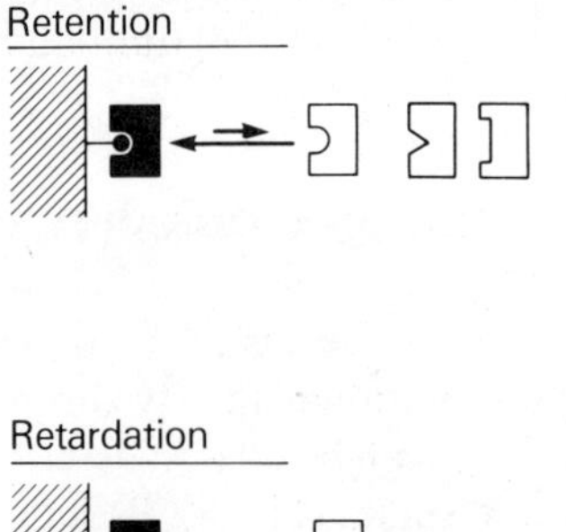

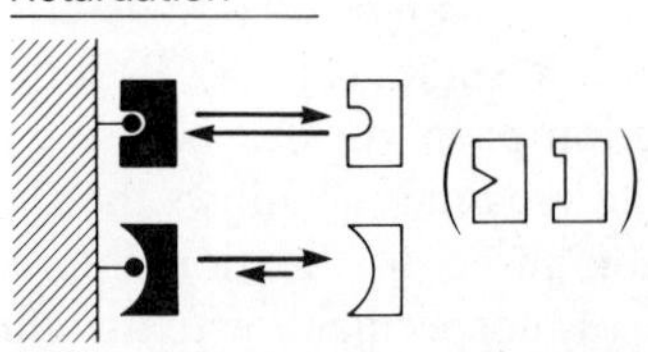

Fig. 2. Schematic representation comparing (*a*) preparative and (*b*) analytical affinity chromatography. (*a*) A sought-after mobile macromolecule binds tightly enough to the affinity matrix to be retained. It is removed in purified form by abrupt change to a chaotropic elution condition. (*b*) The mobile macromolecule is eluted on the affinity matrix at column and buffer conditions, including nature of immobilized interactor, designed to effect retardation but not retention. Isocratic elution allows resolution of interacting molecules from noninteracting ones as well as quantitation of matrix interaction with each mobile interactor resolved by differential affinity.

Here, M is the matrix-immobilized ligand, P is the mobile macromolecule, L is the competing soluble ligand, and $K_{M/P}$ *and* $K_{L/P}$ are the dissociation constants of the MP and LP complexes, respectively. For porous gels, the variation of elution volume V of the zone of P can be related to dissociation constants and amounts of M and L by

$$\frac{1}{V - V_0} = \frac{K_{M/P} + [P]_T}{(V_0 - V_m)\,[M]_T} + \frac{K_{M/P}\,[L]}{K_{\overline{L/P}}(V_0 - V_m)\,[M]_T} \tag{2}$$

and, for $[L] = 0$ (no soluble competitive ligand present), by

$$\frac{1}{V - V_0} = \frac{K_{M/P} + [P]_T}{(V_0 - V_m)\,[M]_T} \tag{3}$$

Here, V_0 is elution volume of unretarded molecule (of the same size range as P), V_m is the volume of the mobile phase (determined by elution of a noninteracting molecule large enough to be fully excluded from the pores), $[M]_T$ is the total concentration of immobilized interactant (determined as the capacity or, less ideally, as total immobilized molecule), and $[P]_T$ is the effective total concentration of P during elution. Normally, for soft affinity gels (e.g., agarose) of high capacity, $[P]_T$ is relatively small and can be neglected. Under such conditions, Eqs. 2 and 3 simplify to

$$\frac{1}{V - V_0} = \frac{K_{M/P}}{(V_0 - V_m)\,[M]_T} + \frac{K_{M/P}[L]}{(V_0 - V_m)\,K_{L/P}[M]_T} \tag{4}$$

and

$$\frac{1}{V - V_0} = \frac{K_{M/P}}{(V_0 - V_m)\,[M]_T} \tag{5}$$

respectively.

For analytical HPAC involving functionally porous matrices (for which $V_0 > V_m$), Eqs. 2–5 also apply. However with nonporous matrices ($V_0 = V_m$), a rather common occurrence for macromolecular elution on rigid high performance affinity matrices, V_0 can be used as an estimate of the accessible volume and Eqs. 2–5 convert to, respectively,

$$\frac{V_0}{V - V_0} = \frac{K_{M/P} + [P]_T}{[M]_T} + \frac{K_{M/P}[L]}{K_{L/P}[M]_T} \tag{6}$$

$$\frac{V_0}{V - V_0} = \frac{K_{M/P} + [P]_T}{[M]_T} \quad \text{when } [L] = 0 \tag{7}$$

$$\frac{V_0}{V - V_0} = \frac{K_{M/P}}{[M]_T} + \frac{K_{M/P}\,[L]}{K_{L/P}\,[M]_T} \quad \text{when } [P]_T = 0 \tag{8}$$

and

$$\frac{V_0}{V - V_0} = \frac{K_{M/P}}{[M]_T} \quad \text{when } [P]_T = 0 \text{ and } L = 0 \tag{9}$$

For many rigid and functionally nonporous high-performance affinity matrices, $[M]_T$ is not large; thus, $[P]_T$ can be ignored (Eqs. 8 and 9) only when it can be reduced by using sufficiently sensitive detection methods.

3. EXPERIMENTAL TESTS OF HPAC USING THE ANALYTICAL APPROACH

Formulations such as Eqs. 2–9 have been used to examine the biospecificity of high-performance affinity matrices. The results have helped select affinity matrices for microscale molecular characterization and isolation. Some of the systems so investigated are summarized in Table 1.

Several neuroendocrine complexes, involving neurophysins and the hormones oxytocin and vasopressin, have been particularly useful to evaluate

Table 1. Some Interacting Macromolecular Systems Evaluated by Analytical HPAC

Affinity Matrix, M	Mobile Molecule, P	$K_{M/P}$ (M)	Related K_d in Solution (M)	Reference
[Con A]Silica[a]	p-nitrophenyl-mannoside	4.5×10^{-5} 6.2×10^{-5}	11.5×10^{-5} 11.5×10^{-5}	17 18
[ADH]Silica[a]	AMP	13.4×10^{-5}	7.0×10^{-5}	19
[BNPII]NPG	AVP	1.1×10^{-5}	$1.6\text{–}2.0 \times 10^{-5}$	20
	BNPII	1.7×10^{-4}	1.7×10^{-4}	20
[BNPII]Silica	AVP	1.0×10^{-5}	$1.6\text{–}2.0 \times 10^{-5}$	21
[BNPII]HXL-agarose[a]	AVP	1.1×10^{-5}	$1.6\text{–}2.0 \times 10^{-5}$	21

[a] Con A is concanavalin A; ADH is alcohol dehydrogenase; HXL is highly cross-linked.

and develop HPAC systems. The first high-performance affinity supports characterized with this latter system were succinamidopropyl derivatives of glass. Bovine neurophysin II (BNP II) was covalently immobilized on both controlled-pore glass (CPG) and nonporous glass (NPG) and the interaction behavior of [^{3}H]Arg^8-vasopressin (AVP) measured by zonal analysis. In general, the extent of retardation was dependent on the amount of mobile component in the zone, especially for nonporous matrix as shown by the data of Fig. 3. As a rule, this strong dependence has not been found with conventional porous gel supports and appears to become evident with high-performance matrices of relatively low capacity (see Section II). The dependence of V on μg of AVP in the zone injected has been used to determine $K_{M/P}$ values from Eq. 9 and the value of $1/(V - V_0)$ extrapolated to AVP = 0. In the case of NPG, affinity matrix equilibrium binding constants so derived and those determined fully in solution are quite similar (Table 1). Generally, such a similarity is taken as an indication that the affinity matrix (here [BNPII]NPG) interacts biospecifically with the mobile molecule (here AVP) and that the matrix therefore can be used reliably for affinity separations, preparative as well as analytical.

The neurophysin/hormone system also has been used to evaluate a number of matrices other than glass beads for their utility in HPAC, by immobilizing BNP II and measuring the biospecificity of interaction with mobile interactors. As judged by the values of chromatographic equilibrium dissociation constants determined, two matrices with adequate biospecificity were obtained with N-hydroxysuccinimide-activated silica and tresyl chloride-activated highly cross-

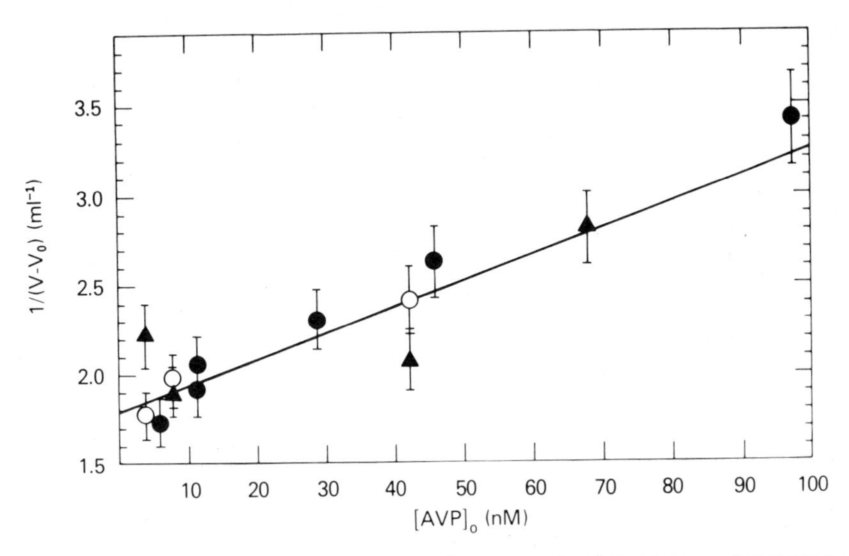

Fig. 3. Zonal elution of tritium-labeled Arg^8-vasopressin (^{3}H-AVP) on [BNPII]NPG showing dependence of the extent of retardation $[1/(V - V_0)]$ on initial concentration of AVP ($[AVP]_0$) in injected zone of 200 μL. The column (25 × 0.46 cm ID) was equilibrated and eluted at room temperature and a flow rate of 0.2 mL/min in 0.4 *M* ammonium acetate, pH 5.7, using a Varian 5000 HPLC system. The symbols ○, ●, and ▲ designate data obtained with different preparations of ^{3}H-AVP. The straight line was obtained by fit to Eq. 7. Figure adapted from Ref. 20.

linked agarose (21). The data obtained with silica, shown in Fig. 4, reflect a general characteristic found for zonal elution on high performance matrices. Here the dependence of $1/(V - V_0)$ on μg AVP again is observed as in Fig. 3. However, by carrying out the elution analysis over a sufficiently wide range of $[AVP]_0$ (initial concentration of AVP in the injected zone), the variation of $1/(V - V_0)$ actually is found to be nonlinear. The upward curvature of $1/(V - V_0)$ vs $[P]_0$ has been found to be prototypic for zonal elution data generally (23). While Eq. 7 predicts a linear relationship between $1/(V - V_0)$ and [AVP], $[AVP]_T$ is the total effective concentration of eluting protein and the value actually plotted in Fig. 4 is μg AVP, a linear function of $[AVP]_0$. Since dilution occurs during elution and the extent of dilution increases as V increases, the relationship of $[AVP]_0$ to $[AVP]_T$ in fact is nonlinear, with $[AVP]_T$ decreasing more sharply than $[AVP]_0$. (For example, halving $[AVP]_0$ will result in an eluted peak of average concentration reduced by greater than 50% because V is increased and the eluted peak is broader.) The upwardly curvilinear variation of $1/(V - V_0)$ vs $[P]_0$ thus is a predictable consequence of the nonlinear relationship in zonal elution between $[P]_0$ and $[P]_T$. Nonetheless, these data can be extrapolated to $1/(V - V_0)$ at $[P]_0 = 0$ to determine $K_{M/P}$. Generally (20–24),

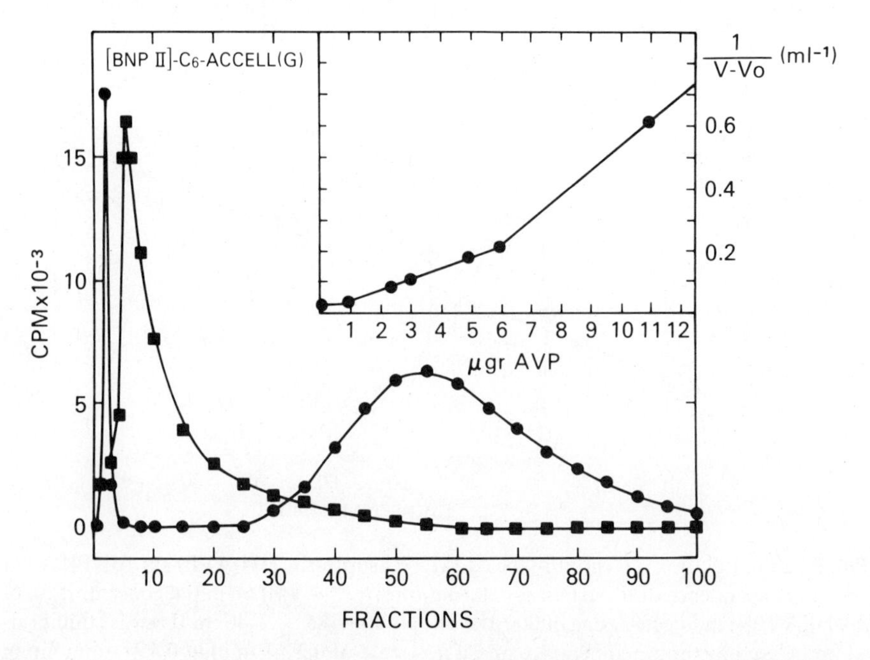

Fig. 4. Zonal elution of ³H-AVP on silica-based [BNPII]C6-ACCELL(G). The column (150 mm × 3 mm ID) was equilibrated and eluted with 0.4 *M* sodium acetate, pH 5.7, at room temperature, using a flow rate of 0.6 mL/min. A 20 μL sample containing 0.1 (●) or 11 (■) μg of ³H-AVP was injected. Elution profiles have an unbound fraction at V_0 = 1.08 mL; for simplicity the peak is shown only for the 0.1 μg elution. Inset: dependence of elution volume on ³H-AVP concentration (expressed as μg AVP in 20 μL injected zone). Values of $1/(V - V_0)$ at lowest and highest μg AVP in inset correspond to ● and ■, respectively, in main figure. Figure taken from Ref. 21 with permission.

analyses at a sufficiently low range of $[P]_0$ or at sufficiently high capacity ($[M]_T$) yield $1/(V - V_0)$ vs $[P]_0$ plots that are essentially linear (as in Fig. 3, making extrapolation to $[P]_0 = 0$ more straightforward.

4. MICROSCALE ISOLATION AND MOLECULAR PROFILING

The impressive microscale fractionation potential of HPAC extends a theme already apparent for conventional affinity chromatography. The latter methodology has always been useful for small-scale isolation owing to the small column size demanded with high-capacity affinity gels and the low nonspecific adsorption encountered with gel supports. Combining this general characteristic of affinity chromatography with the precision of HPLC technology and the potential for increased miniaturization (very small columns packed with increasingly smaller diameter, high-capacity rigid matrices, and consequent shorter analysis times) leads to a technological trend favoring HPAC for microscale isolation.

Beyond simple isolation needs, the development of analytical HPAC broadens the expectations for biomolecular affinity separations from two-step batch procedures, which separate binding from nonbinding molecules (Fig. 2*a*), to more complete chromatographic separations of mixtures of molecular species of different functional binding properties by isocratic, competitive or more elaborate gradient elutions. Such expectations suggest that, as with other HPLC tools (such as reverse phase, ion exchange, or size exclusion), HPAC can be used diagnostically to determine the presence of specific components of biomolecular families in biological fluids and cell and tissue extracts. The potential for this is apparent in the differential retardation of hormones and neurophysins on [BNPII]NPG observed by isocratic elution and shown in Fig. 5. These data show that hormones, neurophysins, and other neurophysin-binding molecules can be separated by isocratic elution and that the positions in the elution profile (V values) identify them by their dissociation constants. Such an analytical separation can show both the number and amounts of binding molecules and their identity as intact molecules versus metabolities (if these possess altered but nonzero interaction affinities); chromatography thus provides a profile of functional biomolecules.

Analytical separation applications such as those for molecular diagnosis seem likely not only to further stimulate the use of HPLC technology for analytical affinity chromatography but also to stimulate instrumental innovation tailored to the needs of this chromatographic mode. Extant pump-injection systems commonly used in HPLC instruments have been quite adequate for analytical HPAC development; and it is likely that most future needs, including miniaturization and biocompatibility, can be satisfied from the natural evolution of HPLC instrumentation generally. Nonetheless, if analytical HPAC continues to be developed for microscale and diagnostic separations, this development would be aided substantially if accompanied by advances in monitoring methods, both on-line and postcollection, especially to make detection more sensitive and more biospecific. As indicated in Fig. 1, the use of post-collector robotic assay (of such properties as enzymatic activity and immunoreactivity) or, better, on-line biosensors would significantly improve the capacity of HPAC for microscale separation and molecular profiling.

5. CONCLUDING COMMENTS

The development of HPAC has helped broaden the use of affinity chromatographic methodology to both microscale and macroscale isolation as well as analytical, multimolecular separations. This range of uses has been assumed generally for such HPLC modes as reversed phase but has not been expected for affinity chromatography owing to the origin of the latter as a largely batchwise, preparative separation method for one molecule at a time. The potential to use

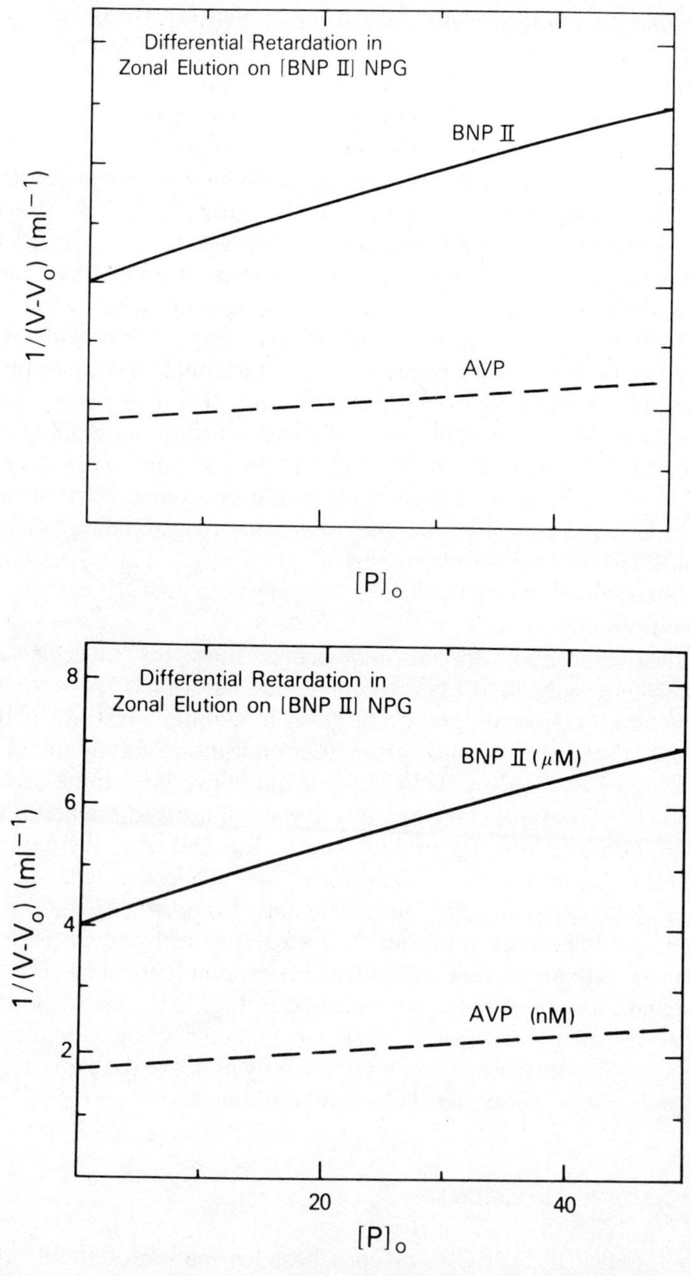

Fig. 5. Comparison of retardation of BNPII and AVP on [BNPII]NPG. AVP profile: redrawn from Fig. 3. BNPII profile: redrawn from Ref. 20. The slightly downward curvilinear trend, which represents data obtained by zonal elution of ^{125}I-BNPII on the same [BNPII]NPG column, was obtained by fit of experimental data to a model, which accounts for both soluble BNPII–matrix BNPII interaction and soluble BNPII self-association.

HPAC more broadly, alone and in combination with other HPLC separation modes, should continue to stimulate technological development, including not only rigid matrices for HPAC but also appropriate instrumentational components. Intriguingly, the expansion of affinity chromatographic usage through HPAC comes at a time when the rules of macromolecular recognition are beginning to be increasingly understood mechanistically. This co-occurrence could stimulate an extension of HPAC separation technology based on the redesign of macromolecular recognition properties and the use of such redesigned macromolecules in affinity chromatographic supports.

Note in Proof. Since this paper was first submitted, high performance affinity chromatography has continued to evolve. New affinity supports are introduced continuously in the search for the high capacity, biocompatability, and low cost needed for isolation applications [see for example the proceedings of the 7th and 8th International Affinity Chromatography Symposia (25, 26)]. Analytical uses also have continued to be examined. With the increased sensitivity to mobile interactor concentration often found with low capacity HPAC supports, frontal elution strategies increasingly have been used since these allow an accurate determination of mobile interactor concentration (27, 28). Molecular profiling, the simultaneous separation and affinity analysis of multiple interacting molecules in the same elution, has been applied successfully (29). And the advantages of weak affinity chromatography for molecular profiling have been emphasized (30). Nonetheless, while progress continues on all applications of HPAC, it still can be observed that, in contrast to other HPLC modes, high performance affinity chromatography is being embraced deliberately. Much preparative affinity chromatography still is accomplished on more traditional soft gels.

REFERENCES

1. P. Cuatrecasas, M. Wilchek, and C.B. Anfinsen, Proc. Natl. Acad. Sci. U.S.A., *61*, 636 (1968).
2. W.B. Jacoby, and M. Wilchek, Eds., *Methods in Enzymology*, Vol. 34, Academic Press, New York, 1974.
3. J. Porath, and T. Kristiansen, in *The Proteins*, 3rd ed., Vol. 1, H. Neurath, and R. Hill, Eds., Academic Press, New York, 1974, p. 95.
4. C.R. Lowe, and P.D.G. Dean, *Affinity Chromatography*, Wiley, London, 1974.
5. W.H. Scouten, *Affinity Chromatography*, Wiley, New York, 1981.
6. I.M. Chaiken, I. Parikh, and M. Wilchek, Eds., *Affinity Chromatography and Biological Recognition*, Academic Press, New York, 1983.
7. P.D.G. Dean, W.S. Johnson, and F.A. Middle, Eds., *Affinity Chromatography—A Practical Approach*, IRL Press, Oxford, 1984.
8. P. O'Carra, in *Chromatography of Synthetic and Biological Polymers, Vol. 2, Hydrophobic, Ion Exchange and Affinity Methods*, R. Epton, Ed., Ellis Horwood, Chichester, 1978, p. 131.

9. O. Hofmann-Ostenhof, M. Breitenbach, F. Koller, D. Kraft, and O. Scheiner, Eds., *Affinity Chromatography*, Pergamon Press, Oxford, 1978.
10. P.O. Larsson, M. Glad, L. Hansson, M.O. Mansson, S. Ohlson, and K. Mosbach, Adv. Chromatogr., *21*, 41 (1983).
11. G. Fassina, and I.M. Chaiken, Adv. Chromatogr., *27*, 247 (1987).
12. I.M. Chaiken, Anal. Biochem., *97*, 1 (1979).
13. D.J. Winzor, in *Affinity Chromatography—A Practical Approach*, P.D.G. Dean, W.S. Johnson, and F.A. Middle, Eds., IRL Press, Oxford, 1984, p. 149.
14. I.M. Chaiken, J. Chromatogr., *376*, 11 (1986).
15. I.M. Chaiken, Ed., *Analytical Affinity Chromatography*, CRC Press, Boca Raton, FL (1987).
16. H.E. Swaisgood, and I.M. Chaiken, in *Analytical Affinity Chromatography*, I.M. Chaiken, Ed., CRC Press, Boca Raton, FL, 1987, p. 65.
17. D.J. Anderson, and R.R. Walters, J. Chromatogr., *376*, 69 (1986).
18. A.J. Muller, and P.W. Carr, J. Chromatogr., *284*, 33 (1984).
19. K. Nilsson, and P.O. Larsson, Anal. Biochem., *134*, 60 (1983).
20. H.E. Swaisgood, and I.M. Chaiken, Biochemistry, *25*, 4148 (1986).
21. G. Fassina, H.E. Swaisgood, and I.M. Chaiken, J. Chromatogr., *376*, 87 (1986).
22. H.E. Swaisgood, and I.M. Chaiken, J. Chromatogr., *327*, 193 (1985).
23. G. Fassina (unpublished data).
24. G. Fassina, Y. Shai, and I.M. Chaiken, Fed. Proc., *45*, 1944 (1986).
25. H.P. Jennissen and W. Muller, Eds. Macromol. Chem. Macromol. Symp. *17* (1988).
26. M. Wilchek, Ed., Proc. 8th International Symp. Affinity Chromatography and Biological Recognition, J. Chromatography, in press.
27. G. Fassina and I.M. Chaiken, J. Biol. Chem., *263*, 13539 (1988).
28. G. Fassina, M. Zamai, M. Brigham-Burke, and I. Chaiken, Biochemistry, *28*, 8811 (1989).
29. P. Caliceti, G. Fassina, and I. Chaiken, Appl. Biochem. Biotechnol., *16*, 119 (1987).
30. S. Ohlson, A. Lundblad, and D. Zopf, Anal. Biochem. *169*, 304 (1988).

CHAPTER 14

Purification of Synthetic Oligodeoxyribonucleotides

GERALD ZON

Applied Biosystems, Foster City, California

1. INTRODUCTION

Much of the importance of synthetic oligodeoxyribonucleotides in biotechnology comes from their increasingly widespread use in diverse areas of both basic and applied research and development (1). The general uses of synthetic oligodeonucleotides, which have been reviewed elsewhere (2–9), include applications as hybridization probes (7, 10–16), primers (7, 17), adaptors for cloning (18), templates for carrying out deletions (19, 20), additions (4), and site-specific mutagenesis (3, 8, 9, 21–30), and the construction of genes (31–39). Examples of such applications that are relevant to the regulatory mission of the Food and Drug Administration include specific probes for detecting genetic diseases (40–42) and organisms with genetic potential for toxin production (43–46), cloning human growth factor (47), the provirus of HTLV-III (48) (a causative agent of AIDS), and human C-reactive (acute phase) protein (49,50), synthesis of genes for hormones (32, 35, 38), human insulin (34), and interferons (36, 37, 39), nonradiolabeled hybridization probes (51–53) for use either in clinics or field studies, vaccine synthesis by recombinant DNA technology (54), new cloning techniques for the diagnosis and treatment of infectious diseases (55), and modified oligonucleotides as targeted chemotherapeutic agents (56, 57).

The rapid progress that has been made in the synthesis of oligonucleotides has led to numerous reviews of this subject (2, 4–6, 8, 13, 58–68), practical guides for laboratory work (69, 70), and increasingly more common use of either commercially available automated DNA synthesizers (68, 71) or simplified manual methods (69, 70, 72–75). The introduction of relatively stable but catalytically activated phosphoamidite reagents (76, 77) and the use of controlled pore glass support having a long-chain alkylamine linker–spacer (78) were notable innovations that, in concert with other technical advances, allow for unattended, rapid synthesis of DNA molecules that have chain lengths in excess of 100 nucleotide units (79). Given the pace of recent chemical developments in this field (80–107), as well as that of RNA synthesis (108–111), it is highly likely that significant breakthroughs will continue to be made.

The most commonly employed versions of solid-phase synthesis of oligodeoxyribonucleotides afford the desired DNA product as one component of a relatively complex mixture, which also contains truncated sequences, base modified full-length molecules, and other contaminants. In some applications of sequencing primers (17), hybridization probes, and primers for site-specific mutagenesis, it has been possible to use the crude synthetic DNA directly, although in the vast majority of cases reported to date, and especially for gene syntheses and physicochemical studies, single or tandem purification procedures have been used to obtain products having a high degree of homogeneity. The specific details for such purification methods may vary according to the chain length of the product, the amount and number of products to be purified, end-use of the desired product, personal preference, available facilities, and other particulars. The basic options to consider for either HPLC or non-HPLC purification of typical length sized (15- to 30-mer), single-sequence, biological amounts ($< 50\ \mu g$) of synthetic DNA have been described elsewhere (5, 8, 69, 70, 112). On the other hand, comparatively little information has been published concerning HPLC purification of relatively long ($>$ 30-mer) products, "mixed-sequence" oligodeoxyribonucleotides, relatively large amounts (10–50 mg) of synthetic DNA, products with self-complementary sequences, and structurally modified oligonucleotides. It was therefore believed that it would be useful to present here selected examples of these latter, less familiar cases, in addition to covering the purification of more or less typical oligodeoxyribonucleotides. Most of the examples and procedures given in this account were taken from the author's data base concerning reversed-phase (RP) HPLC purification, which has been derived from approximately 3000 syntheses that were carried out during the period 1982–1985.

In concluding these introductory remarks, it should be emphasized that this chapter was written with the intention of providing a brief overview of the title subject, with hopefully helpful procedures, observations, and commentary based on personal experience. The opinions expressed here are neither claims nor endorsements of the superiority of a particular procedure or column. References to lead articles or representative studies have been included however, some important citations may have been inadvertently omitted. A review of the chronological development of liquid chromatography in nucleic acid research, which covers improvements in both silica- and polymer-based matrices, has been published very recently (113) together with a short review of the use of HPLC for isolation, purification, and analysis of oligodeoxyribonucleotides (114). For the sake of simplicity, oligodeoxyribonucleotides are hereafter referred to as "oligonucleotides," and are designated in the $5' \rightarrow 3'$ direction by line-formula, such as d(AGCT . . .), where dA = 2′-deoxyadenosine, dG = 2′-deoxyguanosine, and so forth.

1.1. General Comments

Efficient purification of synthetic oligonucleotides can presently be achieved by either polyacrylamide gel electrophoresis (PAGE) (69, 70, 112) or HPLC (112, 115, 116), or combinations of these methods. The continued popularity of PAGE is undoubtedly associated with the fact that electrophoresis of either ^{32}P-labeled or unlabeled oligonucleotides (UV imaging) (117) is a relatively simple and inexpensive method for isolating the desired product in the highest degree of purity that is presently possible, regardless of chain length. In addition to being the current standard method for evaluating the purity of final products with regard to DNA content, PAGE can also provide both confirmation of chain length and diagnostically useful qualitative information concerning the synthesis, that is, relative intensities of full-length and shorter sequences (Fig. 1) (112). On the other hand, PAGE can be relatively time-consuming, labor-intensive, and low-yielding in terms of the recovery of usable DNA from the gel. Moreover, PAGE does not lend itself to automation, nor does it lend itself to large-scale (multimilligram) purification of synthetic oligonucleotides for physicochemical studies [e.g., X-ray crystallography (118, 119) and nuclear magnetic resonance (NMR) spectroscopy (120)] and biotechnological applications (e.g., commercialization of specific hybridization probes as diagnostic devices).

Relatively rapid purification of oligonucleotides, with easy scale-up or automation or both, is possible using an appropriate mode of HPLC: (1) silica-based ("bonded-phase") ion-exchange (e.g., Partisil 10 SAX, Permaphase AAX, PEI (121–129)); (2) polymer-based ("nonbonded-phase") ion-exchange [e.g., RPC-5, NACS (130)]; (3) silica-based reversed-phase [mostly C_{18} (131–134) with some C_4 (135)]; and (4) polymer-based reversed-phase [e.g., PRP-1 (136, 137), PICS (114)]. Variations of these four basic modes of chromatography include tandem ion-exchange/reversed-phase (138–147), reversed-phase ion-pairing (148–153), and simultaneously combined ("mixed") modes (154). HPLC methods are also applicable to the purification of synthetic oligoribonucleotides (116) as well as the determination of the sequence (155, 156) chain length (157, 158), and base composition (159) of oligonucleotides, as alternatives to conventional analytical procedures (160–162).

A discussion and comparison of the aforementioned HPLC purification methods, which go beyond the scope of this chapter, have been presented elsewhere (114); however, a few points should be mentioned here. Ion-exchange HPLC has the advantage of separating oligonucleotides primarily according to chain length (163), with newly developed bonded phases such as polyethyleneimine (PEI) (126–128) allowing separations of oligonucleotides having chain lengths up to 50 residues, and Nucleogen-DEAE affording fractionation of 25- to 1500-basepair DNA restriction fragments (129). This mainly chain-length-determined mode for separation of oligonucleotides is exemplified by the HPLC trace for a commercially available sample of "$(dT)_{12-18}$" shown in Fig. 2 (left). That oligonucleotides having the same length but slightly different base composition can nevertheless show markedly different mobility upon silica-based

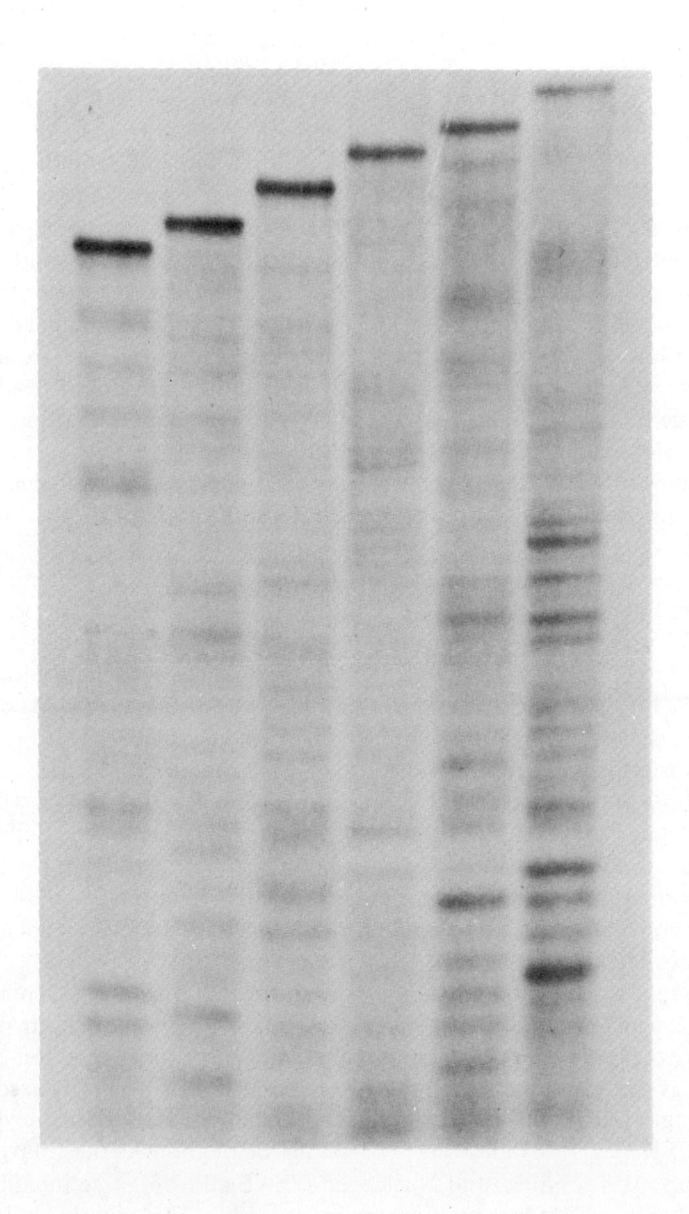

Fig. 1. Autoradiogram of ^{32}P-labeled oligonucleotides in crude mixtures of chemically synthesized 51- to 75-mers, as labeled at the top of each lane. A 12% polyacrylamide gel with 7 *M* urea was used for the separation. Reprinted from Ref. 112 with permission.

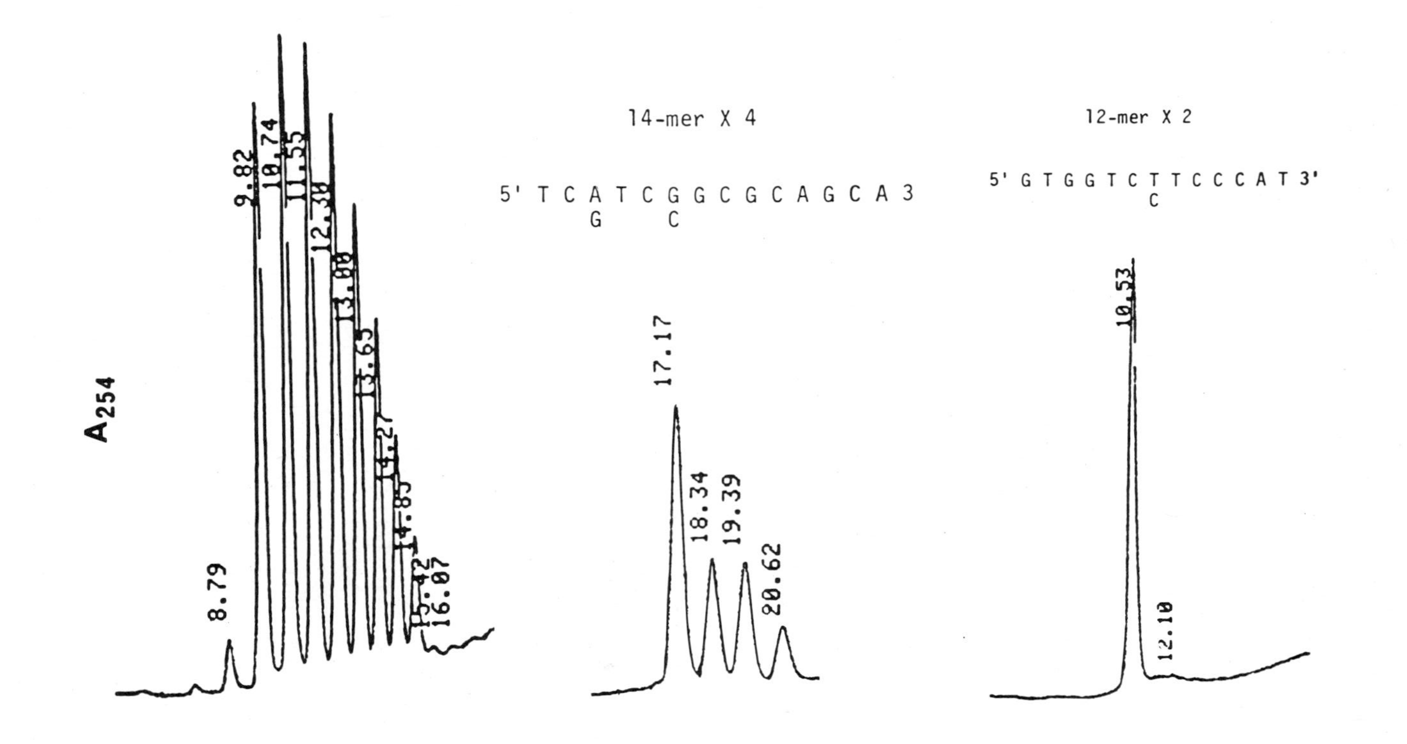

Fig. 2. Ion-exchange HPLC traces obtained using a Whatman Partisil PXS 10/25 SAX column (3.9 mm × 20 cm) and a 2%/min linear gradient of 0.4 *M* potassium phosphate (pH 6.8, 30% v/v EtOH, buffer A) vs 1 m*M* potassium phosphate (pH 6.8, 30% v/v EtOH, buffer B) starting at A:B = 20:80, flow rate = 2 mL/min, with detection at 254 nm and elution times given in minutes. (Left) Commercially available "$(dT)_{12-18}$," which was found by PAGE to actually contain $(dT)_{8-19}$. The peak at 10.74 min was assigned to $(dT)_{12}$. (Middle) Separation of four 14-mers having either A-or-G, or G-or-C at the positions shown. Peak assignments were not made. (Right) Co-elution of two 12-mers having either T or C at the position shown.

ion-exchange HPLC is apparent from Fig. 2 (middle) wherein complete separations were obtained for a mixture of four 14-mers, which differ by only one or two bases at the indicated positions in the chain. On the other hand, the apparently single peak seen in Fig. 2 (right) for two 12-mers, which differ by only a single base at the same position in the chain, indicates that differences in base composition are not necessarily reflected by measurable differences in elution time. Thus, constitutional isomerism of the type illustrated by d(. . . CAT . . .), d(. . . ACT . . .), d(. . . ATC . . .), and so forth, may or may not lead to detectable separation by ion-exchange HPLC (147).

Oligonucleotides are generally eluted from a strong-anion-exchange (SAX) column by use of nonvolatile salts such as potassium phosphate. This may necessitate purification of the salt buffer, taxes equipment and columns, and requires postchromatographic "desalting" by either dialysis, gel filtration, or ion-exchange liquid chromatography with volatile salts (Elutip-D, NACS, etc.) or reversed-phase liquid chromatography (Sep-Pak, Nonsorb, PICS, etc.). Problems with low recovery of oligonucleotides from silica-based ion-exchange HPLC columns and limited lifetimes of these ion-exchange columns (30–50 injections) were noted in early studies (143) and are still present. Reduced column lifetimes are especially observed under typical conditions (140, 142) used to disrupt intra- and intermolecular association of the nucleotide: phosphate buffer containing ethanol (30%) and urea (3–7 *M*) at elevated column temperatures (50–70°C). On the other hand, desalting as well as improvement in purity can be achieved by RP-HPLC, after ion-exchange HPLC, while PICS (113, 114) offers high sample recovery.

Early studies by Khorana and co-workers (131, 132) on the use of a silica-based (μ-Bondapak) C_{18} column to separate oligonucleotides and protected oligonucleotides demonstrated that 5′- and 3′-terminal protective groups that were highly hydrophobic (e.g., monomethoxytrityl and *t*-butyldiphenylsilyl) and easily removed could be utilized as added "handles" for selectivity to improve the purification of synthetic oligonucleotides. These observations and the subsequent development of solid-phase synthetic methods that can provide oligonucleotide products with a hydrophobic 5′-dimethoxytrityl (DMT or DMTr, Fig. 3) protective group have led over the years to increasing use of the so-called "trityl-on"option for product purification by reversed-phase (RP)-HPLC. A significant advantage of this method is the applicability of relatively fast isocratic or gradient elution with relatively "volatile" (i.e., easily removable) mobile-phase components, which invariably include acetonitrile as the organic modifier, and usually include triethylammonium acetate as the ion-pairing agent, although other "organic" (hydrophobic) salts are sometimes employed. Samples that are thus collected as 5′-DMT derivatives can be detritylated with acetic acid, a volatile acid, and then dried *in vacuo* for use as is, without further processing.

Reported variations of the "trityl-on" strategy for RP-HPLC purification include syntheses that provide alternative hydrophobic 5′-termini, such as 4-hexadecyloxy-(4′-methoxy-)trityl (HMT) and 4-(phenylazo-)phenyloxycar-

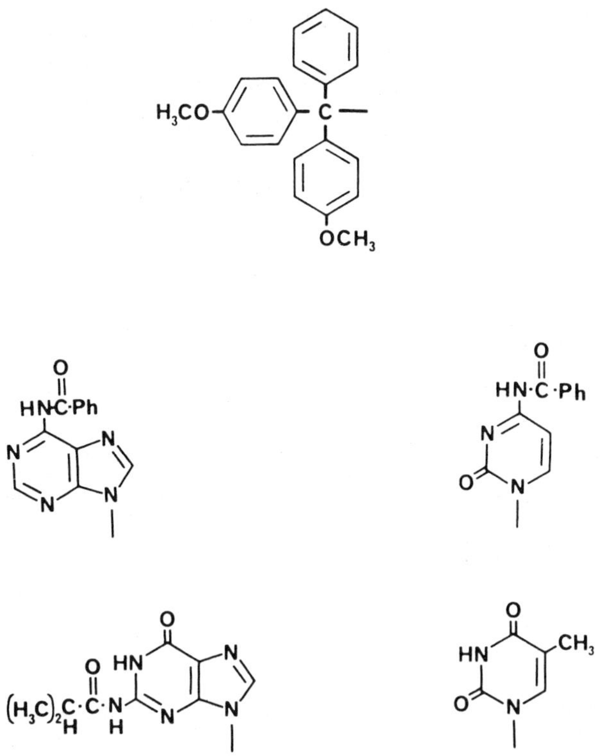

Fig. 3. The structure of the 4,4′-dimethoxytrityl (DMT or DMTr) group, and partial structures of N^6-benzoyl dA, N^4-benzoyl dC, N^2-isobutyryl dG, and dT.

bonyl (164, 165). The former can be incorporated in the last coupling cycle, following the use of DMT-bearing reagents during all previous cycles, in order to provide a 5′-HMT product that has increased hydrophobicity relative to contaminating, shorter 5′-DMT sequences. In contrast to this approach, incorporation of the 4-(phenylazo-)phenyloxycarbonyl group during each coupling cycle allows for acid-free synthesis and, therefore, should lead to crude products that have less contamination by DNA sequences that arise from depurination. The 5′-azo moiety can also serve as visually detectable chromophore for more reliable (e.g., dual wavelength) selection of the product peak during HPLC.

That the elution of 5′-hydroxyl oligonucleotides from RP columns is influenced by chainlength, base composition, and sequence was noted in early studies (131) and has been examined in more detail in a recent report (147) that compares SAX columns, Partisil 10 SAX, and MicroPak AX-10, with Nucleosil C_{18}, and in a comparative study (163) of RP-HPLC using Zorbax ODS and ion pairing with LiChrosorb RP-8 vs Partisil 10 SAX. The mechanism for retention of 5′-hydroxyl oligonucleotides during RP-HPLC involves, at least in part, an ion-pair adsorption process represented by the following equilibrium, which is taken from the report of Schill and co-workers (148):

$$\mathrm{ON}_m^{-z} + z\mathrm{Q}_m^{+} \overset{K_{\mathrm{eq}}}{=} \mathrm{ON}\cdot\mathrm{Q}_{z,s}$$

where ON_m^{-z} is the anionic oligonucleotide in the mobile phase, Q_m^{+} is the counter-ion present in the mobile phase, and $\mathrm{ON}\cdot\mathrm{Q}_{z,s}$ is the ion-pair adsorbed to the stationary phase. By changing the hydrophobicity of the counterion and its concentration, it is possible to alter the elution time, that is, the distribution of the oligonucleotide between the stationary mobile phases, which is given by the equilibrium constant, K_{eq}:

$$K_{\mathrm{eq}} = \frac{[\mathrm{ON}\cdot\mathrm{Q}_z]_s}{[\mathrm{ON}^{-z}]_m[\mathrm{Q}^+]_m^z} \tag{1}$$

The magnitude of K_{eq} depends on the nature of the ion-pair components and the properties of the phases. From Eq. 1 it is possible to derive an expression for the distribution ratio of the oligonucleotide (D),

$$\mathrm{D} = \frac{[\mathrm{ON}\cdot\mathrm{Q}_z]_s}{[\mathrm{ON}^{-z}]_m} = K_{\mathrm{eq}}\cdot[\mathrm{Q}^+]_m^z \tag{2}$$

which shows that the higher the numerical charge of the sample, the stronger is the effect of a change of kind and concentration of the counterion. However, in adsorption processes it is also necessary to account for the capacity of the adsorbent. Here, the ion pairs of the sample will compete with other mobile phase components for the limited number of accessible adsorption site, thus giving the following equation for the capacity ratio (K'):

$$K' = \frac{q\cdot K^0\cdot K_{\mathrm{eq}}\cdot[\mathrm{Q}^+]_m^z}{A + K_{\mathrm{eq}}\cdot[\mathrm{ON}^{-z}]_m\cdot[\mathrm{Q}^+]_m^z} \tag{3}$$

where q is the phase ratio in the column, K^0 is the capacity of the adsorbent, and A is a term that includes the influence of the competing agents.

Schill and co-workers (148) have shown that the effect of the concentration of tetrapentylammonium ion on the retention of d(TT), d(CGp), d(AAp), d(TTp), and d(TTTTp) follows a relationship of this kind. They also observed that, at a given concentration of tetrapropyl-, tetrabutyl-, or tetrapentylammonium ion, the retention of dAp or d(AAAp) increased with the hydrophobicity (hydrocarbon content) of the counterion. These investigators concluded that other effects are operative, since using an eluent without tetraalkylammonium ion led to a retention volume for the oligonucleotide that was lower than the hold-up volume of the column. It was suggested that the influence of a size-exclusion mechanism may counteract the effect of the ion-pair retention mechanism.

In a more recent study (153) of the use of C_{18} column to purify oligonucleotides, it was emphasized that this RP-HPLC separation method should be more properly referred to as RP ion-pair chromatography; however, for the purposes of the present account, RP-HPLC has been used as a relatively short mechanism-independent descriptor.

The ion-pairing mechanism must also be operative during RP-HPLC of DMT-bearing oligonucleotides, although all evidence clearly indicates that the elution of such compounds is markedly influenced by attractive interactions between the RP matrix and hydrophobic "appendages" on DNA (166). This feature is especially important in HPLC of backbone-modified 5′-hydroxyl oligonucleotides that are difficult if not impossible to separate by an exclusively ion-exchange mechanism, as in the case of stereoisomers of a given sequence (see Section 7).

Other sources of information may be consulted regarding theoretical and practical aspects of HPLC (167), ion pairing (168), or application of HPLC to biological compounds (169), in general, or oligonucleotides (115, 116). The remainder of this chapter deals mostly with RP-HPLC separations of oligonucleotides and analogs thereof using methods that were developed on an "as-needed" basis for various applications, and were neither optimized nor systematically studied in detail. Consequently, some of the reported observations require further investigation in order to draw firm conclusions. Examples of relatively simple separations of the type needed for molecular biological research and biotechnological development are covered first, in detail, while subsequent examples of large-scale purifications and relatively "exotic" oligonucleotide analogues will, it is hoped, be of interest as demonstrations of the versatility and resolving power of RP-HPLC. This wide range of applications attests to the foresight of early investigators who expressed their belief that the "HPLC method undoubtedly offers many possibilities for separation in the nucleic acid field" (131).

2. SINGLE-SEQUENCE OLIGONUCLEOTIDES

Synthetic oligonucleotides having a single sequence are now routinely used for specific hybridization, sequencing, site-specific mutagenesis, and construction of genes or regulatory elements. These oligonucleotides, which are generally 15- to 30-mers, are oftentimes required in large numbers and as soon as possible, although usually in relatively small amounts (<1 OD_{260}-unit, $<30\,\mu g$). The comparatively fast and simple postchromatographic processing of 5′-DMT derivatives of oligonucleotides that are rapidly purified by RP-HPLC have led to increasingly widespread adoption of this strategy as the sole method of purification. The chemical synthesis of these derivatives can be achieved by either the phosphotriester method or the newer, much more popular phosphoramidite method, which was recently reviewed in connection with the current status of "gene machines" (68). The phosphoramidite method proceeds with

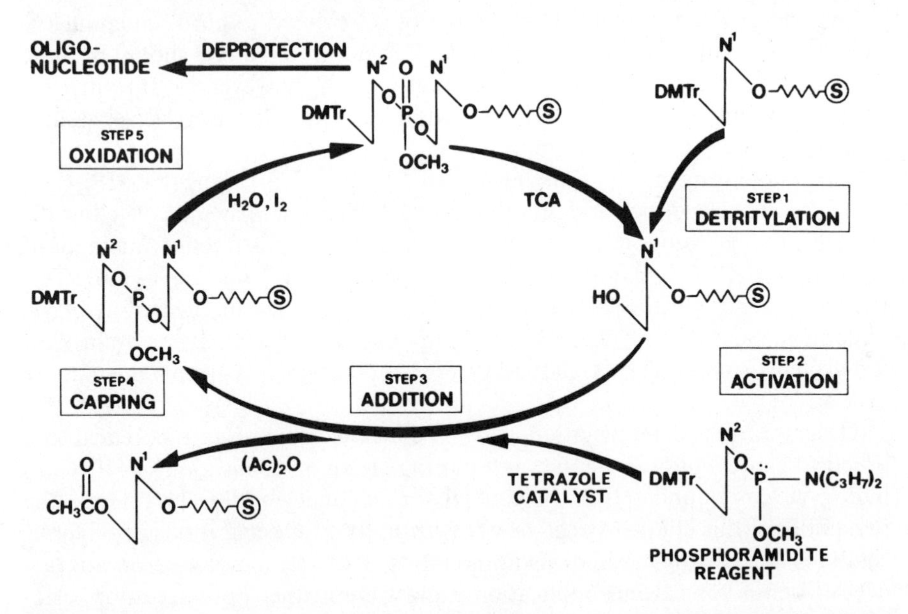

Fig. 4. Schematic representation of the phosphoramidite method (64) for DNA synthesis, which starts (step 1) with either zinc bromide-, trichloroacetic acid- (TCA), or dichloroacetic acid-catalyzed detritylation of 5′-DMTr-N-protected deoxynucleoside N^1 (cf. Fig. 3) that is attached at the 3′ position to a solid support (S). The incoming *O*-methyl-*N*,*N*-diisopropylphosphoramidite derivative of protected deoxynucleoside N^2 is activated (step 2) by mixing with 1*H*-tetrazole, which catalyzes the addition (step 3) to the 5′-HO group of support-bound N^1, thus giving a phosphite triester, $(RO)_2POCH_3$ [use of an *O*-β-cyanoethyl phosphoramidite reagent gives $(RO)_2POCH_2CH_2CN$]. Unreacted 5′-HO groups are capped (step 4) with acetic anhydride and *N*,*N*-dimethylaminopyridine. Oxidation (step 5) of phosphorus with I_2–H_2O completes cycle 1, and is followed by detritylation of the N^2 residue to begin cycle 2 for coupling of the next phosphoramidite reagent bearing N^3. After the last coupling cycle, thiophenol-triethylamine is used for *O*-demethylation. Ammoniolytic cleavage from the support and base-deprotection affords the crude, "trityl on," 5′-DMT product.

significantly higher coupling yields (97–99%), and allows relatively simple and "clean" cleavage from support, backbone-, and base-deprotection, which are features that greatly facilitate purification of the product by RP-HPLC.

2.1. General Procedures and Points to Consider

The crude products referred to in this chapter were obtained by the use of Applied Biosystems Model 380A, 380B, and 381A DNA synthesizers that employed either *O*-methyl or *O*-β-cyanoethyl phosphoramidite reagents and either 10-μmol, 1-μmol, or 0.2-μmol scale columns packed with long-chain alkylamine controlled-pore-glass (CPG) support. Other synthesizers (71) which use essentially the same chemistry cycle (Fig. 4) can be expected to give crude products having analogous quality. The 10-μmol scale column provides "physicochemical amounts" (multimilligrams) of DNA, while the 0.2-μmol scale column is nicely suited for "biological amounts" (micrograms) of DNA. In the author's laboratory, the standard work-up procedure for a 1-μmol scale synthesis is as follows. The machine-delivered or manually obtained ammonia solution of the *crude "trityl-on" deprotected material is kept basic by addition of triethylamine (3–5% v/v) to prevent detritylation* during either storage or removal of most of the ammonia under a stream of N_2 or air, which is followed by dilution with water to a convenient, standard volume (2 mL). A standard analytical-size injection (10 μL, 0.5%) of the sample solution onto a μBondapak C_{18} column that is eluted with a standard gradient of acetonitrile versus 0.1 *M* triethylammonium acetate (TEAA) generally affords adequate signal detection (0.1 AUFS, 254 nm) and resolution, as shown in Fig. 5 for a crude 29-mer. The shorter sequences with 5′-hydroxyl groups, which are inaccurately but widely referred to as "failure sequences," are not retained significantly under the standard operating conditions specified in Fig. 5 and are therefore eluted rapidly (2–3 min). These are followed by elution of benzamide, $PhC(O)NH_2$ (5.5 min), which is the expected by-product from ammoniolysis of benzoyl [PhC(O)] protecting groups on dA and dC residues (Fig. 3); the similarly formed isobutyramide, $(CH_3)_2CHC(O)NH_2$, derived from dG residues (Fig. 3) is not detected by monitoring at 254 nm. The subsequently eluted major peak (8.8 min) contains the 5′-DMT derivative of the product, which is collected during a repeat run with more sample (200 μL, 10%). Since the 5′-DMT derivatives of 15- to 30-mer oligonucleotides are generally eluted at 10 $\pm$ 4 min, time and eluent can be saved by either foregoing the analytical injection or using automated collection devices or both with or without peak-sensing. The collected fraction is taken to *dryness* on either a rotary evaporator at 50°C or in a vacuum centrifuge (e.g., Savant "Speed Vac Concentrator"), and the 5′-hydroxyl group is liberated by detritylation under mild conditions: dissolution of the 5′-DMT DNA in dilute aqueous acetic acid (1 mL of 3% v/v, pH 2.5–2.7) followed by standing at room temperature for 5–10 min. The sample is neutralized with concentrated ammonium hydroxide and then lyophilized. Additional details for postchromatographic sample work-up are given in Section 2.2, while some short-cuts from the synthesizer to the HPLC-purified product are described in Section 2.3. With regard to the aforementioned procedure for detritylation, it should be noted that measurement by RP-HPLC of the pseudo-first-order rate constant for detritylation of a mixture of 14-mers, 5′-DMT-d[CC(A/G)TA(A/G)TCCAT(A/G)TC], under the specified conditions gave $\tau_{1/2} \cong 1.5$ min. If the removal of triethylamine is incomplete, the resultant elevated pH of the solution during attempted detritylation can lead to incomplete reaction. This problem can be avoided by either reconcentration (to dryness) of the 5′-DMT DNA sample from water or use of more concentrated aqueous acetic acid (5% v/v). The conventional procedure using 80% acetic acid is not used, as it has occasionally led to extensive depurination.

For the 29-mer discussed previously, detritylation afforded a 5′-hydroxyl

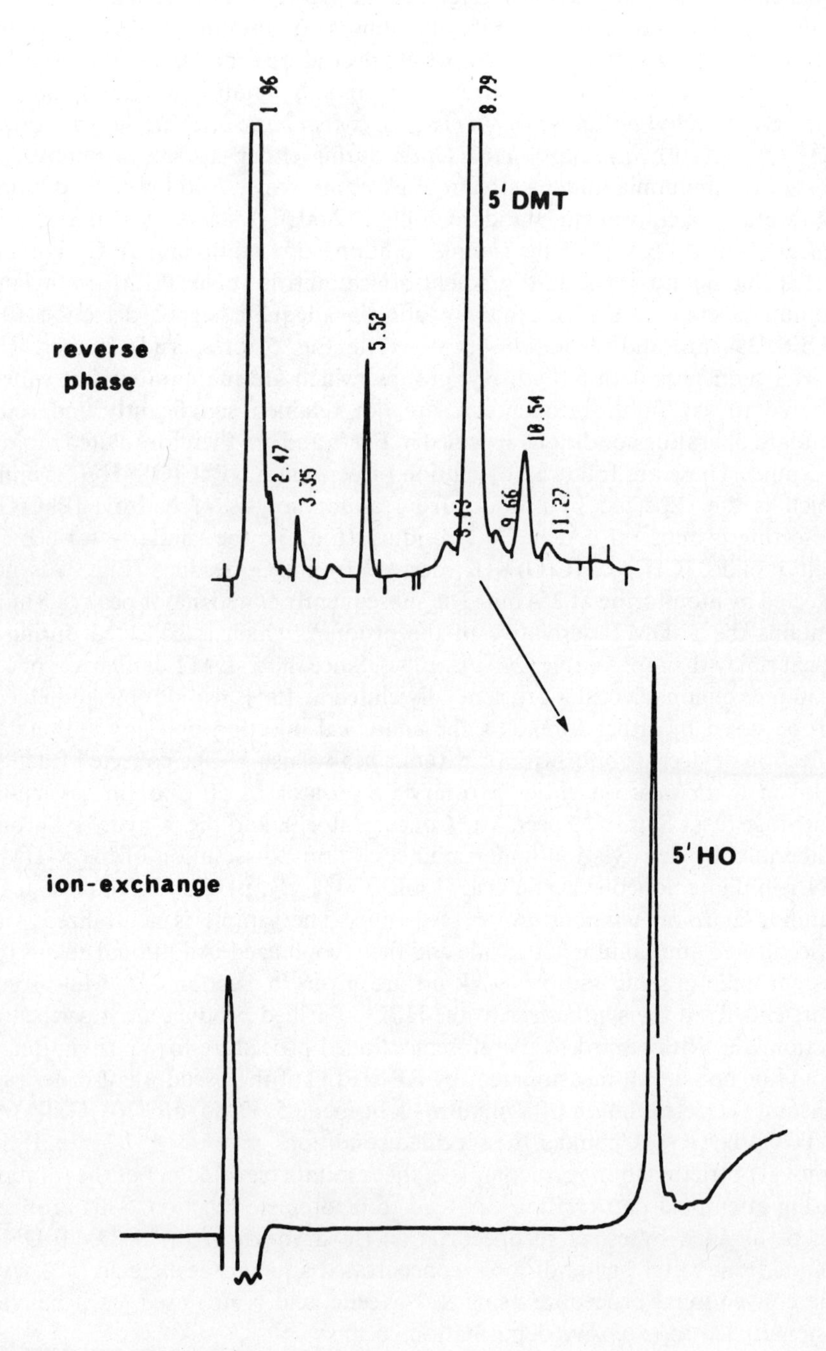
1.96
8.79
5'DMT
reverse phase
5.52
10.54
2.47
3.35
9.13
9.66
11.27
5'HO
ion-exchange

product (4 OD_{260}-units), which gave rise to essentially one peak when analyzed by ion-exchange HPLC (Fig. 5, bottom) under conditions used to separate $(dT)_{8-19}$ and $(dA)_{12-30}$ as standards for checking column performance. The HPLC-derived 5′-hydroxyl 29-mer was used without further purification as a primer for triple-site mutagenesis (K. McKennney, unpublished results).

None of the more than 1000 oligonucleotide probes or primers that have been purified in the author's laboratory using this RP-HPLC method have failed due to insufficient purity. Nevertheless, it is important to recognize that the successful use of probes and primers is not a direct index of the homogeneity of oligonucleotides, and that there are limitations for purification of synthetic DNA by RP-HPLC (and other techniques), as discussed in Section 2.1.2. These are other factors which one should be aware of when using the RP-HPLC method are as follows.

2.1.1. Theoretical versus Isolated Yield

The theoretical yield of an *n*-mer oligonucleotide synthesized by solid-phase methods can be calculated by the continuous product of the individual cycle yields for $n - 1$ cycles of coupling. The results of such calculations (W. Egan, unpublished) for various constant cycle yields (y) are shown in Fig. 6 as a family of curves for the percentage theoretical yield (equal to fractional yield × 100) vs the number of cycles. It can be seen that the synthesis of a 29-mer (28 cycles of coupling) that proceeds with $y = 99\%$ should give a ~75% theoretical yield of product, whereas somewhat less efficient coupling with $y = 95\%$ would give only a ~20% yield of product. In practice one finds that the amount of isolable product is substantially less than that expected based on the theoretical yield. For example, the average coupling yield determined by the conventional "trityl assay" was 98.7% for the 29-mer in Fig. 5, which indicated a 69% theoretical yield $[(0.987)^{28} \times 100]$, or ~174 OD_{260}-units. The molar extinction coefficient (ε) of an oligonucleotide can be roughly estimated by summing the absorptivity of the nucleotide residues (~14,000 for dA and dG, ~7,000 for dC and dT) without further correction; since 1 μmol/1 mL = 1 m*M*, one can then simply

Fig. 5. (Top) RP-HPLC trace obtained for the crude 29-mer product, 5′-DMT-d(GTGTTTTTCTTTCGATCCATTTCTTAC), using a Waters μBondapak C_{18} column (7.8 mm × 30 cm, 10-μm particle size) and a 1%/min linear gradient of acetonitrile vs 0.1 *M* TEAA (pH 7.0) that began at acetonitrile-TEAA = 20:80 and remained at 30:70 after 10 min; flow rate = 4 mL/min, detection at 254 nm, elution time in minutes. The 5′-DMT 29-mer that eluted at 8.8 min was collected and detritylated to give 5′-HO DNA. (Bottom) Ion-exchange HPLC trace obtained for the 5′-HO 29-mer using a Waters SAX cartridge (8 mm × 10 cm, 10-μm particle size) and a 2%/min linear gradient of 1 *M* potassium phosphate (pH 6.8, 30% v/v EtOH, 5 *M* urea, buffer A) vs the same eluent at 1 m*M* (buffer B) starting at A:B = 10:90; flow-rate = 2 mL/min, 5′-HO DNA at 16.8 min.

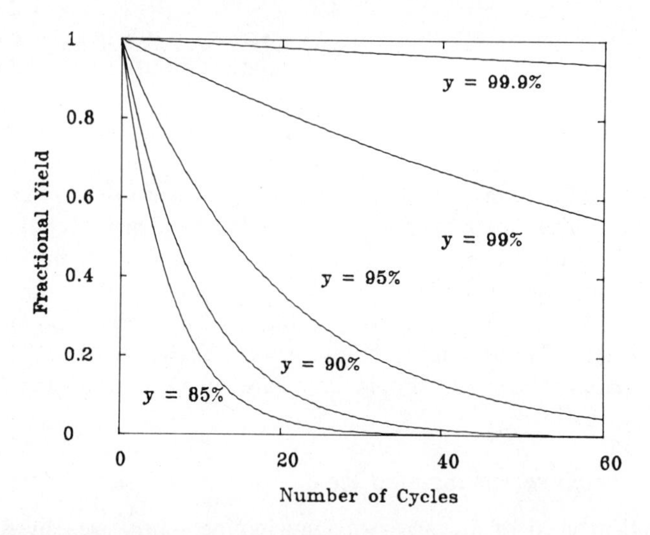

Fig. 6. Smoothed plots of calculated theoretical yields (as mole-fraction) of synthetic oligonucleotides versus the number of coupling cycles for various cycle yields (y). Published with permission of W. Egan.

multiply $\varepsilon \times 10^{-3}$ by the theoretical yield to obtain the OD_{260}-units of expected product. The amount of HPLC-isolable 29-mer product was 40 OD_{260}-units [4 OD_{260} − units × (2 mL/300 μL)], which represents a corrected, isolated yield of 23%; the usual range has been found to be 20–80%. "Hold up" during HPLC purification (see Section 2.1.5) accounts for some of the difference between theoretical and isolated yields, although various types of side reactions (see Section 2.1.2) during synthesis are probably more important contributing factors. For syntheses that proceeded with high-yield couplings, the peak-area ratio for DMT-DNA product to benzamide should be relatively high, at least qualitatively (cf. Fig. 5); calculating an accurate ratio is risky due to the ill-defined fate of the benzamide. An abnormally low ratio can indicate the loss of DMT-DNA product due to detritylation during either work-up or chromatography; it can also indicate column "overload" (see Section 2.1.7). To prevent detritylation during chromatography, it is advisible to have columns and syringes that are "dedicated" for HPLC purification of DMT-DNA products.

2.1.2. *Multiple DMT-Bearing Species*

While methods for DNA synthesis have been steadily improving during the past decade, it is important to be aware of various problems that can complicate the purification of the product by RP-HPLC. Incomplete detritylation (possibly due to blocked pores in CPG support), acid-catalyzed or other reactions leading to depurination, base alkylation (e.g., dT → N^3-methyl-dT via phosphate demethylation), and secondary phosphitylation reactions [e.g., O^6 position in dG

(106)] may occur during synthesis. In addition, strand cleavage induced by PhS^- or elimination at apurinic sites may occur during deprotection and cleavage from support. These processes not only lead to lower than expected yields of isolated product but also result in the formation of DMT-bearing oligonucleotides other than the desired product. That thymine residues in an oligonucleotide are subject to methylation at N^3 by the internucleotide methyl phosphotriester linkages (cf. Fig. 4) was recently demonstrated by Jones (170) and Urdea (171) and their co-workers by the use of HPLC. For example, RP-HPLC of 5′-DMT-d(GCTTTCG) with a μBondapak column afforded material that upon detritylation and re-HPLC led to separation of d(GCTTTCG) and a relatively slow-eluted impurity (170). Both compounds were isolated and then enzymatically digested with snake venom phosphodiesterase followed by alkaline phosphatase. HPLC analysis of the digests showed the presence of dG, dC, and dT from the sample of d(GCTTTCG), while the same deoxynucleosides plus N^3-methyl-dT were derived from the slow-eluted impurity. It was suggested (170) that the use of either ion-exchange HPLC or PAGE is unlikely to resolve such methylated sequences, since they do not differ in either length or charge from the unmodified compound, and that RP-HPLC purification will therefore be limited to 10-mers through 15-mers depending on the methylated sequences. Alternative phosphate protecting groups such as *O*-β-cyanoethyl apparently provide a remedy for this problem (170). Indirect evidence which suggests that even PAGE-purified oligonucleotides contain contaminants that lead to imperfect recombinant clones has recently been reported (172) in connection with the chemical synthesis and expression in yeast of a gene encoding connective tissue activating peptide-III. In this construction, manual synthesis on a silica support using *O*-methyl-*N*,*N*-diisopropyl phosphoramidite reagents followed by PAGE was employed to obtain 20 oligonucleotide products with chain lengths ranging from 14 to 32. A single annealing and ligation step afforded a 280-base-pair fragment that was then partially purified by PAGE and cloned. DNA sequencing of the clones revealed, on average, one transitional or transversional point mutation per recombinant at apparently random loci. These anomalies were tentatively attributed (172) to chemical modification(s) or degradation of nucleotides during chemical synthesis, such as deamination of cytosine, depurination, or condensation of enol tautomers of bases or from $(n - 1)$-mer contaminents in the n-mer preparations or both, each possessing a single random deletion.

Such base-modified and shorter sequences with DMT groups may not be resolved by RP-HPLC; consequently, the isolated products are perhaps most accurately referred to as being only "enriched" in the full-length perfect sequence. The degree of enrichment relative to shorter contaminants can be assessed by PAGE following detritylation; however, the determination of the amount and nature of full-length contaminants requires the use of other analytical methods.

DMT-bearing oligonucleotide impurities account for essentially all of the relatively low intensity peaks that are eluted near (or under) the product peak (cf. Fig. 5) during RP-HPLC purification, as is readily demonstrated by comparative analysis of an aliquot of the crude reaction mixture that has been detritylated. An investigation (173) of mixed-sequence oligonucleotides has shown that RP-HPLC separations of DMT-DNA's can be strongly influenced by structural differences near the hydrophobic 5′-DMT group. Examples of such separations for the two-component mixtures 5′-DMT-d[(A/G)TTTTT] and 5′-DMT-d[(A/G)TTTTTTTT] are given in Fig. 7. For these and other separations of closely related sequences, the longer oligonucleotides generally elute faster than their shorter counterparts, although there is a leveling-off effect with increasing chain length. In connection with the investigation of "trityl on" purification of synthetic oligonucleotides, Becker and co-workers (174) found that RP-HPLC-purified oligonucleotides were contaminated mainly with DMT-bearing sequences that were usually one to three nucleotides shorter than the full-length product. The elution times obtained for compounds **1–4** and **5** was compared to check the degree of separation of each hypothetical 19-mer "product" (**1–4**) and their resolution versus the 18-mer "failure" sequence (**5**):

		t_R (min)	Δt_R (relative to **5**) (min)
1	DMT-d(GTCACAGTCTGGTCTCACC)	6.70	2.02
2	DMT-d(CTCA . . . . . . C)	7.61	1.11
3	DMT-d(ATCA . . . . . . C)	7.94	0.78
4	DMT-d(TTCA . . . . . . C)	9.81	−1.09
5	DMT-d(TCA . . . . . . C)	8.72	0

Three of the 19-mers, compounds **1–3**, had retention times (t_R) that were shorter than that observed for the 18-mer, whereas 19-mer **4** had a longer retention relative to **5**. The values of Δt_R indicated that the degree of separation was markedly dependent on the nature of the "extra" 5′-residue. By collecting a wide versus narrow "cut" of a 5′-DMT product peak, it was shown (174) by PAGE that one respectively obtains more or less contaminating truncated sequences of the type $n - 1$, $n - 2$, and so forth. Similar results have been obtained in the author's laboratory, where the standard protocol calls for taking a narrow "cut" of the main 5′-DMT peak in order to achieve maximum enrichment of the product. The autoradiogram in Fig. 8 shows the level of $n - 1$, $n - 2$, and shorter-length contaminants that are typically for a *wide* "cut" (i.e., virtually all) of the main 5′-DMT peak under the standard RP-HPLC conditions. In these examples, the two preparations of 20-mers (lanes 4 and 6) and the two preparations of 21-mers (lanes 8 and 10) were used without further purification to create single site mutations in the *E. coli* gene for dihydrofolate reductase cloned into a derivative of pBR322 (W.J. Stec and S.J. Benkovic, unpublished results).

An impressive example of RP-HPLC separation of oligonucleotides based on compositional differences at the 5′ terminus was reported by Becker et al. (174), who resolved a random sequence 28-mer (t_R = 28 min) and its 5′-extended

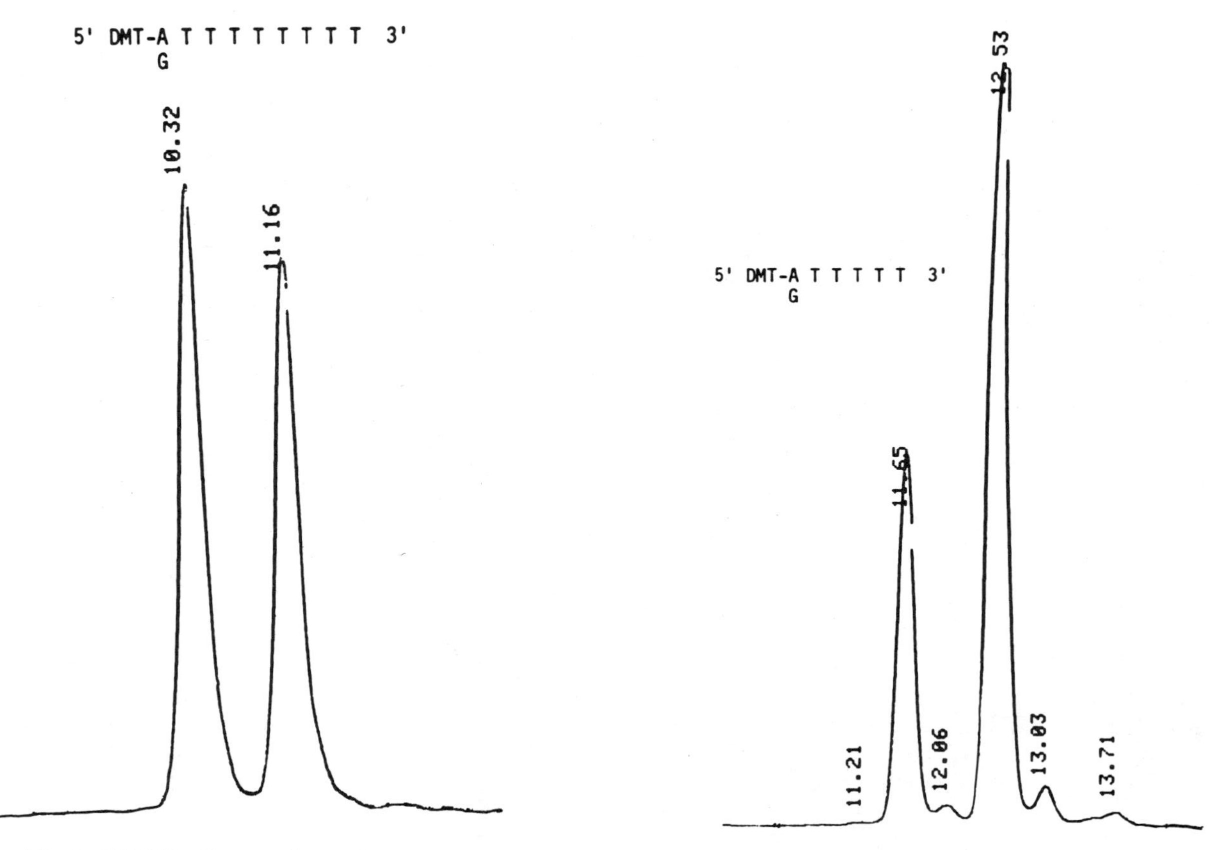

Fig. 7. RP-HPLC separations of 5′-DMT derivatives of nonamers and hexamers that have either A or G at the 5′-end. The HPLC conditions are given in Fig. 5. In each case the G-containing sequence was eluted faster.

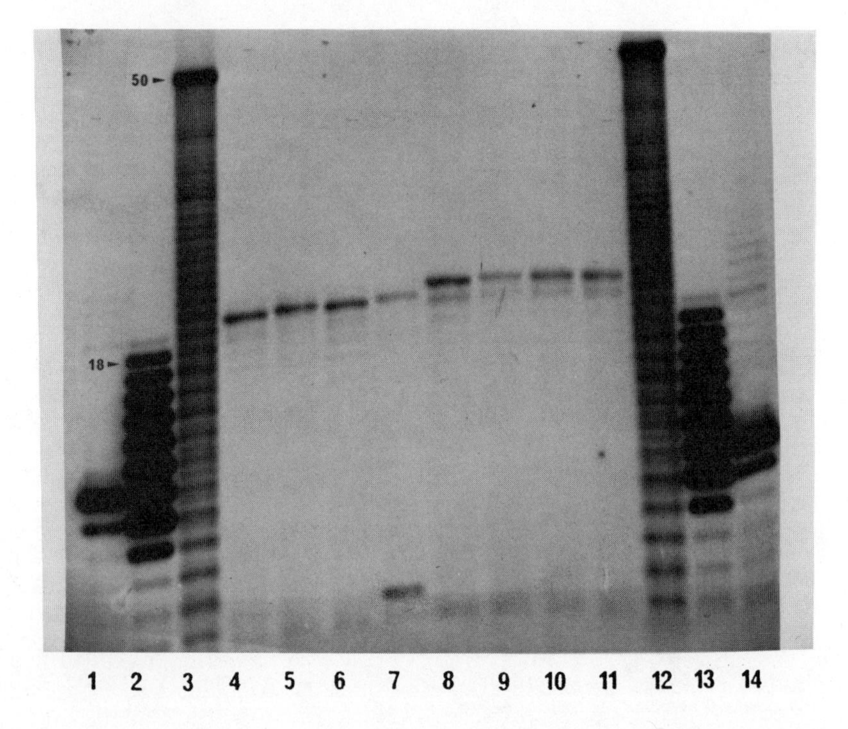

Fig. 8. Autoradiogram of ^{32}P-labeled oligonucleotides after separation by PAGE under typical conditions (112). Lane 1, commercial sample of "$(dT)_{12}$"; lane 2, commercial sample of "$(dT)_{12-18}$"; lane 3, crude mixture of a synthetic 50-mer of dT; lane 4, RP-HPLC-derived 20-mer, d(TGTATCTGGTGCATATCGAC); lane 5, same as lane 4 but different "cut" of 5′-DMT peak; lane 6, same as lane 4 but with a d(. . . $GT_{PS}GC$ . . .) backbone modification, PS = phosphorothioate linkage; lane 7, same as lane 6 but different "cut" of 5′-DMT peak; lane 8, RP-HPLC-derived 21-mer, d(ATTATGGGTCGCCAAACCTGG); lane 9, same as lane 8 but different "cut" of 5′-DMT peak; lane 10, same as lane 8 but with a d(. . . $AA_{PS}AC$ . . .) backbone modification; lane 11, same as lane 10 but different "cut" of 5′-DMT peak; lane 12, same as lane 3; lane 13, same as lane 2; lane 14, same as lane 1.

29-mer counterpart (t_r = 21 min) having the identical sequence through 28 nucleotide units and an additional 5′-terminal dA residue.

2.1.3. Structural Effects on Elution Time

In addition to the previously mentioned primary influence of the strongly hydrophobic DMT group, and the secondary influence of 5′-terminal bases on the elution time for DMT-DNA's, the effects of total base composition and chain length must be considered as well. Quantitative treatments of these various different effects are conceivable and useful, from the viewpoint of making *a priori* predictions about the desired product; however, there appear to be no such reports for 5′-DMT oligonucleotides. Ikuta et al. (137) have studied RP-HPLC behavior of 5′-hydroxyl oligonucleotides and concluded that the retention time seems to be directly proportional to the hydrophobicity of the DNA oligomers. The net hydrophobicity of an oligomer was defined (137) as a sum of two terms, the first being the hydrophobicity due to the bases, and the second being the hydrophobicity due to the polar groups (which may be assigned negative values), mainly the phosphodiester linkages. The second term was considered to be constant for all nucleotide residues in the oligomer, i.e., base independent, whereas the first term is presumably base dependent. The net hydrophobicity was thus regarded as a sum of the terms for all the constituent nucleotides

$$H = \sum_{i=1}^{n} B_i + (n - 1)P$$

where B_i is the hydrophobicity of base i and P is the effect of the phosphodiester linkages of which there are $n - 1$ in an n-mer oligonucleotide. It was assumed that for homopolymers, $(dB)_n$, B = A, G, C, or T, the various B_i's are identical; hence, $H = nB + (n - 1)P$. The difference in net hydrophobicity given by H for $(dB)_{n+1}$ minus H for $(dB)_n$ equals $(n + 1)B + nP - nB - (n - 1)P$, or simply $B + P$. Thus, depending on the sign of $B + P$, either the shorter or the longer homopolymer will be more slowly eluted from the RP column.

Ikuta et al. (137) noted that the order (175) of hydrophobicity for nucleotides at pH 7 is C < G < T < A, and that C < T may be explained by the added methyl group in T, while G < A may be due to more polar groups (CO and NH_2) in G compared to A (only NH_2). Although P is negative, A and T have large B values; therefore, $B + P$ is still positive. Hence, longer oligomers of $(dA)_n$ or $(dT)_n$ are eluted later from a RP column, and it has been possible to use HPLC to resolve components in the series $p(dA)_n$ and $p(dT)_n$ for $n = 2, 4, 6, 8$, and 10 (175). It was reasoned that longer oligomers are eluted closer together compared to shorter ones because although $H^{n+1} - H^n$ is constant, the percentage of difference in H decreases with increasing values of n.

Ikuta et al. (137) concluded that the magnitude for $B + P$ is smaller for C and G, relative to A and T, but still positive. Thus, when they compared components in the series $d(CG)_n$, the order of reversed-phase elution times was 2 < 3 < 4.

The order of elution times for 2′-deoxynucleotides from a C_{18} column using a gradient of acetonitrile in TEAA is dC < dG < dT < dA. To the extent that this order reflects the order of hydrophobicity, it is interesting that replacement of N^7 in dG with a CH moiety (7-deaza-dG) leads to slower elution, as might be expected due to fewer polar groups, whereas replacement of NH_2 in dG with a CH moiety (2′-deoxyinosine) leads to faster elution, which is contrary to what is expected based simply on the relative number of polar groups. Another caveat obtains from the elution orders found by the present author for the series d(TBT) (Fig. 9) and d(TBTT), which were G < C < A < T in both of these

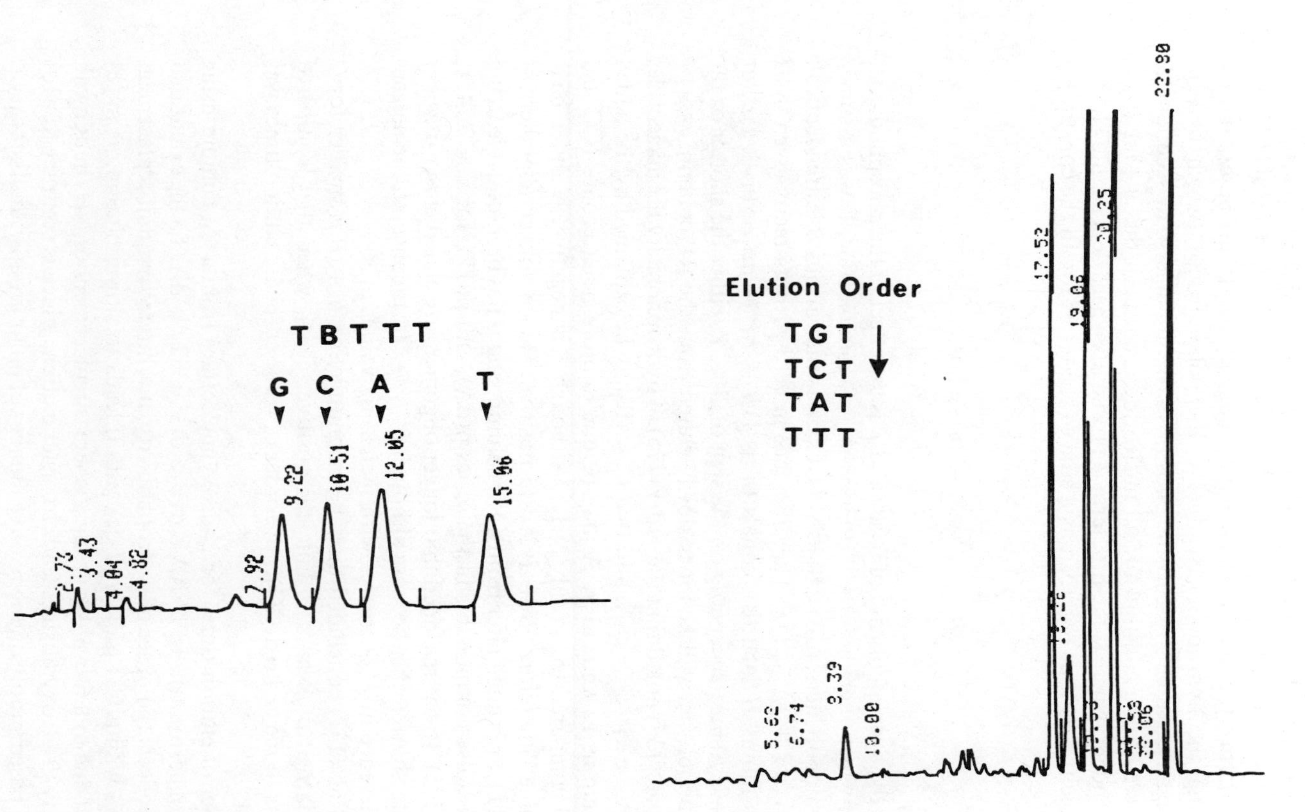

Fig. 9. RP-HPLC separations of four-component mixtures of crude d(TBTTT) and d(TBT), B = A, G, C or T. The column and conditions used for the pentamers are given in Fig. 5. For the trimers, the Waters Nova-Pak C_{18} column (7.8 mm × 30 cm) was eluted with a 0.33%/min gradient of acetonitrile vs TEAA (0.1 *M*, pH 7.0) starting at acetonitrile-TEAA = 5:95; flow rate = 3 mL/min. Peak detection was at 254 nm and elution time is given in minutes.

cases. Reversal of the orders for C and G, and A and T, relative to the 2′-deoxynucleosides, with preservation of the order for C/G and A/T suggests that secondary structural factors (e.g., sequence or nearest neighbors) may need to be taken into account by adding empirically derived terms to the aforementioned equation for H. With a large enough experimental database, it may thus be possible to accurately calculate relative retention times for 5′-hydroxyl oligonucleotides. Similar calculations for 5′-DMT derivatives would, of course, have to account for the role of the strongly hydrophobic 5′-terminal region of the molecule. While the elution order for the aforementioned series of 5′-DMT 19-mers, **1–4**, is the same as that found for the series of 5′-hydroxyl tetranucleotides, d(TBTT), namely G < C < A < T, the elution order for the series 5′-DMT-d(BTT) is G < A < C < T, which clearly indicates that simple qualitative rationalizations of elution times for 5′-DMT derivatives of oligonucleotides may be misleading. Elution times for DMT-bearing oligonucleotides usually (but not always) decrease with increasing chain length, as seen in Fig. 7 for the related 6-mers and 9-mers, and as evidenced by the 42-mer and 50-mer examples in Fig. 10. The HPLC trace obtained for the 5′-DMT 60-mer shown in Fig. 11 likewise indicates relatively fast elution from a C_{18} column under our standard conditions. The 60-mer was collected and then detritylated, ^{32}P-labeled, analyzed by PAGE, and thus compared with the crude material. An overexposed autoradiogram (Fig. 11) revealed that the HPLC-isolated product contained shorter sequences, down to ~35-mers, which evidently coeluted with the product. No attempt was made to improve this degree of enrichment by use of a different gradient.

Recent unpublished studies of C_{18} RP-HPLC purification of 5′-DMT oligonucleotides by Iwabuchi have led to remarkable separations (Fig. 12, PAGE analyses not available) on Nucleosil columns packed with either 5 μm particles that have 100 Å pore diameter, or 7 μm particles that have 300 Å pore diameter. The former column is routinely used for separations that involve products with chain lengths <40–50, while the latter column having a larger pore diameter is routinely used for separations of relatively longer product molecules (see Section 2.1.4). The sharper peaks obtained with these C_{18} Nucleosil columns, relatively to a 10 μm particle C_{18} μBondapak column, result at least in part from the smaller particle size. The significantly longer retention time for the 30-mer (20 min) compared to the 21-mer (17 min) would appear to be too large of a difference to attribute to the somewhat higher AT-content of the 30-mer (57% AT) vs the 21-mer (52% AT), and probably reflects the different gradient conditions. The same can be said for the virtually identical elution times (~20 min) obtained for the 42-mer (48% AT) and the 80-mer (56% AT) on the 7 μm Nucleosil column. It is also reasonable to assume that secondary structure (hairpin formation) and effective size are factors which govern the elution characteristics of such relatively long 5′-DMT derivatives of oligonucleotides during RP-HPLC.

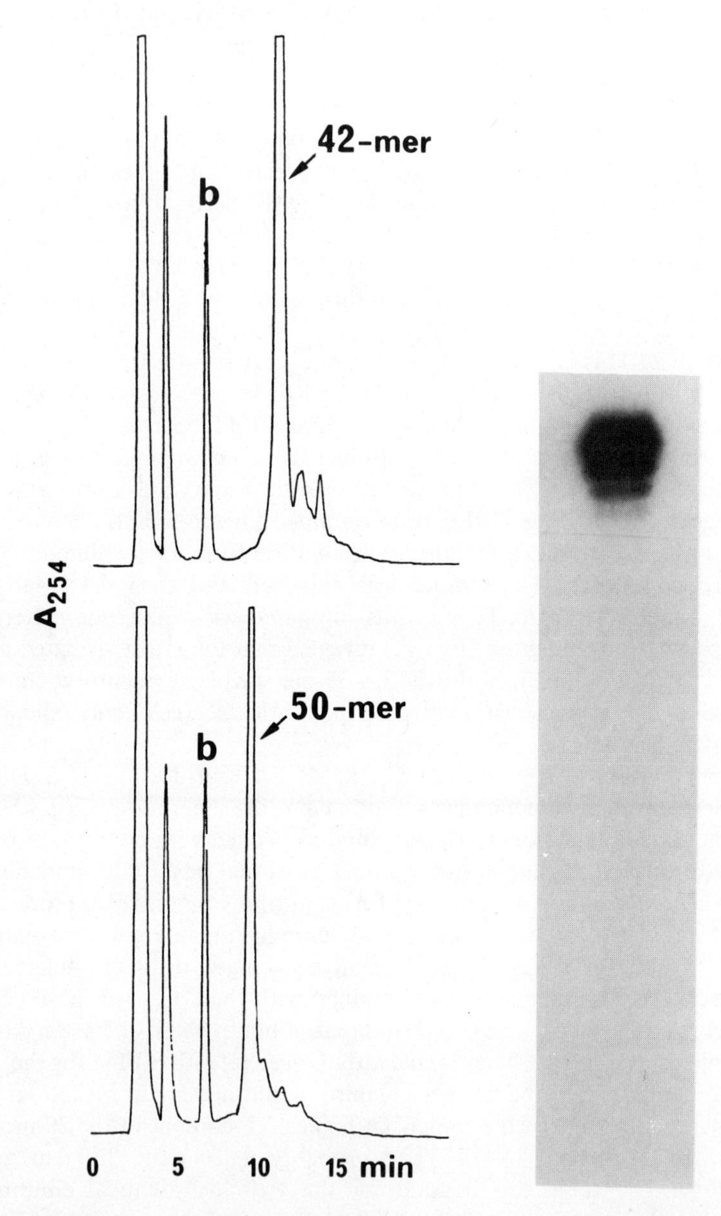

Fig. 10. RP-HPLC separations of a crude 5′-DMT 42-mer (base composition $A_3G_{11}C_{15}T_{13}$) and a crude 5′-DMT 50-mer (base composition $A_{10}G_{13}C_{14}T_{13}$) under the conditions given in Fig. 5; "b" = benzamide. The inset is an overexposed autoradiogram obtained after PAGE of the ^{32}P-labeled material derived from a center-"cut" of the 5′-DMT 50-mer.

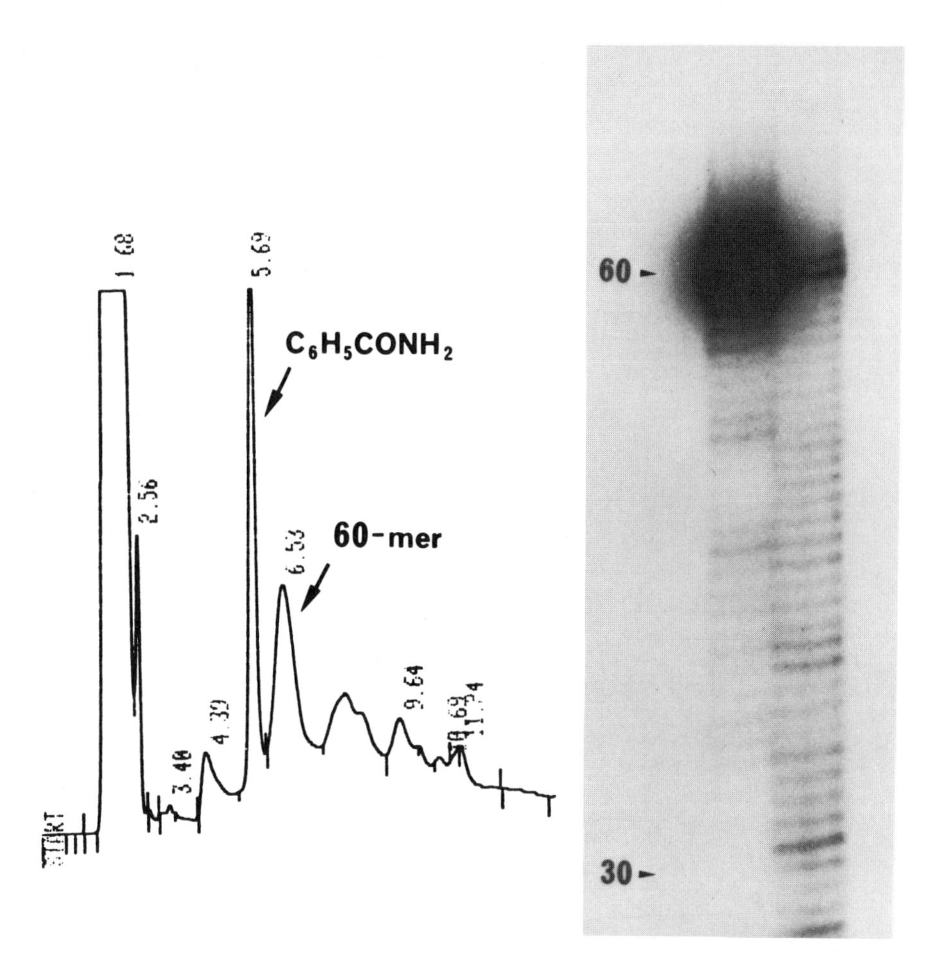

Fig. 11. RP-HPLC separation of a crude 5′-DMT 60-mer (base composition $A_{23}G_{10}C_{14}T_{13}$) using the column and conditions described in Fig. 5. Also shown is an overexposed autoradiogram obtained after PAGE of the detritylated, ^{32}P-labeled crude product (right lane) and HPLC-purified material (left lane).

2.1.4. C_{18} versus C_4 columns

Becker et al. (174) have found that RP-HPLC with a large-pore 300 Å Vydac C_4 matrix allows fast and efficient purification of 5′-DMT derivatives of oligonucleotides up to 104 bases in length. This report prompted the present author to compare the performance of a 300 Å pore size, 10 μm particle, Vydac C_4 column (10 mm × 25 cm) with that of 125 Å pore size, 10 μm particle, μBondapak C_{18} column (7.8 mm × 30 cm), which had similar bed volumes, and were used under identical conditions. The separations shown in Fig. 13 for test samples of

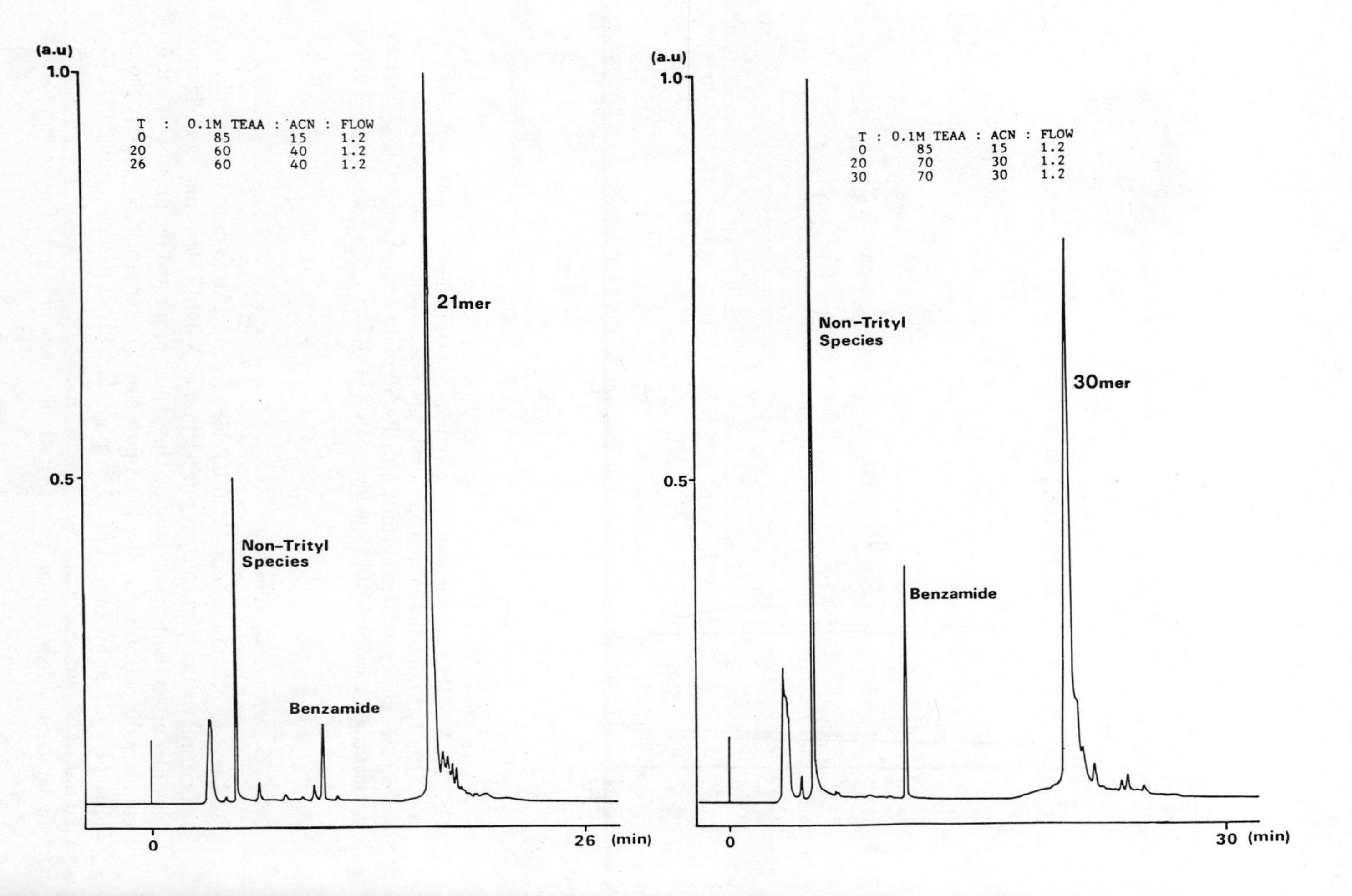
(a.u)
1.0
0.5
T : 0.1M TEAA : ACN : FLOW
0 85 15 1.2
20 60 40 1.2
26 60 40 1.2
21mer
Non-Trityl
Species
Benzamide
0
26 (min)
(a.u)
1.0
0.5
T : 0.1M TEAA : ACN : FLOW
0 85 15 1.2
20 70 30 1.2
30 70 30 1.2
Non-Trityl
Species
30mer
Benzamide
0
30 (min)

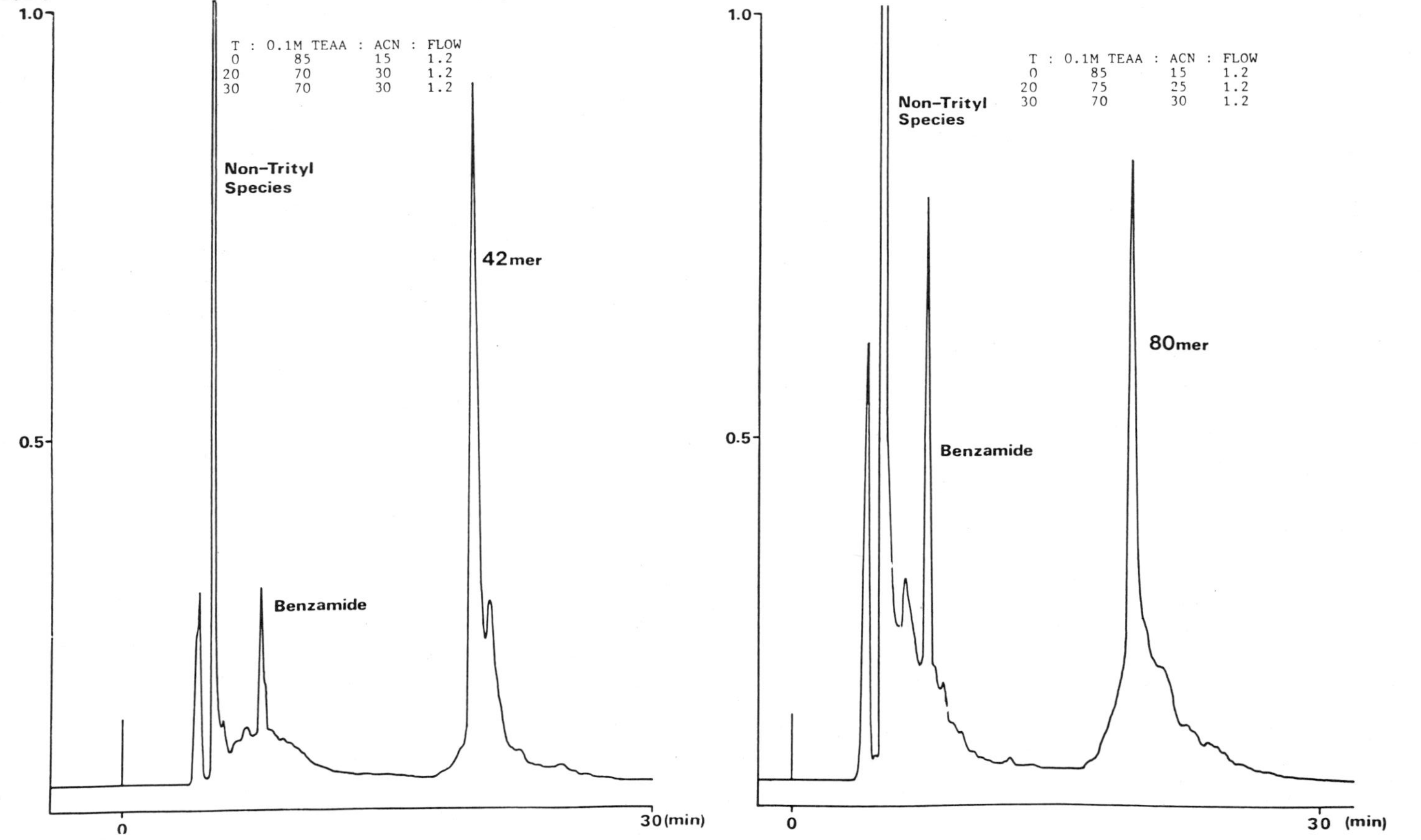

Fig. 12. RP-HPLC separations of a crude 5′-DMT 21-mer (base composition $A_9G_7C_3T_2$), 5′-DMT 30-mer (base composition $A_{11}G_9C_4T_6$), 5′-DMT 42-mer (base composition $A_{13}G_{13}C_9T_7$), and 5′-DMT 80-mer (base composition $A_{25}G_{19}C_{16}T_{20}$). The 21-mer and 30-mer were eluted from a 5 μm particle, 100 Å pore diameter Nucleosil C_{18} column (6 mm × 25 cm, Yamamura Chemical Co., Ltd.) with a 1.2 mL/min gradient of acetonitrile (ACN) vs TEAA (0.1 *M*, pH 7.5). The 42-mer and 80-mer were eluted from a 7 μm particle, 300 Å pore diameter Nucleosil C_{18} column (6 mm × 25 cm). Syntheses were performed on a 1-μmol scale with an Applied Biosystems Model 381A DNA synthesizer that employed *O*-methyl-*N*,*N*-diisopropylphosphoramidite reagents. Published with permission of T. Iwabuchi, Japan Scientific Instrument Co., Ltd.

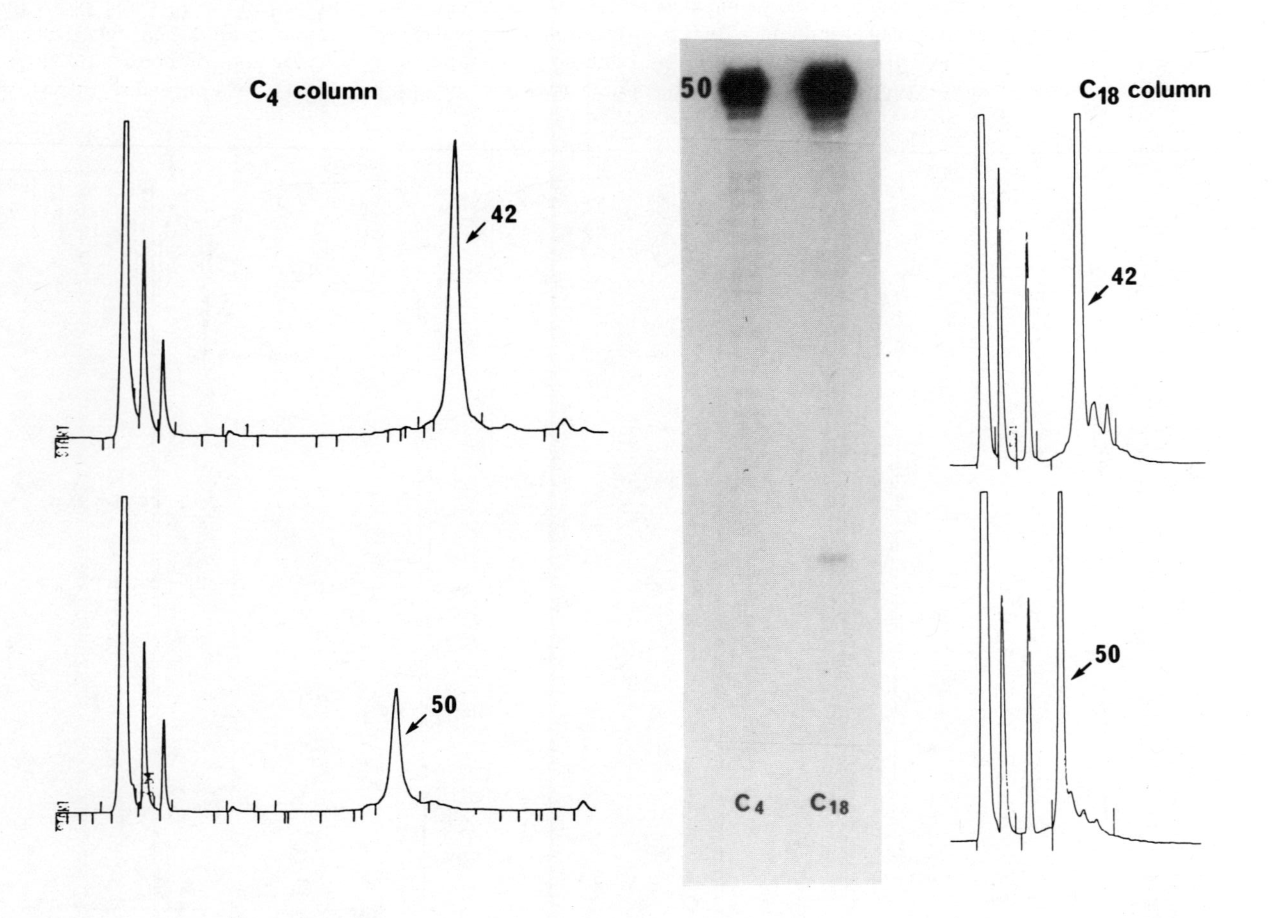

C_4 column
42
50
50
C_4
C_{18}
C_{18} column
42
50

a crude 5′-DMT 42-mer and a crude 5′-DMT 50-mer suggested that the degree of resolution afforded by these columns was more or less the same. However, comparison of the PAGE-derived autoradiogram (Fig. 13) of narrow "center cuts" of the 50-mer revealed that the isolate from the C_{18} column was substantially purer, that is, less contamination by shorter sequences. The generality of these findings, which pertain to nonoptimized HPLC conditions, has yet to be established.

2.1.5. C_{18} versus PRP-1 Columns

The porous, "reversed-phase" matrix in PRP-1 columns does *not* have a discrete, chemically bonded component such as C_{18} or C_4 alkyl moieties, and is simply a hydrophobic copolymer of cross-linked styrene and divinylbenzene in the form of spherical, 10 μm particles having 20–600 Å pores (136). By contrast to silica-based bonded-phase columns, HPLC purification of oligonucleotides using this relatively chemically inert copolymer can be performed under highly alkaline conditions (up to pH 13, 0.5 N buffer) and as elevated temperatures (50–70°C). While these features are especially important for the purification of oligonucleotides that tend to associate intra- or intermolecularly [e.g., complementary sequences, hairpin segments, or presence of $(dG)_n$], the reported (137) additional advantages of PRP-1 columns over bonded-phase columns include higher capacity (10–25 mg DNA/injection), higher recovery (> 95%), and longer operational lifetime (~ 1 year). These factors and the substantially lower cost of PRP-1 vs C_{18}/C_4 columns are of considerable interest and practical importance for analysis and purification of oligonucleotides in general. The author has recently begun a comparative study of PRP-1 and μBondapak C_{18} columns, and the separations achieved to date have been roughly comparable. This is exemplified by the chromatograms for an 8-mer given in Fig. 14. The chromatograms shown in Fig. 15 were obtained by use of a PRP-1 column to purify 5′-DMT products derived from 0.2 μmol scale syntheses of a 17-mer, two 25-mers, a 28-mer, a 41-mer, and a 50-mer. In each case 25% of the entire crude product was injected. The significantly faster elution of 25-mer A (9.5 min) compared to that of 25-mer B (12.0 min) is opposite to what is expected based

---

Fig. 13. (*Opposite*) RP-HPLC separations of a crude 5′-DMT 42-mer (base composition $A_3G_{11}C_{15}T_{13}$) and a crude 5′-DMT 50-mer (base composition $A_{10}G_{13}C_{14}T_{13}$) on μBondapak C_{18} and Vydac C_4 columns (see text for details). The conditions for elution from the C_{18} column are described in Fig. 5. Elution from the C_4 column employed a 1%/min gradient of acetonitrile vs TEAA (0.1 M, pH 7.0) for 15 min starting at acetonitrile-TEAA = 15:85; flow rate = 5 mL/min and peak detection at 254 nm. The elution times for the 5′-DMT product peaks are as follows: C_{18}, 42-mer, 10.8 min; 50-mer, 9.3 min; C_4, 42-mer, 17.0 min; 50-mer, 14.6 min. Also shown is an overexposed autoradiogram obtained after PAGE of the detritylated, ^{32}P-labeled 50-mers: left lane, C_4-isolate; right lane, C_{18}-isolate.

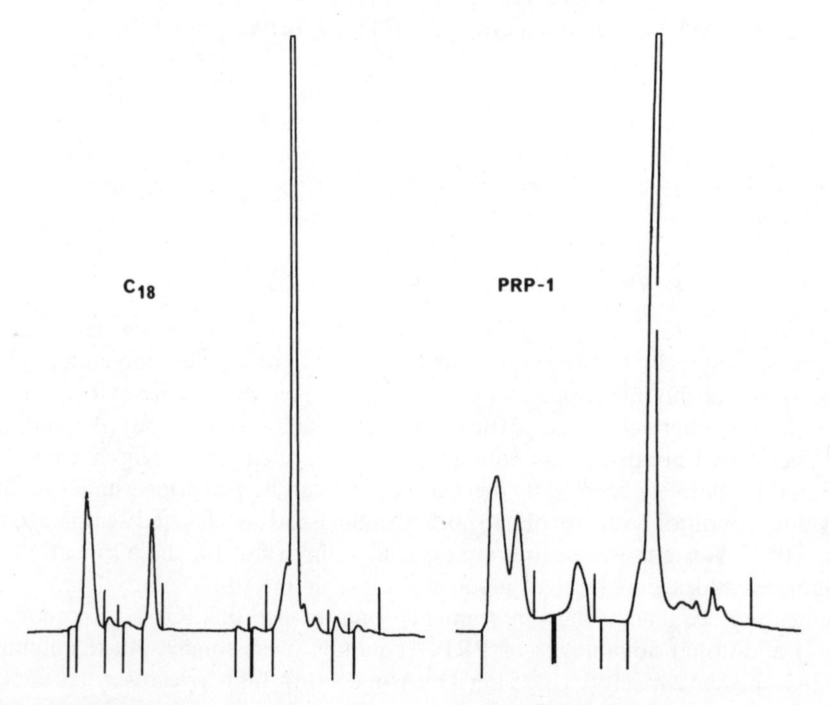

Fig. 14. RP-HPLC separations of crude 5′-DMT-d(GTGTACAC) on Waters μBondapak C_{18} and Hamilton PRP-1 columns. The conditions for elution from the C_{18} column (10 μm particle, 7.8 mm × 25 cm) are described in Fig. 5. Similar conditions were used for elution from the PRP-1 column (10 μm particle, 7 mm × 30 cm), except for a decrease in the flow rate from 4 to 2 mL/min. Peaks were detected at 254 nm; the 5′-DMT product peaks were recorded at 13.8 min for C_{18} and 10.0 min for PRP-1.

on net hydrophobicity, since *A* and *B* have 56% and 32% AT-content, respectively; however, the relative elution times are consistent with the relative hydrophobicity of the base residues near the 5′-DMT group. The ~ 15% 2′-deoxyinosine-content in both the 41-mer and the 50-mer did not cause any unusual chromatographic behavior, as is also true for purification of 2′-deoxyinosine-containing oligonucleotides on a C_{18} μBondapak column.

The recovery of oligonucleotides from a C_{18} column, which is dependent on chain length (114), has been found (114) to range from 50 to 80%, which is lower than the > 95% recovery reported (137) for PRP-1 columns but is nevertheless more than adequate for providing "biological amounts" (microgram quantities) of oligonucleotides derived from syntheses that are carried out on either 1 μmol or 0.2 μmol scale syntheses. On the other hand, the author has found significant "ghost" peaks with large-scale collections on a PRP-1 column, which necessitates the use of several postcollection column washes with the eluting gradient. For biophysical studies that require large amounts of DNA, and for syntheses using precious reagents with ^{13}C or ^{15}N labels, recovery of the product during purification by RP-HPLC is a critical factor in selecting methods and columns; consequently, one should consider the use of newly developed, high-recovery PICS columns (113, 114).

The operational or usable lifetime of "semipreparative" μBondapak C_{18} columns in this author's laboratory has averaged ~2–3 months, either with or without a C_{18} precolumn, and with nightly storage in 100% acetonitrile to prevent hydrolysis of the bonded phase. PRP-1 columns have been used (and abused!) for 6 months, to date; consequently, the advantage afforded by the relatively rugged, cheaper PRP-1 column speaks for itself. It is also worthwhile to mention here that "analytical size" PRP-1, μBondapak C_{18} and Zorbax ODS columns (~5 mm ID) have given inferior separations of 5′-DMT-bearing products, compared to their "semipreparative" counterparts, which presumably reflects differences in the number of theoretical plates.

2.1.6. Repurification

RP-HPLC can be used to further purify 5′-hydroxyl oligonucleotides after detritylation of their corresponding, RP-HPLC-isolated 5′-DMT derivatives, or after isolation by either ion-exchange HPLC or PAGE. In the author's laboratory, RP-HPLC analysis after detritylation has indicated that, in general, the separated impurity peaks have an integrated absorbance, which is ~1–10% of the absorbance of the apparent single product peak (Fig. 16). Consequently, RP-HPLC is used to check for this level of purity of the DMT-derived 5′-hydroxyl oligonucleotides, but a second purification step by RP (or other) HPLC is not usually warranted.

Occasionally, one may wish to use RP-HPLC to either analyze or repurify the HPLC-collected DMT-DNA. In these situations it is necessary to add a liberal amount of triethylamine (3–5% v/v) to the collected fraction to prevent loss of product by detritylation that will occur upon either prolonged storage of the collected material, or concentration of the collected material *in vacuo*.

2.1.7. Overloading

Under typical conditions used for RP-HPLC purification of "biological amounts" of oligonucleotides as their DMT derivatives (cf. Fig. 5), the capacity of a 7.8 mm × 25 cm μBondapak C_{18} column is usually not exceeded upon injection of 10% of the crude product that is obtained from a 1 μmol scale synthesis. The capacity of the column under these conditions is however, occasionally exceeded when scaling-up from analytical size (0.5%) to preparative-size (10%) injections of crude mixtures that contain relatively long oligonucleotides. This column "overloading" effect is evident by comparison of the traces shown in Fig. 17 for 10 μL (0.5%) and 100 μL (5%) aliquots of a 2 mL solution that contained a mixed-sequence 20-mer. The decreased intensity (~50%) of the peak (6.8 min) for the major DMT-bearing species, relative to the peak (4.6 min) for benzamide, together with the appearance of the "lost absorbance"

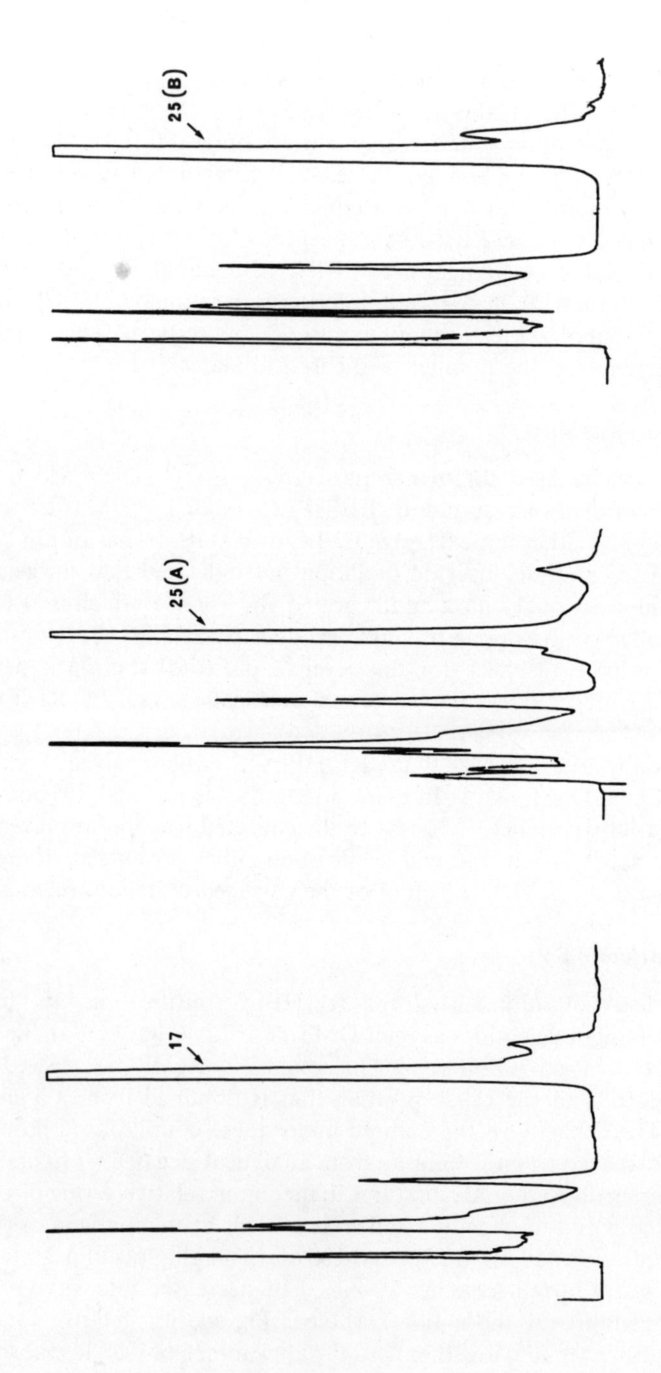
25(B)
25(A)
17

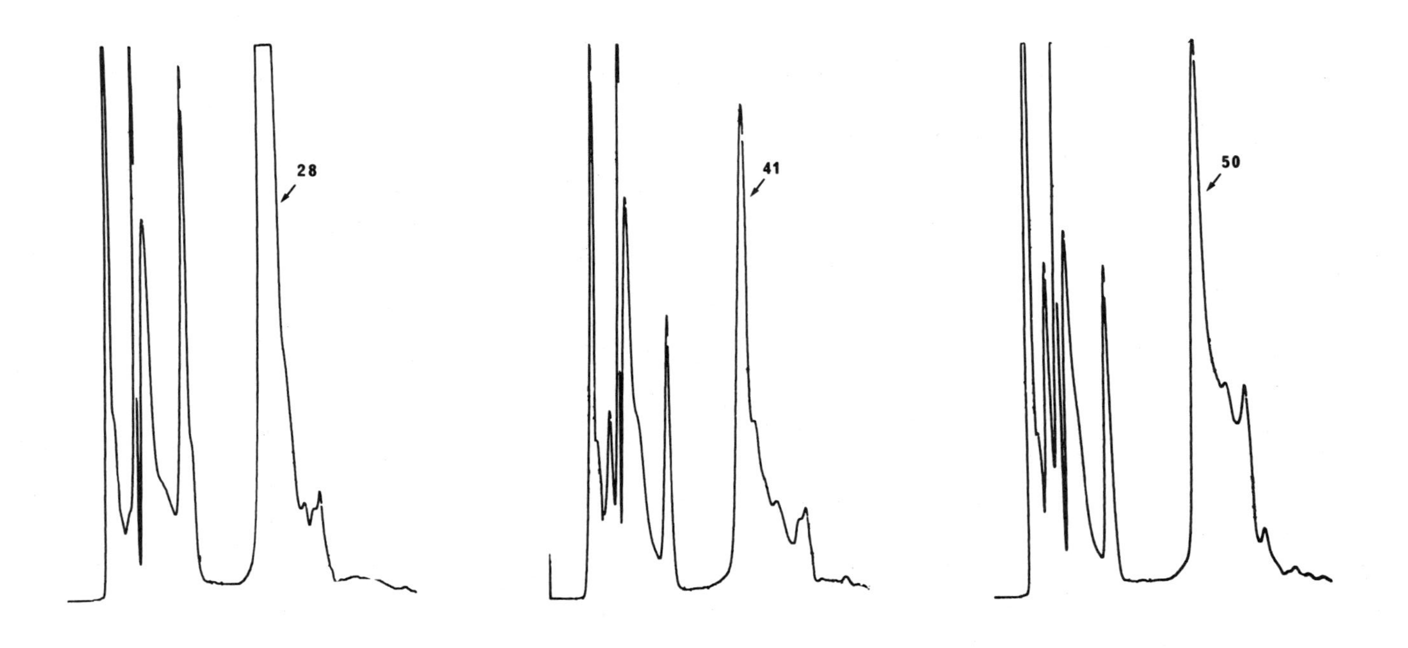

Fig. 15. RP-HPLC separations of crude 5′-DMT products (0.2 μmol scale syntheses). The PRP-1 column (cf. Fig. 14) was eluted with a 1%/min linear gradient of acetonitrile vs TEAA (0.1 *M*, pH 7.0) for 10 min starting at acetonitrile-TEAA = 20:80; flow-rate = 3 mL/min and peak detection at 254 nm. 17-mer = 5′-DMT-d(GTCAACAATATGTGGGC), 11.7 min; 25-mer *A* = 5′-DMT-d(GCCCGCAAAAAAAAACCACTCTTAC), 9.5 min; 25-mer *B* = 5′-DMT-d(ACAAGTCTTGCGGGCCAACGGCGGC), 12.0 min; 28-mer base composition = $A_{13}G_7C_7T_1$, 10.3 min; 41-mer base composition = $A_7G_9C_{11}T_7I_7$, 10.1 min; 50-mer base composition = $A_{11}G_{14}C_9T_9I_7$.

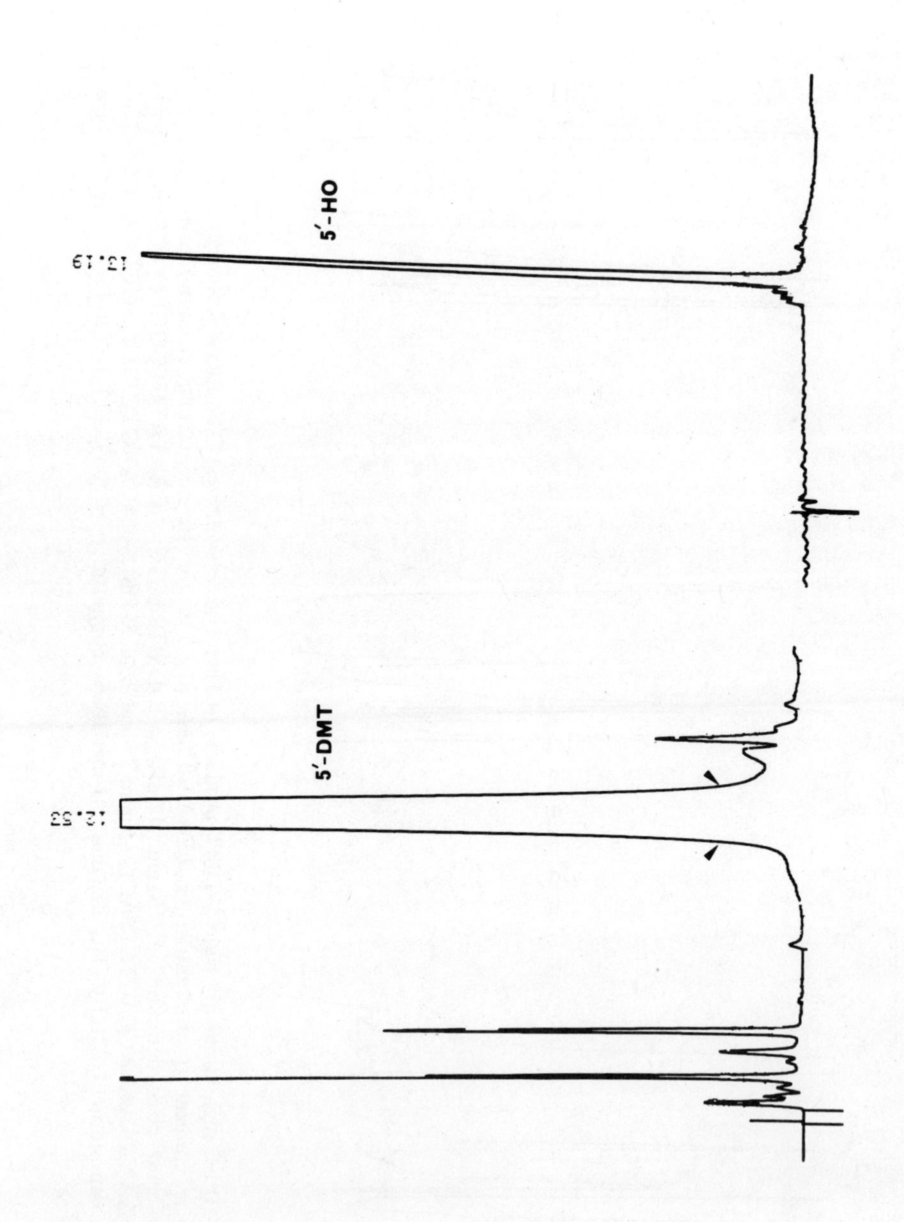
5′-HO
13.19
5′-DMT
12.53

as a new, fast-eluted peak (3.7 min) are good indicators of column overloading. Re-injection of a small portion of the collected, fast-eluted, DMT-product which was not retained led to a normal elution time for this material. In other cases the unretained DMT-product has coeluted with the shorter 5′-hydroxyl oligonucleotides. It remains to be established whether the greater capacity of PRP-1 versus C_{18} columns will lessen the frequency of this problem, which must be avoided when dealing with mixed-sequence products to avoid adventitious fractionation resulting from possible sequence-dependent "overloading."

2.1.8. Column Degradation

The HPLC peak for benzamide, the by-product of deprotection of dA and dC residues, not only serves as a useful internal reference for the relative absorbance of DMT-bearing species but also provides a reliable "benchmark" for gauging the extent of the degradation of a C_{18} column, over a period of time, due to hydrolytic loss of the hydrophobic bonded-phase. The elution of benzamide from a new "semipreparative" μBondapak C_{18} column (7.8 mm × 25 cm) occurs at ~6–6.5 min, under typical operating conditions (cf. Fig. 5). Gradual degradation of the column leads to a corresponding diminution of the elution time for benzamide. When a ~5 min elution time is observed for benzamide, there is usually some broadening and possibly tailing of the peaks, which likewise elute relatively faster, and there is a loss of resolution, relative to that found with a newer column. These differences can be seen by comparison of the chromatograms in Fig. 18 for a crude 5′-DMT 18-mer that was eluted on relatively old and new C_{18} columns under otherwise identical conditions. The peak doubling observed for the DMT-bearing species upon elution from the older column is another indicator of column degradation.

Contamination of RP-HPLC-purified oligonucleotides with small amounts of C_{18}- and $(CH_3)_3Si$-containing material has been found by the author using mass spectrometric and NMR spectroscopic analyses; however, these contaminants, which possibly can be removed by either gel filtration or precipitation, do not appear to interfere with either 5′-end labeling or subsequent biological uses of the oligonucleotides. Larger amounts of these contaminants and possibly silica are present in the products as water-insoluble particulates when a column begins to show significant signs of degradation. These particles can be removed by filtration without appreciable loss of the oligonucleotide products.

Fig. 16. (*Opposite*) RP-HPLC separations of crude 5′-DMT-d(ATATCGATAT) and the 5′-hydroxyl DNA derived from collection and detritylation of the major 5′-DMT peak. The C_{18} column and conditions used at the 5′-DMT stage are described in Fig. 5; arrowheads indicate the "cut." The same column was employed at the 5′-hydroxyl stage with a 1%/min linear gradient of acetonitrile vs TEAA (0.1 *M*, pH 7.0) starting at acetonitrile-TEAA = 5:95; flow rate = 4 mL/min, peak detection at 254 nm.

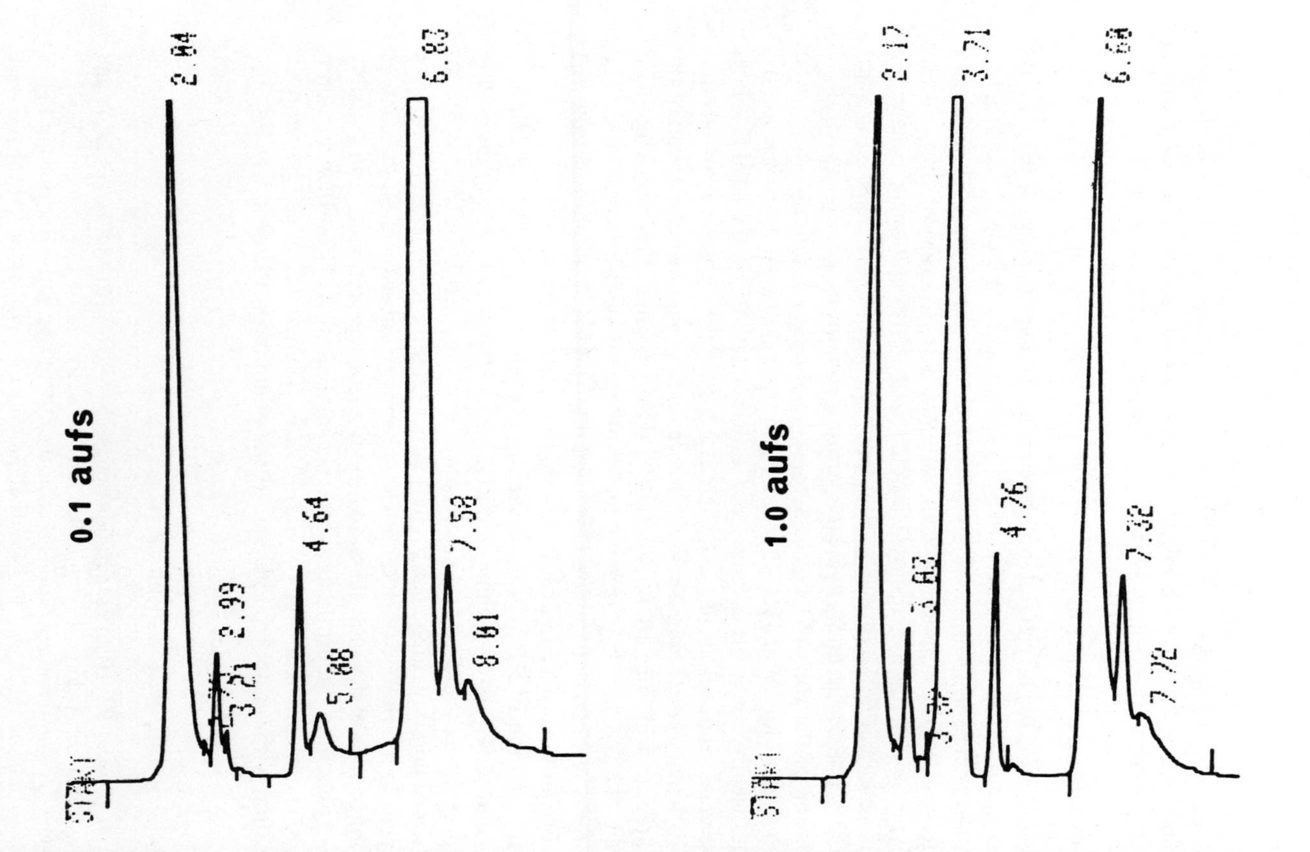

Fig. 17. RP-HPLC separations of a crude, mixed-sequence 20-mer, 5′-DMT-d[GTCAT(I/C)AG(I/C)GT(I/C)CCGTTIAT]. The μBondapak C_{18} column and conditions are the same as those described in Fig. 5, except that the gradient was 2.5%/min for 4 min. The left and right traces were respectively obtained by injections of 0.5% and 5% of the total crude material from a 1 μmol scale synthesis.

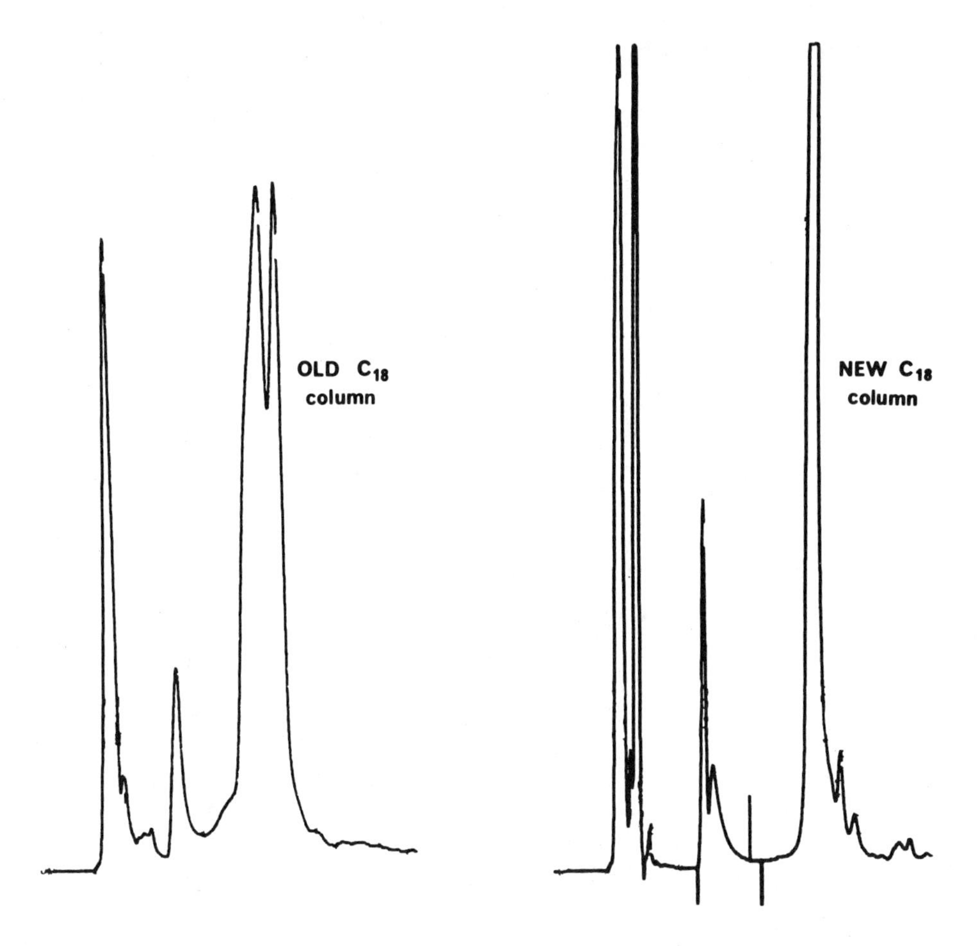

Fig. 18. RP-HPLC separations of a crude 18-mer, 5′-DMT-d(CGGGATTTGTTAGG-CACG). The μBondapak C_{18} column and conditions are the same as those described in Fig. 5. The left and right chromatograms were obtained by injection of 5% of the crude product (1 μmol) on relatively old and new columns, respectively.

2.2. Step-by-Step Post-Chromatographic Sample Work-Up

Purification of crude 5′-DMT oligonucleotides by RP-HPLC (see Section 2.1) leads to the collection of an approximately 0.1 to 2 mL fraction in acetonitrile-TEAA (~ 30:70), depending on the size of the "cut" that is taken. The following procedures are routinely employed in the present author's laboratory for concentration of the collected sample, detritylation, extraction of non-DNA organics, filtration, and then measurement of the optical density of the final 5′-hydroxyl oligonucleotide product, which is either used directly or stored at low temperature. Obvious modifications in this protocol can be made if desired or needed.

1. Concentrate to dryness (or a gum) on a rotary evaporator or in a vacuum centrifuge; lyophilization is risky due to possible melting and "bumping" of the sample, which has a depressed freezing point. Reconcentrate from water if there is an odor of triethylamine.

2. Briefly vortex or sonicate the residue with 1 mL of 3% v/v acetic acid in water at room temperature, preferably in a Pyrex or Kimax test tube. After 10 min neutralize (or make weakly basic) with concentrated NH_4OH (1–2 drops), vortex briefly, add an equal volume of ethyl acetate, vortex briefly, and then discard the separated, upper layer, after centrifugation if necessary (this also allows removal of any solids derived from the HPLC column).

3. From the resultant 1 mL aqueous solution of DNA, which is saturated with ethyl acetate, remove a 50 μL aliquot and then concentrate both the aliquot and the stock solution to dryness. Dissolve the residue from the aliquot in 1 mL water and measure the absorbance at 260 nm, A_{260}. Multiply A_{260} by 19 to determine the number of OD_{260}-units of the dried stock sample.

2.3. Short-Cuts from Synthesizer to Purified Product

In situations where a purified oligonucleotide is needed as soon as possible, one or more of the following shortcuts have been used, regardless of whether *O*-methyl or *O*-β-cyanoethyl phosphoramidite reagents or both were employed in the synthesis. The time required for cleavage from support and backbone/base deprotection is significantly decreased by omitting treatment with thiophenol, in the case of *O*-methyl amidites, and transferring the solid support (from 1 μmol or 0.2 μmol scale syntheses) to a vial that contains ethylenediamino-ethanol (1:1 v/v, 0.5 mL) for heating at 80°C for 15 min (Teflon-lined screw-cap). The cooled mixture is then briefly vortexed with water (1.5 mL). A portion (100–200 μL, or more) of the resultant aqueous solution is injected onto a PRP-1 column, without obtaining a prior analytical trace, and the "cut" containing the DMT-protected product is placed under a stream of N_2 to quickly remove most of the acetonitrile. The resultant solution is adjusted to pH $\sim$2.8 with glacial acetic acid, and then further processed as described in Section 2.2.

In connection with the replacement of a restriction fragment in the structural gene for bacterioopsin by synthetic DNA fragments containing altered codons, Lo and coworkers (176) have reported an extremely rapid, efficient, and general method for purification of synthetic oligonucleotides, which utilizes C_{18} Sep-Pak cartridges in place of a RP-HPLC column for removal of "failed sequences." The Sep-Pak was connected to a 10 mL syringe and was washed successively with 10 mL of acetonitrile, 5 mL of 30% acetonitrile in 0.1 *M* triethylammonium bicarbonate (TEAB), and 25 m*M* TEAB. The solution of crude 5′-DMT DNA in 25 m*M* TEAB was passed through the Sep-Pak (20 sec, total capacity 50–100 A_{260}-units), and the 5′-hydroxyl DNA was eluted with 10–15 mL of 10% acetonitrile in 25 mM TEAB. 5′-DMT DNA was then eluted with 5 mL of 30% acetonitrile in 0.1 *M* TEAB, and the resultant, detritylated material was further purified by PAGE, using a recovery procedure that employed a Sep-Pak. To evaluate the degree of purification of the 5′-DMT DNA that is afforded by this method, the present author applied the Sep-Pak method (with TEAA) to a typical 24-mer, and used RP-HPLC to analyze the fractions. Elution with 10% acetonitrile in 25 m*M* TEAA gave an 80:20 absorbance-ratio for the 5′-hydroxyl:5′-DMT DNA peaks, while elution with 30% acetonitrile in 0.1 *M* TEAA gave a 35:65 absorbance ratio for these peaks, thus indicating significant enrichment of the product, albeit far less than that achieved by RP-HPLC (typically 1:99 to 10:90).

Another non-HPLC method for rapid purification of synthetic oligonuclotides has been reported more recently (177), and utilizes PAGE, UV imaging, and ethanol-precipitation of the excised product.

2.4. Analysis of Oligonucleotides

It is quite evident from comparison of earlier versus current studies dealing with the synthesis and use of oligonucleotides that there is less of a need or concern for unambiguously establishing the structure of a synthetic oligonucleotide by means of size analysis, sequencing, and determination of base composition. There are various reasons for this situation, the most important of which is the established reliability of the synthetic methods. On the other hand, one must still address the problem of how to best establish the structure of the DNA in cases where there is no biological experiment or other maneuver that employ the product and thereby serve as a basis for product quality control. The same is true for cases in which analysis of the oligonucleotide is an integral part of the intended study or application. Conventional non-HPLC methods for these structural analyses include PAGE, chemical sequencing (160–162, 178), and "wandering spot" analysis (160, 161). Alternative methods include the use of enzymatic degradation of oligonucleotides and HPLC analysis of the digests. For example, while the determination of chain lengths by SAX-HPLC versions of PAGE is not practical because of the secondary influence of base composition upon elution time, chain length and base composition can be determined by snake venom phosphodiesterase (SVPDE)-catalyzed hydrolysis of the oligonucleotide followed by RP-HPLC using a C_8 column to separate the eight possible dN and pdN products (158). Total analysis times involved a 1.5-hr digestion and a 40-min HPLC analysis. Chain lengths up to 36 base residues could thus be determined with less than 0.5 base-residue error, and with a sensitivity of ~50 pmole of oligonucleotide.

Tandem digestion (179, *vide infra*) of ~1 nmole of oligonucleotide with SVPDE and alkaline phosphatase (AP) gives dN products for analysis by RP-HPLC using a C_{18} column, as shown in Fig. 19 for the digest derived from a base-modified 16-mer that contained a 2′-deoxyinosine residue as a probe for specific interactions between DNA and a monoclonal antinative DNA autoantibody (180). This same method is useful for studies of backbone-modified oligonucleotides that yield undigested fragments, d[N(X)N′], (X = phosphotriester, phosphorothioate, and so on), which can be quantified, collected, and

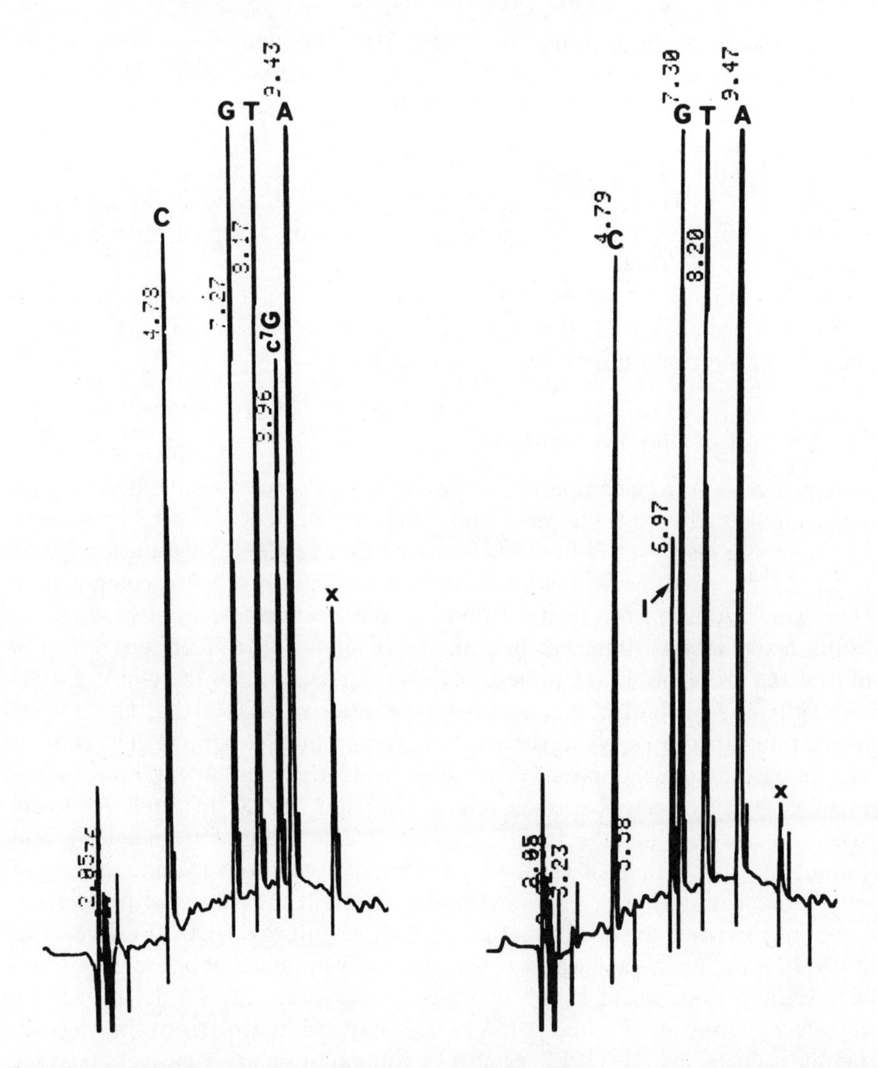

Fig. 19. RP-HPLC separations of 2′-deoxynucleosides formed by tandem digestion of the 18-mer, d(ATATAGCc^{7}GCc^{7}GCGCTATAT), and the 16-mer, d(ATATAICG-CGCTATAT), with snake venom phosphodiesterase and then alkaline phosphatase; dc^{7}G = 7-deaza-dG, dI = 2′-deoxyinosine. The order of elution from the Waters Nova-Pak C_{18} column (7.8 mm × 30 cm, 5 μm particle) is dC < dI < dG < -dT < dc^{7}G < dA using a 1%/min gradient of acetonitrile vs TEAA (0.1 *M*, pH 7.0) starting at acetonitrile-TEAA = 1:99, flow rate = 4 mL/min; "X" is a background peak.

characterized (181). Compositional data can also be obtained by formic acid degradation of oligonucleotides followed by RP-HPLC analysis of the released nucleobases (182, 183).

The following protocol for tandem digestions with SVPDE and AP is used in this author's laboratory to obtain 2′-deoxynucleosides for quantification by RP-HPLC using a 5 μm particle Nova-Pak C_{18} column, which gives superior resolution compared to 10 μm particle μBondapak C_{18} and PRP-1 columns.

1. Prepare Tris-acetate buffer (0.1 *M*): dissolve 1.21 g of Tris and 0.30 g of $MgCl_2 \cdot 6H_2O$ in 80 mL of H_2O, adjust the pH to 8.8 with glacial acetic acid, and then dilute to 100 mL. The buffer can be stored for several months at 5°C.

2. Add approximately 0.1–0.2 OD_{260}-units of the DNA sample, as a solution, to an Eppendorf tube, and then concentrate to dryness in a vacuum centrifuge. Repeat the procedure for replicate samples, and also prepare replicate standards either as of a mixture of 2′-deoxynucleosides [0.05 OD_{260}-units of each using stock solution(s) containing weighed material], or as an oligonucleotide of known base composition.

3. Add 30 μL of SVPDE (Sigma P-6761) solution (made to 1 unit/1 mL with Tris-acetate buffer) to each Eppendorf, vortex, and then heat at 37°C for 3 hr. The incubation time can be decreased to 1.5 hr by use of 60 μL of SVPDE solution.

4. Add 10 μL of AP (Sigma P-9761) solution (made to 20 units/mL with Tris-acetate buffer) to each Eppendorf, vortex, and heat at 37°C for 2 hr. The incubation time can be decreased to 40 min by use of 30 μL of AP solution.

5. Dilute each tandem digestion mixture to 220 μL by the addition of water to the Eppendorf, heat at 100°C for 3 min, cool in an ice-water bath, and then centrifuge.

6. Inject 200 μL of each supernatant and a "blank" (everything but DNA) onto a Nova-Pak C_{18} column (7.8 mm × 30 cm) and elute for 20 min with a 1%/min gradient of acetonitrile versus TEAA (0.1 *M*, pH 7.0), starting at acetonitrile-TEAA = 1:99, flow rate = 4 mL/min, with peak detection at 254 nm, 0.05 AUFS. The injection can also be made on a PRP-1 column with a flow rate = 3 mL/min.

7. Use the values of peak area-% found for either the authentic mixture of 2′-deoxynucleosides or the reference compound to calculate correction factors to convert the values of peak area-% to mol-% for the samples of interest.

Although the precision of these HPLC determinations is quite good ±0.2%), the larger cumulative error limits for accuracy (±5–10%) of the method restrict its use to that of providing supportive evidence for the identification of relatively short oligonucleotides, in conjunction with either NMR spectroscopic or other evidence for chain length and purity. On the other hand, the method is well suited for analyses of modified bases, such as the determination of the amount of N_3-methyl-dT (170, 171) or 7-deaza-dG (dc^7G) (180). In the latter case, the found and calculated (in parentheses) values of mol-% for the 18-mer, d(ATATAGCc^7GCc^7GCGCTTATAT), were as follows: dC, 22.8 (22.2); dG, 11.1 (11.1); dT, 27.3 (27.8); dc^7G, 9.8 (11.1); dA, 29.9 (27.8).

An automated HPLC method for sequence analysis (155, 156) utilized SAX-HPLC to quantify, as a function of time, the pdN products obtained by stepwise degradation of an oligonucleotide with SVPDE, which proceeds in the $3' \rightarrow 5'$ direction. For example, 20 aliquots of a digestion mixture containing ~ 10 nmoles of the decanucleotide, d(TATCAAGTTG), and 0.1 μg of SVPDE were automatically removed, as a function of time, and then analyzed by HPLC to obtain data for individual plots of mole-fraction pdN vs time, from which the sequence was determined. At the end of the digestion, following release, collection, and quantification of the 5′-terminal deoxynucleoside, base composition was also determined.

Horn and Urdea (184) have developed a new strategy for enzymatic purification of chemically synthesized oligonucleotides prior to removal of the DNA from the solid support. Their approach involves complete deprotection of support-bound DNA, except for a 5′-blocking group on the product, followed by $5' \rightarrow 3'$ exonuclease digestion of truncated sequences and then cleavage of essentially pure product from the support. These investigators believed that automation could be easily achieved. If so, it may be possible to automatically sample the digestion reaction for HPLC analysis to determine the sequence of the oligonucleotide still attached to the support. In principle, polarized attachment of any oligonucleotide (including double-stranded DNA restriction fragments) to a suitable solid support (185, 186) would allow for automated sequencing by digestion in either the 5′ or 3′ direction.

3. MIXED-SEQUENCE OLIGONUCLEOTIDES

Following the proposal by Wallace et al. (10) that mixtures of oligonucleotides representing all possible coding sequences for a peptide might be useful as specific hybridization probes (Fig. 20), Wallace and co-workers (187) were the first to use such "mixed-sequence" probes to isolate a cloned cDNA gene (β_2-microglobulin) from a "shotgun library." This method of using protein sequences to design oligonucleotide pools to specifically detect or manipulate mRNA/DNA sequences is now widely employed in either its original version or with the introduction of dG·dT (188) and/or dI·dN (189–191) base pairing (dN = dC, dA, dT, or dG), as options for decreasing the required number of oligonucleotide sequences (lower "redundancy"). Some operational guidelines for probe design with dG·dT and dI·dN base pairing have been published very recently (191), although the success rate for applications of these options versus the original method has yet to be established.

Ideally, the synthesis of mixed-sequence oligonucleotides by competitive coupling reactions of mixtures of either trimer, dimer, or monomer units should occur with negligible selectivity and thus provide an essentially equimolar ("unbiased") distribution of all the possible sequences. Analytical data (173) obtained recently for the competitive coupling of methyl and β-cyanoethyl phosphoramidite reagents was consistent with virtually nonselective couplings,

```
                 1   2   3   4   5   6
         NH
           2    Lys Val Glu Gln Ala Val

                 A   A   A   A   A
mRNA  (5')    AA  GU  GA  CA  GC  GU   (3')
                 G   G   G   G   G
                     C           C
                     U           U

                 T   T   T   T   T
DNA   (3')    TT  CA  CT  GT  CG  CA   (5')
                 C   C   C   C   C
                     G           G
                     A           A
```

Fig. 20. *N*-Terminal amino acid sequence and all of the possible mRNA sequences which encode this hexapeptide; there are $2 \times 4 \times 2 \times 2 \times 4 = 128$ possibilities. The oligodeoxyribonucleotides that are complementary to the mRNA sequences constitute a "mixed-sequence" hydribization probe.

although the gradual and different rates of decomposition of the reagents was identified as a potential problem to avoid (173). A relatively easy way to check for the possibility of biased mixed couplings involves the synthesis of d[T(A/G)T], d[T(C/T)T], d[T(A/G/C/T)T], and so on, and the use of RP-HPLC to separate the resultant trimers, TAT, TGT, TCT, and TTT, for measurement of their relative molar ratios by use of appropriate correction factors which convert peak area-% into mol-%.

The use of PAGE for isolation of a set of mixed-sequence oligonucleotides is complicated by the sometimes marked influence of base composition on electrophoretic mobility (192). While the same problem obtains for purification by use of either ion-exchange of reversed-phase HPLC, experience has shown that the latter method can be employed with a gradient that minimizes the likelihood of inadvertently "losing" the product(s) having perfect (high) complementarity.

A series of mixed-sequence 5′-hydroxyl oligonucleotides representing 2-, 4-, and 8-fold redundancy were synthesized by competitive coupling of *O*-methyl phosphoramidite reagents, and each pool of *n*-mers was then isolated as an apparent single band by PAGE (W. Efcavitch, unpublished studies). SAX-HPLC of the 2-fold redundant 12-mer showed (Fig. 21) only one peak at 10.5 min, while two different 4-fold redundant 14-mers gave rise in one case to completely separated peaks at 17.2–20.6 min, and in the other case three of the four product peaks were separated at 10.4–11.9 min. The 8-fold-redundant 15-mer gave rise to three major peaks and three minor peaks at 15.6–19.2 min. The various components were not analyzed with regard to base composition, but it was clear that the use of SAX-HPLC to purify crude mixtures of mixed-sequence 5′-hydroxyl oligonucleotides can lead to possible overlap of peaks for

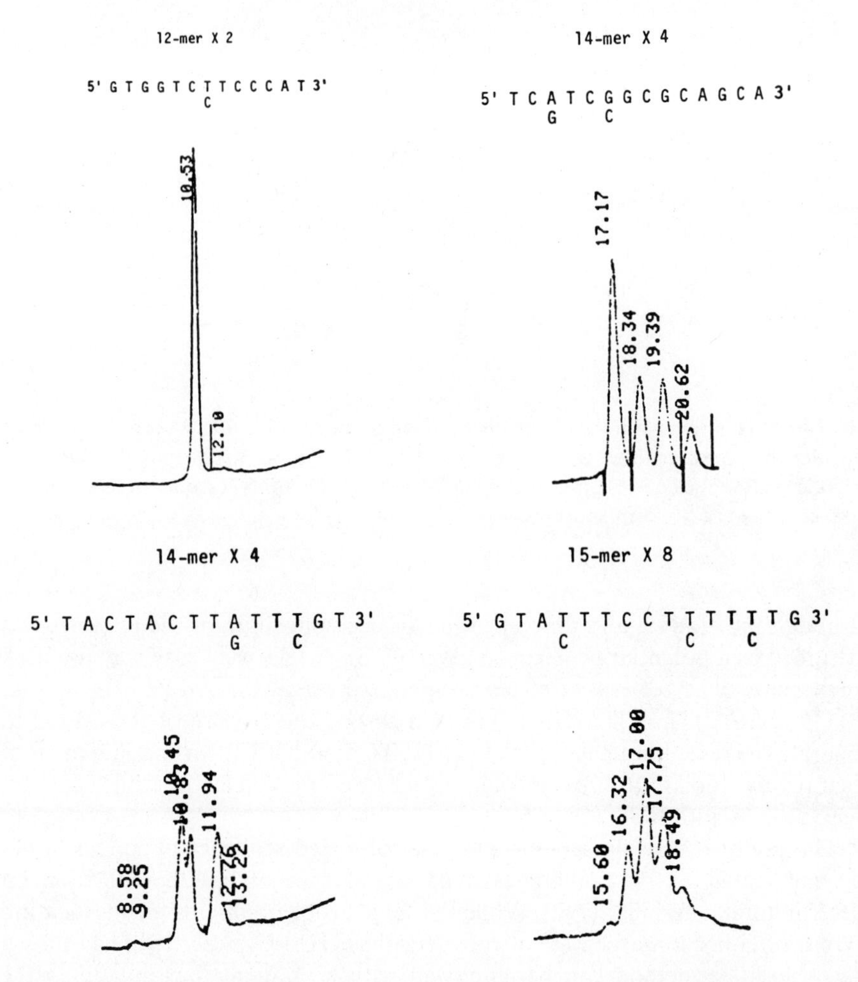

Fig. 21. Ion-exchange HPLC separations of 2-, 4-, and 8-fold redundant, mixed-sequence 5′-hydroxyl oligonucleotides after their isolation by PAGE (J.W. Efcavitch and G. Zon, unpublished studies). The column and conditions are described in Fig. 2. Elution times are given in minutes; 0.025 AUFS at 254 nm.

the desired *n*-mers and shorter sequences. This necessitates taking a relative wide "cut" that begins at a more or less arbitrarily chosen elution time and ends with the slowest-eluted material. RP-HPLC of mixed-sequence 5′-DMT oligonucleotides does not generally lead to separation of the different product sequences (192), except in cases having different bases at or near the 5′-end (cf. Fig. 7). This makes is easier, relative to SAX-HPLC, to collect the pool of products, although there is still some arbitrariness or guesswork as to deciding what minor peaks might be products and what peaks might be impurities. For example, a typical

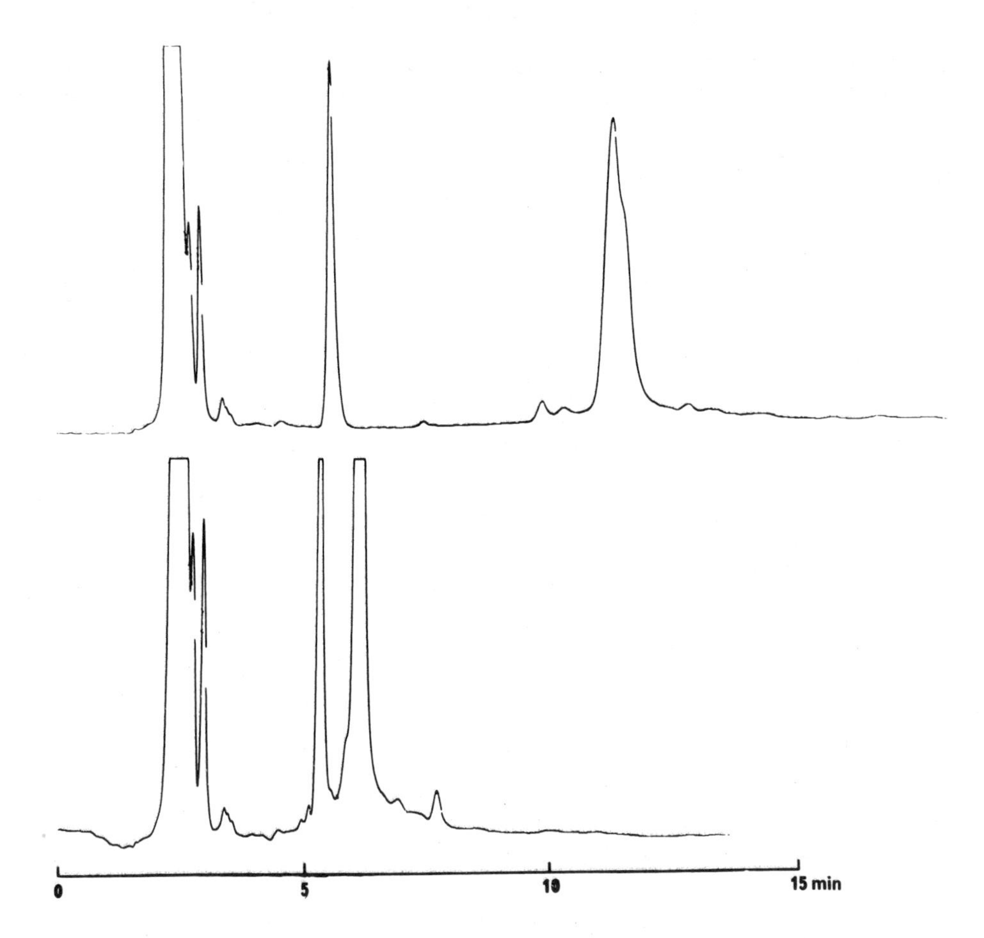

Fig. 22. RP-HPLC separations of the 32-fold redundant, crude 12-mer, 5′-DMT-d[CC(N)AA(A/G)AT(N)AAT], N = A, G, C, or T. The top trace was obtained with a gradient of 1%/min, as described in Fig. 5. The bottom trace was obtained with a "flash" gradient of 2.5%/min acetonitrile vs TEAA (0.1 *M*, pH 7.0) for 4 min, starting at acetonitrile-TEAA = 20:80, flow rate = 4 mL/min.

RP-HPLC gradient separates (Fig. 22) the crude, 32-fold redundant, 5′-DMT 12-mer, d[CC(N)AA(A/G)ATNAAT], into one major peak (11.2 min) with a trailing shoulder, and minor fast- and slow-eluted peaks (~ 10 min and ~ 13 min, respectively). In the best case scenario, wherein the synthesis occurred in an unbiased manner, each product represents $\frac{1}{32}$-nd or 3.3 mol-% of the entire product pool and thus gives rise to 3.3% of the total product absorbance (ignoring differences in molar extinction coefficients). Consequently, the minor DMT-bearing components seen in Fig. 22 could be either impurities or legitimate 12-mer products. In view of the fact that single-sequence oligonucleotide probes can be used successfully as crude synthetic products (typically ⩾75% pure), one can opt for either use of the crude mixed-sequence probe, or use of RP-HPLC to achieve enrichment of the product under conditions that favor collection of the entire product pool at the expense of size homogeneity. The latter option is carried out in this author's laboratory by means of a single change in the standard conditions normally used for single sequences, namely, an increase in the gradient of acetonitrile from 1%/min to 2.5%/min. The effect of this increased gradient on separation is seen in Fig. 22 for the aforementioned mixed-sequence 12-mer, which was collected over the period 5.5 to 10 min. As expected, analyses by PAGE of 5-hydroxyl oligonucleotide-pools derived by this "flash" HPLC method show somewhat higher levels of undesired sequences of the type $n - 1$, $n - 2$, and so on, relative to the results normally obtained with single sequences isolated by use of the slower gradient.

Occasionally, these "flash" HPLC conditions will clearly not lead to coelution of the product pool, as exemplified in Fig. 23 with comparative HPLC traces for the 64-fold redundant, crude 24-mer, ′-DMT-d[(A/G)TC(I/C)ACCAT(T/C)TT(T/C)TT(A/G)TGAAT]. The individual fast- and slow-eluted major peaks were collected (normal gradient), detritylated, and then tandemly digested with SVPDE and AP to obtain base compositions (cf. Fig. 23 legend for values). If one assumes that the stronger hydrophobicity of the 5′-DMT-dA moiety, as compared to the 5′-DMT-dG moiety, in the 24-mers leads to slow- and fast-eluted sets of sequences, respectively, then the calculated versus found values of mol-% for dA (25.5 vs 27.5) and dG (6.4 vs 6.1) in the slow-eluted set, and dA (23.4 vs 23.3) and dG (8.5 vs 9.9) in the fast-eluted set are in reasonably good agreement; in addition, the values of mol-% found for dC, dI, and dT are nearly equivalent in the two sets, as would be expected, and are in reasonably good agreement with the calculated values of mol-%.

Another example of the separation of 5′-DMT products that is presumably caused by differences at the 5′-terminus was found in the case of the crude 12-mer, 5′-DMT-[(A/G)TA(T/C)GC(A/G)TT(A/G)TA]. The HPLC trace (not shown that was obtained by use of the "flash" gradient gave a ~1:1 ratio of peaks at 8.5 and 8.0 min.

The RP-HPLC traces shown in Fig. 24 were obtained for the 16-fold redundant, crude 14-mers, 5′-DMT-d[TCGTC(N)GCTTG(N)GT] and 5′-DMT-d[TCGTC(N)GCCTG(N)GT], which formally differ only at the underlined position. The gradient was essentially a step gradient that went from 20% to 30% acetonitrile in 1 min. The virtually identical elution profiles for the DMT-bearing species, which were collected over a 4–6 min period, demonstrate that a centrally located, single-base change from dT to dC had no measurable effect on the pattern of elution. No attempt was made to identify the fast-eluted sequence(s) at 4.5 min, although this peak was completely resolved when a normal gradient was used (trace not shown).

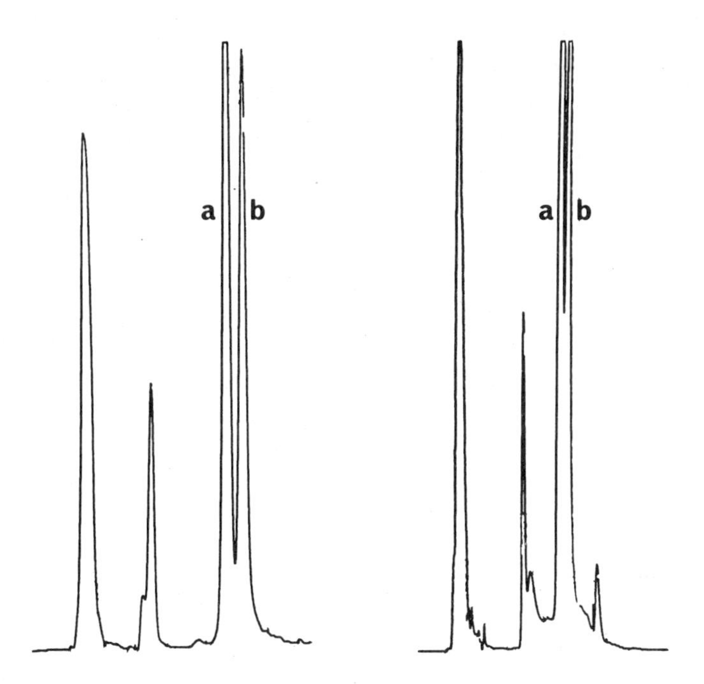

Fig. 23. RP-HPLC separations of the 64-fold redundant, crude 24-mer, 5′-DMT-d[(A/G)TC(I/C)AC(I/C)ACCAT(T/C)TT(T/C)TT(A/G)TGAAT]. The left and right traces correspond to normal and "flash" gradients, respectively, as described in Fig. 22. Under normal conditions, product fractions *a* and *b* were eluted at 9.9 and 10.7 min; with the "flash" gradient, *a* and *b* were eluted at 7.4 and 7.4 min, respectively. Tandem enzymatic digestions of 5′-hydroxyl oligonucleotides derived from *a* indicated the following base composition (mol-%): dC, 23.3; dI, 6.4; dG, 9.9; dT, 37.2; dA, 23.3; the digest from *b* gave dC, 22.0; dI, 6.6; dG, 9.9; dT, 37.2; dA, 23.3.

4. SELF-COMPLEMENTARY OLIGONUCLEOTIDES

Oligonucleotides with self-complementary sequences serve as "adaptors" for use in molecular cloning; these oligonucleotides associate to form double-stranded DNA ("duplexes") having blunt ends and a restriction site in the middle. Self-complementary oligonucleotides are also important as "tailor made" substrates or pseudosubstrates for investigating restriction enzymes and DNA binding proteins, templates for polymerases, and defined-sequence duplexes for physicochemical studies of the structure, dynamics, and sequence-specific properties of DNA.

The tendency for fully and partially self-complementary oligonucleotides to form either duplexes or hairpin-loop structures with themselves, and "heteroduplexes" with shorter fragments, which are present in crude synthetic mixtures, complicates purification by HPLC. This was found during SAX-HPLC in an early study by Gait et al. (142), who noted that, in the case of $d(GC)_5$, the large

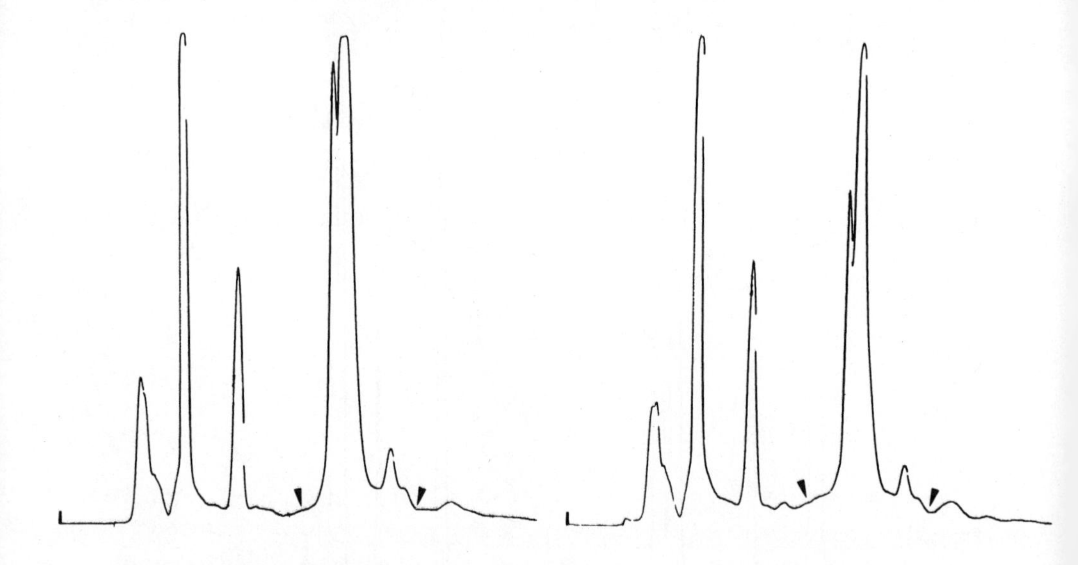

Fig. 24. RP-HPLC isolation of 10% of the crude (1 μmol), 16-fold redundant, 14-mers, 5′-DMT-d[TCGTC(N)GCTTG(N)GT] (left trace) and 5′-DMT-d[TCGTC(N)GCCTG(N)GT] (right trace), N = A, G, C, or T. The μBondapak C_{18} column (3.9 mm × 30 cm) was eluted for 1 min with a 10%/min gradient of acetonitrile vs TEAA (0.1 *M*, pH 7.0) starting at acetonitrile-TEAA = 20:30; flow rate = 2 mL/min. A 4–6 min "cut" was taken in each case.

tendency for self-association in salt solution necessitated the use of 30% ethanol in the eluting buffers and elevated column temperatures to ensure disaggregation. Such "denaturation" during SAX-HPLC was also achieved by use of 7 *M* urea (140).

This author has synthesized numerous self-complementary oligonucleotides in relatively large amounts (20–50 mg), and chose to use RP-HPLC for purification based on its well-known suitability for scale-up applications in organic chemistry, and, moreover, the relatively easy postchromatographic processing of the collected product, as compared to SAX-HPLC. The crude, self-complementary, 5′-DMT oligonucleotides are first analyzed using standard conditions: elution from a μ-Bondapak C_{18} column (7.8 mm × 25 cm) at ambient temperature with a 1%/min gradient of 20–30% acetonitrile versus TEAA (0.1 *M*, pH 7.0) at a flow rate of 4 mL/min, or elution from a PRP-1 column under similar conditions. This has generally led to normal elution patterns (i.e., one major product peak) for various self-complementary oligonucleotides composed of only dA and dT residues (100% AT), such as 5′-DMT-d$(AT)_n$ with n = 5, 6, 9, and 10, 5′-DMT-d$(AATT)_3$, 5′-DMT-d(ATTT-

TTTAAAAAAT), 5′-DMT-d(AAAATTTTAAAATTTT), and 5′-DMT-d(AAAAAAATTTTTT). Normal elution patterns have also been obtained for the following self-complementary oligonucleotides that have 50–80% AT bases: 5′-DMT-d(ATATCGATAT), 80%; 5′-DMT-d(TATAGCTAGCTATA), 71%; 5′-DMT-d(ATTCGTACGAAT), 71%; 5′-DMT-d(TATATG-CCGCATATA), 71%; 5′-DMT-d(TATGCCGCATA), 60%; 5′-DMT-d(TATGGGTACCCATA), 57%; 5′-DMT-d(CCTTAAGG), 50%; 5′-DMT-d(GGAATTCC), 50%; and 5′-DMT-d(GGTATACC), 50%. By contrast, significant amounts of DMT-bearing species, which are eluted *faster* than the major DMT-bearing species, are usually observed for self-complementary oligonucleotides that have ⩾55–60% of GC bases. This characteristically different pattern of elution for associated oligonucleotides is evident by comparison of the reversed-phase HPLC traces (Fig. 25) obtained for the crude 8-mer, 5′-DMT-d(GGAATTCC), which has a T_m at (or below) room temperature and shows one major product peak, and the higher melting crude 12-mer, 5′-DMT-d(CGCGAATTCGCG), which shows multiple, relatively fast-eluted, DMT-bearing species in addition to the major peak. Denaturation of the crude 12-mer in 1:1 formamide–water at 70°C and then injection of the hot solution onto the column caused only minor changes in the appearance of the HPLC trace. Decreasing the concentration of TEAA from 0.1 to 0.01 *M* to favor disaggregation during elution from the RP-HPLC column led only to shorter retention times. Replacement of TEAA at pH 7 with tetraalkylammonium hydroxide at pH to maintain ion pairing but disrupt hydrogen bonding during elution from a chemically inert PRP-1 column is a reasonable approach to investigate.

The use of a heated RP-HPLC column under otherwise standard conditions has in some cases apparently led to thermal denaturation and a normal pattern of elution. This is shown in Fig. 26 for separations on C_{18} μBondapak of the crude 6-mer, 5′-DMT-d(GCGCGC), at room temperature and at 70°C, and in Fig. 27 for separations of the crude 18-mer, 5′-DMT-d(AUAUAGC-GCGCGCUAUAT), at room temperature and at 70°C. Collection of the major 18-mer product peak at elevated temperature followed by detritylation and then tandem digestions with SVPDE and AP to 2′-deoxynucleosides gave a base composition which was in good agreement with that expected for the desired product. Similar results have been obtained with a hot (70°C) PRP-1 column, as was also found by Ikuta et al. (137) in the purification of 5′-DMT-d(CGCGCG) (60°C).

It is important to remember that elution of an apparently single peak under denaturing conditions could sometimes be due to loss of resolution rather than disaggregation. Direct evidence for association of the DMT-bearing product and shorter sequences can, however, be easily obtained by PAGE analysis of collected fractions. Such was the case for ^{32}P 5′-end labeled fractions that had been derived from RP-HPLC separation of the crude 18-mer, DMT-d(ATATAGCGm^{5}CGCGCTATAT) (m^{5}C = 2′-deoxy-5-methylcytosine), under normal HPLC conditions at room temperature [Fig. 28, (A), peaks a–g],

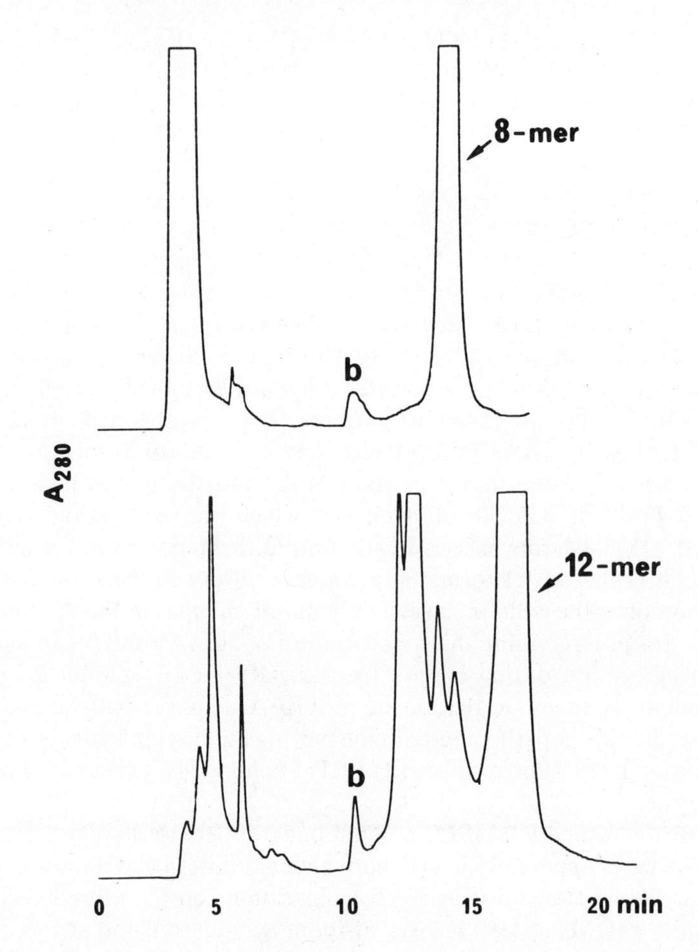

Fig. 25. Preparative-scale (5 μmol) RP-HPLC separations of a crude 8-mer 5′-DMT-d(GGAATTCC), and a crude 12-mer, 5′-DMT-d(CGCGAATTTCGCG). The DuPont Zorbax ODS column (21.5 mm × 25 cm) was eluted for 10 min with a 1%/min gradient of acetonitrile vs TEAA (0.1 *M*, pH 7.0) starting at acetonitrile-TEAA = 20:30, flow-rate = 11.25 mL/min. The *b* peaks are benzamide, which has a low A_{280}/A_{260} ratio compared to that for oligonucleotides.

and under disaggregating conditions [Fig. 28, (B), peaks 1–4], which employed a column temperature of 60°C and 50% v/v formamide in both the acetonitrile and TEAA. An autoradiogram (Fig. 29) revealed the presence of various sizes and amounts of shorter sequences in fractions (lanes) a–g, whereas fraction (lane) 1 contained relatively pure 18-mer. Findings such as these have not, to this author's knowledge, been previously reported.

Substitution of either 2′-deoxyinosine (dI) or 7-deaza-2′-deoxyguanosine (dc^7G) for dT at the underlined position in the 18-mer, 5′-DMT-d(ATATAGC-

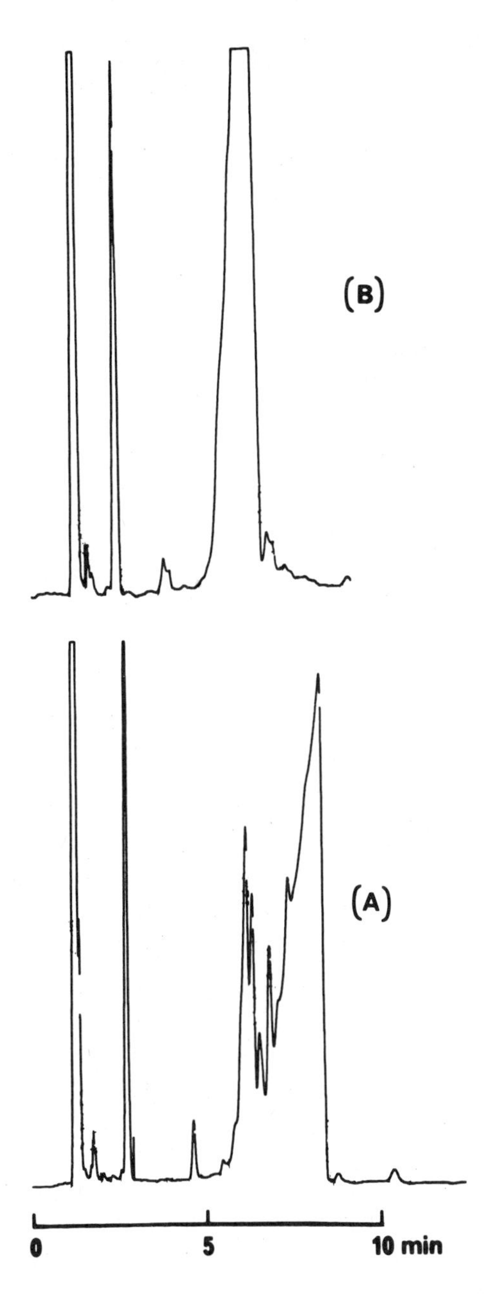

Fig. 26. RP-HPLC separations of the crude 6-mer, 5′-DMT-d(GCGCGC), at room temperature (A) and 70°C (B). The column and other conditions are described in Fig. 5.

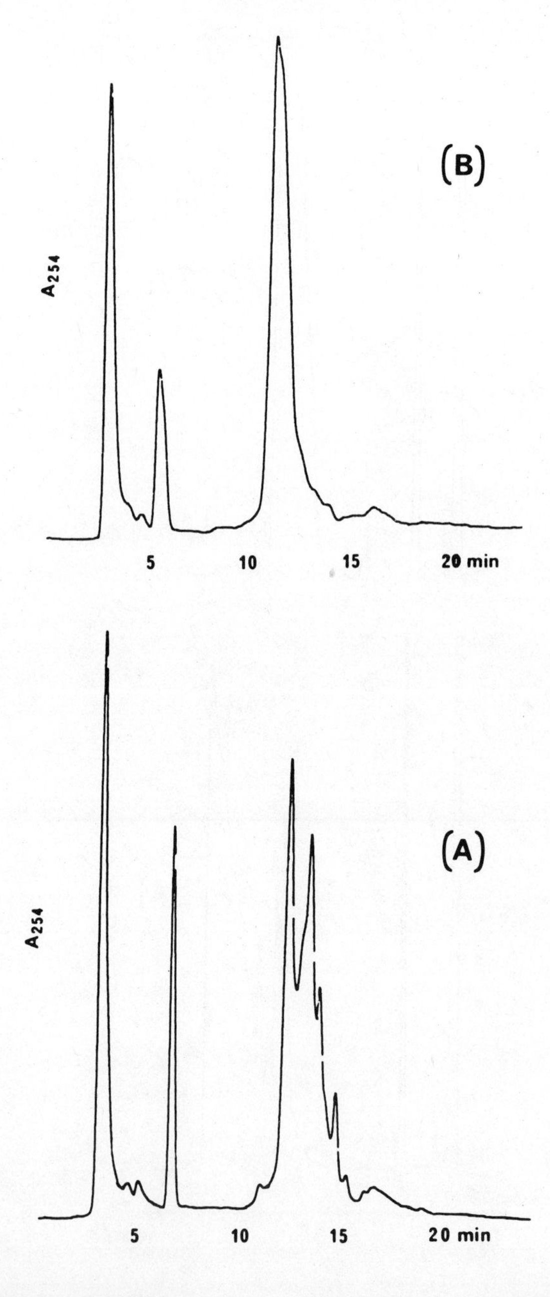

Fig. 27. RP-HPLC separations of the crude 18-mer, 5′-DMT-d(AUAUAGC-GCGCGUAUAT), at room temperature (A) and 70°C (B). The column and conditions for (A) and (B) are described in Fig. 5, with the exception that the gradient for (B) was decreased to 0.5%/min.

GCGCGCTATAT), led to essentially normal patterns of elution from the RP-HPLC column at room temperature, which thus differed from the behavior of the aforementioned m^5C-bearing analogue. This indication of mainly dissociated eluate suggested that I·C and c^7G·C led to a lowered T_m, relative to the case with m^5C·G.

The "parent" 18-mer, d(ATATAGCGCGCGCTATAT), and analogs modified with one or more dU, dm^5C, dI, dc^7G, and phosphorothioate moieties were purified by RP-HPLC and used (180) to define the binding site in native DNA for a murine monoclonal anti-DNA autoantibody by measurement of their competitive binding versus ^{3}H-labeled *E. coli* native DNA. Based on the relative competition factors, it was concluded that the major contacts with the antibody are in the major groove, within a sequence of bases that, in this case, must include a $d(GC)_3$ or $d(GC)_4$ core: the antibody apparently binds to portions of C and G bases in the major groove, a limited region of the backbone, and the 2-amino group of one guanine in the minor groove.

5. LARGE-SCALE PURIFICATION

Relatively large amounts (5–50 mg) of synthetic oligonucleotides are generally required for X-ray crystallographic and NMR studies of the structure and dynamics of DNA, "recognition" of DNA by proteins and enzymes, and binding of drugs or mutagens to DNA. These amounts of DNA, which are roughly 10^6 times the quantities normally used for biological work, may also be needed in the future as specific hybridization probes for medical diagnosis, detection of potentially pathogenic organisms (43–46), and so on. In view of the significance of such biophysical and biotechnological research and development, it is important to have relatively simple, reliable, and effective methods for large-scale purification of oligonucleotides by, for example, RP-HPLC. This author has used RP-HPLC methodology which is essentially identical to that described in Section 2 for the purification of 0.1–0.2 μmol of the crude DMT-DNA; however, scale-up to accommodate 10–30 μmol of crude material requires the use of increased flow-rates and larger columns.

5.1. Column Comparisons

The capacity of a "semipreparative", 7.8 mm × 25 cm, μBondapak C_{18} column is usually exceeded by injections that contain more than approximately 0.5–1 μmol of the crude DMT-DNA, although this limiting amount is dependent on both chain length (lower capacity for longer molecules) and the degree of resolution that must be maintained in the scale-up to achieve purification. Time-consuming repetitive collections on this "semipreparative" column can be avoided by the use of a "preparative," 21.1 mm × 25 cm, Zorbax ODS column which can usually accommodate 5 to 10 μmol scale injections of the crude DMT-DNA. This column is operated at a flow-rate of 11–13 mL/min with the

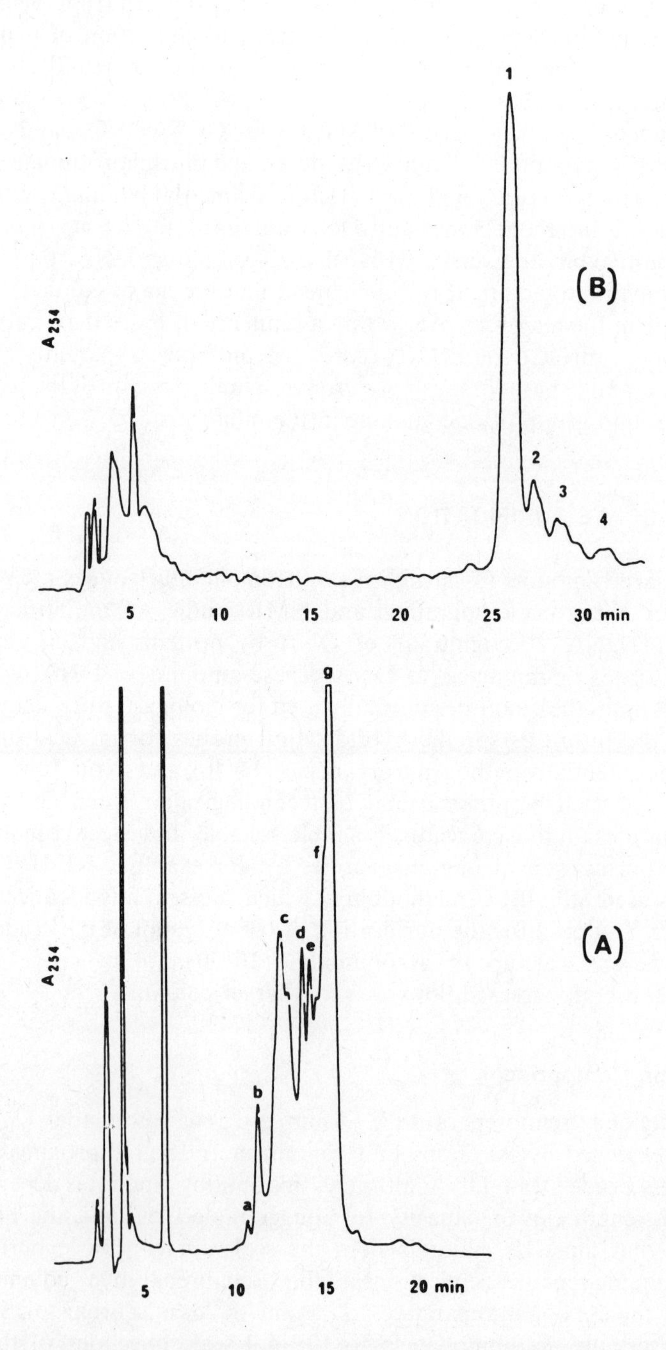

(B)
A254
1
2
3
4
5
10
15
20
25
30 min
(A)
A254
a
b
c
d
e
f
g
5
10
15
20 min

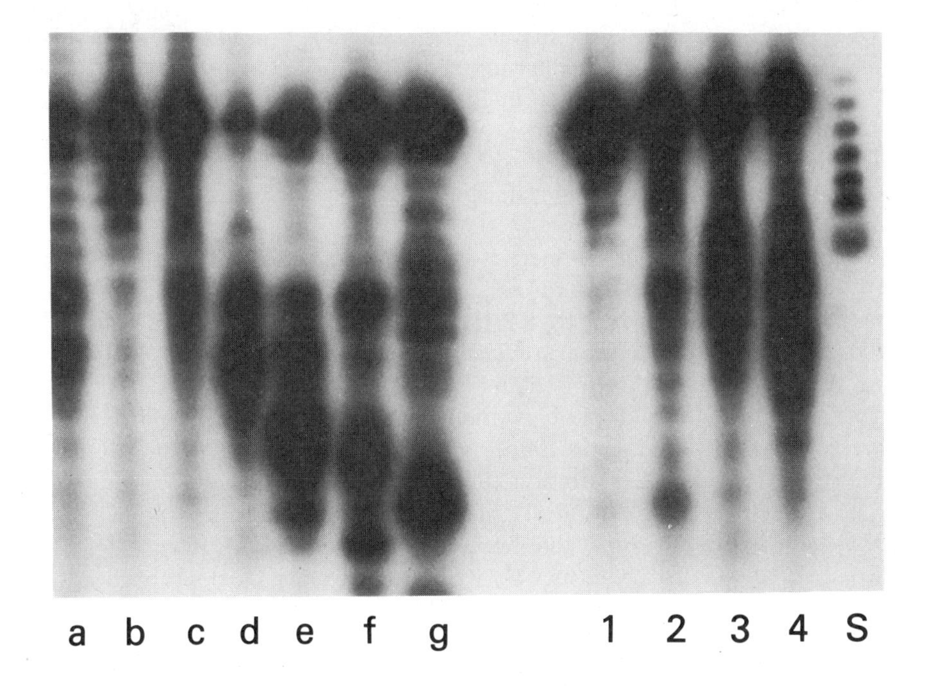

Fig. 29. Overexposed autoradiogram following PAGE of material derived from peaks *a–g* and 1–4 in Fig. 28.

typical gradient: 1%/min acetonitrile versus TEAA (0.1 *M*, pH 7) for 10 min, starting at acetonitrile-TEAA = 20:80. For dimers through hexamers, the gradient is applied for 20–30 min to elute these more retained materials, whereas the gradient for longer molecules is applied for only 10 min, which is then followed by isocratic elution. Comparative separations obtained for crude

---

Fig. 28. (*Opposite*) RP-HPLC separations of the crude 18-mer, 5′-DMT-d(ATATAGC-Gm^{5}CGCGCTATAT), (dm^{5}C = 2′-deoxy-5-methylcytosine), at room temperature (A) and at 60°C with added formamide (B). The column and other conditions for (A) are described in Fig. 5; the peaks at 10–16 min (*a–g*) were collected, detritylated, 5′-end labeled with ^{32}P, and then analyzed by PAGE (autoradiogram in Fig. 29, lanes *a–g*, respectively). The same column as in (A) was eluted with a 1%/min gradient of 0–40% acetonitrile-formamide (1:1 v/v) against TEAA (0.1 *M*, pH 7.0)-formamide (1:1 v/v) at a flow rate = 4 mL/min. Spectral grade formamide from Eastman Kodak was used as received without deionization. Peaks 1–4 were collected and processed as in (A) (autoradiogram in Fig. 29, lanes 1–4, respectively).

5′-DMT-d$(AT)_5$, 5′-DMT-d(ATATTGGATAT), and 5′-DMT-d(CGCGAATTCGCG) on the "semipreparative" μBondapak and "preparative" Zorbax columns are shown in Figs. 30 and 31. The relative intensities of the fast-eluted 5-hydroxyl DNA's, benzamide, and DMT-product peaks are different, owing to detection of peaks at 254 and 280 nm, respectively; however, resolution is maintained fairly well during these large-scale separations.

The self-complementary 10-mer, 5′-DMT-d$(AT)_5$, was collected and then detritylated to give the 5′-hydroxyl product (482 OD_{260}-units, 15.6 mg). Analysis of the final product by RP-HPLC showed (Fig. 30) the presence of several relatively fast- and slow-eluted contaminants that had a combined peak area only 3% of that for the main product. Detailed ^{1}H NMR studies of this decamer by Suzuki and co-workers (193) subsequently provided direct evidence for purity and, more important, the conformation of the duplex: wrinkled D-form with a hydration tunnel in the minor groove.

The aforementioned A·T-rich 11-mer also gave analytical and preparative HPLC traces (Fig. 31, top), that were similar, whereas the trace (Fig. 31, bottom) for the G·C-rich 12-mer showed a somewhat greater proportion of fast-eluted DMT-bearing peaks during the preparative run, which was attributed to more association at higher concentrations of the oligonucleotides. In the latter case, only the slowest-eluted peak contained pure 12-mer; no attempt was made to increase the yield of this product by use of denaturing conditions (see Section 4).

A relatively small PRP-1 column with a bed volume of only 1.4 mm × 15 cm has been reported (137) to have a capacity of 10–25 mg of DNA, which is comparable to that afforded by the much larger (and far more expensive) 21.1 mm × 25 cm, Zorbax ODS column that was discussed above. The present author has used a PRP-1 column with dimensions of 7 mm × 30 cm to purify a variety of crude DMT-DNA's, and found that the capacity of this "semipreparative" column is much better than that of a comparable size μBondapak C_{18} column but does not exceed that of the "preparative," 21.5 mm × 25 cm, Zorbax ODS column, and does not maintain adequate resolution at the reported (137) 10–25 mg loads.

A problem to be aware of when using a PRP-1 column is the reappearance of peaks from a previous, large-scale injection. Thus, if a preparative collection of crude DMT-DNA is followed by repeated gradient elutions using injections of water, the DMT-DNA product peak will be seen with gradually diminished intensity. The same results are obtained even if the column is eluted with acetonitrile-TEAA = 30:70 for an extended period of time after the preparative collection. These observations, which also hold for less hydrophobic 5′-hydroxyl oligonucleotides, may be due to the nonrigid nature of the PRP-1 matrix as compared to silica-based C_{18} columns that do not give rise to a similar problem with "ghost" peaks.

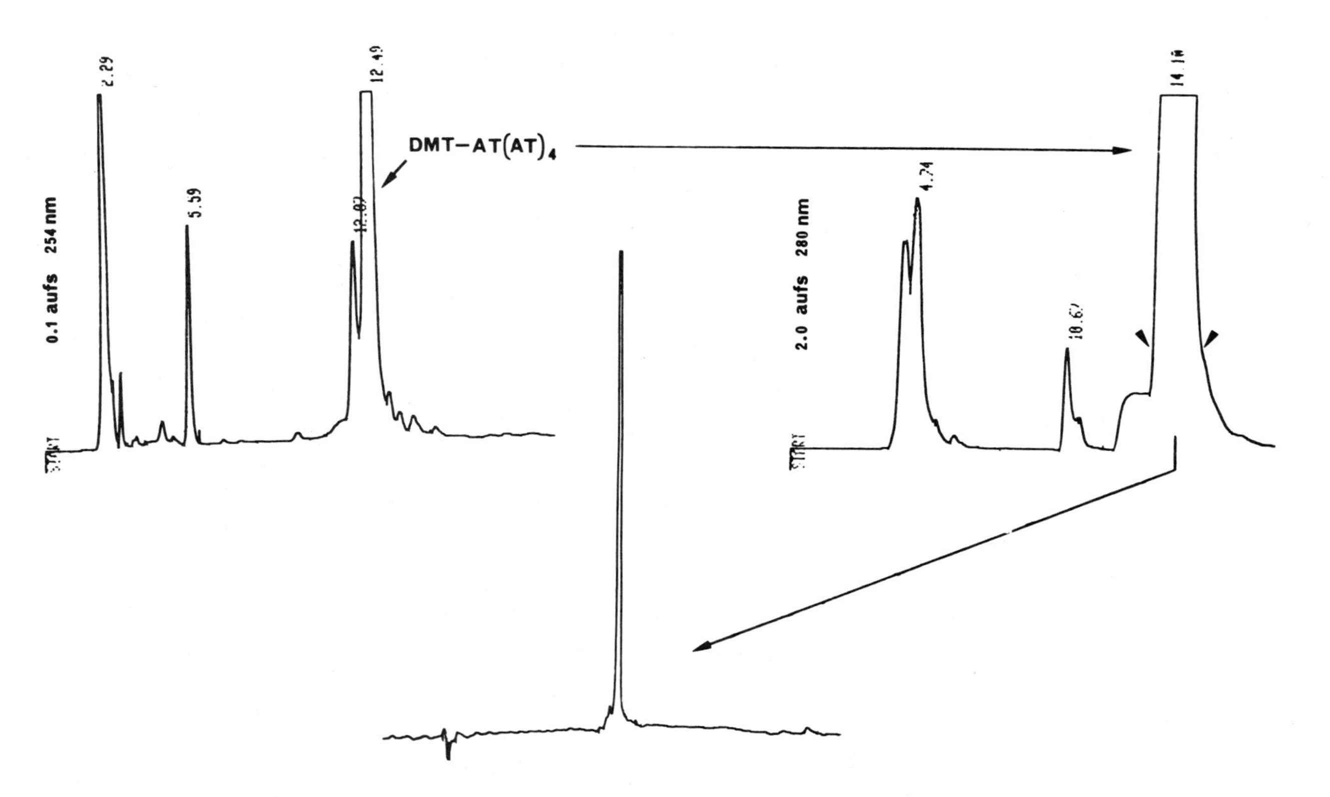

Fig. 30. RP-HPLC separations of the crude 10-mer, 5′-DMT-d(ATATATATAT), under analytical conditions, as described in Fig. 5, and under preparative (5 μmol injection) conditions as described in Fig. 25; the arrowheads indicate the "cut" that was collected, detritylated, and then analyzed as described in Fig. 16.

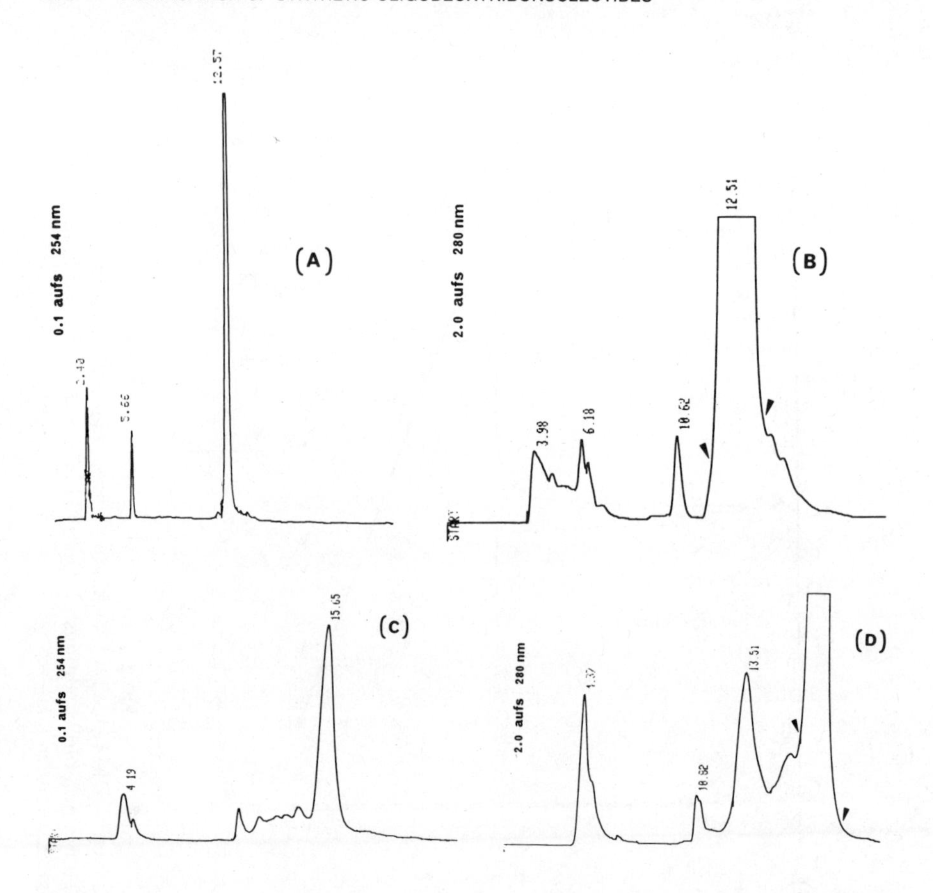

Fig. 31. RP-HPLC separations of a crude 11-mer, 5′-DMT-d(ATATTGGATAT) (top traces), and 12-mer, 5′-DMT-d(CGCGAATTCGCG) (bottom traces), under analytical conditions (A and C), as described in Fig. 5, and under preparative conditions (B and D) as described in Fig. 25. The arrowheads (B and D) indicate the "cuts" that were collected and further processed as described in the text to provide ~15 mg samples for ^{1}H-NMR studies.

5.2. Postchromatographic Sample Work-Up

The majority of this author's large-scale HPLC purifications of crude 5′-DMT oligonucleotides have involved elutions of 4-mer to 14-mer products from a preparative-size, 21.5 mm × 25 cm, Zorbax ODS column at flow rates of ~11–13 mL/min. The collected eluate from one or more injections (200–500 OD_{260}-units) is generally contained in 20–60 mL of acetonitrile-TEAA (0.1 *M*, pH 7.0), which can be concentrated to dryness either on a rotary evaporator at 40–50°C or in a vacuum centrifuge; partial detritylation will usually occur during this concentration step. The resultant residue is vortexed with water (2–5 mL) and then treated with enough glacial acetic acid to obtain a solution pH of 2.8–3.0. A "milky" suspension or opaque solution is formed by precipitation of trityl alcohol, which is extracted with ethyl acetate, after allowing ~15 min for detritylation, and neutralization with concentrated NH_4OH. The aqueous solution is filtered if necessary, and then taken to dryness either in a vacuum centrifuge or by lyophilization.

Residual triethylammonium acetate can be removed by various standard techniques, such as dialysis, ion exchange, gel filtration, or precipitation. In this author's laboratory, dialysis with boiled and washed Spectrapor tubing (1000 or 2000 MW cutoff) sometimes led to the introduction of significant amounts of an unknown, ^{1}H NMR-observable (δ 1.4) contaminant, as did ion exchange with Chelex-100 (^{1}H NMR, apparent singlet, δ 2.0). Consequently, we routinely use the following procedure for precipitation of the product, which is applicable to duplexes with $T_m > \sim 25$–30°C and noncomplementary n-mers with $n \geqslant 8$–10, and can be scaled-up or scaled-down in terms of OD_{260}-units/mL, depending on the recovery from the first precipitation. The 5′-hydroxyl oligonucleotide (~100–500 OD_{260}-units) is dissolved in 0.6 mL of 1 *M* aqueous NaCl, and absolute ethanol (1.4 mL) is then added dropwise with vortexing. If no precipitate forms, then either more absolute ethanol (1.4 mL) is added or the sample is reconcentrated for precipitation by addition of absolute ethanol (0.7–1.4 mL) to the reconstituted DNA solution in 0.3 mL water. The precipitate is cooled in an ice bath, centrifuged briefly, and the cold supernatant is then removed by pipet; the supernatant generally contains no more than 5–10% of the original optical density. The precipitate, which sometimes is an oil, is dried *in vacuo* and then reprecipitated two more times.

Relatively short oligonucleotides and longer sequences that do not precipitate well are obtained as their sodium salt, free of triethylammonium acetate, by gel filtration using prepacked columns of PD-10 Sephadex G-25M. A solution of the oligonucleotide in 1 mL of 0.1–0.5 *M* NaCl is loaded onto the column with 1 mL of water wash, and the DNA is then eluted with water, taking 1 mL fractions. Fractions 3–5, which generally contain most of the oligonucleotide (Fig. 32), are pooled and then lyophilized. If the ^{1}H-NMR spectrum shows the presence of residual triethylammonium acetate ($CH_3CH_2N^+$, δ 1.2, t and δ 3.1, q; $CH_3CO_2^-$, δ 1.8), then the gel filtration procedure is repeated. This gel-filtration method also removes trimethylsilyl-bearing impurities, which derive from end-capped, silica-based reversed-phase columns and have sometimes been difficult to remove by precipitation of oligonucleotides from solutions of ethanol–aqueous NaCl. One does not have to cope with these impurities if a PRP-1 column is used to isolate the DMT-DNA.

Reversed-phase HPLC purification of the crude self-complementary 12-mer, 5′-DMT-d(GCGTACGTACGC), on the preparative Zorbax ODS column followed by detritylation, extraction, and precipitation from ethanol–aqueous NaCl, as described previously, afforded 300 OD_{260}-units of materials that was crystallized in the presence of $Co(NH_3)_6Cl_3$ (Fig. 32, bottom). The X-ray data to assign the symmetry and crystal system are now being collected, with the aim

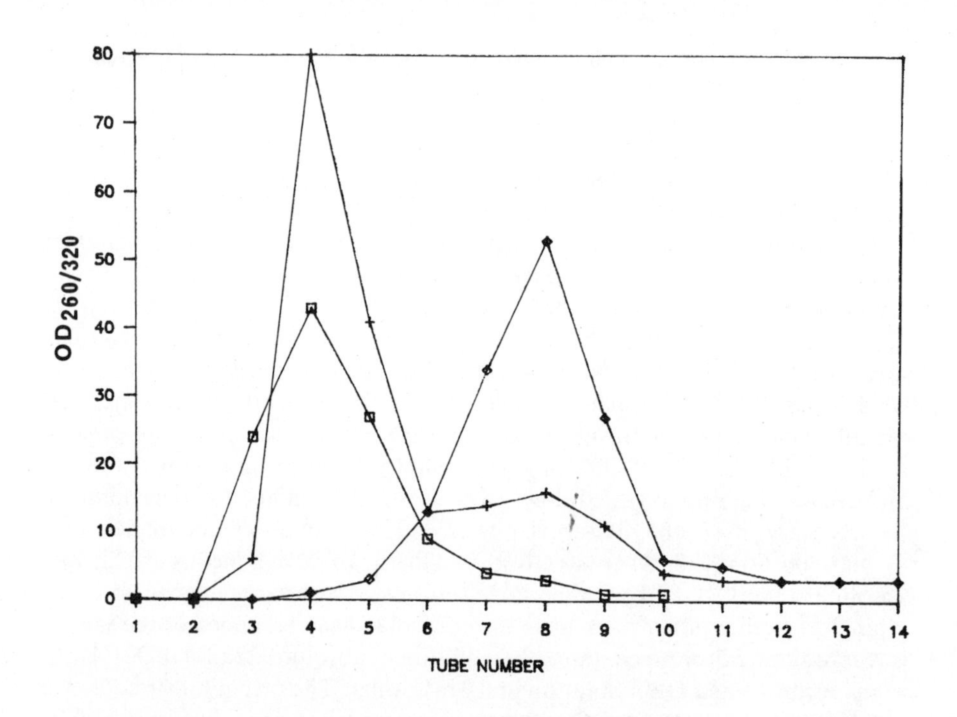
80
70
60
50
40
30
20
10
0
OD260/320
1
2
3
4
5
6
7
8
9
10
11
12
13
14
TUBE NUMBER

of performing a structure analysis (M. Sundaralingam and co-workers, unpublished studies).

6. CHEMICAL SYNTHESIS AND HPLC PURIFICATION OF 5′-PHOSPHORYLATED AND 5′-THIOPHOSPHORYLATED OLIGONUCLEOTIDES

Chemical syntheses of 5′-phosphorylated oligonucleotides by means of the phosphoramidite method have been developed independently by several investigators (194; M. Urdea, unpublished studies; J.A. Thompson and P. Groody, Unpublished studies; G. Zon, unpublished studies). Such syntheses, which can be easily adapted for use with automated DNA synthesizers, provide relatively cheap and convenient alternatives to enzymatic procedures, and can be applied to large-scale preparations that utilize RP-HPLC (194). Aside from the obvious applications in molecular biological studies, 5′-phosphorylated oligonucleotides can be used for various types of affinity-based isolation procedures, for labeling with fluorescent groups, dyes, and other "reporter" moieties, and for attachment of reactive functional groups (195–199). 5′-Thiophosphorylated oligonucleotides can likewise be used for coupling, labeling, and further functionalization (194, 200).

The author's synthetic method for preparing 5′-phosphorylated and 5′-thiophosphorylated oligonucleotides, which is essentially the same as that recently reported by Uhlmann and Engels (194), uses tetrazole-catalyzed coupling of *O*-methyl-*O*-*p*-nitrophenylethyl-*N*,*N*-diisopropylphosphoramidite in an otherwise conventional cycle of chemistry (cf. Fig. 4); replacement of the normal I_2–H_2O oxidant by a hot solution of elemental sulfur in 2,6-lutidine provides the 5′-thiophosphate endgroup. The *p*-nitrophenylethyl (NPE) phosphoramidite reagent is used regardless of whether one employs *O*-methyl or *O*-β-cyanoethyl phosphoramidites for construction of the chain.

O-Demethylation with thiophenol-triethylamine, cleavage from the support with concentrated NH_4OH, and backbone/base-deprotection by heating with concentrated NH_4OH, all in the usual manner, affords the crude oligonucleotide as a *p*-nitrophenylethyl phosphodiester (NPEp) derivative. Compared to other synthetic methods (M. Urdea, unpublished studies; J.A. Thompson and P.

Fig. 32. (*Opposite*)(Top) Gel filtration of 5′-hydroxyl oligonucleotides with prepackaged columns (16 mm × 50 mm) of PD-10 Sephadex G-25M. The square- and plus-shaped symbols refer to d(CGACCAG) and d(CGCGCGCGCG), respectively, the latter of which contained added *p*-nitrophenol. Gel filtration of *p*-nitrophenol alone, which is indicated by the diamond-shaped symbols, was monitored at 320 nm. Each tube contained 1 mL. (Bottom) Crystals of the duplex, $[d(GCGTACGTACGC)]_2$, with dimensions up to 0.15 mm × 0.3 mm, which were grown in the presence of $Co(NH_3)_6Cl_3$ (M. Sundaralingam and co-workers, unpublished studies).

Groody, unpublished studies), which lead directly to either a 5′-phosphate or a 5′-thiophosphate group, the presence of the hydrophobic NPE moiety can be used as a "handle" for convenient isolation of the product by RP-HPLC (194). The HPLC trace shown in Fig. 33 for the NPEp derivative of the crude 8-mer, d(NPEpGGAATTCC), is qualitatively similar to that which was obtained for the analogous 5′-DMT derivative, although the comparatively weaker lipophilicity of the NPE group required less acetonitrile in the gradient for separation of the product from the shorter 5′-hydroxyl sequences.

The corresponding *p*-nitrophenylethyl thiono phosphodiester (NPEps) derivative of the crude 8-mer, d(NPEpsGGAATTCC), gave rise to the expected (179) pair of diastereomers with R_p and S_p absolute configurations at phosphorus, which were separated by RP-HPLC as shown in Fig. 33. This chromatogram also revealed the presence of 18% d(NPEpGGAATTCC), which resulted from incomplete sulfurization. The effect of sulfur substitution on the elution time, and the ability to separate diastereomeric oligonucleotides by reversed-phase HPLC is discussed in a following section that deals with phosphorothioate-containing modified DNA's.

The NPEp and NPEps derivatives of the 8-mer were collected and identified by ^{31}P-NMR spectroscopy, and the NPEp compound was digested with nuclease P1 to give NPEpdG, which was also independently synthesized. Each of the samples was reacted with 1,8-diazobicyclo[5.4.0]undec-7-ene (DBU) in formamide (1:1 v/v) at 70°C for 30 min to remove the NPE group. The reaction mixtures were injected onto a μBondapak C_{18} column to isolate the respective products in essentially quantitative yield, and the products were unambiguously identified by ^{31}P NMR spectroscopy. The 5′-thiophosphorylated material was coupled (200) with monobromobimane (Thiolyte) to give an essentially quantitative yield of the hydrolytically stable, fluorescent-labeled 8-mer shown below. These synthetic methods were successfully applied to a 20-mer (Fig. 34), which could be visualised at ~1-pmol levels by long-wavelength irradiation of its Thiolyte derivative.

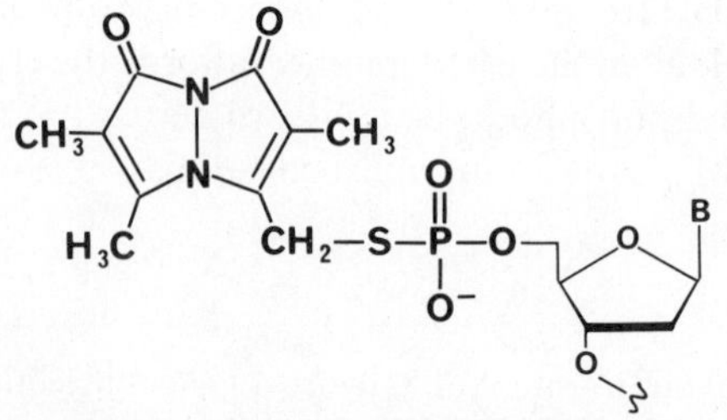

O O N CH3 CH3 N O H3C CH2—S—P—O— O B O⁻ O

7. BACKBONE-MODIFIED OLIGONUCLEOTIDES

Oligonucleotides that have either isotopic (^{17}O, ^{18}O), isoelectronic (S, Se), or other structural modifications (alkoxy, alkyl) at one or more phosphorus positions in the "backbone" of DNA are useful for the investigation of specific

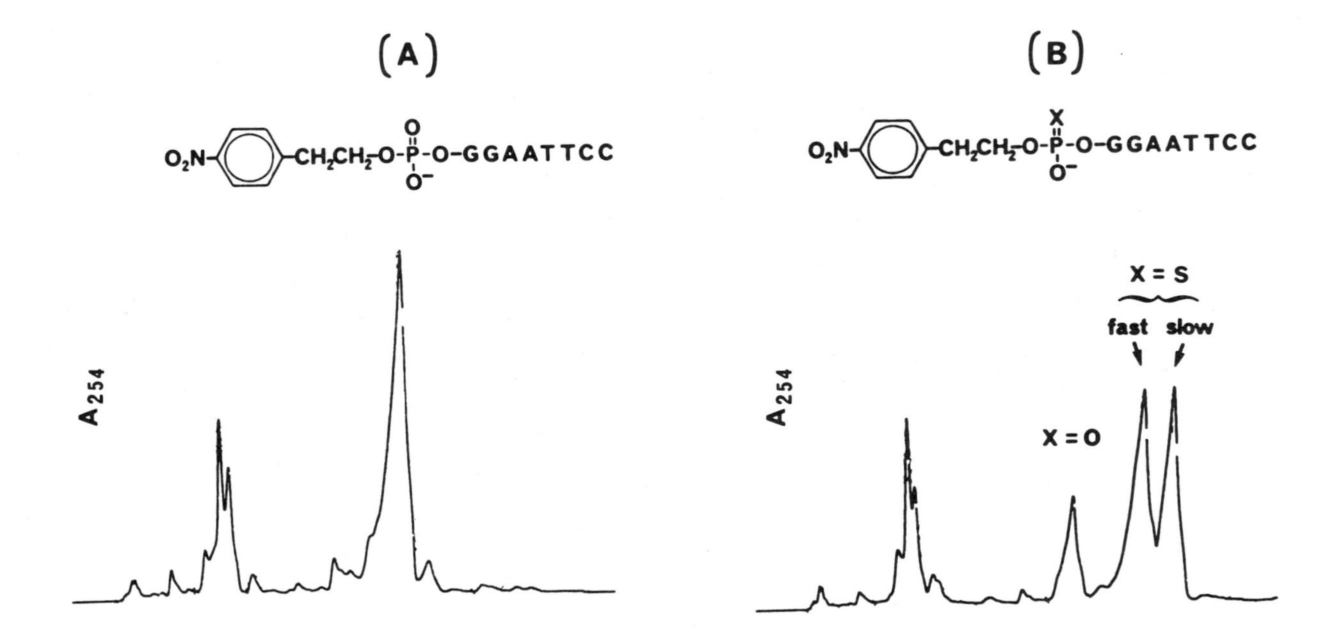

Fig. 33. RP-HPLC separation of the crude 5′-*p*-nitrophenylethyl phosphodiester (A, 18 min) and diastereomeric (B, 21 and 23.5 min) 5′-*p*-nitrophenylethyl thiono phosphodiester derivatives of the 8-mer, d(GGAATTCC). In each case the μBondapak C_{18} column (7.8 mm × 30 cm) was eluted for 30 min with a 0.33%/min gradient of acetonitrile vs TEAA (0.1 *M*, pH 7.0) starting at acetonitrile-TEAA = 10:90; flow rate = 4 mL/min.

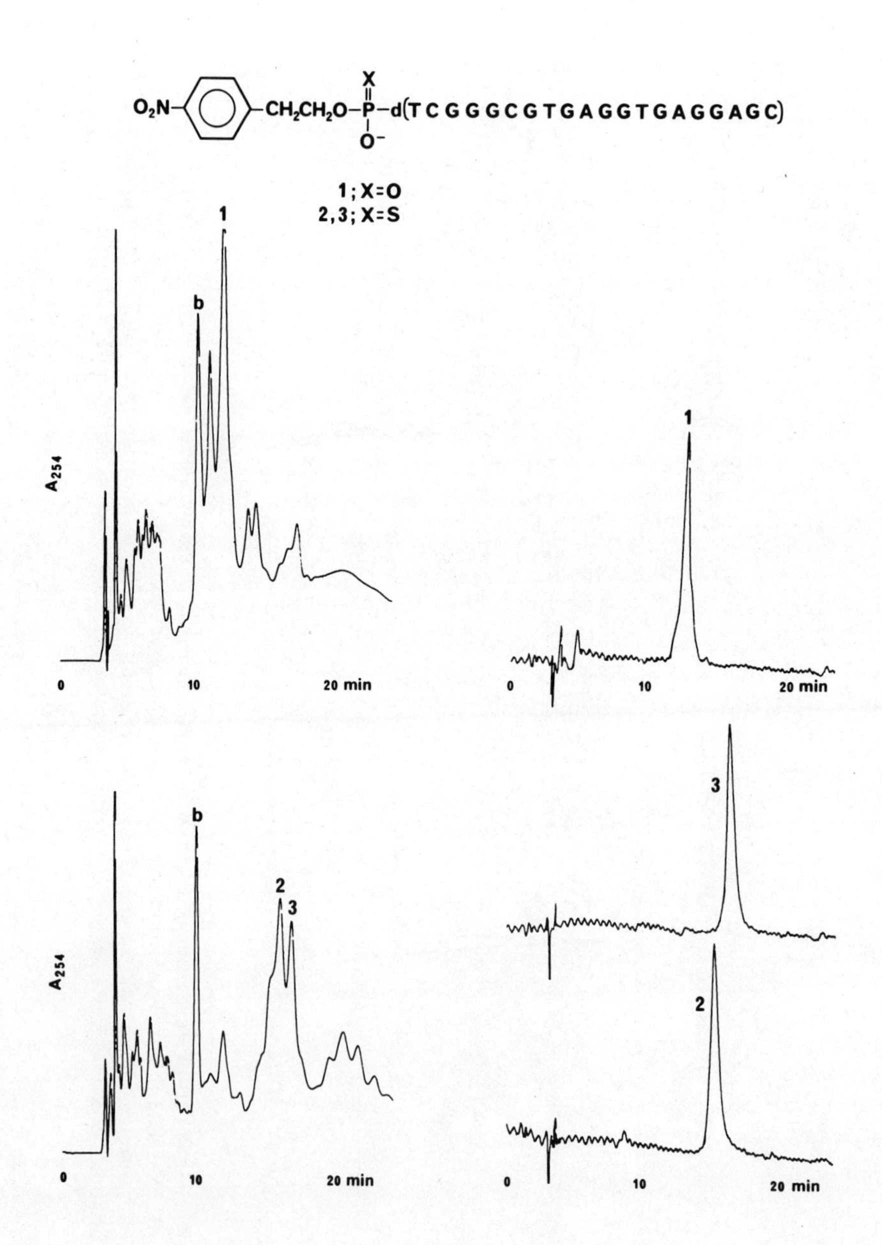

X
$O_2N-C_6H_4-CH_2CH_2O-P(=X)(O^-)-d(TCGGGCGTGAGGTGAGGAGC)$
1; X=O
2,3; X=S
1
b
A_{254}
0
10
20 min
1
0
10
20 min
b
2
3
A_{254}
0
10
20 min
3
2
0
10
20 min

interactions of DNA with proteins (180, 201) and enzymes (202, 203), and have potential applications as DNA/RNA-targeted reagents (204, 205), nondegradable diagnostic probes, and therapeutic agents (206). Substitution of one of the nonbridging phosphoryl oxygens at any internucleotide phosphate group generates an asymmetric (chiral) center with either the R_p or S_p absolute configuration at phosphorus, and thus leads to chemically nonequivalent (diastereomeric) chains. The total number of diastereomers (N) is exponentially related to the number of chiral phosphorus centers (n): $N = 2^n$.

Diastereomerically pure R_p or S_p dimer-blocks (201, 207) can be used to synthesize the corresponding diastereomers of such oligonucleotides, in which case purification of the product by HPLC is more or less routine. By contrast, the use of diastereomerically pure, monomeric phosphoramidite reagents has, to date, resulted in the formation of 2^n diastereomeric products (208), in which case HPLC can be used to both purify and separate the diastereomers (181). Sections 7.1–7.7 describe how RP-HPLC methods have been utilized for these purposes.

7.1. Phosphorothioates

Replacement of the normal I_2–H_2O oxidant used in the synthesis of DNA (cf. Fig. 4) with a hot solution of elemental sulfur (S_8) in 2,6-lutidine affords a thiono phosphotriester group, which in turn gives rise to a phosphorothioate (PS) linkage in the backbone-deprotected oligonucleotide product. Investigations (179, 203) of the separation and stereochemistry of these PS-containing analogs of oligonucleotides have demonstrated that μBondapak C_{18} columns could be used to separate the R_p and S_p diastereomers of a wide variety of mono-phosphorothioates, and in some cases it was possible to separate all four diastereomers of bis-phosphorothioate analogs, such as $G_{PS}GC_{PS}C$ and $T_{PS}ATA_{PS}A$ (Table 1). These separations were generally achieved by tandem RP-HPLC, first

---

Fig. 34. (*Opposite*) RP-HPLC separations of crude synthetic mixtures that contained 20-mers having 5′-*p*-nitrophenylethyl phosphodiester (*1*; X = 0, top left trace) and 5′-*p*-nitrophenylethyl thiono phosphodiester (*2,3*; X = S, bottom left trace) termini, which were introduced by 1*H*-tetrazole-catalyzed double-coupling (15 min) of *O*-*p*-nitrophenylethyl-*O*-methyl-*N,N*-diisopropylphosphoramidite (0.2 *M*) followed by oxidation with either I_2–H_2O-lutidine (30 sec) or sulfur-lutidine (60°C, 30 min), respectively. 5′-Terminal *O*-demethylation with PhSH-Et_3N was followed by cleavage from the support with conc. NH_4OH (25°C, 1 hr) and then ammoniacal decyanoethylation/base-deprotection (55°C, 6 hr). The *b* peaks are benzamide. Collection of *1* (13.4 min) followed by re-HPLC gave the top right trace. Collection of the fast- (16.2 min) and slow-eluted (17.1 min) diastereomers, *2* and *3*, respectively, followed by re-HPLC gave the correspondingly numbered tracings shown at the bottom right. The Waters μBondapak C_{18} column (7.8 mm × 30 cm) was eluted in each case with a gradient of acetonitrile (A) vs TEAA, 0.1 *M*, pH 7.0 (B). The times and corresponding A:B ratios were as follows: 0 min, 10:90; 7 min, 14:86; 20 min, 16:84; 21 min, 30:70; 30 min, 30:70; flow rate = 4.0 mL/min.

Table 1. Separation of Diastereomers of Mono- and Bis-phosphorothioate (PS) Analogs of Di- to Octameric Oligodeoxyribonucleotides by Reversed-Phase HPLC[a]

Formula, 5′ → 3′	Elution Time (min)		Absolute Configuration at Phosphorus
	5′-DMT	5′-HO	
$T_{PS}T$	12.5*	13.4**	R
		14.2	S
$T_{PS}A$	12.3*	12.3**	R
		13.1	S
$T_{PS}G$	–	10.6**	R
		11.3	S
$A_{PS}T$	22.0	13.3**	R
		14.1	S
$A_{PS}A$	16.0***	16.5	R
		18.3	S
$G_{PS}G$	10.1*	12.5	R
		14.0	S
$G_{PS}C$	10.4*	12.7	R
		13.9	S
$G_{PS}A$	10.6*	10.8**	R
		11.6	S
$C_{PS}C$	16.8***	9.9	R
		11.6	S
$C_{PS}G$	10.9*	11.0	R
		12.0	S
$T_{PS}TT$	20.5	19.4	S
	21.5	19.1	R
$TT_{PS}T$	18.7	17.8	R
		19.4	S
$AT_{PS}T$	12.5	14.8	R
		16.2	S
$T_{PS}T_{PS}T$	21.9	21.5	S,R
		22.5	S,S
		20.0	R,R
		22.2	R,S
$A_{PS}A_{PS}A$	18.5	29.5	R,R
		31.3	R,S
		31.7	S,R
		33.3	S,S
$G_{PS}G_{PS}G$	23.5	22.2	S,R
		23.9	S,S
	24.6	20.4	R,R
		22.5	R,S
$G_{PS}G_{PS}A$	17.5	13.2**	S,R
	18.2	13.6	S,S
		12.6	R,R
	19.1	13.1	R,S

Table 1. (*Continued*)

Formula, 5′ → 3′	Elution Time (min) 5′-DMT	5′-HO	Absolute Configuration at Phosphorus
$C_{PS}C_{PS}C$	18.0	10.7§	R,R
		12.0	S,R
		13.1	R,S
		14.2	S,S
$TTT_{PS}T$	18.4	27.2	R
		29.3	S
$TAT_{PS}A$	14.2***	19.0	R
		19.8	S
$T_{PS}ATA$	15.6***	20.4	S
	16.3	19.9	R
$T_{PS}TT_{PS}T$	16.0***	24.1	S,R
		24.9	S,S
	17.0	23.7	R,R
		24.6	R,S
$T_{PS}AT_{PS}A$	15.3***	22.0	S,R
		22.8	S,S
	16.0	21.5	R,R
		22.3	R,S
$G_{PS}GC_{PS}C$	18.0*	17.0	S,R
		18.1	S,S
	16.0	16.0	R,R
		17.5	R,S
$TTT_{PS}T_{PS}T$	17.2	32.1	R,R
		33.3	S,R
		33.6	R,S
		34.5	S,S
$T_{PS}ATA_{PS}A$	15.3	20.9§§	S,R
		22.0	S,S
	16.0	20.2	R,R
		21.3	R,S
$TTT_{PS}TTT$	16.5	14.1**	R
		14.4	S
$GC_{PS}GCGC$	10.9	20.7	R
		21.5	S

[a]HPLC of 5′-DMT and 5′-HO analogs on a μBondapak C_{18} column (30 cm × 7.8 mm) with a linear gradient of acetonitrile in 0.1 *M* TEAA (pH 7.0) at a flow rate of 4 mL/min, starting at acetonitrile-TEAA (20:80) for 5′-DMT (gradient 1%/min), and acetonitrile-TEAA (5:95) for 5′-HO (gradient 0.5%/min), unless specified otherwise. With entries such as $T_{PS}T_{PS}T$, it should be understood that the 5′-DMT component at 21.9 min gave, after ditritylation, a pair of 5′-HO diastereomers at 21.5 and 22.5 min, while the 5′-DMT component at 22.9 min similarly afforded a pair of diastereomers at 20.0 and 22.2 min.

*Gradient 0.33%/min.
**Gradient 1%/min.
***Gradient 1.33%/min.
§Gradient 0.25%/min.
§§Gradient 0.17%/min.

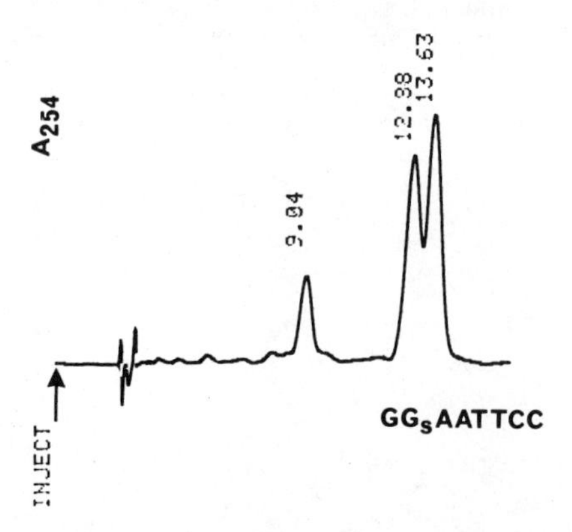

Fig. 35. RP-HPLC separation of d(GGAATTCC) (9.0 min) and the fast- (12.9 min) and slow-eluted (13.6 min) diastereomers of d(GG$_S$AATTCC). The μBondapak C_{18} column (7.8 mm × 30 cm) was eluted with a 0.0625%/min gradient of acetonitrile vs TEAA (0.1 *M*, pH 7.0) starting at acetonitrile-TEAA = 10:90; flow rate = 5 mL/min.

as the 5′-DMT derivatives and then as the fully deprotected 5′-hydroxyl compounds. Unseparable diastereomers of the latter type could not be resolved by SAX-HPLC, at least in the few cases which were studied by the present author. While these findings can be understood in terms of the different mechanisms of separation which are operative in RP- and SAX-HPLC, the reason(s) for the remarkably slower elution that results, in a cumulative fashion, from substitution of P=O with P=S is (are) unknown at this time (181). An example of this effect is seen in Fig. 35; here, replacement of a single P=O with P=S led to a ~4 min increase in the elution time.

The difference in elution times (Δt_R) for diastereomeric mono-phosphorothioate oligonucleotide analogs was found (179) to be greatest when the chiral PS moiety was adjacent to the 5′-DMT group. This relation is exemplified by the series of constitutionally isomeric octanucleotides given in Table 2, wherein the best separation of a pair of diastereomers was obtained with 5′-DMT-d(G$_{PS}$GAATTCC). That this relation is largely independent of chain length was evidenced by the data in Table 3. For example, the 8-mer 5′-DMT-d(G$_{PS}$GAATTCC) has Δt_R = 1.4 min, and the 19-mer 5′-DMT-d(T_{PS}ATA-TACGATATATA) has Δt_R = 1.0 min. The data given in Table 3 also indicated that the values of Δt_R were significantly reduced upon removal of the 5′-DMT group.

7.2. Stereodifferentiation: A Working Hypothesis

Various lines of evidence have suggested that the elution times for 5′-DMT oligonucleotides from a RP-HPLC column are primarily dependent on the hydrophobicity of the DMT group, although there is a marked secondary

Table 2. Separation of Diastereomers of Mono- and Bis-Phosphorothioate (PS) Analogs of the Octamer GGAATTCC By Reversed-Phase HPLC[a]

	Elution Time (min)		Absolute Configuration
Formula, 5′ → 3′	5′-DMT	5′-HO	at Phosphorus
$G_{PS}GAATTCC$	18.4*	18.2	R
	19.8	18.9	S
$GG_{PS}AATTCC$	19.2*	14.7**	R
	21.8	14.7	S
$GGA_{PS}ATTCC$	11.4	17.7	R + S
$GGAA_{PS}TTCC$	11.2	18.2	R + S
$GGAAT_{PS}TCC$	10.9	17.4	R
		17.9	S
$GGAATT_{PS}CC$	11.2	16.9	R
		17.7	S
$GGAATTC_{PS}C$	11.9	17.1	R
		17.6	S
$G_{PS}G_{PS}AATTCC$	13.2	15.6	S, R
	13.6	15.8	S, S
	14.0	15.3	R, R
		15.6	R, S

[a]See Table 1 for details of the standard HPLC conditions for 5′-DMT and corresponding 5′-HO analogs.
*Gradient 0.5%/min, flow-rate 5 ml/min.
**Gradient 1%/min.

influence of the flanking base residue(s). In addition, the chromatographic results (e.g., Tables 2 and 3) obtained for diastereomeric mono-phosphorothioate oligonucleotide analogs have indicated that the best separation of diastereomers (stereodifferentiation) is generally found when the added chiral center with R_p or S_p stereochemistry is located immediately adjacent to the DMT group, and that increasing the distance between the DMT group and the chiral PS moiety leads to loss of stereodifferentiation. One way to rationalize these observations is to assume that the dominant attractive interactions between the hydrophobic 5′-DMT group of an oligonucleotide and the hydrophobic C_{18} moieties of the RP matrix lead not only to longer residence times in the stationary phase but also more "contact" between DMT-proximate groups and the C_{18} layer. This can either increase or decrease the residence time of one diastereomer relative to another depending on the attractive or repulsive nature of the "contact," which must be dependent on stereochemistry. Expressed more succinctly, it can be assumed that nonequivalent, stereochemically dependent interactions between the 5′-DMT end of the oligonucleotide and the stationary phase, as opposed to the mobile phase, leads to the separation of diastereomers of the general structure 5′-DMT-d($N^1_{PS}N^2N^3$ . . .) and 5′-DMT-d($N^1N^2_{PS}N^3$ . . .). The same rationale applies to RP-HPLC stereodifferentiation

Table 3. Separation of Diastereomers of Deca- to Hexadecameric 5′-Terminal Mono-phosphorothioate (PS) Analogs of Oligodeoxyribonucleotides by Reversed-Phase HPLC[a]

Formula, 5′ → 3′	Elution Time (min)		Absolute Configuration at Phosphorus
	5′-DMT	5′-HO	
G_{PS}GGAATTCCC	23.5*	17.7**	R
	24.6	18.1	S
T_{PS}GCATACGAC	15.4***	20.6	S
	16.3	20.7	R
T_{PS}ATATCGATATA	11.2§	23.2	S
	12.2	22.4	R
T_{PS}ATATATCCGATATATA	10.1§	23.2	S
	11.1	22.9	R

[a]See Table 1 for details of the standard HPLC conditions for 5′-DMT and corresponding 5′-HO analogs.
*Gradient 0.33%/min.
**Flow rate 5 mL/min.
***Gradient 0.5%/min for 10 min, then isocratic.
§Gradient 2*/min for 5 min, then isocratic.

of *p*-nitrophenylethyl thiono phosphodiesters (NPE_{PS}) of the general structure $d(NPE_{PS}N_1N_2N_3 \ldots)$.

Extension of this admittedly simplistic analysis to oligonucleotide analogs with hydrophobic substituents attached to phosphorus led to several expectations regarding stereodifferentiation of R_p and S_p diastereomers by RP-HPLC. The first was that hydrophobic backbone substituents that were separated from the 5′-hydroxyl terminus would interact with the RP-matrix, in the absence of competition by the more hydrophobic 5′-DMT group, and thus favor separation of diastereomers. Conversely, such hydrophobic backbone substituents would be relatively unimportant "contact" groups in a DMT-bearing backbone-modified oligonucleotide, thus leading to less effective separation of diastereomers. It was also expected that the location of a given hydrophobic backbone substituent along the oligonucleotide chain would not be especially important with regard to separation of diastereomers, although the relative hydrophobicity of flanking bases in otherwise comparable sequences might be important. The final possibility was that an increase in the hydrophobicity of a given type of backbone substituent could lead to better separation of diastereomers, other things being equal.

These various factors, which are important from the viewpoint of designing new or improved strategies for the synthesis and separation of diastereomers of backbone-modified analogs of DNA, are exemplified and discussed in Sections 7.3–7.7.

7.3. Alkyl Phosphotriesters

The effects of alkylating agents on the structure and function of DNA constitute an important part of the molecular-level mechanisms of carcinogenesis, mutagenesis, and cytotoxicity. Traditional strategies for studying DNA alkylation have employed HPLC to separate and identify alkylated fragments that are obtained by either nuclease or chemical degradation of treated DNA. New constructional strategies are now possible through the stepwise chemical synthesis of specific, site-modified oligonucleotides. Recent examples of synthetic oligonucleotides with altered bases have included the incorporation of O^6-methylguanine (209), N^6-methyladenine (210), and N^3-methylthymine residues (G. Zon, unpublished studies). The R_p and S_p configurations of several nucleoside methyl phosphotriesters, d(N_{OMe}N′), have been recently assigned (211–214); however, there have been fewer stereochemical studies reported for ethyl phosphotriester modified oligonucleotides, d(. . . N_{OEt}N′ . . .) (215, 216).

Ethyl phosphotriester modified oligonucleotides with R_p and S_p stereochemistry have been nonselectively generated by alkylation of poly[d(AT)]·-poly[(AT)] with *N*-ethyl-*n*-nitrosourea, which is a carcinogen and an ethylating reagent that is generally used for ethylation-interference (217) experiments to deduce phosphate contacts in DNA-protein/enzyme recognition. The stepwise synthesis of backbone-ethylated oligonucleotides, by Miller et al. (215) has led to the individual diastereomers of the modified 10-mer, d(CCAAG_{OEt}ATTGG), via diastereomerically pure dimer blocks, d(G_{OEt}A). Although the R_p and S_p configurations were not assigned, these 10-mers were found (215) to exhibit different efficiency as templates for *E. coli* DNA polymerase I.

The desirability of improved synthetic routes to alkyl phosphotriester modified oligonucleotides has prompted recent investigations (218–221) of the phosphite triester method, which employ protected 2′-deoxynucleoside *O*-ethyl- and *O*-isopropyl-*N*,*N*-diisopropylphosphoramidite reagents in an otherwise conventional cycle of chemistry (cf. Fig. 4). The current version of this synthetic procedure affords both R_p and S_p alkyl phosphotriester linkages in the product. Consequently, in cases where it was important to obtain the individual diastereomers, separation was achieved by RP-HPLC. The relative elution times for these diastereomers provide a convenient physical parameter to tentatively characterize the compounds during the assignment of their absolute configuration by means of enzymatic, chemical, and spectroscopic methods. As a representative example, consider the 8-mer shown in Fig. 36. The RP-HPLC trace obtained for the crude DMT-bearing material indicated a 70:30 ratio of 5′-DMT-d(GGAATTCC) : 5′-DMT-d(GGAA_{OEt}TTCC), which were eluted as overlapped peaks at 17.2 and 17.3 min, respectively. The relatively large amount of the unmodified 8-mer in this particular preparation resulted from de-ethylation during ammonolytic deprotection of the base residues. Detritylation of the same crude material followed by re-HPLC also indicated a 70:30 ratio of the 8-mers, d(GGAATTCC) : d(GGAA_{OEt}TTCC); the former compound was eluted at 8.0 min and the desired product was separated as fast- and slow-eluted

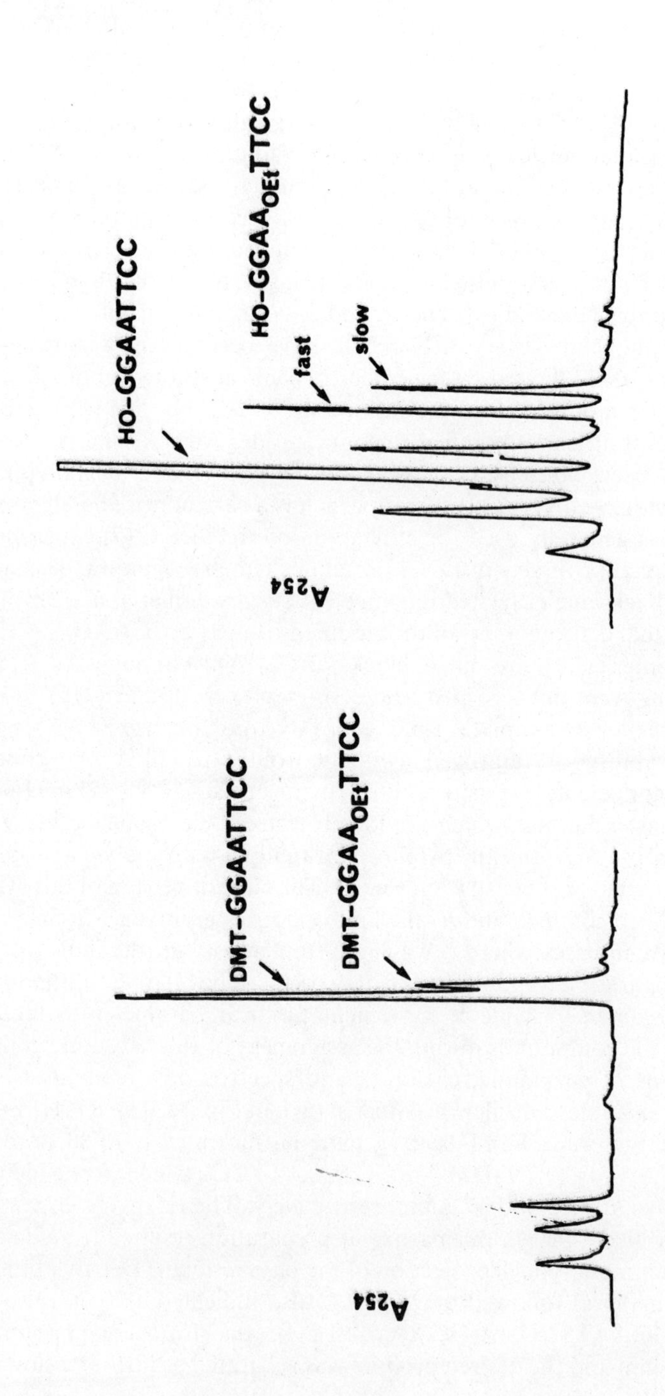
A254
DMT-GGAATTCC
DMT-GGAA$_{OEt}$TTCC
A254
HO-GGAATTCC
HO-GGAA$_{OEt}$TTCC
fast
slow

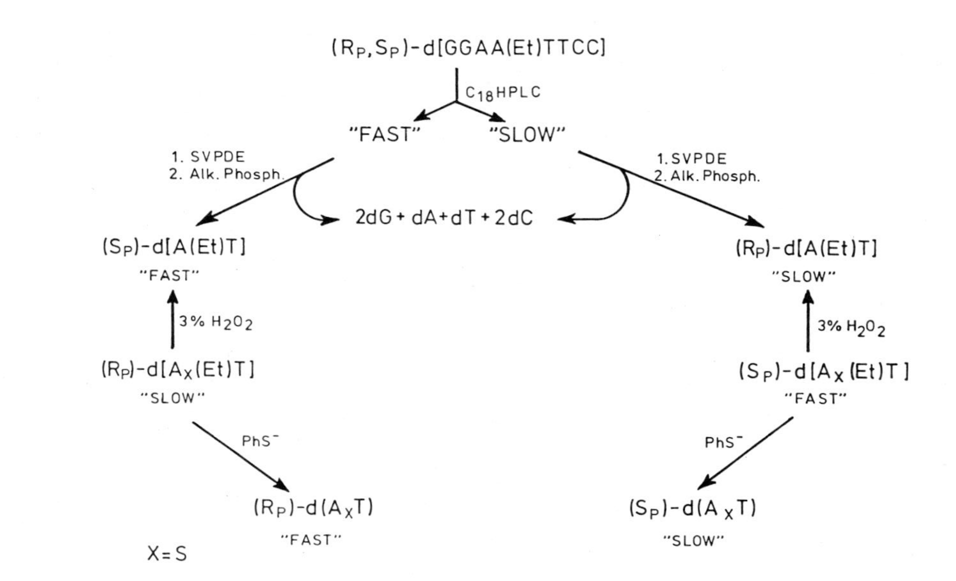

Fig. 36. RP-HPLC separations of the crude, ethyl phosphotriester (OEt) modified 8-mers, 5′-DMT-d($GGAA_{OEt}$TTCC) and d($GGAA_{OEt}$TTCC). Also shown is the scheme used to assign S_p and R_p absolute configurations to the latter, fast-, and slow-eluted, 8-mers, respectively; SVPDE = snake venom phosphodiesterase and Alk. Phosph. = alkaline phosphatase.

diastereomers at 11.5 and 12.7 min, respectively. Resolution of these diastereomers as the 5′-hydroxyl species but not as the 5′-DMT derivatives was consistent with the expectation (see Section 7.2) that stereodifferentiation by RP-HPLC of diastereomers having a hydrophobic backbone-substituent is more effective in the absence of "competition" by the strongly hydrophobic 5′-DMT group.

As indicated in Fig. 36, which summarizes the configurational correlation scheme (218, 221) for ethyl phosphotriester modified oligonucleotides, tandem digestions of the individual fast- and slow-eluted diastereomers of d(GGAA$_{OEt}$TTCC) with SVPDE and alkaline phosphatase followed by RP-HPLC analysis of the resultant mixtures of 2′-deoxynucleosides gave, in each case, 2:1:1:2 molar ratios of dG:dA:dT:dC. In addition, the digest from the fast-eluted 8-mer was found to contain 1 mol-equiv of the undigested dimer fragment, (S_p)-d(A$_{OEt}$T), which eluted at 19.4 min, while the digest from the slow-eluted 8-mer was found to contain 1 mol-equiv of (R_p)-d(A$_{OEt}$T), which eluted at 19.7 min. These results established that the fast- and slow-eluted diastereomers of d(GGAA$_{OEt}$TTCC) had the S_p and R_p absolute configurations, respectively. The configurations at phosphorus in the dimers, d(A$_{OEt}$T), were determined by use of the independently synthesized (*vide infra*) ethyl thiono phosphotriesters, d(A$_{P(S)OEt}$T), which were separated by RP-HPLC and then (1) oxidized with hydrogen peroxide (retention of configuration) to give d(A$_{OEt}$T) and (2) deethylated with thiophenol-triethylamine (retention of configuration) to give d(A$_{PS}$T) of known absolute stereochemistry at phosphorus (203). These new procedures represent improvements compared to the use of *m*-chloroperbenzoic acid as the oxidant and KSeCN as the dealkylating agent (218). An alternative procedure for dealkylation of phosphotriester-modified oligonucleotides having sterically hindered alkyl substituents, which are relatively resistant toward thiophenolate anion or KSeCN, involves hydrolysis in borate buffer (G. Zon, unpublished studies).

The combined enzymatic and chemical procedures discussed above constitute a general method for the assignment of absolute configuration at phosphorus in alkyl-phosphotriester-modified oligonucleotides, the validity of which has been confirmed by two-dimensional nuclear Overhauser effect (2D-NOE) ^{1}H NMR measurements with the individual R_p–R_p and S_p–S_p duplexes of d(GGAA$_{OEt}$TTCC) (M.F. Summers and G. Zon, unpublished studies). Similar analysis of the isopropyl (iPr) analogue required the separation of the diastereomers of d(GGAA$_{OiPr}$TTCC) by RP-HPLC, as shown in Fig. 37. As in the case of the ethyl compound (Fig. 36), there was no separation of the DMT-bearing isopropylated diastereomers, which were therefore cocollected, detritylated, and then separated by RP-HPLC as the 5′-hydroxyl species. The other compounds listed in Table 4 were also separated into pairs of fast- and slow-eluted diastereomers, either at the 5′-DMT stage or as the 5′-hydroxyl species. Octamers with phosphotriester groups either at the 5′-terminal linkage, for example, 5′-DMT-d(G$_{OR}$GAATTCC) R = Et and iPr, or adjacent to the 5′-terminal linkage, for example, 5′-DMT-d(GG$_{OiPr}$AATTCC), were separable

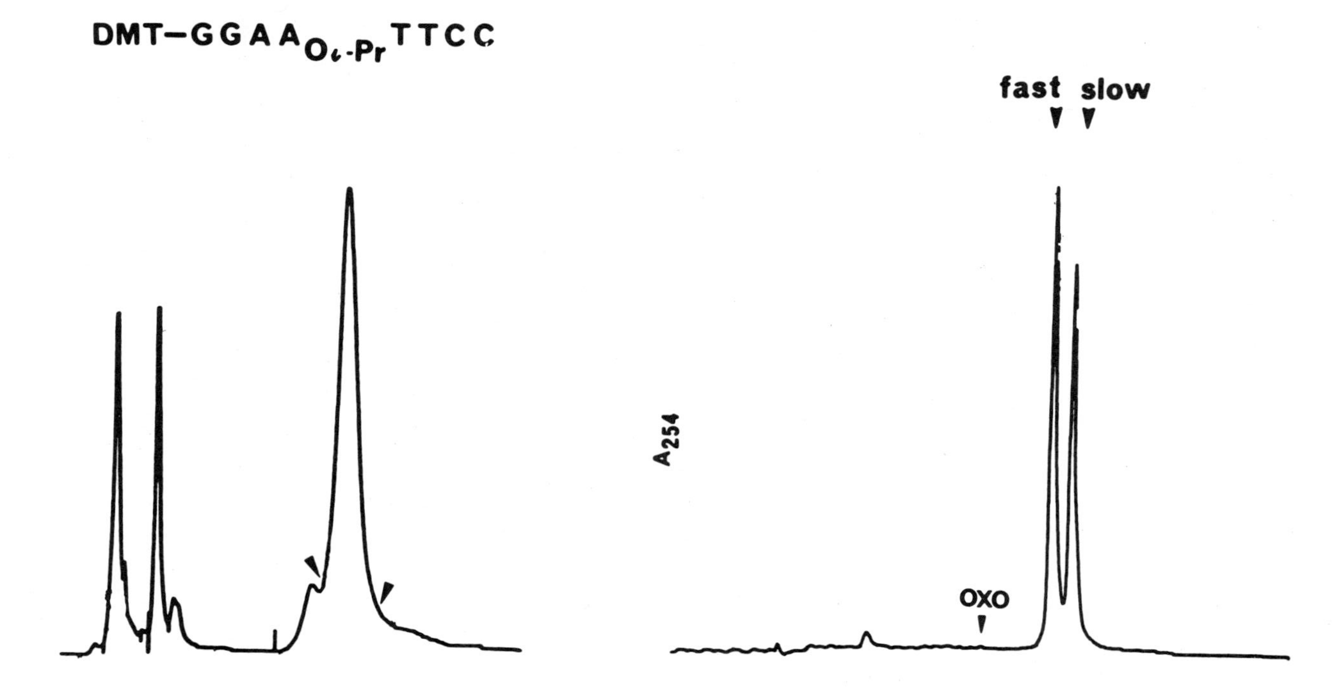

Fig. 37. RP-HPLC isolation of 5′-DMT-d(GGAA$_{OiPr}$TTCC) (7 min) by elution from a μBondapak C_{18} column with a 0.26%/min gradient of acetonitrile vs TEAA (0.1 *M*, pH 7.0) at 5 mL/min starting at acetonitrile-TEAA = 25:75. Ditritylation of the collected material followed by RP-HPLC using the same column and a 1%/min gradient, starting at acetonitrile-TEAA = 5:95, led to collection of the fast- (14.9 min) and slow-eluted (15.7 min) diastereomers; "oxo" refers to the elution time for d(GGAATTCC).

Table 4. Separation of Diastereomers of Ethyl (OEt) and Isopropyl (OiPr) Phosphotriester Analogs of Oligonucleotides by Reversed-Phase HPLC[a]

	Elution Time (min)	
Formula, 5′→ 3′	5′-DMT	5′-HO
$G_{OEt}GAATTCC$	23.5*	10.8
	25.5	10.8
$GGA_{OEt}ATTCC$	17.2**	11.5***
	17.3	12.7
$G_{OiPr}GAATTCC$	16.7	16.0§
	22.2	16.6
$GG_{OiPr}AATTCC$	12.7	14.3§
	13.4	13.9
$GA_{OiPr}ATTCC$	8.4§§	13.1§
		13.7
$GGAA_{OiPr}TTCC$	8.0§§	14.1§
		14.9
$CGCG_{OiPr}CG$	—	13.3§§§
		17.9
$TATAC_{OiPr}ATAT$	14.0	17.7#
		18.7

[a]HPLC on a μBondapak C_{18} column (30 cm × 7.8 mm) with a linear gradient of 1%/min acetonitrile in 0.1 *M* TEAA (pH 7.0) for 10 min at a flow rate of 4 mL/min, starting at acetonitrile-TEAA (20:80), unless specified otherwise.

*μBondapak C_{18} column (30 cm × 4.6 mm), gradient 1.25%/min acetonitrile in 0.1 *M* triethylammonium bicarbonate (TEAB, pH 7.5) at a flow rate of 1.5 mL/min, starting at acetonitrile-TEAB (5:95).

**Gradient 0.5%/min.

***Gradient 0.5%/min, starting at acetonitrile-TEAA (10:90).

§ Initial acetonitrile-TEAA composition is 5:95.

§§ Gradient 0.25%/min, starting at acetonitrile-TEAA (25:75).

§§§ Zorbax ODS column (25 cm × 21.5 mm), gradient 0.167%/min acetonitrile in 0.1 *M* TEAA (pH 7.0) containing 0.5 *M* NaCl at a flow rate of 11.25 mL/min, starting at acetonitrile-TEAA/NaCl (10:90).

Initial acetonitrile-TEAA composition is 5:95, 15 min linear gradient to 12:88 followed by 10 min linear gradient to 13:87.

for presumably the same reasons as those which were discussed in Section 7.2 for the analogous phosphorothioate-containing oligonucleotides. The phosphotriester group in the other DMT derivatives is further removed from the strongly hydrophobic 5′-end, and separation of the diastereomers was either barely detectable or not observed, whereas removal of the DMT group led to comparatively good separations, which was in accord with the working hypothesis for stereodifferentiation by RP-HPLC.

The self-complementary hexamer, d($CGCG_{OiPr}CG$), was interesting in that separation of its diastereomers under normal RP-HPLC conditions was complicated by the elution of multiple species, which were assumed to result from association and, possibly, a conformational equilibrium between B and Z forms. Addition of NaCl (0.5 *M*) to the TEAA eluent led to the separation of configurationally pure forms of fast- and slow-eluted diastereomers, as shown by ^1H NMR analysis and digestion to $G_{OiPr}C$.

The marked improvement of separation upon changing from a relatively slow, linear gradient (0.25%/min) to a tailored, virtually isocratic elution is evident in Fig. 38, which shows chromatograms that were obtained for the diastereomers of the 9-mer, d(TATAC$_{OiPr}$ATAT). Similar improvements in the separation of diastereomers have been found with methanephosphonate analogues of oligonucleotides (see Section 7.5).

A major advantage of the use of the phosphoramidite method to synthesize alkyl phosphotriester modified oligonucleotides is that it affords support-bound phosphite triester linkages, 5′O-P(OR)-O3′, which can be reacted with S_8 in 2,6-lutidine to give final products that have thiono phosphotriester linkages, 5′O-P(S)(OR)-O3′. Compounds of this type not only serve as key "relay" compounds in configurational correlations (cf. Fig. 36), but can also be used to synthesize diastereomerically pure phosphorothioate-containing oligonucleotides that are difficult if not impossible to separate by RP-HPLC. An example of this application concerns the 8-mer, d(GGAA$_{PS}$TTCC), whose diastereomers could not be separated by HPLC (Table 2). By contrast, the diastereomers of d(GGAA$_{P(S)OiPr}$TTCC) were easily separated by RP-HPLC under the same conditions as those listed in Table 4 for d(GGAA$_{OiPr}$TTCC). The individual fast- and slow-eluted diastereomers of d(GGAA$_{P(S)OiPr}$TTCC) were collected and then hydrolyzed in borate buffer (1 *M*, pH 7.4) at 70°C to give the corresponding diastereomers of the desired dealkylated product, d(GGAA$_{PS}$TTCC), in essentially quantitative yield. RP-HPLC was also used to measure the kinetics of the hydrolysis, which obeyed a pseudo-first-order rate law that gave $\tau_{1/2}$ = 40 hr (K.-L. Shao, unpublished studies). Evidence to support the assumption that this hydrolysis, and the hydrolysis of d(. . . N$_{OiPr}$N′ . . .) proceed by an S_N1-like mechanism at carbon comes from RP-HPLC kinetic measurements with d(GGAA$_{OR}$TTCC), R = $CH(CH_3)_2$ and $CD(CD_3($_2$, which gave the expected (222) kinetic isotope effect, k_{7H}/k_{7D} = 1.36. In addition, it was found that $\tau_{1/2}$ = 148 hr for hydrolysis of d(GGAA$_{OEt}$TTCC). The longer half life for the ethyl versus isopropyl derivative was consistent with the instability of $CH_3CH_2^+$ compared to $(CH_3)_2CH^+$, which are intermediates formed by formal hydrolytic ionization of the carbon–oxygen bond in the P(O)–O–C moiety.

Initial attempts by the author to use more hydrophobic phosphotriester groups to improve stereodifferentiation by RP-HPLC have included replacement of $(CH_3)_2CH$ with $Cl_3C(CH_3)_2C$ in the test-sequences d(T$_{OR}$T) and d(CGCG$_{OR}$CG). Unfortunately, the degree of separation of diastereomers was not significantly improved, and the negative inductive effect of the Cl_3C moiety led to marked resistance to hydrolytic dealkylation.

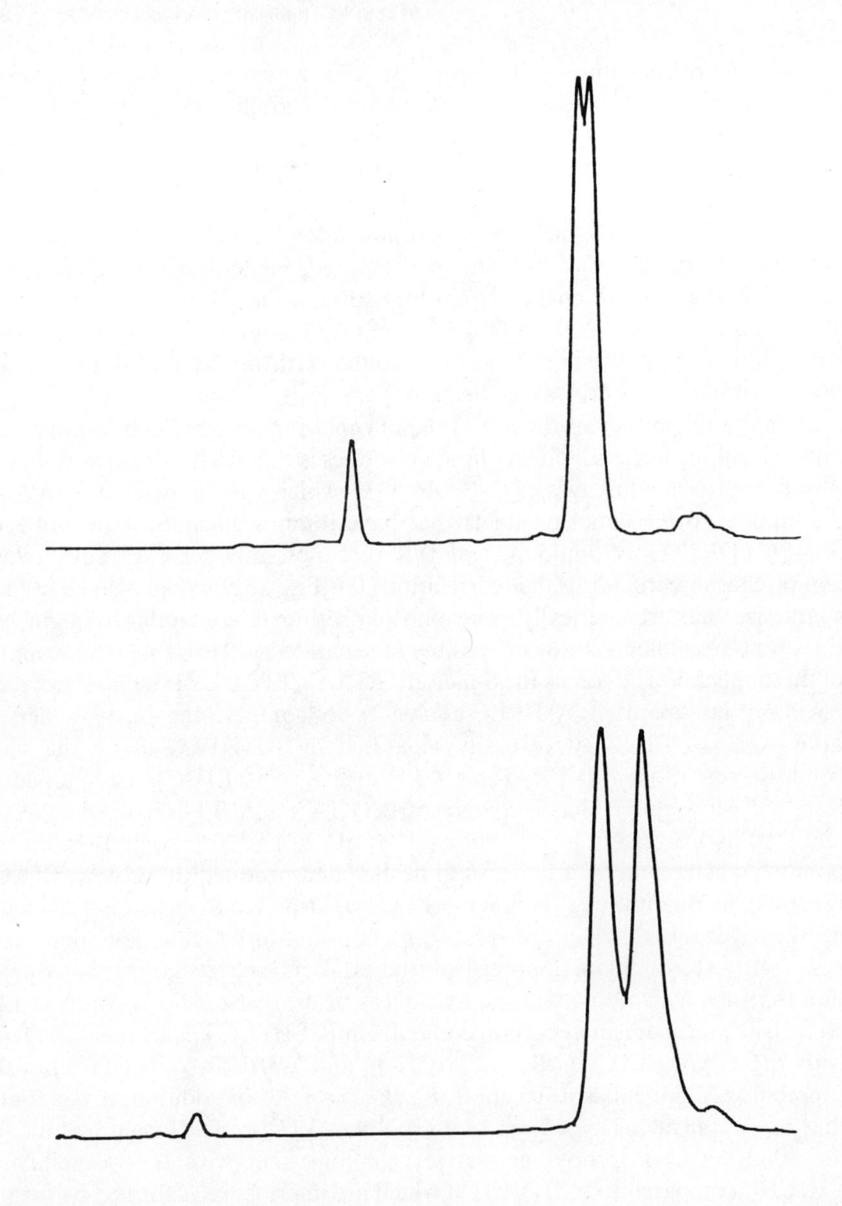

Fig. 38. RP-HPLC separations of diastereomers of the 9-mer, d(TATAC$_{OiPr}$ATAT). In the top trace, the μBondapak C_{18} column (7.8 mm × 30 cm) was eluted with a 0.25%/min gradient of acetonitrile vs TEAA (0.1 *M*, pH 7.0) starting at acetonitrile-TEAA = 10:90; flow rate = 5 mL/min. The diastereomers were eluted at 17.4 and 17.6 min; d(TATACATAT) was eluted at 11.6 min. In the bottom trace, the initial acetonitrile-TEAA composition was 5:95; a 15 min linear gradient to 12:88 was followed by a 10 min linear gradient to 13:87 [the d(TATACATAT) contaminant had been removed from this sample].

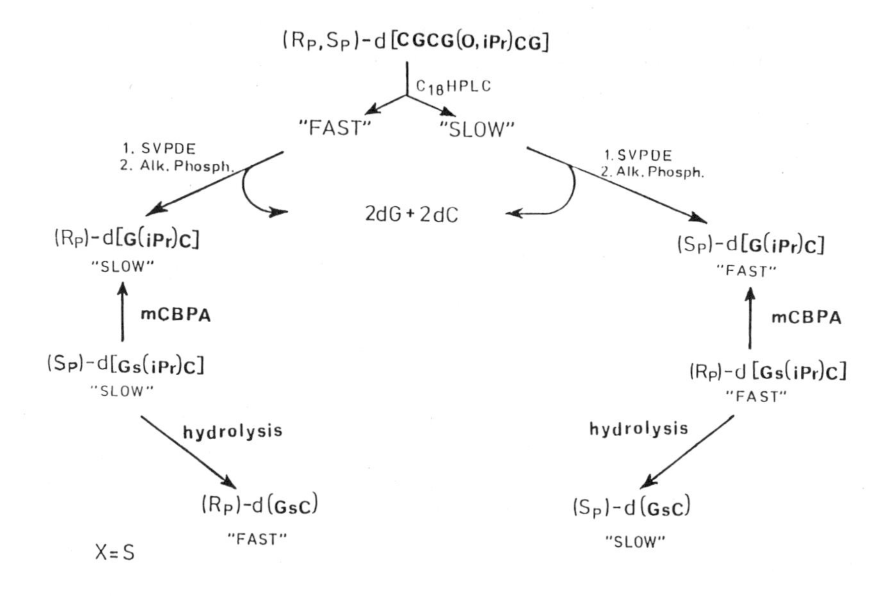

Fig. 39. Scheme used to assign R_p and S_p absolute configurations to the ^{17}O-labeled, isopropylated precursors of the isotopomers, d[CGCG(^{17}O)CG]. SVPDE = snake venom phosphodiesterase, Alk. Phosph. = alkaline phosphatase, mCPBA = *m*-chloroperbenzoic acid, hydrolysis = 1 *M* borate buffer, pH 7.4, 80°C. The R_p and S_p diastereomers of d(G_SC) are known (203).

7.4. Isotopomers

Oligonucleotides that are diastereomeric by virtue of isotopic substitution at an internucleotide linkage, as in 5′O–P(X)O$^-$–O3′ (X = ^{17}O or ^{18}O), can serve as useful probes for stereochemical studies of DNA. While these "isotopomers" can not be separated by currently available techniques, diastereomeric alkyl phosphotriester and thiono phosphotriester modified oligonucleotides can be fractionated by RP-HPLC, and may serve as convenient precursors of isotopomeric DNA molecules. Moreover, the absolute configuration of these triester groups can be unambiguously assigned by either chemoenzymatic or NOE methods, which in turn leads to the assignment of absolute configuration at phosphorus in isotopomers derived from these phosphotriesters.

The scheme in Fig. 39 summarizes the procedures that were used to obtain the individual R_p and S_p ^{17}O-labeled isotopomers, d[CGCG(^{17}O)CG], and to assign their absolute configurations at phosphorus. A mixture of the R_p and S_p diastereomers of the ^{17}O-labeled isopropyl phosphotriesters, d[CGCG(^{17}O, iPr)CG], was synthesized by employing I_2 and [^{17}O]–H_2O as the oxidant in the coupling cycle with an *O*-isopropyl G-phosphoramidite reagent. The mixture was separated into fast- and slow-eluted diastereomers by RP-HPLC (Table 4), and the absolute configuration of each diastereomer was assigned by correlation with the known R_p and S_p dinucleoside phosphorothioates, d(G_sC) (203). Hydrolytic removal of the isopropyl group with retention of configuration at phosphorus converts each isopropylated octamer into its corresponding isotopomer.

Uznanski and co-workers (220) have similarly synthesized and separated the fast- and slow-eluted, ^{18}O-labeled ethyl phosphotriesters, d[G(^{18}O, Et)GAATTCC], which were then deethylated with thiophenol-triethylamine in dioxane (50°C, 6 hr) to give the corresponding isotopomers, d[G(^{18}O)GAATTCC]. Experiments directed at the assignment of the absolute configurations of the ethyl phosphotriester precursors were reported to be in progress.

7.5. Methanephosphonates

Methanephosphonate-modified analogs of DNA have internucleotide linkages of the type 5′O–P(O)CH_3–O3′, which are not too dissimilar to phosphodiester linkages but are nonionic and are centers of chirality. Members of this class of compounds, the simplest of the alkanephosphonate modified oligonucleotides, have been used to study DNA structure (223), interactions between proteins and nucleic acids (201), and specific inhibition of mRNA translation (206, 224–226). Mono-methanephosphonates are considered in this section, while polymethanephosphonates are discussed in Section 7.6.

As with other types of backbone modifications, such as phosphorothioate and phosphotriester linkages, methanephosphonate linkages with a given stereochemistry at phosphorus have been incorporated into synthetic oligonucleotides by the use of diastereomerically pure dimer-blocks that are appropriately functionalized and protected (223). More recently, Stec et al. (183) investigated the possibility of introducing methanephosphonate linkages without control of stereochemistry, by use of Arbusov-type reactions with electrophilic halides (R′X, Fig. 40) during synthesis by the phosphite triester method. Several examples of methanephosphonate (Me) modified dimers and decamers prepared in this manner are given in Table 5; however, there were serious complications due to side reactions. One of these side reactions was identified as detritylation followed by addition of the DMT group to the phosphite triester to afford a 4,4′-dimethoxytriphenylmethanephosphonate linkage, for example, d(T_{DMT}T). On the other hand, by use of DMTCl as the electrophilic halide, it was possible to synthesize various oligonucleotides (Table 5) that have this rather unusual DMT-phosphonate moiety. All of the compounds listed in Table 5 could be separated into fast- and slow-eluted diastereomers. Of particular interest were the separations for d($TTTTT_{Me}TTTTT$), and d($TTTTTTTTT_{Me}T$), which were achieved with the 5′-hydroxyl species but not with the 5′-DMT derivatives. These observations were consistent with working hypothesis for stereodifferentiation of diastereomers during RP-HPLC (see Section 7.2).

In view of the limited utility of the aforementioned Arbusov-type reaction, the present author has recently begun an investigation of the suitability of protected 2′-deoxynucleoside methylphosphonamidites (227) as reagents for the

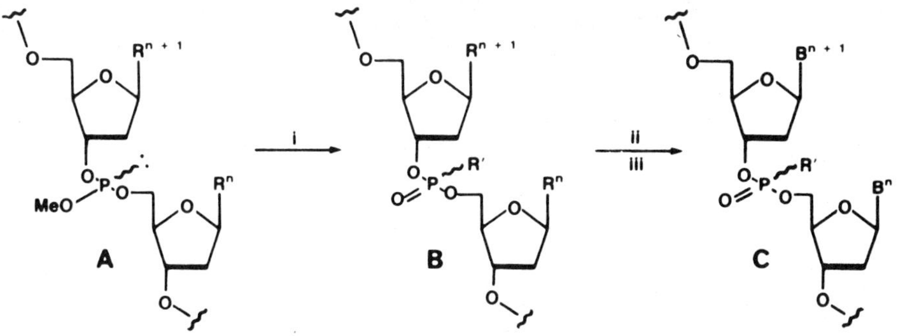

R = A^{Bz}, G^{iBu}, C^{Bz}, T

(i) R′X–CH_3CN
(ii) PhSH–Et_3N
(iii) conc. NH_4OH–heat

B = A, G, C, T

Fig. 40. Generalized scheme for an Arbusov-type reaction (i) of an electrophilic halide (R′X) with a support-bound phosphite (*A*) to give a support-bound phosphonate intermediate (*B*), which would have a phosphonate modified oligonucleotide upon demethylation of the backbone with (ii) and then cleavage from support and base deprotection with (iii).

incorporation of methanephosphonate linkages during otherwise conventional synthesis with either *O*-methyl- or *O*-β-cyanoethyl-*N*,*N*-diisopropylphosphoramidites. Initial results (J.W. Efcavitch, C. McCollum, and G. Zon, unpublished studies) indicated that methylphosphonamidite reagents gave good coupling yields ($\geqslant 96\%$) when used with an automated DNA synthesizer for 1-μmol scale preparations up to 21-mers. Cleavage from support and backbone/base-deprotection were carried out with either ethylenediamine–ethanol (228) or ammonia–methanol under mild, anhydrous conditions to minimize chain cleavage. Examples of some of the methanephosphonate modified oligonucleotides which were thus prepared for stereochemical studies are listed below:

d(A_{Me}T), d(T_{Me}T), d(T_{Me}A)
d(GGAA_{Me}TTCC)
d(GGAAT_{Me}TCC)
d(CCTT_{Me}AAGG)

The dimer, d(A_{Me}T), had been prepared earlier by two other synthetic methods (183, 229), and the absolute configuration at phosphorus in the fast-eluted (RP-HPLC) diastereomer has been established as R_p by X-ray crystallography (183, 230). The RP-HPLC trace obtained in the purification and separation of the diastereomers of the 8-mer, d(GGAA_{Me}TTCC), is shown in Fig. 41. Similar separation was obtained with the constitutional isomer, d(GGAAT_{Me}TCC); however, separation of the diastereomers of the reversed

Table 5. Separation of Diastereomers of Methyl (Me) and 4,4-Dimethyltrityl (DMT) Phosphonate Analogs of Oligodeoxyribonucleotides[a]

	Elution Time (min)		Absolute Configuration		Elution Time (min)		Absolute Configuration
Formula, 5′ → 3	5′-DMT	5′-HO	at Phosphorus	Formula 5′ → 3′	5′-DMT	5′-HO	at Phosphorus
$T_{Me}T$	—	14.1	R (?)	$T_{DMT}C$	—	16.1**	—
		14.8	S (?)			16.6	—
$A_{Me}T$	—	15.0	R	$C_{DMT}C$	—	14.4**	—
		16.2	S			15.0	—
$TTTTT_{Me}TTTTT$	23.3*	4.5	—	$A_{DMT}A$	—	17.1**	—
		4.8	—			18.0	—
$TTTTTTTTT_{Me}T$	24.2*	15.4	R	$GG_{DMT}AATTCC$	—	15.2***	—
		15.6	S			15.8	—
$T_{DMT}T$	—	18.8**	—	$GGAA_{DMT}TTCC$	—	15.1***	—
		20.0				16.5	

[a]HPLC on a μBondapak C_{18} column (30 cm × 7.8 mm) with a linear gradient of 1%/min acetonitrile in 0.1 *M* TEAA (pH 7.0) at a flow rate of 4 mL/min, starting at acetonitrile-TEAA (5:95), unless specified otherwise.

*Gradient 0.5%/min, flow rate, 3 mL/min, starting at acetonitrile-TEAA (20:80).

**Gradient 1%/min for 10 min, then isocratic, starting at acetonitrile-TEAA (30:70).

***Gradient 2%/min for 10 min, followed by gradient 1%/min, starting at acetonitrile-TEAA (5:95).

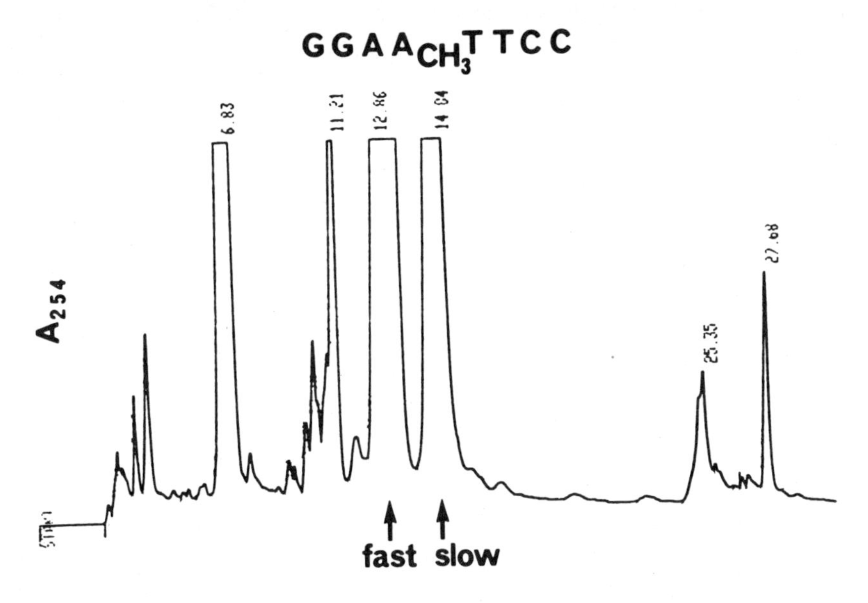

Fig. 41. RP-HPLC separation of the crude, methanephosphonate (Me) modified 8-mer d(GGAA$_{Me}$TTCC). The μBondapak C_{18} column (7.8 mm × 30 cm) was eluted with acetonitrile-TEAA (0.1 *M*, pH 7.0) as follows: initial composition acetonitrile-TEAA = 5:95, a linear gradient for 7 min to 12:88, isocratic up to 18 min, a linear gradient for 2 min to 30:70, and then isocratic up to 30 min. The peaks for the fast- and slow-eluted diastereomers are at 12.9 and 14.8 min, respectively.

sequence, d(CCTT$_{Me}$AAGG), was much more difficult, and required small-scale injections (to avoid overloading) and isocratic elution from the μBondapak C_{18} column at acetonitrile-TEAA (0.1 *M*, pH 7.0) = 10:90. Since the degree of separation of the diastereomers of d(A$_{Me}$T) and d(T$_{Me}$A) by C_{18} HPLC are roughly comparable, the markedly different behavior found for d(GGAA$_{Me}$TTCC) and d(CCTT$_{Me}$AAGG) suggests that the nature of the base residues that flank the hydrophobic methanephosphonate moiety can have a significant influence on whether the separation of diastereomers will be relatively easy or difficult.

The structure and stereochemical purity of methanephosphonate-modified oligonucleotides can be checked by tandem digestions with SVPDE and alkaline phosphatase to give 2′-deoxynucleosides and the undigested dinucleoside methanephosphonate(s), which is analogous to the procedure described for phosphotriester modifications (Figs. 36 and 39), with the exception that there are no chemical or spectroscopic methods for assigning the absolute configuration at phosphorus in dinucleoside methanephosphonates and, hence, in precursor oligonucleotides of interest. One example of the potential utility of such methods concerns the diastereomerically pure mono-methanephosphonate analogs of segments of the *lac* operator that were recently synthesized (201) by the use of diastereomerically pure, appropriately protected and functionalized $T_{Me}T$. Experiments (201) with these segments provided evidence for operator–repressor contact at a specific internucleotide phosphate group. The affinity of *lac* repressor for the modified operators was shown to be dependent on the relative stereochemistry of the methanephosphonate linkage, but the absolute configuration at phosphorus at the presumed internucleotide contact was not determined. To solve this stereochemical problem, and at the same time develop general methods for assignment of chirality at methanephosphonate modifications in DNA, various strategies are being tested using compounds such as the 8-mer, d(GGAAT_{Me}TCC) (M.F. Summers, W.D. Wilson, and G. Zon, unpublished studies). Tandem digestions of the individual fast- and slow-eluted (RP-HPLC) diastereomers with SVPDE and alkaline phosphatase gave the expected ratios of 2′-deoxynucleosides and the fast- (16.4 min) and slow-eluted (17.2 min) diastereomers of the undigested dimer, d(T_{Me}T), respectively, as determined by RP-HPLC under the conditions given in Fig. 19. Since these diastereomers can be enzymatically exercised from the $T_{Me}T$-containing segments of the *lac* operator in question, assignment of the absolute configuration at phosphorus in d(GGAAT_{Me}TCC) completes the stereochemical correlation. This assignment will be made by NOE measurements for the P–CH_3 protons and the upstream H_3' proton in the B-form duplex, [d(GGAAT_{Me}TCC)]$_2$, as the distance (NOE) between these protons is shorter (stronger) when phosphorus has the *S* configuration.

The previously mentioned relation of distance, NOE, and absolute stereochemistry should be general for methanephosphonate-modified B-helices, regardless of sequence, and has been recently confirmed in the case of [d(GGAA_{Me}TTCC]$_2$. The HPLC-separated (Fig. 41) fast- and slow-eluted diastereomers of d(GGAA_{Me}TTCC) were individually digested with SVPDE and alkaline phosphatase to afford the fast- (19.2 min) and slow-eluted (20.4 min) diastereomers of d(A_{Me}T), respectively, by use of RP-HPLC as described in Fig. 19. It has been shown (183) previously that fast- and slow-eluted d(A_{Me}T) have the R_p and S_p configurations, by RP-HPLC comparisons with authentic material studied by X-ray crystallography (230); consequently, fast- and slow-eluted d(GGAA_{Me}TTCC) must have R_p and S_p stereochemistry. This finding was in accord with the NOE cross-peak observed for P–CH_3 and the upstream H_3' in the $S_p \cdot S_p$ but not the $R_p \cdot R_p$ duplex, [d(GGAA_{Me}TTCC)]$_2$.

This approach to the assignment of absolute configuration at phosphorus in any enzymatically excised dimer, d($N^1_{Me}N^2$), can in principle be generalized by the synthesis and separation of the diastereomers of, for example, d(GG$N^1_{Me}N^2N^3N^4$CC), wherein the basepairs N^1–N^4 and N^2–N^3 are each complementary, which leads to diastereomeric duplexes for configurational assignments of NOE measurements.

7.6. Poly-Methanephosphonates

Miller and co-workers have shown that oligodeoxyribonucleoside methanephosphonates ("poly-methanephosphonates"), which are nonionic nucleic acid analogs, are resistant to hydrolysis by nucleases and are able to penetrate the membranes of living cells (206, 228–230). These properties and the ability of poly-methanephosphonates to form more stable hydrogen-bonded complexes with complementary polynucleotides (206, 223) enable one to selectively inhibit the function of cellular nucleic acids. For example, Miller and co-workers have recently reported (226) the effects of 4- to 12-unit poly-methanephosphonate analogs on translation of rabbit globin mRNA in rabbit reticulocyte lysates and in rabbit reticulocytes. That these compounds also represent a novel class of potential chemotherapeutic agents (206) has been demonstrated (P. Miller, unpublished studies) with the analogue, $d(T_pC_{Me}C_{Me}T_{Me}C_{Me}T_{Me}G)$, which is complementary to the acceptor splice junction of immediate early pre-mRNA 4 and 5, and caused a dose-dependent inhibition of herpes simplex virus (HSV-1): 99% reduction of virus titers in cells treated with 150–300 μM concentrations. Compounds of this sort, including P-aryl phosphonates, have been patented (231), as have P–C_{1-10} alkyl versions (232). In this connection, it is worthwhile to mention that "intercalating agents specifying nucleotides" have also been recently patented (233). Phenanthridinium compounds having P–NHR linkages to dimers of dT were prepared (233) as site-specific inhibitors for enzymatic processes involving polynucleotides, as reagents for selective cleavage or modification of polynucleotide chains, and as agents for introducing markers, for example, fluorescent moieties (204), at specific regions in polynucleotides. Thus, one of these prototypes was shown to bind strongly to poly(A) at 0°C and weakly or not at all to poly(G), poly(C), poly(U), and poly(I).

Two methods have been previously employed (228) to purify poly-methanephosphonate analogs of DNA derived from condensation of protected 2′-deoxynucleoside 3′-methanephosphonate triethylammonium salts, depending on the structure. For oligomers which contain only methanephosphonate linkages, the 5′-DMT derivative was isolated by C_{18} RP-HPLC: shorter, nontritylated species were first eluted with a gradient of 0–25% (or 30%) acetonitrile in water gradient, and the DMT-bearing material was then eluted with a step gradient of 50% acetonitrile in water. The latter material was detritylated with 80% acetic acid at room temperature for subsequent RP-HPLC, which led to separation of the desired product from shorter oligomers that had coeluted with the product at the DMT-stage. Recoveries from the C_{18} column were found to vary from 50–90%, and it was suggested this could have been due to irreversible adsorption of the oligomers to the column, depending on the base composition of the oligomer.

The second method for purification reported by Miller and co-workers (228) was applicable to poly-methanephosphonate oligomers that have a normal, negatively charged phosphodiester linkage, which was necessary for ^{32}P 5′-end labeling with T4-polynucleoside kinase. RP-HPLC of the crude synthetic mixture containing the 9-mer, 5′-DMT-d($A_pA_{Me}A_{Me}A_{Me}G_{Me}C_{Me}A_{Me}A_{Me}G$), led to elution of the major peak at 22 min in 50% acetonitrile–water, whereas the detritylated crude material showed the corresponding major peak at 15 min. The detritylated crude material was partially purified by ion exchange on a DEAE cellulose column, and the product fraction was further purified by RP-HPLC using a gradient of 0–30% acetonitrile in 0.1 *M* ammonium acetate. The resultant product fraction was finally desalted on a Bio-Gel P-2 gel filtration column.

Recent investigations (K. Shinozuka and G. Zon, unpublished studies) of poly-methanephosphonate modified oligonucleotides derived from methylphosphonamidite reagents have also utilized monovalent sequences of the general type d($N_pN'_{Me}N''_{Me}$ . . .), primarily for reasons of better solubility in biological media. However, the method of purification is based solely on RP-HPLC. The product is first isolated from the crude synthetic mixture as its 5′-DMT derivative by elution from either a C_{18} or PRP-1 column with a gradient of acetonitrile in water. The 5′-DMT derivative thus obtained is then detritylated for either use "as is" or further purification by RP-HPLC to remove shorter contaminating sequences, which can be detected by PAGE (225) or RP-HPLC or both. This methodology for purification of poly-methanephosphate analogs of DNA is relatively convenient, as the elutions are performed with acetonitrile–water, and there are no anion-exchange and desalting steps.

As a typical example of the conditions of HPLC and the results obtained, elution of the crude 19-mer, 5′-DMT-d[$T_p(T_{Me})_{17}T$], from a "semipreparative" PRP-1 column with a gradient of acetonitrile in water led to collection of the major peak seen at 24.8 min in Fig. 42. Ditritylation of the eluate followed by analysis under the same HPLC conditions showed a major peak at 13.2 min for the product, d[$T_p(T_{Me})_{17}T$], and a series of minor fast-eluted peaks at 6–12 min due to shorter oligomers, which had been co-collected with the major DMT-bearing peak, and were also evident in autoradiograms after PAGE.

Based on what has been said in Section 2.1.3, it was not surprising to find that the elution times for these charge-neutralized oligomers were influenced by base composition (no ion-pairing mechanism possible). For example, the 19-mers of d[$A_p(A_{Me})_{17}A$] with 5′-DMT and 5′-hydroxyl groups were eluted at 19.4 and 11.7 min, respectively, under conditions that were essentially identical to those described in Fig. 42 for the corresponding, more retained, dT-containing oligomers ($\Delta t_R = 5.4$ and 1.5 min, respectively). This relation of elution time and base composition was also evident from comparative HPLC data obtained for the T-rich 12-mer, 5′-DMT-d($T_pT_{Me}T_{Me}T_{Me}T_{Me}T_{Me}C_{Me}T_{Me}C_{Me}C_{Me}A_{Me}T$), which was eluted at 21.1 min, and the A-rich 12-mer, 5′-DMT-d($A_pA_{Me}A_{Me}$-$G_{Me}A_{Me}T_{Me}G_{Me}T_{Me}A_{Me}T_{Me}T_{Me}C$), which was eluted at 15.2 min (Fig. 42, $\Delta t_{sR} = 5.9$ min). Both of these DMT-bearing 12-mers were collected, detritylated, ^{32}P-labeled, and then shown by autoradiography after PAGE to be largely free of shorter contaminating oligomers.

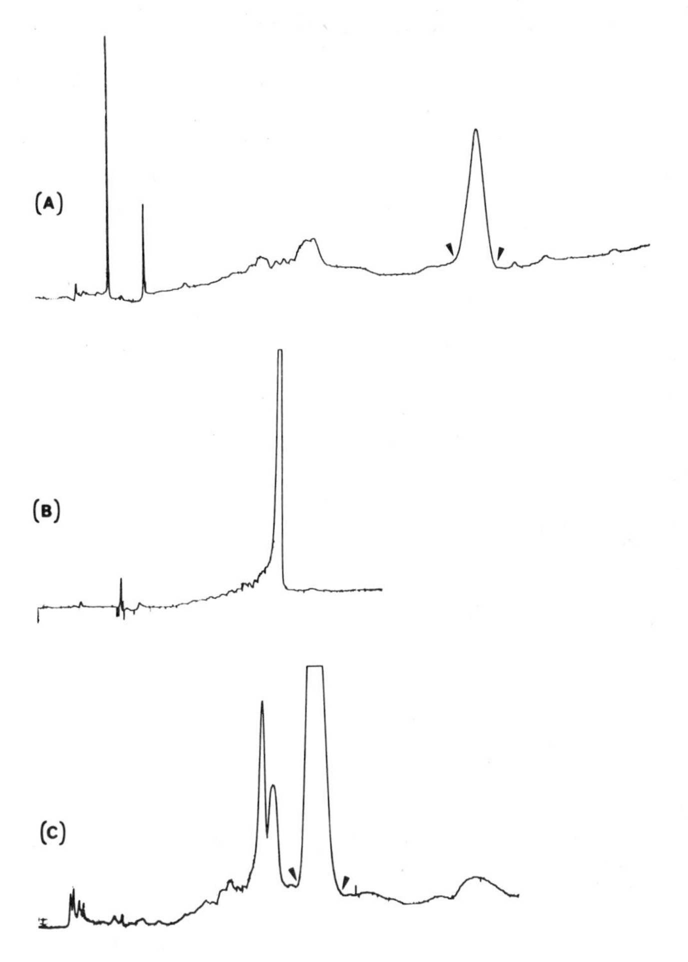

Fig. 42. RP-HPLC separations of poly-methanephosphonate (Me) modified oligonucleotides having a 5′-terminal phosphodiester linkage (p). In each case, the PRP-1 column (7 mm × 25 cm) was eluted at 3 mL/min with a gradient of acetonitrile versus water. The times and acetonitrile–water ratios were as follows: 0 min, 10:90; 7.5 min, 25:75; 20 min, 27:73; 30 min, 48:52. (A) The crude 19-mer, 5′-DMT-d[$T_p(T_{Me})_{17}T$], 24.8 min. (B) The 19-mer, d[$T_p(T_{Me})_{17}T$], 13.2 min, derived from collection and then detritylation of the "cut" indicated by the arrowheads in (A). (C) The crude 12-mer, 5′-DMT-d($A_pA_{Me}A_{Me}G_{Me}A_{Me}T_{Me}G_{Me}T_{Me}A_{Me}T_{Me}T_{Me}C$), 15.2 min; the arrowheads indicate the "cut" which was taken.

7.7. Absolute Configuration by HPLC: A Caveat

For a series of structurally related, modified dinucleoside phosphates, d(N_XN'), one might expect to find a correlation between relative elution time of diastereomers from a RP-HPLC column and stereochemistry at phosphorus. The series of phosphorothioate-containing compounds (X = PS) listed in Table 1 has provided evidence for such a correlation: in each case the R_p diastereomer was eluted from a μBondapak C_{18} column faster than the S_p diastereomer (similar results were obtained from a PRP-1 column). The same type of correlation may hold for ethyl phosphotriesters, X = OEt, as it has been noted (215) that diastereomers with the S_p configuration were eluted faster than their R_p counterparts upon RP-HPLC with an ODS-2 column in all of the cases studied. On the other hand, (R_p)- and (S_p)-d(T_{OiPr}T) appear as fast- and slow-eluted peaks, respectively, on a μBondapak C_{18} column (216), while the diastereomers of d(G_{OiPr}C) show the *opposite* pattern of elution (cf. Fig. 39) under HPLC conditions that were identical to those used to study d(T_{OiPr}T). Since the methodology for assignment of the absolute configurations of the diastereomers of d(N_{OiPr}N′) is firmly established (216, 220, 221), the HPLC characteristics for the diastereomers of d(T_{OiPr}T) versus d(G_{OiPr}C) represent a caveat for the use of empirical correlations between relative elution time from a RP-HPLC column and stereochemistry at phosphorus as a basis for assigning unknown absolute configurations. On the other hand, diastereomers with known absolute configurations can be reliably identified on the basis of their elution pattern under a specified set of HPLC conditions.

ACKNOWLEDGEMENTS

I thank Bill Hancock, who suggested this account, and I am very grateful to my collaborators and co-workers whose contributions have been referenced or discussed herein. Wojciech J. Stec deserves special thanks for his indefatigable pursuit of ever more challenging separations of diastereomers by HPLC, and his continued conceptual contributions to our collaborative work. The editorial assistance of Ellen Kirshbaum is also appreciated.

ADDENDUM

An investigation of possible donor–acceptor pi-pi electron interactions leading to "charge-transfer HPLC" of oligonucleotides has been reported for a multifunctional stationary phase made by covalent bonding of 3,6-diamino-10-methylacridinium ion to epoxysilylated silica (234).

Several different approaches to HPLC of oligonucleotides under strongly alkaline denaturing conditions have been described. In one procedure, concentrated NH_4OH solutions of oligomers up to 160 bases in length were first purified on a polymeric ion-exchange resin in the presence of 0.01 *M* NaOH as the denaturant, and the collected fractions were then desalted, detritylated (80% acetic acid), and further purified under the same HPLC conditions (235). In another procedure, 0.05 *M* NaOH was used as the denaturant for purification of 5′-DMT product by reversed-phase (PRP-1) HPLC (236). Alternatively, a mixture of poly(tetrafluoro ethylene) and DEAE-cellulose can reportedly be used to absorb 5′-DMT product directly from the concentrated NH_4OH solution of crude material (237). This can be followed by on-column detritylation (3% dichloroacetic acid in CH_2Cl_2) to further minimize the time required for product purification (237).

The oligonucleotide purification cartridge (OPC™, Applied Biosystems) was recently introduced for rapid, high-quality purification of synthetic oligonucleotides within 10–15 min, manually (238). The OPC™ method has a number of desirable features: (1) direct loading of the concentrated NH_4OH solution of crude synthetic DNA; (2) specific binding of 5′-DMT material, and (3) elution of 5′-HO failure sequences in the presence of NH_4OH to maintain denaturing conditions, which is not possible with silica-based (e.g., C_{18} Sep-Pak™, Waters) cartridges; (4) on-column detritylation (2% trifluoroacetic acid) and (5) elution of the 5′-HO product in a volatile buffer, which eliminates the need for a final desalting step; (6) retention of the DMT alcohol by-product on the resin. The capacity of OPC™ and the use of denaturing conditions are significant advantages relative to Nensorb™ (DuPont), as are the use of OPC™ for desalting 5′-HO oligonucleotides, which have been isolated by either ion-exchange HPLC or PAGE, and very rapidly purifying 5′-labeled fluorescent primers for automated sequencing of DNA by the Sanger method (239). It should be emphasized that the degree of purification obtained with any of the previously mentioned cartridges is largely determined by the homogeneity of the crude synthetic DNA with regard to DMT-bearing species, since the step gradients and particle-size distributions cannot afford additional resolution of the type achieved by HPLC.

A five-part series of articles entitled "A Review of HPLC in Nucleic Acids Research" has been published. Part I (240) focused on the historical development of chromatography supports for liquid chromatography applications in nucleic acids research, and covered parallel improvements in silica- and polymer-based HPLC resins, and in chemical coupling techniques. In Parts II (241), III (242), and IV (243) of this series, the use of these new HPLC resins for the isolation, purification and analysis of oligodeoxyribonucleotides, supercoiled plasmid DNA, and DNA restriction fragments was described, featuring the PICS (paired-ion chromatography systems) resin that has recently been commercialized (Applied Biosystems) as a HPLC column. Part V (244) describes the use of chromatography supports and purified nucleic acids for creating solid supports for DNA affinity chromatography.

By comparison with typical base composition analyses of oligodeoxyribonucleotides (245) that use HPLC, a significantly improved HPLC system for this enzymatic method has been reported. Experimental data for oligomers which range from 18 to 140 bases in length were found to be in excellent agreement with theoretical compositions, and the method is applicable to high-molecular-weight genomic DNA (246). This method has also been used to identify and quantify the extent of undesirable dT (247) and dG (248) modifications that can possibly occur during chemical synthesis of oligodeoxyribonucleotides.

REFERENCES

1. P.J. Abelson, Science, *219*, 611–613 (1983).
2. K. Itakura and A.D. Riggs, Science, *209*, 1401–1405 (1980).
3. M. Smith and S. Gillam, "Constructed Mutants Using Synthetic Oligodeoxyribonucleotides as Site-Specific Mutagens," in *Genetic Engineering Principles and Methods*, Vol. 3, J.K. Setlow and A. Hollaender, Eds., Plenum Press, New York, 1981, pp. 1–32.
4. M.D. Edge and A.F. Markham, "Applications of Oligonucleotide Synthesis to Interferon Research," in Biochim. Biophys. Acta, *695*, 35–48 (1982).
5. R.B. Wallace and K. Itakura, "Solid Phase Synthesis and Biological Applications of Polydeoxyribonucleotides," in W.H. Scouten, Ed., Wiley, New York, 1983, pp. 631–663.
6. R.B. Wallace and K. Itakura, "Solid-Phase Synthesis of Polydeoxyribonucleotides for Biological Applications," in *Nucleic Acid Research Future Development*, K. Mizobuchi, I. Watanabe, and J.D. Watson, Eds., Academic Press, New York, 1983, pp. 227–245.
7. M. Smith, "Synthetic Oligodeoxyribonucleotides as Probes for Nucleic Acids and as Primers in Sequence Determination," in *Methods of DNA and RNA Sequencing*, S.M. Weissman, Ed., Praeger Publishers, New York, 1983, pp. 23–68.
8. K. Itakura, J.J. Rossi, and R.B. Wallace, "Synthesis and Use of Synthetic Oligonucleotides," Ann. Rev. Biochem., *53*, 323–356 (1984).
9. C.S. Craik, BioTechniques, *3*, 12–19 (1985).
10. R.B. Wallace, J. Shaffer, R.F. Murphy, J. Bonner, T. Hirose, and K. Itakura, Nucleic Acids Res., *6*, 3543–3557 (1979).
11. J.W. Szostak, J.I. Stiles, B.-K. Tye, P. Chiu, F. Sherman, and R. Wu, Methods Enzymol., *68*, 419–428 (1979).
12. K. Itakura, T. Miyake, E.H. Kawashima, Y. Ike, H. Ito, C. Morin, A.A. Reyes, M.J. Johnson, M. Schold, and R.B. Wallace, "Chemical Synthesis and Application of Oligonucleotides of Mixed Sequence," in *Recombinant DNA*, A.G. Walton, Ed., Elsevier, New York, 1981, pp. 273–289.
13. K.L. Agarwal, J. Brunstedt, and B.E. Noyes, J. Biol. Chem., *256*, 1023–1028 (1981).
14. M. Jaye, H. de la Salle, F. Schamber, A. Balland, V. Kohli, A. Findeli, P. Tolstoshev, and J.-P. Lecocq, Nucleic Acids Res., *11*, 2325–2335 (1983).
15. E. Ohtsuka, S. Matsuki, M. Ikehara, Y. Takahashi, and K. Matsubara, J. Biol. Chem., *260*, 2605–2608 (1985).
16. Y. Takahashi, K. Kato, Y. Hayashizaki, T. Wakabayashi, E. Ohtsuka, S. Matsuki, M. Ikehara, and K. Matsubara, Proc. Natl. Acad. Sci. USA, *82*, 1931–1935 (1985).
17. R. Sanchez-Pescador and M.S. Urdea, DNA, *3*, 339–343 (1984).
18. R.J. Rothstein, L.F. Lau, C.P. Bahl, S.A. Narang, and R. Wu, Methods Enzymol., *68*, 98–109 (1979).
19. R.B. Wallace, P.F. Johnson, S. Tanaka, M. Schold, K. Itakura, and J. Abelson, Science, *209*, 1396–1400 (1980).
20. V.-L. Chen and M. Smith, Nucleic Acids Res., *12*, 2407–2419 (1984).
21. G. Dalbadie-McFarland, L.W. Cohen, A.D. Riggs, C. Morin, K. Itakura, and J.H. Richards, Proc. Natl. Acad. Sci. USA, *79*, 6409–6413 (1982).

22. K. Norris, F. Norris, L. Christiansen, and N. Fill, Nucleic Acids Res., *11*, 5103–5112 (1983).
23. K.A. Osinga, A.M. Van der Bliek, G. Van der Horst, M.J.A. Groot Koerkamp, H.F. Tabak, G.H. Veeneman, and J.H. Van Boom, Nucleic Acids Res., *11*, 8595–8608 (1983).
24. M.J. Zoller and M. Smith, Methods Enzymol., *100B*, 468–500 (1983).
25. H.U. Goringer, R. Wagner, W.F. Jacob, A.E. Dahlberg, and C. Zwieb, Nucleic Acids Res., *12*, 6935–6950 (1984).
26. K.-M. Lo, S.S. Jones, N.R. Hackett, and H.G. Khorana, Proc. Natl. Acad. Sci. USA, *81*, 2285–2289 (1984).
27. M. Schold, A. Colombero, A.A. Reyes, and R.B. Wallace, DNA, *3*, 469–477 (1984).
28. M.J. Zoller and M. Smith, DNA, *3*, 479–488 (1984).
29. C.S. Craik, C. Largman, T. Fletcher, S. Roczniak, P.J. Barr, R. Fletterick, and W.J. Rutter, Science, *228*, 291–297 (1985).
30. T. Grundstrom, W.M. Zenke, M. Wintzerith, H.W.D. Matthes, A. Staub, and P. Chambon, Nucleic Acids Res., *13*, 3305–3316 (1985).
31. H.G. Khorana, K.L. Agarwal, M. Buchi, M.H. Caruthers, N.K. Gupta, K. Kleppe, A. Kumar, E. Ohtsuka, U.L. RajBhandary, J.H. van de Sande, V. Sgaramella, T. Terao, H. Weber, and T. Yamada, J. Mol. Biol., *72*, 209–217 (1972).
32. K. Itakura, T. Hirose, R. Crea, A.D. Riggs, H.L. Heyneker, F. Bolivar, and H.W. Boyer, Science, *198*, 1056–1063 (1977).
33. E.L. Brown, R. Belagaje, M.J. Ryan, and H.G. Khorana, Methods Enzymol., *68*, 109–151 (1979).
34. R. Crea, A. Kraszewski, T. Hirose, and K. Itakura, Proc. Natl. Acad. Sci. USA, *75*, 5765–5769 (1978).
35. J. Smith, E. Cook, I. Fotheringham, S. Pheby, R. Derbyshire, M.A.W. Eaton, M. Doel, D.M.J. Lilley, J.F. Pardon, T. Patel, H. Lewis, and L.D. Bell, Nucleic Acids Res., *10*, 4467–4482 (1982).
36. M.D. Edge, A.R. Greene, G.R. Heathcliffe, V.E. Moore, N.J. Faulkner, R. Camble, N.N. Petter, P. Trueman, W. Schuch, J. Hannam, T.C. Atkinson, C.R. Newton, and A.F. Markham, Nucleic Acids Res., *11*, 6419–6435 (1983).
37. S. Tanaka, T. Oshima, K. Ohsuye, T. Ono, A. Mizono, A. Veno, H. Nakazato, M. Tsujimoto, N. Higashi, and T. Noguchi, Nucleic Acids Res., *11*, 1707–1723 (1983).
38. M.S. Urdea, J.P. Merryweather, G.T. Mullenbach, D. Coit, U. Heberlein, P. Valenzuela, and P.J. Barr, Proc. Natl. Acad. Sci. USA, *80*, 7461–7465 (1983).
39. E. Jay, J. Rommens, L. Pomeroy-Cloney, D. MacKnight, C. Lutze-Wallace, P. Wishart, D. Harrison, W.-Y. Lui, V. Asundi, M. Dawood, and F. Jay, Proc. Natl. Acad. Sci. USA, *81*, 2290–2294 (1984).
40. M. Piratsu, Y.W. Kan, A. Cao, B.J. Conner, R.L. Teplitz, and R.B. Wallace, N. Engl. J. Med., *309*, 284–287 (1983).
41. S.H. Orkin, A.F. Markham, and H.H. Kazazian, J. Clin. Invest., *71*, 775–779 (1983).
42. V.J. Kidd, R.B. Wallace, K. Itakura, and S.L.C. Woo, Nature, *304*, 230–234 (1983).
43. C.W. Finn, Jr., R.P. Silver, W.H. Habig, M.C. Hardegree, G. Zon, and C.F. Garon, Science, *224*, 881–884 (1984).

44. W.E. Hill, W.L. Payne, G. Zon, and S.L. Mosely, Appl. Environ. Microbiol., *50*, 1187–1191 (1985).
45. W.E. Hill, B.A. Wentz, J.A. Jagow, W.L. Payne, and G. Zon, J. Assoc. Off. Anal. Chem., *69*, 531–536 (1986).
46. M. Nishibuchi, W.E. Hill, G. Zon, W.L. Payne, and J.B. Kaper, J. Clin. Microbiol., *23*, 1091–1095 (1986).
47. A. Ullrich, C. Berman, T.J. Dull, A. Gray, and J.M. Lee, EMBO J., *3*, 361–364 (1984).
48. B.H. Hahn, G.M. Shaw, S.K. Arya, M. Popovic, R.C. Gallo, and F. Wong-Staal, Nature, *312*, 166–169 (1984).
49. K.-J. Lei, T. Liu, G. Zon, E. Soravia, T.-Y. Liu, and N.D. Goldman, J. Biol. Chem., *260*, 13377–13383 (1985).
50. N.Y. Nguyen, A. Suzuki, S.-M. Cheng, G. Zon, and T.Y. Liu, J. Biol. Chem., *261*, 10450–10455 (1986).
51. M. Renz and C. Kurz, Nucleic Acids Res., *12*, 3435–3444 (1984).
52. A. Chollet and E.H. Kawashima, Nucleic Acids Res., *13*, 1529–1541 (1985).
53. T. Kempe, W.I. Sundquist, F. Chow, and S.-L. Hu, Nucleic Acids Res., *13*, 45–57 (1985).
54. D.G. Kleid, "Vaccine Synthesis by Recombinant DNA Technology," Ann. Rep. Med. Chem., *19*, 223–230 (1984).
55. N.C. Engleberg and B.I. Eisenstein, N. Engl. J. Med., *311*, 892–901 (1984).
56. P.O.P. Ts'o, P.S. Miller, and J.J. Greene, "Nucleic Acid Analogs with Targeted Delivery as Chemotherapeutic Agents," in *Development of Target-Oriented Anticancer Drugs*, Y.-C. Cheng, B. Goz, and M. Minkoff, Eds., Raven Press, New York, 1983, pp. 189–206.
57. A.S. Levina and E.M. Ivanova, Bioorg. Khim., *11*, 231–238 (1985).
58. R.I. Zhdanov and S.M. Zhenodarova, "Chemical Methods of Oligonucleotide Synthesis," Synthesis, *3*, 222–245 (1975).
59. C.B. Reese, "The Chemical Synthesis of Oligo- and Poly-Nucleotides by the Phosphortriester Approach," Tetrahedron, *34*, 3143–3179 (1978).
60. S.A. Narang, H.M. Hsiung, and R. Brousseau, Methods Enzymol., *68*, 90–98 (1979).
61. S.A. Narang, R. Brousseau, H.M. Hsiung, and J.J. Michniewicz, Methods Enzymol., *65*, 610–620 (1980).
62. G. Alvarado-Urbina, G.M. Sathe, W.-C. Liu, M.F. Gillen, P.D. Duck, R. Bender, and K.K. Ogilivie, Science, *214*, 270–274 (1981).
63. M.H. Caruthers, "The Design and Synthesis of Gene Control Regions Useful for Genetic Engineers," in *Recombinant DNA*, A.G. Walton, Ed., Elsevier, New York, 1981, pp. 261–272.
64. M.H. Caruthers, S.L. Beaucage, C. Becker, W. Efcavitch, E.F. Fisher, G. Galluppi, R. Goldman, R. deHaseth, F. Martin, M. Matteucci, and Y. Stabinsky, "New Methods for Synthesizing Deoxyoligonucleotides," in *Genetic Engineering Principles and Methods*, Vol. 4, J.K. Setlow and A. Hollaender, Eds., Plenum Press, New York, 1982, pp. 1–17.
65. M.H. Caruthers, "New Methods for Chemically Synthesizing Deoxyoligonucleo-

tides," in *Methods of DNA and RNA Sequencing*, S.M. Weissman, Ed., Praeger Publishers, New York, 1983, pp. 1–22.

66. R.L. Letsinger, "Chemical Synthesis of Oligodeoxyribonucleotides: A Simplified Procedure," in *Genetic Engineering Principles and Methods*, Vol. 5, J.K. Setlow and A. Hollaender, Eds., Plenum Press, New York, 1983, pp. 191–207.
67. T.C. Atkinson, BioTechniques, *1*, 6–10 (1983).
68. M.H. Caruthers, Science, *230*, 281–285 (1985).
69. Various authors in *Chemical and Enzymatic Synthesis of Gene Fragments: A Laboratory Manual*, H.G. Gassen and A. Lang, Eds., Verlag Chemie, Weinheim, 1982, pp. 1–249.
70. Various authors, in *Oligonucleotide Synthesis: A Practical Approach*, M.J. Gait, Ed., IRL Press, Washington, DC, 1984, pp. 1–217.
71. J. Van Brunt, Bio/Technology, *3*, 775–782 (1985).
72. T. Tanaka and R.L. Letsinger, Nucleic Acids Res., *10*, 3249–3260 (1982).
73. H. Seliger, C. Scalfi, and F. Eisenbeiss, Tetrahedron Lett., *24*, 4963–4966 (1983).
74. R.L. Letsinger, E.P. Groody, N. Lander, and T. Tanaka, Tetrahedron, *40*, 137–143 (1984).
75. M.E. Schott, Am. Biotechnol. Lab., *3*, 20–23 (1985).
76. S.L. Beaucage and M.H. Caruthers, Tetrahedron Lett., *22*, 1859–1862 (1981).
77. L.J. McBride and M.H. Caruthers, Tetrahedron Lett., *24*, 245–248 (1983).
78. S.P. Adams, K.S. Kavka, E.J. Wykes, S.B. Holder, and G.R. Galluppi, J. Am. Chem. Soc., *105*, 661–663 (1983).
79. B.D. Warner, M.E. Warner, G.A. Karns, L. Ku, S. Brown-Shimer, and M.S. Urdea, DNA, *3*, 401–411 (1984).
80. T.M. Cao, S.E. Bingham, and M.T. Sung, Tetrahedron Lett., *24*, 1019–1020 (1983).
81. T. Dorper and E.-L. Winnacker, Nucleic Acids Res., *11*, 2575–2584 (1983).
82. J.-L. Fourrey and J. Varenne, Tetrahedron Lett., *24*, 1963–1966 (1983).
83. R. Frank, W. Heikens, G. Heisterberg-Moutsis, and H. Blocker, Nucleic Acids Res., *11*, 4365–4377 (1983).
84. B.C. Froehler and M.D. Matteucci, Nucleic Acids Res., *11*, 8031–8036 (1983).
85. B.C. Froehler and M.D. Matteucci, Tetrahedron Lett., *24*, 3171–3174 (1983).
86. L.J. McBride and M.H. Caruthers, Tetrahedron Lett., *24*, 2953–2956 (1983).
87. N.S. Sinha, J. Biernat, and H. Koster, Tetrahedron Lett., *24*, 5843–5846 (1983).
88. A.D. Barone, J.-Y. Tang, and M.H. Caruthers, Nucleic Acids Res., *12*, 4051–4061 (1984).
89. S.L. Beaucage, Tetrahedron Lett., *25*, 375–378 (1984).
90. C. Claesen, G.I. Tesser, C.E. Dreef, J.E. Marugg, G.A. van der Marel, and J.H. van Boom, Tetrahedron Lett., *25*, 1307–1310 (1984).
91. J.L. Fourrey and J. Varenne, Tetrahedron Lett., *25*, 4511–4514 (1984).
92. H. Koster, J. Biernat, J. McManus, A. Wolter, A. Stumpe, C.K. Narang, and N.D. Sinha, Tetrahedron Lett., *40*, 103–112 (1984).
93. G. Kumar and M.S. Poonian, J. Org. Chem., *49*, 4905–4912 (1984).
94. J.E. Marugg, C.E. Dreef, G.A. van der Marel, and J.H. van Boom, Recl. Trav. Chim. 103, 97–98 (1984).

95. J. Katsuzaki, H. Hotoda, M. Sekine, and T. Hata, Tetrahedron Lett., *25*, 4019–4022 (1984).
96. J. Ott and F. Eckstein, Nucleic Acids Res., *12*, 9137–9142 (1984).
97. N.D. Sinha, J. Biernat, and H. Koster, Nucleosides & Nucleotides, *3*, 157–171 (1984).
98. N.D. Sinha, J. Biernat, J. McManus, and H. Koster, Nucleic Acids Res., *12*, 4539–4557 (1984).
99. H. Takaku, S. Veda, and Y. Tomita, Chem. Pharm. Bull. *32*, 2882–2885 (1984).
100. M.H. Caruthers, L.J. McBride, L.P. Bracco, and J.W. Dubendorff, Nucleosides & Nucleotides, *4*, 95–105 (1985).
101. C.A.A. Claesen, R.P.A.M. Segers, and G.I. Tesser, Recl. Trav. Chim. *104*, 119–122; *104*, 209–214 (1985).
102. J.W. Efcavitch and C. Heiner, Nucleosides & Nucleotides, *4*, 267 (1985).
103. J.-L. Fourrey and J. Varenne, Tetrahedron Lett., *26*, 1217–1220 (1985).
104. M.F. Moore and S.L. Beaucage, J. Org. Chem., *50*, 2019–2025 (1985).
105. W. Pfleiderer, F. Himmelsbach, R. Charubala, H. Shirmeister, A. Beiter, B. Schulz, and T. Trichtinger, Nucleosides & Nucleotides, *4*, 81–94 (1985).
106. R.T. Pon, M.J. Damha, and K.K. Ogilvie, Nucleic Acids Res., *13*, 6447–6465 (1985).
107. H. Seliger and K.C. Gupta, Angew. Chem. *97*, 711–713 (1985).
108. D.G. Norman, C.B. Reese, and H.T. Serafinowska, Tetrahedron Lett., *25*, 3015–3018 (1984).
109. K.K. Ogilvie, M.J. Nemer, and M.F. Gillen, Tetrahedron Lett., *25*, 1669–1672 (1984).
110. R.T. Pon and K.K. Ogilvie, Nucleosides & Nucleotides, *3*, 485–498 (1984).
111. R.T. Pon and K.K. Ogilvie, Tetrahedron Lett., *25*, 713–716 (1984).
112. Anonymous, in *Evaluation and Purification of Synthetic Oligonucleotides*, Applied Biosystems, Inc., Foster City, CA, User Bulletin Issue No. 13, November 9, 1984, pp. 1–27.
113. J.A. Thompson, "A Review of High-Performance Liquid Chromatography in Nucleic Acids Research, I. Historical Considerations," in BioChromatogr., see ref. 242.
114. G. Zon and J.A. Thompson, "A Review of High-Performance Liquid Chromatography in Nucleic Acids Research, II. Isolation, Purification, and Analysis of Oligodeoxyribonucleotides," in BioChromatogr., *1*, 22–32 (1986).
115. L.W. McLaughlin and J.U. Krusche, "Applications of High Performance Liquid Chromatography to Oligonucleotide Separation and Purification," in *Chemical and Enzymatic Synthesis of Gene Fragments: A Laboratory Manual*, H.G. Gassen and A. Lang, Eds., Verlag Chemie, Weinheim, 1982, pp. 177–198.
116. L.W. McLaughlin and N. Piel, "Chromatographic Purification of Synthetic Oligonucleotides," in *Oligonucleotide Synthesis: Practical Approach*, M.J. Gait, Ed., IRL Press, Washington, DC, 1984, pp. 117–133.
117. P. Clarke, H.-C. Lin, and G. Wilcox, Anal. Biochem., *124*, 88–91 (1982).
118. H.R. Drew, R.M. Wing, T. Takano, C. Broka, S. Tanaka, K. Itakura, and R.E. Dickerson, Proc. Natl. Acad. Sci. USA, *78*, 2179–2183 (1981).

119. J. Grable, C.A. Frederick, C. Samadzi, L. Jen-Jacobson, D. Lesser, P. Greene, H.W. Boyer, K. Itakura, and J.M. Rosenberg, J. Biomol. Struct. Dyn., *1*, 1149–1160 (1984).
120. D.J. Patel, S.A. Kozlowski, S. Ikuta, and K. Itakura, Fed. Proc., *43*, 2663–2670 (1984).
121. K.E. Norris, F. Norris, and K. Brunfeldt, Nucleic Acids Res., Symp. Ser. No. 7, 233–241 (1980).
122. K. Miyoshi and K. Itakura, Tetrahedron Lett., *20*, 3635–3638 (1979).
123. R. Crea and T. Horn, Nucleic Acids Res., *8*, 2331–2348 (1980).
124. K. Miyoshi and K. Itakura, Nucleic Acids Res., Symp. Ser. No. 7, 281–291 (1980)
125. K. Miyoshi, T. Miyake, T. Hozumi and K. Itakura, Nucleic Acids Res., *8*, 5473–5489 (1980).
126. J.D. Pearson and F. Regnier, J. Chromatogr., *255*, 137–149 (1983).
127. T.G. Lawson, F.E. Regnier, and H.L. Weith, Anal. Biochem., *133*, 85–93 (1983).
128. R.R. Drager and F.E. Regnier, Anal. Biochem., *145*, 47–56 (1985).
129. R. Hecker, M. Colpan, and D. Riesner, J. Chromatogr., *326*, 251–261 (1985).
130. J.A. Thompson, R.W. Blakesley, K. Doran, C.J. Hough, and R.D. Wells, Methods Enzymol. *100B*, 368–400 (1983).
131. H.-J. Fritz, R. Belagaje, E.L. Brown, R.H. Fritz, R.A. Jones, R.G. Lees, and H.G. Khorana, Biochemistry, *17*, 1257–1267 (1978).
132. R.A. Jones, H.-J. Fritz, and H.G. Khorana, Biochemistry, *17*, 1268–1278 (1978).
133. K. Makino, H. Ozaki, H. Wada, T. Tateuchi, M. Ikehara, T. Fukui, H. Sasaki, and Y. Kato, Nucleic Acids Res., Symp. Ser. No. 16, 45–48 (1985).
134. A.M. Delort, R. Derbyshire, A.M. Duplaa, A. Guy, D. Molko, and R. Teoule, J. Chromatogr., *283*, 462–467 (1985).
135. C.R. Becker, J.W. Efcavitch, C.R. Heiner, and N.F. Kaiser, J. Chromatogr., *326*, 293–299 (1984).
136. D.P. Lee and J.H. Kindsvater, Anal. Chem., *52*, 2425–2428 (1980).
137. S. Ikuta, R. Chattopadhyaya, R.E. Dickerson, Anal. Chem., *56*, 2256–2257 (1984).
138. E.E. Leutzinger, P.S. Miller, and P.O.P. Ts'o, "Application of High-Pressure Liquid Chromatography in the Synthesis of Oligonucleotides," in *Nucleic Acid Chemistry: Improved and New Synthetic Procedures, Methods and Techniques*, Part Two, L.B. Townsend and R.S. Tipson, Eds., Wiley, New York, 1978, pp. 1037–1043.
139. R. Arshady, E. Atherton, M.J. Gait, K. Lee, and R.C. Sheppard, J. Chem. Soc., Chem. Commun., 423–425 (1979).
140. M.J. Gait and R.C. Sheppard, Nucleic Acids Res., *6*, 1259–1268 (1979).
141. P.S. Miller, D.M. Cheng, N. Dreon, K. Jayaraman, L.-S. Kan, E.E. Leutzinger, S.M. Pulford, and P.O.P. Ts'o, Biochemistry, *19*, 4688–4698 (1980).
142. M.J. Gait, S.G. Popov, M. Singh, and R.C. Titmas, Nucleic Acids Res., Symp. Ser. No. 7, 243–257 (1980).
143. M.J. Gait, M. Singh, and R.C. Sheppard, Nucleic Acids Res., *8*, 1081–1096 (1980).
144. A.F. Markham, M.D. Edge, T.C. Atkinson, A.R. Greene, G.R. Heathcliffe, C.R. Newton, and D. Scanlon, Nucleic Acids Res., *8*, 5193–5205 (1980).

145. M.L. Duckworth, M.J. Gait, P. Goelet, G.F. Hong, M. Singh, and R.C. Titmas, Nucleic Acids Res., *9*, 1691–1706 (1981).

146. F. Chow, T. Kempe, and G. Palm, Nucleic Acids Res., *9*, 2807–2817 (1981).

147. H. Schott and H. Eckstein, J. Chromatogr., *296*, 363–368 (1984).

148. A. Sokolowski, N. Balgobin, S. Josephson, J.B. Chattopadhyaya, and G. Schill, Chem. Scr., *18*, 189–191 (1981).

149. W. Jost, K. Unger, and G. Schill, Anal. Biochem., *119*, 214–223 (1982).

150. Z. El Rassi and C. Horvath, "Effect of Mobile Phase Composition on the Retention Behavior of Oligonucleotides in Reversed Phase Chromatography," in *Practical Aspects of Modern High Performance Liquid Chromatography*, I. Molnar, Ed., W. de Gruyter & Co., Berlin, 1983, pp. 1–14.

151. B. Allinquant, C. Musenger, and E. Schuller, J. Chromatogr., *326*, 281–291 (1985).

152. C.Y. Ip, D. Ha, P.W. Morris, M.L. Puttemans, and D.L. Venton, Anal. Biochem., *147*, 180–185 (1985).

153. M. Kwiatkowski, A. Sandstrom, N. Balgobin, and J. Chattopadhyaya, Acta. Chem. Scand., *B38*, 721–733 (1984).

154. J.B. Crowther, S.D. Fazio, and R.A. Hartwick, J. Chromatogr., *282*, 619–628 (1983).

155. J.H. van Boom and J.F.M. de Rooy, J. Chromatogr., *131*, 169–177 (1977).

156. J.F.M. de Rooij, W. Bloemhoff, and J.H. van Boom, J. Chromatogr., *177*, 380–384 (1979).

157. J.B. Crowther, R. Jones, and R.A. Hartwick, J. Chromatogr., *217*, 479–490 (1981).

158. J.B. Crowther, J.P. Caroni, and R.A. Hartwick, Anal. Biochem., *124*, 65–73 (1982).

159. H.-J. Fritz, D. Eick, and W. Werr, "Analysis of Synthetic Oligodeoxyribonucleotides," in *Chemical and Enzymatic Synthesis of Gene Fragments: A Laboratory Manual*, Verlag Chemie, Weinheim, 1982, pp. 199–223.

160. R. Wu, N.-H. Wu, Z. Hanna, F. Georges, and S. Narang, "Purification and Sequence Analysis of Synthetic Oligonucleotides," in *Oligonucleotide Synthesis: A Practical Approach*, IRL Press, Washington, DC, 1984, pp. 135–151.

161. R. Frank and H. Blocker, "The 'Wandering Spot' Sequence Analysis of Oligodeoxyribonucleotides," in *Chemical and Enzymatic Synthesis of Gene Fragments: A Laboratory Manual*, Verlag Chemie, Weinheim, 1982, pp. 225–246.

162. S. Chandrasegaran, S.-B. Lin, and L.-S. Kan, BioTechniques, *3*, 6–8 (1985).

163. W. Haupt and A. Pingoud, J. Chromatogr., *260*, 419–427 (1983).

164. H. Seliger, U. Kotschi, and G. Schmidt, 4th Intl. Symp. on HPLC of Proteins, Peptides, and Polynucleotides, December 10–12, 1984, Baltimnore, MD, abstr. no. 310.

165. H. Seliger and U. Kotschi, Nucleosides & Nucleotides, *4*, 153 (1985).

166. H. Seliger and H.-H. Gortz, Angew. Chem. Int. Ed. Engl., *20*, 683–684 (1981).

167. L.R. Snyder and J.J. Kirkland, *Introduction to Modern Liquid Chromatography*, 2nd ed., Wiley, New York, 1979, pp. 1–863.

168. P.A. Perrone and P.R. Brown, in *Ion-Pair Chromatography: Theory and Biological and Pharmaceutical Applications*, M.T.W. Hearn, Ed., Marcel Dekker, New York, 1985, pp. 259–282.

169. W.S. Hancock and J.T. Sparrow, *HPLC Analysis of Biological Compounds: A Laboratory Guide*, Marcel Dekker, New York, 1984, pp. 1–361.
170. X. Gao, B.L. Gaffney, M. Senior, R.R. Riddle, and R.A. Jones, Nucleic Acids Res., *13*, 573–584 (1985).
171. M.S. Urdea, L. Ku, T. Horn, Y.G. Gee, and B.D. Warner, Nucleic Acids Res., Symp. Ser. No. 16, 257–260 (1985).
172. G.T. Mullenbach, A. Tabrizi, R.W. Blacher, and K.S. Steiner, J. Biol. Chem., *261*, 719–722 (1986).
173. G. Zon, K.A. Gallo, C.J. Samson, K.-L. Shao, M.F. Summers, and R.A. Byrd, Nucl. Acids Res. *13*, 8181–8196 (1985).
174. C.R. Becker, J.W. Efcavitch, C.R. Heiner, and N.F. Kaiser, J. Chromatogr., *326*, 293–299 (1985).
175. D.P. Lee and J.H. Kindsvater, Anal. Chem., *52*, 2425–2428 (1980).
176. K.-M. Lo, S.S. Jones, N.R. Hackett, and H.G. Khorana, Proc. Natl. Acad. Sci. USA, *81*, 2285–2289 (1984).
177. R.S. Lloyd, A. Recinos, and S.T. Wright, BioTechniques, *4*, 8–10 (1986).
178. A. Rosenthal, S. Schwertner, V. Hahn, and H.-D. Hunger, Nucleic Acids Res., *13*, 1173–1183 (1985).
179. W.J. Stec and G. Zon, Tetrahedron Lett., *25*, 5275–5278 (1984).
180. B.D. Stollar, G. Zon, and R.W. Pastor, Proc. Natl. Acad. Sci. USA, *83*, 4469–4473 (1986).
181. W.J. Stec and G. Zon, J. Chromatogr., *236*, 263–280 (1985).
182. U. Gunthert, M. Schweiger, M. Stupp, and W. Doerfler, Proc. Natl. Acad. Sci. USA, *73*, 3923–3927 (1976).
183. W.J. Stec, G. Zon, W. Egan, R.A. Byrd, L.R. Phillips, and K.A. Gallo, J. Org. Chem., *50*, 3908–3913 (1985).
184. T. Horn and M.S. Urdea, Nucleic Acids Res., Symp. Ser. No. 16, 153–156 (1985).
185. R.W. Blakesley and J.A. Thompson, Fed. Proc., *42*, 1955 (1983).
186. J. Ruth, R.D. Smith, and R. Lohrmann, Fed. Proc., *44*, 1622 (1985).
187. S.V. Suggs, R.B. Wallace, T. Hirose, E.H. Kawashima, and K. Itakura, Proc. Natl. Acad. Sci. USA, *78*, 6613–6617 (1981).
188. K.L. Agarwal, J. Brunstedt, and B.E. Noyes, J. Biol. Chem., *256*, 1023–1028 (1981).
189. G. Felsenfeld and H.T. Miles, Ann. Rev. Biochem., *31*, 407–448 (1967).
190. R.C. Grant and S.J. Harwood, J. Am. Chem. Soc., *90*, 4474–4476 (1968).
191. F.H. Martin, M.M. Castro, F. Aboul-ela, and I. Tinoco, Nucleic Acids Res., *13*, 8927–8938 (1985).
192. R. Frank and H. Koster, Nucleic Acids Res., *6*, 2069–2087 (1979).
193. E.P. Suzuki, N. Nattabiraman, G. Zon, and T.L. James, Biochemistry, *25*, 6854–6865 (1986).
194. E. Uhlmann and J. Engels, Tetrahedron Lett., *27*, 1023–1026 (1986).
195. B.C.F. Chu and L.E. Orgel, DNA, *4*, 327–331 (1985).
196. B.C.F. Chu and L.E. Orgel, Proc. Natl. Acad. Sci. USA, *82*, 963–967 (1985).
197. R.S. Youngquist and P.B. Dervan, J. Am. Chem. Soc., *107*, 5528–5529 (1985).
198. G.B. Dreyer and P.B. Dervan, Proc. Natl. Acad. Sci. USA, *82*, 968–972 (1985).

199. G. Lancelot, U. Asseline, N.T. Thuong, and C. Helene, Biochemistry, *24*, 2421–2529 (1985).
200. R. Cosstick, L.W. McLaughlin, and F. Eckstein, Nucleic Acids Res., *12*, 1791–1810 (1984).
201. S.A. Noble, E.F. Fisher, and M.H. Caruthers, Nucleic Acids Res., *12*, 3387–3404 (1984).
202. B.A. Connolly, F. Eckstein, and A. Pingoud, J. Biol. Chem., *259*, 10760–10763 (1984).
203. W.J. Stec, G. Zon, W. Egan, and B. Stec, J. Am. Chem. Soc. *106*, 6077–6079 (1984).
204. K. Yamana and R.L. Letsinger, Nucleic Acids Res., Symp. Ser. No. 16, 169–172 (1985).
205. C. Helene, T. Montenay-Garestier, T. Saison, M. Takasugi, J.J. Toulme, U. Asseline, G. Lancelot, J.C. Maurizot, F. Toulne, and N.T. Thuong, Biochimie, *67*, 777–783 (1985).
206. P.S. Miller, C.H. Agris, K.R. Blake, A. Murakami, S.A. Spitz, P.M. Reddy, and P.O.P. Ts'o, "Nonionic Oligonucleotide Analogs as New Tools for Studies on the Structure and Function of Nucleic Acids Inside Living Cells," in *Nucleic Acids*: *The Yectors of Life*, B. Pullman and J. Jortner, Eds., D. Reidel Publishing Co., Hingham, MA, 1983, pp. 521–535.
207. R. Cosstick and F. Eckstein, Biochemistry, *24*, 3630–3638 (1985).
208. W.J. Stec and G. Zon, Tetrahedron Lett., *25*, 5279–5282 (1984).
209. B.L. Gaffney, L.A. Marky, and R.A. Jones, Biochemistry, *23*, 5686–5691 (1984).
210. G.V. Fazakerley, R. Teoule, A. Guy, H. Fritzsche, and W. Guschlbauer, Biochemistry, *24*, 4540–4548 (1985).
211. B.V.L. Potter, F. Eckstein, and B. Uznanski, Nucleic Acids Res., *11*, 7087–7103 (1983).
212. W. Hertering and F. Seela, J. Org. Chem., *50*, 5314–5323 (1985).
213. M. Weinfeld, A.F. Drake, J.K. Saunders, and M.C. Paterson, Nucleic Acids Res., *13*, 7067–7077 (1985).
214. M.R. Hamblin and B.V.L. Potter, FEBS Lett., *189*, 315–317 (1985).
215. P.S. Miller, S. Chandrasegaran, D.L. Dow, S.M. Pulford, and L.S. Kan, Biochemistry, *21*, 5468–5474 (1982).
216. D.E. Jensen and D.J. Reed, Biochemistry, *17*, 5098–5107 (1978).
217. U. Siebenlist and W. Gilbert, Proc. Natl. Acad. Sci., *77*, 122–126 (1980).
218. W.J. Stec, G. Zon, K.A. Gallo, R.A. Byrd, B. Uznanski, and P. Guga, Tetrahedron Lett., *26*, 2191–2194 (1985).
219. L.R. Phillips, K.A. Gallo, W.J. Stec, B. Uznanski, and G. Zon, Org. Mass. Spectrom., *20*, 781–783 (1985).
220. B. Uznanski, M. Koziolkiewicz, W.J. Stec, G. Zon, K. Shinozuka, and L.G. Marzilli, Chem. Scr., *26*, 221–224 (1986).
221. M. Koziolkiewicz, B. Uznanski, G. Zon, and W.J. Stec, Chem. Scr., *26*, 251–260 (1986).
222. I.H. Williams, J. Chem. Soc., Chem. Commun., 510–511 (1985).
223. P.S. Miller, N. Dreon, S.M. Pulford, and K.B. McParland, J. Biol. Chem., *255*, 9659–9665 (1980).

224. A. Murakami, K.R. Blake, and P.S. Miller, Biochemistry, *24*, 4041–4046 (1985).
225. K.R. Blake, A. Murakami, and P.S. Miller, Biochemistry, *24*, 6132–6138 (1985).
226. K.R. Blake, A. Murakami, S.A. Spitz, S.A. Glave, M.P. Reddy, P.O.P. Ts'o, and P.S. Miller, Biochemistry, *24*, 6139–62 (1985).
227. A. Jager and J. Engels, Tetrahedron Lett., *25*, 1437–1440 (1984).
228. P.S. Miller, C.H. Agris, A. Kurakami, P.M. Reddy, S.A. Spitz, and P.O.P. Ts'o, Nucleic Acids Res., *11*, 6225–6242 (1983).
229. P.S. Miller, Y. Yano, E. Yano, C. Carrol, K. Jayaraman, and P.O. Ts'o, Biochemistry, *18*, 5134–5143 (1979).
230. K.K. Chako, K. Lindner, W. Saenger, and P.S. Miller, Nucleic Acids Res., *11*, 2801–2813 (1983).
231. P.O.P. Ts'o and P.S. Miller, Chem. Abstr., *101*, 230961x (1984).
232. H. Koester, Chem. Abstr., *101*, 230954x (1984).
233. R.L. Letsinger and M.E. Schott, Chem. Abstr., *101*, 211660g (1984).
234. P.A.D. Edwardson, C.R. Lowe, and T. Atkinson, J. Chromatogr., *368*, 363–369 (1986).
235. H. Sawai, Nucleic Acids Res. Symp. Ser. No. 17, 113–116 (1986).
236. M.W. Germann, R.T. Pon, and J.H. van de Sande, Anal. Biochem., *165*, 399–405 (1987).
237. A. Meyerhaus, G. Heisterberg-Moutsis, G. Kurth, H. Blocker, and R. Frank, Nucleosides Nucleotides, *4*, 235 (1985).
238. L.J. McBride, C. McCollum, S. Davidson, J.W. Efcavitch, A. Andrus, and S.J. Lombardi, BioTechniques, *6*, 362–367 (1988).
239. C. Connell, S. Fung, C. Heiner, J. Bridgham, V. Chakerian, E. Heron, B. Jones, S. Menchen, W. Mordan, M. Raff, M. Recknor, L. Smith, J. Springer, S. Woo, and M. Hunkapiller, BioTechniques, *5*, 342–348 (1987).
240. J.A. Thompson, BioChromatogr., *1*, 16–20 (1986).
241. G. Zon and J.A. Thompson, BioChromatography, *1*, 22–32 (1986).
242. J.A. Thompson, BioChromatography, *1*, 68–80 (1986).
243. J.A. Thompson, BioChromatography, *2*, 4–18 (1986).
244. J.A. Thompson, BioChromatography, *2*, 68–79 (1986).
245. M.E. Swartz and J.E. Oberholtzer, BioChromatogr., *2*, 98–100 (1987).
246. J.S. Eadie, L.J. McBride, J.W. Efcavitch, L.B. Hoff, and R. Cathcart, Anal. Biochem. *165*, 442–447 (1987).
247. L.J. McBride, J.S. Eadie, J.W. Efcavitch, and W.A. Andrus, Nucleosides Nucleotides, *6*, 297–300 (1987).
248. J.S. Eadie and D.S. Davidson, Nucleic Acids Res., *15*, 8333–8349 (1987).

CHAPTER 15

Design of Chromatographic Matrices For the Purification and Analysis of Antibodies

DAVID R. NAU

Research Specialty Products Division, J.T. Baker Inc., Phillipsburg, New Jersey

1. INTRODUCTION

Antibody purification is one of the oldest challenges in protein biochemistry. Even today, despite a long history of use as tools for analysis, identification, characterization, and purification, the development of rapid and economical purification schemes for antibodies remains a major problem. Recently, this dilemma has become even more acute, since the development of monoclonal antibody technology, by providing an enormous impetus for progress in numerous fields of science and technology (57), has created a number of new and more demanding purification problems.

Until recently, antibody purification was performed in the research laboratory on a relatively small scale, using conventional open column chromatography on soft gels. For many years, this type of chromatography was quite adequate for conducting basic research aimed at understanding the functions and properties of antibodies and other proteins. However, with the development of monoclonal antibody technology and the commercialization of a host of antibody-based diagnostic and therapeutic products, the preparative purification of antibodies has become an area of major interest. Today, the economics of protein purification are of paramount importance, not only in the biotechnology and pharmaceutical industries to increase profitability, but also, in academic laboratories. For the academic and industrial research scientist, it is highly desirable to reduce the time, effort, and cost of immunoglobulin purification, in order to facilitate the conquest of more important scientific endeavors.

With the increasing use of monoclonal antibodies in the health care field, large quantities of high-purity, nonpyrogenic immunoglobulins are now required for in vivo diagnostics (imaging) and immunotherapeutics (14, 24, 56,

79), the purification of pharmaceuticals by affinity chromatography (18, 61, 91, 99), in vitro diagnostics (46, 58, 98), and research purposes (immunotitration, cytological labeling, immunoassays, probes, and so on).

Prior to the advent of monoclonal antibodies, the universal source of immunoglobulins was serum/plasma. Since IgG antibodies typically represent a major portion of the total protein complement of serum, rigorous purification was often not essential for the intended use, since the major concern was usually the elimination of nonspecific interactions rather than absolute product purity. Similar criteria also apply to the purification of monoclonals from ascites fluid, since the antibody level is often rather high and the intended use is typically not one which necessitates extremely high product purity. However, current trends favor using hybridoma cell cultures to produce large amounts of antibody, owing to the ease of scaling-up production and other economic considerations (4, 29, 68, 79), as well as the desirability of using hybrid human monoclonals as immunotherapeutics and in vivo diagnostics (14, 24, 28, 56, 79), even though the levels of antibody are two to three orders of magnitude lower in culture supernatants than in ascites fluid or serum. Furthermore, most cells in culture require fetal calf serum or horse serum which contain a diverse mixture of proteins, many of which are present at high levels (62, 67–71). Even in serum-free tissue culture media, where the protein content is not extremely complex, the purification may be nonetheless difficult, since the monoclonal is often present as a minor component and numerous nonimmunoglobulin proteins may also be secreted from the cells (62, 69–71). In the case of ascites fluids, contaminating proteins may also consist of host polyclonal antibodies, while fetal bovine sera may also contain low levels (40–400 μg/mL) of bovine serum immunoglobulins (4, 10, 37, 62, 67–71).

In addition, antibody purification is no longer limited to the IgG subclasses, since today, monoclonals from other immunoglobulin classes, particularly IgM and IgA monoclonals, are of both commercial and academic interest (29, 62, 67–71). Furthermore, as techniques for hybridoma manipulation and immunochemistry become increasingly sophisticated, new purification procedures are needed for cell lines secreting multiple immunoglobulins, active and inactive monoclonals, biospecific hybrid antibodies, protein-antibody conjugates, antibody fragments, monoclonals with diverse heavy and light chains, hybrid fragments, and chemically modified monoclonals (28, 67–71). Therefore, modern antibody purification is not one problem, but many.

The successful large-scale use of monoclonals is dependent on practical solutions to a number of technical problems, including the development of fast and efficient purification methods. Without a doubt, the inability to purify immunoglobulins from cell cultures in an economical manner has limited their usefulness, since the cost of the separation is typically greater than that of the actual production. As a partial solution, elaborate cell immobilization, encapsulation, and other fermentation procedures, and numerous serum-free media supplements have been developed in order to reduce the levels of nonimmunoglobulin proteins and the related purification problems (62, 69, 70). As a result of the disadvantages that are inherent to many conventional separation techniques, most modern antibody purification schemes consist of an elaborate series of steps; this greatly reduces the cost effectiveness of the entire production procedure. Indeed, it has been suggested that protein purification is the Achilles heel of modern biotechnology.

For protein chemists or process engineers faced with the task of purifying numerous antibodies from a number of sources, the most practical approach to antibody purification would involve the use of one or two highly selective chromatographic matrices that have universal applicability, with minor adjustments in purification protocols, to attain any level or purity from laboratory to process scale.

The requirement for an economical chromatographic matrix able rapidly to purify large quantities of antibodies to homogeneity compelled us to investigate synthetic approaches in an attempt to construct a chromatographic surface that would bind antibodies more selectively than conventional ion exchange matrices. Using mixed-mode interactions as the basis, silica gel as the most advantageous support, and a hydrophilic polymeric coverage to increase stability, eliminate nonspecific protein binding, and maximize recovery of mass and immunological activity, we developed BAKERBOND ABx* (antibody exchanger), a unique chromatographic matrix that behaves like an ion exchanger in that immunoglobulins may be resolved via the manipulation of buffer pH and ionic strength, but also, exhibits an affinity-like sensitivity toward all immunoglobulins, indicative of more complex interactions (67–71). ABx may be used to purify pyrogen-free antibodies of any class or subclass (IgG_1, IgG_2, IgG_3, IgG_4, IgA_1, IgA_2, IgM, IgD, and IgE), from any source (serum/plasma, ascites fluids, cell culture supernatant, or chicken egg yolk) from virtually any species (human, mouse, rat, rabbit, horse, cow, pig, sheep, goat, guinea pig, and so on), as well as antibody fragments and protein-antibody conjugates (62–71). The patented ABx surface chemistry (weak cation exchange, mild anion exchange, and mild hydrophobic interaction) facilitates the binding of all immunoglobulins, without binding most albumins, transferrins, proteases or other common protein contaminants, nor the common pH indicator dyes in cell culture media (e.g., phenol red, phenolpthalein).

This chapter focuses upon the use of ABx in the analysis and the preparative-scale purification of immunoglobulins from a number of diverse sources, and the optimization of chromatographic conditions in order to increase capacity and resolution and to facilitate the purification. Methods will be presented for the preparative purification of polyclonal antibodies form the serum/plasma of the most commonly immunized animals, monoclonal antibodies from ascites fluid and cell culture supernatants supplemented with fetal bovine serum, the purification of antibody classes, subclasses, fragments, and conjugates, and, the use of ABx in the analytical mode, to monitor and "quality control" hybridoma cell lines and fermentation broths, clinical samples, and purified antibody preparations.

2. MATERIALS AND METHODS

All reagents, buffer salts and solvents were products of J.T. Baker Inc. (Phillipsburg, NJ). (2[*N*-morpholino]ethanesulfonic acid] was purchased from Sigma Chemical Company (St. Louis, MO). SDS PAGE (sodium dodecyl sulfate polyacrylamide gel electrophoresis) was conducted under reducing conditions as detailed previously (67).

Serum samples, ascites fluids, cell culture supernatants, purified antibodies, fragments, classes, subclasses and conjugates, and serum-free media supplements were purchased from Sigma Chemical Company (St. Louis, MO), Jackson Immunoresearch Laboratories, Inc. (Avondale, PA), GIBCO (Grand Island, NY), Pel-Freez Biologicals (Rogers,AK), Cooper Biomedical (Malvern, PA), Boehringer Mannheim (Indianapolis, IN), Protogen AG (Laufelfingen, Switzerland), DuPont (Wilmington, DE), Walgene R & D Laboratories (Arcadia, CA), or were received as gifts from various pharmaceutical and biotechnology companies. Endotoxins were detected and quantitated by the Limulus Amebocyte Lysate test from Whittaker Bioproducts Inc. (Walkersville, MD).

HPLC (high-performance liquid chromatography) was conducted on equipment from Waters Associates (Bedford, MA), LDC Milton Roy Corporation (Riviera Beach FL), Beckman Instruments (Fullerton, CA), Pharmacia-LKB Biotechnology (Piscataway, NJ), Gilson Company, Inc. (Worthington, OH) and Varian Associates, Inc. (Palo Alto, CA). Chromatography was conducted on columns which were not tested by any quality control procedures.

3. BAKERBOND ABx (ANTIBODY EXCHANGER): GENERAL PROPERTIES

The major advantage of BAKERBOND ABx is that it binds all immunoglobulins rather selectively while exhibiting little or no affinity for the majority of the protein contaminants present within serum/plasma, ascites fluid, or serum-supplemented cell culture media (Figs. 1–3, and Table 1). In general, more than 90% of the nonimmunoglobulin proteins, including most albumins, transferrins, proteases, and insulin, as well as pH indicator dyes, elute in the void volume, leaving the antibody(s) and a relatively small number of these contaminating proteins bound to the matrix. Therefore, ABx may be used to separate immunoglobulins of any class and species from any crude antibody preparation in a single step, at purities greater than 75% (Figs. 1–3).

Because most of the proteins present in crude antibody samples are not bound to ABx, the majority of the ABx ligands are available to bind immunoglobulins; this factor substantially increases the "real life" binding capacity of ABx. For instance, analytical columns containing about 4 mL of ABx may be used in some cases to purify up to 10 mL of crude ascites fluid, 200 mL of serum-supplemented cell culture fluid and several milliliters of serum/plasma, with greater than 90% recovery (see Tables 2–4a). The fact that ABx does not bind

Table 1. Retention Times of Polyclonal and Monoclonal Immunoglobulins on BAKERBOND ABx

Source	Antibody or Protein	Antigen if Known**	Retention Time (Minutes)
mouse ascites	$IgG_{1,k}$		36
mouse ascites	$IgG_{2a,k}$		26
mouse ascites	$IgG_{2a,k}$		26
mouse ascites	$IgG_{2b,k}$		26
mouse ascites	$IgG_{2b,kk}$		26
mouse ascites	IgG_{3}		30
mouse ascites	IgA		16
mouse ascites	IgA		25
mouse ascites	IgM		28
mouse ascites	IgG	BSA	29
mouse ascites	IgG	SRBC	40
mouse ascites	IgG		35
mouse ascites	IgG		45
mouse ascites	IgG		22
mouse ascites	IgD		40
mouse ascites	IgG		60/67
mouse ascites	IgG	HCG	25
mouse ascites	IgG	HCG	27
mouse ascites	IgG	HCG	27
mouse ascites	IgM		52/60
mouse ascites	IgM		60
mouse ascites	IgM		68
mouse ascites	IgM		68
mouse ascites	IgG	I1-2	30
mouse ascites	IgG	I1-2	42
tissue culture	IgG		42
tissue culture	IgG	Int	40
tissue culture	IgG	Ren	31
tissue culture	IgG		45
tissue culture	IgG		50
tissue culture	IgM		30
tissue culture	IgM	ODC	27
tissue culture	IgG		30
human serum	IgG		22-35
equine serum	IgG		22-31
bovine serum	IgG		22-31
guinea pig serum	IgG		22-31
mouse serum	IgG		25-37
rabbit serum	IgG		24-32
bovine serum	albumin		3
human serum	albumin		3
mouse serum	albumin		3
bovine serum	transferrin		3
human serum	transferrin		3
mouse serum	transferrin		3
various	proteases		3

*All retention times were determined on a BAKERBOND ABx analytical column (4.6 × 250 mm); 250 μL of sample was chromatographed over a 60 min linear gradient of 10 mM MES, pH 6.0, to 250 mM KH_2PO_4, pH 6.8, with a flow rate of 1.0 mL/min.
**Abbreviations are BSA, bovine serum albumin; SRBC, sheep red blood cells; HCG, human chorionic gonadotropin; IL-2, interleukin 2; Int, interferon; Ren, renin; ODC, ornithine decarboxylase.

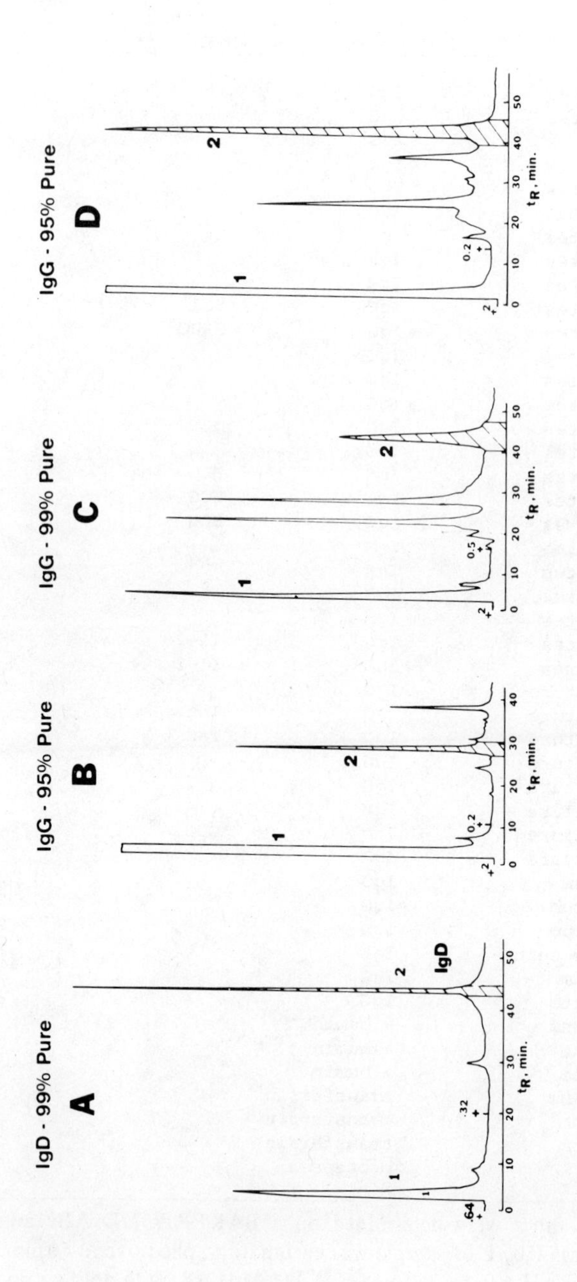

IgD - 99% Pure
A
IgG - 95% Pure
B
IgG - 99% Pure
C
IgG - 95% Pure
D
IgD
1
2
.64
.32
0.2
0.5
t_R, min.

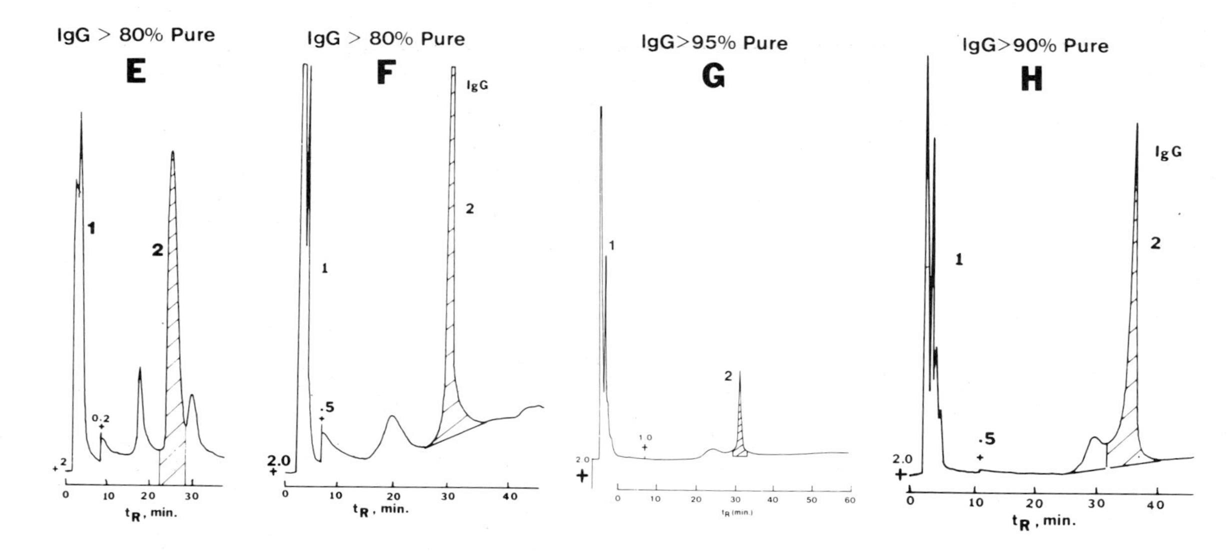

Fig. 1. Purification of monoclonal antibodies from eight different mouse ascites fluids on BAKERBOND ABx. Chromatography was conducted on columns (4.6 × 250 mm) containing 5 micron ABx (chromatograms A, B, C, and D), 15 micron ABx (chromatogram E), or 40 micron ABx (chromatograms F, G and H). The mobile phase consisted of an initial (A) buffer of 10 m*M* KH_2PO_4, pH 6.0, and a final (B) buffer of 250 m*M* KH_2PO_4, 6.8, with a linear gradient from 100% A to 50% B buffer over 1 hr. The flow rate was 1.0 mL/min and this produced back pressures of 1000, 200, or 30 psi for the 5, 15, and 40 micron particles, respectively. At least 0.4 mL of ascites fluid (diluted to 1.2 mL with buffer A) was loaded onto the column in each case. Peak 1 was composed of albumins, transferrins, and proteases, and peak 2 contained immunoglobulin (cross-hatched area). The proteins were detected by UV absorbance at 280 nm and the attenuation (absorbance units full scale; AUFS) was decreased following the elution of the void volume peak (1), as indicated in each chromatogram (see +), in order to better visualize the antibody peak.

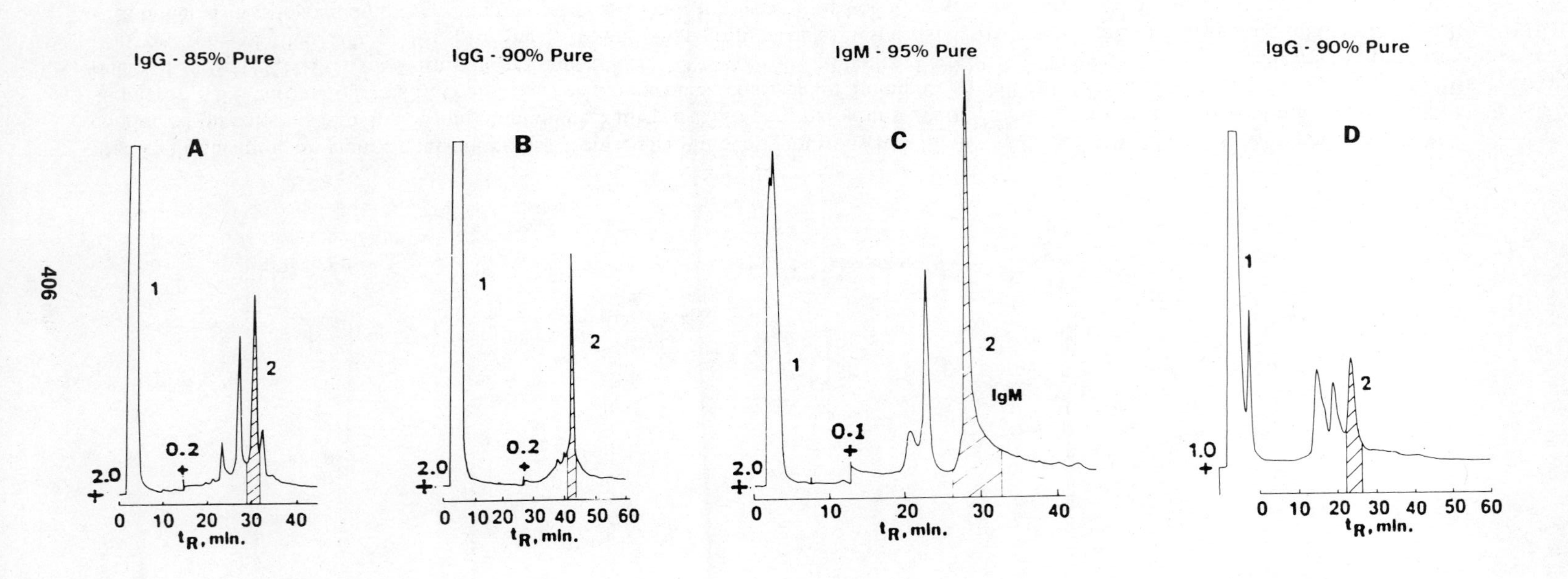

IgG - 85% Pure
A
1
2
0.2
2.0
0 10 20 30 40
tR, min.
IgG - 90% Pure
B
1
2
0.2
2.0
0 10 20 30 40 50 60
tR, min.
IgM - 95% Pure
C
1
2
IgM
0.1
2.0
0 10 20 30 40
tR, min.
IgG - 90% Pure
D
1
2
1.0
0 10 20 30 40 50 60
tR, min.

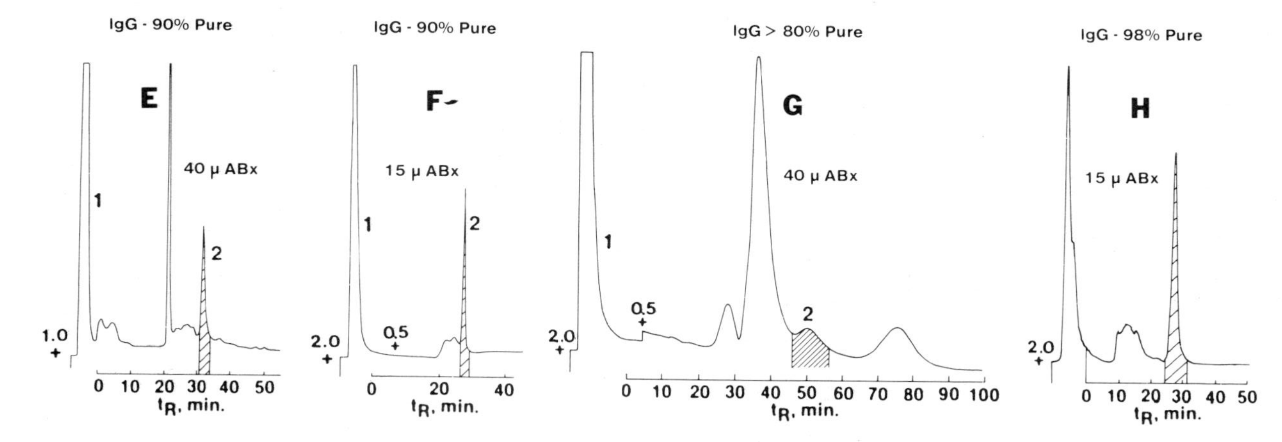

Fig. 2. Purification of monoclonal antibodies from various fetal bovine serum-supplemented cell culture fluids on BAKERBOND ABx. Chromatography was conducted on columns (4.6 × 250 mm) packed with either 5 micron ABx (chromatograms A, B, C, and D), 15 micron ABx (chromatograms E and F), or 40 micron ABx (G and H), as in Fig. 1, except, in chromatogram D, the starting (A) buffer was 10 m*M* MES, pH 5.6, the elution (B) buffer was 250 m*M* $(NH_4)_2SO_4$ plus 10 m*M* NaOAc, pH 5.6, and the gradient was linear from 100% A to 100% B buffer over 1.5 hr; in chromatogram E, the starting (A) buffer was 25 m*M* MES, pH 5.6, the elution (B) buffer was 500 m*M* NaOAc, pH 7.0, and the gradient was linear from 100% A to 100% B buffer over 1 hr; in chromatogram F, the starting (A) buffer was 10 m*M* MES, pH 6.15, and the gradient was linear from 100% A to 100% B over 1 hr; in chromatogram G the starting (A) buffer was 10 m*M* MES, pH 5.6, the elution (B) buffer was 500 m*M* NaOAc, 7.0, with a linear gradient from 100% A to 100% B over 2 hr and the flow rate was 0.5 mL/min; in chromatogram H, the starting (A) buffer was 25 m*M* MES, pH 5.6, the elution (B) buffer was 500 m*M* NaOAc, pH 7.0, the gradient was isocratic at 100% A buffer for 15 min, followed by a step gradient to 50% B buffer and a linear gradient from 50% B to 100% B over 1 hr, and the column size was 7.75 × 100 mm. The samples in chromatograms A–G were from conventional (batch) cell cultures containing fetal bovine serum (some were concentrated by tangential flow ultrafiltration prior to chromatography); the sample in chromatogram H was from an encapsulation broth (35% pure initially). At least 0.4 mL of neat cell culture fluid or ultrafiltrate (diluted to 1.2 mL with buffer A) was loaded onto the column in each case.

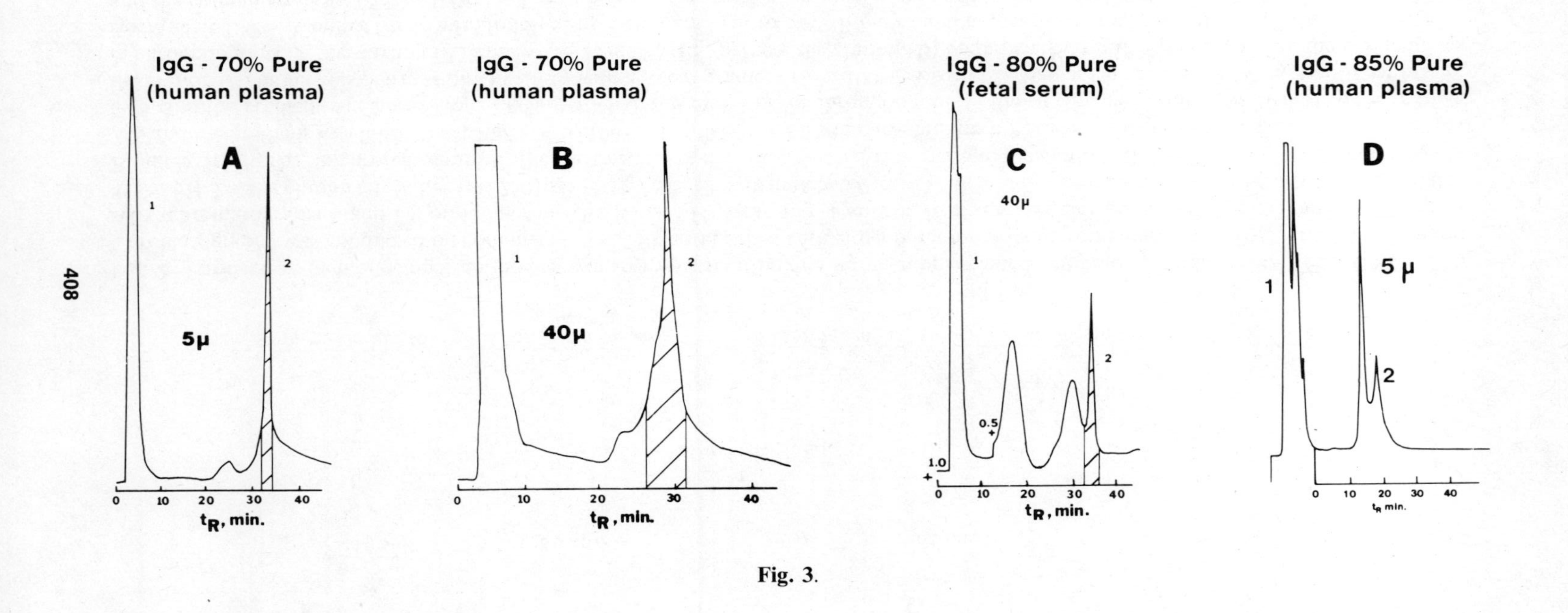

IgG - 70% Pure
(human plasma)
A
5μ
1
2
0
10
20
30
40
t_R, min.
IgG - 70% Pure
(human plasma)
B
40μ
1
2
0
10
20
30
40
t_R, min.
IgG - 80% Pure
(fetal serum)
C
40μ
1
2
0.5
1.0
0
10
20
30
40
t_R, min.
IgG - 85% Pure
(human plasma)
D
5 μ
1
2
0
10
20
30
40
t_R min.

Fig. 3.

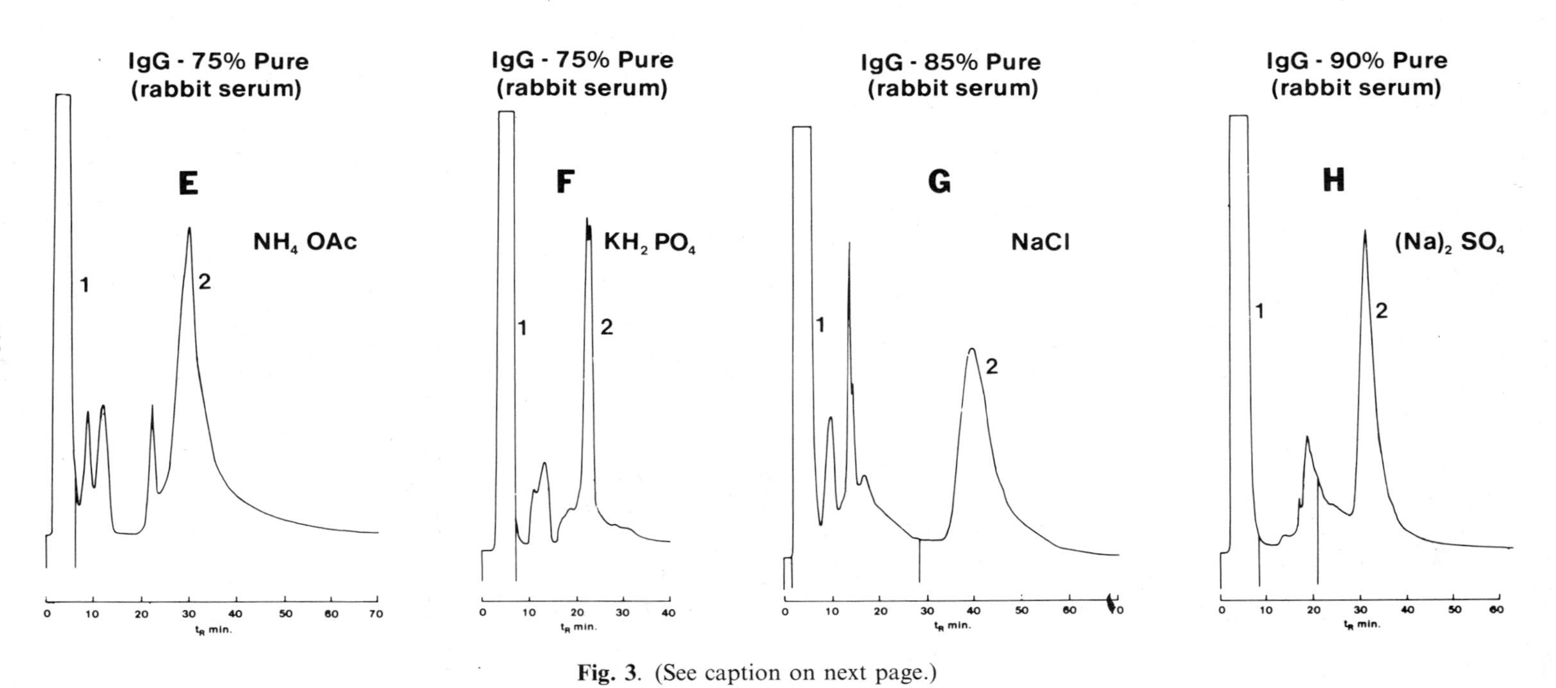

Fig. 3. (See caption on next page.)

Fig. 3. Purification of polyclonal antibodies from human plasma, rabbit serum, and fetal calf serum on BAKERBOND ABx. Chromatography was conducted as in Fig. 1, except, in chromatogram A, of human plasma, the column was packed with 5 micron ABx, and the initial (A) buffer was 10 m*M* MES, pH 6.0, in chromatogram B, of human plasma, a 6.0 × 200 mm column packed with 40 micron ABx was used and the gradient was 100% A to 100% B buffer over 30 min; in chromatogram C, of fetal calf serum, the column contained 40 micron ABx, the initial (A) buffer consisted of 10 m*M* MES, pH 5.6, and the final (B) buffer was 250 m*M* KH_2PO_4, pH 5.8; and in chromatogram D, of plasma from an immunodeficient human, a 7.75 × 100 mm column packed with 5 micron ABx was used; the initial buffer (A) was 10 m*M* MES, pH 5.6, the elution (B) buffer was 400 m*M* $(NH_4)_2SO_4$ plus 20 m*M* NaOAc, pH 6.0, and elution was achieved with a step gradient from 100% A to 10% B buffer, 12 min after the injection, held at 10% B for 14 min, and followed by a linear gradient from 10% B to 100% B over 5 min. The purification of polyclonal immunoglobulins from rabbit serum (chromatograms E, F, G, and H) was conducted on a column (7.75 × 100 mm) containing 5 micron ABx, at a flow rate of 1 mL/min with a pressure drop of 180 psi. The starting (A) buffer was 25 m*M* MES, pH 5.2, and the elution (B) buffer was either, 500 m*M* NH_4OAc, pH 7.0 (E), 250 m*M* KH_2PO_4, pH 7.0 (F), 500 m*M* NaCl plus 12 m*M* KH_2PO_4, pH 7.0 (G), or 400 m*M* $(Na)_2SO_4$ plus 10 m*M* KH_2PO_4, pH 7.0 (H). The vertical lines in each chromatogram refer to the injection, a step gradient, or the initiation of the linear gradient or all three. In chromatogram E, a linear gradient from 100% A to 100% B buffer over 1 hr was begun 7 min after the injection. In chromatogram F, a step gradient from 100% A to 20% B buffer was used 7 min after the injection, followed directly by a linear gradient from 20% B to 100% B over 1 hr. In chromatogram G, a step gradient from 100% A to 13% B was initiated 2 min after the injection (held at 13% B buffer for 26 min) and followed by a linear gradient from 13% B to 100% B buffer over 30 min. In chromatogram H, an initial isocratic gradient at 100% A buffer for 8 min was followed by a step gradient from 100% A to 10% B (held for 14 min) and a linear gradient from 10% B to 100% B buffer over 20 min. In chromatogram E, the purity of the entire IgG peak was approximately 75% (fractions 1 through 11 were 50, 60, 65, 65, 70, 80, 75, 75, 75, 70, and 60% pure, respectively, and 12, 13, and 14 were below the levels of accurate quantitation). In chromatogram F, the IgG peak was approximately 75% pure (fractions 1 through 7 were 70, 75, 80, 85, 80, 70, and 60% pure, respectively). In chromatogram G the purity of the entire IgG peak was about 85% (fractions 1 through 10 were 70, 75, 80, 90, 90, 85, 80, 80, 75, and 70% pure, respectively, and 11 and 12 were below detection limits). In chromatogram H, the IgG peak was approximately 90% pure (fractions 1 through 6 were 65, 80, 90, 95, 90, and 85% pure, respectively, and 7 was below the detection limit). In chromatograms A–C at least 0.2 mL of serum, diluted to 1.0 mL, was loaded onto the column; in D–H at least 1.0 mL of crude serum (diluted with 2.0 mL of buffer A) was loaded. Proteins were detected by absorbance at 280 nm at 2.0 AUFS unless indicated otherwise on the chromatogram.

Table 2. Effect of Various Equilibration and Dilution Buffers on the Binding of Rabbit Polyclonal Immunoglobulins (IgG, IgA, and IgM) to 40 micron BAKERBOND ABx[a]

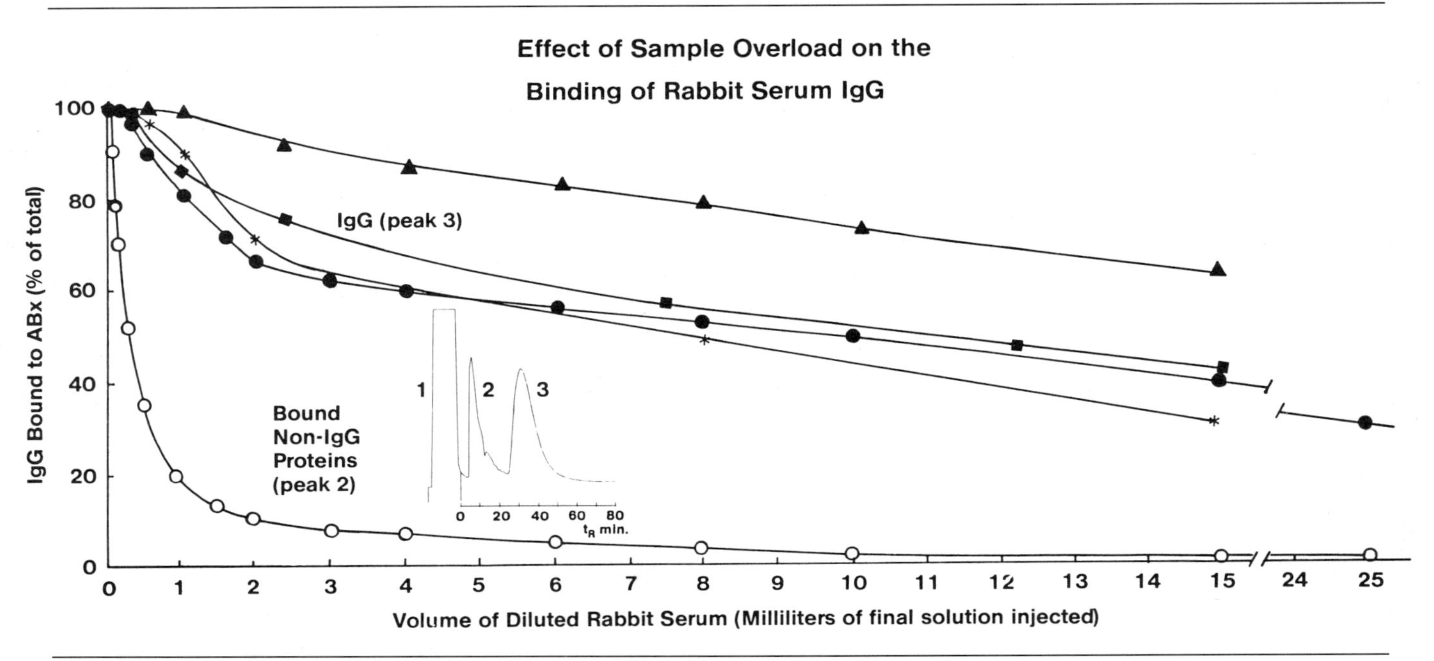

[a]Chromatography was conducted on a column (10 × 100 mm) packed with 40 micron ABx, as detailed in Fig. 18. Antibody binding was determined as a function of sample load, in presence of various initial equilibration and dilution buffers, including either, (●) 25 mM MES, pH 5.6 (1 volume rabbit serum plus 2 volumes buffer), or, (□) 25 mM MES, pH 5.6 (1 volume neat serum plus 3 volumes buffer), or, (*) 10 mM MES, pH 5.6 (1 volume serum plus 3 volumes buffer), or, (△) 10 mM MES, pH 5.6 (1 volume dialyzed serum plus 3 volumes buffer; dialysis was against 20 mM MES plus 15 mM KH_2PO_4, pH 5.8). The displacement of proteins from the bound nonimmunoglobulin peak (peak 2 in the insert) was also monitored (O) for the samples which were diluted with 2 volumes of 25 mM MES, pH 5.6.

Table 3. Effect of Column Overload with Ascites Fluid on the Binding and Resolution of a Monoclonal Antibody on BAKERBOND ABx and BAKERBOND MAb Matrices[a]

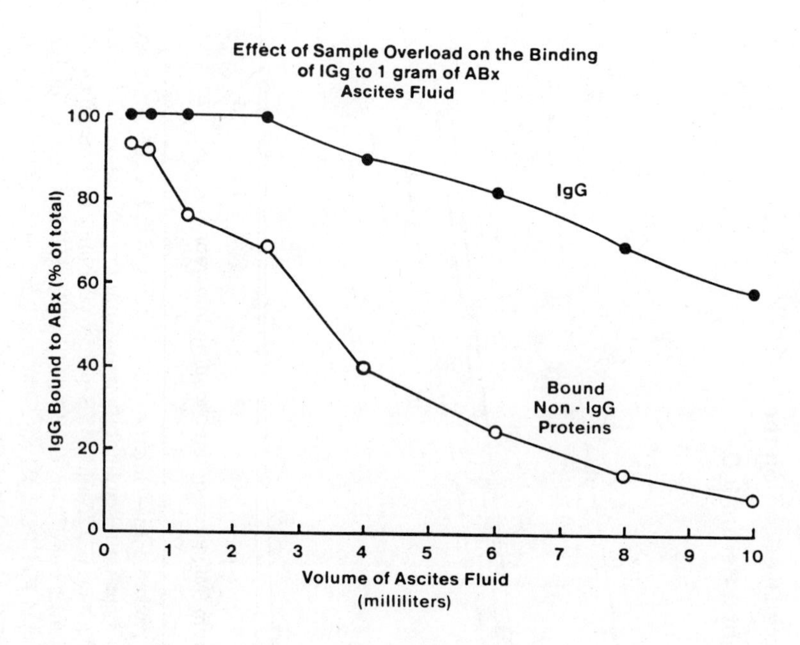

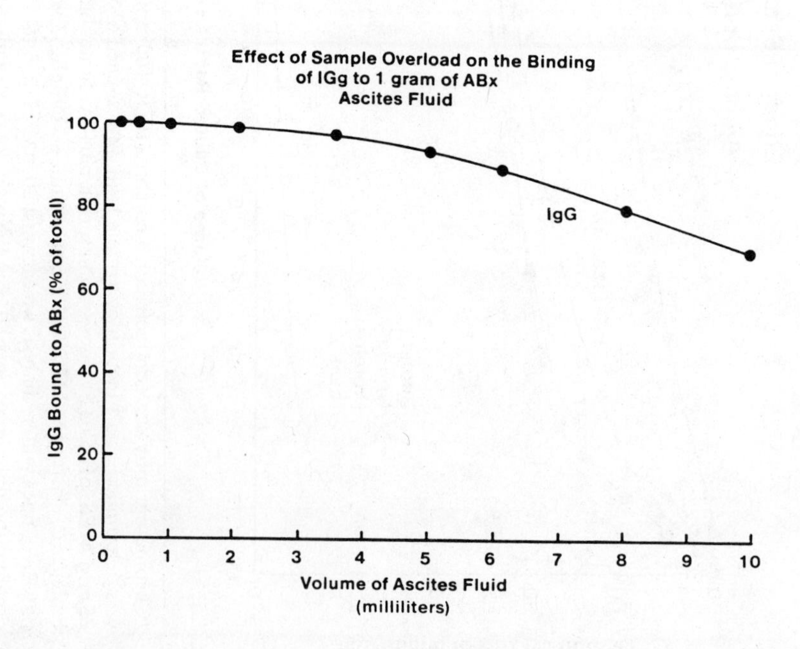

Table 3. (*Continued*)

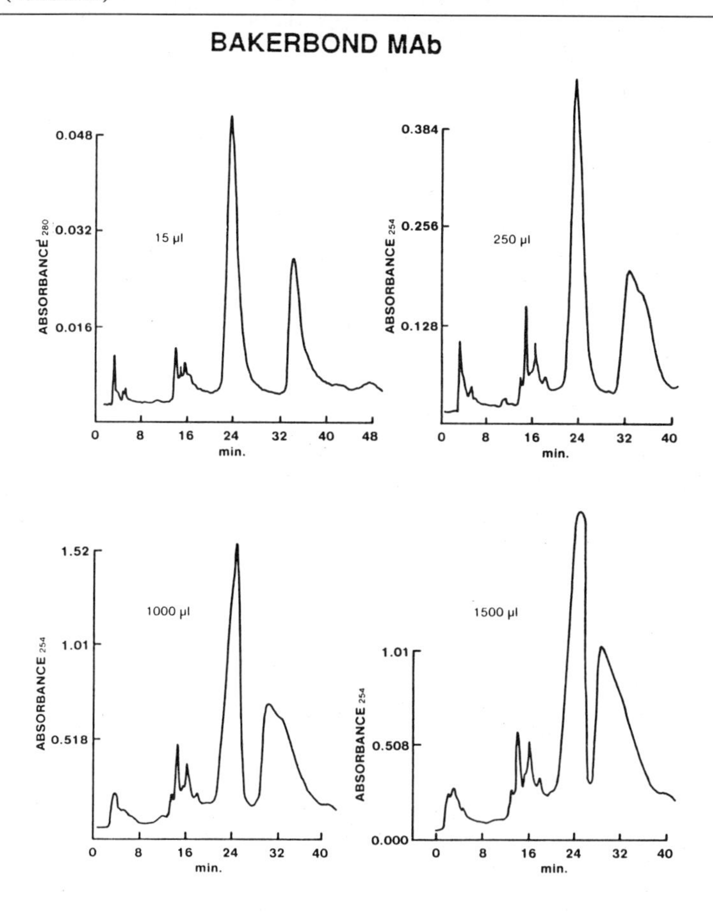

[a]Increasing amounts of a crude mouse ascites fluid containing a monoclonal antibody at approximately 10 mg/mL were loaded onto columns (4.6 × 250 mm) packed with either 15 micron ABx or 5 micron MAb. ABx chromatography was conducted on a column (7.75 × 100 mm) packed with 15 micron ABx, with a starting (A) buffer of 10 mM MES, pH 5.2, and an elution buffer of 1 M NaOAc, pH 7, at a flow rate of 0.7 mL/min; the samples were equilibrated to 10°C prior to injection and consisted of 1 part ascites fluid plus 3 parts buffer (A). Chromatographic analysis of the ABx void volume containing the monoclonal antibody molecules which failed to bind was conducted as detailed in Table 4. High-performance anion-exchange chromatography was conducted on a column packed with 5 micron MAb, with an intial (A) buffer of 10 mM KH_2PO_4, pH 6.8, and a final (B) buffer of 500 mM KH_2PO_4, pH 6.8, with a linear gradient from 100% A to 50% B buffer over 1 hr, at a flow rate of 1.0 mL/min, with a back pressure of 1000 psi. The loading capacity for ascites fluids on ABx is significantly higher than that on anion exchangers, despite the relatively low ligand density on the batch used in these early experiments (seed Table 4G).

Table 4. Effect of Sample Loading Conditions on the Binding of a Monoclonal Antibody and Other Proteins to 40 micron BAKERBOND ABx[a]

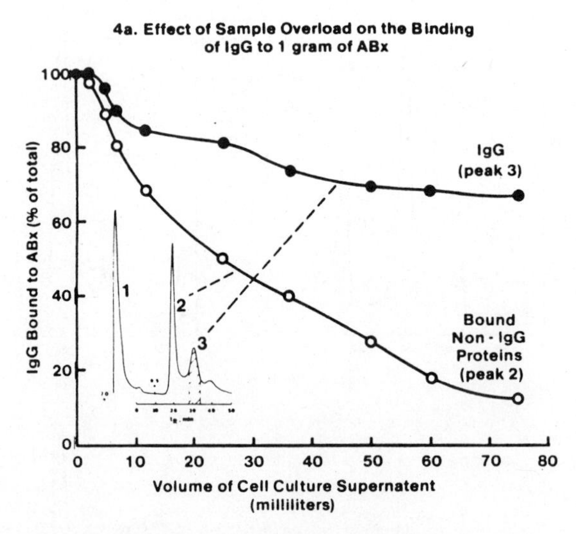

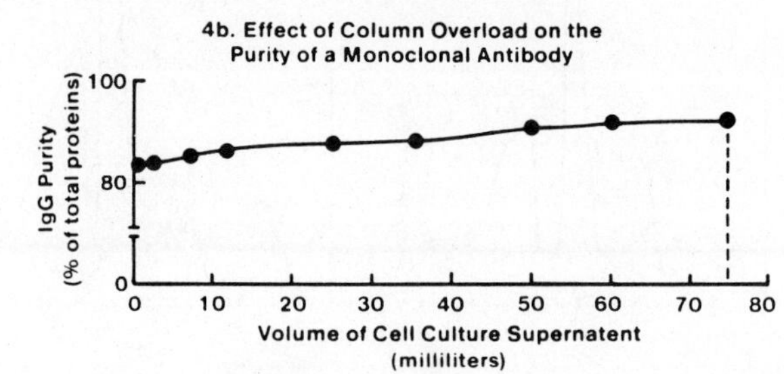

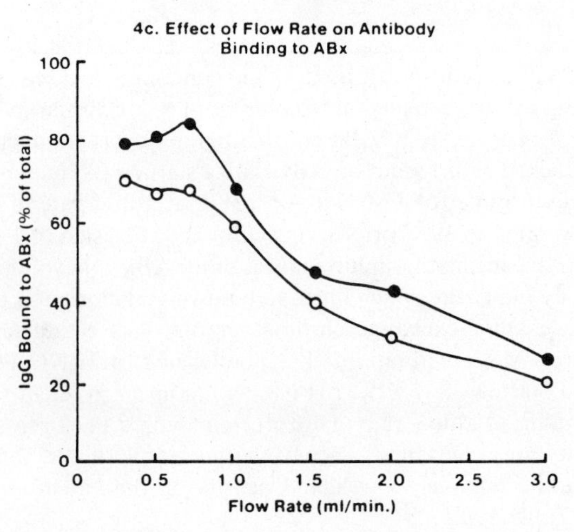

Table 4. (*Continued*)

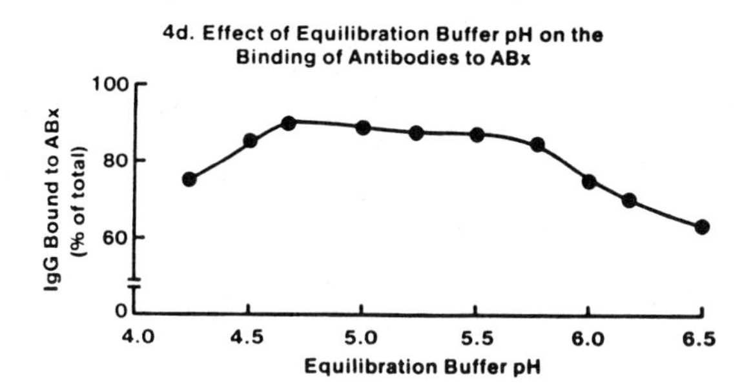

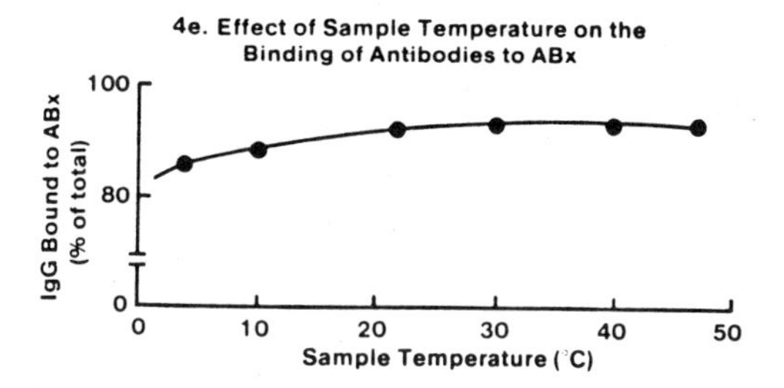

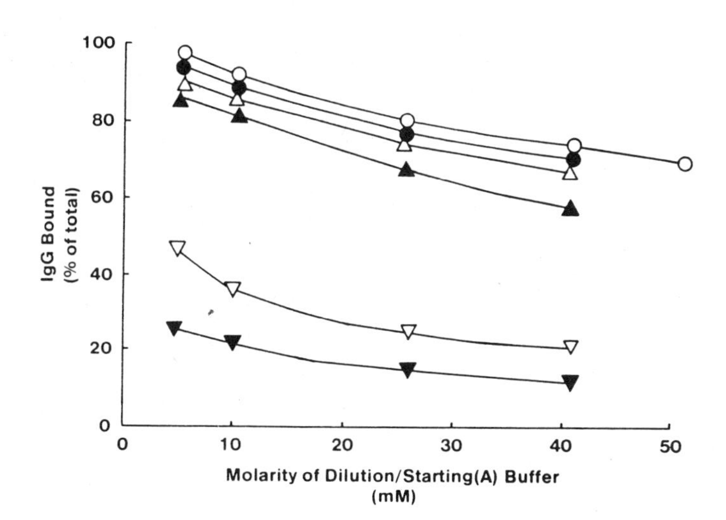

Table 4. (*Continued*)

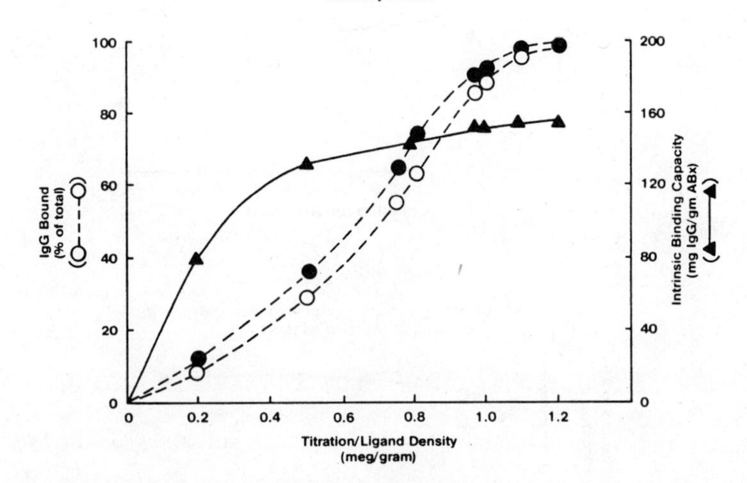

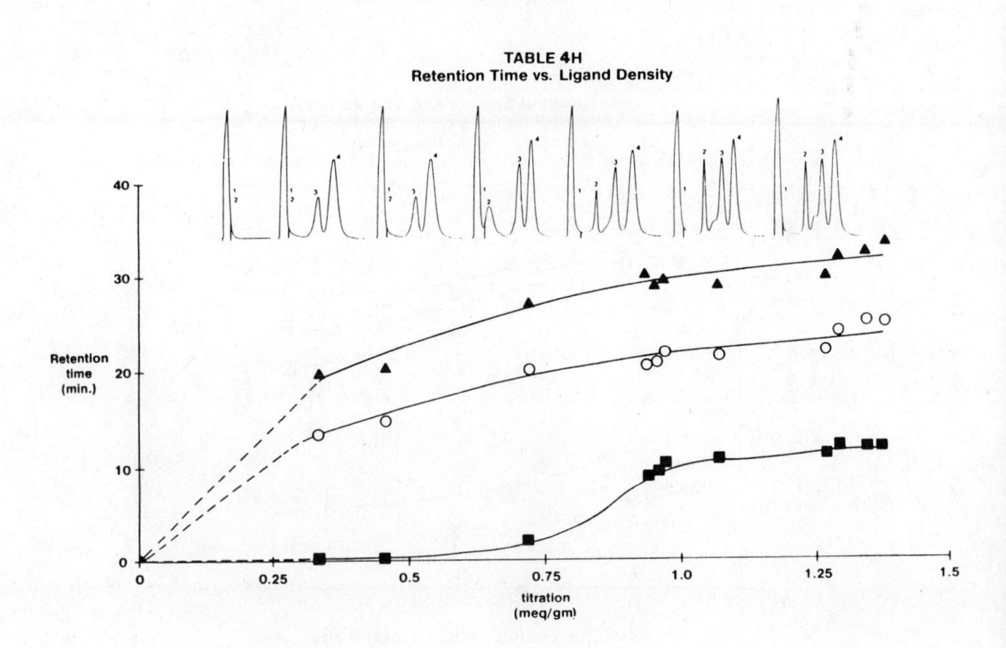

Table 4. (*Continued*)

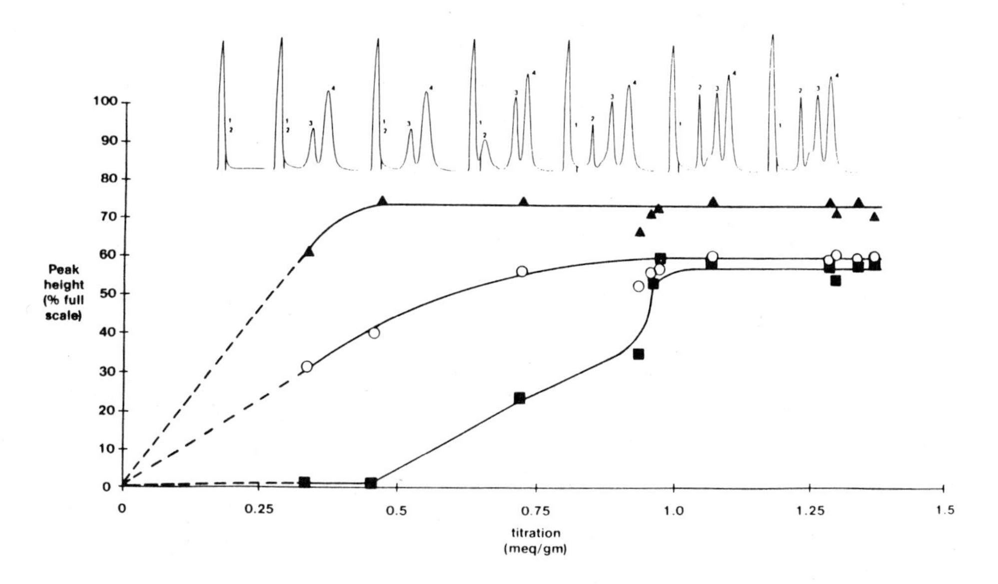

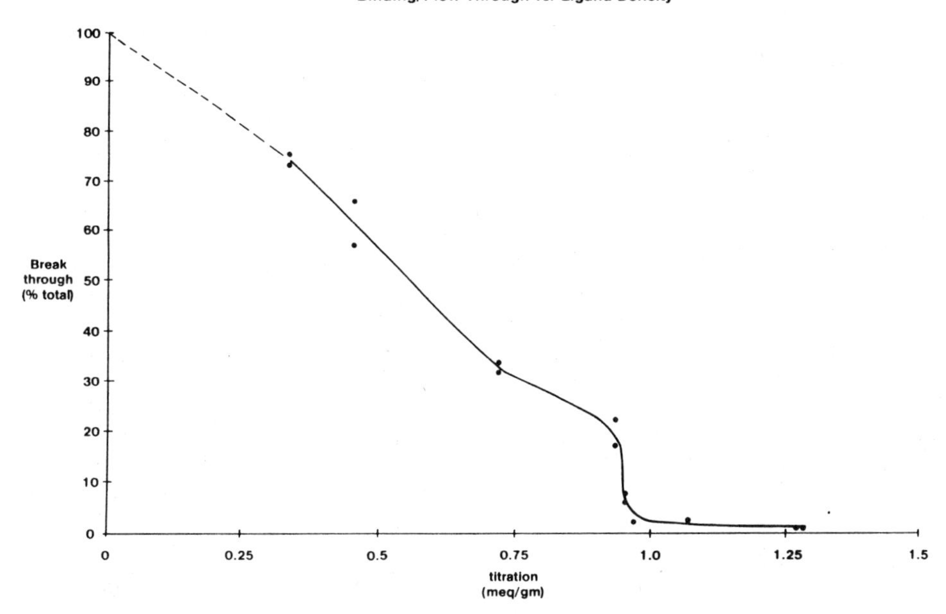

Table 4. (*Continued*)

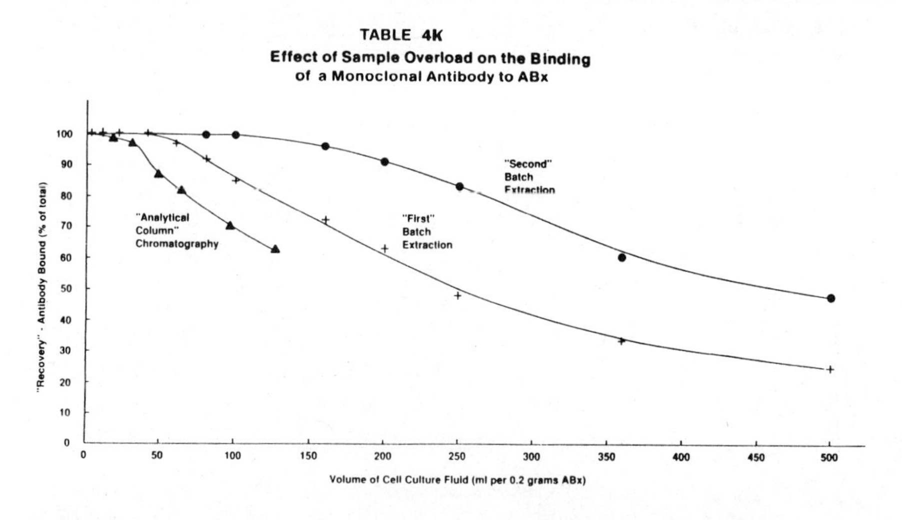

[a]In Tables 4A–4G and 4K the monoclonal antibody was present in a cell culture supernantant containing 10% fetal bovine serum and represented less than 1% of the total protein. Prior to chromatography the sample was concentrated 20-fold by ultrafiltration (against 50 mM Tris, pH 7.0) and diluted (1:6) with 25 mM MES buffer, pH 5.6. Chromatographic conditions are the same as those given in Fig. 9, with the minor changes noted below. The column (4.6 × 250 mm) contained approximately 1.5 g of ABx that had a relatively low ligand density; and the initial (A) buffer was 25 mM MES, pH 5.6 the sample load was 1.0 mL of concentrate (in 6 mL) containing approximately 55 mg of protein (unless stated otherwise). Analysis was conducted on a 5 micron ABx column (7.75 × 10 mm) as detailed in Fig. 9, by injecting 1.0 mL of the void volume from each of the 40 micron ABx runs, and measuring the height and the area of the monoclonal peak relative to the neat sample, taking into account changes in volume, dilution factors, or detection sensitivity, or all three. (Table 4A) The effect of sample overload on the binding of IgG to Abx in the column chromatographic mode. The volume values given in the graph refer to mL of original cell culture fluid. As the sample load is increased substantially, most of the bound nonimmunoglobulin proteins (○, bound, non-IgG contaminants; peak 2) are displaced by the antibody molecules (●), IgG; peak 3). As a result, the purity of the antibody peak actually increasess as the sample load increases (Table 4B). A number of conditions used in this experiment were later found to be less than optimal; the sample temperatue was 4°C (Table 4E), the equilibration buffer was pH 5.6 (Table 4D), the flow rate was 0.8 mL/min (Table 4C) and the ligand density (Tables 4G, 4H, 4I, 4J) of the ABx used in this study (0.80 meq/g) was significantly below the present product specification (0.95 meq/g). The optimization of these parameters should further increase binding and loadability. (Table 4B) Effect of ABx column overload on the purity of a monoclonal antibody. The monoclonal antibody peak from the loading experiments (Table 4A) was analyzed (by SDS PAGE); the purity of the IgG peak was actually found to increase with increasing column overload due to the preferential displacement of the bound, contaminating proteins (peak 2), relative to the monoclonal (peak 3). (Table 4C) Effect of flow rate on antibody binding to ABx. The binding of antibodies to ABx is faciltated by moderate flow rates, presumably because of hindered binding kinetics of large IgG molecules at rapid linear velocities; the reduced sample

Table 4. (*Continued*)

temperature (6 mL of sample at 4°C) probably affected binding kinetics (see Table 4E). Unexpectedly, slower flow rates also produced a decrease in binding affinity. The equilibration (A) buffer was either 20 mM MES, pH 5.60(●) or 20 mM MES, pH 6.15 (○). (Table 4D) Effect of equilibration buffer pH on the binding of antibodies to ABx. Optimal binding for this particular monoclonal was at low ionic strength in zwitterionic buffers (e.g., MES) at pH values between 4.7 and 5.7. However, for monoclonals with higher isoelectric points, higher pH values are typically more effective. (Table 4E) Effect of sample temperature on the binding of antibodies to ABx. The optimal temperatures for loading large samples onto ABx are those above 10°C. (Table 4F) Effect of the molarity of the sample dilution and equilibration (A) buffer on the binding of a monoclonal antibody to ABx. The cell culture ultrafiltrate was diluted (1:4) with and bound to the column in various equilibration buffers (MES, pH 5.6 (○), MOPSO, pH 5.6 (●), acetate, pH 5.6 (△), KH_2PO_4, pH 5.6 (▲), succinate, pH 5.6 (▽), or citrate, pH 5.6 (▼)) of different molarities, and chomatographed as detailed above. (Table 4G) Effect of the ABx ligand density on the binding of a monoclonal antibody to ABx. Eight columns were packed with experimental 40 micron ABx bonded phases in which the ligand densities were varied (as measured by elemental analysis, colorimetric assay and titration), in order to determine the effect on the binding of a monoclonal antibody from a crude sample. The flow rate was either 1.0 (○) or 0.5 (●) mL/min. In a separate set of experiments, the intrinsic binding capacity for IgG was determined on these same bonded phases (▲), in the batch or static mode. Briefly, 0.1 g of ABx was equilibrated in 25 mM MES, pH 5.4, and added to 20 mg of mouse polyclonal IgG with stirring. After 10 min, the amount of IgG that was present in the supernatant (that which was not absorbed) was measured spectrophotometrically at 280 nm against a standard curve. (Tables 4H–4J) The effects of ABx ligand density on protein binding in the preparative mode was also investigated using a more sensitive test system. A series of columns (4.6 × 250 mm) containing ABx bonded phases with varying degrees of ligand density were overloaded with a synthetic sample mixture. This sample contained 10 mg/mL ovalbumin (peak 1) and 1.5 mg/mL each of three proteins which bind to ABx to a different degree, peaks 2 (■), 3 (○), and 4 (▲). This sample mixture also contained 100 mM NaCl plus 100 mM KH_2PO_4, pH 6.7, in order to hinder binding and promote breakthrough of these three proteins. Large amounts (8.0 mL) of this sample were injected onto each column at a high flow rate (3.0 mL/min). Following the elution of the void volume peak (at the tic marks) the gradient was intiated from 100% A to 100% B buffer over 45 min, at a flow rate of 3.0 mL/min, with an initial (A) buffer of 25 mM MES, pH 5.8, and a final (B) buffer of 25 mM MES, pH 5.8, and a (B) buffer of 1 M NaOAc, pH 5.8. The data obtained were processed by three separate methods as detailed below. (Table 4H). In the first method, the relative retention times of each of the three bound proteins were measured (from the intitiation of the gradient) and plotted against the ABx ligand density; a good correlation was found. (In contrast, at low sample loads no significant difference was found.) (Table 4I) In the second method, the peak heights of the individual peaks were plotted against ligand density; again, a good correlation was observed. (Table 4J) In the third method, the void volume peak (about 20 mL) was collected and absorbance was measured spectrophotometrically at 400 nm a wavelength at which only the first two bound, chromophore-containing proteins (peaks 2 and 3) abosrb to any significant extent. The absorbance (color) of the void volume was used to quantitate the severity of the split peak, or amount of these proteins which failed to bind; again, this method gave a good correlation between the ABx ligand density and protein binding. (Table 4K) Effect of increasing sample volumes on the ability of ABx to bind a monoclonal antibody from

Table 4. ***(Continued)***

a cell culture ultrafiltrate by column chromatrography and by batch extraction. The sample consisted of a 10% fetal bovine serum-supplemented cell culture fluid that was concentrated by ultrafiltration (as in Table 4A) and diluted with 1 part of 25 mM MES, pH 4.2. Various volumes of this sample were added to conditioned 40 micron ABx and the monoclonal antibody was purified by batch extraction (+), as detailed in Fig. 20. A second batch extraction (●) was also performed using the supernatents from this first extraction. ABx column chromatography was conducted with a column (7.75 × 100 mm) containing the same ABx lot used in the batch extraction, using chromatographic conditions as detailed above (▲). Better binding was obtained in the batch mode, as well as on this higher load ABx (Table 4K) relative to experimental batches of ABx which had a lower ligand density (Table 4A).

the majority of nonimmunoglobulin proteins not only increases capacity, but also eliminates the need for an initial ammonium sulfate, salt fractionation.

To date, literally hundreds of antibodies have been purified on ABx in this and other laboratories (25, 29, 35, 62, 69, 70, 72, 80). Research in a number of laboratories indicates that the recoveries of immunological activity from ABx is either quantitative, or at least greater than those obtained by other high-performance or state-of-the-art methods (25, 29, 35, 62, 69, 70, 72, 80). These high recoveries reflect the mild elution conditions used for ABx chromatography.

As indicated previously, each antibody preparation represents a unique purification problem. Furthermore, not every antibody is destined for the same use, and as such, purity requirements may differ substantially. Although many immunoglobulins elute from ABx as electrophoretically pure peaks, the development of a purification protocol may involve the optimization of several parameters. These include choosing the correct buffer conditions, using step gradients, and manipulating flow rates in conjunction with reduced gradient steepness or step gradient elution in order to improve resolution.

4. EFFECTS OF MOBILE PHASE ON RESOLUTION

Substantial experimental evidence has been gathered in recent years demonstrating the effects of mobile phase composition on selectivity and the ability to manipulate protein elution profiles on reversed phase matrices. However, a review of the literature reveals little work on the optimization of buffer species and pH for the separation of proteins on ion-exchange supports (62, 65). Even those studies that have been conducted on high-performance ion-exchange supports have demonstrated only minor effects of mobile phase components on the basic chromatographic profile. In some cases, the lack of significant effects were attributable to the minimal differences in the salt species employed, while in other cases, these minor effects appeared to be the result of an inherent inability to manipulate selectivity on conventional ion exchangers, in general (62, 65). While manufacturers have made substantial efforts in developing literally hundreds of different chromatographic matrices for antibody purification, relatively little attention has been paid to the optimization of the mobile phase conditions required for protein and antibody separations on these supports.

Owing to their complex structure, individual proteins have numerous sites at which physical interactions may occur. Therefore, bonded phases such as ABx that consist of mixed-modes of binding ligands offer multiple chromatographic adsorption mechanisms. These multiple ligands allow for changes in selectivity in the presence of various buffer systems that are not typically encountered with more conventional chromatographic matrices (62, 63, 65). Physical interactions between any given protein and the ABx surface chemistry may differ substantially in the presence of various mobile phases due to the differences in buffering capacity, pH, ionic strength, and ion composition of various elution and binding buffers.

4.1. Elution Buffers

Dramatic changes in selectivity and resolution have been achieved on ABx by manipulating buffer conditions to optimize resolution in a one-step purification

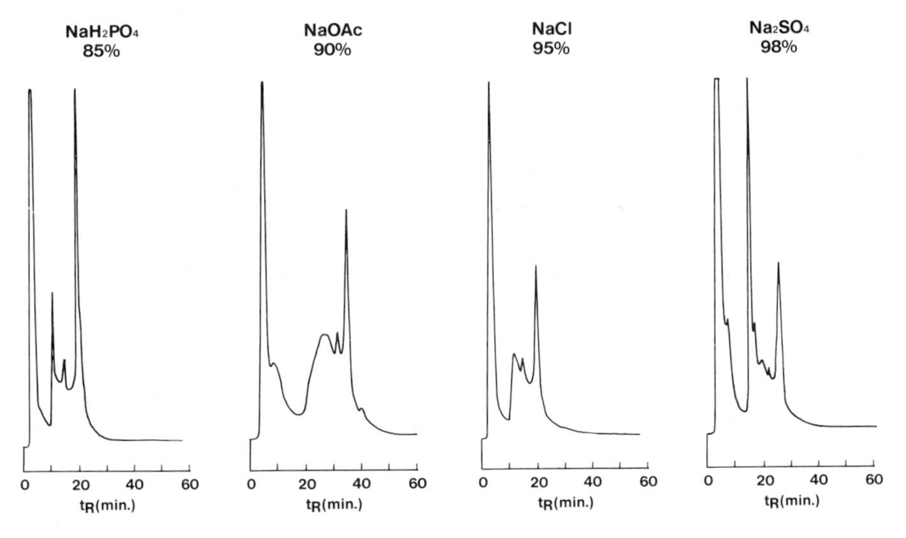

Fig. 4. Effects of elution buffer on the purification of a monoclonal antibody from mouse ascites fluid with 5 micron BAKERBOND ABx. Chromatography was conducted on a column (7.75 × 100 mm) containing 5 micron ABx. The mobilie phase consisted of an initial (A) buffer of 25 m*M* MES, pH 5.4, and a final elution (B) buffer of either 500 m*M* KH_2PO_4, pH 6.8, or, 1 *M* NaOAc, pH 7.0, or, 1 *M* NaCl plus 20 m*M* NaOAc, pH 5.7, or, 500 m*M* $(NH_4)_2SO_4$ plus 20 m*M* NaOAc, pH 6.7, with a linear gradient of 100% A to 50% B buffer over 1 hr, at 1.0 mL/min and 200 psi. Proteins were detected by UV absorbance at 280 nm at an attenuation of 2.0 AUFS. The sample consisted of 0.4 mL of crude mouse ascites fluid in 1.6 mL of buffer A. The monoclonal antibody ("MAb") was in the last peak in each chromatogram.

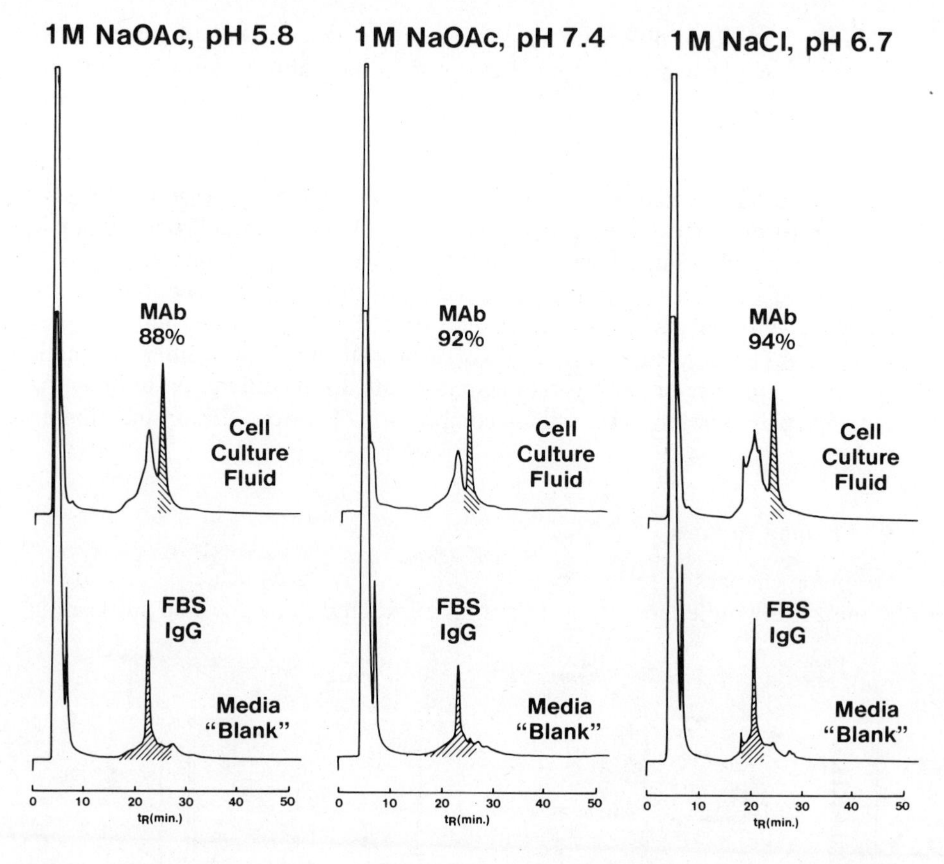

Fig. 5. Effect of elution buffer on the purification of a mouse IgG monoclonal antibody from a cell culture fluid ultrafiltrate with 5 μ BAKERBOND ABx. Chromatography was conducted on a column (7.75 × 100 mm) containing 5 micron ABx. The mobile phase consisted of an initial (A) buffer of 10 m*M* MOPSO plus 15 m*M* MES, pH 5.6, and a final elution (B) buffer of either, 1 *M* NaOAc, pH 5.8, or, 1 *M* NaOAc, pH 7.0, or, 1 *M* NaCl plus 20 m*M* NaOAc, pH 6.7, or, 500 m*M* KH_2PO_4, pH 5.2, or, 500 m*M* KH_2PO_4, pH 7.4, or, 500 m*M* $(NH_4)_2SO_4$ plus 20 m*M* NaOAc, pH 6.7. The gradient consisted of an initial hold at 100% A buffer for 4 min, followed by a linear gradient from 100% A to 100% B buffer over 26 min. The initial flow rate was 0.7 mL/min for 4 min, then changed to 1.0 mL/min; the back pressure was less than 200 psi. Proteins were detected by UV absorbance at 280 mn at an attenuation of 2.0 AUFs. The sample consisted of either, 0.4 mL of fetal bovine serum containing a hormone/peptide supplement ("media blank"), or 0.3 mL of cell culture ultrafiltrate ("culture fluid") diluted to 1.0 mL with buffer A. The monoclonal antibody (cross-hatched peaks labeled "MAb") eluted as the last major peak and was resolved from most protein contaminants, including the host (fetal bovine serum polyclonal) IgG (cross-hatched peaks labeled "FBS IgG"). The purity of the monoclonal (determined by SDS PAGE and high performance size exclusion chromatography) is given in each chromatogram.

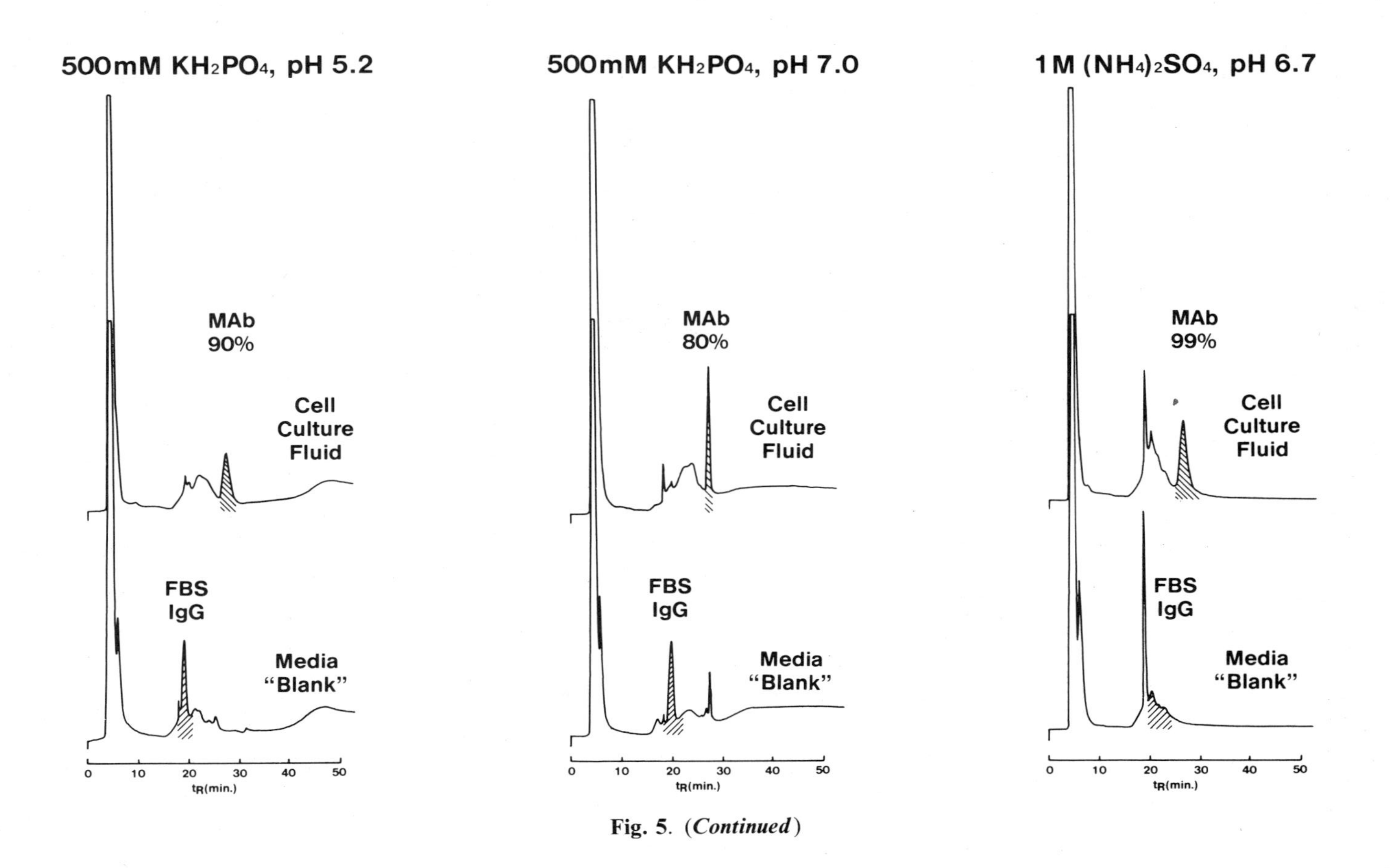

Fig. 5. (*Continued*)

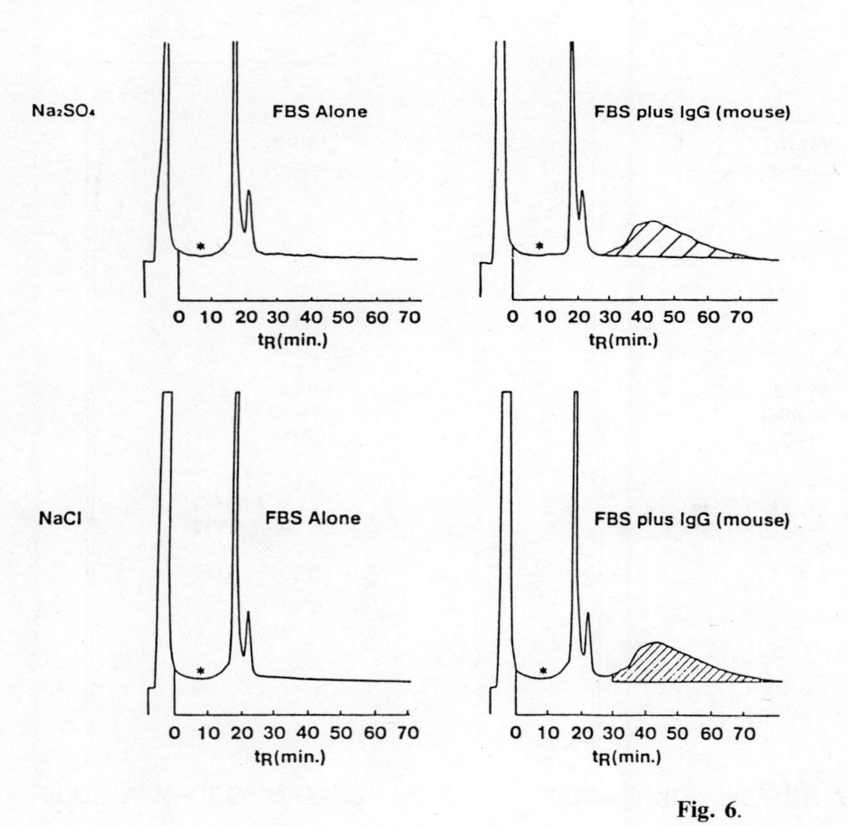

Na₂SO₄
FBS Alone
0 10 20 30 40 50 60 70
tR(min.)
FBS plus IgG (mouse)
0 10 20 30 40 50 60 70
tR(min.)
NaCl
FBS Alone
0 10 20 30 40 50 60 70
tR(min.)
FBS plus IgG (mouse)
0 10 20 30 40 50 60 70
tR(min.)

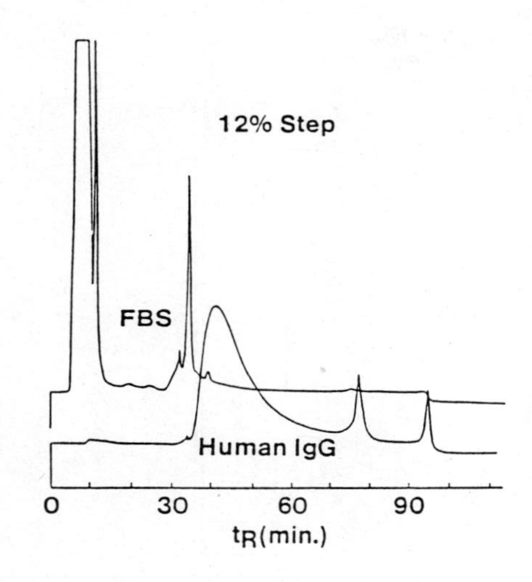

12% Step
FBS
Human IgG
0 30 60 90
tR(min.)

Fig. 6.

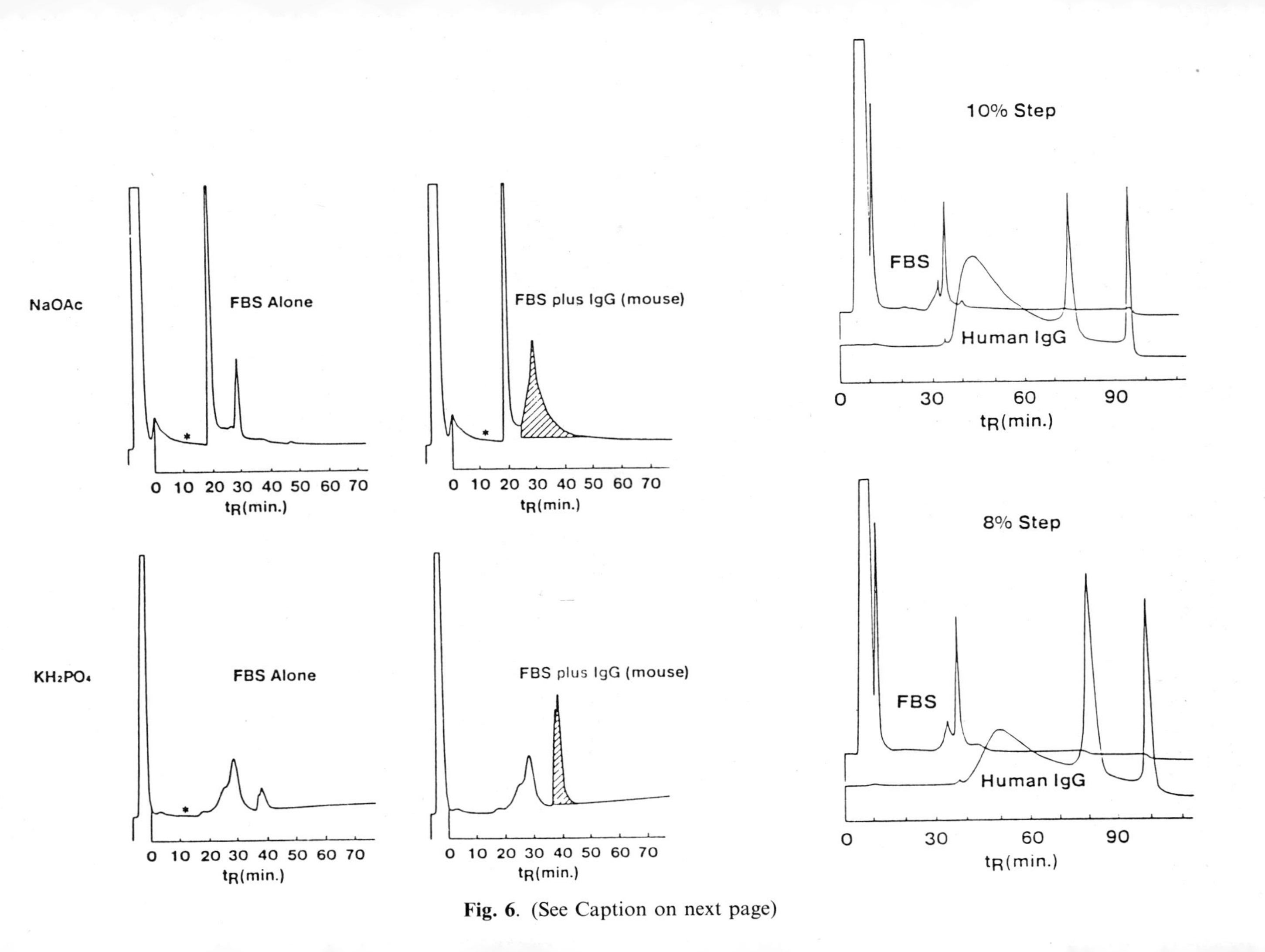

Fig. 6. (See Caption on next page)

Fig. 6. Effect of various elution buffers and step gradients on the separation of mouse polyclonal IgG from fetal bovine serum proteins on BAKERBOND ABx. Chromatography was conducted on a column (4.6 × 250 mm) packed with 5 (top left and top middle), 15 (right), or 40 micron ABx. The mobile phase consisted of an initial (A) buffer of 25 m*M* MES, pH 5.8, or, pH 5.5 (right), or 10 m*M* MES, pH 5.6 (top left) and a final elution (B) buffer of either, 250 m*M* $(NH_4)_2SO_4$ plus 10 m*M* NaOAc, pH 5.6, or, 1 *M* NaCl plus 20 m*M* NaOAc, pH 5.7, or, 1 *M* NaOAc, pH 7.0, or, 500 m*M* KH_2PO_4, pH 6.8, or, 500 m*M* $(NH_4)_2SO_4$ plus 20 m*M* NH_4OAc, pH 7.0 (right). In the gradient elution runs (left and middle) the flow rate was 0.5 mL/min, with a back pressure of 500 psi and 20 psi for the 5 and 40 micron runs, respectively, with a linear gradient from 100% A to 100% B buffer over 90 min. In the step gradient chromatograms (right), the flow rate was 0.7 mL/min for 7 min, and then 1.0 mL/min, at a back pressure of less than 200 psi, with an initial isocratic gradient at 100% A buffer for 7 min, followed by a step gradient to either 12% B, 10% B or 8% B (held for 56 min), and a third step to 100% B. Proteins were detected by UV absorbance at 280 nm at an attenuation of 2.0 AUFS with or without a change (see +) to 1.0 AUFS. The sample consisted of 0.5 mL of crude fetal bovine serum with or without mouse polyclonal IgG, diluted in 1.5 mL of buffer A. The cross-hatched peaks correspond to the retention times of purified mouse polyclonal IgG (left and middle) or, human serum IgG (right). Similar results were obtained with polyclonal IgG and IgM from other species. Elution with ammonium sulfate allows any antibody to be resolved from all fetal bovine serum proteins. The chromatograms on the right demonstrate the ability of step gradients to improve resolution.

scheme (Figs. 3–7 and 11). Alternatively, rechromatography of the immunoglobulin-containing peak under a second set of buffer conditions may be used to produce homogeneous antibody (Figs. 11 and 12).

Early experiments indicated that elution with phosphate, acetate, sulfate, or halide mobile phases produced distinct elution profiles, and often, the monoclonal and other bound proteins actually reverse their relative order of elution (Figs. 3–6 and 11). In the presence of sulfate or halide elution buffers, immunoglobulins are bound by ABx more strongly than virtually any other protein present in mouse ascites fluid, plasma, fetal bovine serum, or serum-based hybridoma cell culture media (Figs. 3–6). Although sulfate- or halide-based mobile phases accentuate the separation, similar, but often less pronounced results are obtained in other buffer systems (Figs. 3–7 and 9–11).

Fig. 7. Effect of initial equilibration buffer pH and ionic strength on the purification of a monoclonal antibody from a serum-supplemented cell culture ultrafiltrate. Chromatography was conducted on a HPLC column (4.6 × 250 mm) containing 15 micron ABx. The mobile phase consisted of an initial (A) buffer of 10 or 25 m*M* MES, at a pH of either 5.6, 5.9, or 6.1 and a final elution (B) buffer of 500 m*M* KH_2PO_4, pH 6.3, with a linear gradient of 100% A to 50% B buffer over 60 min, at a flow rate of 1.0 mL/min, with a back pressure of 200 psi. Proteins were detected by UV absorbance at 280 nm at an attenuation of 1.0 AUFS. The sample was 0.5 mL of crude serum-supplemented cell culture ultrafiltrate (10 ×) plus 1.5 mL of the appropriate A buffer. The cross-hatched peak contains the monoclonal antibody purities of 80, 90, 93, 95, and 97%, from top left to bottom right, respectively (determined by SDS PAGE).

10 mM MES

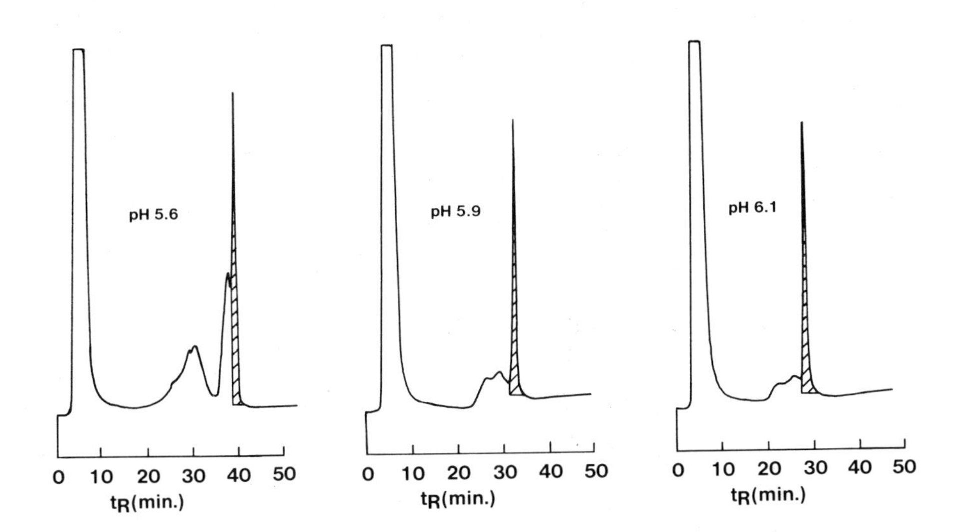

25 mM MES

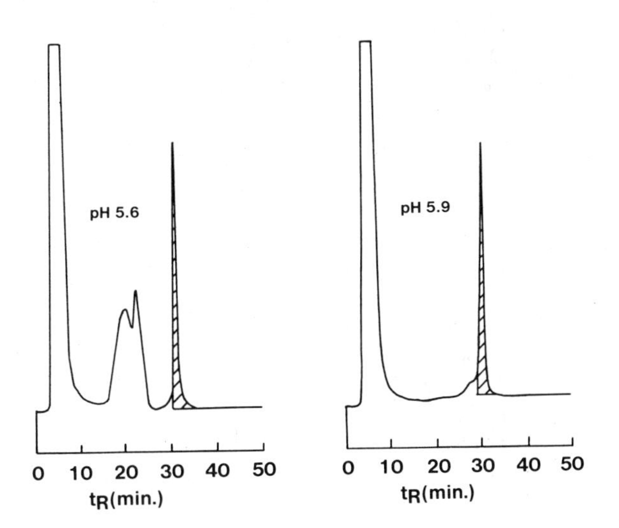

Fig. 7.

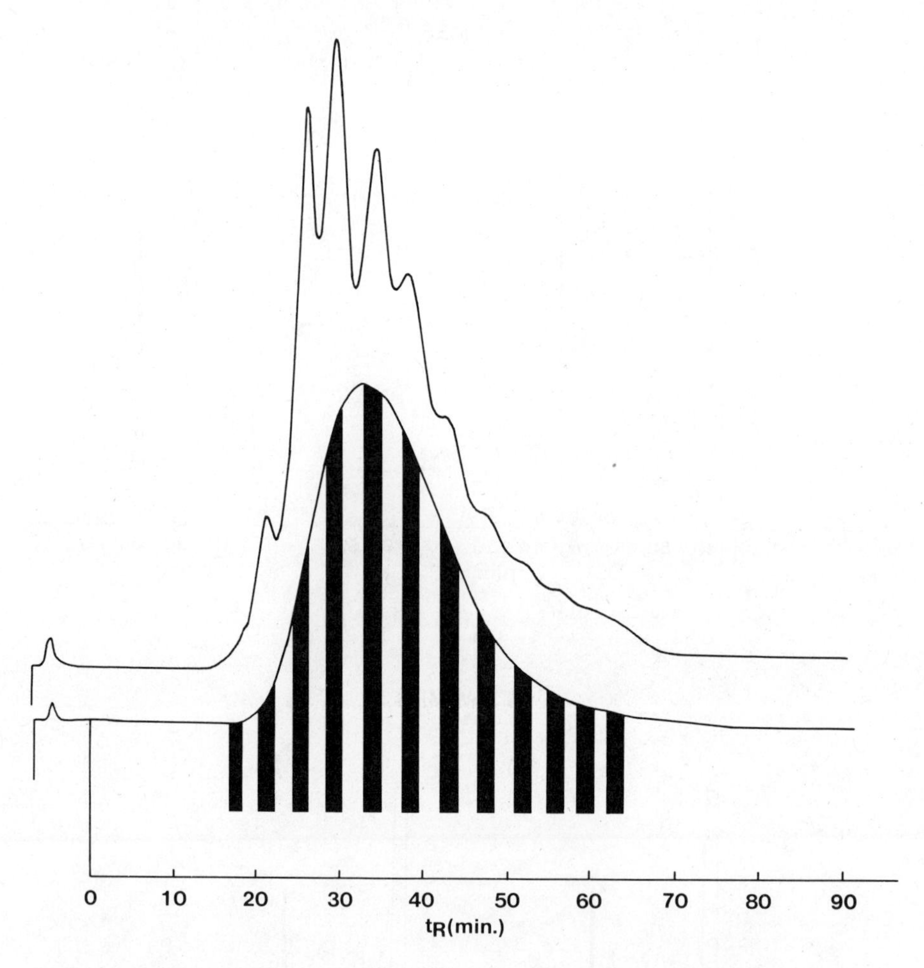

Fig. 8. Rechromatography of antibody fractions from human polyclonal IgG on 5 micron BAKERBOND ABx, and the resolution of various immunoglobulin species. Chromatography was conducted on a column (7.75 × 100 mm) containing 5 micron ABx. The mobile phase consisted of an initial (A) buffer of 25 m*M* MES, pH 5.4, and a final elution (B) buffer of 500 m*M* $(NH_4)_2SO_4$ plus 20 m*M* NaOAc, pH 6.7, with a linear gradient of 100% A to 25% B buffer over 1 hr, at a linear flow rate of 1.0 mL/min, with a back pressure of 200 psi. Proteins were detected by UV absorbance at 280 nm at an attenuation of 2.0 AUFS (below) or 0.02 AUFS (top). The sample consisted of 25 mg of purified human IgG (below), or 0.04 mL of each of the 12 shaded fractions (top) diluted in 4.0 mL of 25 m*M* MES, pH 5.0. In the lower chromatographic run every other fraction (shaded areas) was collected, diluted, and reinjected under identical conditions, to produce the chromatogram on top. These chromatograms illustrate the high resolution on ABx, and the ability to separate immunologically active polyclonal fractions in order to enhance specificity, reduce nonspecific binding, and to produce a preparation which is more similar to a monoclonal antibody.

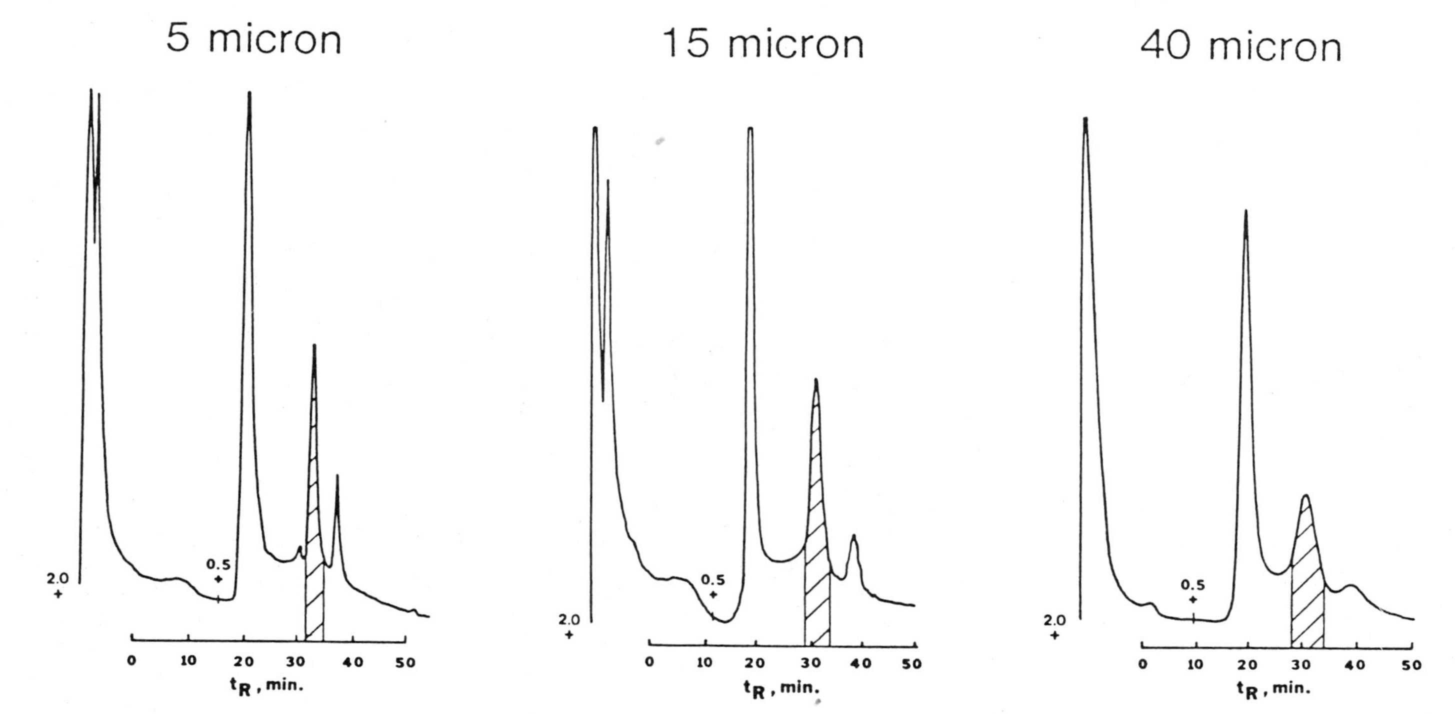

Fig. 9. Chromatographic profile of a serum-supplemented cell culture ultrafiltrate chromatographed on 5, 15, and 40 micron BAKERBOND ABx. Chromatography was conducted on columns (7.75 × 100 mm) packed with 5, 15, or 40 micron ABx. The mobile phase consisted of an initial (A) buffer of 10 m*M* MES, pH 5.6 and a final (B) buffer of 1 *M* NaOAc, pH 7.0, with a 1 hr gradient from 100% A to 100% B buffer, at a flow rate of 1.0 mL/min and a back pressure of 200, 15, and 3 psi for 5, 15, 40 micron ABx, respectively. The sample was 0.5 mL of tissue culture ultrafiltrate, diluted to 1.2 mL with buffer A. Proteins were detected by UV absorbance at 280 mm, and the attenuation was changed from 2.0 to 0.2 AUFS (see +). The purity of the monoclonal (cross-hatched peaks) was greater than 95, 92, and 87% (by SDS PAGE) for the 5, 15, and 40 micron ABx runs, respectively. Note the similarity in elution profiles due to the identical surface chemistry. This facilitates scale-up (see Fig. 10 and Table 4).

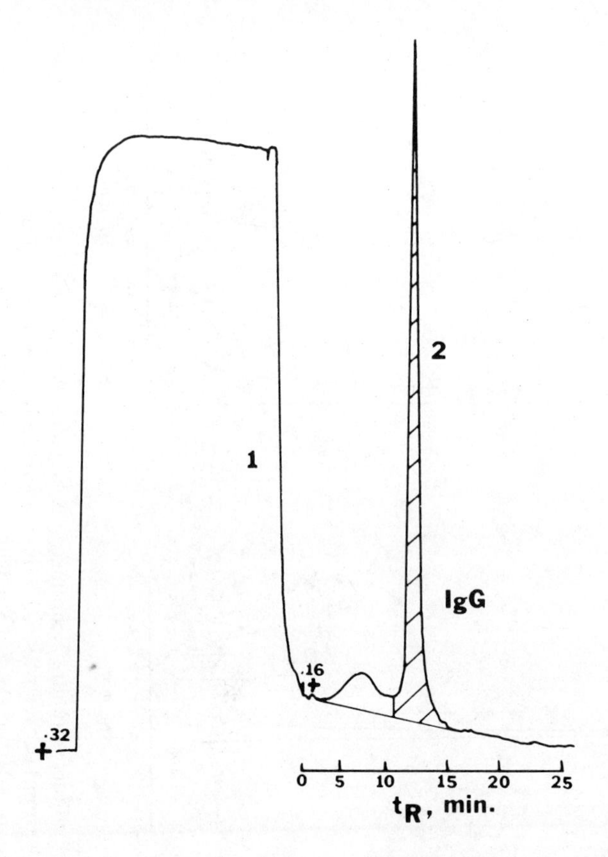

Fig. 10. Preparative chromatography of serum-based cell culture supernatant on 40 micron BAKERBOND ABx. Chromatography was conducted on a column (21.2 × 150 mm) containing 15 g of 40 micron ABx. The mobile phase consisted of an initial (A) buffer of 10 m*M* MES, pH 5.6, and a final (B) buffer of 250 m*M* KH_2PO_4, pH 6.8. The gradient was linear from 100% A to 100% B buffer over 40 min (beginning at t_0). The flow rate was 20 mL/min and the back pressure was 20 psi, and the sample consisted of 1.5 L of cell culture fluid that was concentrated 20-fold by ultrafiltration and diluted 5-fold with buffer A (375 mL final volume). The detector (UV at 280 nm) sensitivity was increased from 0.32 AUFS to 0.16 AUFS following the elution of the void volume peak (1) which contained albumins, transferrins, and other proteins. Peak (2) contained greater than 95% of the original IgG at a purity greater than 90% (by SDS PAGE). Method development for this sample is shown in Fig. 9 and in Table 4.

An extension of this work indicates that for rabbit serum, most ascites fluids, and most cell culture fluids, the purity of the immunoglobulin fractions and the resolution between the antibody(s) and the major bound, nonimmunoglobulin peak(s) progressively increases as the anionic species of the elution buffer is changed from phosphate, to acetate, to halide (chloride), and then, to sulfate (Figs. 3–6). That is, elution with phosphate often produces antibodies of the lowest purity, while elution with sulfate produces higher purities. In some cases,

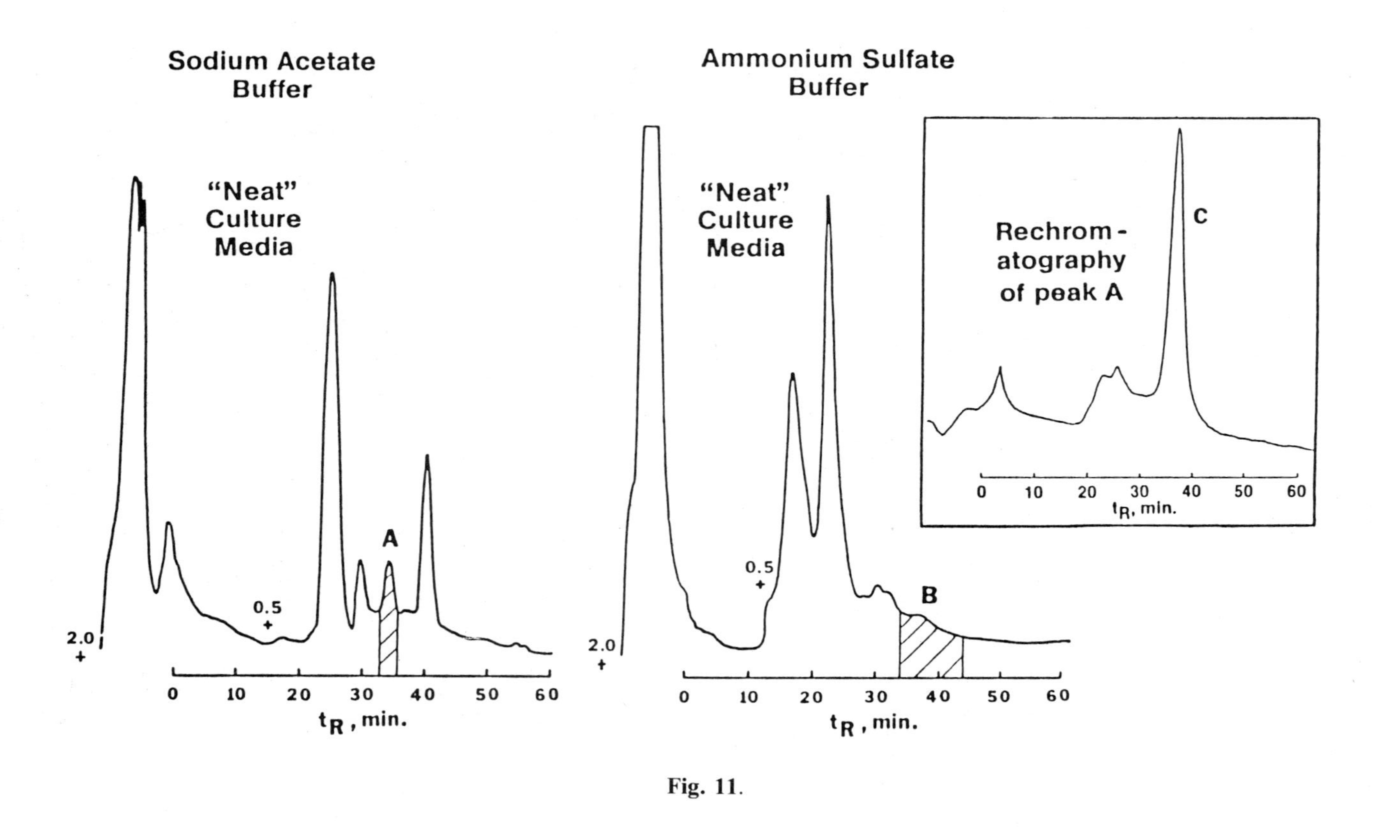
Sodium Acetate Buffer
"Neat" Culture Media
2.0
0.5
A
0 10 20 30 40 50 60
t_R, min.
Ammonium Sulfate Buffer
"Neat" Culture Media
2.0
0.5
B
0 10 20 30 40 50 60
t_R, min.
Rechrom-atography of peak A
C
0 10 20 30 40 50 60
t_R, min.

Fig. 11.

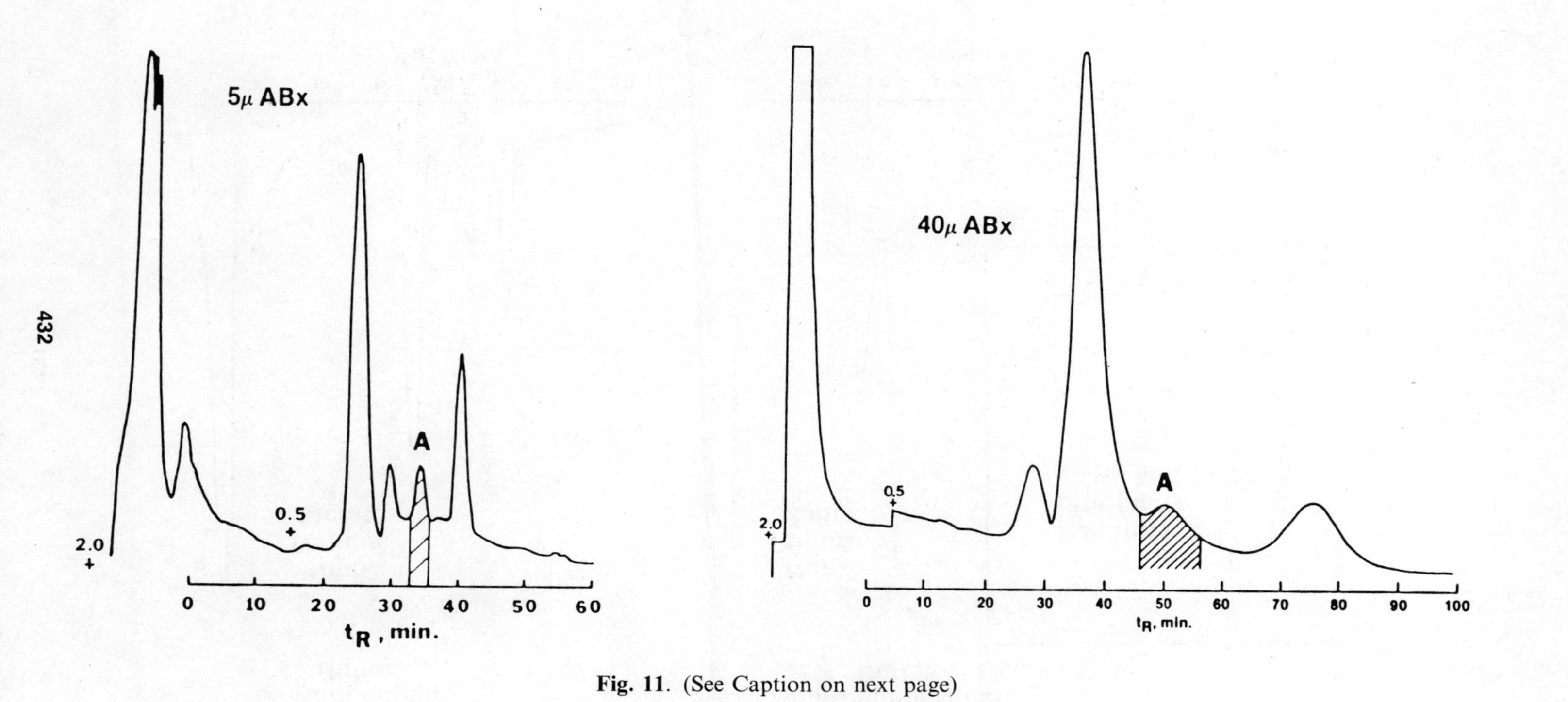

Fig. 11. (See Caption on next page)

Fig. 11. Method development and method transfer of a two-step ABx purification of a monoclonal antibody from a serum-supplemented cell culture fluid on BAKERBOND ABx. Chromatographic method development (upper row) was conducted on a column (7.75 × 100 mm) packed with 5 micron ABx. The mobile phases consisted of an initial (A) buffer of 10 m*M* MES, pH 5.6, and a final (B) buffer of either 500 m*M* NaOAc, pH 7.0 (A, above left), or 250 m*M* $(NH_4)_2SO_4$ plus 10 m*M* NaOAc, pH 5.6 (B, above right, and C in the insert). The gradient was linear from 100% A to 100% B buffer over 1.5 hr and the flow rate was 0.7 mL/min, with a back pressure of 150 psi. The sample load was 0.4 mL of neat cell culture fluid (diluted to 1.2 mL with A buffer). The detection was at 280 nm at 2.0 and 0.5 AUFS (see +). Although adequate purity (greater than 85%) was achieved with either elution buffer alone, rechromatography of peak A in the presence of ammonium sulfate buffer produced homogeneous antibody (99+ % by SDS PAGE). Elution with the acetate mobile phase alone resulted in the separation of the monoclonal from more than 80% of the bovine serum immunoglobulins; the second step further reduced these levels by more than 95%. The lower chromatograms represent the transfer of the initial step to 40 micron ABx. Chromatography was conducted as detailed in chromatogram A (top left), except the column was packed with 40 micron ABx, the gradient was increased to 2 hr following a 15 min loading period at 100% A buffer, the flow rate was reduced to 0.5 mL/min, the back pressure was below 20 psi, and the purity of the cross-hatched peak (A) was 78% (by SDS PAGE analysis). (Further method transfer and scale-up are shown in Fig. 12.)

acetate buffer also produces excellent purity with both mouse and rat ascites fluids (25, 29, 62, 69, 70, 80).

In the presence of ammonium sulfate and chloride elution buffers, ABx resolves virtually all the nonimmunoglobulin proteins present within fetal bovine serum from mouse or bovine serum polyclonal IgG (Fig. 6). Because polyclonal antibodies consist of many different monoclonals, these results suggest that any monoclonal antibody will also be resolved from all fetal bovine serum components. In contrast, some coelution of the polyclonal antibodies does occur in the presence of both acetate and phosphate elution buffers (Fig. 6). For this reason, sulfate- or halide-based elution buffers are suggested for the purification of monoclonals from cell culture supernatants on ABx.

A common misconception is that sharp peaks always correlate with increased resolution and purity. Although this generalization often holds, many times the opposite is actually true (Figs. 4 and 5). For example, the broad peaks that result from the elution of polyclonal immunoglobulins from ABx with sulfate buffers are due to high resolution and the separation of a number of polydisperse immunoglobulin species (Fig. 8). Likewise, immunoglobulins present in fetal bovine serum are virtually undetectable following elution with sulfate buffers, eluting over a 40 min period (Fig. 6); in contrast, elution with phosphate produces low resolution of these multiple antibody species which coelute as an extremely sharp peak (Fig. 3). With phosphate buffers, sharp peaks may also result in a decrease in resolution of a monoclonal antibody from contaminating proteins (Figs. 3–6).

Despite the fact that the elution of monoclonal antibodies from ABx with

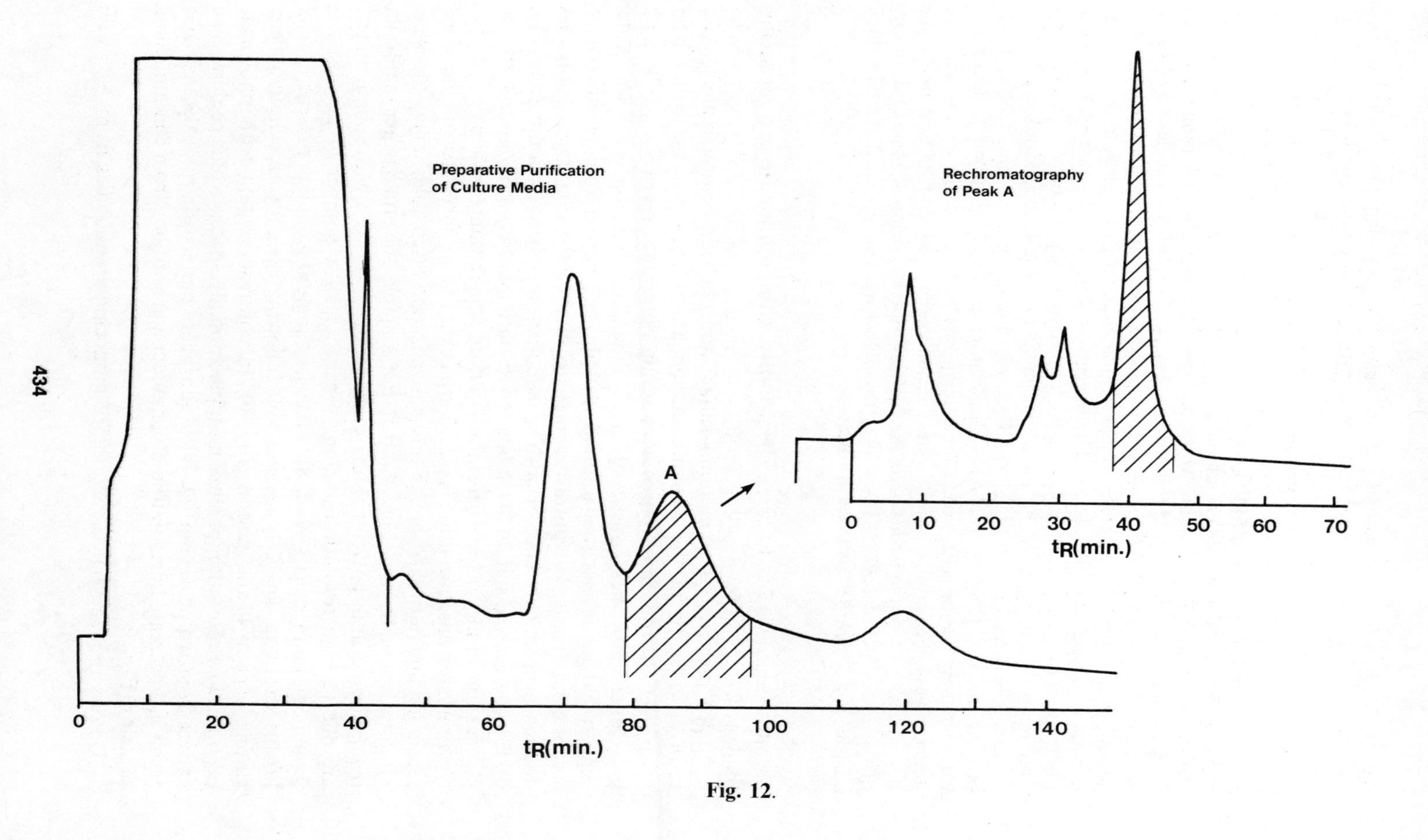
Preparative Purification of Culture Media
Rechromatography of Peak A
A
t_R(min.)
0
20
40
60
80
100
120
140
t_R(min.)
0
10
20
30
40
50
60
70

Fig. 12.

phosphate buffers tends to produce the lowest purity of all mobile phases, Corleone and Swaminathan (25) have recently shown that elution of monoclonals from 80 different ascites fluids with phosphate buffer produced average purities of greater than 90%; however, even higher purities were obtained with acetate elution buffers. With either buffer, both the purity and the recovery of antibodies from these ascites was higher on ABx than on high-performance anion exchange chromatography (25). Epstein et al. (29) have also recently found that the acetate elution of monoclonals antibodies from nine mouse ascites and cell culture fluids on ABx gave nearly homogenous IgM; a purity higher than that achieved following a two-step purification by either high-performance anion-exchange plus size exclusion chromatography, or, salt precipitation plus size exclusion chromatography. Ross et al. (80) have also used acetate elution of mouse and rat ascites fluids on ABx to obtain purities greater than 99% for more than half the monoclonals tested; of the remaining, most were 95% pure, while one was only 80% pure.

The particular cation species (i.e., Na^+, K^+, Li^+, NH_4^+, and so on) present within the ABx elution buffer is not of major significance. Comparable elution profiles and antibody purities have been obtained with a series of cationic salts of either phosphate [KH_2PO_4, NaH_2PO_4, $(NH_4)_2HPO_4$], acetate (KOAc, NaOAc, NH_4OAc), sulfate [$(NH_4)_2SO_4$, Na_2SO_4] or chloride (KCl, NaCl, NH_4Cl). However, minor increases in peak sharpness have been observed in the presence of ammonium phosphate relative to sodium phosphate, and in the presence of ammonium acetate relative to sodium acetate (43, 62, 63, 69–71). Furthermore, for sodium chloride or ammonium sulfate mobile phases, neither

---

Fig. 12. Preparative two-step purification of a monoclonal antibody from a serum-supplemented cell culture fluid on BAKERBOND ABx. The initial preparative chromatographic step was conducted on a 1 in. column (22 × 250 mm) packed with 40 micron ABx (left), while the second step was on a half inch column (10 × 250 mm) containing 15 micron ABx (right). The mobile phase consisted of an initial (A) buffer of 25 m*M* MES, pH 5.4, and a final elution (B) buffer of either 500 m*M* $(NH_4)_2SO_4$ plus 20 m*M* NaOAc, pH 6.7 (right), or, 2 *M* NaOAc, pH 7.0 (left). The gradient for the initial run consisted of an isocratic gradient at 100% A buffer for 45 min, followed by a linear gradient from 100% A to 50% B buffer over 1 hr, while on the semipreparative column the gradient was linear from 100% A to 50% B buffer over 90 min, with a flow rate of 25.0 mL/min on the preparative column and 4 mL/min on the semipreparative column, with back pressures of 20 and 150 psi, respectively. Proteins were detected by UV absorbance at 280 mm at an attenuation of 2.0 AUFS. The initial sample consisted of 200 mL of crude serum-supplemented cell culture media (dialyzed for 8 hr against 25 m*M* MES plus 15 m*M* KH_2PO_4, pH 6.2), plus 400 mL of 25 m*M* MES, pH 5.0. The cross-hatched peak from the initial run was collected and dialyzed (as above), and a 20 mL aliquot was diluted in an equal volume of 25 m*M* MES, pH 4.0, rechromatographed on the semi-prep column and eluted with NaOAc. The purity of the final monoclonal antibody peak was greater than 97% (by SDS PAGE) and contained less than 10% of the immunoglobulins from the fetal bovine serum. (Method development is detailed in Fig. 11.)

the acutal identity of the buffer species which is added at low ionic strengths (e.g., 20 mM KH_2PO_4, or 20 mM NaOAc, and so on), nor the final pH (provided it is between 5.0–7.8) is of major importance in determining the chromatographic profile and final product purity (62, 63, 69–71). In contrast, the manipulation of pH of the acetate or phosphate buffer may have substantial effects on the final purity of the monoclonal. In general, phosphate buffers at lower pH (5.6) give better resolution than those at neutral pH, while acetate buffers at neutral pH tend to produce higher purities than those at a lower pH (62). These results appear to be due to the sharp increases in elution pH that accompanies elution of antibodies in neutral phosphate buffers, which causes proteins to pile up, versus, the gradual increase in eluent pH characteristic of the neutral acetate buffers, which actually facilitates resolution (62, 63, 69–71).

Although specific elution buffers are recommended, virtually any biological or assay buffer [phosphate buffered saline (PBS), and so on] may be used as an exchange buffer, in order to rapidly elute the antibody fraction from ABx, provided it is at or above physiological ionic strength (62, 63). Simply by changing from an ammonium sulfate gradient to such an exchange buffer, following the elution of all nonimmunoglobulin proteins, the antibody(s) may be eluted from the column as a sharp peak of high purity in a buffer suitable for immediate use (62, 63, 69–71).

4.2. Binding Buffers

In addition to changing the elution buffer conditions, significant changes in elution profile and selectivity can be achieved on ABx with various equilibration or binding buffers. An example is the effect of the starting buffer pH and ionic strength on the elution profile and the purity of a monoclonal antibody present in a serum-supplemented cell culture ultrafiltrate (Fig. 7). The use of phosphate as an elution buffer results in the coelution of the monoclonal and nonimmunoglobulin contaminants in the presence of equilibration buffers of low ionic strength and pH. However, increasing the pH or ionic strength of the equilibration buffer or both causes the coeluting, nonimmunoglobulin proteins to be selectively displaced, so that the purity of the monoclonal antibody increases substantially (Fig. 7). However, it should be noted that increasing the pH or the ionic strength of the starting buffer or both can reduce the binding of some antibodies in the preparative mode, particularly those with a low isoelectric point (Appendix V). Therefore, in some instances the choice of the optimum starting buffer may represent a compromise between resolution and capacity. However, higher pH values can be used with some monoclonals (particularly those with a high pI) to increase binding capacity, as well as resolution (62, 63, 69–71).

Two techniques that represent alternatives to manipulating the binding buffer conditions are changing the starting buffer species or pH or both directly after sample loading and antibody binding, or, operating the ABx in the non-equilibrated mode in which the column is not fully equilibrated or titrated prior to sample loading (62, 63, 69, 70). Although the ability to change elution profile and antibody purity on ion exchangers by manipulating inital equilibration buffer conditions (Fig. 7) is rarely discussed in the literature, these techniques should receive serious consideration during method development.

5. METHOD DEVELOPMENT AND SCALE-UP WITH LINEAR AND STEP GRADIENTS

Despite a long tradition of antibody purification on ion exchangers, and the variability in retention of different monoclonals on these supports, few studies have focused on the optimization of such separations. In part, the lack of such a systematic approach to method development may be associated with the low resolution and long chromatographic run times that are inherent to conventional gels. Long chromatographic run times make method development a long and tedious task, while poor resolution makes it difficult to assess peak identity and purity, and, less likely that antibodies will be resolved from contaminants via rechromatography.

To facilitate method development and scale-up, the ABx surface chemistry has been bonded to three silica particle sizes producing bonded phases with remarkably similar chromatographic properties (Fig. 9). These silica sizes include, 5 micron for high-performance column chromatography, 15 micron for high resolution, medium pressure, preparative chromatography, and 40 micron bulk silica (BAKERBOND PREPSCALE matrices) for large-scale batch extraction and traditional, open column chromatography (62, 63). The differences in resolution on these three particle sizes are minimized due to the use of an identical surface chemistry (Fig. 9).

Although the scale-up of a chromatographic method may be accomplished in a number of ways, the use of HPLC columns containing high performance liquid chromatographic matrices (e.g., 5 or 15 micron ABx) to conduct initial method development has a number of distinct advantages. High-performance packings provide fast chromatographic run times and high resolution, so that proteins elute as sharp peaks at high specific activities (Fig. 9). These factors facilitate the identification of the antibody via electrophoresis or immunological assay, increase sensitivity and permit the use of extremely small amounts of precious sample. Because high-performance matrices also provide more rapid separations, detailed investigations may be conducted into the factors that influence chromatographic selectivity, such as mobile phase composition and gradient profile (Figs. 3–5, 7, 13, 15); because resolution is better, small but significant changes in elution profile can be easily detected (e.g., impurities eluting as shoulders on the main peak can be identified) and the resolution from major contaminating proteins can be ascertained.

These factors also facilitate method development because the antibody peak eluted in the presence of one mobile phase can be used to identify active fraction in other buffer systems; this reduces the need to analyze all fractions by tedious

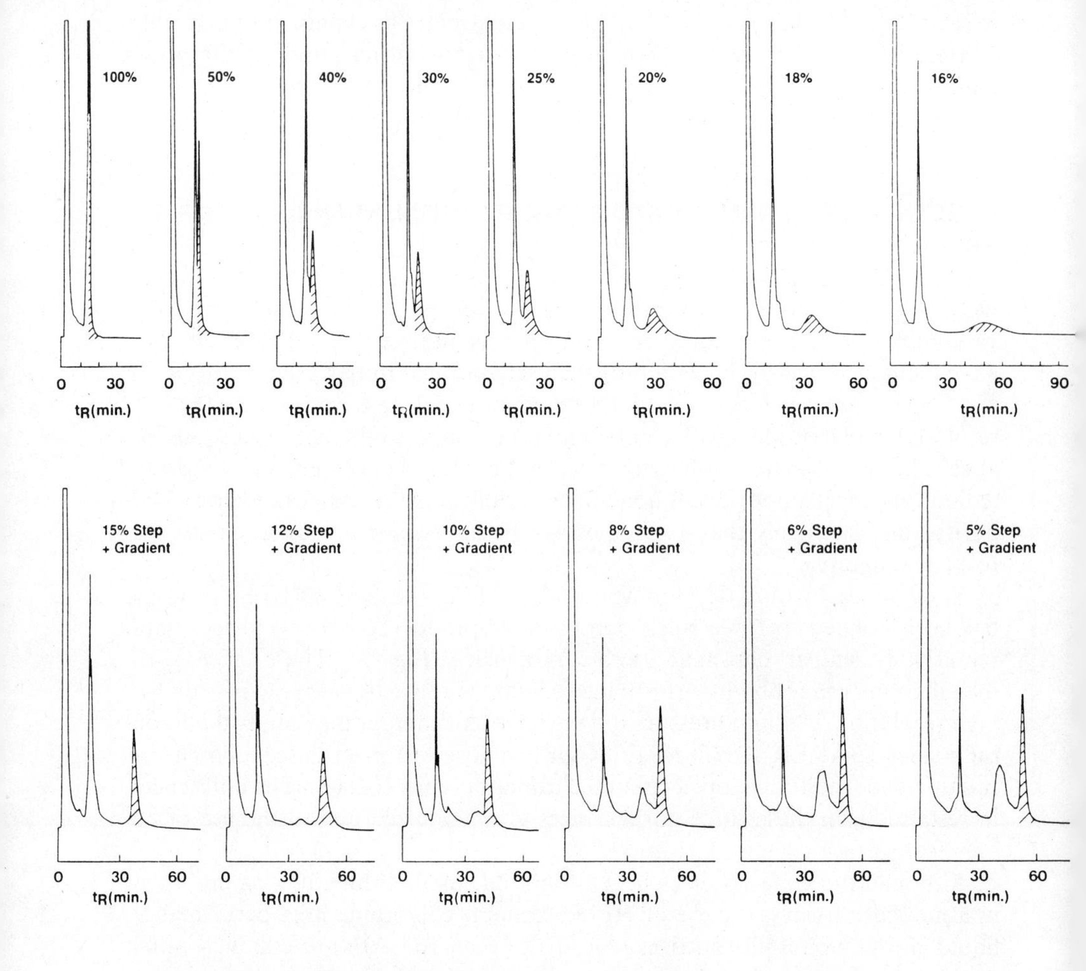

Fig. 13. Effects of step gradients on the resolution of a monoclonal antibody from a fetal bovine serum-supplemented cell culture fluid on 15 micron BAKERBOND ABx. Chromatography was conducted on a column (7.75 × 100 mm) packed with 15 micron ABx. The equilibration (A) buffer was 10 m*M* MOPSO plus 15 m*M* MES, pH 5.6 and the elution buffer was 500 m*M* $(NH)_2SO_4$ plus 20 m*M* NaOAc, pH 6.7. The flow rate was 1.0 mL/min (except as noted below) at a back pressure of 50 psi, and proteins were detected by UV absorbance at 280 nm at an attenuation of 1.0 AUFS. The sample consisted of 0.4 mL of a serum-supplemented cell culture ultrafiltrate diluted to 1.0 mL with buffer A. An initial isocratic gradient at 100% A buffer was held for 4 min, followed directly by either a single step gradient to various percentages of the B buffer (top; 100% B through 16% B), or (second row) with lower ionic strength step gradients in conjunction with a concomitant linear gradient to 100% B buffer over 30 min (i.e., 100% A, step to 15% B, then 15% B to 100% B over 30 min, etc.), or (third and fourth rows) either a linear gradient (see Fig. 5 for details), or, an initial step gradient to 16% B buffer, with or without a second step gradient to 50% B buffer or changes in flow rate or both. In

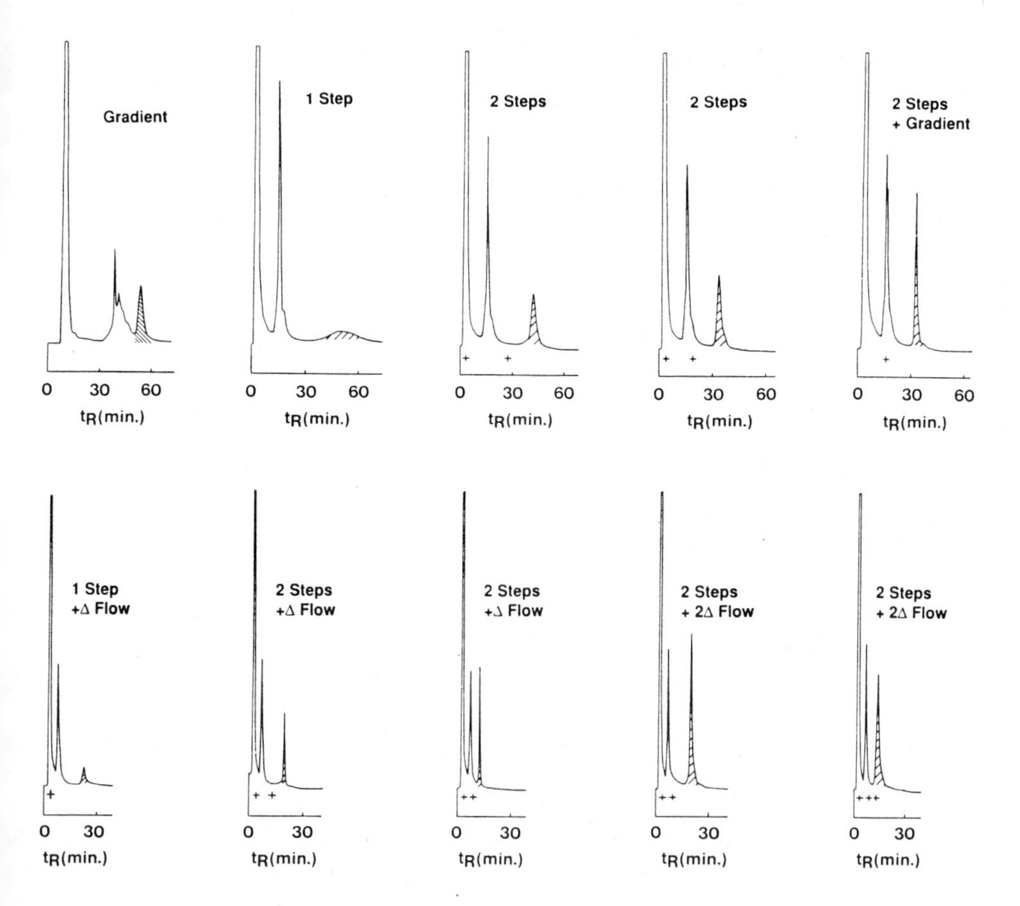

these chromatograms in the lower two rows, the first + denotes the first step gradient to 16% B buffer, and in the bottom row, a concomitant increase in flow rate to 2.5 mL/min; the second + denotes a second step gradient to 50% B buffer; and, the second and third + signs in the last two chromatograms, respectively, denote a decrease in flow rate (back to 1.0 mL/min); in the chromatogram labeled "2 Steps + Gradient," a linear gradient from 50% B to 100% B over 5 min was begun immediately after the initiation of the second step gradient. The upper set of chromatograms illustrate the failure of excessively high salt step gradients (i.e., above 20% B buffer) to enhance resolution (relative to the linear gradient). In contrast, in the second row, the step gradients are becoming excessively low; a linear gradient is employed after these steps in order to visualize impurities and to illustrate the inadequacy of these low ionic strength steps in fully separating contaminating proteins from the monoclonal. The two lower sets of chromatograms illustrate that, once a near optimal step gradient has been found (e.g., 16% B buffer), increased flow rates and second step gradients can be used to sharpen the antibody peak, reduce elution volume, and run time.

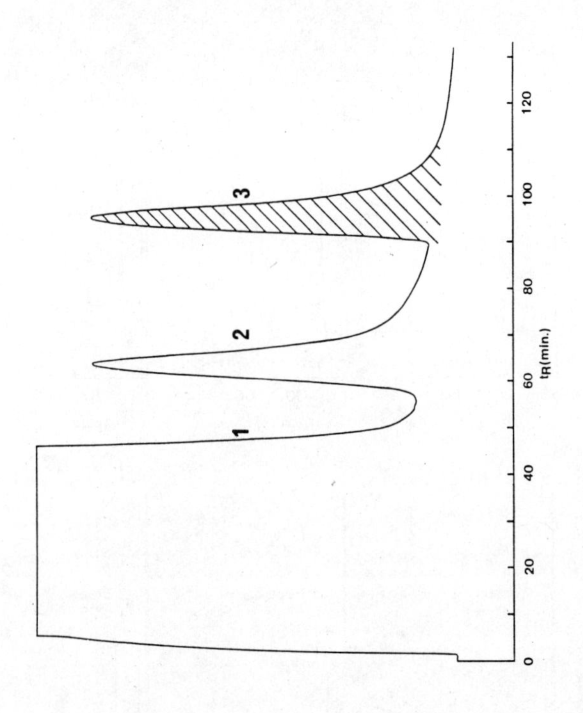
1
2
3
0
20
40
60
80
100
120
tR(min.)

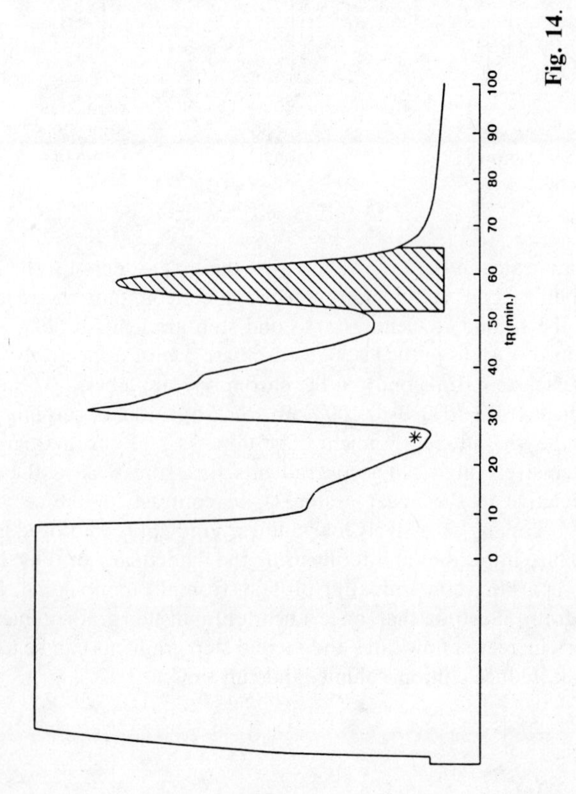
*
0
10
20
30
40
50
60
70
80
90
100
tR(min.)

Fig. 14.

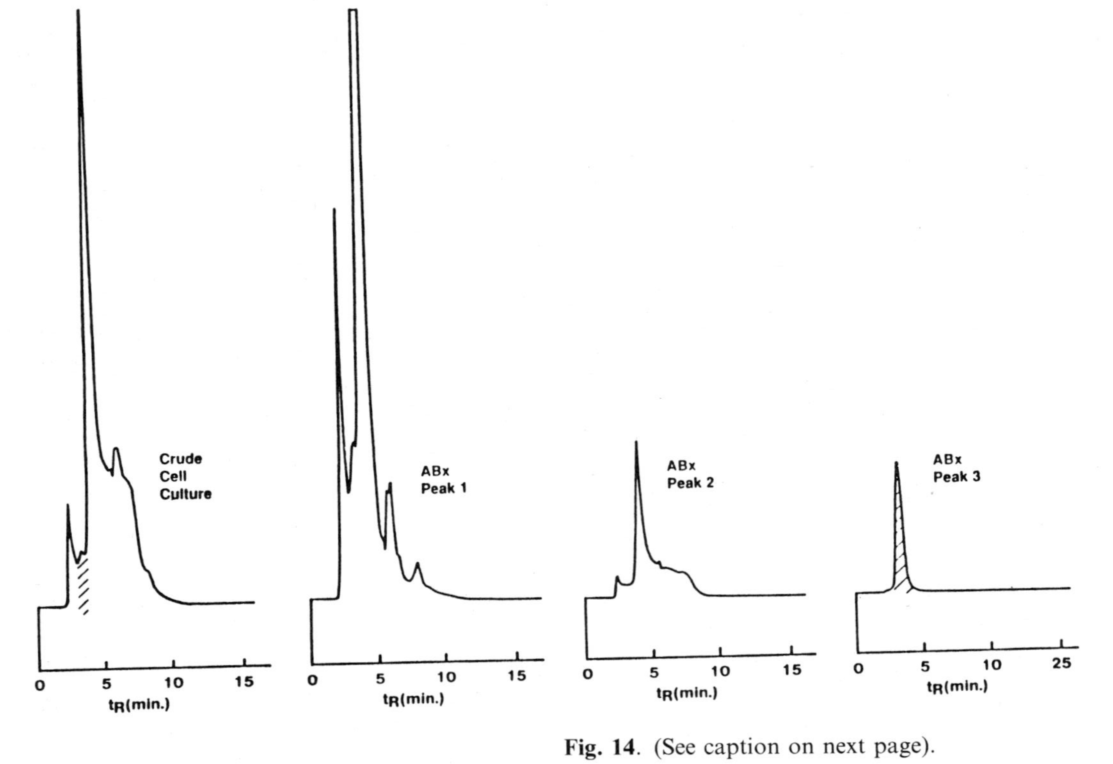

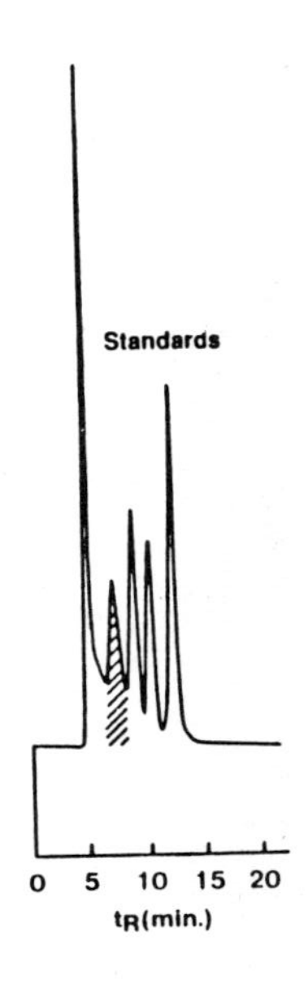

Fig. 14. (See caption on next page).

Fig. 14. Preparative purification of a monoclonal antibody from a serum-supplemented cell culture ultrafiltrate using either linear or step gradient elution from BAKERBOND ABx. Linear gradient elution was conducted on a column (10 × 250 mm) containing approximately 11 g of 15 micron ABx (left). The mobile phase consisted of an initial (A) buffer of 25 m*M* MES, pH 5.4, and a final elution (B) buffer of 500 m*M* $(NH_4)_2SO_4$ plus 20 m*M* NaOAc, pH 6.7, with an initial isocratic gradient at 100% A buffer for 43 min, followed by a linear gradient from 100% A to 100% B buffer over 120 min, at a linear flow rate of 5.0 mL/min, with a back pressure of 200 psi. Proteins were detected by UV absorbance at 280 nm and the attentuation was changed from 2.0 to 1.0 AUFS (see +). The sample consisted of 75 mL of ultrafiltrate (1500 mL of original cell culture fluid, containing 180 mg of monoclonal antibody) diluted in 150 mL of buffer A. The cross-hatched peak contains the monoclonal antibody at a purity of greater than 90% (by SDS PAGE and size exclusion chromatography) with less than 15% of the original fetal bovine serum immunoglobulins. Step gradient elution was conducted on a column (21.2 × 150 mm) packed with 15 g of 40 micron ABx (right). The chromatographic conditions were the same as detailed above, except, the gradient consisted of an initial isocratic gradient at 100% A buffer for 42 min followed by a step to 17% buffer (held for 36 min) and a second step to 50% B buffer [1 *M* $(NH_4)_2SO_4$ plus 40 m*M* NH_4OAc, pH 7.0]; the flow rate was 8.0 mL/min and the back pressure was less than 20 psi; detection was at 280 nm at 2.5 AUFS. The sample consisted of 160 mL of ultrafiltrate (3.2 L of original supernatant) diluted in 160 mL of buffer A; the cross-hatched peak contains 370 mg of the monoclonal antibody at a purity of greater than 95% (by SDS PAGE and size exclusion chromatography) with virtually no fetal bovine serum immunoglobulins. The chromatograms on the bottom represent the size exclusion chromatographic analysis of the neat sample and various peaks eluted from the preparative ABx run (right top). Chromatography was conducted on a column (7.75 × 250 mm) packed with an experimental 5 micron size exclusion chromatographic matrix. The mobile phase consisted of 200 m*M* KH_2PO_4, pH 7.2 (isocratic), at a flow rate of 0.7 or 0.3 mL/min (for the protein standards); the sample injection volume was 0.05 mL. Proteins were detected by UV absorbance at 280 nm at various sensitivities (AUFS). Analysis of ABx peak number 2 indicates that relatively little α_2-macroglobulin (the first protein peak in the size exclusion chromatogram) has been bound by the ABx because of displacement and exclusion of this large (700 kD) protein from the matrix. In the chromatogram on the right, protein standards were separated to check the column efficiency; these included bovine thyroglobulin (670 kD), bovine IgG (150 kD), ovalbumin, (44 kD), horse myoglobin (17 kD) and vitamin B-12 (1.3 kD) which eluted at 5, 7, 9, 11 and 13 min, respectively (the IgG peaks are cross-hatched.)

immunological or electrophoretic methods. High-performance supports also provide valuable information during the method development stage as to whether an adequate separation might be obtained on a less expensive medium performance support, or, whether a high-performance matrix is required (Figs. 4, 5, 31).

Although a significant amount of literature is available concerning scale-up, several basic points should be emphasized. In order to ensure reproducible results during scale-up, a similar column length should be used for both method development and preparative chromatography; linear velocities (relative flow rates) should also be kept the same. Furthermore, although purification methods which were developed on 5 micron ABx HPLC columns can be transferred directly to larger 15 micron ABx HPLC columns or 40 micron ABx open columns (Figs. 3 and 27), it is often desirable to lengthen the gradient and reduce the flow rates on the preparative run (Figs. 11 and 12). Finally, when step gradient elution techniques are being used, method development experiments should also focus on transfer of the procedure from the 5 or 15 micron ABx column to a similar column containing 40 micron ABx, because higher ionic strength steps are typically required on 40 micron ABx due to its higher surface area.

Extensive method development was conducted on the purification of one particular monoclonal antibody which was present in a fetal bovine serum-supplemented cell culture ultrafiltrate (Appendices III–V, Figs. 9, 10, and Tables 4A–4G). The information obtained in these investigations should be of general usefulness for similar samples. Initial method development conducted on 5 micron ABx and was transferred directly to analytical columns containing 15 and 40 micron ABx (Fig. 9). Additional method development on the 40 micron ABx column included the effects of sample load, binding buffer species, ionic strength and pH, flow rate, and temperature, on antibody binding, recovery and resolution (Table 4). It was found that at higher sample loads, during frontal chromatography, the amount of bound, nonimmunoglobulin protein was reduced substantially due to displacement chromatography, leaving mostly just antibody bound to the column (Table 4 and Appendices III–V). As such, even with phosphate elution buffer, the purity of the monoclonal antibody peak was excellent; furthermore, the antibody eluted as a sharp peak, in a small volume of low ionic strength buffer. These optimized conditions were employed in a preparative chromatographic separation of 1.5 L of the serum-supplemented cell culture media on a column containing 15 g of 40 micron ABx (Fig. 10). Electrophoretic analysis indicated the purity of the monoclonal was greater than 90% with a recovery above 95%. (Details are given in Figs. 9 and 10, Table 4, and Appendices III-V).

In another scale-up study, it was found that a procedure comprised of two steps on ABx were necessary to achieve the purification of a monoclonal antibody that was present at extremely low levels within a serum-supplemented cell culture supernatant (Figs. 11 and 12). Initial method development on a 5 micron ABx HPLC column focused on the effect of mobile phase composition on resolution of the antibody from the large amounts of fetal bovine serum proteins and on the second step ABx rechromatography of the antibody peak in the presence of an alternate mobile phase (Fig. 11). Next, the method was transferred on the analytical level from this 5 micron ABx column to a small 40 micron ABx column using a longer gradient time in order to maintain adequate resolution (Fig. 11). Finally, the purification was scaled-up on a larger, preparative column which was dry packed with 40 micron ABx, and the active fractions were rechromatographed in the second buffer system on a semipreparative column containing 15 micron ABx (Fig. 12).

As indicated in the previous examples, gradient elution often provides

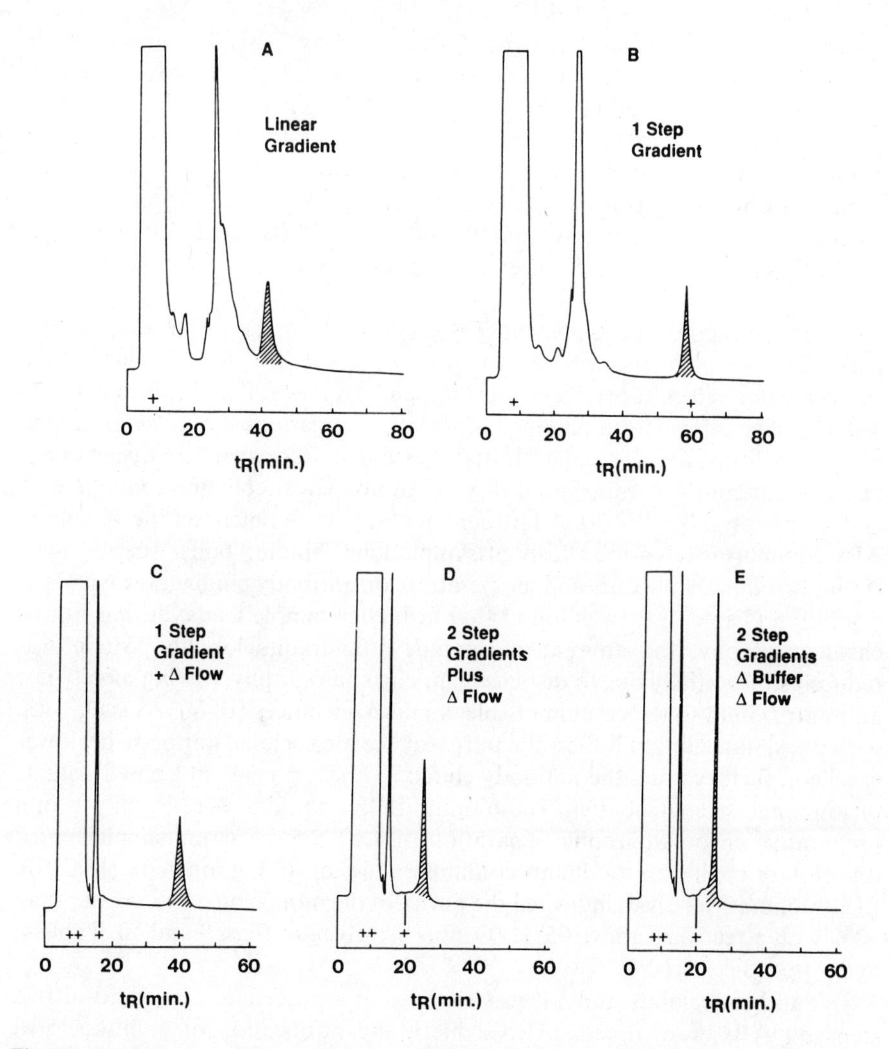

Fig. 15. Use of step gradient elution and increased flow rates to improve resolution and purity, and reduce run times and elution volumes with the BAKERBOND ABx seperation of a monoclonal antibody from a serum-supplemented cell culture ultrafiltrate. Method development was conducted on a column (7.75 × 100 mm) packed with 15 micron ABx (chromatograms A–E). The equilibration (A) buffer was 25 m*M* MES, pH 5.5 and the elution buffer was 500 m*M* $(NH_4)_2SO_4$ plus 20 m*M* NaOAc, pH 7.0. The flow was 1.0 mL/min (except as noted below) at a back pressure of 50 psi, and proteins were detected by UV absorbance at 280 mn at an attenuation of 1.0 AUFS. The sample consisted of 1.0 mL of a serum-supplemented cell culture ultrafiltrate (20 ×), diluted to 5 mL with buffer A. In each chromatogram, an initial isocratic gradient at 100% A buffer was held for 7 min during sample loading; in chromatogram A, a linear gradient from

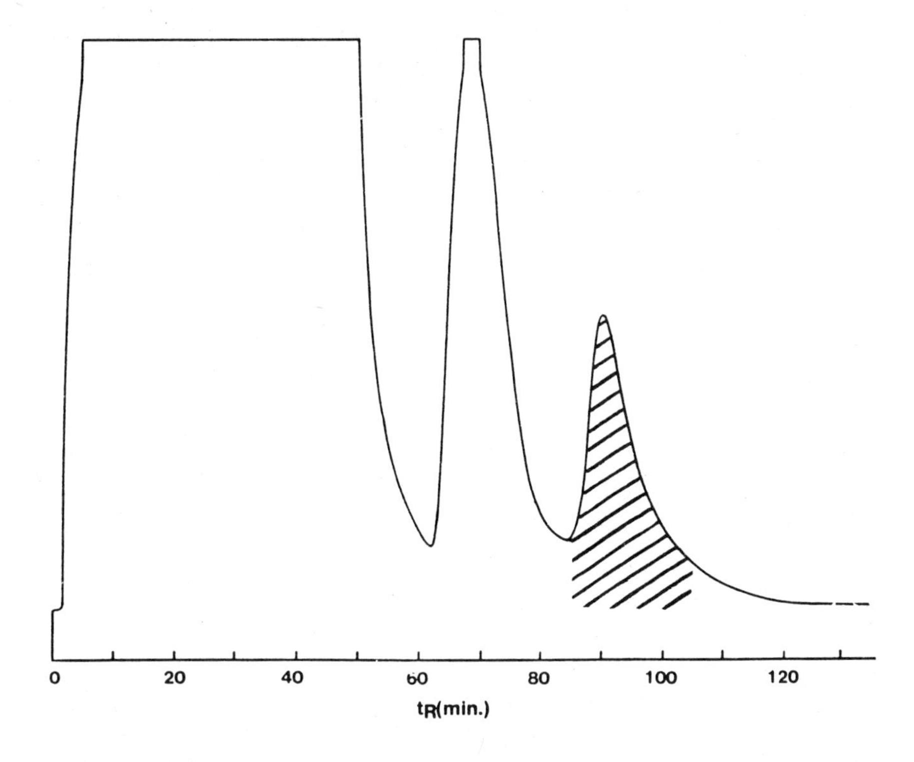

100% A to 100% B buffer over 60 min was begun at 7 min; in chromatogram B, a single step gradient to 17% B buffer was held isocratically for 53 min, and followed by a 20 min linear gradient to 100% B buffer; in chromatogram C, a step gradient to 17% B buffer was held isocratically for 33 min, followed by a 20 min linear gradient to 100% B buffer, and the flow rate was changed to 4.0 mL/min between 10 min and 16 min. In chromatogram D, a step gradient to 17% B was held isocratically for 13 min, followed by a second step to 100% B buffer, and the flow rate was changed to 4.0 mL/min between 10 and 20 min; in chromatogram E, the conditions were the same as those in chromatogram D, except the step gradient at 20 min was to 100% C buffer (1 *M* NaOAc, pH 7.0). The cross-hatched peak in chromatograms A–E contains the monoclonal antibody at a purity of greater than 99% (by SDS PAGE and size exclusion chromatography). Preparative chromatography using two step gradients was conducted on a 1 in. (21.2 × 150 mm) column packed with 40 micron ABx (bottom). Chromatography was conducted as above, except that elution was with an isocratic gradient at 100% for 48 min, followed by an initial step gradient to 17% B buffer (held for 27 min) and a final step gradient to 30% B buffer at 75 min. The sample consisted of 180 mL of the cell culture ultrafiltrate (3.6 L of original supernatant) diluted in 180 mL of buffer A; the cross-hatched peak contains 220 mg of the monoclonal antibody at a purity greater than 93% (by SDS PAGE and size exclusion chromatography), with virtually no fetal bovine serum immunoglobulins.

protein products of adequate purities. However, in most cases, a more effective, more highly resolving method for purifying monoclonals on ABx is the use of step gradients in conjunction with optimized mobile phase conditions (62, 64, 71). Step gradients represent one of the oldest methods for eluting proteins from ion exchange matrices. In many production environments stepwise elution is used extensively, not only for economic reasons (simpler equipment and reduced buffer costs), but more important, resolution can be increased dramatically (Figs. 13–19). However, most chromatographers make little use of these more traditional elution methods, perhaps because gradient elution is a simpler technique. Indeed, one of the disadvantages of step elution is that more extensive method development is typically required to optimize a given separation (Figs. 13, 15, 16). However, if initial method development aimed at examining the effects of mobile phase on resolution is conducted (Figs. 3–7), it should be rather easy to extend this work toward the optimization of a step elution method (Figs. 13, 15, 16), particularly since most of the basic optimization experiments for ABx have already been conducted.

Another problem with step gradients is the possible appearance of false peaks or split peaks (i.e., either changes in absorbance which arise from sudden shifts in salt concentration, or, the appearance of the protein/antibody of interest in several different peaks). False peaks and split peaks (69, 70, 97) typically arise from excessive, rapid, or multiple changes in ionic strength (or pH or both), or, when methods are developed with lower sample loads and then scaled-up directly to preparative separations which employ significantly higher loads or excessive column overload (frontal chromatography). These problems can be circumvented by conducting the proper experiments during method development, including loading studies and scale-up runs (62–64, 71).

In order to determine the proper step gradient to employ, a first approximation is to run a long gradient, determine the ionic strength at which the monoclonal elutes, and in subsequent chromatographic run, use a step which is

---

Fig. 16. Use of step gradient elution of a monoclonal antibody from a mouse ascites fluid to improve resolution and purity and reduce run times and elution volumes on 15 micron BAKERBOND ABx. Chromatography was conducted on a column (7.75 × 100 mm) packed with 15 micron ABx. The mobile phase consisted of an initial (A) buffer of 25 m*M* MES, pH 5.5, and a final elution (B) buffer of 1 *M* NaCl plus 20 m*M* NaOAc, pH 7.0. In each chromatogram, an initial isocratic gradient of 100% A buffer was used for 12 min; in chromatogram A, a linear gradient from 100% A to 100% B buffer over 30 min was used; in chromatogram B, a step gradient to 20% B was held isocraticly; in chromatogram C, a step gradient to 15% B was used; and, in chromatogram D, a step gradient to 15% B buffer was held for 25 min, and followed by a second step to 35% B buffer. The flow rate was 1.0 mL/min with a back pressure of 50 psi. Proteins were detected by UV absorbance at 280 nm at an attenuation of 2.0 AUFS. The samples contained 1.0, 1.3, 1.7, and 1.7 mL of neat mouse ascites fluid (diluted in 2 volumes of buffer A) and the monoclonal antibody was contained within the cross-hatched peak at purities of greater than 93, 95, 99, and 99% (by SDS PAGE and size exclusion chromatography), in chromatograms A, B, C and D, respectively.

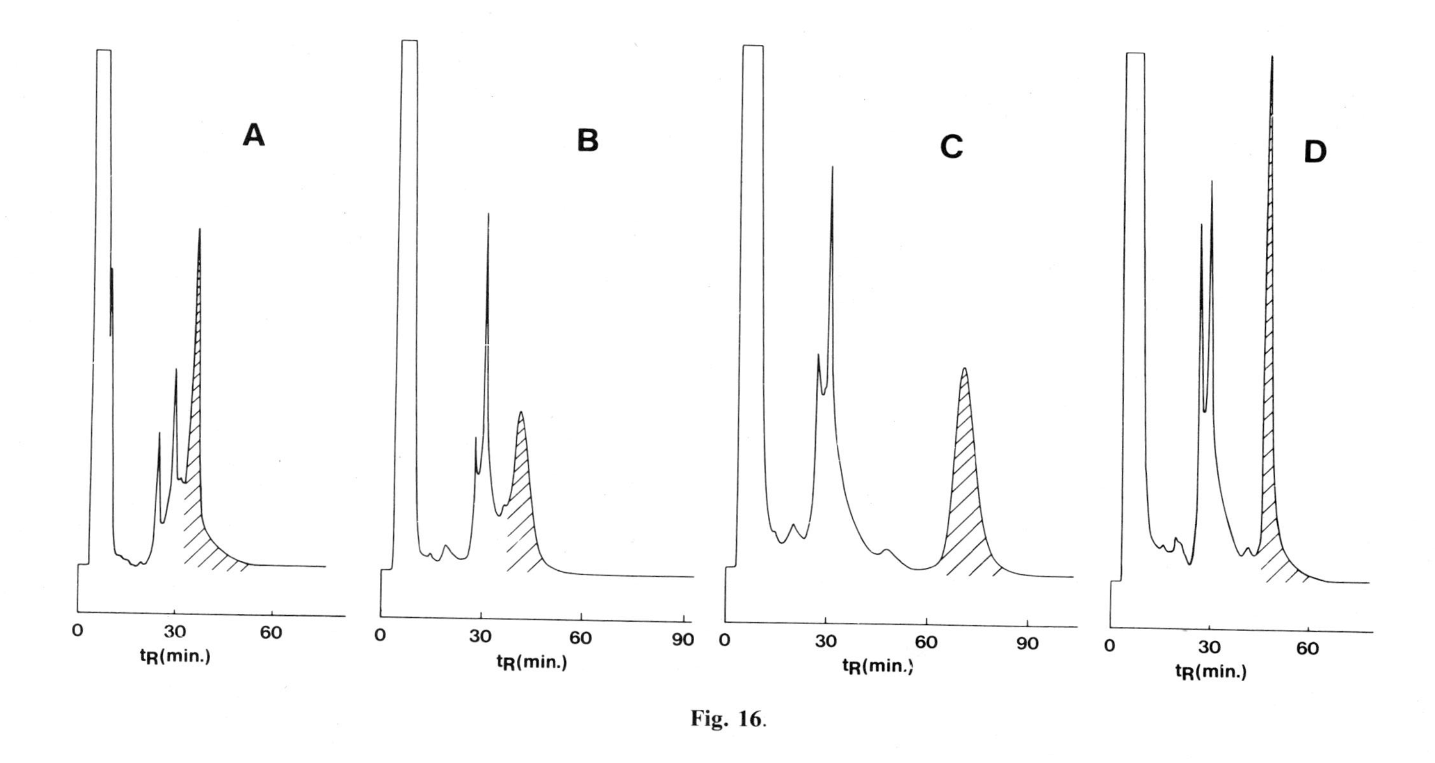

Fig. 16.

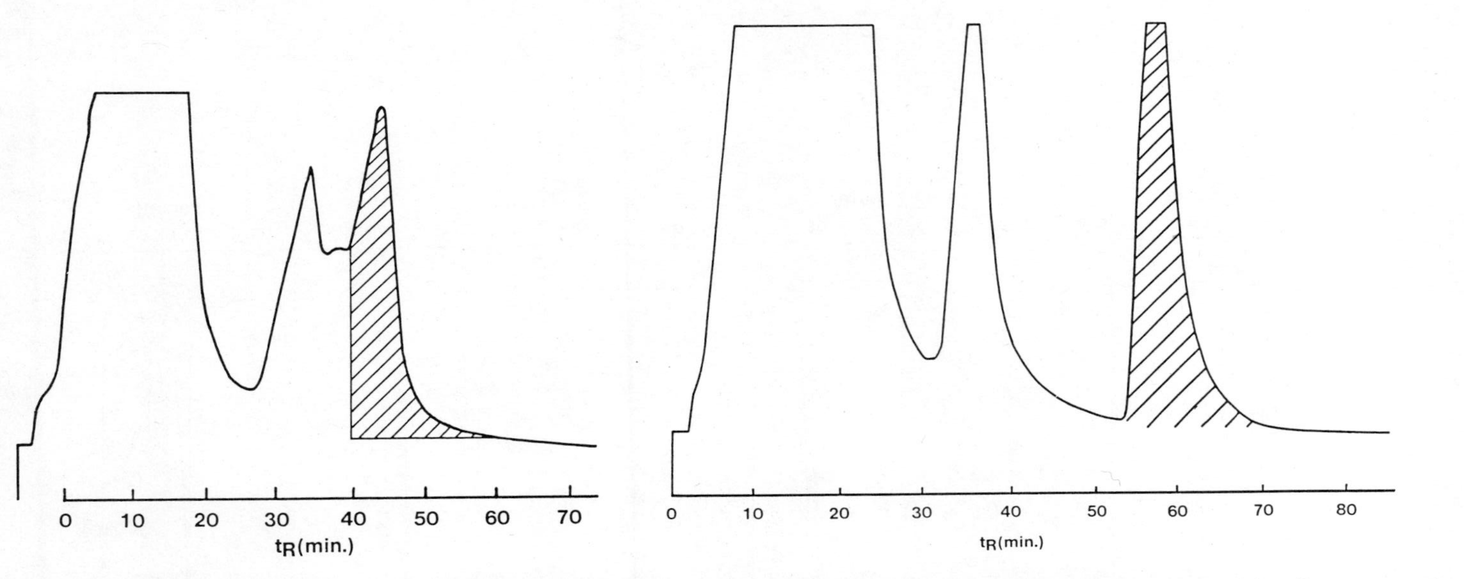

Fig. 17. Preparative purification of a monoclonal antibody from a mouse ascites fluid on BAKERBOND ABx using either linear or step gradient elution. Preparative chromatography with a linear gradient was conducted on a column (10 × 250 mm) containing 15 micron ABx (left). The mobile phase consisted of an initial (A) buffer of 25 m*M* MES, pH 5.4, and a final elution (B) buffer of 500 m*M* $(NH_4)_2SO_4$ plus 20 m*M* NaOAc, pH 6.7, with an initial isocratic gradient of 100% A buffer for 18 min, followed by a linear gradient of 100% A to 100% B buffer over 1 hr, at a flow rate of 5.0 mL/min, with a back pressure of 150 psi. Proteins were detected by UV absorbance at 280 nm at an attenuation of 3.0 AUFS. The sample consisted of 25 mL of crude ascites fluid plus 50 mL of 25 m*M* MES, pH 5.4. The monoclonal antibody was contained within the cross-hatched area at a purity of greater than 95% (by SDS PAGE) and the recovery of antibody mass was greater than 96%. Preparative chromatography using step gradient elution (right) was conducted as detailed above, except the column was packed with 40 micron ABx and the elution buffer (B) was 2 *M* NaCl plus 20 m*M* NaOAc, pH 7.0, with an initial isocratic gradient at 100% A buffer for 25 min, followed by a step gradient to 13% B buffer (held for 15 min) and a second step gradient to 30% B buffer. The back pressure was 20 psi, and the monoclonal antibody was contained withn the cross-hatched area at a purity of greater than 98% (by SDS PAGE and size exclusion chromatography), and the recovery of antibody mass was greater than 98%. (Method development experiments are detailed in Figs. 4 and 16.)

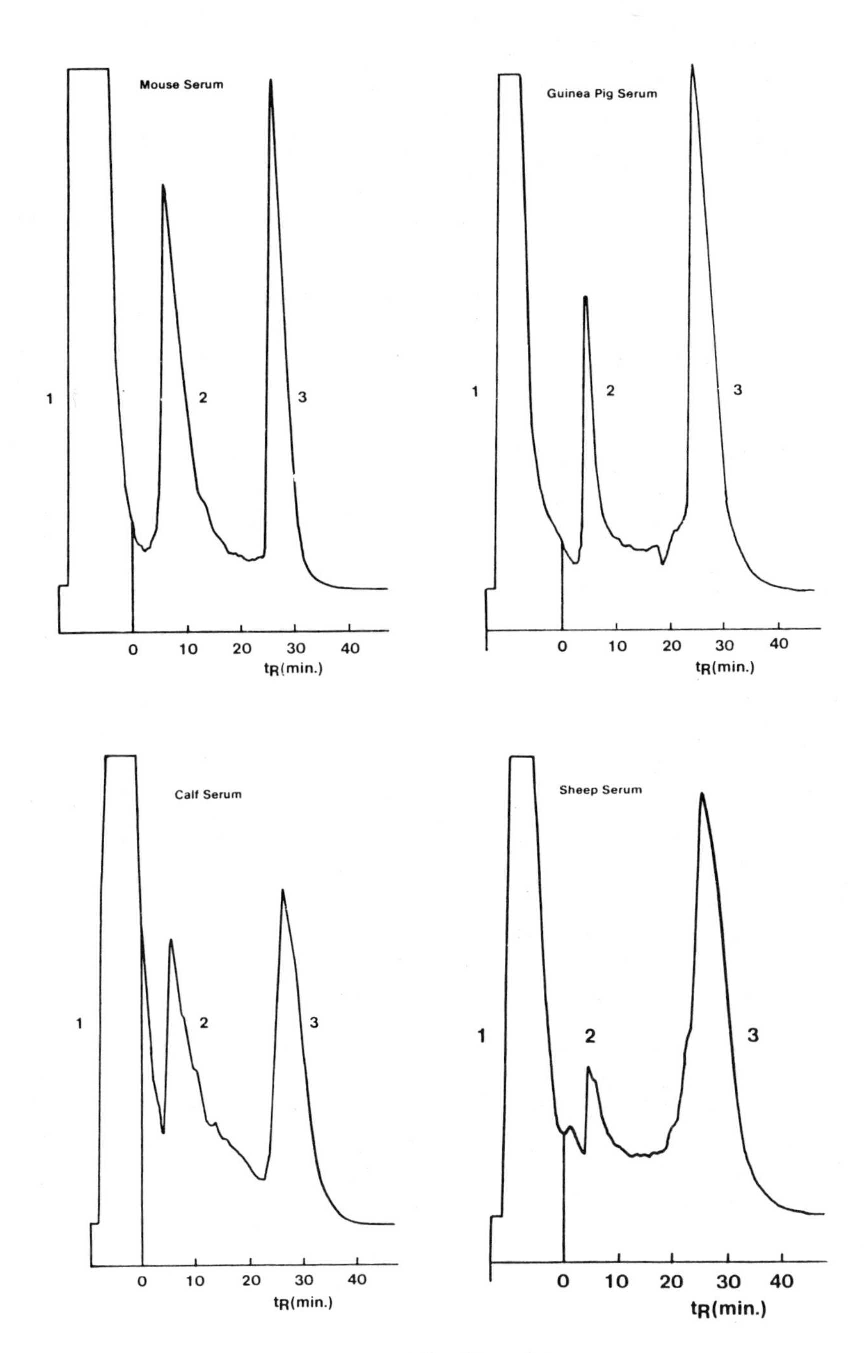
Mouse Serum
1
2
3
0
10
20
30
40
tR(min.)
Guinea Pig Serum
1
2
3
0
10
20
30
40
tR(min.)
Calf Serum
1
2
3
0
10
20
30
40
tR(min.)
Sheep Serum
1
2
3
0
10
20
30
40
tR(min.)

Fig. 18.

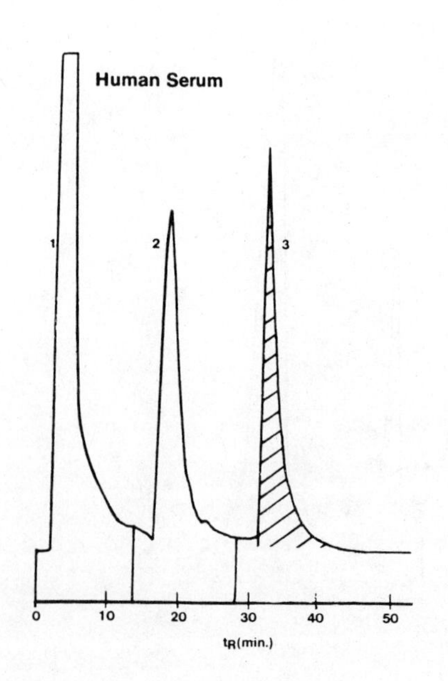

Human Serum
1
2
3
0
10
20
30
40
50
tR(min.)

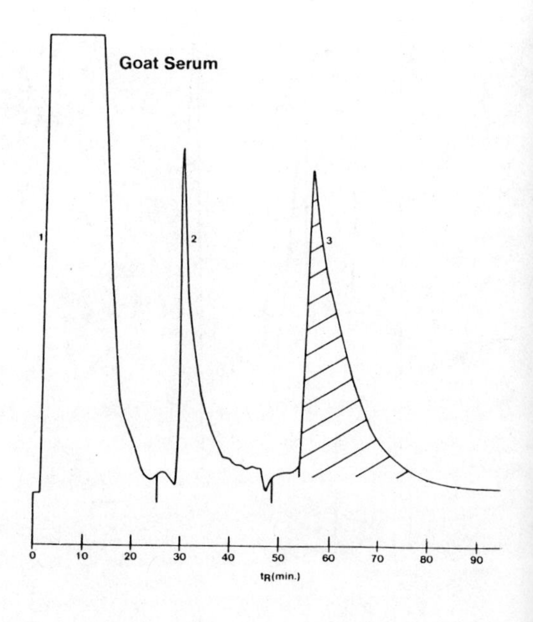

Goat Serum
1
2
3
0
10
20
30
40
50
60
70
80
90
tR(min.)

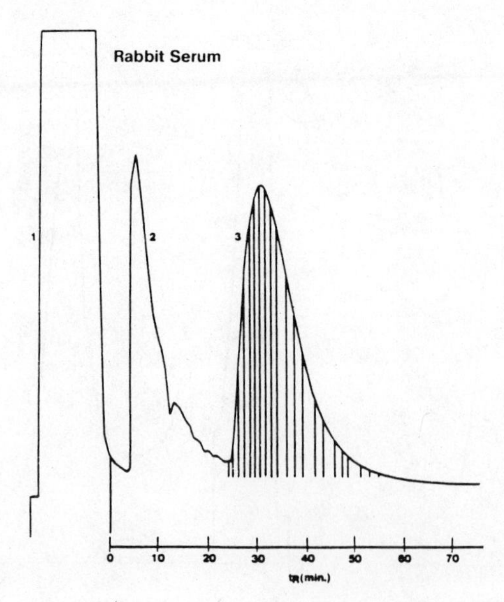

Rabbit Serum
1
2
3
0
10
20
30
40
50
60
70
tR(min.)

Fig. 18.

Fig. 18. Purification of polyclonal immunoglobulins (IgG, IgA, and IgM) from serum on 40 micron BAKERBOND ABx with the use of step gradient elution. Chromatography was conducted on a column (10 × 100 mm) packed with 40 micron ABx. The mobile phase consisted of an initial (A) buffer of 25 m*M* MES, pH 5.5, and a final elution (B) buffer of either 1 *M* NaCl plus 20 m*M* NaOAc, pH 5.8 (for the mouse, guinea pig, bovine, and sheep serum chromatograms) or 500 m*M* $(NH_4)_2SO_4$ plus 20 m*M* NaOAc, pH 6.7 (for the human, goat, and rabbit serum chromatograms). Various step gradients were employed to elute the immunoglobulins, these included; for the mouse serum, an initial isocratic gradient at 100% A buffer for 14 min, followed by a step to 10% B (at t_0), which was held for 20 min, and linear gradient from 10% B to 100% B over 10 min, at an initial flow rate of 1.2 mL/min, and 2.0 mL/min after the initial step, with a back pressure of 20 psi, for the guinea pig serum, an initial isocratic gradient at 100% A buffer for 14 min, followed by a step to 10% B (held for 20 min) and a linear gradient from 10% B to 100% B over 20 min, at an initial flow rate of 1.4 mL/min, and 2.0 mL/min following the initial step, with a back pressure of 20 psi; for the bovine serum, an initial isocratic gradient at 100% A buffer for 9 min, followed by a step to 7% B (held for 20 min) and a linear gradient from 7% B to 100% B over 20 min, at an initial flow rate of 1.5 mL/min, and 2.0 mL/min after the initial step, with a back pressure of 30 psi; for the sheep serum, an initial isocratic gradient at 100% A buffer for 14 min, followed by a step to 7% B (held for 20 min) and a linear gradient from 7% B to 100% B over 20 min, at an initial flow rate of 1.8 mL/min, and 2.0 mL/min during the run, with a back pressure of less than 30 psi; for human serum, an initial isocratic gradient at 100% A buffer for 13 min, followed by a step to 7% B (held for 15 min) and another step to 100% B, at a flow rate of 2.0 mL/min, with a back pressure of 20 psi; for goat serum, an initial isocratic gradient at 100% A buffer for 25 min, followed by a step to 10% B (held for 23 min) and another step to 50% B, at a linear flow rate of 2.0 mL/min, with a back pressure of 20 psi; and for rabbit serum, an initial isocratic gradient at 100% A buffer for 17 min, followed by a step to 10% B (held for 20 min) and a linear gradient from 10% B to 100% B over 30 min, at a flow rate of 2.0 mL/min, with a back pressure of 20 psi. Proteins were detected by UV absorbance at 280 nm at an attenuation of 0.5 AUFS (for the mouse, guinea pig, bovine calf, or rabbit serum) or 1.0 AUFS (for the sheep, goat, or human serum). The sample injection volumes were 4.5 mL (1.5 mL of mouse, guinea pig, bovine calf, or sheep serum diluted with 25 m*M* MES, pH 5.0), 4.0 mL (1.0 mL of human serum plus buffer A), 12 mL (4 mL of goat serum in 25 m*M* MES, pH 5.0), or 10 mL (4 mL of rabbit serum, in 25 m*M* MES, pH 5.0). Peak 1 contained albumins, transferrins and proteases, peak 2 contained the bound, nonimmunoglobulin proteins and peak 3 contained the serum polyclonal immunoglobulins. The purity of the immunoglobulin peaks were determined by SDS PAGE analysis to be greater than 75% ($>75\%$ for guinea pig serum; $>80\%$ for mouse, bovine calf serum, or sheep serum; $>85\%$ for human serum; $>90\%$ for goat or rabbit serum).

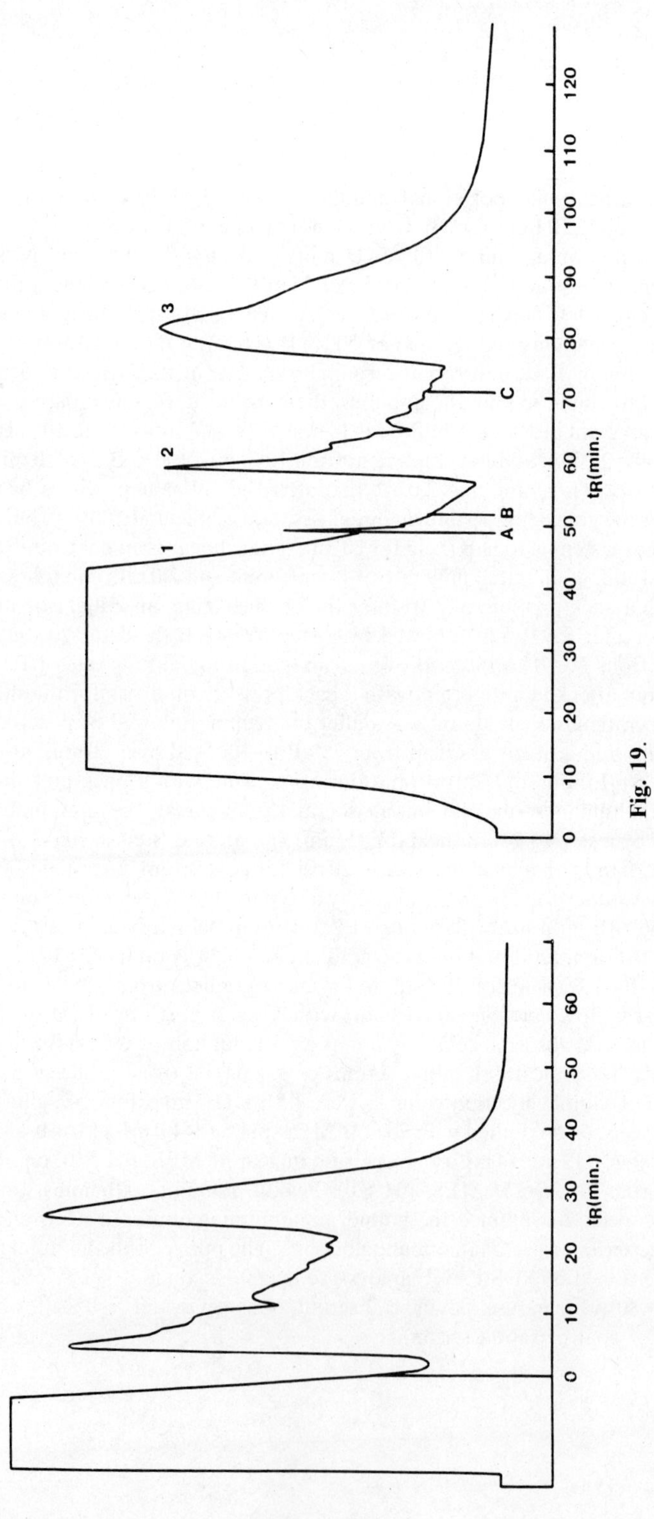

0
10
20
30
40
50
60
t_R(min.)
0
10
20
30
40
50
60
70
80
90
100
110
120
t_R(min.)
1
2
3
A B
C

Fig. 19.

one-half of this ionic strength (Fig. 6). This step can be held constant for a short period to elute the contaminating proteins, and then, followed either by a linear gradient or by a second step gradient to elute the monoclonal (Figs. 13, 15, 16).

Step gradients of a wide range of ionic strengths are often capable of providing adequate resolution between a monoclonal antibody and the bound, nonimmunoglobulin proteins (Fig. 13). However, the most efficient step gradient lies at an ionic strength that is between two extremes. At one extreme, the ionic strength is too low to enable the full elution of the bound, contaminating proteins in a timely manner (Fig. 13). (Although one of the main purposes of employing step gradients is to increase resolution and purity, another is to reduce chromatographic run time, since, in many cases, recovery as well as process economics are proportional to chromatographic residence time). On the other hand, if the ionic strength of the step gradient is too high, the separation is rapid, but the monoclonal partially coelutes with the contaminating proteins (Fig. 13). Again, this defeats one of the purposes of employing step gradients; that is, to enhance product purity. Obviously then, the ionic strength of the step gradient should be between these two extremes; high enough to achieve the desired goals of quickly eluting all of the nonimmunoglobulin proteins, but neither at the expense of creating false peaks or split peaks, nor by eluting the monoclonal so quickly that it coelutes with contaminants, and reduces product

←

Fig. 19. Preparative purification of polyclonal immunoglobulins (IgG, IgA, and IgM) from goat and rabbit serum on 40 micron BAKERBOND ABx. Preparative chromatography of goat serum (left) was conducted on a column (22 × 150 mm) packed with 40 micron ABx. The mobile phase consisted of an initial (A) buffer of 25 m*M* MES, pH 5.4, and a final elution (B) buffer of 1 *M* NaCl plus 30 m*M* NaOAc, pH 6.7, with an initial isocratic gradient at 100% A buffer for 18 min, followed by a step to 10% B buffer (held for 20 min) and a linear gradient from 10% B to 100% B over 20 min, at a linear flow rate of 10.0 mL/min with a back pressure of 30 psi. Proteins were detected by UV absorbance at 280 nm at an attenuation of 1.0 AUFS. The sample consisted of 45 mL of crude goat serum diluted in 90 mL of 25 m*M* MES, pH 5.0. The peak eluting between 22 and 35 min contained the goat serum polyclonal immunoglobulins at a purity of 90% (by SDS PAGE). Preparative chromatography of dialyzed rabbit serum (right) was conducted on a column (10 × 250 mm) packed with 40 micron ABx. The mobile phase consisted of an initial (A) buffer of 25 m*M* MES, pH 5.4, and a final elution (B) buffer of 500 m*M* $(NH_4)_2SO_4$ plus 20 m*M* NaOAc, pH 6.7, with an initial isocratic gradient at 100% A buffer for 52 min, followed by a step gradient (at "B") to 10% B buffer (held for 20 min) and a linear gradient (at point "C") from 10% B to 100% B buffer over 30 min at an initial flow rate of 3.0 mL/min that was changed (at point "A") to 5.0 mL/min, with a back pressure of 150 psi. Proteins were detected by UV absorbance at 280 nm at an attenuation of 2.0 AUFS. The sample consisted of 28 mL of crude rabbit serum dialyzed against 20 m*M* MES plus 15 m*M* KH_2PO_4, pH 6.0, and diluted with 82 mL of 25 m*M* MES, pH 5.4. Peak 1 contained albumins, transferrins and proteases, peak 2 contained the bound, non-immunoglobulin proteins, and peak 3 contained the rabbit immunoglobulins (IgG, IgA, and IgM) at a purity (by SDS PAGE) and a recovery of greater than 90%.

purity. Ideally, there is an ionic strength at which all the contaminating proteins elute rather quickly, yet the monoclonal antibody is still tightly bound (Fig. 13, 15, 16). In fact, because the bound contaminating proteins tend to elute at a rather constant position, despite relatively large changes in the ionic strength of the step (Figs. 6, 13), rather low salt concentrations may be used. Typically, it is advantageous to employ a second step gradient or a rapid linear gradient following the elution of the contaminating proteins, in order to quickly elute the monoclonal (Figs. 13, 15, 16). This second step gradient has the advantage of not only reducing the chromatographic run time, but also, eluting the product into a smaller volumne of a more appropriate buffer (e.g., ammonium acetate for lyophilization, concentrated ammonium sulfate for binding to a HIC matrix, PBS, an assay buffer, a formulation buffer, or any buffer which enhances the stability of the monoclonal). Finally, it should be mentioned that the manipulation of flow rate can be used in conjunction with step gradients in order to further reduce run time, and, to increase the specific activity of the final product (Figs. 13, 15, 16). As a result of the high degree of resolution which can be obtained with step gradients, 40 micron ABx may be used to achieve large scale, high performance separations in an economical manner (Figs. 14, 17, 19).

Linear and step gradient elution methods were developed for the preparative purification of monoclonal antibodies that were present at relatively low levels within an ascites fluid and a cell culture fluid (Figs. 4, 16, 17). Initial method development experiments on 5 micron ABx suggested that even with rapid gradients, ammonium sulfate was the optimal mobile phase for both samples (Figs. 4 and 5). Further method development showed that step gradients were able to enhance resolution and antibody purity (Figs. 13, 14, 15, 16). Therefore, two separate methods, employing either linear or step gradients, were transferred to larger columns which were dry packed with 15 or 40 micron ABx (Figs. 14, 17). Details of these experiments are given in the figure legends.

For these types of challenging samples this approach facilitates the optimization of the entire preparative purification process (62–64, 71). This is essential for monoclonals that must be purified on a routine basis, because small amounts of time spent on method development can result in substantial savings over the long term. However, in most cases, scale-up may be much easier, particularly when quantity or purity requirements are not extremely high. In these cases, method development may simply consist of several initial runs on an analytical column, followed by direct scale-up to a larger column (Figs. 18, 19). This is particularly true in the case of serum samples, due to the extensive method development that has been conducted in this laboratory (Figs. 3, 18, 19, and Table 2).

The purification of serum immunoglobulins on ABx may be easily accomplished with the use of two step gradients using ammonium sulfate or sodium chloride elution buffers. Because plasma immunoglobulins from various species display slightly different binding strengths/affinities toward ABx, different chromatographic conditions have been employed for each type of serum (Fig. 18). An initial step gradient is used to selectively elute the majority of the bound, nonimmunoglobulin proteins, while the polyclonal antibodies remain bound. The immunoglobulins may be eluted either with another step gradient or with a linear gradient (Fig. 18).

6. BATCH EXTRACTION

One of the oldest methods of protein purification is batch or solid phase extraction. Batch extraction is a simple and efficient procedure and can be conducted in any laboratory because it requires no columns, pumps, or other chromatographic equipment. Furthermore, in contrast to column chromatography, batch extractions typically allow for higher capacities (Table 4K), and, more reproducible scale-up (62, 63, 69 70) and require less buffer. ABx is well suited for the batch extraction of monoclonal or polyclonal antibodies because, under a specific set of buffer conditions, it is possible to selectively remove immunoglobulin from crude samples. ABx also has excellent physical properties for batch extraction; it separates quickly from the aqueous phase by gravity, is easily sterilized and regenerated, and does not support microbial growth (62, 63, 66–70). These factors make batch extractions with ABx suitable for large-scale applications where aseptic conditions and ease of handling are essential. The exact conditions required for aqueous batch extraction of proteins on ABx media, as well as the kinetics of the process may be conveniently monitored and maximized using ABx HPLC analysis (Figs. 20, 22).

One of the major advantages of the batch extraction method is that it provides extremely high capacities; higher than those that can be achieved in a chromatographic column containing the same amount of ABx matrix (Table 4K). Quantitative recovery/binding of a monoclonal antibody from 250 mL of a serum supplemented cell culture fluid has been achieved on as little as 1 g of ABx in the batch mode (Table 4K).

Another major advantage of the batch extraction method is the ease and accuracy of scale-up. Unlike column chromatography, which does not always lend itself to accurate method transfer for scale-up, batch extraction can be readily scaled-up to the large scale from method development that was conducted in a small test tube (62, 63, 69, 70). Accurate scale-up of batch extractions simply involves maintaining a constant ratio between the amounts of ABx, sample volume and buffer volume, while holding buffer ionic strength and pH, dilution factors and extraction times constant. Virtually identical results have been obtained when the optimized conditions for the batch extraction of rabbit serum immunoglobulins were scaled-up 1000-fold, or, with the 200-fold scale-up of the batch extraction of monoclonal antibodies from an ascites fluid or a fetal bovine serum-supplemented cell culture supernatant (62, 63, 69, 70).

Prior to scale-up, method development was conducted in order to optimize the conditions for the batch extraction of polyclonal immunoglobulins (IgG, IgM, and IgA) from rabbit serum, and monoclonal antibodies from ascites fluids and cell culture fluids. As mentioned previously, elution from ABx with

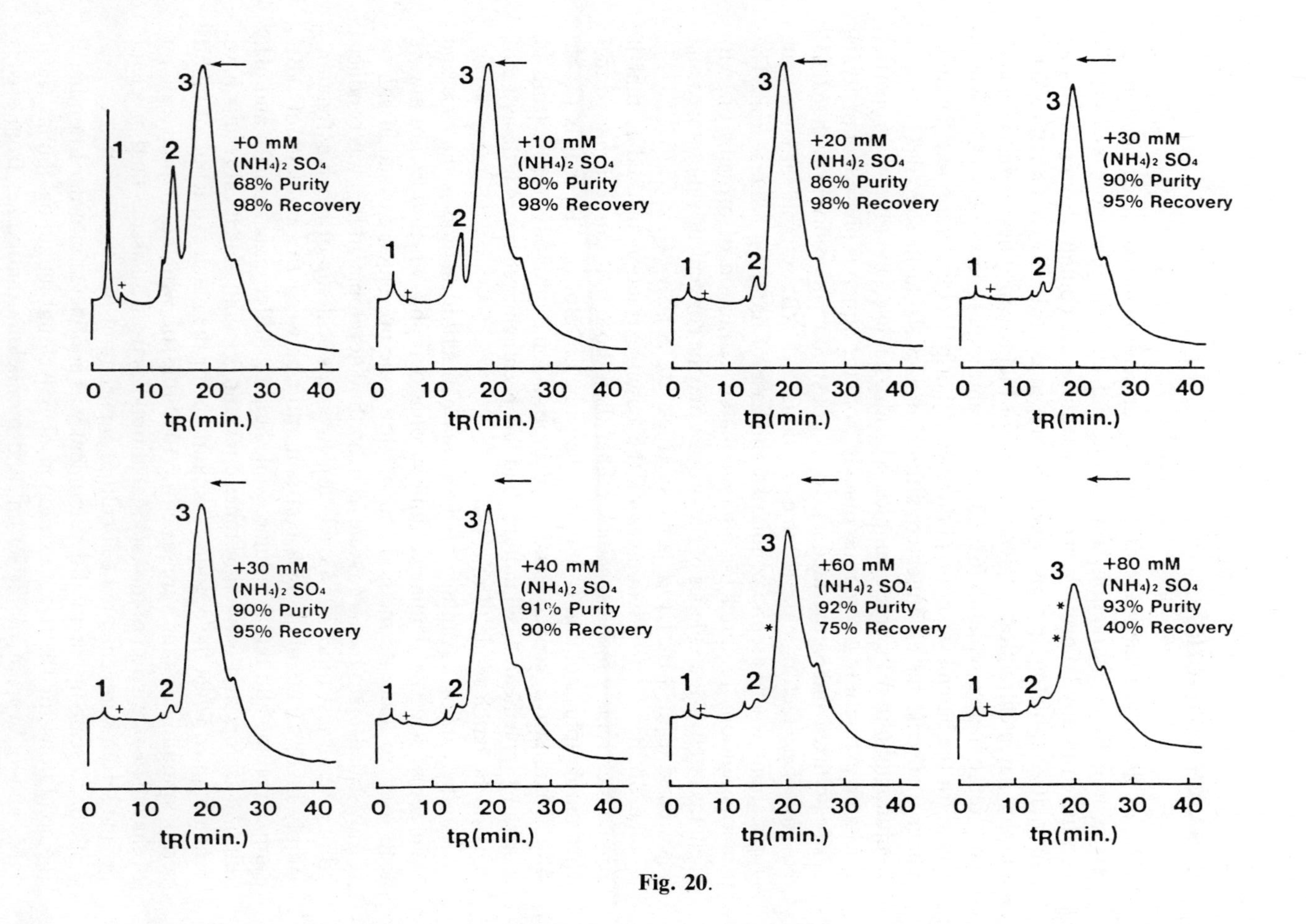

+0 mM
(NH₄)₂ SO₄
68% Purity
98% Recovery
+10 mM
(NH₄)₂ SO₄
80% Purity
98% Recovery
+20 mM
(NH₄)₂ SO₄
86% Purity
98% Recovery
+30 mM
(NH₄)₂ SO₄
90% Purity
95% Recovery
+30 mM
(NH₄)₂ SO₄
90% Purity
95% Recovery
+40 mM
(NH₄)₂ SO₄
91% Purity
90% Recovery
+60 mM
(NH₄)₂ SO₄
92% Purity
75% Recovery
+80 mM
(NH₄)₂ SO₄
93% Purity
40% Recovery
1
2
3
0
10
20
30
40
tR(min.)

Fig. 20.

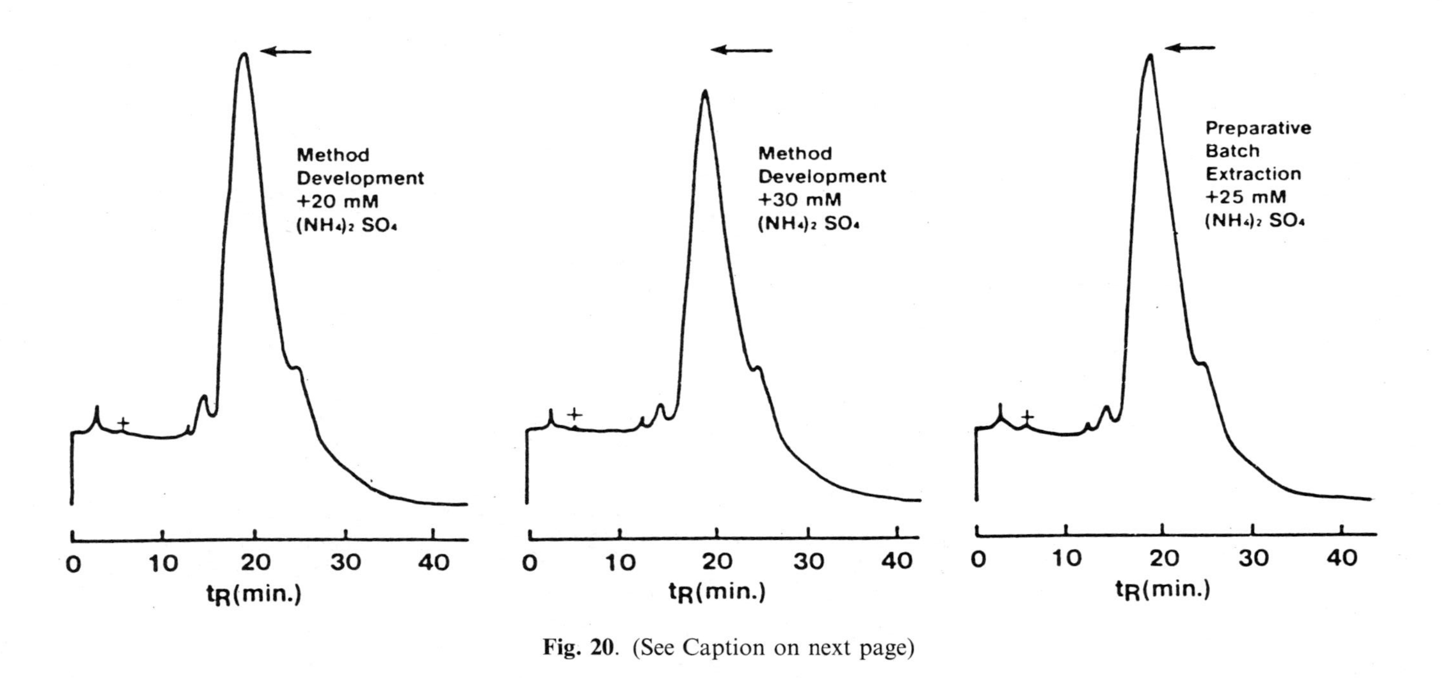

Fig. 20. (See Caption on next page)

Fig. 20. Effects of ammonium sulfate on recovery and purity in the batch extraction of immunoglobulins from rabbit serum with 40 micron BAKERBOND ABx. Chromatography was conducted on a column (4.6 × 250 mm) containing 15 micron ABx. The mobile phase consisted of an initial (A) buffer of 25 m*M* MES, pH 5.4, and a final elution (B) buffer of 500 m*M* $(NH_4)_2SO_4$ plus 20 m*M* NaOAc, pH 6.7, with an initial isocratic gradient at 100% A buffer, followed by a linear gradient from 100% A to 100% B buffer over 30 min, at a flow rate of 1.0 mL/min with a back pressure of 150 psi. Proteins were detected by UV absorbance at 280 mm at an attenuation of 1.0 AUFS and 0.5 AUFS (see +). The sample consisted of 0.2 mL of the batch extracted immunoglobulin fraction (see protocol below) diluted in 1.8 mL of 25 m*M* MES, pH 5.4. Peak 1 contained albumins, transferrins and proteases, peak 2 nonimmunoglobulin proteins, and peak 3 contained the immunoglobulins. Purity and recovery data are given with each chromatogram. The horizontal arrow indicates the peak height for 100% recovery during extraction. The three chromatograms on the bottom represent the ABx analysis of the 1000-fold scale-up of the optimized conditions (500 mL of rabbit serum were processed on 20 g of ABx). Although the optimum conditions for any given sample may vary, the addition of approximately 20–30 m*M* ammonium sulfate typically give the maximum recovery and purity. Conditioning new ABx column prior to use is suggested for cell culture fluids due to the low levels of antibody. To condition ABx, 3 mg of albumin or 0.1 mL of fetal bovine serum were diluted in 2 mL of 25 m*M* MES, pH 5.5, and added to 200 mg of 40 micron ABx. After 10 min of mixing, 5 mL of 500 m*M* $(NH_4)_2SO_4$ plus 20 m*M* NaOAc, pH 5.8, were added with stirring for 2 min. The ABx was allowed to settle, and after 10 min, the supernatant was removed; 8 mL of 50 m*M* MES, pH 5.0 were added with stirring and removed after 5 min. Next, 9 mL of 25 m*M* MES, pH 5.5, were added with stirring for 5 min and removed after the ABx settled. The following is the method development protocol used to optimize batch extraction on ABx. First, 24 mL of 25 m*M* MES, with a pH less than 4.0, was added to 8 mL of sample (the final pH was less than 6.0). Next, 2 mL volumes of this diluted sample were added to 14 test tubes containing 200 mg of conditioned, pH equilibrated (in 25 m*M* MES, at pH 5.5) ABx. [For cell culture fluids, the amount of ABx required for quantitative binding (see Table 4K) can be reduced 5- to 50-fold.] These tubes were capped, mixed for 5 min, and allowed to settle for 5 min. Next, 3 mL volumes of ammonium sulfate buffer were added to the 2 mL of the diluted sample to give various final concentrations of MES and $(NH_4)_2$ SO_4 from 0 to 80 m*M*. These samples were capped and gently mixed for 20 min, and the ABx was allowed to settle for 10 min. Next, 5 mL volumes of supernatant were decanted off the top of each sample, 8 mL of 25 m*M* MES, pH 5.5, were added to each tube with stirring, and the ABx was allowed to settle for 10 min. Next, 8 mL of supernatant were removed, and 1 mL of 500 m*M* $(NH_4)_2SO_4$ plus 20 m*M* NaOAc, pH 6.8, was added to each tube with stirring for 10 min. The ABx was allowed to settle for 10 min, and 1 mL of the supernatant containing the desorbed IgG was decanted. This last step was repeated in order to recover the liquid associated with the interstitial spaces of ABx. This protocol was used to optimize ABx batch purification of antibodies from serum, ascites fluid and cell culture fluid (Figs. 20–22).

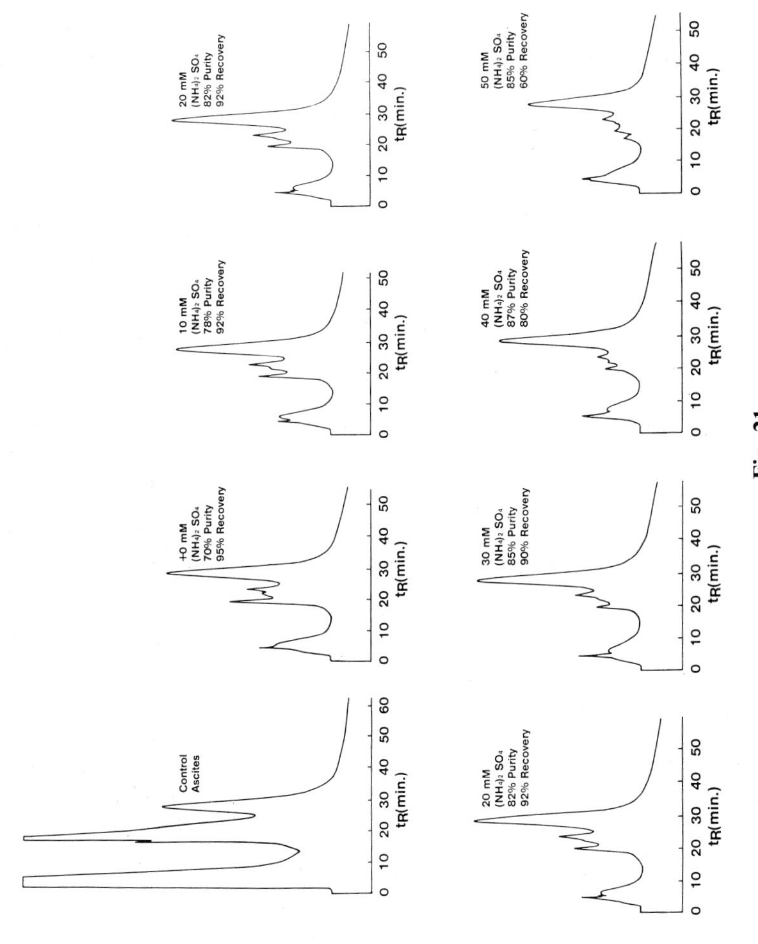
Control
Ascites
t_R(min.)
+0 mM
$(NH_4)_2\ SO_4$
70% Purity
95% Recovery
t_R(min.)
10 mM
$(NH_4)_2\ SO_4$
78% Purity
92% Recovery
t_R(min.)
20 mM
$(NH_4)_2\ SO_4$
82% Purity
92% Recovery
t_R(min.)
20 mM
$(NH_4)_2\ SO_4$
82% Purity
92% Recovery
t_R(min.)
30 mM
$(NH_4)_2\ SO_4$
85% Purity
90% Recovery
t_R(min.)
40 mM
$(NH_4)_2\ SO_4$
87% Purity
80% Recovery
t_R(min.)
50 mM
$(NH_4)_2\ SO_4$
85% Purity
60% Recovery
t_R(min.)

Fig. 21.

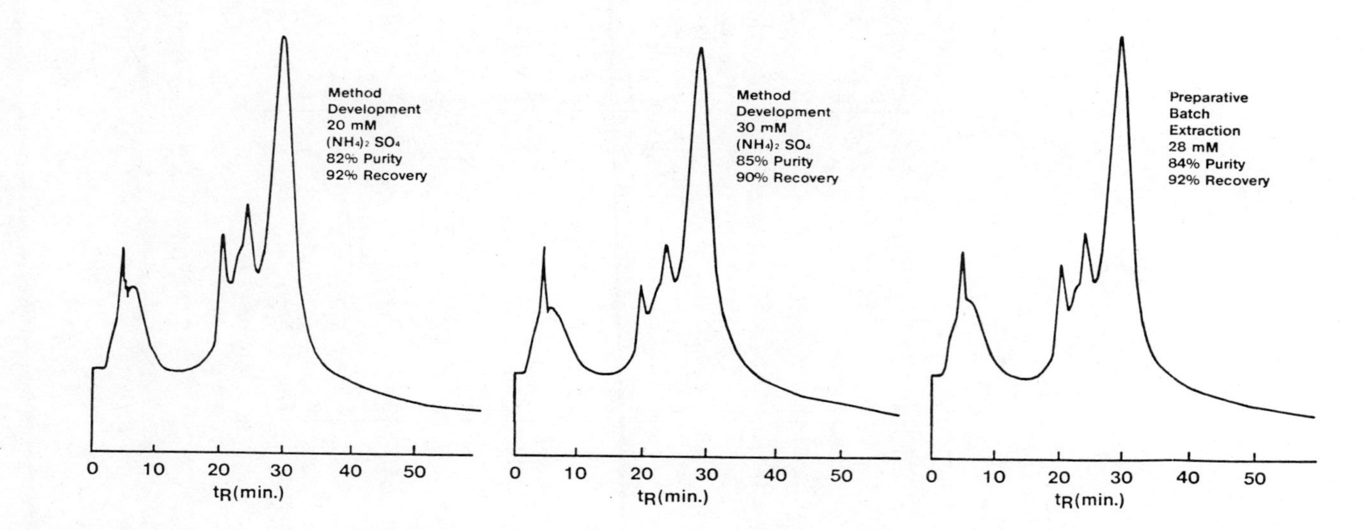

Fig. 21. Effects of ammonium sulfate on the selective batch extraction of a monoclonal antibody from mouse ascites fluid with 40 micron BAKERBOND ABx. Chromatography was conducted on a column (4.6 × 250 mm) containing 15 micron ABx. The mobile phase consisted of an initial (A) buffer of 25 m*M* MES, pH 5.4, and a final elution (B) buffer of 500 m*M* $(NH_4)_2SO_4$ plus 20 m*M* NaOAc, pH 6.7, with an initial isocratic gradient of 100% A buffer for 2 min, followed by a step gradient from 100% A to 15% B buffer, and a linear gradient from 15% B to 50% B buffer over 60 min, at a flow rate of 1.0 mL/min, with a back pressure of 150 psi. Proteins were detected by UV absorbance at 280 nm at an attenuation of 0.2 AUFS. The sample consisted of 0.4 mL of neat ascites fluid or 0.2 mL of batch extracted monoclonal antibody, diluted to 2.0 mL with 25 m*M* MES, pH 5.4. The monoclonal antibody is the last peak to elute (at 30 min). Details for the experiment are given in Fig. 20. Purity and recovery data are given with each chromatogram. The purification factor for the samples with 20 and 30 m*M* $(NH_4)_2SO_4$ was approximately 20-fold. The lower set of three chromatograms represent the ABx analysis of the 200-fold scale-up of this batch extraction, in the presence of 25 m*M* $(NH_4)_2SO_4$ (100 mL of ascites fluid were processed with 4 g of ABx).

ammonium sulfate results in the separation of all immunoglobulins from virtually all fetal bovine serum proteins, as well as the majority of nonimmunoglobulin proteins present within various adult sera (Figs. 3, 15–25, 27–28). Therefore, various concentrations of sulfate salts were added to diluted serum in the batch mode, in an attempt to selectively desorb or inhibit the adsorption of the nonimmunoglobulin proteins, while at the same time, allow all or most of the immunoglobulins to be adsorbed (Figs. 20–22). The addition of ammonium sulfate at a final concentration up to 20 or even 30 mM did not prevent the quantitative binding of the rabbit polyclonal immunoglobulins to ABx (Fig. 20). More important, the purity of the antibodies in the absorbed fraction increased significantly as higher concentrations of ammonium sulfate

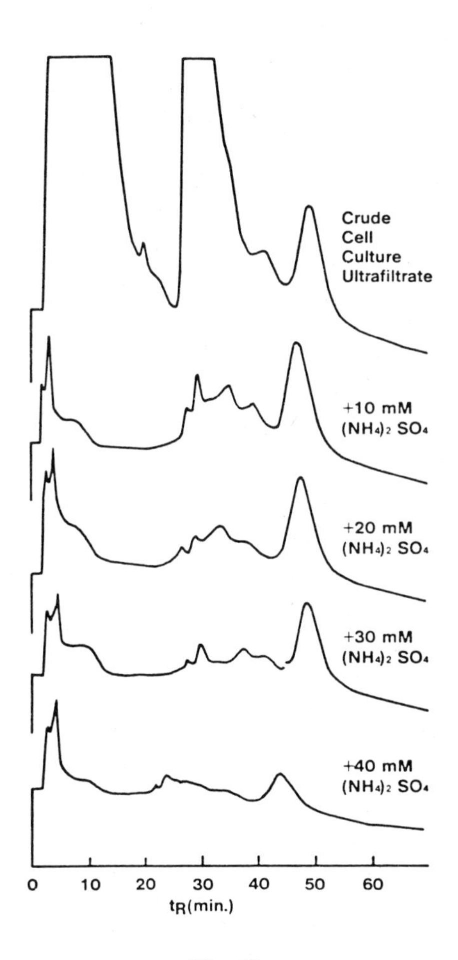

Fig. 22.

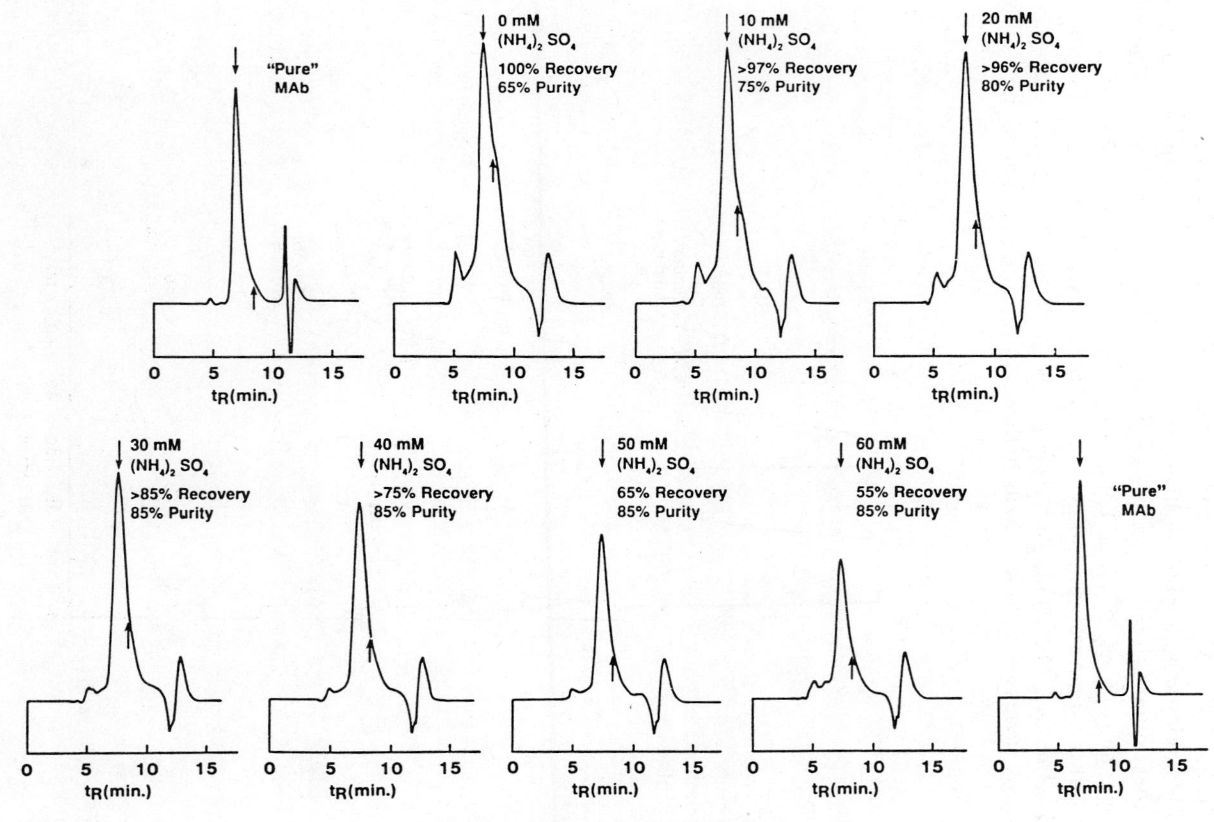

Fig. 22. (See Caption on next page).

Fig. 22. ABx and size exclusion chromatographic analysis of the batch extraction of monoclonal antibodies from two serum-supplemented cell culture ultrafiltrates with 40 micron BAKERBOND ABx. ABx chromatography was conducted on a column (4.6 × 250 mm) containing 15 micron ABx (above). The mobile phase consisted of an initial (A) buffer of 25 m*M* MES, pH 5.4, and a final elution (B) buffer of 500 m*M* $(NH_4)_2SO_4$ plus 20 m*M* NaOAc, pH 6.7, with a linear gradient of 100% A to 100% B buffer over 1 hr, at a flow rate of 1.0 mL/min, with a back pressure of 200 psi. Proteins were detected by UV absorbance at 280 nm at an attenuation of 0.05 AUFS. The sample consisted of either neat cell culture ultrafiltrate or batch extracted immunoglobulins, diluted to 8.0 mL in 25 m*M* MES, pH 5.4. The ABx void volume peak contained albumins, transferrins, proteases, and small molecules, and the peak eluting at 45 min was the monoclonal antibody. The purification factor for the 20 m*M* $(NH_4)_2SO_4$ sample was about 50-fold. Batch extraction was conducted as detailed in Fig. 20. Size exclusion chromatographic analysis of the batch extraction of another cell culture ultrafiltrate was conducted on a column (7.75 × 250 mm) packed with an experimental size exclusion chromatographic matrix (below). The mobile phase was 200 m*M* KH_2PO_4, pH 7.2, with isocratic elution at a linear flow rate of 0.3 mL/min. Proteins were detected by UV absorbance at 280 nm. The sample contained the batch extracted monoclonal antibody at the indicated purities and recoveries (by SDS analysis and size exclusion chromatography). The downward arrow indicates the retention time of the monoclonal and the peak height for quantitative recovery during extraction. The upward arrow denotes the retention time for bovine transferrin and the height of this arrow and the broadness of the monoclonal peak suggests the approximate level of contamination by proteins with molecular weights of 50 to 70 kD.

were added and the protein contaminants were selectively desorbed from, or prevented from binding to the ABx matrix. Ideally, an intermediate salt concentration can be found which provides the optimum recovery and purity, without compromising either (Figs. 20–22). Chromatographic and electrophoretic analysis and comparison of the neat serum and the batch extracted samples indicated that the recovery of immunoglobulins from this method was nearly quantitative, provided that either a second extraction was conducted in order to remove the extracted and desorbed immunoglobulin from the interstitial spaces of the bonded phase, or provided that all the supernatant was decanted or filtered in the first extraction step. The efficacy of this procedure is illustrated by the high capacities (Table 4K), large purification factors (up to 100-fold), good recoveries and high purities (greater than 70%) obtained with most samples.

Although batch extraction with ABx provides the advantage of higher capacities relative to those which can be obtained via column chromatography (Table 4K), purity is typically lower (Figs. 20–22). However, batch extracted monoclonals can be purified to near homogeneity by a second purification step with either ABx (Figs. 20–22), anion exchange (Fig. 23), hydrophobic interaction (Fig. 24), or even size exclusion (Fig. 22) chromatography.

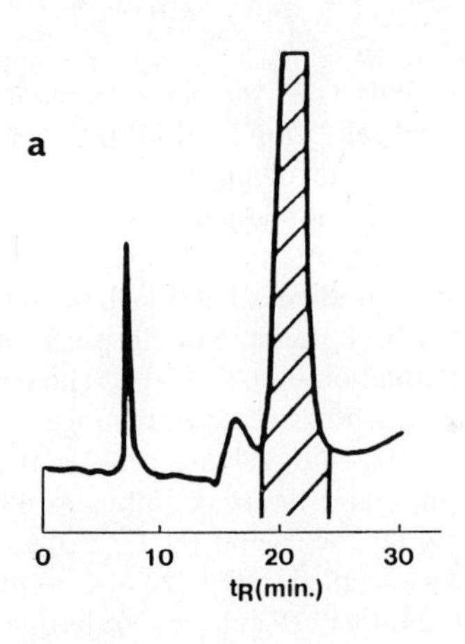
a
0
10
20
30
tR(min.)

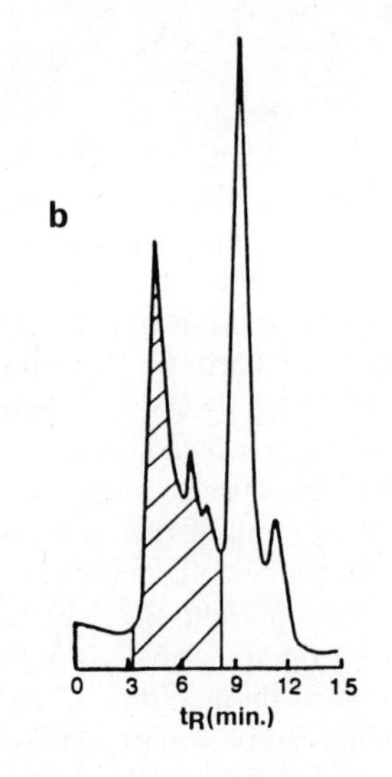
b
0
3
6
9
12
15
tR(min.)

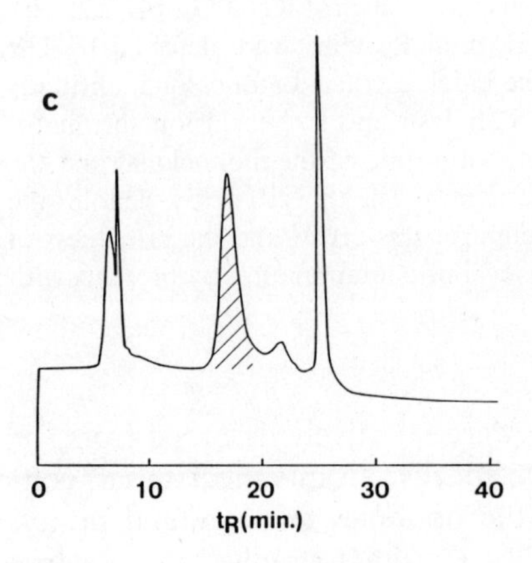
c
0
10
20
30
40
tR(min.)

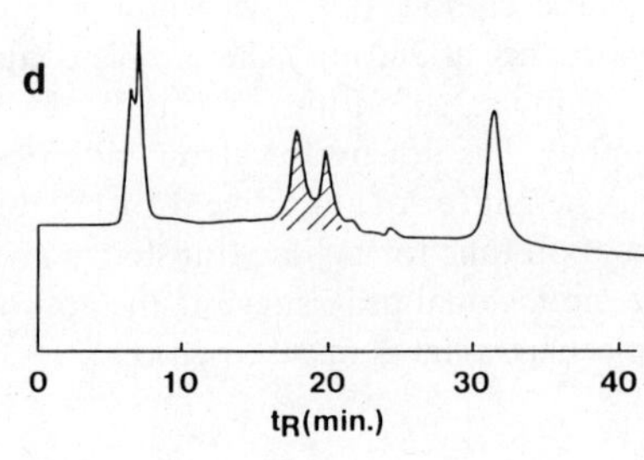
d
0
10
20
30
40
tR(min.)

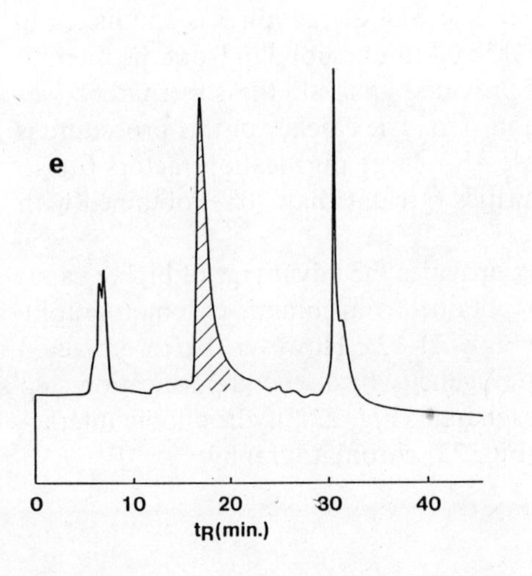
e
0
10
20
30
40
tR(min.)

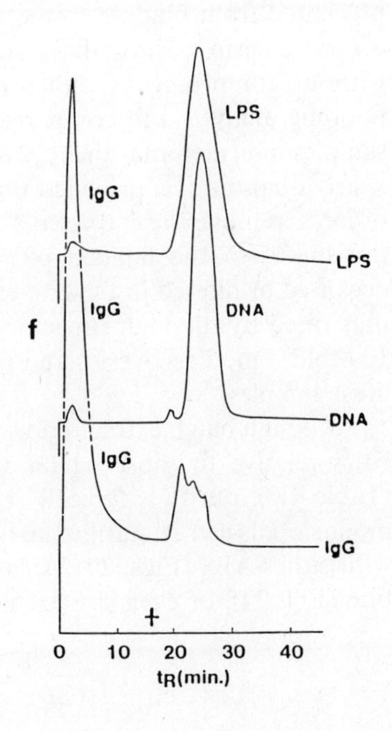
f
LPS
IgG
LPS
IgG
DNA
DNA
IgG
IgG
+
0
10
20
30
40
tR(min.)

7. SECOND STEP PURIFICATION

In cases where extreme product purity is required or large sample volumes are being processed, a purification scheme comprised of several steps on ABx may be possible (Figs. 23, 24, 33, 35). Initial purification may be conducted on a large column containing 40 micron ABx using traditional, open column chromatographic equipment, or, in the batch mode. Since greater than 96% of the

---

Fig. 23. Second step purification of antibodies by anion exchange chromatography on BAKERBOND MAb following initial purification with ABx. Chromatogram A represents the second step purification to homogeneity of a monoclonal antibody from a cell culture supernatant on MAb following initial purification by 5 micron ABx (see Fig.2*a*). Anion exchange chromatography was conducted on a column (4.6 × 500 mm) packed with 15 micron MAb, with an initial (A) buffer of 10 m*M* KH_2PO_4, pH 6.6, and a final (B) buffer of 250 m*M* KH_2PO_4, pH 6.5, with a linear gradient from 100% A to 50% B buffer over 1 hr, at a flow rate of 1.0 mL/min, with a back pressure of 200 psi. Chromatogram B represents the second step purification (< 95%) of polyclonal antibodies from human plasma on BAKERBOND MAb following an initial purification of ABx. Anion exchange chromatography was conducted on a column (4.6 × 250 mm) packed with 5 micron MAb, with an initial (A) buffer of 10 m*M* KH_2PO_4, pH 6.6, and a final (B) buffer of 250 m*M* KH_2PO_4, pH 6.5, with a linear gradient from 100% A to 50% B buffer over 1 hr, at a flow rate of 1.0 mL/min, with a back pressure of 200 psi. Chromatograms C, D, and E all represent second step purifications of monoclonal antibodies on MAb following batch extractions with ABx (C and D are from serum-supplemented cell culture ultrafiltrates, and E is from a mouse ascites fluid). Batch extractions with ABx were conducted as detailed in Fig. 20, except that no ammonium sulfate was added with the sole purpose of reducing antibody purity. Anion exchange chromatography was conducted on a column (7.75 × 100 mm) packed with 15 micron MAb. The mobile phase consisted of an initial (A) buffer of 50 m*M* Tris-OAc, pH 7.5, and a final (B) buffer of 2 *M* NaOAc, pH 5.8, at an initial flow rate of 0.5 mL/min and 1.0 mL/min following the tic mark, with a back pressure of less than 50 psi. Elution was conducted with a linear gradient from 100% A to 100% B buffer over 60 min (chromatogram C) or 90 min (chromatograms D and E). The batch extracted samples (0.5 mL) were diluted with 1.5 mL of buffer A and injected. The void volume peak and the final peaks contained contaminating proteins and the cross-hatched peaks contained the monoclonal antibodies at a purity of greater than 98%, 96%, and 97% in chromatograms C, D, and E, respectively (by SDS PAGE and size exclusion chromatography). Chromatogram F represents the relative elution profiles of nucleic acid (human spern DNA), pyrogen (*E. coli* lipopolysaccharide), and antibodies (human polyclonal IgG) on MAb, suggesting the ability to separate and strip DNA and pyrogens from monoclonal antibodies. Chromatography was conducted on a SCOUT* column (4.6 × 50 mm) containing 15 micron MAb, with an initial (A) buffer of 50 m*M* Tris-OAc, pH 6.4, and an elution (B) buffer of 2 *M* NaOAc, pH 5.8, at a flow rate of 1.0 mL/min, with a step gradient from 100% A to 100% B buffer at 17 min. DNA was detected at 260 nm, while the IgG and pyrogen were detected at 280 nm.

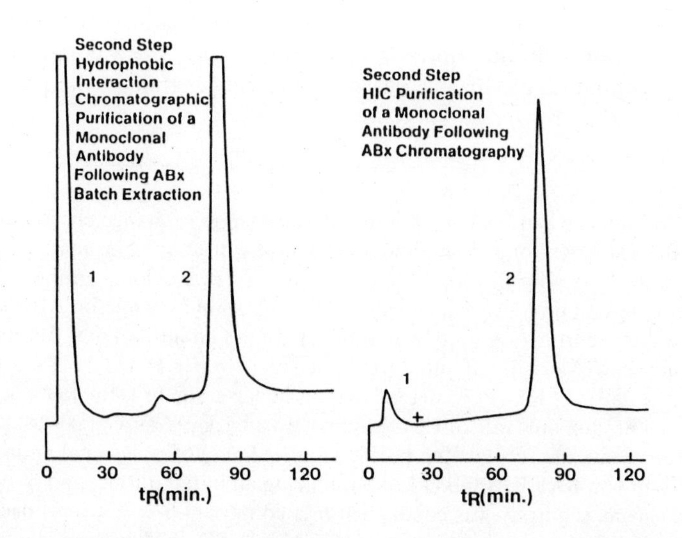
Second Step
Hydrophobic
Interaction
Chromatographic
Purification of a
Monoclonal
Antibody
Following ABx
Batch Extraction
1
2
0
30
60
90
120
tR(min.)
Second Step
HIC Purification
of a Monoclonal
Antibody Following
ABx Chromatography
1
2
0
30
60
90
120
tR(min.)

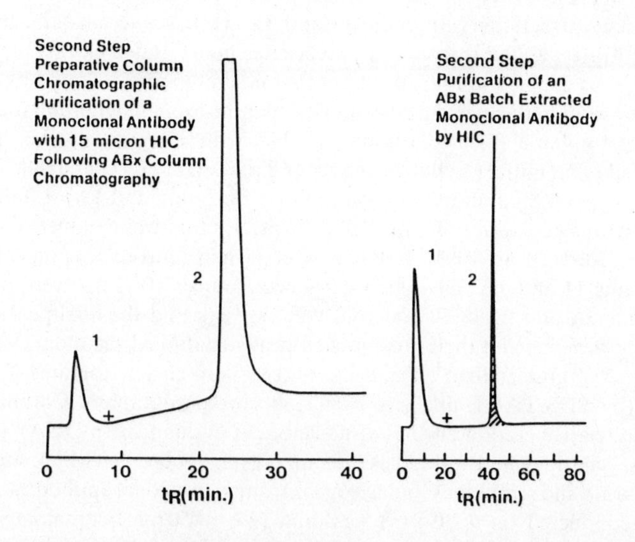
Second Step
Preparative Column
Chromatographic
Purification of a
Monoclonal Antibody
with 15 micron HIC
Following ABx Column
Chromatography
1
2
0
10
20
30
40
tR(min.)
Second Step
Purification of an
ABx Batch Extracted
Monoclonal Antibody
by HIC
1
2
0
20
40
60
80
tR(min.)

nonimmunoglobulin proteins will be removed in this first step, the second step is typically conducted on a much smaller HPLC column containing high resolution 5 or 15 micron ABx with a second buffer system (Figs. 33, 35). The active fractions may also be diluted, dialyzed, or ultrafiltrated, loaded in low ionic strength binding buffer, and rechromatographed with the use of a step gradient into NaOAc buffer at pH 7.0 in order to maximize resolution (62, 63, 69–71).

As an alternative to rechromatography on ABx, those immunoglubilin fractions that contain minor contaminants may also yield homogenous antibody by a second chromatographic step on a high-performance anion-exchange matrix, such as BAKERBOND MAb*, or a high-performance hydrophobic interaction chromatographic (HIC) support, such as BAKERBOND HI-Propyl* (62–65, 68–71). Since the mechanisms of separation on these complementary bonded phases are each entirely different, ABx plus MAb or HI-Propyl offer selectivity and resolution which is greatly enhanced relative to any single chromatographic matrix alone. Second step purifications of ABx fractions on MAb or HI-Propyl have been shown to produce near homogeneous monoclonal antibodies in every case tested, either in the batch mode or in the colunm chromatographic mode (Figs. 23–26).

Two-dimensional chromatography using ABx followed by MAb has been carried out with cell culture supernatants, mouse ascites fluids, and, human plasma (Fig. 23). MAb chromatography has also been used successfully to purify monoclonals following ABx batch extraction (Fig. 23). Briefly, the technique involves collecting the immunoglobulin(s) eluted from ABx, diluting

Fig. 24. Second step purification of monoclonal antibodies by high performance hydrophobic interaction chromatoraphy on BAKERBOND HI-Propyl following an initial purification step with ABx. Hydrophobic interaction chromatography (HIC) was conducted on a column (7.75 × 100 mm) packed with 15 micron HI-Propyl. The mobile phase consisted of an initial (A) buffer of 700 m*M* $(NH_4)_2SO_4$ plus 35 m*M* NaOAc plus 25 m*M* KH_2PO_4, pH 7.2, and an elution (B) buffer of 25 m*M* KH_2PO_4, pH 7.2, with either a linear gradient from 100% A to 100% B buffer over 1 hr, at a flow rate of 0.5 mL/min, with a back pressure of 25 psi (chromatograms A and B), or, a linear gradient of 100% A to 100% B buffer over 15 min or 30 min, at a flow rate of 1.0 mL/min, with a back pressure of 50 psi (chromatograms C and D, respectively). The sample injection volume was 2.0 mL, and consisted of either 1.0 mL or 0.5 mL of an ABx batch extracted cell culture supernatant (in chromatograms A and D, respectively), or, 0.4 mL or 1.0 mL of fractions containing monoclonal antibodies from preparative column chromatographic runs of serum-supplemented cell culture supernatants (chromatograms B and C, respectively). These samples were added to buffer A and a high salt buffer to give a final salt concentration equal to buffer A. Proteins were detected by UV absorbance at 280 nm and the attenuation was changed (see +) from 1.0 to 2.0 AUFS. The monoclonal antibody was contained in the last peak; the recovery of mass was quantitative, and the purity was greater than 99% (by SDS PAGE and size exclusion chromatography). (See Fig. 26 for more details of this method.)

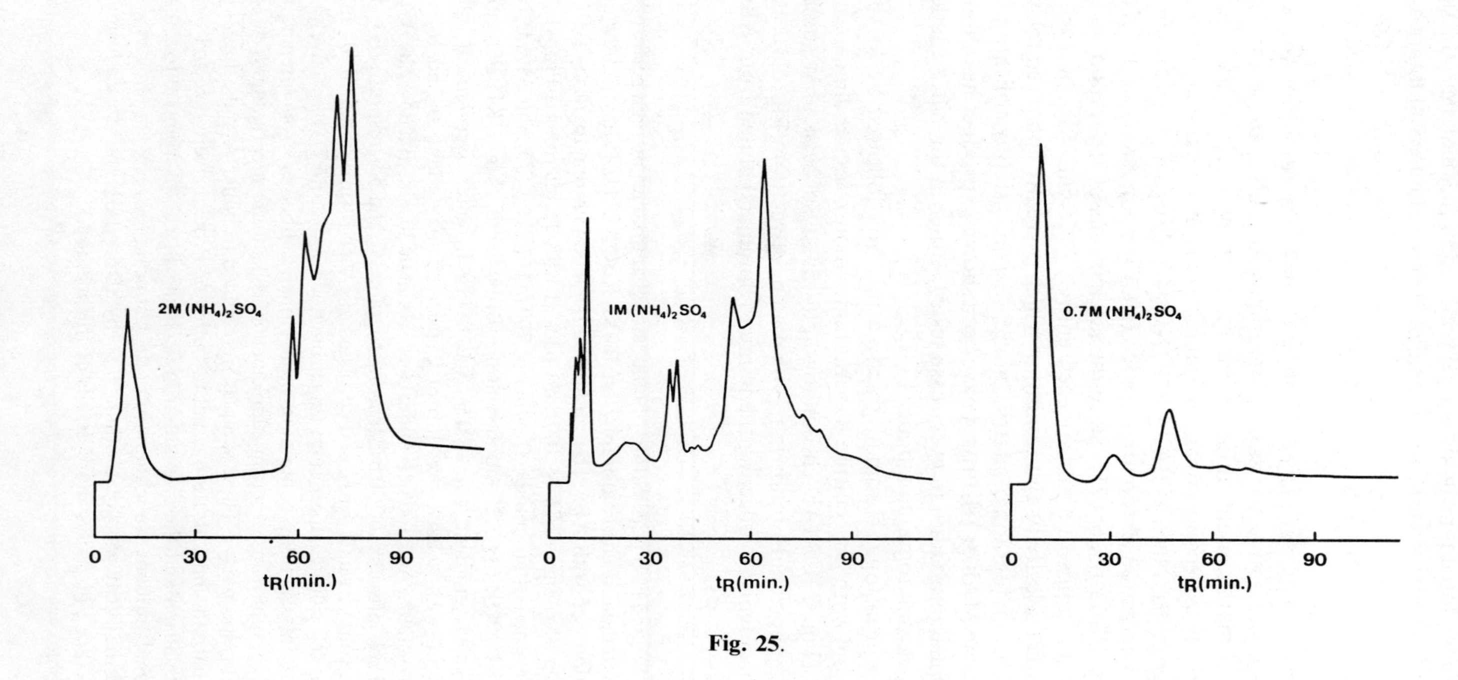
2M $(NH_4)_2SO_4$
IM $(NH_4)_2SO_4$
0.7M $(NH_4)_2SO_4$
0
30
60
90
t_R(min.)
0
30
60
90
t_R(min.)
0
30
60
90
t_R(min.)

Fig. 25.

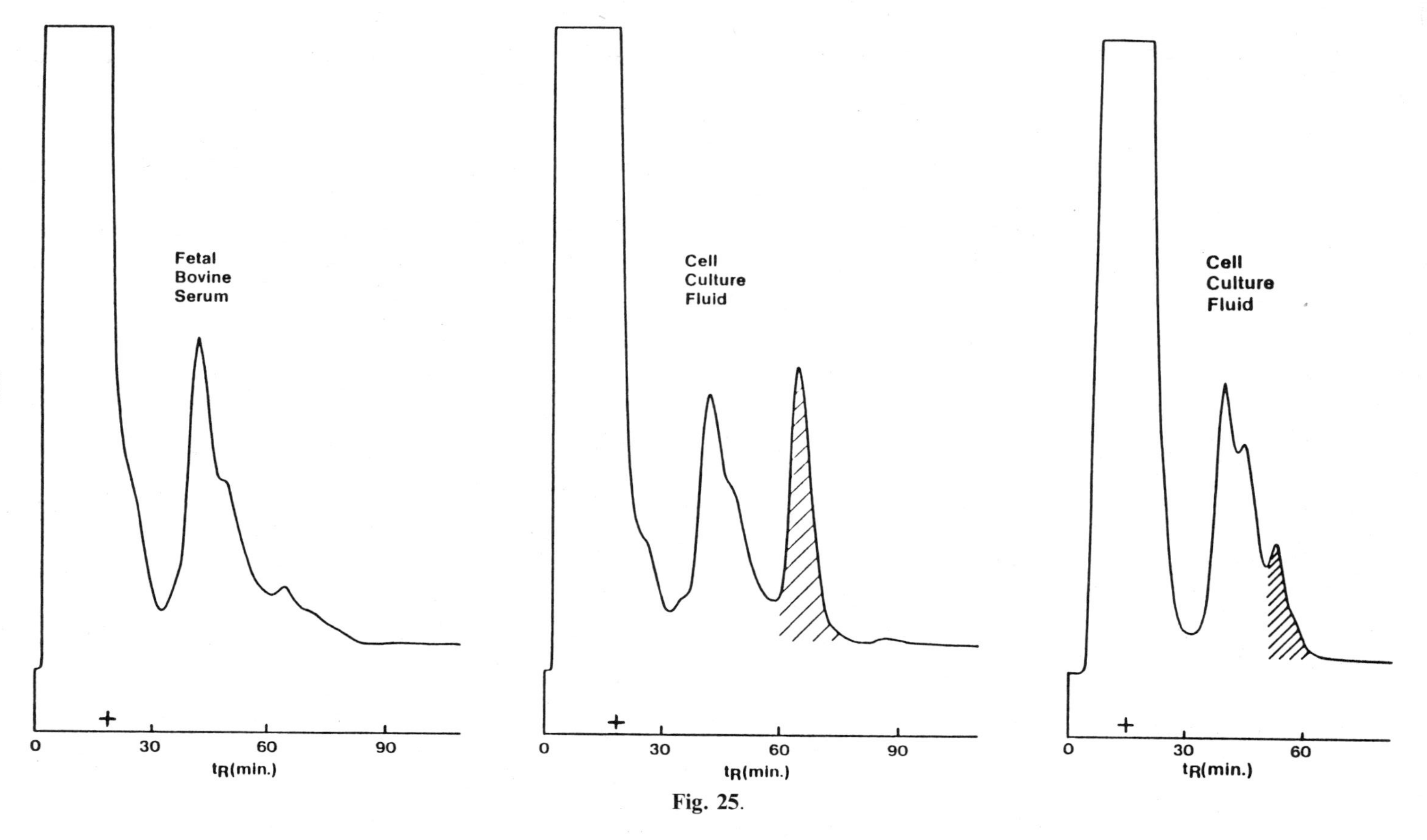
Fetal
Bovine
Serum
Cell
Culture
Fluid
Cell
Culture
Fluid
0
30
60
90
t_R(min.)
0
30
60
90
t_R(min.)
0
30
60
t_R(min.)

Fig. 25.

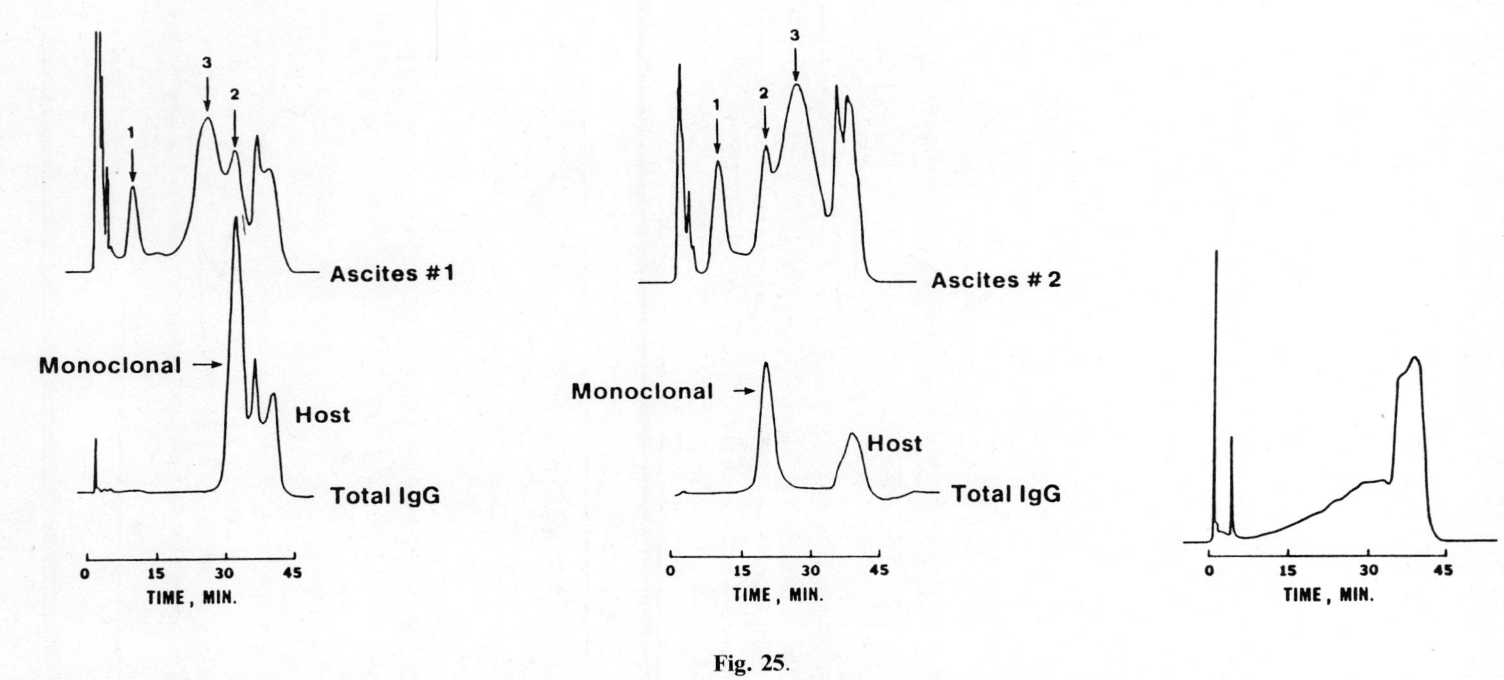

3
2
1
Ascites #1
Monoclonal →
Host
Total IgG
0
15
30
45
TIME, MIN.
3
2
1
Ascites # 2
Monoclonal →
Host
Total IgG
0
15
30
45
TIME, MIN.
0
15
30
45
TIME, MIN.

Fig. 25.

with 2 to 4 parts of MAb equilibration buffer, and re-injecting this lower level of protein onto a smaller MAb column.

Hydrophobic interaction chromatography (HIC) with BAKERBOND HI-Propyl is also an excellent complement to ABx as a second chromatographic step when homogenous antibody preparations are required (62, 63, 68–71). Antibodies eluted from ABx may often be loaded directly onto HIC matrices (Fig. 24). Furthermore, with the proper buffer conditions, virtually all of the contaminating fetal bovine serum proteins that bind to ABx do not bind to HI-Propyl, and flow through in the void volume, while all mouse or human IgG

Fig. 25. Effect of reduced salt concentration on the binding of fetal bovine serum proteins and the purification of monoclonal antibodies from cell culture and ascites fluids on BAKERBOND HI-Propyl. Hydrophobic interaction chromatography in the upper and middle sets of chromatograms was conducted on a column (7.75 × 100 mm) packed with 15 micron HI-Propyl. In the upper row, the sample was fetal bovine serum and the mobile phase consisted on an initial (A) buffer of either 2 *M* $(NH_4)_2SO_4$ plus 35 m*M* NaOAc plus 25 m*M* KH_2PO_4, pH 7.2, or 1 *M* $(NH_4)_2SO_4$ plus 35 m*M* NaOAc plus 25 m*M* KH_2PO_4, pH 7.2, or, 700 m*M* $(NH_4)_2SO_4$ plus 35 m*M* NaOAc plus 25 m*M* KH_2PO_4, pH 7.2, and an elution (B) buffer of 25 m*M* KH_2PO_4, pH 7.2, with a linear gradient from 100% A to 100% B buffer over 1 hr, at a flow rate of 0.5 mL/mn, with a back pressure of 50 psi. In the middle set of chromatograms, elution of monoclonal antibodies from serum-supplemented cell culture ultrafiltrates was conducted with an initial isocratic gradient at 100% A buffer [700 m*M* $(NH_4)_2SO_4$ plus 35 m*M* NaOAc plus 25 m*M* KH_2PO_4, pH 7.2] for 15 or 18 min (see +), followed by a linear gradient from 100% A to 100% B buffer over 40 min or 30 min (for the chromatogram on the right), at a flow rate of 1.0 mL/min, with a back pressure of 50 psi. The sample was 0.5 mL of fetal bovine serum (top row) or 3.0 mL of a cell culture ultrafiltrate (middle) diluted (to 2 mL or 12 mL) with buffer A and 2.1 *M* $(NH_4)_2SO_4$ buffer to a final salt concentration equal to buffer A. Proteins were detected by UV absorbance at 280 nm at an attenuation of 2.0 AUFS. The recovery of monoclonal antibody mass was quantitative and the purity (cross-hatched peaks) was greater than 90% (center) or 80% (right) (by SDS PAGE, size exclusion, and ABx chromatography). In the presence of 2 *M* $(NH_4)_2SO_4$ virtually all fetal bovine serum proteins are bound by HI-Propyl, while in the presence of 0.7 *M* salt, predominantly immunoglobulin G is bound (see also Fig. 26); this more selective binding buffer provides high resolution of monoclonal antibodies. In contrast, intermediate selectivity is obtained in the presence of a 1 *M* $(NH_4)_2SO_4$ binding buffer. In the bottom row, hydrophobic interaction chromatography of two mouse ascites fluids and purified mouse serum polyclonal IgG (bottom right) was conducted on a column (4.6 × 250 mm) containing 5 micron HI-Propyl; the starting (A) buffer was 1 *M* $(NH_4)_2SO_4$ plus 25 m*M* KH_2PO_4, pH 7.0, the elution (B) buffer was 25 m*M* KH_2PO_4, pH 7.0, and the gradient was 100% A to 25% B buffer over 30 min, then 25% B to 100% B over 5 min. The flow rate was 1.0 mL/min, with a back pressure of 1200 psi and injection volume was 0.15 mL. Proteins were detected by UV absorbance at 280 nm at 0.5 AUFS and identified by SDS PAGE. Peak 1 was transferrin, peak 2 was the monoclonal antibody, and peak 3 was albumin. The purity of the monoclonal antibody peaks was only about 75%, despite the higher level of monoclonal antibody present in these ascites fluids (bottom) relative to the cell culture fluids (middle).

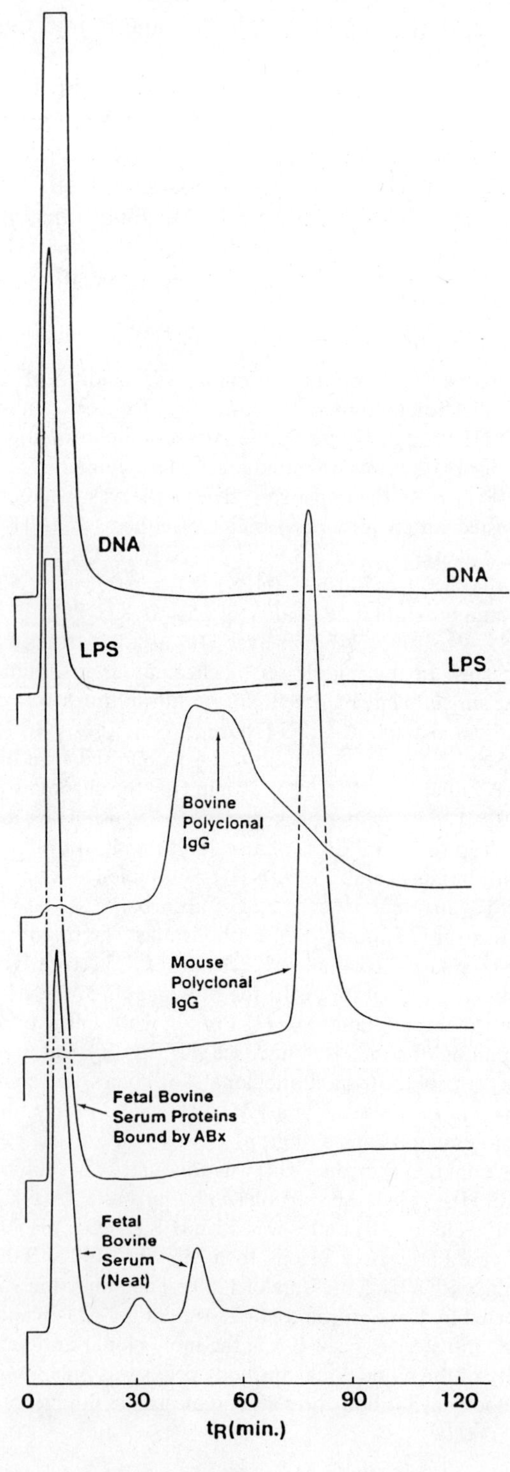
DNA
DNA
LPS
LPS
Bovine
Polyclonal
IgG
Mouse
Polyclonal
IgG
Fetal Bovine
Serum Proteins
Bound by ABx
Fetal
Bovine
Serum
(Neat)
0
30
60
90
120
tR(min.)

monoclonal antibodies bind rather strongly (Figs. 24, 25, 26). Therefore, antibodies eluted from ABx may be bound rather selectively by HIC supports, either in the column mode or even in the batch mode, and eluted in a substantially purified form, free of bovine immunoglobulins and other fetal bovine serum proteins (Figs. 24, 26). (More details of these methods are provided in Appendix I-5.)

One of the major advantages of downstream purification steps with BAKERBOND MAb and/or BAKERBOND HI-Propyl is the ability of these matrices to depyrogenate and strip nucleic acid contaminants from monoclonal antibodies. Although bacterial endotoxins (lipopolysaccharides) do not bind to ABx, some DNA and RNA coelute with immunoglobulins (62, 63, 68–71); furthermore, nucleic acids and pyrogens do exhibit a significant amount of nonspecific binding to antibodies. However, on anion exchangers with a high ligand density (e.g., BAKERBOND MAb), nucleic acids and lipopolysaccharides can be bound strongly and stripped from the antibody in the presence of an equilibration buffer at pH 6.4, while the antibody is eluted in the void volume, free from adsorbed DNA and pyrogens (Fig. 23). A significant amount of nucleic acids and pyrogens can also be stripped from monoclonals by hydrophobic interaction chromatography, particularly since mouse IgG can be bound rather selectively to BAKERBOND HI-Propyl at an ionic strength at which nucleic acids and lipopolysaccharides exhibit little or no binding (Fig. 24). These techniques can bring the levels of pyrogen and nucleic acid contamination in monoclonal antibody preparations well below those required for in vivo use (i.e., 10 pg DNA per dose and 0.05 ng lipopolysaccharide per mL).

Fig. 26. Hydrophobic interaction chromatographic separation of mouse polyclonal IgG from fetal bovine serum proteins, nucleic acids, lipopolysaccharides and bovine polyclonal IgG with BAKERBOND HI-Propyl. Hydrophobic interaction chromatography was conducted on a column (7.75 × 100 mm) packed with 15 micron BAKERBOND HI-Propyl. The mobile phase consisted of an initial (A) buffer of 700 m*M* $(NH_4)_2SO_4$ plus 35 m*M* NaOAc plus 25 m*M* KH_2PO_4, pH 7.2 and an elution (B) buffer of 25 m*M* KH_2PO_4, pH 7.2, with a linear gradient from 100% A buffer to 100% B buffer over 1 hr, at a flow rate of 0.5 mL/mn, with a back pressure of 25 psi. The sample consisted of either 2 mg of commercially purified nucleic acid (human sperm DNA), pyrogen (*E. coli* lipopolysaccharides), bovine or mouse immunoglobulin G, 0.5 mL of neat fetal bovine serum, or 0.5 mL of fetal bovine serum proteins which bind to ABx, diluted to 2 mL with buffer A and 2.1 *M* $(NH_4)_2SO_4$ buffer to give a salt concentration equal to buffer A. Proteins were detected by UV absorbance at 280 nm at various AUFS values. These chromatograms suggest that mouse IgG can be separated from greater than 95% of the fetal bovine serum IgG, and that, while mouse IgG is bound strongly, most nucleic acids, pyrogens and fetal bovine serum proteins are not bound; furthermore, virtually none of fetal bovine serum proteins which bind to ABx are bound to HI-Propyl.

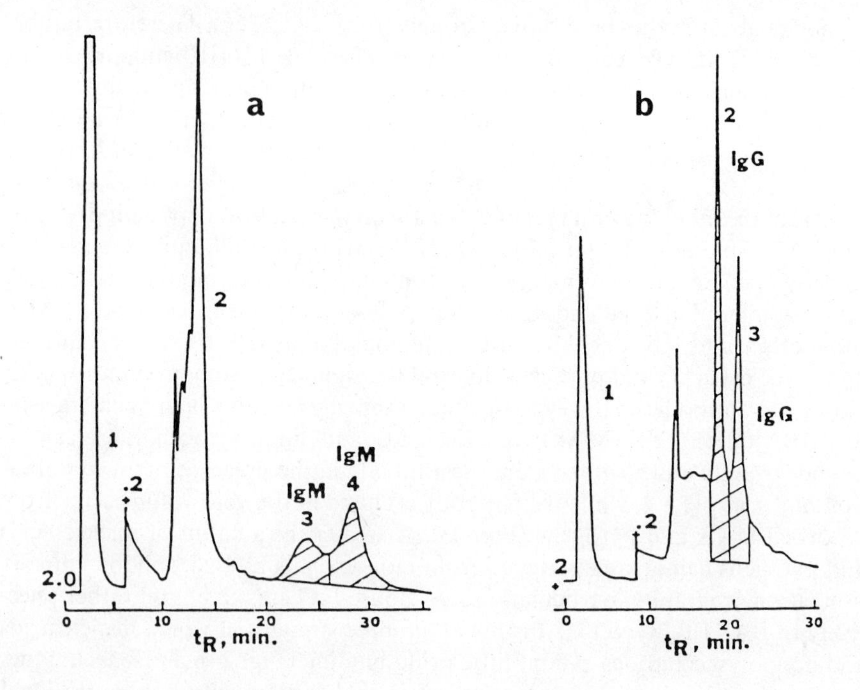
a
1
2
.2
2.0
IgM
3
IgM
4
0
10
20
30
tR, min.
b
2
IgG
1
3
IgG
2
.2
0
10
20
30
tR, min.

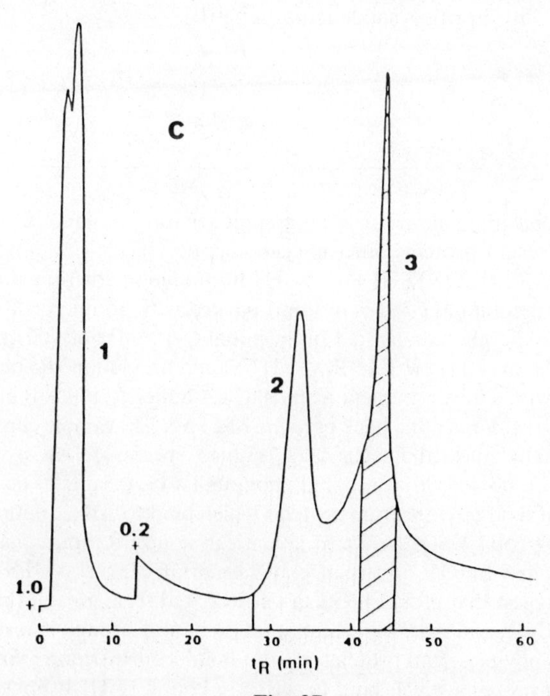
c
1
2
3
0.2
1.0
0
10
20
30
40
50
60
tR (min)

Fig. 27.

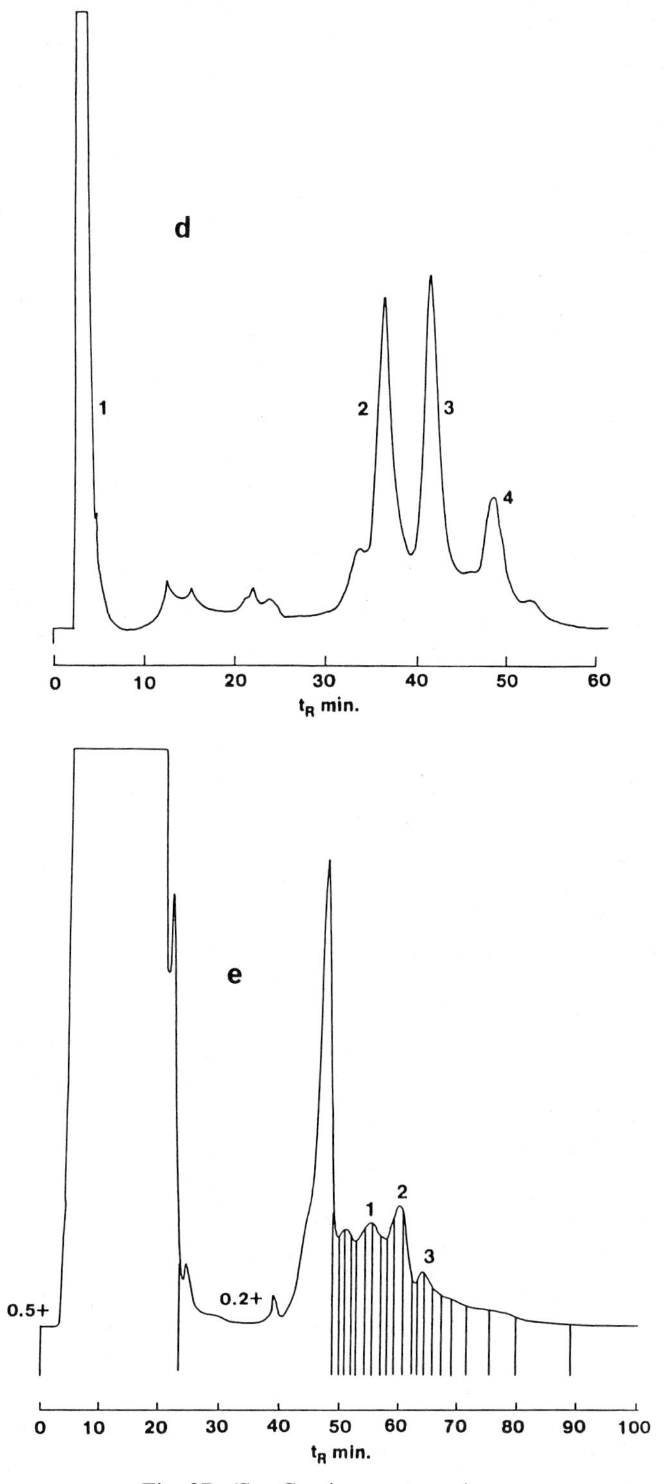

Fig. 27. (See Caption next page)

Fig. 27. Separation of multiple antibody species from ascites fluids and cell culture fluids on BAKERBOND ABx. Chromatography was conducted on either an analytical column (4.6 × 250 mm) containing 5 micron ABx (chromatograms A and B) or 15 micron ABx (chromatogram C), or, on a GOLD column (7.75 × 100 mm) containing 5 micron ABx (chromatograms D and E). In chromatogram A, the initial (A) buffer was 10 m*M* KH_2PO_4, pH 6.0, and the final (B) buffer was 500 m*M* KH_2PO_4, pH 6.8, with a linear gradient from 100% A to 50% B buffer over 1 hr. In chromatogram B, the starting (A) buffer was 10 m*M* MES, pH 6.0, the elution (B) buffer was 250 m*M* KH_2PO_4, pH 6.8, and elution was with 100% A buffer for 10 min, followed by a step to 20% B buffer and linear gradient from 20% B to 50% B over 60 min. In chromatogram C, the initial (A) buffer was 10 m*M* MES, pH 6.0, and the final (B) buffer was 250 m*M* KH_2PO_4, pH 6.8, with a linear gradient from 100% A to 50% B buffer over 1 hr. In chromatogram D, the mobile phase consisted of an initial (A) buffer of 25 m*M* MES, pH 5.6, and a final elution (B) buffer of 1 *M* NaOAc, pH 7.0, with an initial isocratic gradient at 100% A buffer for 3 min, followed by a step gradient to 15% B buffer, and a linear gradient from 15% B to 100% buffer over 1 hr. In chromatogram E, the starting (A) buffer was 25 m*M* MES, pH 5.2, the elution (B) buffer was 1 *M* NH_4OAc, pH 7.0, and the gradient was linear from 100% A to 100% B over 2 hr. The flow rate was 1.0 mL/min at a back pressure of 1000 psi (5 micron ABx, 4.6 × 250 mm column), or 200 psi (5 micron ABx, 7.75 × 100 mm column, or, 15 micron ABx, 4.6 × 250 mm column). Proteins were detected by UV absorbance at 280 mm and the attenuation was varied (see +; in D the attenuation was 2.0 AUFS). The sample consisted of 0.5, 0.5, 0.4, or 0.6 mL of mouse ascites fluid (chromatograms A, B, C, and D, respectively) or 2 mL of a serum-supplemented cell culture ultrafiltrate (chromatogram E). Peak 1 consisted of albumin, transferrin and proteases; in chromatogram A, peak 2 contained weakly bound proteins, and peaks 3 and 4 (cross-hatched) were two distinct IgM species, being greater than 95% pure (by SDS PAGE). In chromatogram B, peaks 2 and 3 (cross-hatched areas) were both IgG monoclonal antibodies at greater than 95% purity (by SDS PAGE). In chromatogram C, peak 2 contained the host IgG, and peak 3 was the monoclonal antibody. In chromatogram D, peaks 2, 3, and 4 were all IgG_1 monoclonal antibodies with variable light chain compositions at greater than 98% purity (by SDS PAGE); peaks 2 and 4 were divalent with two fast and two slow migrating light chains, respectively, and peak 3 was a monovalent hybrid containing both fast and slow light chains. Chromatogram E represents the ABx chromatographic separation of monospecific from bispecific monoclonal antibodies secreted by a hybrid hybridoma (quadroma). Fractions 1–20 all contained immunological activity and were rather pure by SDS PAGE analysis (fractions 1 through 12 were 40, 70, 85, 90, 95, 95, 95, 95, 95, 80, 70, and 70% pure, respectively, and fractions 13–20 were below the levels of accurate detection). Immunological assays and IgG subclass-specific affinity chromatography indicated that the anti-horseradish peroxidase monospecific IgG_1 eluted between 52 and 58 min, the anti-somatostatin monospecific IgG_{2b} eluted between 62 and 68 min, and the anti-horseradish peroxidase/anti-somatostatin bispecific hybrid antibody (IGG_1/IgG_{2b}) eluted between 56 and 64 min. The recovery of total antibody mass was greater than 97% and the recovery of immunological acitivity for the bispecific monoclonal antibody was greater than 90%.

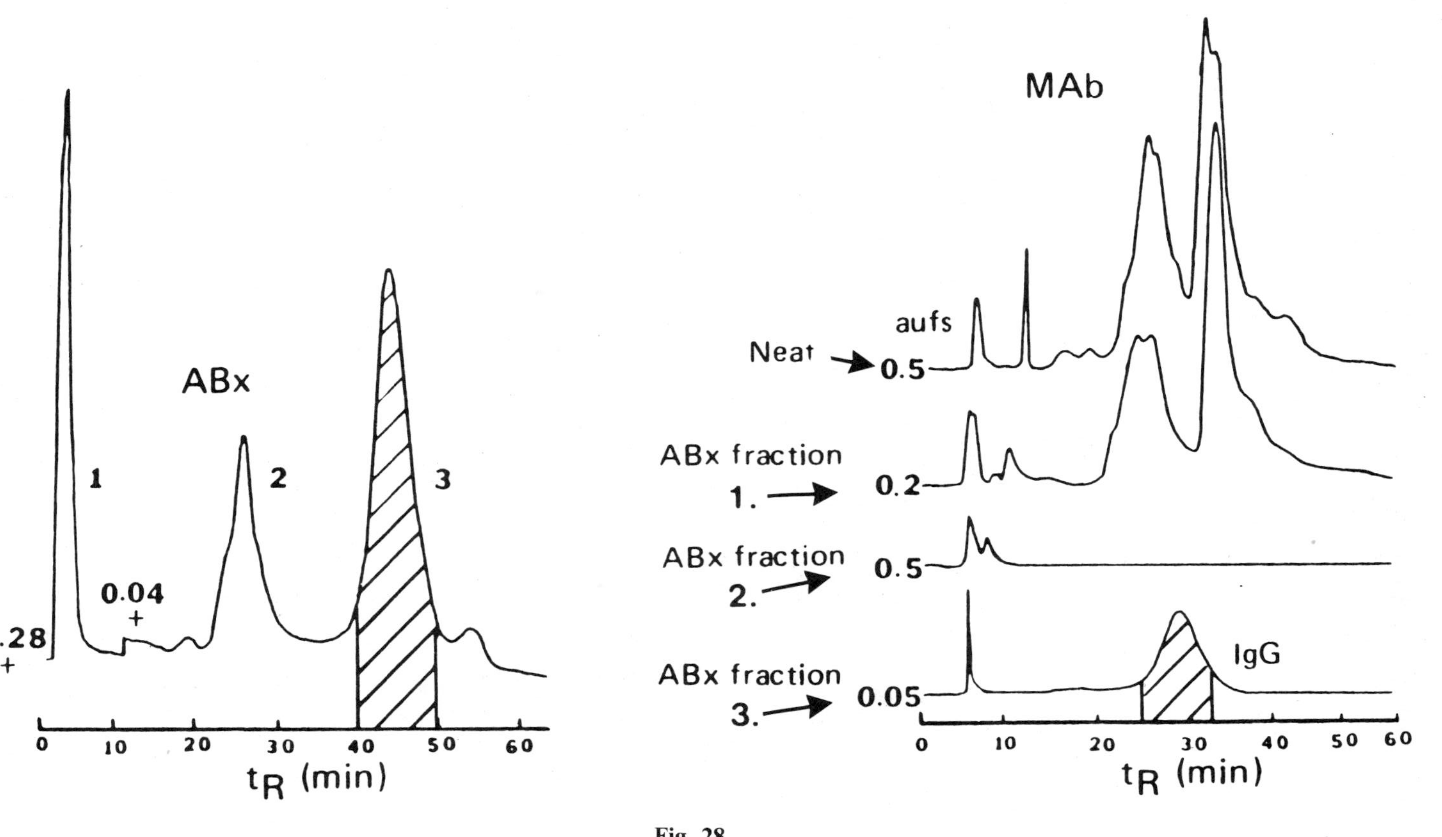
ABx
1
2
3
1.28
+
0.04
+
0 10 20 30 40 50 60
tR (min)
MAb
aufs
Neat
0.5
ABx fraction
1.
0.2
ABx fraction
2.
0.5
ABx fraction
3.
0.05
IgG
0 10 20 30 40 50 60
tR (min)

Fig. 28.

8. RESOLUTION OF MULTIPLE IMMUNOGLOBULIN SPECIES

Under similar chromatographic conditions antibodies have been found to elute from ABx at a wide range of retention times (Table 1). Variations in elution times among different antibodies are apparently the result of subtle differences in antibody structure and molecular diversity, including differences in isoelectric point, surface charge distribution, antibody class and subclass, light chain variations, surface glycosylation, idiotype, epitope, source, species, and other types of immunoglubulin molecular diversity which are not yet fully understood, nor totally defined.

Whereas the variability in antibody elution times may be a disadvantage on other chromatographic support due to possible coelution of the antibody with albumins, transferrins, or proteases or all three, it represents a distinct advantage on ABx, because most of these nonimmunoglobulin contaminants do not bind (Table 1). This variation in antibody binding affinity facilitates the resolution of different monoclonals with minimal possibility of contamination. An unexpected benefit of such diversity in antibody binding strength is that ABx is capable of fractionating the heterogeneous population of polyclonal antibodies present within serum/plasma (Fig. 8). Further resolution has been achieved by rechromatography of active fractions in the presence of a second buffer system (data not shown). This technique faciltates the separation of a highly purified epitope-specific immunological activity from serum with minimal contamination by other polyclonal antibodies.

In an analagous manner, ABx is able to separate monoclonal antibodies from serum polyclonal (host) immunoglobulins in mouse ascites fluid or in fetal bovine serum (Figs. 4, 11, 12, 14, 27, 33, 34). Any monoclonal antibody can be

Fig. 28. Rechromatography of ABx fractions (peaks) from a serum-based cell culture fluid on BAKERBOND MAb, high performance anion exchange matrix. Neat hybridoma tissue culture media was chromatographed on a column containing 40 micron BAKERBOND ABx, and the 3 major ABx peaks (1, 2, 3) were rechromatographed on a 15 micron BAKERBOND MAb column. More than 95% of the total proteins were not bound by ABx (fraction 1), while these nonimmunoglobulin proteins were bound and even coeluted with the monoclonal antibody on the MAb and other anion exchangers. The purity of the monoclonal fraction (3) from the ABx column was greater than 90% IgG (by SDS PAGE). Note the 32-fold change in recorder sensitivity (AUFS, see +), and the low levels of monoclonal present. ABx chromatography was conducted on a column (4.6×250 mm) containing 40 micron BAKERBOND ABx with a starting (A) buffer of 10 m*M* MES, pH 5.6, and a final elution (B) buffer of 250 m*M* KH_2PO_4, pH 6.8, with a linear gradient from 100% A to 50% B over 1 hr, at a flow rate of 1.0 mL/min, with a back pressure of 30 psi. Anion exchange chromatography was conducted on a column (4.6×500 mm) packed with 15 micron BAKERBOND MAb matrix. The mobile phase consisted of an initial (A) buffer of 10 m*M* KH_2PO_4, pH 6.6, and a final (B) buffer of 500 m*M* KH_2PO_4, pH 6.6, with a linear gradient (100% A to 50% B) over 1 hr. The flow rate was 1.0 mL/min with a back pressure of 150 psi.

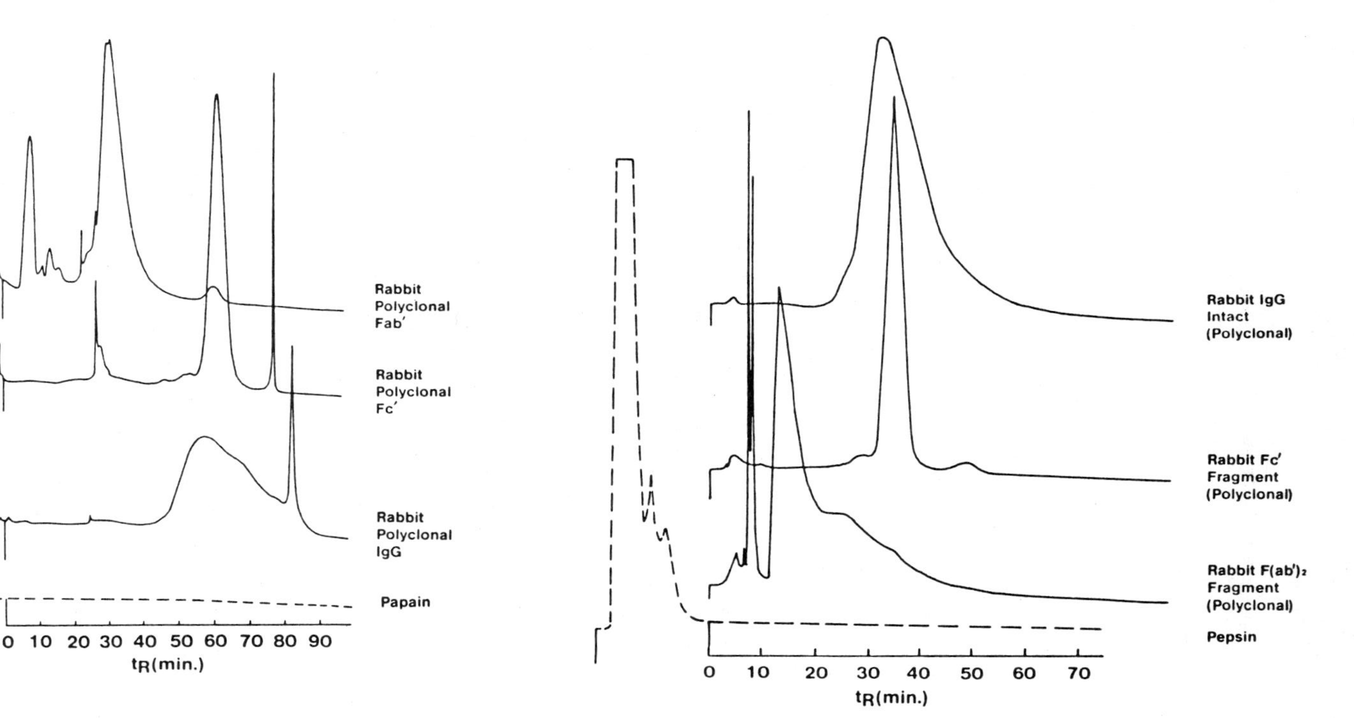
Papain Digest of Rabbit Polyclonal IgG on 5μ BAKERBOND ABx
Rabbit Polyclonal Fab′
Rabbit Polyclonal Fc′
Rabbit Polyclonal IgG
Papain
0 10 20 30 40 50 60 70 80 90
tR(min.)
Pepsin Digest of Rabbit Polyclonal IgG on 5μ BAKERBOND ABx
Rabbit IgG Intact (Polyclonal)
Rabbit Fc′ Fragment (Polyclonal)
Rabbit F(ab′)2 Fragment (Polyclonal)
Pepsin
0 10 20 30 40 50 60 70
tR(min.)

Fig. 29.

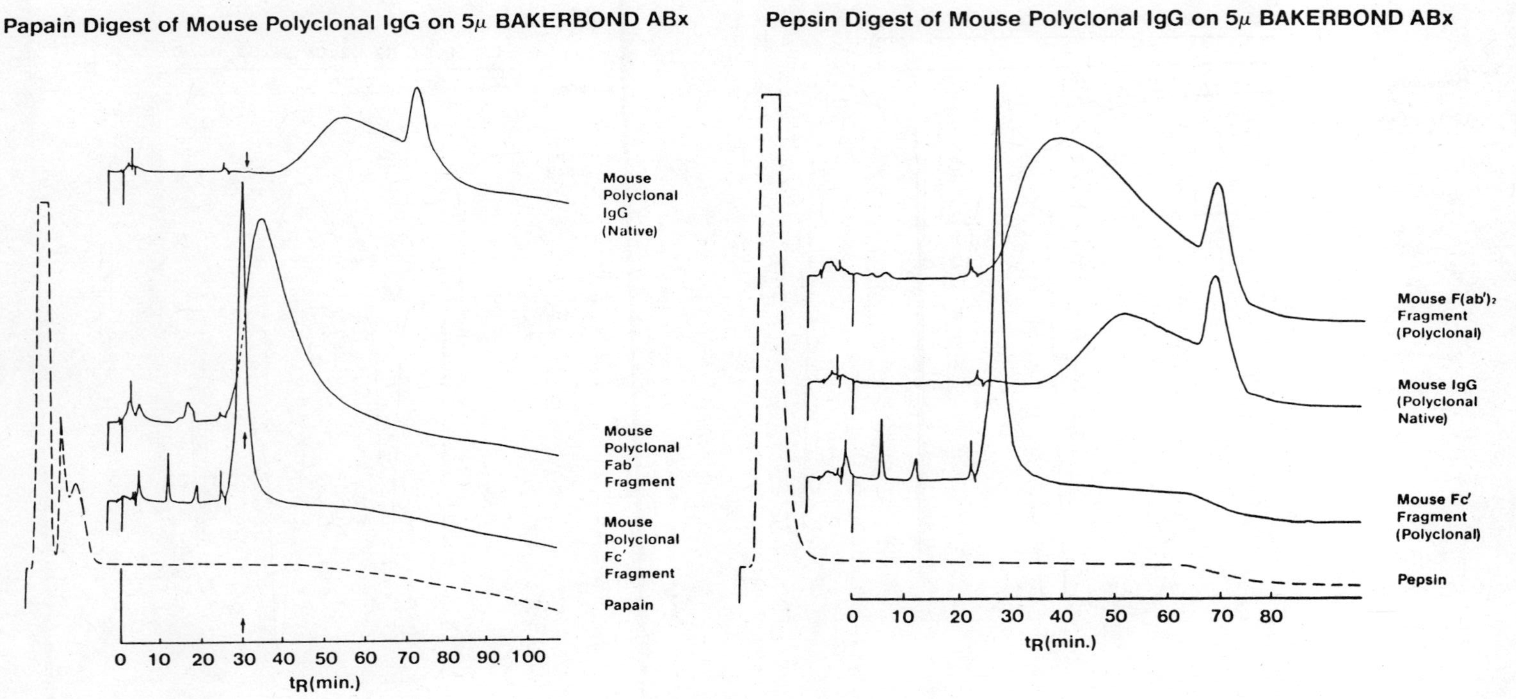

Papain Digest of Mouse Polyclonal IgG on 5μ BAKERBOND ABx
Mouse Polyclonal IgG (Native)
Mouse Polyclonal Fab′ Fragment
Mouse Polyclonal Fc′ Fragment
Papain
0 10 20 30 40 50 60 70 80 90 100
tR(min.)
Pepsin Digest of Mouse Polyclonal IgG on 5μ BAKERBOND ABx
Mouse F(ab′)2 Fragment (Polyclonal)
Mouse IgG (Polyclonal Native)
Mouse Fc′ Fragment (Polyclonal)
Pepsin
0 10 20 30 40 50 60 70 80
tR(min.)

Fig. 29.

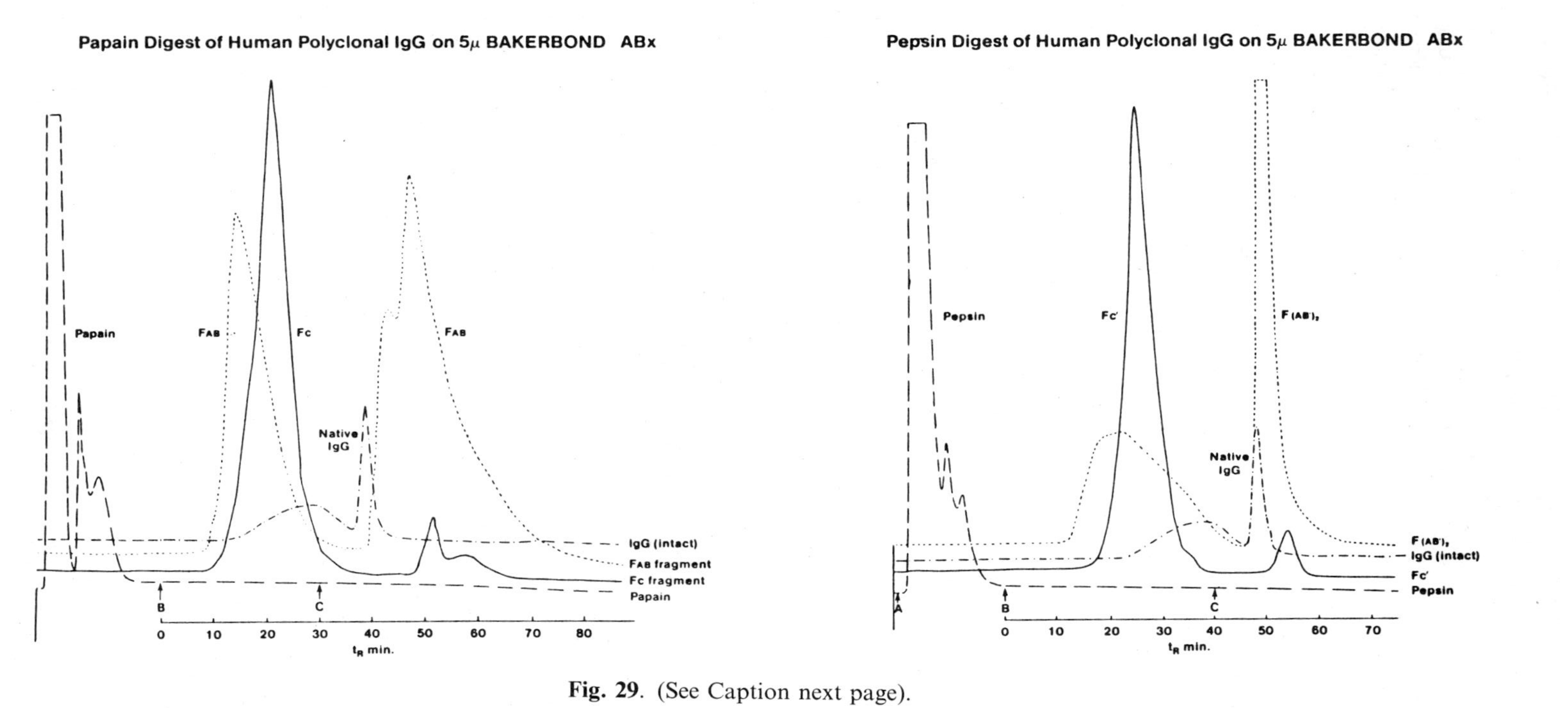

Fig. 29. (See Caption next page).

successfully separated from greater than 80% of the serum polyclonal contaminants present within either equine or fetal bovine serum-supplemented cell culture media via elution with ammonium sulfate or sodium chloride buffers (Figs. 4, 6, 27). Further removal of host immunoglobulins may be accomplished with a second ABx chromatographic step with an alternate elution buffer (Figs. 11, 12), or, by HIC (Figs. 25, 26).

ABx is also able to resolve multiple antibody species (idiotypes) secreted from various forms of double producing hybridomas (25, 35, 62, 63, 67–71, 80). In initial studies, multiple forms of IgG and IgM were separated in a highly purified state from ascites fluids and cell culture media (Figs. 27, 34). More recently, three forms of monoclonal antibody have been resolved from single hybridomas secreting active and inactive light chain variant antibodies in this and other laboratories (25, 35, 62, 63, 67–71, 80). Several IgG_1 species with variable light chain compositions $[(K_1)_2$, $(K_1)_1(K_2)_1$ and $(K_2)_2)]$ have been purified to homogeneity in a single step from mouse ascites fluids (Fig. 27). Ross et al. (80) have also recently identified several mouse and rat ascites fluids which contain three distinct peaks of immunological activity by ABx chromatography.

Fig. 29. Separation of human, rabbit and mouse polyclonal IgG antibody fragments of proteolytic digestion with BAKERBOND ABx. Chromatography was conducted on a column (7.75 × 100 mm) containing 5 micron ABx. The mobile phase consisted of an initial (A) buffer of either 25 m*M* MES, at a pH of either 5.4 (chromatograms A and C), 5.0 (chromatogram B) or 5.2 (chromatogram D), or 10 m*M* MES, pH 5.6 (chromatograms E and F), and a final elution (B) buffer of 500 m*M* or 250 m*M* (as in chromatograms E and F) $(NH_4)_2SO_4$ plus 20 m*M* NaOAc, pH 6.7. In chromatogram A, the rabbit IgG (papain) fragments were eluted with a linear gradient from 100% A to 25% B buffer over 75 min, followed by a step gradient to 100% B buffer. In chromatogram B, the rabbit IgG (pepsin) fragments were eluted with an initial isocratic gradient of 100% A for 2 min, followed by a step gradient up 10% B buffer (held for 11 min), and a linear gradient to 50% B buffer over 1 hr. In chromatogram C, the mouse IgG (papain) fragments were eluted with an initital isocratic gradient at 100% A buffer, followed by a linear gradient from 100% A to 25% B buffer over 1 hr, and followed by another linear gradient from 25% B to 100% B buffer over 10 min. In chromatogram D, the mouse IgG (pepsin) fragments were eluted with an initial isocratic gradient at 100% A buffer for 2 min, followed by a linear gradient from 100% A to 25% B buffer over 1 hr, and a linear gradient from 25% B to 100% B over 10 min. In chromatogram E, the human IgG (papain) fragments were eluted with an initial isocratic gradient at 100% A buffer for 23 min, followed by a step gradient (at "B") from 100% A to 21% B buffer (held isocratically for 30 min), and a linear gradient (at "C") from 21% B to 100% B over 10 min. In chromatogram F, elution of the human IgG (pepsin) fragments was conducted as in chromatogram E, except that the initial hold at 100% A buffer was for only 21 min, while the step gradient to 21% B buffer was held for 40 min. Proteins were detected by UV absorbance at 280 nm at an attenuation of 2.0 AUFS. In all cases, the flow rate was 1.0 mL/min, with a back pressure of 200 psi, and the sample contained at least 1 mg of protein diluted in 2 mL of 25 m*M* MES, at a pH equal to buffer A. In each case protein standards were used to identify the elution profiles of the various components.

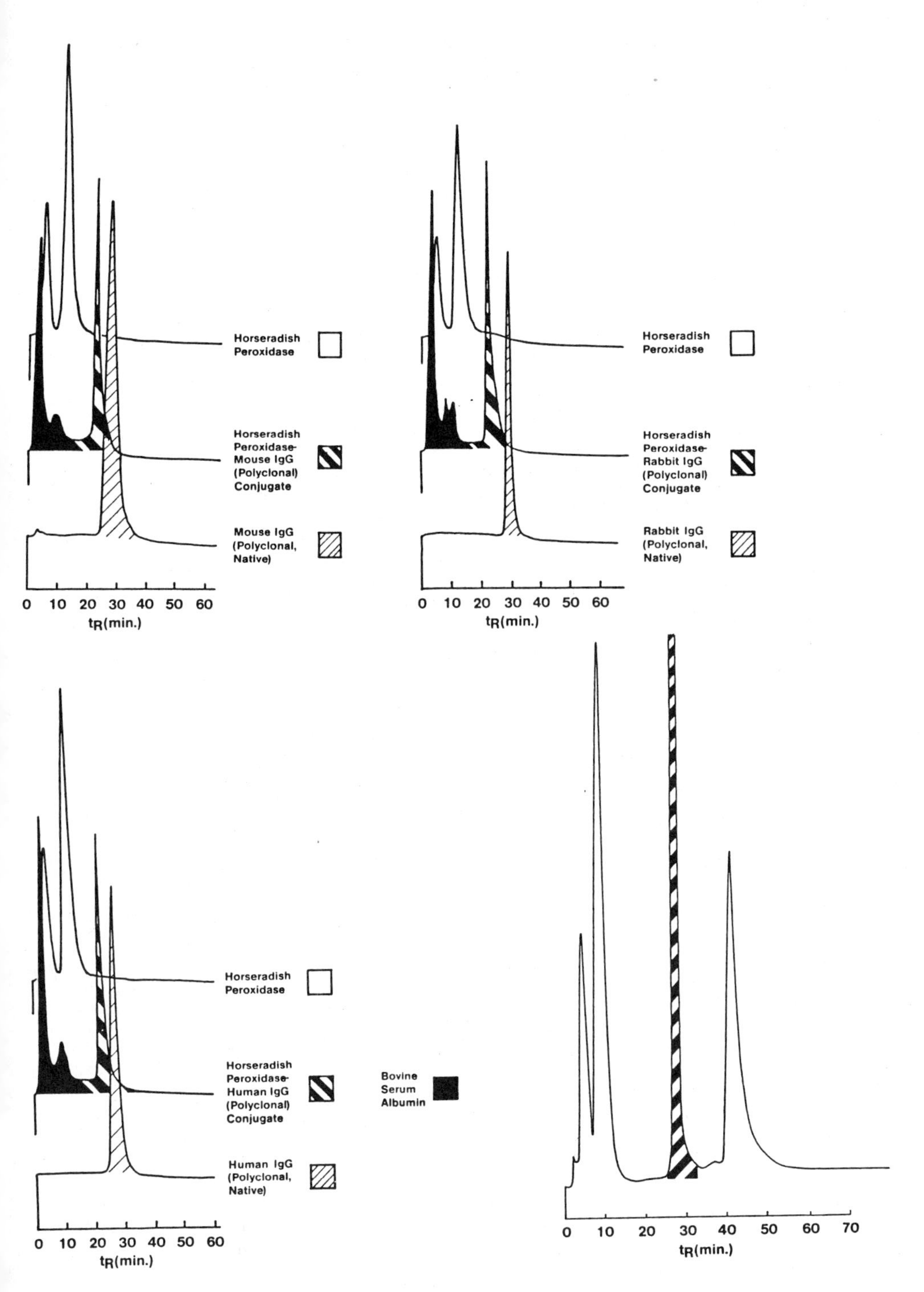
Horseradish Peroxidase
Horseradish Peroxidase-Mouse IgG (Polyclonal) Conjugate
Mouse IgG (Polyclonal, Native)
0 10 20 30 40 50 60
tR(min.)
Horseradish Peroxidase
Horseradish Peroxidase-Rabbit IgG (Polyclonal) Conjugate
Rabbit IgG (Polyclonal, Native)
0 10 20 30 40 50 60
tR(min.)
Horseradish Peroxidase
Horseradish Peroxidase-Human IgG (Polyclonal) Conjugate
Human IgG (Polyclonal, Native)
0 10 20 30 40 50 60
tR(min.)
Bovine Serum Albumin
0 10 20 30 40 50 60 70
tR(min.)

Fig. 30.

In that study (80), all three antibodies from one mouse ascites fluid were shown to have identical heavy chains, although their light chains were different, with the first and last eluting peaks containing LCa and LCb, respectively, while the middle peak contained a mixture of these.

Immunological assays indicated that the first antibody peak contained the vast majority of the immunological activity, the middle peak only slight activity, and the last peak virtually none (80). Although in this case (80), and others (Figs. 27, 34) the monoclonal antibody with the fast migrating light chain eluted first, in other instances, this slow light chain variant may elute first on ABx (62, 63, 69, 70), suggesting that there is no real correlation between light chain type and ABx chromatographic retention. However, the monovalent, hybrid monoclonal antibody, which represents a mixture of the two light chains (and has an intermediate immunoreactivity), does appear to consistently elute between the two bivalent antibodies (62, 63, 69, 70), as its physicochemical properties are presumably intermediate between the latter two.

In other cases, multiple forms of immunoglobulins, including two myeloma IgG_1 and two monoclonal IgG_{2a} antibodies from a single hybridoma have also been resolved on ABx. These four immunoglobulins exhibited no noticeable differences in their light chain compositions by SDS PAGE. Previously, several hybridomas were identified in this laboratory that produced only two forms of immunoglobulin that were also indistinguishable by SDS PAGE (Figs. 27, 33, 34). Similar results have recently been obtained with a number of other samples in this and other laboratories (62, 63, 68–71). More recently, ABx has been used

Fig. 30. Separation of horseradish peroxidase, polyclonal IgG and horseradish peroxidase-polyclonal IgG conjugates from three species, and a mouse monoclonal IgG-HRP conjugate on 5 micron BAKERBOND ABx. Chromatography was conducted on a column (7.75 × 100 mm) containing 5 micron ABx. The mobile phase consisted of an initial (A) buffer of 25 m*M* MES, pH 5.2, and a final elution (B) buffer of 500 m*M* $(NH_4)_2SO_4$ plus 20 m*M* NaOAc, pH 6.7, and (for the top three chromatograms) with an initial isocratic gradient at 100% A buffer for 10 min, followed by a linear gradient from 100% B buffer over 30 min, at a flow rate of 1.0 mL/min, at a back pressure of 200 psi. Proteins were detected by UV absorbance at 280 nm at an attenuation of 2.0 AUFS. The sample consisted of at least 1.0 mg of protein diluted in 2.0 mL of 25 m*M* MES, pH 5.2 (either mouse, rabbit or human polyclonal IgG, horseradish peroxidase or HRP-IgG conjugates). The ABx chromatogram on the bottom represents the purification of a monoclonal antibody-horseradish peroxidase conjugate from a reaction mixture. Chromatographic conditions are the same as those detailed above, except that elution was with an initial isocratic gradient at 100% A buffer for 10 min, followed by a step gradient to 10% B, and a linear gradient from 10% B to 20% B over 30 min. The sample consisted of 0.5 mL of conjugation reaction mixture plus 1.5 mL of 25 m*M* MES, pH 5.4. The cross-hatched peak corresponds to the conjugated product, the initial peaks (up to 10 min) correspond to the horseradish peroxidase isozymes, and the final peak is the native monoclonal antibody. The conjugate peak was greater than 98% pure (by SDS PAGE).

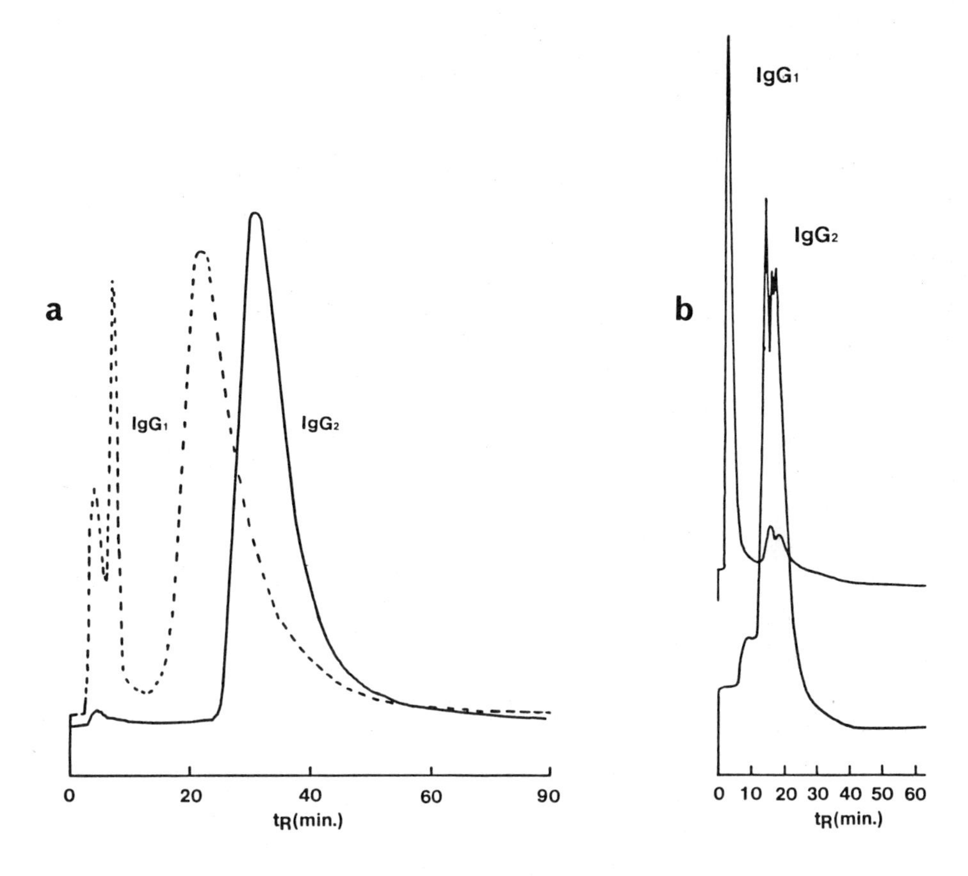

Fig. 31. Separation of bovine, goat and sheep polyclonal IgG_1 and IgG_2 on 5 micron BAKERBOND ABx. Chromatography was conducted on a column (7.75 × 100 mm) containing 5 micron ABx. In chromatogram A (bovine polyclonal IgG_1 and IgG_2), the mobile phase consisted of an initial (A) buffer of 25 m*M* MES, pH 6.3, and a final elution (B) buffer of 500 m*M* $(NH_4)_2SO_4$ plus 20 m*M* NaOAc, pH 6.7, with a linear gradient from 100% A to 25% B buffer over 75 min, followed by a step gradient to 100% B at 75 min, at a linear flow rate of 1.0 mL/min, with a back pressure of 200 psi. In chromatogram B (goat IgG_1 and IgG_2), the mobile phase consisted of an initial (A) buffer of 10 m*M* MES plus 10 m*M* Tris-OAc, pH 7.2, and a final elution (B) buffer of 500 m*M* $(NH_4)_2SO_4$ plus 20 m*M* NaOAc, pH 6.7, at a linear flow rate of 1.0 mL/min and back pressure of 200 psi. In chromatograms C and D (sheep IgG_1 and IgG_2), the mobile phase consisted of an initial (A) buffer of 10 m*M* MES plus 10 m*M* Tris-OAc, at a pH of either 6.8 (chromatogram C) or 7.2 (chromatogram D), and a final elution (B) buffer of 500 m*M* $(NH_4)_2SO_4$ plus 20 m*M* NaOAc, pH 6.7, with a linear gradient from 100% A to 100% B buffer over 1 hr, at a flow rate of 1.0 mL/min and a back pressure of 200 psi. Proteins were detected by UV absorbance at 280 nm at an attenuation of 1.0 AUFS. The sample contained at least 1.0 mg of affinity-purified serum polyclonal IgG_1 or IgG_2 diluted in 2.0 mL of the appropriate buffer A.

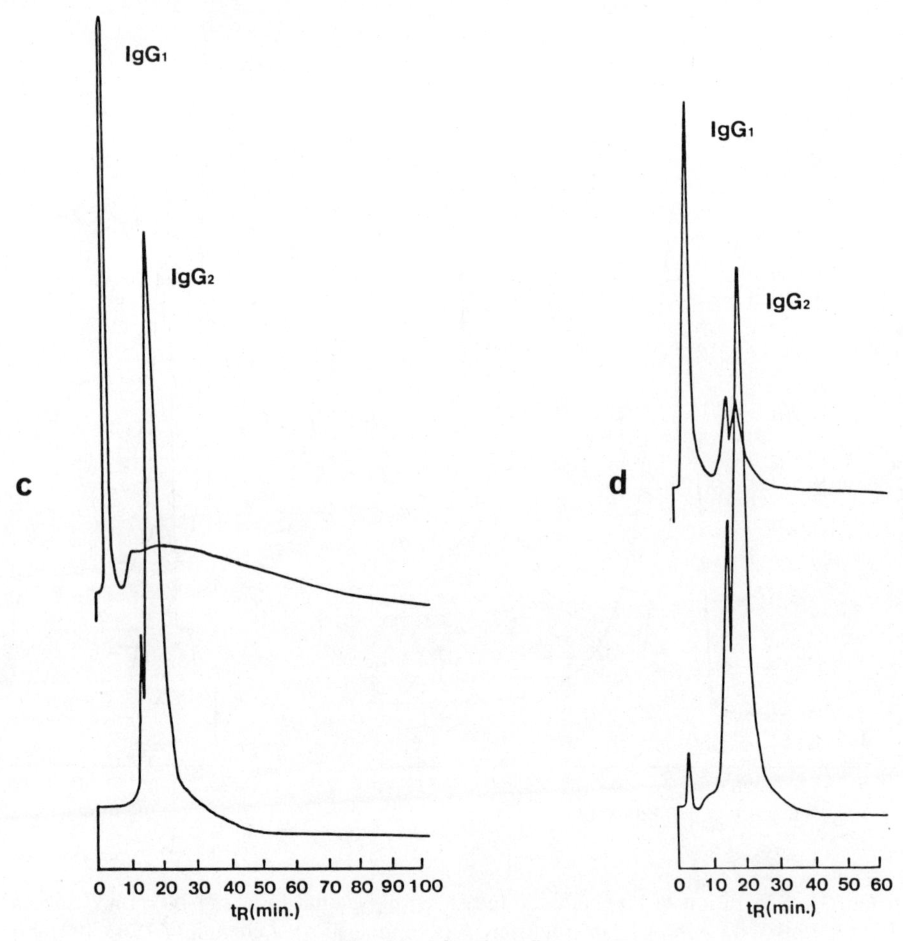

Fig. 31. (*Continued*)

to separate and identify three distinct monoclonal antibodies in a purified hybridoma cell culture which also exhibited no noticeable difference by SDS PAGE analysis (Figs. 33, 34). Although no differences in the electrophoretic patterns have been observed among any of the multiple monoclonals from these particular types of hybridomas, there is often a difference in the relative amounts, as well as the subclass or immunoreactivities or both of these antibodies. Thus, ABx appears to be capable of resolving monoclonal (and polyclonal) immunoglobulins not only on the basis of differences in their class and subclass, but also on the basis of differences in their light chain compositions, as well as other, more subtle differences in immunoglobulin molecular diversity.

Finally, ABx has been used to resolve a bispecific hybrid monoclonal antibody (28) from two monospecific monoclonal antibodies which were secreted by the same hybrid hybridoma or quadroma (Fig. 27). Despite the low

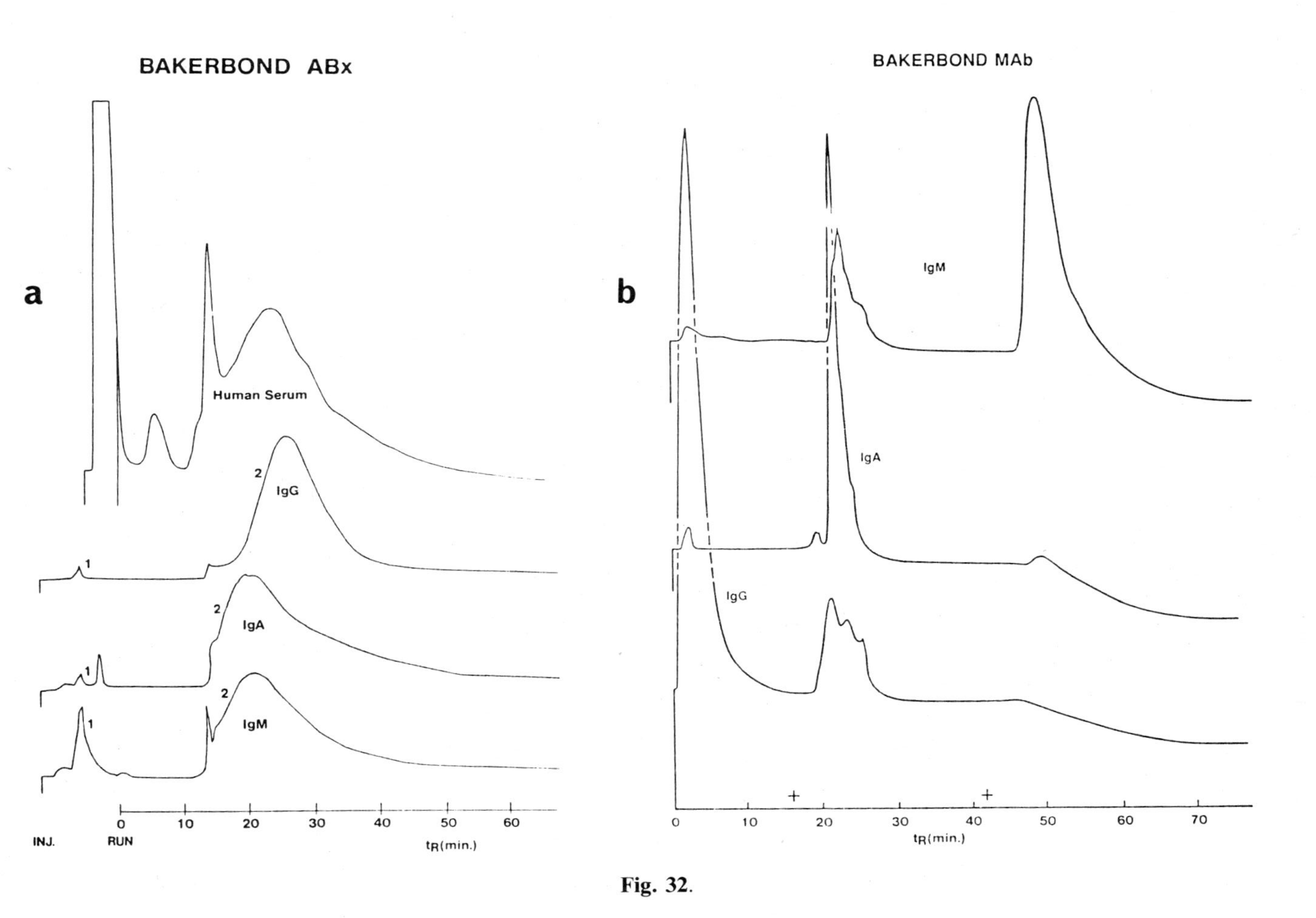

BAKERBOND ABx
a
Human Serum
IgG
IgA
IgM
1
2
INJ.
RUN
0
10
20
30
40
50
60
t_R(min.)
BAKERBOND MAb
b
IgM
IgA
IgG
0
10
20
30
40
50
60
70
t_R(min.)

Fig. 32.

levels of each antibody and the presence of large amounts of fetal bovine serum, the bispecific hybrid IgG_1/IgG_{2a} (anti-somatostatin/anti-horseradish peroxidase) was resolved not only from the monospecific IgG_1 (anti-horseradish peroxidase) and the monospecific IgG_{2b} (anti-somatostatin), but also from the nonimmunoglobulin proteins as well; all the immunologically active fractions contained highly purified antibody (greater than 90% by SDS PAGE analysis).

Although it is obvious that detailed investigations are required in this area, it is anticipated that the ability to resolve multiple forms of immunoglobulins will facilitate basic research and development in the field of immunology, as well as the purification of truly homogeneous antibody preparations.

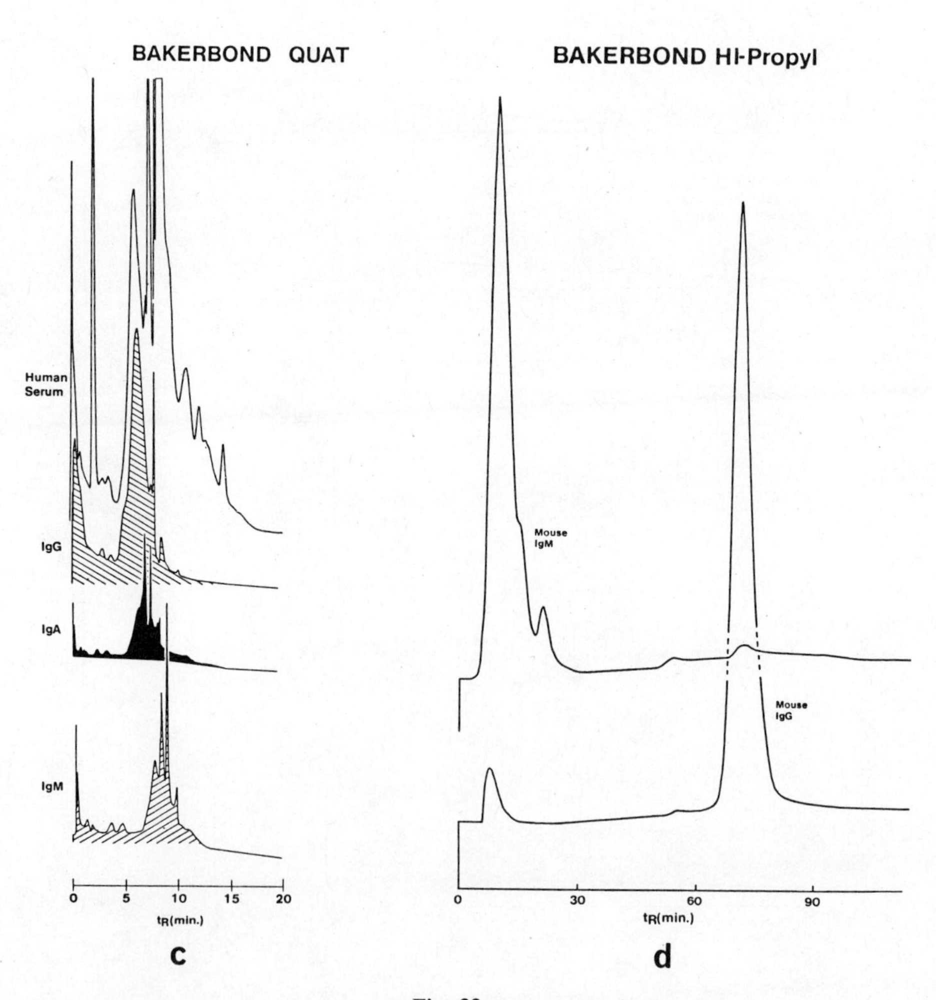

Fig. 32.

9. SUMMARY

Several factors are important in the preparative-scale purification of antibodies. These include capacity, recovery, resolution, specificity, speed, durability, hygiene and economy. ABx has been designed with these factors in mind.

As new types of monoclonal antibodies are purified and as purity requirements increase, it becomes increasingly apparent that the molecular diversity inherent within the immunoglobulin family makes a universal single step purification method less likely. However, the development of dedicated chromatographic matrices such as ABx, which utilize composite surface interactions to maximize selective antibody binding and resolution, represent a major step in the right direction.

The selectivity and resolving power of the ABx surface chemistry, plus the rigid polymer-coated silica support with its high ligand density and high

---

Fig. 32. Separation of immunoglobulin classes with BAKERBOND ABx (mixed-mode), BAKERBOND MAb (weak anion exchange), BAKERBOND QUAT (strong anion exchange), or BAKERBOND HI-Propyl (hydrophobic interaction) chromatography. In chromatogram A, the partial separation of the immunoglobulin classes (IgG, IgA, and IgM) from human plasma was conducted on a column (7.75 × 100 mm) containing 5 micron ABx. The mobile phase consisted of an initial (A) buffer of 25 m*M* MES, pH 5.4, and elution (B) buffer of 500 m*M* $(NH_4)_2SO_4$ plus 20 m*M* NaOAc, pH 6.3, with a linear gradient (at t_0) from 100% A to 100% B buffer over 1 hr with a flow rate of 1.0 mL/min, and a back pressure of 200 psi. The sample injection volume was 2.0 mL, consisting of 0.5 mL of neat human plasma, or purified immunoglobulin class standards, diluted in 25 m*M* MES, pH 5.4. Peak 1 contained albumins, transferrins, and proteases, peak 2 contained human immunoglobulin classes (IgA, or IgM). In chromatogram B, the separation of human immunoglobulin classes was conducted on a SCOUT column (4.6 × 50 mm) containing 15 micron MAb. The mobile phase consisted of initial (A) buffer of 50 m*M* Tris-OAc, pH 6.4, and a final (B) buffer of 2 *M* NaOAc, pH 5.8, with an initial isocratic gradient of 100% A buffer for 16 min, followed by a step gradient to 15% B buffer (held for 26 min), and a second step to 50% B buffer at 42 min, with an initial flow rate of 0.7 mL/min for 16 min, changed to 1.0 mL/min, at a back pressure of less than 50 psi. The sample consisted of purified human immunoglobulin classes diluted in 2.0 mL of buffer A. In chromatogram C, the separation of the immonoglobulin classes from human serum was conducted on a SCOUT column (4.6 × 50 mm) containing 5 micron QUAT. The mobile phase consisted of an initial (A) buffer of 25 m*M* Tris-OAc, pH 7.9, and a final elution (B) buffer of 2 *M* NaOAc, pH 5.8, with a linear gradient from 100% A to 100% B buffer over 1 hr, with a flow rate of 1.0 mL/min, and a back pressure of 200 psi. The sample consisted of 0.1 mL of neat human plasma, or purified immunoglobulin standards diluted in 2.0 mL of 25 m*M* Tris-OAc, pH 7.9. In chromatogram D, mouse immunoglobulin classes were separated by hydrophobic interaction chromatography on a column (7.75 × 100 mm) packed with 15 micron HI-Propyl. The mobile phase consisted of an initial (A) buffer of 700 m*M* $(NH_4)_2SO_4$ plus 35 m*M* NaOAc plus 25 m*M* KH_2PO_4, pH 7.2, with a linear gradient from 100% A to 100% B buffer over 1 hr, at a flow rate of 0.5 mL/min, with a back pressure of 25 psi. The sample consisted of purified mouse immunoglobulin class standards diluted in 2.0 mL of buffer A.

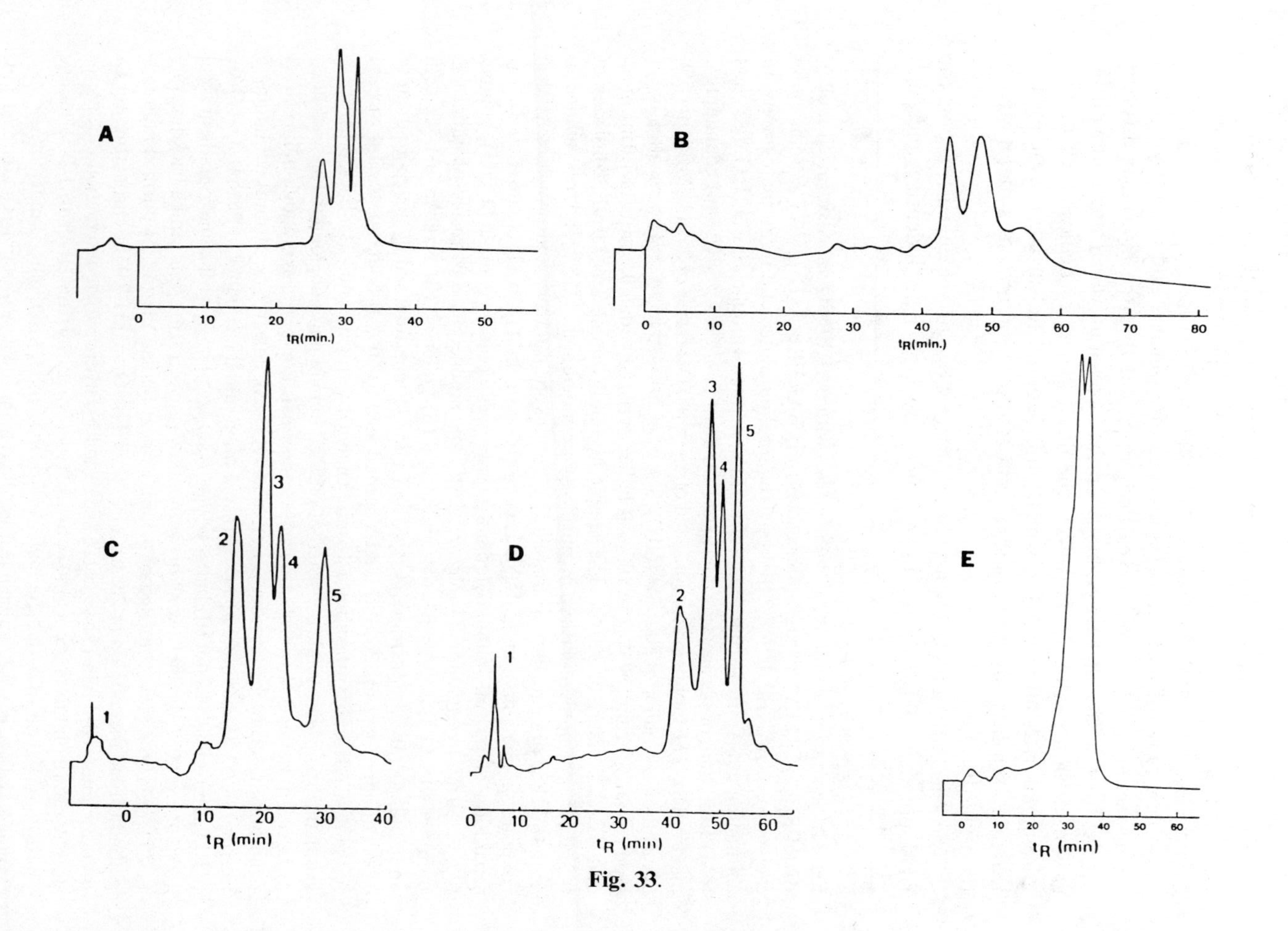

Fig. 33.

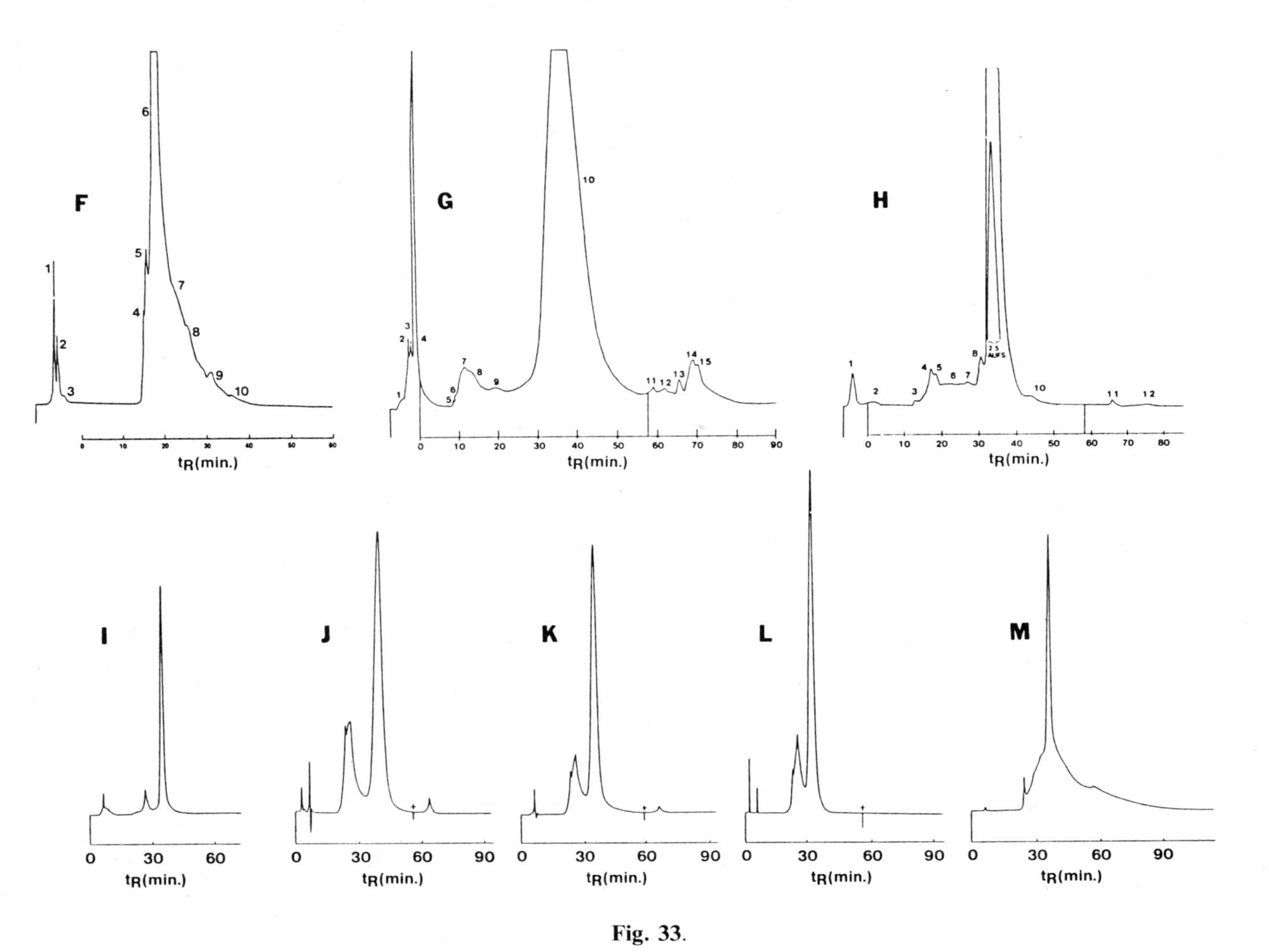

Fig. 33.

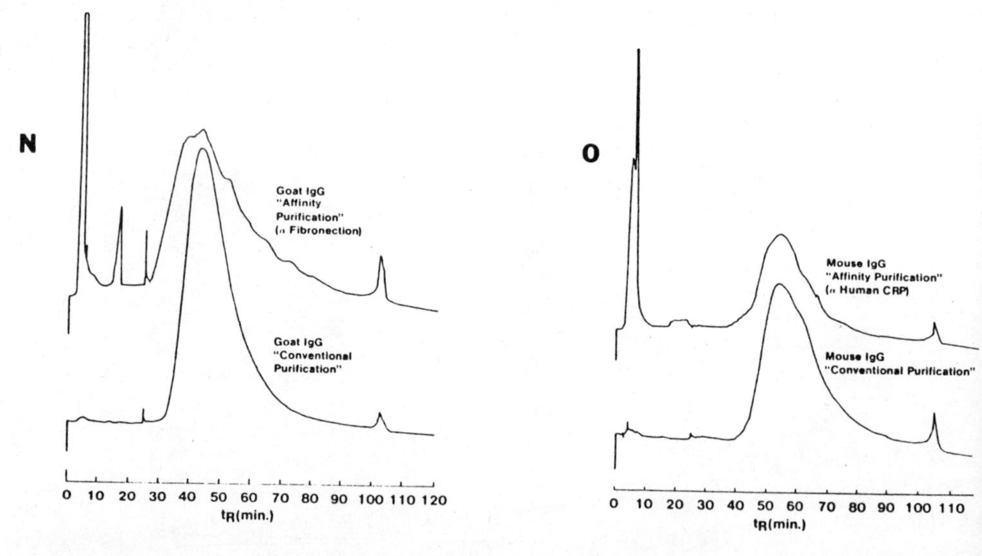

Fig. 33. Quality control analysis of purified monoclonal antibody preparation containing multiple immunoglobulins on 5 micron BAKERBOND ABx. Chromatography was conducted on columns (7.75 × 100 mm in chromatograms A–C and E–O, 4.6 × 250 mm in chromatogram D) containing 5 micron ABx. The mobile phase consisted of an initial (A) buffer of either 10 m*M* MES, pH 5.6 (chromatograms C, D, E, and F) or 25 m*M* MES, pH 5.6 (chromatograms A, B, and G–O), and a final elution (B) buffer of either 500 m*M* $(NH_4)_2SO_4$ plus 20 m*M* NaOAc, pH 6.7 (as in chromatograms B, G, H, N, and O), or, 1 *M* NaOAc, pH 7.0 (as in chromatograms A, C, D, F, and I–M), or, 250 m*M* KH_2PO_4, pH 5.5 (chromatogram E). Elution was conducted with a linear gradient from 100% A (held for several minutes) followed either by a linear gradient (at t_0) from 100% A to 50% B buffer over 1 hr (as in chromatograms A, B, E, F, and H), or, over 90 min (chromatogram C), or, 100% A to 35% B buffer over 90 min followed by a second gradient from 35% B to 100% B over 10 min (chromatograms N and O), or, with step gradients to 25% B buffer just after the injection (chromatogram D), or, to 18% B for 58 min, plus a linear gradient from 18% B to 100% B over 20 min (chromatogram G), or, to 20% B for 60 min, plus a second step to 100% B at 60 min (in chromatograms I, J, K, L, and M); in chromatogram H the gradient was interrupted at 58 min by a step to 100% B buffer. The flow rate was 1.0 mL/min or 0.5 mL/min (in chromatogram E), with a back pressure below 200 psi. Proteins were detected by UV absorbance at 280 nm at an attenuation of 0.5 AUFS (this was changed to 2.5 AUFS during the elution of the monoclonal antibody peak in chromatogram H) or 1 AUFS (in chromatograms N and O). The sample consisted of 0.2–2.0 mg of the purified antibody diluted in 2.0 ml of the appropriate A buffer. In chromatograms A and B, the three major peaks are three distinct immunoglobulins with similar light chain compositions, at purities of at least 95 and 98%, respectively (by SDS PAGE). In chromatograms C and particularly D, ABx was able to resolve small amounts of contamination (peak 1) and multiple forms of immunoglobulin (peaks 2, 3, 4, and 5) at a purity greater than 99% (by SDS PAGE). Immunological methods later revealed the presence of both myeloma IgG_1 (65% of the

capacity make ABx a versatile tool not only for purification, but also for analysis and quality control purposes. The following appendix focuses upon the use of ABx in the preparative mode, as well as in the analytical mode to monitor and quality control clinical samples and antibody preparations, and, in the purification of antibody fragments, conjugates, classes, and subclasses.

10. TRADEMARK AND PATENT INFORMATION

BAKERBOND, BAKERBOND ABx, BAKERBOND MAb, BAKERBOND HI-Propyl, PREPSCALE, SCOUT, and BAKERBOND spe are all trademarks of J. T. Baker Inc., Phillipsburg, NJ. BAKERBOND ABx is registered under U. S. Patent No. 4,606,825.

---

total IgG) and monoclonal IgG_{2a} (35% of the total IgG). In chromatogram E, the major peaks are IgG light chain variant monoclonal antibodies which were purified from a mouse ascites fluid by high-performance anion-exchange and size-exclusion chromatography; the early eluting peaks contained nonimmunoglobulin proteins, and the shoulder eluting between 25 and 30 min contained host, mouse polyclonal IgG. Chromatogram F is the ABx quality control of a monoclonal antibody from an encapsulation broth following an extensive purification procedure on conventional chromatographic matrices. Analysis of collected fractions by SDS PAGE indicated that peaks 1, 2, 3, 4, and 5 contained albumins, transferrins and several other proteins and peaks 7, 8, 9, and 10 contained mostly antibody. Rechromatography of peak 6 under identical conditions produced a single sharp peak, suggesting that the immunoglobulins in peaks 7, 8, 9, and 10 were derived from horse serum (which was used as a supplement). Chromatograms G–M are the ABx analysis of IgG monoclonal antibodies which were purified by multistep procedures on conventional chromatographic media. In G, peaks 1–3 and 11–15 consisted of nonimmunoglobulin proteins, while peaks 9–13 also contained immunoglobulins; peak 10 was the monoclonal antibody at a purity greater than 99% (by SDS PAGE). In H, peaks 1–8 and 10–12 contained mostly nonimmunoglobulin proteins, while peaks 6–10 also contained some immunoglobulin; peak 9 was the monoclonal antibody at a purity greater than 99% (by SDS PAGE analysis). In chromatograms I–L, the monoclonals were purified from mouse ascites fluids, and in M, the monoclonal was from a fetal bovine serum-supplemented cell culture fluid. In each case, the peaks eluting before 30 min contained protein contaminants, and the major peak eluting between 30 and 40 min was homogeneous monoclonal antibody (by SDS PAGE). Some host polyclonal IgG was also present, particularly in M, which contained a significant amount of bovine polyclonal IgG (eluting between 30 and 90 min). In chromatograms N and O, affinity-purified polyclonal antibodies were analyzed by ABx chromatography; the ABx void volume peaks contained some non-immunoglobulin proteins, as well as IgG which was devoid of immunological activity.

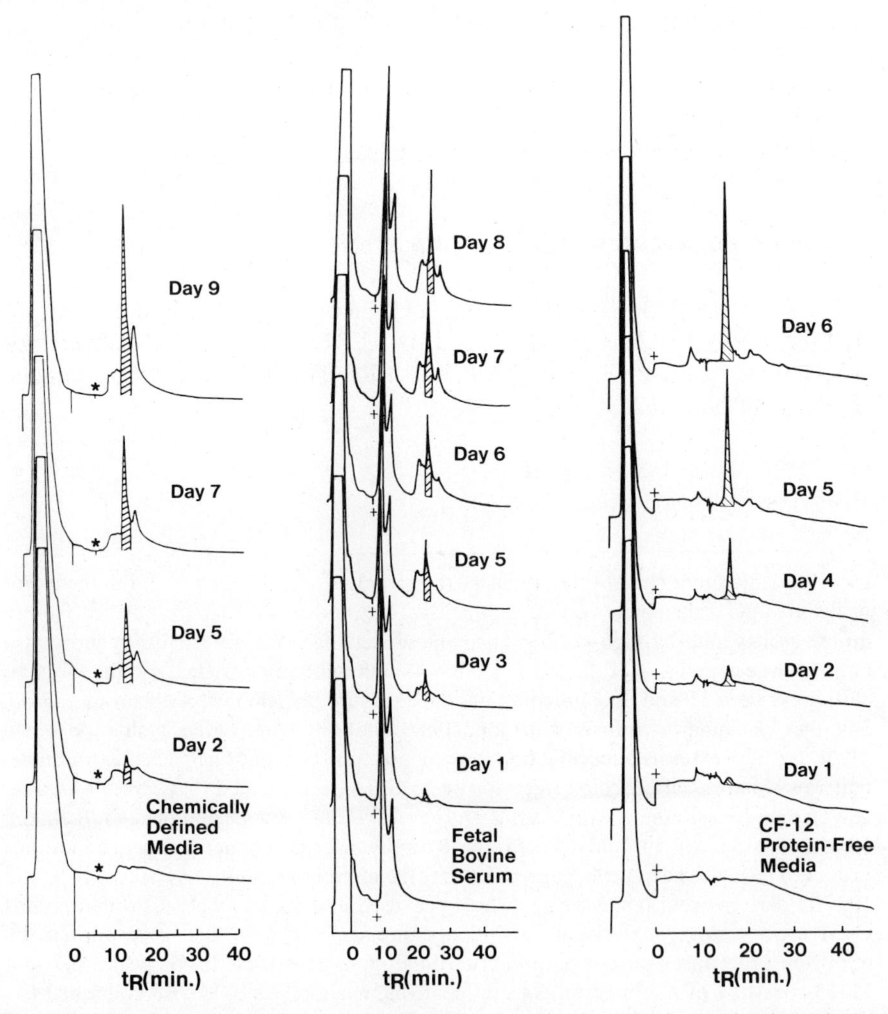

Fig. 34. ABx chromatographic monitoring of monoclonal antibody production (time-course) by three hybridoma cell lines grown in various culture media supplements. Chromatography was conducted on a column (7.75 × 100 mm) containing 5 micron ABx. The mobile phase consisted of an initial (A) buffer of 25 m*M* MES, pH 5.4, and elution (B) buffer of 500 m*M* $(NH_4)_2SO_4$ plus 20 m*M* KH_2PO_4, pH 6.7. In chromatogram A, elution was conducted with an initial isocratic gradient at 100% A for 12 min, followed by a linear gradient from 100% A to 100% B buffer over 30 min. In chromatogram B, elution was with an isocratic gradient at 100% A buffer for 8 min, followed by a step gradient to 8% B buffer and linear gradient from 8% B to 100% B over 30 min. In chromatogram C, elution was conducted with an initial isocratic gradient at 100% A buffer for 5 min, followed by a step gradient to 8% B (held for 10 min) and a linear gradient from 8% B to 100% B buffer over 30 min. In all cases, elution was conducted at a flow rate of 1.0 mL/min, with a back pressure of 200 psi. Proteins were detected by UV absorbance at 280 nm at an attenuation of 0.05 AUFS and was changed to 0.01 AUFS (see +). The sample consisted of 1.0 mL of neat hybridoma cell culture super-

ACKNOWLEDGMENTS

Special thanks to Dr. Ko for the sheep IgG_2, Dr. Fred Darfler for the CF-12 supplemented hybridoma cell time-course samples, Dr. Steven Rosenthal for the myeloma and diseased-state plasma samples, Dr. Claudio Cuello for the hybrid hybridoma (quadroma) cell culture supernatant, and those who kindly provided various crude and purified monoclonal antibody preparations. Special thanks as well to Dr. Laura J. Crane and Mrs. Joyce Guenther for helping proof the final manuscript. Dr. Mike Flashner for the ascites fluid loading study on the MAb matrix (chromatograms on the bottom of Table 3), and Dr. Steven Berkowitz for the HIC chromatograms of ascites fluid (Fig. 25, bottom row) and the publishers of *BioChromatography* and Marcel Dekker, Inc., for permission to print previously published material.

APPENDIX I. DESIGN OF CHROMATOGRAPHIC MATRICES FOR THE PURIFICATION AND ANALYSIS OF ANTIBODIES: REVIEW OF CHROMATOGRAPHIC METHODS AND DETAILS OF ABx USE

I.1. Chromatographic Matrices for Antibody Purification

I.1.1. Traditional Antibody Purification

Anion-exchange chromatography has traditionally been the most widely used method in antibody purification schemes. Since the conditions for the gradient elution of proteins from ion-exchange matrices are relatively gentle, antibodies may be recovered with high activity. Since the mechanism of separation on ion exchangers is based on reversible surface charge interactions between the chromatographic matrix and the protein itself, the technique may be highly resolving, owing to the array of surface charge distributions and isoelectric points of the various proteins present in antibody preparations.

---

natant from various days after subculture plus 2.0 mL of 25 m*M* MES, pH 5.0. In all cases, the cross-hatched peaks correspond to the fraction containing more than 90% of the immunological activity at a purity greater than 95% (by SDS PAGE) after day 5 subculture. In chromatogram A, the cells were grown for various times in PC-1 protein-defined, serum-free media. The peak eluting at 15 min also contained immunoglobulin, but was immunologically inactive. In chromatogram C, the cells were grown for various times in a fetal bovine serum-supplemented cell culture media. The peaks eluting at 20 and 25 min also contained immunoglobulin, but were immunologically inactive. In chromatogram B, the cells were grown in of CF-12 chemically defined, protein-free culture media supplement for various times. Protein concentrations were determined on each day to be 0, 2.0, 4.4, 11.0, 15.4, 27.2, and 54.0 mg/L on days 0–6, respectively. The monoclonal concentrations were determined by electrophoresis and densitometric tracing to be 0, 1, 2, 4, 7, 12 and 16 mg/L on days 0–6 respectively, while ABx analysis indicated that the monoclonal levels were 0, 0.7, 2, undetermined, 6, 13 and 17 mg/L on days 0–6, respectively. Cell viability on days 1–6 was 90%, 90%, 90%, 85%, 80% and 40%, respectively.

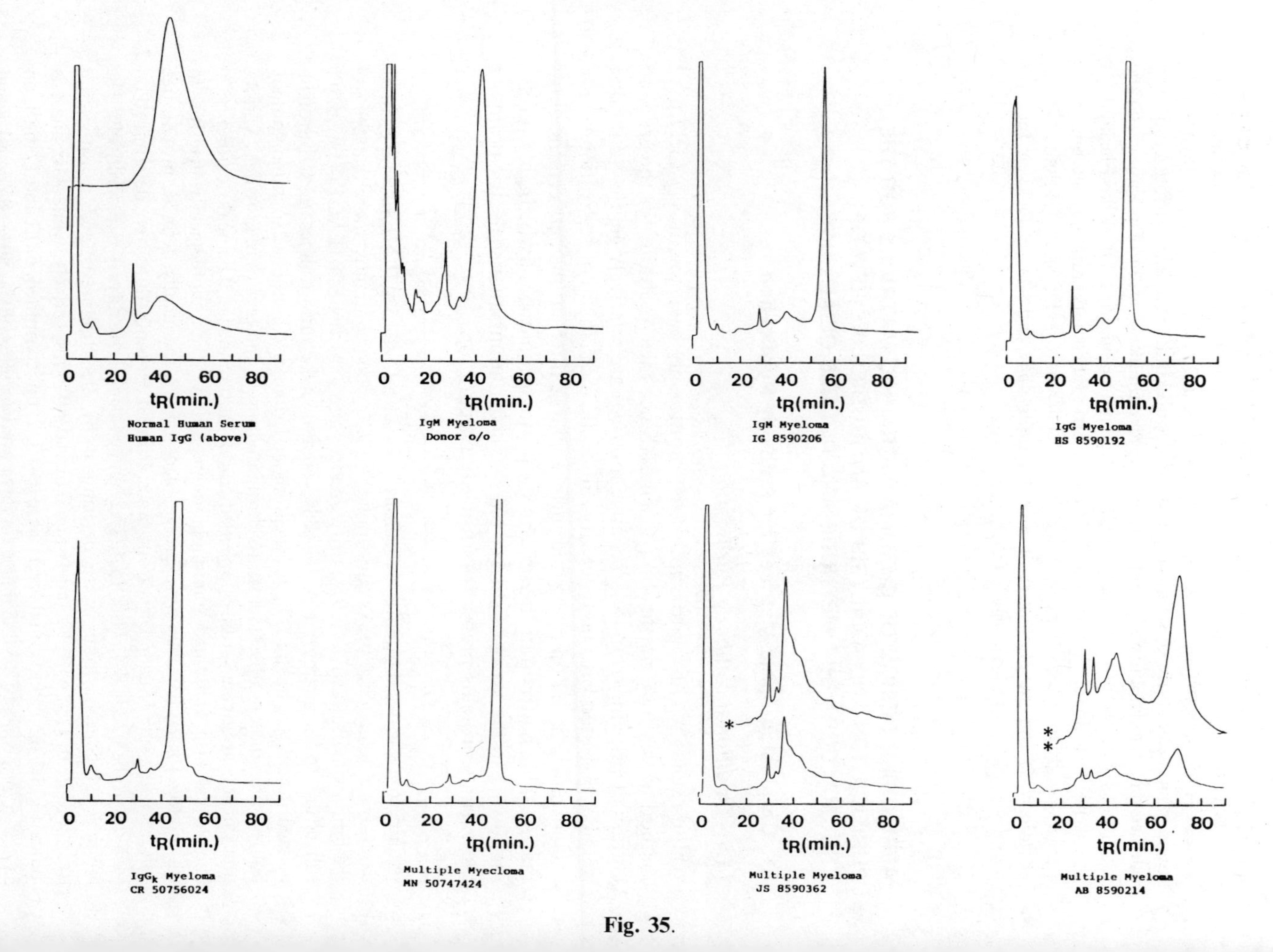
0 20 40 60 80
t_R(min.)
Normal Human Serum
Human IgG (above)
IgM Myeloma
Donor o/o
IgM Myeloma
IG 8590206
IgG Myeloma
HS 8590192
IgG_k Myeloma
CR 50756024
Multiple Myecloma
MN 50747424
Multiple Myeloma
JS 8590362
Multiple Myeloma
AB 8590214

Fig. 35.

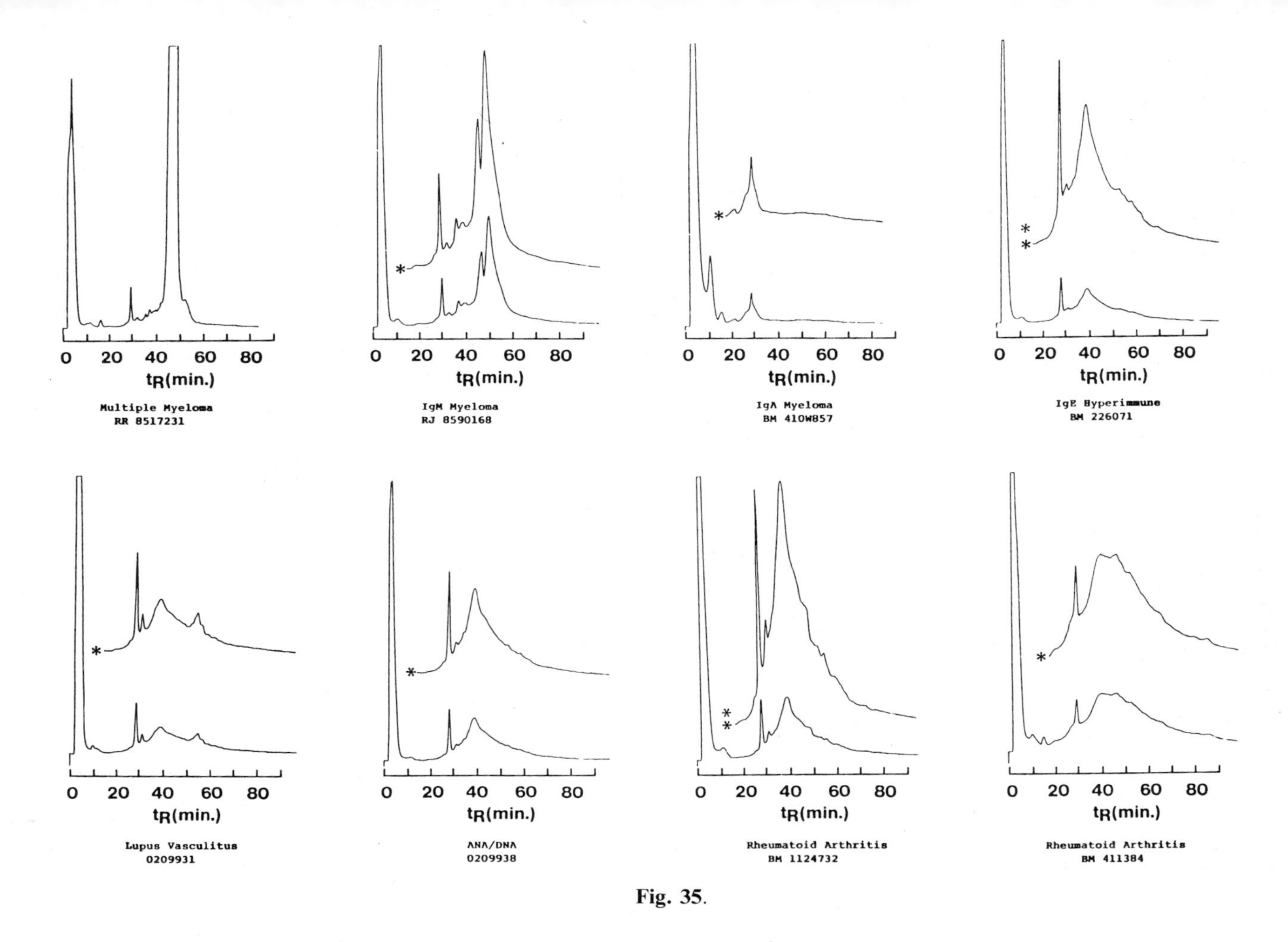

Multiple Myeloma
RR 8517231
IgM Myeloma
RJ 8590168
IgA Myeloma
BM 410W857
IgE Hyperimmune
BM 226071
Lupus Vasculitus
0209931
ANA/DNA
0209938
Rheumatoid Arthritis
BM 1124732
Rheumatoid Arthritis
BM 411384
tR(min.)
0
20
40
60
80

Fig. 35.

Preparative-scale antibody purification has traditionally been carried out using classical liquid chromatography with open columns packed with DEAE (diethylaminoethyl) soft gels which are based upon porous, insoluble polysaccharides (11, 31, 32, 36, 45). Although traditional anion exchange chromatography on soft gels is easy to use, low in cost, requires simple equipment, can be scaled-up to some extent, and produces antibodies at a purity which is adequate for a number of end uses, traditional soft gels suffer from a number of serious disadvantages for preparative-scale protein purification. Owing to the large particle sizes and broad particle size distribution (50 to 150 microns), soft gels provide extremely low resolution between antibodies and contaminating proteins, and chromatography must typically be combined with an initial ammonium sulfate, salt fractionation step (11, 31, 32, 36, 45). Therefore, extremely long chromatographic run times and large quantities of both gel and buffer are required to achieve adequate separations. The inability to conduct rapid and multiple chromatographic runs also makes method development a tedious process, or one which is simply omitted (51, 62, 63, 69–71). Because the optimization of a separation via the manipulations of the mobile phase conditions, gradient profile and other conditions which alter chromatographic selectivity, is rarely achieved or even attempted, the quality and effectiveness of the purification step is severely compromised (51, 62, 63, 69–71). The concomitant reduction in the degree of purification at any given step results in the need for further downstream purification steps, which increases costs and labor, and decreases product recovery. In addition to long run times, rapid flow rates which are desirable in preparative chromatography for high throughput and rapid re-equilibration and regeneration, are not possible on soft gels due to the high back pressures, poor mechanical strength and the possibility of column collapse (51, 62, 63, 69–71). Furthermore, owing to large changes in bed volume that occur with changes in ionic strength and pH, it is often neither practical, nor possible, to repeat the separation without repacking the column (51, 62, 63, 69–71). Since the life expectancy of soft gels is rather short, it is tempting to discard the matrix after each use, because of the cumbersome maintenance and

Fig. 35. Chromatographic analysis of plasma samples on 5 micron BAKERBOND ABx. Chromatography was conducted on a column (7.75 × 100 mm) containing 5 micron ABx. The mobile phase consisted of an initial (A) buffer of 25 m*M* MES, pH 5.4, and a final elution (B) buffer of 500 m*M* $(NH_4)_2SO_4$ plus 20 m*M* NaOAc, pH 6.7, with a linear gradient from 100% A to 50% B buffer over 1 hr and a flow rate of 1.0 mL/min, at a back pressure of 200 psi. Proteins were detected by UV absorbance at 280 nm at an attenuation of 1.0 AUFS, with inserts above several of the individual chromatograms at higher sensitivities (*, 0.5 AUFS; or **, 0.25 AUFS) in order to better visualize the changes in antibody profile. The sample consisted of 0.1 mL of neat human plasma, diluted to 2.0 mL with 25 m*M* MES, pH 5.4. The first chromatogram, on the top left, depicts normal human plasma and purified, normal human polyclonal IgG that are shown to demonstrate the relative retention times and the curvilinear nature of a normal serum immunoglobulin peak.

regeneration procedures and the fact that the polysaccharide-based gels themselves support microbial growth (62, 66). This biodegradability, as well as chemical instability and the need for rigorous sterilization lead to the possibility of endotoxin and degradation product contamination of the final antibody preparation (62, 66). Finally, soft gels often bind immunoglobulin irreversibly, further decreasing yields (11, 31, 32, 36, 45, 51, 61, 62). These factors reduce the overall economy of soft gel use and make them less attractive than the state-of-the-art chromatographic supports which are now available for preparative-scale antibody purification.

I.1.2. Modern Antibody Purification

I.1.2.1. Anion-Exchange Chromatography. Recent work in this and other laboratories indicates that chromatography on high performance anion exchange matrices may provide a useful approach to monoclonal antibody purification, since these materials overcome many of the problems inherent to soft gels (1, 8, 9, 12, 13, 15–17, 23, 25–27, 29, 32, 34, 37, 40, 43, 51, 52, 54, 55, 67–72, 75, 78, 85, 92, 93). These HPLC supports, as well as many of the new medium performance supports (30 to 100 micron silica-based, cross-linked soft gel or polymeric supports) produce much higher resolution than traditional matrices (62, 63, 68–71). The capacity of many of these matrices is often higher than those of soft gels, binding as much as 150 mg of purified IgG per gram of packing (62, 63, 68–71). These factors, combined with the ability to use higher flow rates, enable substantially higher purities and throughputs on much smaller columns. Finally, modern chromatographic supports are more rugged than soft gels, allowing longer column lifetimes and increased chemical stability (62, 63, 66–72). The combination of these factors increases the cost-effectiveness of antibody purification, particularly in the case of the medium performance matrices (62, 63, 67–72).

However, despite this recent progress, it has become increasingly apparent that the purification of monoclonal antibodies by high-performance anion-exchange chromatography also has a number of serious limitations. Not only do the background profiles of contaminating nonimmunoglobulin proteins tend to vary considerably, depending upon species, strain, individual hybridoma, or culture media supplement, but more important, individual monoclonals may also elute at a wide range of retention times within the gradient, from the extremes of not even binding, to eluting after the albumin peak (13, 23, 27, 29, 68–72). These factors not only reduce the possibility of achieving a homogenous antibody preparation, but also hinder method development and identification of the antibody peak. Although some investigators have suggested that this variability in retention is related to the subclass of the monoclonal (27), others have clearly demonstrated that this view is a gross oversimplification (25, 26, 69–72).

Anion exchangers are particularly unsuitable for use with serum-based tissue culture media (Fig. 28), since the monoclonal antibody inevitably coelutes with a complex mixture of other proteins and dyes which are present in substantial amounts. The binding of dyes (i.e., phenol red, phenolphthalein, and so on), decreases resolution by broadening the antibody and albumin peaks, changes the elution characteriestics of the immunoglobulin peaks, and actually changes the mechanism of the separation (37, 62, 69–72). These dyes also reduce capacity and the amount of sample that can be purified (29, 37, 62, 69–72), and in some cases, the binding of these dyes to polymer-based supports may be irreversible (9, 29, 62, 69–72). As a solution to this problem some investigators have suggested the use of salt fractionation (4, 11, 13, 19, 29, 32) or even reversed phase chromatography (9) for sample preparation; steps which can reduce recovery and process economy.

In fact, anion exchange chromatography is, *a priori*, undesirable as an initial step for large-scale antibody purification; since virtually all of the proteins and dyes are bound, the capacity of the matrix that is available to bind immunoglobulin is drastically reduced (37, 62, 69–72). This decrease in capacity is a particular disadvantage when attempting method development, since analytical chromatography on these matrices often does not correctly predict the chromatographic profile in the presence of the overloaded conditions (Table 3) which are typically used for preparative chromatography (23, 27, 29, 37, 67–72). This is particularly true if one considers that, with very few exceptions, totally different surface chemistries must be used if preparative chromatography on traditional low-pressure, open columns is to be scaled-up from method development which was conducted on high-performance matrices (62, 67–71).

High-performance anion-exchange supports also typically produce low product purities with immunoglobulin classes other than IgG, as IgM and IgA monoclonals often coelute with albumin, while IgD and IgE monoclonal antibodies often coelute with transferrin and α_2-macroglobulin (1, 8, 15, 26, 29, 62, 69, 70). One major problem is that the level of these monoclonals is often rather low, even in ascites fluids. In contrast, monoclonals of the IgG subclasses may often comprise more than half of the total protein content of ascites fluids (26, 29, 67–72, 92) and as such, the partial coelution of minor contaminating proteins does not drastically reduce the purity of the monoclonal antibody peak.

Despite inherent disadvantages, however, anion exchange chromatography is an extremely valuable tool, particularly as a second chromatographic step following ABx batch extractions, and in the removal of endotoxins and nucleic acids which are bound to immunoglobulins (35, 62, 63, 69, 70). On anion exchangers with a high ligand density (e.g., BAKERBOND MAb), nucleic acids and lipopolysaccharides can be bound strongly and stripped from the antibody, which may be eluted in the void volume, free from adsorbed DNA and pyrogens (Fig. 23). This technique can reduce the levels of pyrogen and nucleic acid contamination in monoclonal antibody preparations below those required for in vivo use (i.e., 10 pg DNA per dose and 0.05 ng lipopolysaccharide per mL).

I.1.2.2. Affinity Chromatography. Recently, the purification of monoclonal and polyclonal antibodies with protein A, protein G (for IgG), Jacalin (for IgA), concanavalin A (for IgM), and anti-immunoglobulin, specific antigen and other immuno-affinity or affinity-like matrices has gained wide acceptance (4, 5, 10, 18, 20, 22, 30, 31, 38, 45, 49, 50, 59–63, 77, 89, 91, 97, 99). The development of these materials represents a major breakthrough in the field of antibody purification, since in many cases, monoclonals may be separated in a substantially purified form, with minimal protease contamination (5, 10, 18, 30, 31, 38, 45, 59, 63, 89, 99). However, a major misconception is that affinity supports bind only the protein of interest. Typically, a series of chromatographic steps on nonaffinity matrices may be required following the affinity step in order to remove other bound proteins and leached affinity ligands. Furthermore, most affinity supports, including protein A, require a unique method development in order to maximize each affinity purification (62, 63, 69, 70). One of the major drawbacks of protein A is its inability to bind all antibody classes, some IgG subclasses, numerous mouse and rat antibodies, F_{ab} and $F_{(ab')2}$ fragments and many antibody conjugates, and, to separate host immunoglobulins from the monoclonal or other multiple antibody species (18, 30, 31, 38, 45, 59–63, 88, 89). Furthermore, protein A and immuno-affinity matrices exhibit high binding constants and tend to bind immunoglobulins tightly or even irreversibly. As such, they require harsh elution conditions including detergents, organic solvents, chaotropic agents, or extremes of pH and ionic strength; these drastic, non-physiological conditions may denature, alter tertiary or quaternary structure or even chemically-modify the monoclonal and reduce immunological activity (5, 10, 11, 18, 30, 31, 38, 45, 59–63, 77, 88, 89, 91, 97, 99). Protein-based affinity matrices are also susceptible to biodegradation, proteolytic digestion, bacterial ligand and endotoxin leakage, are difficult to sterilize and require refrigeration; this is particularly true for affinity supports which are based on polysaccharide soft gels (62, 66–70). These factors make the use of affinity chromatography less desirable in cases where the product is destined for in vivo use, as an injectable therapeutic or diagnostic.

Affinity matrices also require large quantities of either highly purified antigen, affinity ligand, or antibodies raised against the monoclonal of interest, and the production, purification or cost of these proteins alone or both may be prohibitive. In addition, protein A and affinity matrices suffer from low capacity, short column lifetime, and high cost, and as such, are typically uneconomical for most preparative or process-scale purifications (62, 63, 69–70).

Recently, a number of silica- and polymer-based affinity matrices have been developed with increased mechanical and biochemical stability, which enable higher flow rates and throughputs, longer column lifetimes, higher product recoveries, and minimal leaching of the ligand from the support (4, 10, 18, 22, 45, 49, 61–63, 77, 89, 97). These factors are of paramount importance for supports which contain antigens or other ligands which are expensive or difficult to purify, or both, and, for products destined for therapeutic use. Although many of these newer supports are based on smaller high performance particles, resolution is typically not a critical issue on immuno-affinity matrices, since in theory, only the protein of interest is bound. Affinity supports based on larger silica particles have been shown to give excellent resolution, retaining the advantage of low elution volumes and durability, but enabling more economical separations which can be conducted on conventional chromatographic equipment, or, via solid phase, batch extraction (62, 63, 99–70).

I.1.2.3. Hydroxyapatite Chromatography. Recent evidence suggests that hydroxyapatite may be a useful matrix for the chromatographic purification of monoclonal antibodies (6, 7, 10, 21, 42, 43, 47–49, 53, 75, 88). The major advantage of hydroxyapatite over conventional chromatographic media is the fact that antibodies are often bound more strongly than other common protein contaminants (10, 21, 43, 47–49, 53, 62, 75). However, recent studies suggest that even anion exchange or hydrophobic interaction chromatography provide higher resolution of monoclonal antibodies than hydroxyapatite chromatography (75). As such, rather long chromatographic run times are required in order to resolve the monoclonal antibody from the numerous bound proteins (10, 21, 42, 47, 49, 53, 75). Furthermore, the capacity of hydroxyapatite is from 10 to 20 times lower than that of chromatographic materials which are based on silica or soft gels (67–71). These problems, combined with broad peaks, produce a purified antibody preparation which is extremely dilute. Hydroxyapatite also has intrinsic limitations for routine use, and particularly for scale-up; it produces high back pressures and low flow rates, tends to bind immunoglobulin non-specifically and irreversibly, and because of its fragile nature and its solubility in inappropriate buffers, it exhibits an extremely short column lifetime (6, 7, 40, 43, 53, 67–71, 75, 90).

The fact that numerous nonimmunoglobulin proteins and pH indicator dyes also bind to hydroxyapatite (21, 43, 47–49), along with its inherently low capacity and ligand density, place serious limitations on the relative volume of ascites fluid, and particularly, serum-based cell culture media which can be purified on a given hydroxyapatite column. Therefore, preparative hydroxyapatite chromatography is typically conducted only after an initial salt fraction or DEAE or affinity chromatography, or both (10, 43, 47, 49, 75). This is apparently because, at moderate sample loads most of the proteins present in antibody preparations do bind to the support (21, 22, 75), and these compete with the antibody molecules for the limited number of available ligands, thereby inhibiting the binding of the antibody (Appendix III).

I.1.2.4. Size-Exclusion Chromatography. Size-exclusion chromatography has traditionally been used as an integral part of multi-step antibody purification schemes (4, 5, 10, 19, 40, 45, 59, 86). Since the relative molecular weights of the common contaminating proteins (50–70 kD) are substantially lower than those of the immunoglobulins ($\geqslant$150 kD), appropriately designed size exclusion columns may yield rather pure antibody preparations.

However, size-exclusion chromatography is inherently a low resolution technique; this makes it difficult or impossible to separate antibodies from contaminating proteins with similar molecular weights or from other immunoglobulin species of the same class, and necessitates the use of extremely long chromatographic run times in order to enhance the separation. As such, an initial purification step is typically required prior to a sizing step (4, 5, 10, 19, 40, 45, 49, 86), particularly in the case of cell culture media, since the antibody is present at such low levels that the large amounts of contaminating proteins often obscure the antibody peak (Fig. 14). High protein concentrations can also hinder the purification of monoclonals on size exclusion supports, since contaminants can often clog the pores and destroy the separation, making column hygiene and regeneration imperative (13).

Perhaps the major disadvantage of size exclusion chromatography is its extremely low capacity (more than 100 times less than ABx). This necessitates the use of large columns or multiple chromatographic runs or both (although the need for long chromatographic run times further limits throughput). Since elution is isocratic and there are no binding ligands, proteins are not concentrated on the support; therefore, as the injection volume increases, the peaks become proportionally broader, and this reduces resolution and limits sample load. Therefore, a chromatographic concentration step may be desirable prior to the size exclusion step. In fact, size exclusion chromatography can be an extremely useful technique provided that initial purification steps are used to concentrate the antibody (Figs. 13–17). Furthermore, as the purity of the sample being loaded onto a size exclusion column increases, so too does the useful capacity of that column. Therefore, despite inherent drawbacks, the increasing use of other, more highly resolving and discriminating, high capacity chromatographic matrices make the use of size exclusion chromatography more attractive, particularly as a downstream step in multistep procedures in which extremely high final product purity is required.

I.1.2.5. Hydrophobic Interaction Chromatography. Traditional reversed-phase chromatographic matrices tend to irreversibly bind and/or denature immunoglobulins, and as such, are unsuitable for either analytical or preparative antibody separations (62–65, 69, 70). However, recent work in this and other laboratories indicates that hydrophobic interaction chromatography (HIC) may be a useful alternative to traditional reversed phase materials, because it allows for quantitative recovery of immunoglobulin mass and activity (2, 44, 54, 62–65, 68–71, 75, 76, 94).

Because hydrophobic interaction supports separate proteins via mechanisms which are totally different from those of ion exchange, HIC not only complements more traditional separation modes, but also adds a new dimension for the biochemist to explore while developing immunoglobulin purification schemes (Figs. 23, 25, 26). Furthermore, since proteins are bound to HIC supports in the presence of high salt concentrations, HIC is a logical second step for fractions eluted from ion exchangers (Fig. 24). This is a particularly convenient approach, since the direct application of samples with high ionic strength is possible, eliminating the need for dialysis, dilution, ultrafiltration, or buffer exchange.

Previously, we had reported that hydrophobic interaction chromatography may be useful in the purification of some monoclonal antibodies from mouse ascites fluids (64, 67–71). Although subsequent research in a number of laboratories has supported this contention, these initial investigations suggested that, in general, HIC is not a sufficiently specific method for the one-step purification of monoclonal antibodies (2, 44, 54, 62–64, 75, 76, 94). These studies suggested that, like anion exchange chromatography, traditional hydrophobic interaction chromatography is, *a priori*, not particularly desirable as an initial purification step; virtually all the proteins present in crude antibody preparations are bound, and as such, the number of ligands which are available to bind antibodies are dramatically reduced (2, 44, 54, 62–65, 67 71, 75, 76, 94). Furthermore, individual antibodies were shown to elute as broad peaks and at a wide range of retention times within the gradient, hindering method development and identification of the antibody peak, and decreasing the possibility of achieving a homogeneous antibody preparation (2, 44, 54, 62–65, 67, 71, 75, 76, 94). These considerations, as well as the apparent need to use extremely large amounts of ammonium sulfate in order to bind immunoglobulins, were believed to limit the effectiveness of traditional HIC as an economical initial step in preparative antibody purification schemes.

However, recent work in this laboratory (64, 65) suggests that the lack of specificity for antibody binding on HIC supports which was observed in these preliminary investigations may have simply resulted from the use of sub-optimal binding buffer conditions (Figs. 24–26). With BAKERBOND HI-Propyl, it is possible to use mild binding conditions [700 mM $(NH_4)_2SO_4$)] to purify human or mouse IgG monoclonal antibodies from fetal bovine serum-supplemented cell culture fluids to near homogeneity (and 95% free of bovine IgG) with near quantitative recovery of immunoglobulin mass (Figs. 24–26). BAKERBOND HI-Propyl is also a simple, efficient, economical, and direct, second purification step for IgG following an initial purification step with BAKERBOND ABx (Fig. 24). Antibodies eluted from ABx may be loaded directly onto HIC matrices (Fig. 24). Furthermore, with these optimized buffer conditions, virtually all of the contaminating fetal bovine serum proteins which bind to ABx do not bind to HI-Propyl, and flow through in the void volume, while all mouse IgG monoclonal antibodies bind rather strongly (Figs. 24–26). Therefore, antibodies eluted from ABx, either in the column mode or even in the batch mode, may be bound rather selectively by HIC supports, and eluted in a substantially purified form, free of bovine immunoglobulins and other fetal bovine serum proteins (Figs. 24, 26). Another advantage of downstream purification steps with BAKERBOND HI-Propyl is that a significant amount of nucleic acids and pyrogens can also be stripped from monoclonals, particularly since mouse IgG can be bound rather selectively at an ionic strength at which nucleic acids and lipopolysaccharides exhibit little or no binding (Fig. 24).

Recently, Porath and co-workers (76) discovered a new kind of chromatographic interaction which exists between immobilized thioether-sulfone ligands and some specific types of proteins, including immunoglobulins. These thiophilic adsorbants or T-gels bind almost exclusively immunoglobulins and α_2-macroglobulin in the presence of both low and high salt concentrations (76).

Although thiophililc interaction chromatography and hydrophobic interaction chromatography are both promoted by high salt concentrations, the separation mechanism are apparently not the same, because the two matrices exhibit different affinities and absorption capacities for several serum proteins in the presence of various salts of the Hofmeister series (76). Porath et al. (76) has suggested that T-gels may represent an almost universal adsorbent for immunoglobulins, because in many cases, these matrices are capable of producing near homogeneous immunoglobulin preparations with good recoveries (75–92%). In many respects, the results obtained on T-gels are reminiscent of those obtained in this laboratory with BAKERBOND HI-Propyl in the presence of low ionic strength binding buffer condtions; these striking similarities are currently being investigated (62, 63).

I.1.2.6. Cation Exchange Chromatography. Several investigators have recently demonstrated the utility of strong cation exchange chromatography in multistep antibody purification schemes (12, 13, 19, 26, 32, 33, 35, 55, 62, 63, 65, 82, 85). In many cases, conventional cation exchange matrices can be used to elute relatively pure monoclonal antibodies from crude ascites fluid or cell culture media (12, 13, 19, 26, 32, 33, 35, 55, 62, 63, 65, 82, 85). However, the inability of conventional cation exchangers to bind most immunoglobulins strongly, without creating a split peak (see Appendix III), appears to represent a major drawback for their general use in antibody purification schemes. Recent studies in our laboratories suggest that even those antibodies that should bind strongly to cation exchangers may be inhibited from binding by other components in antibody preparations (Appendix III). While this breakthrough behavior and the concurrent decrease in recovery is a problem with serum and ascites fluid, it is a major dilemma with cell culture media and often places strict limitations on the amount of sample which can be loaded onto a column (13, 55, 65, 82).

Another disadvantage of cation exchangers is that these supports tend to bind numerous contaminating ascites proteins more strongly than the antibody itself, presumably due to the low pH values (5.0 or less) which are often required to bind the antibody (13, 19, 32, 35, 45, 55, 62, 82, 85). Even following ammonium sulfate, salt fractionation a significant proportion of these bound proteins often coelute with the monoclonal antibody (13, 19, 32, 35, 45, 55, 62, 82, 85). These factors necessitate not only the optimization of buffer conditions for the purification of each monoclonal, but also the use of ammonium sulfate, salt fractionation or other chromatographic steps prior to the cation exchange step (13, 19, 32, 35, 45, 55, 62, 82, 85). Indeed, high performance strong cation exchange chromatography has been used as a final step for the purification of a monoclonal antibody from a serum-free cell culture supernatant, but only after anion exchange and size exclusion chromatography; the final recovery of antibody was 50% (32). Strong cation exchange chromatography has also been used as an initial purification step prior to size exclusion chromatography for a series of hybridoma cells grown in serum-free culture media (19); the sizing step was required because in each case a number of contaminating proteins coeluted with the monoclonal antibody, despite the fact that the levels of the exogeneous non-immunoglobulin proteins were actually lower than those of the monoclonal antibodies, indicating that the final increase in product specific activity was typically less than 2-fold following both steps. Furthermore, the recovery of antibody mass was typically only 50%, possibly due to the low pH (4.0–5.0) required to bind the antibody to the strong cation exchanger. Sivakoff et al. (85) have recently suggested that although cation exchange chromatography may have applicability in the purification of some monoclonals from mouse ascites fluids, it is more often an ineffective method, due to the similar isoelectric points of the major protein contaminants and most monoclonal antibodies. In a recent investigation, hydrophobic interaction, anion exchange and cation exchange chromatography were all found to be equally effective in the purification of mouse monoclonal antibodies from ascites fluids, despite the high isoelectric points of all the IgG monoclonals tested (26).

Despite a number of limitations, however, there may be great utility for cation exchange chromatography in multistep purification schemes where high product purity is required, particularly in cases in which the monoclonal has a high isoelectric point and is tightly bound, and/or antibody recovery/capacity is not a major concern. In many process-scale antibody purification schemes, strong cation exchange chromatography plays a vital role as an initial clean-up step (62, 63, 69, 70).

APPENDIX II. EFFECT OF MATRIX STRUCTURE AND SURFACE MODIFICATION ON CAPACITY AND RECOVERY ON ION EXCHANGERS

Significant differences exist between the protein binding capacities of various commercially-available ion exchange supports (1–3, 8–18, 21–23, 26, 32, 33, 37, 51, 52, 62–71, 73, 75, 78, 81, 82, 85, 92, 94). In general, high-performance supports (1–10 micron) tend to have much lower intrinsic binding capacities than do the medium performance (15–60 micron) or soft gel (40–160 micron) matrices. The BAKERBOND MAb high-performance anion-exchange matrix has an intrinsic binding capacity for albumins and immunoglobulins which is as much as twice as high as other polymer-coated, silica based anion exchangers (1, 2, 54, 73, 81, 94) and from 4 (1, 16, 51, 52, 78) to 20 (17, 92) times higher than the most widely used commercially-available polymer-based high-performance column packings. This higher capacity permits the MAb matrix to accommodate significantly higher sample loads without reducing resolution (Table 3). Significant differences in binding capacities also exist among the other chromatographic matrices for antibody purification (62–71). For instance, immunoaffinity supports and hydroxyapatite have the lowest intrinsic binding capacities for immunoglobulins, absorbing about 20 times less protein than ABx or MAb (10, 18, 31, 38, 47–49, 62, 67–70, 89). However, it must be kept in mind that, like

ABx, affinity supports do not bind the majority of the nonimmunoglobulin proteins, and this factor may increase the tolerated sample load significantly. This is particularly true in the case of serum-supplemented cell culture supernatants in which the antibody typically represents only a small portion of the total protein content.

For instance, the capacity of ABx to purify a monoclonal from one cell culture fluid containing 10% fetal bovine serum was found to be more than 20 times that of conventional anion exchangers (62, 68–71). In contrast, ascites fluids typically contain much more antibody than cell culture media, and as a result, the relative real-life capacity of ABx is only approximately 5–10 times higher than that of the high capacity anion exchangers. For instance, an analytical MAb column is typically capable of purifying up to 1.5 mL of crude ascites, while the same size ABx column is capable of purifying up to 8.0 mL of crude ascites fluid (Table 3).

Owing to its high surface area and chemical reactivity macroporous silica provides an excellent base for chromatographic supports. Of the silica-based high performance ion exchangers, the highest capacities and recoveries are typically obtained when the support is encapsulated within a layer of hydrophilic polymer (1, 2, 54, 62–71, 73, 77, 81, 94). This polymeric coating may either be absorbed to the silica surface and then cross-linked, or, it may be covalently linked to the silica, alleviating the need for cross-linking (62, 63). In cases where the polymer is noncovalently absorbed to the silica surface, extensive cross-linking is required to enhance stability (2, 77). However, cross-linking reactions may create several problems, including a reduction in both binding capacity and protein recovery (62, 65). Several recent studies (54, 62, 94) indicate that the extensive cross-linking of PEI and other polymeric supports may produce bonded phases which have strong hydrophobic characteristics, and depending upon the mobile phase composition, these supports may be used either as ion exchangers or for hydrophobic interaction chromatography (65). However, recent work in our laboratories indicates that proteins which are both acidic and hydrophobic (e.g., ovalbumin, calmodulin, and many immunoglobulins) may be bound irreversibly by either anion exchangers with strongly hydrophobic ligands or by HIC supports that have anion exchange characteristics, while proteins which are both basic and hydrophobic (e.g., lysozyme, ribonuclease) may give low recoveries either on cation exchangers with excessive hydrophobic areas or HIC supports containing negative ligands (62, 65). The fact that immunoglobulins contain surface regions which are rather acidic as well as those which are also hydrophobic may explain the low recoveries (nonspecific binding) which are often obtained on some anion exhangers (62). In contrast, mildly hydrophobic areas on ion exchangers, such as those present on the ABx and MAb matrices, or relatively nonpolar surfaces on HIC matrices such as those on BAKERBOND HI-Propyl (Figs. 14, 17, 19, 24, 26), facilitate quantitative recovery of protein mass and activity (25, 29, 35, 62–72). In contrast to most polymer-based anion exchangers, these supports have been used for over 1000 h under preparative conditions with little change in chromatographic performance and protein binding capacity, suggesting that nonspecific binding and protein losses are minimal (62–71).

APPENDIX III. LIGAND DENSITY, CAPACITY AND BINDING KINETICS

Chromatographic matrices, such as hydroxyapatite, cation exchangers, ABx, and affinity matrices, which do not to bind the majority of the proteins in crude antibody preparations, share in common certain, rather unique features (62, 69, 70). The major problem inherent to these affinity-like supports is that, during frontal chromatography in the presence of highly overloaded conditions, a substantial percentage of antibody may fail to bind (Tables 2, 3, and 4). Under these or other suboptimal chromatographic loading conditions a split peak (97) is obtained, in which a given homogeneous antibody species is split into two separate peaks; one being the void volume peak containing those antibodies which fail to bind, and the other, being a bound peak containing those antibody molecules which did bind to the matrix. Although these phenomena are similar to displacement chromatography (44,62,96), they are dissimilar in that the antibodies which flow through (elute in the void volume) are not displaced from the support by a more strongly bound, competitor or displacer molecule, as in true displacement chromatography. Rather, the antibodies are actually inhibited from binding due to a combination of interactions with other proteins (similar protein–protein interactions have been shown to produce split peaks on affinity matrices), and the ability of the nonbinding contaminating proteins to swamp out the surface interaction between the antibody and the support (62, 97). Restricted diffusion and suboptimal binding conditions can also be partially responsible for this type of behavior, and will be discussed in detail elsewhere (62).

The fact that some of these non-immunoglobulin proteins which are not strongly bound in the overloaded or preparative, frontal chromatographic mode do indeed interact, albeit weakly, with these bonded phases (hydroxyapatite, cation exchangers and ABx), is made evident when binding conditions are optimized and the sample load is reduced, numerous peaks elute several minutes after the void volume, while other proteins may bind rather strongly (21, 62). For example, several recent investigations have shown that the majority of the proteins present in antibody preparations are, in fact, bound by hydroxyapatite at low sample loads. This suggests that these contaminating proteins are being displaced from these supports by the antibody at high sample loads (in contrast to the contaminating proteins displacing or preventing the binding of the antibody).

As the amount of these bound nonimmunoglobulin proteins, which may occupy ligands which would otherwise be available for binding the antibody, and, their affinity toward the support increases, so too does the relative amount of antibody which will be eluted in the void volume. This will reduce the amount of sample which can be bound to the column in a single pass, without recycling the void volume peak. Regardless of how much protein is actually bound to the column, the amount of total protein and other electrolytes in the sample can be important factors in determining the ratio of sample to support, which is required for quantitative binding of the antibody in a single chromatographic run (Tables 2, 3, 4).

Although this capacity/recovery (split peak) problem also occurs on ABx during frontal chromatography (Tables 2, 3, and 4), it is much more acute on conventional cation exchange matrices and hydroxyapatite, which have rather low ligand densities. Early research in this laboratory suggested that, by properly optimizing the ABx surface chemistry (Appendix IX), the severity of these problems might be minimized, or even circumvented. This work also indicated that on ion exchangers, ligand density may be more important in preparative chromatography than intrinsic binding capacity alone (Tables 4G, 4H, 4I, 4J). For instance, although little difference exists between the intrinsic binding capacities (i.e., binding of a purified protein in the static or batch mode) of experimental ABx bonded phases in which the surface chemistry was modulated to produce ligand densities/titrations ranging from 0.2 to 1.4 meq/g (ion exchange groups/mass of support), it was found that under heavily overloaded conditions, ABx supports which contained higher ligand densities bound much more of the total immunoglobulin (Tables 4G–4J). These results also suggested that for chromatographic matrices like ABx which have extremely high ligand densities and binding capacities, there is a certain ligand density above which a maximum amount of protein is bound, and subsequent increases in ligand density, above this threshold value, have little or no effect on further increasing the intrinsic binding capacity (Table 4G). However, maximizing the ligand density is extremely important in the preparative mode, because large amounts of electrolytes (protein, salt, amino acids, and so on) which are present in crude biological samples tend to interact with the antibody(s) and the support, and prevent quantitative binding (Table 4G–4J). This places strict limitations on the amount of sample which may be loaded with quantitative binding of the antibody (Tables 2, 3, 4).

The intrinsic binding capacity of commercially available 5 micron ABx is approximately 150 mg of purified IgG per gram of bonded phase, corresponding to a titration of 0.5 meq/g packing. Similar capacities have been obtained on 5 micron ABx in the column chromatographic mode by frontal analysis or the breakthrough method (this is the level at which 50% of the subsequently injected (purified) IgG fails to bind to the column). In other experiments, more than 100 mg of IgG per gram of ABx have been loaded with quantitative binding (e.g., over 1.1 gram of rather pure horse IgG was bound quantitatively on a column containing 11 grams of 15 micron ABx). The ligand titrations of the 15 micron and 40 micron (PREPSCALE) preparative media are even higher than those on 5 micron ABx (typically at least 0.7 meq/g and 1.0 meq/g, respectively). However, due to the higher surface areas and lower densities of the 15 and 40 micron silica bases, the total capacity per volume and the actual ligand densities of 5, 15 and 40 micron ABx are quite similar, being 0.35, 0.35 and 0.40 meq/mL, respectively. Hence, the elution profiles of the three matrices are remarkably similar with gradient elution (Figs. 31, 34).

Regnier et al. (2, 77) have shown that the intrinsic binding capacity of anion exchangers for proteins smaller than 100 kD (e.g., albumins and transferrins) increases with pore size up to 240 Å, due to steric inaccessibility within smaller pores. However, intrinsic binding capacity decreases dramatically as the pore size increases above 240 Å, due to the concurrent decrease in surface area of the support (2, 62, 63, 77). Therefore, anion exchangers with an average pore diameter of 240 Å may bind two to four times more hemoglobin than those with pore diameters of 500 and 1000 Å, respectively (2, 77). These results may also explain some of the differences in total binding capacity among ion exchangers, because the ABx and MAb bonded phases contain 300 Å pores, while many commercially-available high-performance ion-exchangers contain much larger pores (1, 8, 9, 12–16, 19, 22, r3, 65, 32–35, 37, 40, 43, 51, 52, 55, 72, 78, 82). These 300 Å pores also allow for high binding capacities for larger proteins, as the intrinsic binding capacities for immunoglobulins on ABx and MAb (150 micrograms/gram) are only slightly lower than those for smaller proteins, such as albumins and transferrins (180 mg/g). In fact, the binding capacity for thyroglobulin or IgM (900 kD) on the MAb matrix is actually double that on commercially available cross-linked agarose matrices with 1000 Å pores (62, 63).

The absorption/binding kinetics for proteins into these 300 Å pores also appears to be rather rapid, since substantial increases in flow rate do not significantly reduce binding capacity, as determined by frontal chromatography. For example, frontal uptake experiments (62, 63) indicate that the binding of a purified 70 kD protein to ABx was only reduced by 38% when the flow rate during the injection/loading was increased from 0.5 mL/min to 30 mL/min (a linear velocity of 3 cm/sec), despite the fact that the column was heavily overloaded (50 mg/g) with a concentrated protein solution (50 mg/mL), and that the chromatography was conducted on a column (4.6 × 50 mm) which we have found to be too short to provide optimum protein binding on ABx (62, 63).

In contrast to these results, in the presence of highly overloaded conditions in which an analytical column is challenged with several grams of protein in a cell culture supernatant, the effect of flow rate on the binding of immunoglobulins to ABx and the generation of a split peak is much more dramatic (Table 4C). Rather than simply being the result of restricted diffusion and slow adsorption kinetics (as is apparently disputed above), these effects may be the results of strong protein-protein interactions and the swamping out of the interactions between the antibody and ABx by the vast amounts of fetal bovine serum proteins and other media components. This veiw is supported by the large discrepancies (approximately 30-fold) in flow rates which can be tolerated for the binding of the purified protein relative to those for the binding of proteins from this crude sample (Table 4C). Therefore, poor diffusion/adsorption kinetics are apparently not totally responsible for the split peak observed on ABx at extremely high sample loads. This view is also supported by the similar split peak behavior observed in the batch mode following a 40 min adsorption period (Table 4K).

Finally, the sharp peaks obtained on ABx for larger proteins such as immunoglobulins, particularly IgM (900 kD), also suggest that the adsorption/desorption kinetics are rather rapid into 300 Å pores.

Although tailing does occur with polyclonal antibody populations due to the polydisperse nature of the IgG molecules (Fig. 3), only a minimal amount of tailing occurs with monoclonal IgG, IgA, or IgM antibodies (Figs. 1, 2, 35). However, it should be pointed out that pores smaller than 300 Å may be inaccessible to IgM; recent work in this laboratory suggests that larger pores (~ 500 Å) may provide higher capacities for IgM, provided the surface area of the silica and the ABx ligand density remain high.

APPENDIX IV. RESOLUTION, PURITY, AND DISPLACEMENT CHROMATOGRAPHY IN THE PREPARATIVE MODE

The high degree of selectivity achieved on the ABx is admirably illustrated when a series of fractions collected from ABx are rechromatographed on a BAKERBOND MAb high performance anion exchange column (Fig. 28). When the ABx void volume fraction from a serum-supplemented cell culture media was rechromatographed on the anion exchanger, it produced a chromatographic profile which was virtually identical to the anion exchange profile of the neat culture media itself, indicating that ABx can selectively remove the monoclonal antibody (Fig. 28). Furthermore, while a substantial portion of contaminating proteins coeluted with the monoclonal antibody on the anion exchanger, they were totally resolved from the immunoglobulins on ABx.

ABx also provides higher resolution than anion exchangers in the preparative, frontal chromatographic mode, not only because the matrix has inherently higher selectivity at low sample loads, but also because the ABx matrix binds less of the non-immunoglobulin protein contaminants that may migrate chromatographically toward the immunoglobulin and further reduce resolution in the presence of overloaded conditions (Table 3). This chromatographic migration which occurs on anion exchangers at high loads appears to be the result of simple band broadening, tailing/peak asymmetry, volume and mass overloading effects, and the non-linear nature of the adsorption isotherm, because the relative, as well as the absolute retention times of the albumin and the monoclonal antibody decrease substantially in the presence of large sample loads (Table 3).

As previously mentioned, a form of true displacement ("self"-displacement) chromatography does occur on ABx during frontal chromatography; as the sample load is increased toward the capacity limit of the matrix, many of the less strongly bound, nonimmunoglobulin proteins are displaced (Table 2–4). The binding of these early eluting contaminating proteins decreases due to displacement by the other more strongly bound proteins (including the antibody, which is typically the most strongly bound protein), as well as by the swamping out mechanism and protein–protein interactions previously detailed (Tables 2–4). Although on most matrices, purity typically decreases as the sample load is increased, on ABx, the purity of the antibody peak may actually increase because column overload may increase resolution, and, via this displacement mechanism, reduce the amount of the bound protein contaminants (Table 4B). Similar results have been obtained with serum (Table 2), ascites fluids (Table 3), and serum-supplemented cell culture fluid (Table 4). As the sample load is increased, the percentage of the non-immunoglobulin proteins which bind decreases substantially (Tables 2–4). The antibody is typically bound quantitatively at moderate sample loads, but further overloading the ABx column causes an increasing percentage of the antibody(s) to be displaced as well (Tables 2–4). A dramatic example of this displacement chromatography is made evident by the differences in elution profile and relative peak heights of the bound peaks in analytical and preparative chromatograms of the same sample (Figs. 9, 10). As the sample load increases, the size of the monoclonal peak also increases, while the relative size of the bound non-immunoglobulin peak decreases substantially (the peaks eluting at 20 min in Fig. 9, and 8 min in Fig. 10). Therefore, the ability to conduct frontal and displacement chromatography on ABx can be used to advantage to increase sample loads (Tables 2–4) and to increase product purity (Table 4B).

APPENDIX V. ANTIBODY BINDING AND SAMPLE LOAD OPTIMIZATION

Loading capacities on ABx depend upon the nature of the sample matrix and the antibody itself; low temperatures, high ionic strength and pH, and high levels of protein and immunoglobulin (Tables 4C–4F), may often reduce the amount of sample which can be loaded onto the column for quantitative binding of the antibody (Tables 2, 3, 4). Therefore, dialysis, dilution or particularly ultrafiltration are recommended in order to reduce the amount of antibody which fails to bind due to competitive interactions with proteins and salts present within the sample itself. Tangential flow ultrafiltration is valuable due to its ability to concentrate large volumes of culture media into low ionic strength buffer (25 mM MES plus 15 mM KH_2PO_4 at pH 6.0), and to reduce the total amount of nonimmunoglobulin protein (more than three-fold in many cases).

Zwitterionic buffers such as MES or MOPSO are used as ABx binding buffers because they facilitate the binding of antibodies, presumably by maintaining the charge integrity of the matrix, as well as the protein itself (Table 4F). This is hypothetically due to the ability of buffers with opposite charges on each end of the molecule to preserve the relative proportion of negative and positive group which are present on the matrix. (MES and MOPSO have a positively charged amine on the morpholino end of the molecule, and a negatively charged aliphatic sulfonic acid group on the other end). For instance, during column re-equilibration the sulfonic acids groups on the MES buffer will bind to the positive, anion exchange ligands of ABx leaving a positive morpholino moiety of MES exposed to act as a new ligand on ABx (i.e., substitution of a new positive group for the original positive ligand); likewise, the positive, amine (morpholino) groups on other MES molecules may bind to the cation exchange ligands of ABx, leaving a sulfonic acid moiety of MES exposed to act as strong cation exchange group, working in concert with the rest of the ABx chemistry. The result is that the total number of both cation and anion exchange groups on the ABx surface, and the relative proportion of these, does not change dramatically (i.e., the charge integrity is maintained). Therefore, binding of antibodies is maximized in these zwitterionic buffers (Table 4F). Of course, unless samples are exhaustively dialyzed or ultrafiltered, non-zwitterionic salts present in a sample may reduce binding at high sample loads. This is presumably because simple, non-zwitterionic salts do not form such complex structures to facilitate binding.

Some salts such as divalent cations or dicarboxylic acids may actually prevent chromatographic binding on ABx, presumably via an analogous mechanism (Table 4F). For instance, during column reequilibration dicarboxylic acids in the starting buffer may bind to a positively charged anion exchange ligand on ABx (as well as on the proteins) and temporarily convert it to a weak cation exchange ligand reversing the charge characteristics. In fact, such a strategy has been intentionally employed to convert anion exchangers to cation exchangers (6,26). Likewise, the presence of dicarboxylic acids has a repulsive action towards the similarly charged cation exchange ligands present on ABx, and may construct a barrier between the proteins and ABx (particularly for the more basic ionotopes of proteins, which tend to bind these dicarboxylic acids).

ABx columns should be equilibrated with approximately 20 to 30 column volumes of starting buffer (10 to 25 mM MES, pH 5.4–5.8). This may be accomplished more quickly by using rapid flow rates or by beginning the equilibration with 50 mM MES, pH 5.4 and then switching to lower ionic strength MES (10 or 25 mM, pH 5.4–5.8). In most cases, 25 mM MES is the best binding buffer because of its higher buffering capacity. MES buffer is also an excellent starting buffer, because pH equilibration may be conveniently monitored by following absorbance at 220 nm due to the decrease in absorbtivity (extinction coefficient) of MES which accompanies column pH equilibration (62, 63). Alternatively, the pH of the column effluent may be monitored to determine column equilibration.

To facilitate antibody binding, flow rates for sample loading should be approximately 60 to 80% of those used for actual chromatography (Table 4C). Of course, higher flow rates (linear velocities) may be used with lower sample loads (Figs. 1–3). The flow rate during chromatography will vary depending upon the internal diameter of the column (4.6 mm ID, use 0.6 mL/min; 7.75 mm ID, use 0.8 mL/min; 10 mm ID, use 3.0 mL/min; and so on).

APPENDIX VI. ECONOMIC ASSESSMENT OF ANTIBODY PURIFICATION

The production of monoclonal and even polyclonal antibodies is typically an expensive undertaking, and the cost of purification is often the major expense in this production. Therefore, it is often assumed that, in order to reduce overall production costs, chromatography should be conducted on the least expensive supports. However, it may be valuable to examine the costs of related factors (overhead) relative to that of the chromatographic support. For example, on a given column packed with 40 micron ABx, the cost of buffers can exceed the cost of the matrix after just 3 workdays (62, 63, 69, 70). In the case of semipreparative HPLC columns, the cost of the buffer salts may exceed the cost of columns prepacked with 15 or 5 micron ABx after just 3 weeks or 3 months, respectively (62, 63, 69, 70). Therefore, even 5 micron ABx HPLC columns can typically cost less than the buffer salts used on that column during its lifetime (over 1000 hours).

Process economy is also more than a matter of choosing the least expensive support; capacity, longevity, product recovery and purity, and a host of other factors must also be considered. It should also be emphasized that the capacity of the matrix determines the size of the column required to process a given amount of sample; in turn, this determines the amount of mobile phase used. The use of high-performance materials typically reduce the numbers of downstream steps, and this typically increases recovery, and eliminates the costs associated with these additional chromatographic steps.

Therefore, in some cases it may be beneficial to conduct the entire purification by HPLC, particularly in cases in which the product is extremely valuable, high resolution is required, the sample volume is relatively small, and the sample is relatively clarified. In other cases, the use of less expensive medium performance matrices for initial step purification may be much more economical, and may often result in similar product purity (62, 69–70).

APPENDIX VII. ABx PURIFICATION OF ANTIBODY FRAGMENTS, CONJUGATES, CLASSES AND SUBCLASSES

VII.1. Antibody Fragments

The production and purification of antibody fragments derived from proteolytic digestion of native immunoglobulins represents an important aspect of modern antibody technology. For certain diagnostic and therapeutic purposes, it is often advantageous to use $F_{(ab')2}$ antibody fragments which are bivalent, but lack the $F_{c'}$, fragment, while for other uses, univalent Fab fragments are more desirable (14, 29, 31, 36, 56, 62). An absence of F_c or $F_{c'}$, fragments is desirable in serological studies involving cells which bind immunoglobulins via F_c receptors and in the production of therapeutics and diagnostics destined for in vivo applications (14, 29, 31, 36, 56, 62). Due to the importance of these fragments in modern medicine and biochemistry, methodologies have been developed for the purification and analysis of rabbit, mouse, and human IgG fragments using ABx chromatography. These methods may be used to monitor product accumulation in studies focusing on the optimization of condition for proteolytic digestion/fragmentation (or conjugation) reactions, as well as in the purification of these products (29, 62, 69, 70).

Like most proteases, neither pepsin nor papain are bound by ABx and elute in the void volume (Fig. 29). In contrast, all native, intact antibodies, as well as all proteolytic fragments of IgG bind strongly to ABx. As a result, ABx is capable of separating all the fragments, as well as the native/intact antibody from either protease (Fig. 29). In most cases, the F_{ab} or $F_{(ab')2}$ fragments are well resolved from native immunoglobulins, and the F_c or $F_{c'}$, fragments as well (Fig. 29). Exceptions occur with peptic digest of human, and particularly mouse polyclonal IgG in which a small amount of the native immunoglobulins co-elute with the $F_{(ab')2}$ fragments (Fig. 20). Intact IgG typically has the highest affinity for ABx, suggesting that the binding strength of the native IgG is due to the combined binding effects of the two fragments (Fig. 29). Human IgG is a partial exception, as the majority of the F_{ab} and $F_{(ab')2}$ fragments elute slightly after the intact immunoglobulins (Fig. 29). As expected, the polyclonal F_{ab} and $F_{(ab')}$ fragments give broader peaks due to the polydisperse nature imparted by the hypervariable regions, while the polyclonal F_c and $F_{c'}$, fragments tend to elute as rather sharp peaks (69–74).

VII.2. Antibody Conjugates

Protein–antibody conjugates represent one of the most useful products of modern biotechnology. By covalently coupling antibodies or antibody fragments to various reporter proteins, it is possible to produce highly sensitive and selective therapeutic and diagnostic agents (46, 62, 69, 70, 86, 87). However, in order to achieve this high sensitivity and to reduce non-specific binding or cross-reactivity (and thus, increase the accuracy of diagnostics and the effectiveness of therapeutics), highly purified conjugate preparations of high specific activities are required (62, 69, 70, 46, 86, 87). Due to the importance of conjugates in modern biotechnology and biochemistry, methods have been developed for the separation of horseradish peroxidase (HRP) and calf intestine alkaline phosphatase conjugates of polyclonal IgG using ABx chromatography. Human, rabbit and mouse polyclonal IgG and the protein conjugates derived from these antibodies all gave comparable elution patterns from ABx (Fig. 30). The conjugates of these proteins were typically eluted between the native proteins and the native antibodies, and were well resolved in all cases. The elution of the conjugated product between these proteins suggests that the conjugate has a binding affinity to ABx which represent a mixture of its constituent proteins. This may be due to the fact that some of the chromatographic binding sites on the antibody (sites which are responsible for the binding to ABx) become inaccessible following conjugation, thereby reducing binding. In this regard, the elution profiles of calf intestine alkaline phosphatase–IgG conjugates are particularly striking (62, 69, 70). While the polyclonal IgG elutes as a broad peak, and the various calf intestine isozymes of alkaline phosphatase elute as multiple weakly bound peaks, the conjugate elutes at an intermediate position, as an extremely sharp peak (62, 69, 70). These results suggest that regions of both proteins which are responsible for different binding strengths/chromatographic affinities towards ABx (i.e., hypervariable regions of the antibodies, and variable regions among the alkaline phosphatase isozymes) are possibly involved in the conjugation reaction. In contrast, in the presence of some mobile phases, goat IgG–alkaline phosphates conjugates elute as a series of peaks, which partially coelute with several of the isozymes of the alkaline phosphatase (data not shown). In fact, this type of chromatographic behavior is typical for conjugates in which multiple protein molecules are conjugated to a single antibody molecule. In most cases (e.g, HRP conjugates), the retention time of the conjugate of ABx decreases as a function of the number of reporter proteins which are coupled per antibody molecule (i.e., their chromatographic properties become more and more like that of the native (unconjugated) protein itself) and those conjugated species which contain the most proteins are eluted closest to the void volume (62, 69, 70).

VII.3. IgG Subclasses

Extensive clinical research has shown a strong correlation between various diseases and imbalances in the IgG subclasses (39, 74, 84, 87, 95). Therefore, a rapid and quantitative chromatographic method to determine relative levels of the IgG subclasses might be valuable in clinical and veterinary research. In normal serum, the IgG_1 and IgG_2 subclasses represent the vast majority of the total immunoglobulin G class (making up 60% and 30%, respectively). Due to the the preponderance of these two subclasses, this investigation focuses upon the separation of IgG_1 and IgG_2 with BAKERBOND ABx.

ABx is capable of separating the two major subclasses of IgG from goat, sheep, and bovine plasma (Fig. 31). As with most separations on ABx, resolution is particularly dependent upon mobile phase conditions (62–71). In fact, although little separation of goat IgG_1 and IgG_2 is achieved when the immunoglobulins are eluted with either phosphate or acetate salts (data not shown), various sulfate salts are quite effective in this regard (Fig. 31). Furthermore, the separations in sulfate and other elution buffers can be further optimized by manipulating binding/equilibration buffer species, ionic strength, and/or pH (Fig. 31). Unfortunately, major changes in the pH and ionic strength of the binding buffer tend to reduce capacity on ABx (62, 69, 70); although this presents no problem for analytical purposes, for preparative applications, capacity must be balanced against resolution (Table 4).

At present, it is unknown whether ABx is capable of separating IgA_1 and IgA_2 or IgG subclasses from other species, although investigations in this and other laboratories is also being conducted toward this goal (62).

VII.4. Immunoglobulin Classes

Numerous clinical studies have demonstrated that, like the IgG subclasses, a strong correlation exists between changes in the levels of the immunoglobulin classes and various pathological states (39, 74, 84, 87, 95). Therefore, reliable methods are needed for the accurate identification and quantitation, as well as the purification of immunoglobulin classes. ABx is capable of separating a significant amount of human polyclonal IgG from the other major human serum immunoglobulins, although this separation is not complete (Fig. 32). ABx may be used in conjuction with either hydrophobic interaction chromatography on BAKERBOND HI-Propyl, weak anion exchange chromatography on BAKERBOND MAb, strong anion exchange chromatography on BAKERBOND QUAT, or size exclusion chromatography, in order to obtain rather pure IgM, IgA, and IgG from human, mouse, bovine, sheep or goat serum (Fig. 32). These methods will be examined in detail elsewhere (62, 63, 69, 70).

APPENDIX VIII. ANALYSIS WITH ABx

VIII.1. Introduction

The analysis of biological samples for the quantitation and characterization of immunoglobulins is typically conducted via immunological, electrophoretic, or chromatographic techniques (4, 5, 9, 10, 12–14, 19, 23, 27, 33, 37, 40, 45, 48, 49, 55, 58, 60, 80, 82–84, 86, 87, 92). However, any analytical method suffers from limitations, and these are no exception. Immunological and electrophoretic methods are often laborious, complicated, semiquantitative and/or unable to distinguish between isotypes, while traditional chromatographic methods also suffer from either poor resolution and/or low mass recoveries.

ABx may be useful in the clinical analysis of plasma to identify immunological disorders, including basic immunodeficiencies, autoimmune diseases, overproduction disorders and those characterized by specific changes in class or subclass composition. ABx may also be useful in monitoring the efficacy of various therapeutic regimes and to determine antibody pharmacokinetics following the administration of therapeutics such as alpha-interferon, interleukin-2, cyclosporine, T-cell inhibitors, exogeneous human polyclonals, monoclonals, hybrid antibodies, fragments, conjugates, or magic bullets (62, 63, 69, 70).

Trace levels of monoclonal antibodies in complex mixtures can be chromatographed on ABx for hybridoma selection, screening and subcloning, antibody characterization, subclass determination, identifying hybridomas which produce multiple antibodies, monitoring hybridoma cell lines, tracking genetic drift and class/subclass switching, quality control, monitoring fermentation broths/culture fluids, optimizing harvest time, process monitoring and purification monitoring (62, 63, 69, 70).

ABx can rapidly separate multiple types of immunoglobulins, making it possible to separate and quantitate undesirable immunoglobulins in crude or purified antibody preparations, which are totally indistinguishable from the monoclonal via conventional means (immunological, electrophoretic, or chromatographic methods). As an alternative to column chromatography with ABx, particularly in situations in which accuracy is less critical, solid phase or batch extraction of monoclonals with ABx may be used (62, 63, 67, 69, 70). This method facilitates the rapid spectrophotometric analysis and/or the simultaneous clean-up and concentration of numerous samples for analysis via other techniques.

It is beyond the scope of this investigation to fully explore all possible clinical and analytical applications of ABx. However, it is anticipated that the basic methods and ideas which will be proposed within this initial study will form a solid foundation and spawn a host of other, more detailed investigations of this kind.

VIII.2. Purification/Process Monitoring

ABx chromatography has been used as an analytical method to monitor the efficacy of ABx solid phase, batch extractions of antibodies from serum, ascites fluids or cell cultures (Figs. 20–22). ABx has also been used to monitor the purity of monoclonals purified by ABx column chromatography, and at the same time provide methods development for a second step purification on ABx, when necessary (Figs. 11, 12). These analytical purification monitoring techniques give rapid information about the approximate purity of immunoglobulins extracted by either column or batch methods, and facilitates method development for the optimization of the conditions required for batch extractions or column chromatography (62–64, 67–71).

VIII.3. Quality Control

ABx has also been used for quality control purposes, to check the purity of monoclonal antibody preparations, including those destined for in vivo use (62–64, 67–71). ABx is well suited for the quality control analysis of purified monoclonals, since it does not bind the majority of protein contaminants, reducing the possibility of the monoclonal coeluting with nonimmunoglobulin proteins (62–64, 67–71). In addition, the resolution and peak sharpness on ABx are excellent, particularly in the isocratic mode or with the use of step gradients (62–64, 67–71). Furthermore, the high capacity of ABx may be put to good use in the purification of minor contaminants for identification and/or use in toxicological studies which may be required for therapeutic or in vivo diagnostic preparations. Finally, ABx is able to separate multiple forms of immunoglobulins (Figs. 27, 33, 34) of the same or of different subclasses, slow migrating, inactive and active, light chain variant antibodies, monoclonals from double producers, bispecific hybrids from monospecific monoclonals produced by quadromas or by chemical means, and monoclonal antibodies from polyclonal antibodies present in either mouse ascites fluids or serum-supplemented cell culture supernatants (62–64, 67–71). Therefore, single peak purity on ABx substantially increases the probability of ensuring that the monoclonal antibody preparation is truly homogeneous with respect to antibody idiotype.

Every analytical technique has intrinsic limitations, including ABx, and there may be cases in which impurities coelute with the monoclonal antibody. However, conducting a series of chromatographic quality control analyses with ABx in the presence of several different elution buffer conditons or gradient profiles minimizes the possibilities of coelution. Basing purity assessments on any single criterion, no matter how sensitive or reliable, is a mistake, particularly when a product is destined for in vivo use. This is true, not only in the case of diagnostics, which are typically only administered once, but particularly for therapeutics, in which case the product must be administered in a series of therapeutic regimes and at relatively large doses.

ABx chromatographic analysis has been used to monitor the production of monoclonal antibodies from several hybridoma cell lines grown in a number of different growth media (Fig. 34). ABx chromatography represents a rapid method for the sensitive quantitative analysis of monoclonal antibody levels (less than 1 microgram) and permits the characterization of the antibody species being produced, because all immunoglobulins can be resolved from all the nonimmunoglobulin proteins in fetal bovine serum, as well as serum-free media (Fig. 6). Although a small amount of non-immunoglobulin protein secreted by the hybridoma cells is also bound by ABx, most of this elutes in the void volume; those proteins which do bind are present at low levels and do not typically interfere with the analysis (Fig. 34).

The ABx analysis of the time-course of antibody production from the hybridoma cell lines investigated reveals a steady increase in monoclonal levels following the initial lag phase (Fig. 34). These analyses correlate well with the electrophoretic analysis of the same samples (Fig. 34). Analysis of these cultures after the last time point shown, failed to detect a significant increase in extracellular antibody levels, despite a decrease in cell viability which suggests that the intracellular proteins were being released into the culture media (Fig. 34). In contrast, this decrease in cell viability and concomitant cell lysis was accompanied by a significant increase in the levels of a number of nonimmunoglobulin proteins as detected by electrophoresis and by an increase in the total extracellular protein levels (Fig. 34).

Finally, it should be noted that ABx was able to separate several forms of immunoglobulin produced by these hybridoma cell lines; one hybridoma was later identified as a double producer, while another was identified as a triple producer (Fig. 34). Immunological assays of these peaks revealed that in each case, only one of the peaks which contained the monoclonal antibody exhibited significant immunological activity.

VIII.4. Clinical Analysis

As previously mentioned, ABx may be used in the clinical laboratory to quantitate plasma immunoglobulins and to identify various immunological disorders. The changes in serum immunoglobulin levels which occur in patients with autoimmune diseases (34, 39,74, 84, 87, 95) are easily detectable with ABx (Fig. 35). The elution profiles of serum from patients with various degrees of rheumatoid arthritis suggest that a rather distinct chromatographic pattern, or fingerprint, may be obtained from each sample (Fig. 35). The clinical significance of these findings is presently unknown, although similar research is currently being conducted in a number of laboratories in an attempt to correlate the ABx chromatographic profile with various biochemical and physiological parameters (62, 63). Similar differences were noted among the plasmas of patients with other autoimmune diseases such as Lupus vasculitus or ANA/DNA (Fig. 35).

Myeloma proteins are monoclonal antibodies which are present in the serum of patients with various lymphoproliferative diseases (34, 39, 74, 84, 87, 95). Analysis of a variety of myeloma sera with ABx reveals striking differenes in both the basic elution profiles and the levels of the individual monoclonal(s). In some cases, particularly for samples taken during an advanced stage of the cancer, the myeloma protein makes up the vast majority of the total protein content of the plasma (Fig. 35). Plasmas which contain IgG myelomas of a specific subclass are typically deficient in the other IgG subclasses (34, 39, 74, 84, 87, 95), and ABx is able to detect this reduction in endogenous immunoglobulins levels (Fig. 35). Furthermore, in some cases ABx is also able to detect a reduction in the levels of nonimmunoglobulin (void volume) proteins relative to those of the control (Fig. 35). Many of the samples analyzed by ABx exhibit elution profiles which suggest the presence of multiple forms of myeloma proteins (Fig. 35). ABx is either resolving several forms of the same monoclonal, or the patient has several different activity proliferating clones or one producing several different monoclonals.

Just as ABx may be used to monitor serum or plasma samples for various lymphoproliferative and other diseases characterized by elevated immunoglobulin levels, it may also be used to detect various immunodeficiencies. It is possible to increase the sensitivity of this chromatographic method with the use of a sharp gradient or a step gradient, so that the polyclonal immunoglobulins are eluted as a larger, sharp peak. This type of protocol permits the accurate quantitation of the extremely low antibody levels present within the serum of immunodeficient patients (Fig. 3D).

In contrast to the aforementioned methods which make use of ABx high performance liquid chromatography, it is possible to rapidly screen multiple serum samples for abnormal immunoglobulin levels using ABx batch, or solid phase extraction techniques (62). These methods facilitate the extraction of relatively pure immunoglobulin fractions, and thus, may be used for the rapid determination of antibody levels in multiple samples via spectrophotometric analysis, or, for clean-up prior to immunological assay or electrophoresis (62, 63, 69, 70).

Although the chromatographic analysis of serum samples with ABx may not replace the conventional electrophoretic or immunological methods which are used on a routine basis in clinical laboratories, these techniques may augment other diagnostic tests. A variety of techniques must ultimately be used to fully understand and characterize diseases of the immune system. ABx chromatographic profiles may aid in providing information and insights into the basic nature of these diseases, and additionally help monitor the efficacy of various therapeutic regimes. The techniques presented within this section will probably be of greatest use in the detection and diagnosis of lymphoproliferative and other neoplastic diseases. However, considering the enormous diversity among the various diseases of the immune system (34, 39, 74, 84, 87, 95), there are a number of possible applications for ABx in the clinical and veterinary diagnosis of various pathological states.

APPENDIX IX. DESIGN AND OPTIMIZATION OF ABx

Although one chromatographic mechanism predominates on ABx (weak cation exchange), the presence of other ligands of the mixed-mode chemistry (mild anion exchange and mild hydrophobic interaction) significantly enhances the selectivity of this matrix (62, 63, 65).

The construction of ABx is initiated with silica bases of various particle sizes (5, 15 and 40 micron) with an average pore size of 300 Å. This architectural base, which consists of high quality, closely sized wide-pore chromatographic silicas with a narrow pore size distribution (factors which enhance resolution and efficiency) is then derivatized with a covalently bound hydrophilic polyamine backbone via a patented process; this in turn, is further derivatized with the various functional groups.

Although the functions of the polyamine base are multiple, most of these roles are non-interactive (62, 65). This polymeric backbone serves seven purposes: (1) it covers non-specific interaction sites on the silica surface, leading to quantitative recovery of antibody mass and immunological activity, thereby reducing the need for routine cleaning or regeneration of the matrix; (2) it protects the entire surface of the silica base, thereby increasing chemical and physical stability, and thus, column lifetime; (3) it increases not only intrinsic binding capacity, but more important, ligand density, and thus, the capacity of the matrix to quantitatively bind immunoglobulins in the presence of highly overloaded conditions; (4) it enhances the effects of mobile phase composition on resolution, elution profile and selectivity control; (5) it reduces the pKa values of the adjacent weak cation exchange groups to insure their ionization at lower pH values; (6) it widens the pKa range of these carboxylic acid ligands, as well as the amines themselves, and thus, provides a polyelectrolyte matrix which is ionizable over an extremely wide pH range; and finally; (7) it provides mild anion exchange groups which can work in concert with the rest of the ABx surface chemistry to bind and resolve immunoglobulins.

Obviously, these amines can function as anion exchange ligands which are capable of interacting in cooperation with other ligands of the ABx surface chemistry, to bind to acidic regions of both the F_c and the F_{ab} chains of immunoglobulins. However, it is imperative that the relative proportion of amines is carefully controlled and kept to a minimum, because, as the level of anion exchange groups increase, albumins and other acidic proteins begin to bind to the matrix (62, 65). Another function of this polymeric coating is that it provides amines and carboxylic acids with a wide range of pKa values, which are ionizable over a wide pH range; this facilitates the generation of internal pH gradients in the presence of various mobile phases, which may be used to enhance resolution (62, 63, 65, 67–71).

One of the more important functions of this polyamine base is that the presence of adjacent amines and amides significantly reduces the pKa values of the weak cation exchange moieties. The ability to reduce the pH of the equilibration buffer and still have a major portion of the carboxylic acid residues fully charged facilitates the binding of immunoglobulins to ABx. Furthermore, because of the polyelectrolytic nature of the nitrogenous base, the pKa values of these carboxylic acid groups cover a wide range. One of the disadvantages of conventional cation exchange matrices is that rather low pH values are required to bind immunoglobulins, and as a result, many nonimmunoglobulin proteins also bind strongly, thereby reducing capacity and resolution (Appendix I). In contrast, higher pH values may be used to bind antibodies to ABx; this may result from a number of complex factors, including the partial contributions to immunoglobulin binding which results from (1) anion exchange functionalities; (2) hydrophobic interactions; (3) size exclusion mechanisms; (4) an increased ligand density; (5) shifts in the pKa values of the ion exchange ligands; and (6) the titratable nature of ABx.

Ligand density is of crucial importance in optimizing the binding of immunoglobulins to ABx, not only in terms of balancing the relative proportion of anion and cation exchange groups, but also, in terms of maximizing the number of carboxylic acid and hydrophobic ligands per unit volume of matrix (Appendix III). While the density of cation exchange groups on ABx is well above 0.4 meq/mL, the ligand density on most commercially-available high performance cation exchangers is below 0.1 or 0.2 meq/mL (13, 62).

Another factor which influences the binding of immunoglobulins to ABx is pore size (62, 63). Larger pores facilitate the binding of large nonimmunoglobulin proteins, such as α_2-macroglobulin (700 kD), which in turn, compete for the binding sites which could otherwise be occupied by IgG (62). Because α_2-macroglobulin can be a major component in crude antibody samples, its exclusion (Fig. 14, Tables 2, 3, 4) can increase the real-life capacity of ABx significantly (29, 62, 65, 71).

The role of hydrophobic interaction chromatographic mechanism in the selectivity and binding of immunoglobulins to ABx is not completely understood (62, 63, 65). However, it is possible that hydrophobic ligands might contribute to antibody binding and selectivity because immunoglobulins exhibit a highly specific affinity towards some HIC supports (Figs. 24, 25, 26). Although high salt concentrations are usually required for most proteins to be driven onto hydrophobic ligands, the selectivity and binding strength which some HIC matrices exhibit towards immunoglobulins reach a maximum at moderately low salt concentrations (Fig. 25). Furthermore, since HIC is actually a mild form of reversed phase chromatography, having the same intrinsic physicochemical basis, some hydrophobic interactions might occur on ABx even at low ionic strengths (65). Several groups (35, 62, 65) have recently used the hydrophobic nature of ABx to selectively desorb the bound, contaminating proteins with a step gradient to high $(NH_4)_2SO_4$ concentrations; under these conditions many monoclonals still remain bound, and may be eluted (free of contaminating proteins) with the use of a lower salt concentration (101).

For decades chromatographers have made use of matrices which were specifically designed for the purification of proteins (65). Although these supports were presumably designed to operate via just one chromatographic mechanism, recent studies suggests that many do exhibit mixed-mode behavior (62, 65). Unfortunately, many of these mixed-mode matrices give poor recoveries and suffer from a number of design flaws and related disadvantages (62, 65). Special care must be taken in order to reduce or eliminate the problems which are inherent to this type of support (62, 65).

Considering the problems associated with designing adequate mixed-mode matrices, the labor and fine-tuning which is required, the complex nature of proteins and their interactions with chromatographic surfaces, and, the myriad of different proteins must ultimately be purified, it is doubtful that the future will bring about a significant proliferation in the rational design of application-specific chromatographic matrices. However, it is anticipated that, for proteins of significant importance which will be required in large quantitities and at high purities, such an approach may prove to be advantageous. BAKERBOND ABx certainly represents a step in that direction.

REFERENCES

1. C.L. Allen, and L. Haff, "Preparative scale isolation of monoclonal antibodies." Poster 3705 presented at HPLC '86 — The Tenth International Symposium on Column Liquid Chromatography, May 18–23, 1986, San Francisco, CA.
2. A.J. Alpert,"New materials and techniques for HPLC of proteins." Ph.D. Dissertation, Purdue University, 1980. Conducted under the direction of F. Regnier.
3. P.U. Antle, G.B. Cox, S.I. Sivakoff, and A.P. Goldberg, "Purification of biomolecules by HPLC," BioChromatography *2*(1), 46–55 (1987).
4. P. Bailon, N. Drugazima, and A.H. Nishikawa, "Large scale purification of monoclonal antibodies from intracapsular supernatants and ascites fluids," Paper 210, presented at Biotechnology '86, January 29–30, 1986, New Orleans, LA.

5. H.F. Bazin, F. Cormont, and L. DeClerca, "Purification of rat monoclonal antibodies," *Methods in Enzymology*, Vol. 121, J.J. Langone and H. Von Vanakis, Eds. Academic Press, New York, 1986, pp. 638–652.

6. G. Bernardi, "Chromatography of proteins on hydroxyapatite," in *Methods in Enzymology*, Vol. 22, Jacoby, W.B., Ed., Academic Press, New York, 1971, pp. 325–339.

7. G. Bernardi, M.G. Giro, and G. Gaillord, "Chromatography of polypeptides and proteins on hydroxyapatite columns: Some new developments." Biochim. Biophys. Acta, *278*, 409–420 (1972).

8. C.R. Birdwell, K.G. Burnett, R.M. Bartholomew, C.L. Peterson, J.M. Levertt, T.A. Ferson, and L. Brumwell. "Analytical and semi-preparative purification of human monoclonal IgM antibody by high performance ion exchange chromatography." Paper 3702 presented at HPLC ' 86- The Tenth International Symposium on Column Liquid Chromatography, May 18–23, 1986, San Francisco, CA.

9. R.M. Brooks, P. Strickler, J. Gemski, M. Swartz, and R. Karol, "High resolution preparative separations of biological molecules by ion exchange and reverse phase chromatography." Paper presented at the Second Preparative-Scale Liquid Chromatography Symposium, May 12–13, 1986, Washington, DC.

10. T.L. Brooks, and A. Stevens, "Techniques for purifying monoclonal antibodies," American Lab, October 54–64 (1985).

11. C. Bruck, J.A. Drebin, C. Glineur, and D. Portetelle, "Purification of mouse monoclonal antibodies from ascitic fluid by DEAE Affi-Gel Blue Chromatography," *Methods in Enzymology*, Vol. 121, J.J. Langone and H. Von Vanakis, Eds., Academic Press, New York, pp. 587–596.

12. S.W. Burchiel, J.R. Billman, and T.R. Alber, "Rapid and efficient purification of mouse monoclonal antibodies from ascites fluid using high performance liquid chromatography," J. Immunol. Methods, *69*, 33–44 (1984).

13. S.W. Burchiel, "Purification and analysis of monoclonal antibodies by HPLC," *Methods in Enzymology*, Vol. 121, J.J. Langone and H. Von Vanakis, Eds., Academic Press, New York, pp. 596–615.

14. S.W. Burchiel, and B.A. Rhodes, *Radioimmunoimaging and Radioimmunotherapy*, Elsevier, New York, 1983.

15. R.F. Burgoyne, "HPLC techniques advance monoclonal antibody isolation," Research and Development, *27*, (66), 82–85 (1985).

16. R.F. Burgoyne, R.M. Brooks, M. Swartz, and R. Karol, "High resolution preparative separations of biological molecules by ion exchange and reverse phase chromatography." Paper 4104 presented at the Tenth International Symposium on Column Liquid Chromatography, May 18–23, 1986, San Francisco, CA.

17. D.J. Burke, J.K. Duncan, L.C. Dunn, L. Cummings, C.J. Sieber, and G.S. Ott, "Rapid protein profiling with a novel anion-exchange material." J. Chromatogr. *353*, 425–437 (1987).

18. G.J. Calton, "Immunosorbent separations." *Methods in Enzymology*, W. Jakoby, Ed., Academic Press, New York, 1984, pp. 104, 381–387.

19. M. Carlsson, A. Hedin, M. Inganas, B. Harfast, and F. Blomberg, "Purification in in vitro produced mouse monoclonal antibodies. A two step procedure using cation exchange chromatography and gel filtration," J. Immunol. Methods, *79* (1), 89–98 (1985).

20. P.A. Carr, Presented at HPLC '86 — The Tenth International Symposium on Column Liquid Chromatography, May 18–23, 1986, San Francisco, CA.
21. P.J. Cheng, and R.E. Morris, "New, rapid, high capacity chromatographic purification of IgG, monoclonal antibodies (MAb) directly from mouse ascites." Poster 3704 presented at HPLC '86 — The Tenth International Symposium on Column Liquid Chromatography, May 18–23, 1986, San Francisco, CA.
22. T.M. Chow, G. Ott, M. Clark, and C. Siebert, "Rapid two-step purification of monoclonal antibodies from tissue culture fluids with an automated preparative chromatograph system." Paper 123 presented at Pittsburgh Conference and Exposition on Analytical Chemistry and Applied Spectroscopy, 1987.
23. P. Clezardin, J.L. McGregor, M. Manach, H. Bonkerche, and M. Deschavanne. "One-step procedure for the rapid isolation of mouse monoclonal antibodies and their antigen binding fragments by fast protein liquid chromatography on a Mono-Q anion exchange column." J. Chromatogr. *319*, 67–77 (1985).
24. S.P. Cobbold, and H. Waldman, "Therapeutic potential of monovalent monoclonal antibodies," Nature, *308*, 460–462 (1984).
25. G. Corleone, and B. Swaminathan (unpublished).
26. A. Danielsson, and A. Ljunglof, "One-step purification of monoclonal antibodies from mouse ascites by FPLC. A screening of different techniques." Paper 806 presented at the Seventh International Symposium on HPLC of Proteins and Peptides and Polynucleotides, Nov. 2, 1987, Washington, D.C.
27. J.R. Deschamps, J.E.K. Kildreth, D. Derr, and J.T. August, "A high performance liquid chromatography procedure for the purification of mouse monoclonal antibodies," Analytical Biochem. *147*, 451–454 (1985).
28. R. Duberstein, "Scientists develop new technique for producing bispecific monoclonals," *Genetic Engineering News,* January, 22–25 (1986).
29. A.L. Epstein, G.S. Naeve, and F.M. Chen, "Comparison of Mono Q, Superose-6, and ABx FPLC for purification of IgM monoclonal antibodies," J. Chromatog. *444*, 153–164 (1988); and paper number 721 presented at the Seventh International Symposium on HPLC of Proteins, Peptides and Polynucleotides, Nov. 2–4, 1987, Washington, DC.
30. P.L. Ey, S.J. Prowse, and C.R. Jenkin, "Isolation of pure IgG_1, IgG_{2a}, IgG_{2b} immunoglobulins from mouse serum using protein A Sepharose," Immunochemistry, *15*, 429–436 (1978).
31. J.L. Fahey, and E.W. Terry, *Handbook of Experimental Immunology*, Vol. 1, D.M. Weir, Eds. Blackwell, London, 1978, pp. 8.1–8.16.
32. U-B. Fredriksson, L. Fagerstam, S. Gronlund, and P. Borwell, "Serial chromatographic process to purify monoclonal Ab from hybridoma cell culture: ion exchange on Sepharose^R and MonoBeads^{TM}." Presented at Biotech '85, Europe.
33. A-K. Frej, J-G. Gustafsson, and P. Hedman, "FPLC for monitoring microbial and mammalian cell cultures," Biotechnology, September, 777–781 (1984).
34. P. Gallo, A. Siden, and B. Tavolako, "Anion-exchange chromatography of normal and monoclonal serum immunoglobulins," J. Chromatogr. *416*, 53–62 (1987).
35. V. Garg, "Use of preparative HPLC in large scale purification of therapeutic grade proteins from mamalian cell culture." Paper number 911 presented at the Seventh International Symposium on HPLC of Proteins, Peptides, and Polynucleotides,

Nov. 2–4, 1987, Washington, D.C. and presented at the 1987 annual meeting of the Society for Industrial Microbiology, Baltimore, MD.

36. J.S. Garvey, N.E. Cremer, and D.H. Susodorf, "A laboratory text for instruction and research," *Methods in Immunology*, Addison-Wesley, Reading, MA, 223–226.
37. M.J. Gemski, B.P. Doctor, J. K. Gentry, M. J. Pluskal, and M. P. Strickler, "Single step purification of monoclonal antibody from murine ascites and tissue culture fluids by anion exchange high performance liquid chromatography," BioTechniques, *3*(5), 378–384 (1985).
38. J.W. Goding, "Use of staphylococcal protein A as an immunological reagent," J. Immunol. Methods, *20*, 241–246 (1978).
39. D.C. Heiner, "Significance of immunolobulin G subclasses", Am. J. Med., *76*, 1–6 (1984).
40. E.A. Hill, R.I. Penney, and B.A. Anderton, "Protein HPLC-fast purification of immunoglobulin G-type mouse monoclonal antibodies by FPLC ion exchange chromatography." Abstract C101 at the 14th Int. Symp. on Chromatography, London, England, 1982.
41. H. Hirano, T. Nishimura, and T. Iwamura, "High flow rate hydroxyapatites," Analytical Biochemistry, *150*, 228–234 (1985).
42. S. Hjerten, "Calcium phosphate chromatography of normal human serum and of electrophoretically isolated serum proteins," Biochem. Biophys. Acta, *31*, 216–235 (1959).
43. S.M. Hochschwender, N. Sanchez, J. Ruse, P.R. Vasquez, M.G. Nichols, G. Christiansen, R.M. Bartholomew, and T.L. Brooks, "Applications of hydroxyapatite — HPLC separation of IgG light-chain variants." Poster 3701 presented at HPLC '86 — The Tenth International Symposium on Column Liquid Chromatography, May 18–23, 1986, San Francisco, CA.
44. C. Horvath, "High performance displacement chromatography." Paper presented at the Second Preparative-Scale Liquid Chromatography Symposium, Washington, DC, May 12–13, 1986.
45. S. Ikeyama, S. Nakagawa, M. Arakawa, H. Sugino, and A. Kakinuma, "Purification and characterization of IgE produced by human myeloma cell line, U266," Molecular Immunology, *23*(2), 159–167 (1986).
46. E. Ishikawa, Y. Hamaguchi, and S. Yoshitake, "Enzyme labeling with *N,N'-o*-Phenylenedimalimide," Enzyme Immunoassay, E. Ishikawa, T. Kawai, and K. Miyai, Eds., Igaku-Shoin Ltd., Tokyo, 1981.
47. H. Juarez-Salinas, G.S. Ott, J-C. Chen, T.L. Brooks, and L.H. Stanker, "Separation of IgG idiotypes by high performance hydroxyapatite chromatography," *Methods in Enzymology*, Vol. 121, J.J. Langone and H. Von Vanakis, Eds., Academic Press, New York, 1986, pp. 615–622.
48. H. Juarez-Salinas, S.C. Engelhorn, W.L. Bigbee, M.A. Lowry, and L.H. Stanker, "Ultrapurification of monoclonal antibodies by high performance hydroxyapatite (HPHT) Chromatography," BioTechniques, May/June, 164–169 (1984).
49. H. Juarez-Salinas, W.L. Bigbee, G.B Lamotte, and G.S. Ott, "New Procedures for the analysis and purification of IgG murine MAb's," *American Biotechnology Laboratory,* March/April, 38–43 (1986).

50. L. Kagedal, H. Hansson, and M. Sparrman, "Influence of the flow rate and gradient volume of the elution pattern in metal chelate affinity chromatography." Poster 3106 presented at HPLC '86 — The Tenth International symposium on Column Liquid Chromatography, May 19, 1986, San Francisco, CA.
51. Y. Kato, K. Nakamura, and T. Hashimoto, "Evaluation of conventional and medium-performance anion exchangers for the separation of proteins," J. Chromatogr. *253*, 219–225 (1982).
52. Y. Kato, K. Nakamura, and T. Hashimoto, "New high performance cation exchanger for the separation of proteins," J. Chromatogr. *294*, 207–212 (1984).
53. T. Kawasaki, "Hydroxyapatite HPLC." Paper 503 presented at HPLC '86 — The Tenth International Symposium on Column Liquid Chromatography, May 18, 1986, San Francisco, CA.
54. L.A. Kennedy, W. Kopaciewicz, and F.E. Regnier, "Multimodal liquid chromatography columns for the separation of proteins in either the anion-exchange or hydrophobic-interaction mode," J. Chromatogr. *359*, 73–84 (1986).
55. M. Keynes, "Rapid purification of mouse monoclonal antibodies from ascites fluid by fast protein liquid chromatography," *Chromatography International*, April, 14–21, 1986.
56. A. Klausner, "Taking aim at cancer with monoclonal antibodies," Biotechnology, *4*, 194–195 (1986).
57. G. Kohler, and C. Milstein, "Continuous cultures of fused cells secreting antibody of predefined specifity," *Nature, 256*, 495–497 (1975).
58. J.J. Langone, and H. Van Vunakis "Immunochemical techniques (Part B),"*Methods in Enzymology*, Vol. 73, S.P. Colowick, and N.O. Kaplan, Eds. 21–34.
59. H.B. Levy, and H.A. Sober, "A simple chromatographic method for preparation of gamma-globulin," Proc. Soc. Exp. Biol. Med., 103, 250 (1960).
60. M. Mariani, D. Nucci, L. Bracci, and G. Antoni, "Determination of antigen sepcific immunoglobulin contents of ascitic fluids and antisera, " Hybridoma, *5*(1), 73 (1986).
61. T.H. Maugh, "Chromatography: from here to affinity." *Bio/Technology*, *3* (1984).
62. D.R. Nau, Manuscripts in preparation (see Refs. 100–104).
63. D.R. Nau, (unpublished).
64. D.R. Nau, "Optimization of monoclonal antibody purification," in *Techniques in Protein Chemistry*, T. Hugli, Ed., Protein Society 1988 Meeting Papers, pp. 339–347.
65. D.R. Nau, "Optimization of protein separations on multimodal chromatographic supports," in *HPLC of Proteins, Peptides and Polynucleotides*, M.T.W. Hearn, Ed., VCH, Inc., Boca Raton, FL (in press).
66. D.R. Nau, and J.G. Guenther, "Effective antimicrobials for the sterilization of silica-based chromatographic supports: solution to an ongoing controversy." Presented at the Sixth International Symposium on HPLC of Proteins, Peptides and Polynucleotides, October 20–22, 1986, Baden-Baden, West Germany.
67. D.R. Nau. "A unique chromatographic matrix for rapid antibody purifcation," BioChromatography, *1*(2), 82–94 (1986).

68. D.R. Nau, "ABx — A novel chromatographic matrix for the purification of antibodies," in *Commercial Production of Monoclonal Antibodies: A Guide for Scaling-up Antibody Production*, S. Seaver, Eds., Marcel Dekker, New York, 1987.
69. D.R. Nau, "Chromatographic analysis and characterization for antibodies and antibody preparations," in *HPLC Analysis, Separation, and Purification of Antibodies and Related Substances*, D.R. Nau, Ed., Marcel Dekker, New York (in preparation).
70. D.R. Nau, "Chromatographic analysis and characterization of antibodies and antibody preparations," in *Characterization and Analysis of Antibodies and Antibody Preparations*, D.R. Nau, Ed., VCH Publishers, New York (in preparation).
71. D.R. Nau, "Chromatographic purification of antibodies," in *Monoclonal Antibodies and Nucleic Acid Probes*, B. Swaminathan and G. Prakash, Eds., Marcel Dekker, New York, 1989.
72. A.R. Nazareth, C. Mello, and R.P. McPartland, "Rapid purification of monoclonal antibodies by anion-exchange HPLC." Poster 3703, presented at HPLC '86, The Tenth International Symposium on Column Liquid Chromatoraphy, May 18–23, 1986, San Francisco, CA.
73. P. O'Neil, W. Kopaciewicz, S. Fulton, and F.E. Pfannkoch, "Sample clean-up and depyrogenation of PAE-300 (anion exchange) columns." Presented at Pittsburgh Conference and Exposition on Analytical Chemistry and Applied Spectroscopy.
74. V.A.Oxelius, "Immunoglobulin G (IgG) subclasses and human disease," Am. J. Med., *76*, 7–18 (1984).
75. B. Pavlu, U. Johansson, C. Nyhlen, and A. Wichman, "Rapid purification of monoclonal antibodies by high performance liquid chromatography," J. Chromatogr. *359*, 449–460 (1986).
76. J. Porath, "Thiophilic interaction and the selective adsorption of proteins," Trends in Biotechnology, *15*, 225–229 (1987).
77. F.E. Regnier, "High performance liquid chromatography of biopolymers: the present and future." Paper 657, presented at the Pittsburgh Conference and Exposition on Analytical Chemistry and Applied Spectroscopy, March 10–14, 1986, Atlantic City, NJ.
78. J. Richey, "FPLC: A comprehensive separation techique for biopolymers," American Lab., Oct. 104–129 (1982).
79. J. Ritz, and S.F. Schlossman, "Utilization of monoclonal antibodies in the treatment of leukemia and lymphoma," Blood, *59*, 1–11 (1982).
80. A.H. Ross, D. Herlyn, and H. Koprowski, Journal of Immunological Methods, *102*(2), 227–232 (1987).
81. M.N. Schmuck, D.L. Gooding, and K.M. Gooding, "Comparison of porous silica packing materials for preparative ion-exchange chromatography," J. Chromatogr. *359*, 323–330 (1986).
82. W.E. Schwartz, F.M. Clark, and I.B. Sabran, "Process scale isolation and purification of immunoglobulin G," LC-GC Magazine, *4*(5), 442–448 (1986).
83. S.S. Seaver, J.L. Rudolph, and J.E. Gabriels, Jr., "Amino acid analysis of cell cultures." Biotechniques, Sept./Oct. 254–260 (1984).

84. F. Shakib, and D.R. Stanworth, "Human IgG subclasses in health and disease," La Ricerca Clin. Lab., *10*, 561–580 (1980).
85. S. Sivakoff, "Analytical to preparative-scale purification of biomolecules using novel surface-stabilized ion exchange packing materials." Lecture 347 Presented at Pittsburgh Conference for Analytical and Applied Spectroscopy, Atlantic City, NJ, March, 1987.
86. S.I. Sivakoff, "HPLC analysis and purification of antibody conjugates," BioChromatography, *1* May–June 42–48 (1986).
87. D. Svaril, *Methods of Enzymatic Analysis*, J. Bergmeyer and M. Grabl, Eds., VCH Publishers, Weinheim, 1986, pp. 60–71.
88. G.J. Smith, R.D. McFarland, H.M. Reisner, and G.S. Hudson, "Lymphoblastoid cell-produced immunoglubulins: preparative purification from culture medium by hydroxyapatite chromatography," Anal. Biochem. *141*, 432–436 (1984).
89. D. Southern, and D. Hollis, "Fast affinity chromatography," Bio/Technology, *4*(6), 519–521 (1986).
90. M. Spencer, and M. Grynpas, "Hydroxyapatite for chromatography 1. physical and chemical properties of different preparations," J. Chromatogr. *166*, 423–434 (1978).
91. T. Staehelin, D.S. Hobbs, H.-F. Kung, C.-Y. Lai, and S. Pestka, "Purification and characterization of recombinant human leukocyte interferon (IFLrA) with monoclonal antibodies," J. Biol. Chem. *256*, 9750–9754 (1981).
92. A. Stevens, T. Morrill, and S. Parlante, "Rapid micropreparative purification of proteins and peptides," BioChromatography, *1*(1), 50–54 (1986).
93. M.E. Suart, M.J. Tomany, D.J. Phillips, and T.L. Tarvin, "Isolation and purification methodologies using new silica-based ion exchange media," BioChromatography, *2*(1), 38–45 (1987).
94. A.P. Tice, and F.E. Regnier, "Rapid concentration and desalting of column fractions following multimodal chromatographic purification of immunoglobulin G," Amicon Lab News, Spring 1–6 (1986).
95. S.G. Tompson, *Clinical Chemistry Theory Analysis and Correlation*, L.A. Kaplan and A.J. Pesc, Eds. The C.V. Mosby Co., St. Louis, MO, 1984, pp. 211–239.
96. A.R. Torres, S.C. Edberg, and E.A. Peterson, "Preparative HPLC of proteins on an anion exchanger using unfractionated carboxymethyldextran displacers." Poster 1502 presented at HPLC '86 — The Tenth International Symposium on Column Liquid Chromatography, May 18–23, 1986, San Francisco, CA.
97. R.R. Walters, "Affinity chromatographic rate constant measurements using the split-peak method." Paper 11 presented at the Twenty-fourth Eastern Analytical Symposium, November 19–22, 1985.
98. R. Wang, W.F. Bermudez, R.L. Saunders, W.A. Present, R.M. Bartholomew, and T. Adams, "The TANDEM™ PAP system: A simplified radioimmunometric assay empoloying monoclonal antibodies to prostatic acid phosphatase," Clin. Chem. *27*, 1063–1069 (1981).
99. L. Wofsy, and B. Burr, "A use of affinity chromatography for the specific purification of antibodies and antigens," J. Immunol. *103*, 380–383 (1969).
100. D.R. Nau, "Chromatographic methods for antibody purification and analysis," BioChromatography, *4*(1), 4–18 (1989).

101. D.R. Nau, "Optimization of mobile phase conditions for antibody purification on a mixed-mode chromatographic matrix," BioChromatography, *4*(3), 131–143 (1989).
102. D.R. Nau, "Effects of mobile phase conditions on protein conformation and chromatographic selectivity in ion exchange and hydrophobic interaction chromatography," BioChromatography, *4*(2), 62–68 (1989).
103. D.R. Nau, "Optimization and scale-up of linear and step gradient elution technieques for preparative antibody purification on a mixed-mode matrix," BioChromatography, *4*(5), 1–14 (1989).
104. D.R. Nau, "Optimization of the hydrophobic interaction chromatography purification of monoclonal antibodies," BioChromatography (in press).

CHAPTER 16

Characterization of Synthetic Polypeptides by Mass Spectrometry

DANIEL R. MARSHAK

Cold Spring Harbor Laboratory, Cold Spring Harbor, New York

BLAIR A. FRASER

Chemical Biology Laboratory, Division of Biochemistry and Biophysics, Food and Drug Administration, Bethesda, Maryland

1. INTRODUCTION

Many of the goals of modern biotechnology focus on the production of polypeptides that are physiologically or antigenically active. Such polypeptides can be synthesized by biological or chemical means. Biologically synthesized polypeptides may be isolated as natural products, products expressed in cells containing clonal DNA, or products of enzymatic condensations (73). Chemically synthesized polypeptides may be produced by liquid phase condensations, condensations on a solid support, or combinations of liquid and solid phase methods. Regardless of the synthetic mode, polypeptides that have important biological roles should be fully characterized using a multidisciplinary approach to structural analysis (47).

Depending on the molecular weight of the polypeptide, different strategies need to be employed for purification and characterization (46). However, all of these strategies capitalize on fundamental techniques of purification and structural analysis. The desired polypeptide is usually purified by a combination of chromatographic and electrophoretic methods, although some polypeptides may be crystallized. The purification of large polypeptides biologically synthesized in a prokaryotic or eukaryotic expression systems often involves the scale up of methods such as ultrafiltration, ion exchange chromatography and affinity chromatography. Characterization of these products usually involves fragmentation by enzymatic or chemical means and structural analysis of these fragments. Particular attention must be paid to post-translational modifications including glycosylation, acylation, methylation, and proteolysis. The purification of chemically synthesized polypeptides often involves solvent extractions and column chromatography. Complete characterization of these products must make use of techniques to evaluate amino acid compositin and sequence, chemical composition, chemical modifications and stereochemistry. In characterizing these chemically synthesized peptides, fast atom bombardment mass spectrometry complements currently available chromatographic and analytical methods.

In this chapter, we will illustrate how peptides synthesized by solid phase methods can be characterized throughout purification by using fast atom bombardment mass spectrometry. Once purified, these synthetic peptides can then be more fully characterized using amino acid analysis, sequential Edman degradation, exopeptidase digestions, and mass spectrometry. In many cases, mass spectrometric analyses contribute information that is difficult or impossible to obtain by other methods.

2. PEPTIDE SYNTHESIS

2.1. Perspective

The central problem in the chemical synthesis of polypeptides is the repetitive formation of peptide bonds in high yield while maintaining the integrity of the amino acid side chains. The first major advance in polypeptide synthesis was the formation of peptide bonds by the condensation of acid azides (29) or acid chlorides (32) with primary amines. Such condensations had limited usefulness in the formation of the defined amino acid polymers until the 1930s. At that time, the second major advance in peptide synthesis came as a result of the introduction of urethane protecting groups such as benzyloxycarbonyl (9). Subsequent improvements of coupling methods included the introduction of mixed anhydrides (23–26), activated esters (12, 72), and coupling reagents (2, 62). These advances culminated in the landmark synthesis of oxytocin by du Vigneaud and co-workers (30), and the subsequent development of analogues of this and other peptide hormones.

The third major advance in peptide synthesis was the innovation of insoluble supports for the growing peptide chain, as conceived by Merrifield (49). This technique allows extensive washing of the peptide on the support to remove reagents and solvents, thus obviating the need for precipitation or purification at intermediate steps. Solid phase peptide synthesis does demand very high coupling efficiency at each step ($>99\%$) in order to minimize deletion products in the final peptide. In addition, removal of the peptide from the support requires strong acids that may lead to undesirable products of side reactions. Nonetheless, solid phase synthesis is an efficient and effective method for assembling proteins of substantial size. For example, Kent and colleagues (27) have reported the synthesis of a biologically active lymphokine of 140 residues using automated, solid phase, stepwise condensations.

Future advances in peptide synthesis must include: (1) the synthesis of large proteins (200 to 1000 amino acid residues), (2) the synthesis of modified proteins, such as glycosylated, acylated or phosphorylated proteins, and (3) the assembly of proteins with directed, complex, disulfide bond formation. These issues are essential to the ultimate goal of chemical synthesis of biologically active proteins. Such syntheses may provide a cost-effective and more efficient alternative product to proteins produced from cloned DNA fragments.

2.2. Methodology

The general methodologies that are currently employed for chemical synthesis of proteins are broadly classified as liquid phase methods and solid phase methods. Both strategies are effective for successful syntheses, but each affords certain advantages and disadvantages to the synthetic chemist and product manager.

Liquid phase synthesis affords the advantage of the purification of intermediates. Two general strategies have been employed: segment condensation and stepwise synthesis. Segment condensation methodologies generally employ the activation of short (3 to 10 residue) peptides at the carboxy terminus using, for example, an azide procedure (38). Assembly of these peptide segments and the isolation, purification, and characterization of the intermediates is a labor intensive occupation, but homogeneous, biologically active peptides have been made in high yield (75). An alternative to segment condensation is stepwise synthesis. Protected peptide intermediates must be isolated by precipitation at each step, so this method appears to be useful for peptides up to 30 amino acids (17, 53). Liquid phase synthesis may also be accomplished using soluble polymers, such as polyethylene glycol (53). Such polymeric supports may extend the maximal length of peptides synthesized in the liquid phase, but these methods have limited usefulness for long syntheses. Thus, the major restrictions of liquid phase synthesis are: (1) the length of the polypeptide, (2) the time and labor involved in purification of intermediates, and (3) slow kinetics of bond formation. The main advantages include the high purity of the final product and the possibility of nearly infinite scale up for production of peptides.

Solid phase peptide synthesis has gained wide acceptance as a useful method for both manual and automated procedures. These methods have been extensively reviewed (5), as have the major side reactions in peptide synthesis (16). To accurately and rapidly interpret data obtained from a combined approach of HPLC and mass spectrometry, it is important to understand the chemical basis of the peptide synthesis as well as the potential side reactions during the preparation. As a guide in these interpretations, we have summarized below some of the most widely used methods in solid phase synthesis.

2.2.1. Attachment

The carboxy-terminal amino acid is covalently attached to a solid support, usually cross-linked polystyrene resins functionalized with a benzyl ester handle, such as the phenylacetamidomethyl (PAM) group. Such a benzyl ester can be hydrolyzed in strong acids including HBr, HF, or trifluoromethane sulfonic acid

(TFMSA), to leave a free C-terminal carboxylic acid. For preparation of the C-terminal α-carboxamides, the *p*-methylbenzylhydrylamine resin has been quite useful (54). Acidolytic cleavage results in release of the α-carboxamide linkage. The finding that many physiologically active peptides from the gut and brain have α-carboxamides (69) has increased the general interest in the use of benzylhydrylamine resins.

2.2.2. Protection

N-α-protected amino acids are added to the derivatized resin through activation of the α-carboxylate of the amino acid. Sequential additions of amino acids requires that the α-amino group of the anchored peptide be free, and that the amino acid to be attached have *N*-α-protection as well as side chain protection. The *N*-α-protecting group must be labile to conditions under which the anchorage to the resin as well as the side chain protection are stable. Under ideal conditions, 100% of the *N*-α-protecting group could be removed with no loss of side chain protection or resin attachment. Such ideal conditions have been referred to as an "orthogonal" synthetic scheme by Barany and Merrifield (5). In practice, conditions are not completely orthogonal, and several side reactions affecting yield can occur. The *N*-α-protecting groups are usually acid labile urethanes, such as tert-butyloxycarbonyl (t-Boc) or benzyloxycarbonyl (Z). These groups are labile to acidolytic cleavage under conditions in which the side chains are relatively stable. However, some deprotection of protecting groups, such as benzyl esters, can occur, resulting in side chain modification. In addition, alkylatin of indoles can occur by reaction of the t-Boc group with unprotected tryptophan. Recently, Sheppard and colleagues (4) have introducted FMOC (9-fluorenylmethyl-oxycarbonyl) protected amino acids that allow removal of the *N*-α-protecting group under basic conditions.

2.2.3. Coupling

Three major methods of peptide bond formation are currently used in solid phase synthesis procedures: (1) direct coupling using coupling reagents, (2) preformed symmetric anhydrides, and (3) activated esters. Direct coupling employs a coupling agent, such as a carbodiimide, that is soluble in organic solvents. Such reagents include diisopropyl and dicyclohexyl carbodiimides dissolved in dichloromethane or dimethylformamide. The coupling reagent is added directly to a mixture of the protected amino acid and the deprotected resin at $pH > 8$ in organic solvent. During the condensation reaction, the urea resulting from hydration of the carbodiimide reagent (e.g., diisopropyl urea or dicyclohexyl urea) is formed, and must be removed by extensive washing. Preformation of symmetric anhydrides of protected amino acids can be achieved by incubation of the coupling reagent with the amino acid, followed by removal of the urea by-product. The concentrated symmetric anhydride reacts faster than direct coupling methods, although formation of the anhydride minimally requires a twofold molar excess of amino acid derivative over resin substituent levels. In some cases, symmetric anhydrides can be unstable, rearranging to form mixed anhydrides. For example, amides such as glutamine and asparagine are not reliably coupled through the α-carboxylate by anhydride formation. In these cases, activated esters are quite useful. Reagents for activating the carboxylate include 1-hydroxybenzotriazole (43), aryl esters [e.g., *p*-nitrophenyl esters, (13)] tetrazole (15), and, more recently, pentafluorophenyl esters (4). Activated esters also can be used in place of anhydrides in synthetic schemes for all amino acids. The newer reagents such as hydroxybenzotriazole and tetrazole appear to accelerate the coupling reactions through catalysis of acylation (14). Thus, effective coupling schemes should include both activated ester and anhydride methods, depending on the particular reaction, to optimize the kinetics and yield of peptide bond formation.

2.2.4. Cleavage and Deprotection

Two general methods have been employed for cleavage of peptides from solid support and/or deprotection of amino acid side chains: catalytic reduction and acidolytic cleavage. Catalytic hydrogenolysis is a useful method for deprotection, although care must be taken to protect the methionyl thioether. However, for cleavage and deprotection of peptides in a single, rapid step, acidolytic treatment has become widely accepted. Inorganic acids, often used for these procedures, include HBr or HF doped with scavenger molecules, such as anisole, to protect the peptide from realkylation by protecting groups (5, 59). In addition, aryl compounds may quench the multiple free radical moieties produced from anhydrous HF. An alternative to this procedure is the use of organic acids, particularly trifluoromethane sulfonic acid (TFMSA), introduced by Yajima and Fujii (71). This method has the advantages of nearly infinite scale up potential and the use of normal laboratory glassware for the reaction. In contrast, HF must be contained in fluorocarbon plastic containers to avoid harmful contact with the researchers. TFMSA has become more useful with the introduction of new arginyl and histidinyl protecting groups, such as mesitylene sulfonate, that are quite labile in these acids. Ultimately, optimal cleavage and deprotection schemes may include a combination of catalytic hydrogenation and acidolytic steps for selective deprotection of residues in the peptide. The use of *N*-α-protecting groups that are base labile, such as FMOC, have permitted the development of attachment protocols that are designed to avoid HF or TFMSA at the cleavage step. For example, the *p*-hydroxymethyl phenoxyacetic acid (HMP) resin can be cleaved in 95% aqueous trifluoroacetic acid (4). This and other new developments in resin technology may help avoid some of the problems of side reactions occurring due to strong acid treatments.

2.3. Purification

The major considerations in selecting purification methods for synthetic peptides and proteins are the amount and the purity of product required. Both of these requirements depend on the usage of the peptide, including immunological, physiological, and biochemical applications.

The two main issues in peptide purification are: (1) the removal of nonpeptide material, including the resin, scavengers, and free protecting groups derivatives, and (2) the separation of the desired peptide from other peptidaceous material, including deletion products, chemically modified products, branched products, and termination products. The initial purification can be accomplished by extraction of the peptide with organic solvents, followed by gel filtration chromatography to remove small molecules. The second purification can be accomplished by HPLC on silica supports in a reversed phase, ion exchange, or hydrophobic interaction mode. Large-scale, reverse-phase HPLC is becoming widely accepted as a rapid means of purification of peptides on the gram level. Alternatively, large scale purification can be achieved by differential solvent extraction using a countercurrent apparatus. This procedure is most useful for repeated, large scale (gram to kilogram quantities) isolation of the same peptide.

Synthetic peptides and peptide–protein conjugates can be used effectively as antigens for the production of site-directed antibodies. In most cases, milligram amounts of peptide are more than adequate for immunizations, but additional peptide may be needed for radioimmunoassays or solid phase immunoassays. Peptide variants containing tyrosinyl residues can be radioiodinated and used as ligands in radioimmunoassay (57). As immunogens, unpurified synthetic peptides can be used, but a large number of antibodies may develop that react with impurities in the sample. If antibodies are prepared with crude peptide, then it is imperative to screen these antibodies using a highly purified, homogeneous peptide or protein in order to avoid false positive reactions. Alternatively, it is often feasible to prepare a small amount of highly purifed peptide for use as an immunogen, while larger preparations can be accumulated during the immunization period in the animal.

Synthetic peptides are also used in physiological and biochemical tests. For example, synthetic hormones or analogs of hormones are injected into animals to measure physiological responses. Under these conditions, it is imperative to prepare large quantitites of highly purified products, since slight modifications or deletions may result in the antagonism or agonism of the desired response. Thus, a 1% contaminant that is 1000 fold more potent than the major product could be responsible for the observed response. Similarly, in biochemical assays, peptides that are used as enzyme substrates must be highly purified to avoid antagonistic effects of substrate analogs. Kinetic measurements (K_M, v_{max}) of enzymes using peptide substrates are invalid unless the peptide used has been documented as homogeneous and unmodified.

3. MASS SPECTROMETRY

3.1. Perspective

Mass spectrometry has supplemented traditional methods of peptide structural analysis for over 20 years, but only recently have desorption ionization methods facilitated the analysis. Advances in desorption ionization, most notably plasma desorption (PD) developed by MacFarlane (48), fast-atom-bombardment (FAB) ionization developed by Barber et al. (6) and mass analyzer design have made possible the direct analysis of underivatized polypeptides as large as trypsin (MW = 23,643) (6, 21, 65, 66).

In the past, peptides presented a problem for mass spectrometry because they are polar, labile, involatile polymers. To be analyzed by mass spectrometry, peptides needed to be derivatized. The new desorption ionization methods have afforded mass spectral analysis of underivatized peptides at nanomolar and picomolar concentrations (20). Gained from the mass spectrum can be additional information that may be integrated into a multidisciplinary approach to structurally analyzing a polypeptide, be it from biosynthetic or chemical synthetic origin.

3.2. Strategy

Mass spectral analysis of biosynthetic polypeptides and proteins requires a strategy exploiting techniques that will also be used when analyzing small, chemically synthesized peptides. Depending on the mass limit of the mass spectrometer, the molecular weight for the protonated protein can be obtained (6, 21, 65). Coupled with amino acid analysis and partial Edman degradation, the mass spectrum can provide some indication that the desired polypeptide is present. In most cases, this direct approach may not be possible and an indirect approach will be necessary. Fragmentation of the intact protein by chemical and enzymatic means is necessary to obtain fragment peptides. The mixture of fragment peptides can be directly analyzed by mass spectrometry, (i.e., FAB-mapping), (53), or the digestion mixture can be partially fractionated by HPLC with each fraction then being characterized by mass spectrometry (33, 35), amino acid analysis, and amino acid sequence analysis. Further HPLC purification of the peptide pools may be necessary depending on any redundant amino acid sequences in the native protein. Both of these approaches have been used to characterize protein products of recombinant DNA technology, [e.g., human interleukin-2 (33), insulin (7)], as well as naturally occurring proteins, [e.g., hemoglobin (55), lysozyme (68), and tRNA synthetases (10, 35)].

Choice of the mass spectrometry method desired for the task is important in the overall strategy for characterizing the polypeptide. Plasma desorption mass spectra display recognizable protonated molecular weights of the polypeptides with less clear fragment ion intensities (48, 65, 66). Direct connection of the HPLC to the mass spectrometer through a thermospray interface (11) allows on-line analysis of the HPLC effluent and also provides molecular weight information for each component (42). However, for a multidisciplinary approach, preparative or semipreparative HPLC of the polypeptide with collection of fractions may be more useful. A portion of each fraction that is collected can then be characterized by amino acid analysis (37), mass spectrometry (39, 71), sequential Edman degradation (40), enzymatic digestion (18), and stereochemical methods (44, 45). Each method has its shortcomings but together they will provide a complementary image of the peptide's structure.

For chemically synthesized peptides, FAB-mass spectrometry has rapidly become a standard tool (1, 21, 22, 51, 71, 76). It can be used throughout the entire synthetic process to characterize the protected amino acids (34), the crude peptide mixture (71), chromatographic fractions obtained during purification (71), and the desired peptide or peptide analog (51, 71). Characterization of these materials will afford better understanding of the problems of chemical synthesis and thereby better synthetic strategies can be designed.

3.3. Methodology

To obtain the mass spectrum for a peptide, it is first ionized. The ionized peptide, usually as the protonated molecular ion, then dissipates this excess energy by bond fission giving rise to daughter ions and neutral fragments (21, 56). The protonated molecular ion and these daughter ions are then detected and their masses and relative abundances are displayed as a histogram or table, the mass spectrum. The peptide bond is the most common site of fragmentation. Two fragment ion patterns, important for discerning the amino acid sequence, are then obtained. The first pattern would arise from peptide bond fission with retention of the positive charge on the amino terminal fragment ion and loss of the carboxy terminus as a neutral fragment, yielding predominantly the acylium and aldiminium ions. The second pattern would arise from peptide bond fission with retention of the positive charge on the carboxy terminal fragment ion and loss of the amino terminus as a neutral fragment, yielding predominantly the ammonium ion series. Each peptide bond is a likely site for fission, depending upon the particular peptide bond strength. The two patterns of fragment ion intensities in the mass spectrum can then be used to infer a proposed structure for the peptide. This interpretation is by no means a unique interpretation but should be consistent with the other chemical data obtained for the peptide.

This process of ionization of the peptide, followed by bond fission, to obtain the mass spectrum is a random process. There is no sequential operation that is performed on the peptide, as with sequential Edman degradation (40) or exopeptidase digestion (58), to yield arrangement of the amino acids in the peptide. Efforts are being made to provide sequence data for peptides by mass spectrometry but these efforts need compositional analysis, i.e., amino acid analysis, and some form of sequential operation, for example, sequential Edman degradation or exopeptidase digestion, to be completely successful.

Certain ambiguities arise in the mass spectrum that need to be resolved by other methods. For example, isoleucine, leucine, and hydroxyproline have identical nominal masses. Edman degradation and amino acid analysis are useful in resolving these amino acid residues. Lysine and glutamine have identical nominal masses. Acetylation of the lysine or deuterium exchange (20, 36, 61) is useful for discerning these in the mass spectrum. Other redundant residue masses for either amino acids or dipeptide combinations, as illustrated in Table 1, can confuse spectral interpretation. Most notable are those that may arise during peptide synthesis, such as β-aspartimide, methyl-S-methionine, diketopiperazine formation, and oxidized residues. These and other redundances need to be considered when interpreting the mass spectrum for a synthetic peptide.

Prosthetic groups, introduced intentionally or unintentionally, may not behave as predicted under fast atom bombardment conditions. Loss of sulfate from a peptide under positive ionization conditions necessitates analysis under negative ionization (50). Reduction of sensitive residues, most natably cystines (19, 21), requires careful examination of several sequentially acquired mass spectra. Dehalogenation may occur during analysis (60). Peptide mixtures may display equivocal mass spectra (20, 28). Cyclic peptides may not display sufficient fragment ion intensities to allow interpretation (76).

In immunochemical studies, chemically synthesized peptides can either be used as haptens in eliciting antisera (67) or as competitive inhibitors of antigen binding to antibody (3). The purity requirements for each of these tasks may be different. Incomplete deprotection may not have as much effect on eliciting an antibody response as it may have in inhibiting the binding of a protein antigen to its antibody. Modification of certain amino acid residues at antigenic sites on myoglobin have been shown to decrease the antigenic reactivity (3). A chemically synthesized peptide spanning that antigenic site may not be a useful inhibitor of binding if it is not completely deprotected, subsequently purified, and then characterized. Chemically synthesized peptides used as biological effectors, such as neuropeptides or endocrine peptides, need to be free of chemical modifications since these chemical modifications may impart undesirable agonist or antagonist properties.

4. EXAMPLES

Our goal in the remainder of this chapter is to present examples from our laboratories of synthetic peptides that have been instructive in evaluating biological problems. These examples illustrate the use of mass spectrometry in the characerization of these pepetides. The problems described in the preceding sections are evaluated in these examples using FAB-MS.

4.1. Example 1: Analysis of a Bioactive Peptide by HPLC and FAB-MS

An example of the use of FAB-MS in peptide synthesis is illustrated in Fig. 1. The decapeptide shown corresponds to the structure of the gonadotropin releasing hormone from the brains of lampreys, a primitive vertebrate species of fish

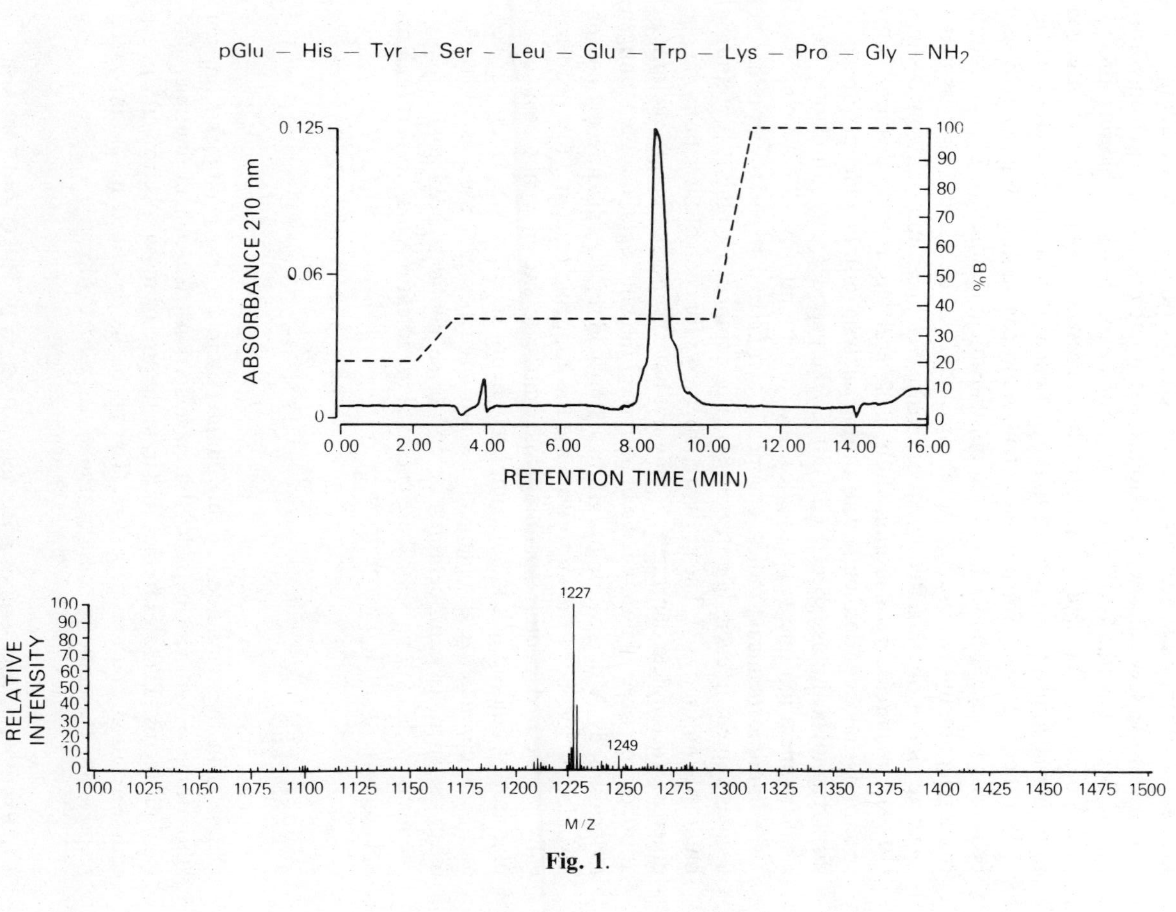
pGlu — His — Tyr — Ser — Leu — Glu — Trp — Lys — Pro — Gly — NH2
ABSORBANCE 210 nm
0.125
0.06
0
%B
100
90
80
70
60
50
40
30
20
10
0
0.00 2.00 4.00 6.00 8.00 10.00 12.00 14.00 16.00
RETENTION TIME (MIN)
RELATIVE INTENSITY
1227
1249
1000 1025 1050 1075 1100 1125 1150 1175 1200 1225 1250 1275 1300 1325 1350 1375 1400 1425 1450 1475 1500
M/Z

Fig. 1.

(63). Following elucidation of the structure of the peptide hormone natural product, we wanted to chemically synthesize the peptide in order to test biological activity. The peptide was synthesized using preformed symmetric anhydrides of t-Boc amino acids on *p*-methyl benzhydrylamine resin using an automated procedure (Applied Biosystems 430A). The tryptophan was not protected as the formyl-derivative. The final purification step employed an isocratic elution step on a reversed phase column. Although some peak asymmetry is evident, the major peak had the amino acid composition and mass of the expected product $(M + H)^+ = 1227$ mu. No evidence was found for any oxidation or alkylation of the indole ring of the tryptophan. A small intensity at $m/z = 1249$ represents the sodium attachment ion of the intact molecular species $(M + Na)^+$. When the product was tested in female lampreys, ovulation and increased plasma steroids were observed at dosages comparable to the natural product (63, 64). Thus, the combined approach of HPLC, FAB-mass spectrometry, and bioassay demonstrated the purity and activity of the synthetic product.

4.2. Example 2: Modifications in Hydrogen Fluoride

The undecapeptide H-Leu-Gln-Val-Glu-Ile-Val-Pro-Ser-Gln-Gly-Glu-OH was synthesized by automated procedurs on PAM-resins and deprotected using HF/anisole (10:1). Upon purification of the peptide by reversed phase HPLC, a complex pattern of peaks was observed (Fig. 2). Fraction 1 appeared to be a major double peak eluting between several smaller contaminants. Fraction 2 contained three major peaks, and eluted at a higher proportion of organic solvent than did fraction 1. FAB-MS analysis of the original mixture indicated that the dominant species had the expected *m*/*z*, 1199. However, several other intensities were noted. After HPLC, fraction 1 contained the expected peptide $(M + H)^+ = 1199$ mu, along with an intensity at $m/z = 1253$. This corresponds to an addition of 54 mu to the molecular ion. In addition, lower level intensities at $m/z = 1217$ and 1235 were observed. Taken together, these intensities arise at intervals of 18 mu above the $(M + H)^+$, and may correspond to substitutions of $F\cdot (M = 19$ mu) for $H\cdot (M = 1$ mu). Such substitutions might be envisioned to occur by a free radical mechanism, perhaps the generation of $F\cdot$ during the HF procedure. Free radical substitution would be favored at tertiary carbon atoms, such as those in valine and isoleucine in the peptide.

Fig. 1. Analysis of synthetic lamprey gonadotropin releasing hormone by HPLC and mass spectrometry. The amino acid sequence of the peptide is shown at the top. The upper graph shows the analysis of the peptide by HPLC on a reversed phase (C_{18}) column using as a mobile phase 0.1% trifluoroacetic acid and increasing proportions of acetonitrile. The absorbance of the effluent at 210 nm is indicated by the solid line and the percentage acetonitrile is indicated by the dashed line. The lower histogram displays a partial FAB-mass spectrum of the peptide from $m/z = 1000$ to 1500. The signals with highest relative intensity are numbered with nominal mass.

Leu — Gln — Val — Glu — Ile — Val — Pro — Ser — Gln — Gly — Glu (M + H⁺ = 1199)

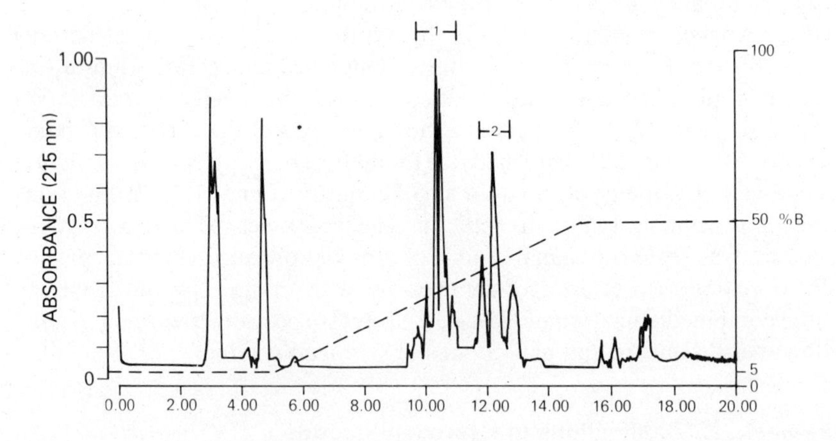
1
2
1.00
0.5
0
ABSORBANCE (215 nm)
100
50 %B
5
0
0.00
2.00
4.00
6.00
8.00
10.00
12.00
14.00
16.00
18.00
20.00

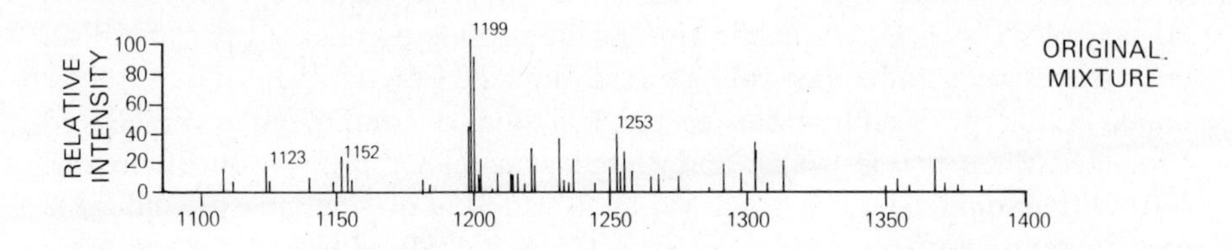
ORIGINAL MIXTURE
RELATIVE INTENSITY
1199
1253
1123
1152
1100
1150
1200
1250
1300
1350
1400

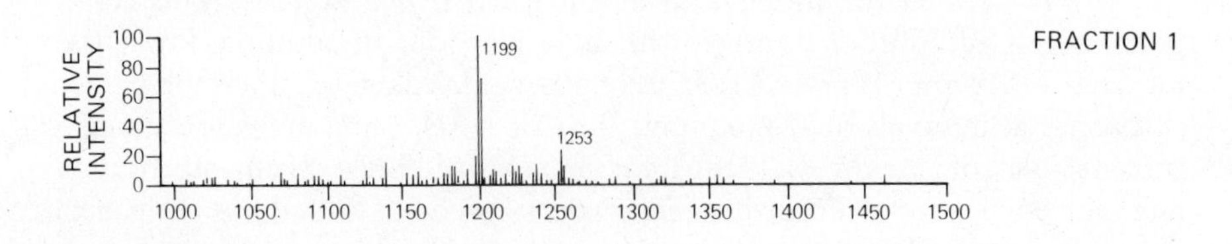
FRACTION 1
RELATIVE INTENSITY
1199
1253
1000
1050
1100
1150
1200
1250
1300
1350
1400
1450
1500

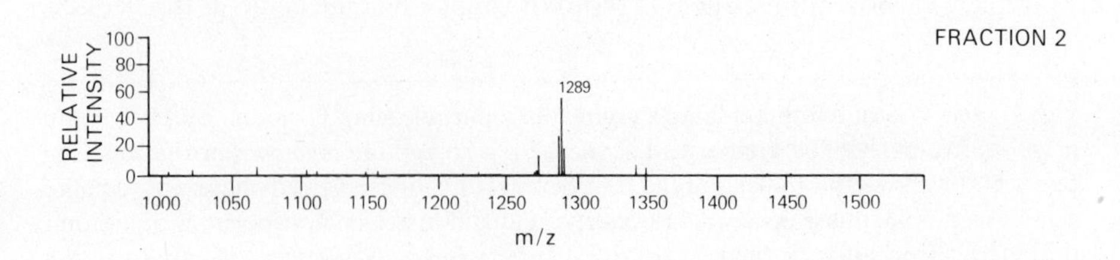
FRACTION 2
RELATIVE INTENSITY
1289
1000
1050
1100
1150
1200
1250
1300
1350
1400
1450
1500
m/z

It is possible, therefore, that substitution of F for H occurs in those peptides containing multiple valinyl, isoleucinyl or leucinyl residues, resulting in small amounts of perfluoridated products. Such products are important to diagnose when peptides are used for immunological procedures, since fluoridated haptens are highly antigenic (31).

Fraction 2 showed an intensity at $m/z = 1289$, an increase of 90 mu over the protonated molecular ion at $m/z = 1199$. This is consistent with the mass of an anisole-peptide adduct in which anisole is esterified to a glutamine or glutamate side chain. The obvervation that there are three species by HPLC, but only a single intensity at $m/z = 1289$ suggests that there are three different adducts formed, all having identical mass. In this case, anisole most likely attached to the glutamine-2, glutamic acid-4, or glutamine-9 side chains. It is imperative that such adducts be removed from peptide mixtures since the anisole adduct may contribute additional immunological or physiological activities.

4.3. Example 3: Modification in Trifluoromethane Sulfonic Acid

In the preceding example, it was clear that cleavage and deprotection of the synthetic peptide in HF/anisole resulted in several products, namely fluoride and anisole attachment products. Acidolytic deprotection by sulfonic acids (74) may be used to avoid these side reactions, but other problems may arise. An example of these difficulties is illustrated in Fig. 3.

The peptide H-Ser-Phe-Glu-Asp-Lys-His-Gly-Leu-Ala-NH_2 was synthesized by solid phase methods using preformed symmetric anhydrides and *p*-methyl benzhydrylamine resin. The peptide was treated with 1M TFMSA in TFA containing thioanisole and *m*-cresol as scavangers, at 0°C for 3 hr. After reaction, the peptide was isolated by precipitation twice in ice cold diethyl ether. The precipitate was lyopholized and further purified by HPLC, as shown in Fig. 3. Using a series of gradient and isocratic steps during the elution, four major products were isolated. Product A had no apparent peptide material by amino acid analysis, while products B, C, and D had identical amino acid compositions that were all consistent with the desired product. Mass spectral analysis revealed that product B contained the desired peptide at $m/z = 1002$. Product C had a major intensity at $m/z = 984$, 18 mu lower than the desired product. Detailed

Fig. 2. Analysis of an undecapeptide treated with HF/anisole by HPLC and mass spectrometry. The amino acid sequence of the peptide is shown at the top. The upper graph shows the HPLC analysis of the crude peptide mixture on a reversed phase (C_{18}) column using as a mobile phase 0.1% trifluoroacetic acid and increasing proportions of acetonitrile. The absorbance of the effluent at 210 nm is indicated by the solid line, and the percentage acetonitrile is indicated by the dashed line. The abscissa list the retention time in minutes following injection. The fractions indicated were collected and subjected to further analysis. The lower histograms display partial FAB-mass spectra from $m/z = 1000$ to 1500 of the original mixture and the two fractions as shown in the HPLC analysis. The signals with highest relative intensity are numbered with nominal mass.

1 5
Ser– Phe– Glu – Asp – Lys – His – Gly – Leu– Ala – NH_2 ($M + H^+ = 1002$)

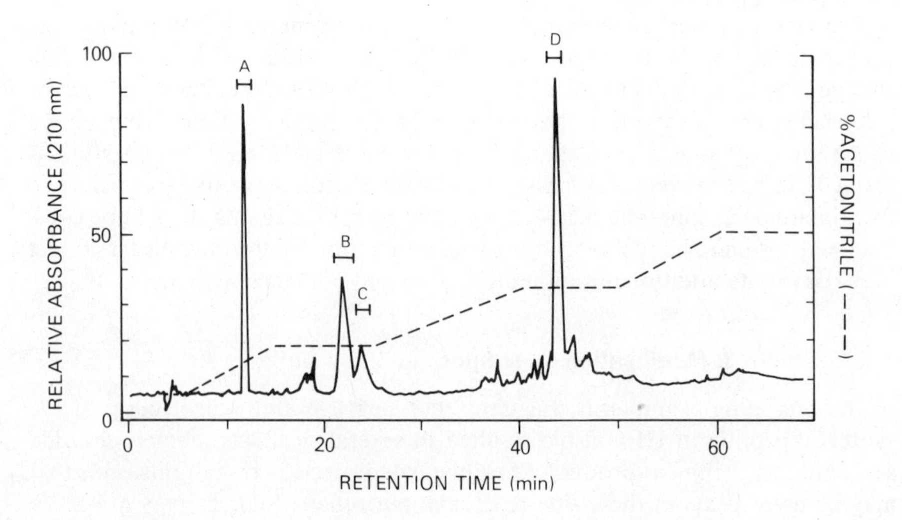

PRODUCT	MASS $(M + H)^+$	STRUCTURE
A	<650	UNKNOWN
B	1002	DESIRED PEPTIDE
C	984	ASPARTIMIDE –HN–HC(H_2C–C(=O))–C(=O)–N–
D	1156	im-Tosyl H_3C–C₆H₄–S(=O)(=O)–N(imidazole)N

Fig. 3. Analysis of a nonapeptide treated with TFMSA/TFA separated by HPLC and characterized by FAB-mass spectrometry. The amino acid sequence of the peptide is shown at the top. The graph shows the HPLC analysis of the peptide on a reversed phase (C_{18}) column using as a mobile phase of 0.1% trifluoroacetic acid and increasing proportions of acetonitrile. The solid line indicates the absorbance of the effluent at 210 nm, and the dashed line indicates the percentage acetonitrile. The fractions labeled A though D were collected and subjected to further analysis. The table summarizes the results of FAB-mass spectrometric analysis of fractions A–D, indicating the nominal mass of the molecular ion and the abbreviated, proposed structure for each fraction.

examination of the complete spectrum indicated the loss of 18 mu occurred at aspartic acid-4, whose side chain had been protected as the benzyl ester. Under the conditions of the reaction, the benzyl ester was apparently displaced by the peptide nitrogen of the Asp-Lys peptide bond, leaving the stable, five membered ring, the β-aspartimide. Such dehydration is known to occur (16) and may be partially avoided by using cyclohexyl esters of aspartic acid rather than benzyl esters. However, in the example presented, it is noteworthy that when such aspartimide forms do occur, it is possible to separate them by isocratic elution on HPLC. In addition, it is essential to identify the products accurately, and FAB-mass spectometry is useful for rapid screening of such products. Separation of aspartimide forms is desirable since these peptides may be physiologically inactive, metabolized at different rates (16) or highly immunogenic compared to the desired peptide containing free aspartic acid.

Product D showed an intensity at $(M + H)^+ = 1156$ mu, 154 mu higher than the desired product. This product corresponds to the tosyl protecting group on the imidazole ring of histidine remaining intact on the peptide. Ultraviolet spectra and the retention time of this product are also consistant with this conclusion. Thus, cleavage of the peptide by TFMSA did not fully remove the tosyl protection of the histidine. This problem is more pronounced with tosyl-arginine (74), but new derivatives such as mesitylene sulfonate (mts) may circumvent these difficulties. Certainly, the tosyl derivatives are not desirable as antigens since the resulting antiserum may be unreactive to the native structure. When this peptide was resynthesized and treated with HF/anisole, product B was the single major species apparent.

4.4. Example 4: Analysis of Peptide Analogues

An example of further diagnostic use of FAB-mass spectrometry is illustrated in Figure 4. Two structural analogues were synthesized by solid peptide synthesis. The first peptide, H-Glu-Asn-Leu-Lys-Asn-Gly-Leu-Phe-OH, was removed from the resin using liquid HF/anisole. A portion of the crude peptide mixture, examined by FAB-MS, did not display an $(M + H)^+$ of 934.5 mu but instead displayed two significant intensities at 336.2 mu and 617.4 mu, corresponding respectively to H-Gly-Leu-Phe-OH and H-Glu-Asn-Leu-Lys-Asn-OH. This peptide was resynthesized and removed from the resin using TFMSA/TFA. The crude peptide mixture, when examined by FAB-MS, displayed the desired $(M + H)^+$ at 934.57 mu but also a significant intensity at 917.55 mu which corresponds to transpeptidation of the peptide to the β-aspartimide form. These two forms of the peptide were separated by reversed phase HPLC on an octadecylsilanyl column using 0.1% trifluoroactic acid and a shallow gradient of acetonitrile. The partial FAB-mass spectrum for the desired peptide exhibited an $(M + H)^+$ at 934.52 mu with little intensity at 917.51 mu (Fig. 4, top). Amino acid analysis and analysis of the entire FAB mass spectrum were consistent with the desired structure.

The second peptide, H-Glu-Asn-Leu-Lys-Asp-Cys(SEt)-Gly-Leu-Phe-OH,

Glu – Asn – Leu – Lys – Asn – Gly – Leu – Phe $(M + H^+ = 934.5)$

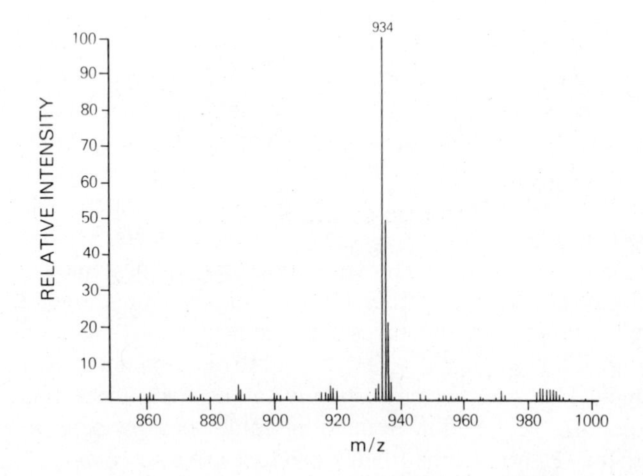

Glu – Asn – Leu – Lys – Asp – Cys(SEt) – Gly – Leu – Phe $(M + H^+ = 1098.5)$

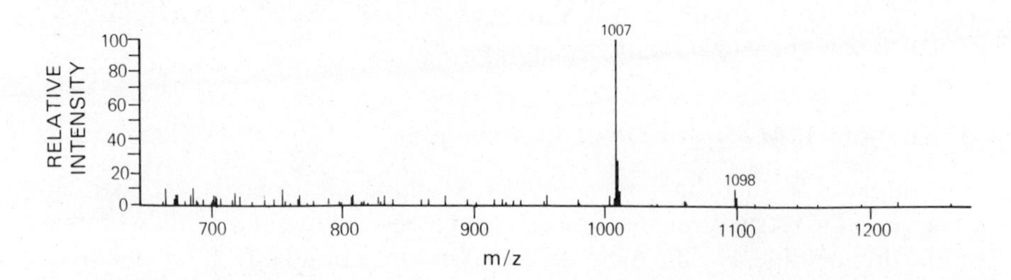

Fig. 4. FAB-mass spectrometric analysis of peptide analogs. The amino acid sequences of the two peptides analyzed are shown at the top of each graph. The graphs show partial FAB-mass spectra of the peptides on a Kratos MS-50 instrument. The signals with highest relative intensity are numbered with nominal mass.

was removed from the resin using liquid HF/anisole. A portion of the crude peptide mixture was examined by FAB-mass spectrometry and exhibited an intensity, among others, at 1098.5 mu, consistent with the $(M + H)^+$ for the desired peptide. This intensity showed an isotopic envelope that indicated the presence of sulfur. The peptide was purified by reverse phase HPLC using 0.1% trifluoroacetic acid and a shallow gradient of acetonitrile. The partial FAB mass spectrum (Fig. 4 bottom) of the purified peptide displayed an unexpected intensity at 1006.6 mu. This intensity did not show the expected isotopic envelope for a sulfur-containing peptide. A lower intensity at 1098.5 mu corres-

Table 1. Redundant Residue Masses[a]

Mass	Amino Acid	Dipeptide
113	Ile, Leu, HyPro	
114	Asn	Gly-Gly
128	Lys, Gln	Gly-Ala
144	β-Aspartimide	Gly-Ser
146	AECys, Met-CH_3	
147	Phe, Met(O)	
156	Arg	Gly-Val
163	Tyr, Met(O_2)	
186	Trp	Gly-Glu, Ala-Asp, Ser-Val
202	Trp + 16	Thr-Thr, Ala-Met

[a] Abbreviations: HyPro hydroxyproline;
AECys: *S*-aminoethyl-cysteine;
Met-CH_3: *S*-methyl-methionine;
Met(O): methionine solfoxide;
Met(O_2): methionine solfone.

ponded to the expected $(M + H)^+$ isotopic envelope for the ethyl-S-cysteine form of the peptide. Amino acid analysis was consistent with the peptide structure. One possible explanation for the intensity at 1006.6 mu could be that the ethyl-S-S is lost due to retro-Michael addition (β-elimination from the ethyl-S-cysteine). The dehydro-alanine so formed could then be hydrogenated. However, alanine is absent from the amino acid analysis of this peptide. Therefore, this chemical reaction could be a consequence of FAB-mas spectral analysis. Analysis of the FAB mass spectrum indicated that certain fragment ions could possibly arise by fragmentation of either the ethyl-S-cysteine or alanine forms of the peptide.

5. SUMMARY

In this chapter we have presented an overview of the chemical methods of peptide synthesis and mass spectrometry as well as several examples of the combined use of HPLC and FAB-mass spectrometry for the characterization of synthetic peptides. These discussions should be useful to purveyors of synthetic peptides for several reasons. First, chemical synthesis is supplanting expression cloning as an easier method for producing peptides below approximately 50 amino acid residues. Thus, many hormones and hormone analogues for human therapeutics will likely be produced by chemical synthesis. Second, characterization of these products by chemical and physical methods is of paramount importance to avoid immunogenic or toxic contamination. Mass spectrometry coupled to a multi-disciplinary approach, including HPLC, is an effective means for characterizing the products. Finally, as the mass ranges of mass spectrometers and of peptide synthesis increase, the methods discussed here will continue to be an important weapon in the arsenal of the modern biotechnologist.

REFERENCES

1. R.T. Alpin, J. Christiansen, and G.T. Young, Int. J. Peptide and Protein Res., *21*, 555–561 (1983).
2. J.F. Arens, Rec. Trav. Chim. Pays-Bas, *74*, 769–770 (1955).
3. M.Z. Atassi, Immunochem., *12*, 423–438 (1975).
4. E. Atherton, A. Dryland, P. Goddard, L. Cameron, J.D. Richards, and R.C. Sheppard, in *Peptides: Structure and Function*, C.M. Deber, V.J. Hruby, and K.D. Kopple, Eds., Pierce Chem. Co., Rockford, IL, 1985, pp. 249–252.
5. G. Barany, and R.B. Merrifield, in *The Peptides: Analysis, Synthesis, Biology*, Vol. 2, E. Gross and J. Meienhofer, Eds), Academic Press, New York, 1979, pp. 1–284.
6. M. Barber, R.S. Bordoli, G.J. Elliott, N.J. Horoch, and B.N. Green, Biochem. Biophys. Res. Commun., *110*, 753–757 (1983).
7. M. Barber, R.S. Bordoli, G.J. Elliott, R.D. Sedgwick, A.N. Tyler, and B.N. Green, J. Chem. Soc., Chem. Commun., 936–938 (1982).
8. M. Barber, R.S. Bordoli, G.V. Garner, D.B. Gordon, R.D. Sedgwick, L.W. Tetler, and A.N. Tyler, Biochem. J., *197*, 401–404 (1981).
9. M. Bergman, and L. Zervas, Ber. Deutsch. Chem. Ges., *65*, 1192–1201 (1932).
10. K. Biemann, Int. J. Mass Spectrom. Ion Physics, *45*, 183–194 (1982).
11. C.R. Blakely, M.L. Vestal, Anal. Chem., *55*, 750–754 (1983).
12. M. Bodanszky, Nature, *175*, 685–686 (1955).
13. M. Bodanszky, in *The Peptides: Analysis, Synthesis, Biology*, Vol. 1, E. Gross and J. Meienhofer, Eds., Academic Press, New York, 1979, pp. 105–196.
14. M. Bodanszky, Int. J. Peptide Protein Res., *25*, 449–474 (1985).
15. M. Bodanszky, and A. Bodanszky, Int. J. Peptide Protein Res., *24*, 563–568 (1984).
16. M. Bodanszky, and J. Martinez, in *The Peptides: Analysis, Synthesis, Biology*, Vol. 5, E. Gross and J. Meienhofer, Eds., Academic Press, New York, 1983, pp. 111–216.
17. M. Bodanszky, and N.J. Williams, J. Am. Chem. Soc., *89*, 685–689 (1967).
18. C.V. Bradley, D.H. Williams, and M.R. Hanley, Biochem. Biophys. Res. Commun., *104*, 1223–1230 (1982).
19. A.M. Buko, and B.A. Fraser, Biomed. Mass. Spectrom., *12*, 577–585 (1985).
20. A.M. Buko, L.R. Phillips, and B.A. Fraser, Biomed. Mass Spectrom., *10*, 324–333 (1983).
21. A.M. Buko, L.R. Phillips, and B.A. Fraser, Biomed. Mass Spectrom., *10*, 408–419 (1983).
22. A.M. Buko, L.R. Phillips, and B.A. Fraser, Biomed. Mass. Spectrom., *10*, 387–393 (1983).
23. H. Chantrenne, Nature, *160*, 603–604 (1947).
24. H. Chantrenne, Biochim. Biophys. Acta, *2*, 286–293 (1948).
25. H. Chantrenne, Nature, *164*, 576–577 (1949).
26. H. Chantrenne, Biochim. Biophys. Acta, *4*, 484–492 (1950).
27. I. Clark-Lewis, R. Aebersold, H. Ziltener, J.W. Schrader, L.E. Hood, and S.B.H. Kent, Science, *231*, 134–139 (1986).

28. M.R. Clench, G.V. Garner, D.B. Gordon, and M. Barber, Biomed. Mass. Spectrom., *12*, 355–357 (1984).
29. T. Curtius, Ber. Deutsch. Chem. Ges., *35*, 3226–3228 (1902).
30. V. du Vigneaud, C. Ressler, J.M. Swan, C.W. Roberts, P.G. Katsoyanni, and S. Gordon, J. Am. Chem. Soc., *75*, 4879–4880 (1953).
31. H.N. Eisen, *Immunology*, Harper and Row, New York, (1974).
32. E. Fischer, Ber. Deutsch. Chem. Ges., *36*, 2094–2106 (1903).
33. K. Fukuhara, T. Tsuji, K. Toi, T. Takao, and Y. Shimonishi, J. Biol. Chem., *260*, 10487–10494 (1985).
34. G.V. Garner, D.B. Gordon, L.W. Tetler, and R.D. Sedgwick, Biomed. Mass Spectrom., *18*, 486–488 (1983).
35. B.W. Gibson, and K. Biemann, Proc. Natl. Acad. Sci. U.S.A., *81*, 1956–1960 (1984).
36. W.R. Gray, J.E. Rivier, R. Galyean, L.J. Cruz, and B.M. Olivera, J. Biol. Chem., *258*, 12247–12251 (1983).
37. R.L. Heinrikson, and S.C. Meredith, Anal. Biochem., *136*, 65–74 (1984).
38. J. Honz, and J. Rudenger, Collect. Czech. Chem. Commun, *26*, 2333–2344 (1961).
39. J. Humphries, E.F. Nurse, S.J. Dunmore, A. Beloff-Chain, G.W. Taylor, and H.R. Morris, Biochem. Biophys. Res. Commun., *114*, 763–766 (1983).
40. M.W. Hunkapiller, and L.E. Hood, Science, *207*, 523–525 (1980).
41. S.B.H. Kent, and R.B. Merrifield, Int. J. Peptide Protein Res., *22*, 57–65 (1983).
42. H.Y. Kim, D. Pilosof, D.F. Dyckes, and M.L. Vestal, J. Amer. Chem. Soc., *106*, 7304–7309 (1984).
43. W. Konig, and R. Geiger, Chem. Ber., *106*, 3626–3635 (1973).
44. W. Konig, and R. Geiger, Chem. Ber., *106*, 3626–3635 (1973).
45. P. Marfey, Carlsberg Res. Commun., *49*, 591–596 (1984).
46. D.R. Marshak, in *Proteins: Structure and Function*, J. L'Italien, Ed. Plenum Press, New York, 1987, pp. 286–298.
47. D.R. Marshak, and B.A. Fraser, in *Brain Peptides Update*, Vol. 1, D.T. Krieger, M.J. Brownstein, and J.B. Martin, Eds. J. Wiley, New York, 1987, pp. 9–36.
48. R.D. McFarlane, Ace. Chem. Res., *15*, 268–275 (1982).
49. R.B. Merrifield, J. Am. Chem. Soc., *85*, 2149–2154 (1963).
50. L.J. Miller, I. Jardine, E. Weissman, V.L.W. Go, and D. Speicher, J. Neurochem., *43*, 835–840 (1984).
51. M.L. Moore, W.F. Huffman, G.D. Roberts, S. Rottschaeffer, L. Sulat, J.s. Stefankiewicz, and F. Stassen, Biochem. Biophys. Res. Commun., *121*, 878–883 (1984).
52. H.R. Morris, M. Panico, and G.W. Taylor, Biochem. Biophys. Res. Commun., *117*, 299–305.
53. M. Mutter, and E. Bayer, in *The Peptides: Analysis, Synthesis, Biology*, Vol. 2, E. Gross and J. Meienhofer, Eds., Academic Press, New York, 1979, pp. 285–332.
54. P.G. Pietta, and G.R.Marshall, J. Chem. Soc. D., 650–651 (1970).
55. P. Pucci, C. Carestia, G. Fioretti, A.M. Mastrobuoni, and L. Pagano, Biochem. and Biophys. Res. Commun., *130*, 84–90 (1985).
56. P. Roepstorff, and J. Fohlman, J. Biomed. Mass Spectrom., *11*, 601 (1984).

57. G. Rougon, and D.R. Marshak, J. Biol. Chem., *261*, 3396–3401 (1986).
58. G.P. Royer, W.E. Schwartz, and F.A. Liberatore, Methods in Enzymology, *47*, 40–45 (1977).
59. S. Sakakibara, Y. Shimonishi, Y. Kishida, M. Okada, and H. Sugihara, Bull. Chem. Soc. Jpn., *40*, 2164–2167 (1967).
60. S.K. Sethi, C.C. Nelson, and J.A. McCloskey, Anal. Chem., *56*, 1977–1979 (1984).
61. S.K. Sethi, D.L. Smith, and J.A. McCloskey, Biochem. Res. Commun., *112*, 126–131 (1983).
62. J.C. Sheehan, and G.P. Hess, J. Am. Chem. Soc., *77*, 1067–1068 (1955).
63. N.M. Sherwood, S.A. Sower, D.R. Marshak, B.A. Fraser, and M.J. Brownstein, J. Biol. Chem., *261*, 4812–4819 (1986).
64. S.A. Sower, J.A. King, R.P. Millar, N.M. Sherwood, and D.R. Marshak, Endocrinol, *120*, 773–779 (1987).
65. B. Sundqvist, I. Kamensky, P. Hakansson, J. Kjellberg, M. Salehpour, S. Widdiyasekera, J. Fohlman, P.A. Peterson, and P. Roepstorff, Biomed. Mass Spectrometry, *11*, 242–257 (1984).
66. B. Sundqvist, P. Roepstorff, J. Fohlman, X. Hedin, P. Hakansson, I. Kamensky, M. Lindberg, M. Salehpour, and G. Sawe, Science, *226* 696–698 (1984).
67. J.G. Sutcliffe, T.M. Shinnick, N. Green, and R.A. Lerner, in *Biotechnology and Biological Frontiers* P.H. Abelson, Ed.) Amer. Assoc. Advance. Science, Washington, DC, 1984, pp. 97–109.
68. T. Takao, M. Yoshida, Y.-M. Hong, S. Aimoto, and Y. Shimonishi, Biomed. Mass Spectrom., *11*, 549–556 (1984).
69. K. Tatemoto, and V. Mutt, Proc. Natl. Acad. Sci. USA, *75*, 4115–4119 (1978).
70. D. Theodoropoulos, P. Cordopatis, D. Dalietos, A. Furst, and T.D. Lee, Biochem. Biophys. Res. Commun., *115*, 653–657 (1983).
71. H. Watanabe, K. Takiguchi, S. Aimoto, and Y. Shimonishi, *Peptide Chemistry*, N. Izumiya, Ed. Protein Research Foundation, Osaka, 1985, pp. 91–96.
72. T. Wieland, W. Schafer, and E. Bokelmann, Liebigs Ann. Chem., *573*, 99–104 (1951).
73. F.P. Widmer, K. Breddam, and J.T. Johansen, Carlsberg Res. Commun., *45*, 453–463 (1980).
74. H. Yajima, and N. Fujii, in *The Peptides: Analysis, Synthesis, Biology*, Vol. 5, E. Gross and J. Meienhofer, Eds., Academic Press, New York, 1983, pp. 66–111.
75. H. Yajima, S. Futaki, N. Fujii, K. Akaji, S. Funakoshi, M. Sakurai, S. Katakura, K. Inoue, R. Hosotani, T. Tobe, T. Segawa, A. Inoue, K. Tatemoto, and V. Mutt, J. Chem. Soc. Chem. Commun., 877–878 (1985).
76. J.D. Young, C.E. Costello, A. Van Langenhove, E. Haber, and G.R. Matsueda, Int. J. Peptide Protein Res., *22*, 374–380 (1983).

INDEX